Metals

Metalloids

Nonmetals

18 8A

13 3A	14 4A	15 5A	16 6A	17 7A	2 **He** 4.00 helium
5 **B** 10.81 boron	6 **C** 12.01 carbon	7 **N** 14.01 nitrogen	8 **O** 16.00 oxygen	9 **F** 19.00 fluorine	10 **Ne** 20.18 neon

10 8B	11 1B	12 2B	13 **Al** 26.98 aluminum	14 **Si** 28.09 silicon	15 **P** 30.97 phosphorus	16 **S** 32.07 sulfur	17 **Cl** 35.45 chlorine	18 **Ar** 39.95 argon
28 **Ni** 58.69 nickel	29 **Cu** 63.55 copper	30 **Zn** 65.39 zinc	31 **Ga** 69.72 gallium	32 **Ge** 72.61 germanium	33 **As** 74.92 arsenic	34 **Se** 78.96 selenium	35 **Br** 79.90 bromine	36 **Kr** 83.80 krypton
46 **Pd** 106.42 palladium	47 **Ag** 107.87 silver	48 **Cd** 112.41 cadmium	49 **In** 114.82 indium	50 **Sn** 118.71 tin	51 **Sb** 121.75 antimony	52 **Te** 127.60 tellurium	53 **I** 126.90 iodine	54 **Xe** 131.29 xenon
78 **Pt** 195.08 platinum	79 **Au** 196.97 gold	80 **Hg** 200.59 mercury	81 **Tl** 204.38 thallium	82 **Pb** 207.2 lead	83 **Bi** 208.98 bismuth	84 **Po** (209) polonium	85 **At** (210) astatine	86 **Rn** (222) radon
110 **Ds** (281) darmstadtium	111 **Rg** (280) roentgenium	112 **Cn** (285)	113 — (284)	114 — (289)	115 — (288)	116 — (292)	117 ** (292)	118 — (294)

64 **Gd** 157.25 gadolinium	65 **Tb** 158.93 terbium	66 **Dy** 162.50 dysprosium	67 **Ho** 164.93 holmium	68 **Er** 167.26 erbium	69 **Tm** 168.93 thulium	70 **Yb** 173.04 ytterbium	71 **Lu** 174.97 lutetium
96 **Cm** (247) curium	97 **Bk** (247) berkelium	98 **Cf** (251) californium	99 **Es** (252) einsteinium	100 **Fm** (257) fermium	101 **Md** (258) mendelevium	102 **No** (259) nobelium	103 **Lr** (260) lawrencium

**Discovered in 2010, element 117 is currently under review by IUPAC.

INTRODUCTORY
CHEMISTRY

NIVALDO J. TRO

INTRODUCTORY CHEMISTRY
FOURTH EDITION

Prentice Hall

Boston Columbus Indianapolis New York San Francisco Upper Saddle River
Amsterdam Cape Town Dubai London Madrid Milan Munich Paris Montréal Toronto
Delhi Mexico City São Paulo Sydney Hong Kong Seoul Singapore Taipei Tokyo

Library of Congress Cataloging-in-Publication Data

Tro, Nivaldo J.
  Introductory chemistry / Nivaldo J. Tro. — 4th ed.
    p. cm.
  ISBN 978-0-321-68793-7
  1. Chemistry. I. Title.
  QD33.2.T76 2012
  540—dc22
                              2010037612

Editor in Chief, Chemistry: Adam Jaworski
Acquisitions Editor: Terry Haugen
Marketing Manager: Erin Gardner
Senior Project Editor: Jennifer Hart
Editorial Assistant: Catherine Martinez
Marketing Assistant: Nicola Houston
VP/Executive Director, Development: Carol Trueheart
Development Editor: Erin Mulligan
Managing Editor, Chemistry and Geosciences: Gina M. Cheselka
Project Manager: Shari Toron
Full-Service Project Management/Composition: Francesca Monaco/Preparé, Inc.
Senior Manufacturing and Operations Manager: Nick Sklitsis
Operations Specialist: Maura Zaldivar
Art Director: Mark Ong
Interior and Cover Designer: LCI Design/Laura C. Ierardi
Senior Technical Art Specialist: Connie Long
Art Studio: Precision Graphics
Cover and Chapter Opening Illustrations: Quade Paul
Photo Research Manager: Elaine Soares
Photo Researcher: Clare Maxwell
Senior Media Producer: Angela Bernhardt
Associate Media Producer: Kristin Mayo
Media Supervisor: Liz Winer
Production Coordinator, Media: Shannon Kong

Printed in the United States of America
10 9 8 7 6 5 4 3

ISBN-10:   0-321-68793-0
ISBN-13: 978-0-321-68793-7

Prentice Hall
is an imprint of

www.pearsonhighered.com

To Annie

ABOUT THE AUTHOR

Niva Tro, *chairman of the Chemistry Department at Westmont College in Santa Barbara, California, has been a faculty member at the college since 1990. He received his B.A. degree in chemistry from Westmont College in 1985 and earned a Ph.D. from Stanford University in 1989, after which he performed postdoctoral research at the University of California at Berkeley. Honored as Westmont's outstanding teacher of the year in 1994, 2001, and 2008, he was also named the college's outstanding researcher of the year in 1996.*

Professor Tro lives in Santa Barbara with his wife, Ann, and their four children, Michael, Alicia, Kyle, and Kaden. For leisure, he enjoys reading, writing, snowboarding, biking, and other outdoor activities with his family.

BRIEF CONTENTS

CONTENTS

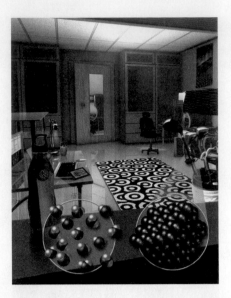

5 Molecules and Compounds 127

4 Atoms and Elements 93

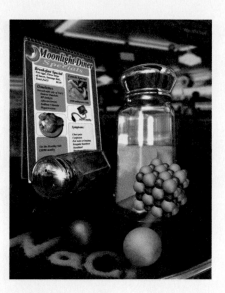

6 Chemical Composition 165

7 Chemical Reactions 205

8 Quantities in Chemical Reactions 249

9 Electrons in Atoms and the Periodic Table 285

10 Chemical Bonding 325

11 Gases 359

15 Chemical Equilibrium   531

16 Oxidation and Reduction   577

17 Radioactivity and Nuclear Chemistry 613

18 Organic Chemistry 643

Problem-Solving Procedures

Solving Unit Conversion Problems	28
Solving Numerical Problems	38
Writing Formulas for Ionic Compounds	136
Obtaining an Empirical Formula from Experimental Data	185
Writing Balanced Chemical Equations	212
Writing Equations for Precipitation Reactions	220
Writing Lewis Structures for Covalent Compounds	331
Predicting Geometry Using VSEPR Theory	338
Balancing Redox Equations Using the Half-Reaction Method	585
Naming Alkanes	655

This book is for *you*, and every text feature has you in mind. I have two main goals for you in this course: to see chemistry as you never have before, and to develop the problem-solving skills you need to succeed in chemistry.

I want you to experience chemistry in a new way. Each chapter of this book is written to show you that chemistry is not just something that happens in a laboratory; chemistry surrounds you at every moment. I have worked with several outstanding artists to develop photographs and art that will help you visualize the molecular world. From the opening example to the closing chapter, you will *see* chemistry. I hope that when you finish this course, you think differently about your world because you understand the molecular interactions that underlie everything around you.

I also want you to develop problem-solving skills. No one succeeds in chemistry—or in life, really—without the ability to solve problems. I can't give you a formula for problem solving, but I can give you strategies that will help you develop the *chemical intuition* you need to understand chemical reasoning.

Look for several recurring structures throughout this book designed to help you master problem solving. The most important ones are (1) a four-step process (Sort, Strategize, Solve, and Check) designed to help you learn how to solve problems; (2) the solution map, a visual aid that helps you navigate your way through a problem; (3) the two-column Examples, in which the left column explains in clear and simple language the purpose of each step of the solution shown in the right column; and (4) the three-column Examples, which describe a problem-solving procedure while demonstrating how it is applied to two different Examples. In addition, since students have specifically asked me to provide connections between Examples and end-of-chapter problems, I have added a For More Practice feature at the end of worked Examples that will guide you to the end-of-chapter problems that will provide more opportunity to practice the skill(s) covered in the Example.

Lastly, know that chemistry is *not* reserved only for those with some superhuman intelligence level. With the right amount of effort and some clear guidance, anyone can master chemistry, including you.

Sincerely,

Nivaldo J. Tro
tro@westmont.edu

To the Instructor

I thank all of you who have used any of the first three editions of *Introductory Chemistry*—you have made this book the most widely selling book in its market, and for that I am extremely grateful. The preparation of the fourth edition has enabled me to continue to refine the book to meet its fundamental purpose: teaching chemical skills in the context of relevance.

Introductory Chemistry is designed for a one-semester, college-level, introductory or preparatory chemistry course. Students taking this course need to develop problem-solving skills—but they also must see *why* these skills are important to them and to their world. *Introductory Chemistry* extends chemistry from the laboratory to the student's world. It motivates students to learn chemistry by demonstrating how it plays out in their daily lives.

This is a visual book. Today's students often learn by seeing, so wherever possible, I have used images to help communicate the subject. In developing chemical principles, for example, I worked with several artists to develop multipart images that show the connection between everyday processes visible to the eye and the molecular interactions responsible for those processes. This art has been further refined and improved in the fourth edition, making the visual impact sharper and more targeted to student learning. For example, I have implemented a hierarchical method of labeling in many of the images: the white-boxed labels are the most important, the tan-tint boxes are second most important, and unboxed labels are the third most important. This allows me to treat related labels and annotations within an image in the same way, so that the relationships between them are immediately evident. My goal is to create an art program that teaches, and that presents complex information clearly and concisely. Many of the illustrations showing molecular depictions of a real-world object or process have three parts: macroscopic (what we can see with our eyes); molecular and atomic (space-filling models that depict what the molecules and atoms are doing); and symbolic (how chemists represent the molecular and atomic world). The goal is for the student to begin to see the connections between the macroscopic world, the molecular world, and the representation of the molecular world with symbols and formulas.

In the fourth edition, I have also refined the problem-solving pedagogy to include four steps: Sort, Strategize, Solve, and Check. The *solution map*, which has been part of this book since the beginning, is now part of the *Strategize* step. This four-step procedure is meant to guide students as they learn chemical problem-solving. Extensive flowcharts are also incorporated throughout the book, allowing students to visualize the organization of chemical ideas and concepts. The color scheme used in both the solution maps and the flowcharts is designed to have pedagogical value. More specifically, the solution maps utilize the colors of the visible spectrum—always in the same order, from violet to red.

Throughout the worked Examples in this book, I use a *two- or three-column* layout in which students learn a general procedure for solving problems of a particular type as they see this procedure applied to one or two worked Examples. In this format, the *explanation* of how to solve a problem is placed directly beside the actual steps in the *solution* of the problem. Many of you have said that you use a similar technique in lecture and office hours. Since students have specifically asked for connections between Examples and end-of-chapter problems, I include a For More Practice feature at the end of each worked Example that lists the review examples and end-of-chapter problems that provide more opportunity to practice the skill(s) covered in the Example.

A successful new feature in the second edition was the Conceptual Checkpoints, a series of short questions that students can use to test their mastery of key concepts as they read through a chapter. Emphasizing understanding rather than calculation, they are designed to be easy to answer if the student has grasped the essential concept but difficult if he or she has not. Your positive remarks on this new feature prompted me to continue adding more of these to the fourth edition, including questions that highlight visualization of the molecular world.

PREFACE

New to This Edition

- **A student-friendly, step-by-step problem-solving approach is presented throughout** (fully introduced and explained in Chapter 2): The format for the majority of the worked examples in the book has been changed to involve four steps: Sort, Strategize, Solve, and Check. "Relationships Used" have been identified and new "Check" step content has been added to most worked examples in the book.
- **In all chapters, figure labels now follow a consistent hierarchy.** Three types of labels appear on the art. The most important are in white shadow boxes; the second most important, in tinted boxes (with no border); and the third appear unboxed.
- **Page numbers have been added to chapter-opening outlines** in all chapters.
- **Approximately 25% more Conceptual Checkpoints** have been added throughout the text.
- **All figures and figure captions have been carefully examined, and images and labels have been replaced or revised when needed** to improve the teaching focus of the art program.
- **Every end-of-chapter question has been carefully reviewed** by the author and editor and revised and/or replaced when necessary.
- **Reading quizzes for every chapter in MasteringChemistry®.**

Some significant improvements have been made to key content areas as well. These include:

- Section 2.6: *Problem Solving and Unit Conversion* is a new section that introduces and explains the new general problem-solving method used throughout the worked Examples in this edition. Problems are organized into Sort, Strategize, Solve, and Check steps. The new category of "Relationships Used" is called out in the Strategize step of many of the worked examples. New "Check" material has also been added.
- Section 2.10 has been revised in conjunction with Section 2.6 *Problem Solving and Unit Conversion* and is now titled *Numerical Problem-Solving Overview*.
- Updated or revised chapter-opening art in Chapters 3, 5, 6, 10, and 17.
- Conversion factor for K to Celsius changed from 273° to 273.15° throughout.
- Introduction to Section 3.8 *Energy* revised to more effectively introduce the concept of work, the law of conservation of energy, and the different forms of energy.
- Time-sensitive data, such as the graph regarding global temperature in Figure 8.2, and the data in Figures 14.20 *Acid rain in the United States* and 14.22 *Emissions of SO_2 from 1980 to 2009,* have been updated throughout.
- Section 5.3: *Chemical Formulas: How to Represent Compounds* has been revised to include the different ways of representing compounds (molecular formulas, structural formulas, ball-and-stick models, and space-filling models.) A new color key for space-filling models has been added to this section.
- Section 5.7: *Naming Ionic Compounds* has been revised. The terms "Type I" and "Type II" ionic compounds have been eliminated, and the subsections and worked Examples dealing with naming binary ionic compounds have been renamed and revised.

- A new solubility flowchart has been added to Section 7.5: *Aqueous Solutions and Solubility: Compounds Dissolved in Water*.
- Every end-of-chapter question has been carefully reviewed and revised or replaced when necessary.

MasteringChemistry is the most effective and widely used online tutorial, homework and assessment system for chemistry. It helps instructors maximize class time with customizable, easy-to-assign, and automatically graded assessments that motivate students to learn outside of class and arrive prepared for lecture. These assessments can easily be customized and personalized by instructors to suit their individual teaching style. The powerful gradebook provides unique insight into student and class performance even before the first test. As a result, instructors can spend class time where students need it most.

I hope the changes in the fourth edition support you in your mission of teaching students chemistry. Ours is a worthwhile cause, even though it requires constant effort. Please feel free to e-mail me with any questions or comments you might have. I look forward to hearing from you as you use this book in your course.

Sincerely,

Nivaldo J. Tro
tro@westmont.edu

The design and features of this text have been conceived to work together as an integrated whole with a single purpose: to help students understand chemical principles and to master problem-solving skills in a context of relevance. Students must be able not only to grasp chemical concepts and solve chemical problems, but also to understand how those concepts and problem-solving skills are relevant to their other courses, their eventual career paths, and their daily lives.

Teaching Principles

The development of basic chemical principles—such as those of atomic structure, chemical bonding, chemical reactions, and the gas laws—is one of the main goals of this text. Students must acquire a firm grasp of these principles in order to succeed in the general chemistry sequence or the chemistry courses that support the allied health curriculum. To that end, the book integrates qualitative and quantitative material and proceeds from concrete concepts to more abstract ones.

ORGANIZATION OF THE TEXT

The main divergence in topic ordering among instructors teaching introductory and preparatory chemistry courses is the placement of electronic structure and chemical bonding. Should these topics come early, at the point where models for the atom are being discussed? Or should they come later, after the student has been exposed to chemical compounds and chemical reactions? Early placement gives students a theoretical framework within which they can understand compounds and reactions. However, it also presents students with abstract models before they understand why they are necessary. I have chosen a later placement for the following reasons:

1. **A later placement provides greater flexibility.** An instructor who wants to cover atomic theory and bonding earlier can simply cover Chapters 9 and 10 after Chapter 4. However, if atomic theory and bonding were placed earlier, it would be more difficult for the instructor to skip these chapters and come back to them later.

2. **A later placement allows earlier coverage of topics that students can more easily visualize.** Coverage of abstract topics too early in a course can lose some students. Chemical compounds and chemical reactions are more tangible than atomic orbitals, and the relevance of these is easier to demonstrate to the beginning student.

3. **A later placement gives students a reason to learn an abstract theory.** Once students learn about compounds and reactions, they are more easily motivated to learn a theory that explains them in terms of underlying causes.

4. **A later placement follows the scientific method.** In science, we normally make observations, form laws, and then build models or theories that explain our observations and laws. A later placement follows this ordering.

Nonetheless, I know that every course is unique and that each instructor chooses to cover topics in his or her own way. Consequently, I have written each chapter for maximum flexibility in topic ordering. In addition, the book is offered in two formats. The full version, *Introductory Chemistry*, contains 19 chapters, including organic chemistry and biochemistry. The shorter version, *Introductory Chemistry Essentials*, contains 17 chapters and omits these topics.

Print and Media Resources

FOR THE INSTRUCTOR

MasteringChemistry (*http://www.masteringchemistry.com*)
MasteringChemistry is the best adaptive-learning online homework and tutorial system. Instructors can create online assignments for their students by choosing from a wide range of items, including end-of-chapter problems and research-enhanced tutorials. Assignments are automatically graded with up-to-date diagnostic information, helping instructors pinpoint where students struggle either individually or as a class as whole.

Instructor Resource and Full Solutions Manual (0-321-73019-4) Prepared by Mark Ott of Jackson Community College, and Matthew Johll of Illinois Valley Community College. This manual features lecture outlines with presentation suggestions, teaching tips, suggested in-class demonstrations, and topics for classroom discussion. It also contains full solutions to all the end-of-chapter problems from the text.

Printed Testbank (0-321-73009-7) Prepared by Michael Hauser of St. Louis Community College. This printed test bank includes more than 1500 questions. A computerized version of the test item file is available on the Instructor's Resource DVD and can be downloaded from the Instructor Resource Center.

Instructor Resource DVD (0-321-73007-0) This resource provides an integrated collection of resources to help instructors make efficient and effective use of their time. This package features the following:

- All the art from the text, including figures and tables in JPG and PDF formats; movies; animations; Interactive Molecules; and the Instructor's Resource Manual files.
- Four PowerPoint™ presentations: (1) a lecture outline presentation for each chapter, (2) all the art from the text, (3) the worked Examples from the text, and (4) CRS (Classroom Response System) questions.
- The TestGen, a computerized version of the Test Item File that allows you to create and tailor exams to your needs.

FOR THE STUDENT

Pearson eText: The integration of Pearson eText within MasteringChemistry gives students, with new books, easy access to the electronic text when they are logged into MasteringChemistry. Pearson eText pages look exactly like the printed text, offering powerful new functionality for students and instructors. Users can create notes, highlight text in different colors, create bookmarks, zoom, view in single-page or two-page view, etc.

Study Guide (0-321-73010-0) by Donna Friedman, St. Louis Community College—Florissant Valley. Each chapter contains an overview, chapter objectives, a chapter review, as well as practice problems for each major concept in the text. This is followed by two or three self-tests with answers located at the end of each chapter so students can check their work.

Student Solution Manual (0-321-73018-6) by Matthew Johll of Illinois Valley Community College. This book provides solutions only to those problems that have a short answer in the text's Answers section (problems numbered in blue in the text).

Acknowledgments

This book has been a group effort, and there are many people whose help has meant a great deal to me. First and foremost, I would like to thank my editor, Terry Haugen, who came on board during this edition. Terry is a good thinker who knows the needs of both students and professors. Thanks, Terry, for your guidance on and commitment to this revision. As always, I am grateful to Paul Corey, the president of the Science Division at Pearson, for his unwavering support.

New to this edition, but not to working with me, is Erin Mulligan, a development editor whose friendship and guidance I continue to cherish. Thanks, Erin, for all your outstanding help and advice. I cannot thank my project editor Jennifer Hart enough. Jennifer, you are always there to guide me, to keep me on task, and to take care of whatever needs to be done. I am so grateful. I would also like to thank Erin Gardner, my marketing manager whose creativity in describing and promoting the book is without equal. Thanks also to Brian Buckley and the MasteringChemistry team who continue to provide and promote the best online homework system on the planet.

I also appreciate the expertise and professionalism of my copy editor, Betty Pessagno, as well as the skill and diligence of Francesca Monaco and her colleagues at Prepare. I am a picky author, and they always accommodated my seemingly endless requests. Thank you, Francesca. Thanks as well to my project manager Shari Toron, managing editor Gina Cheselka, senior technical art specialist Connie Long, and the rest of the Pearson-Prentice Hall team—they are part of a first-class operation. This text has benefited immeasurably from their talents and hard work. I owe a special debt of gratitude to Quade Paul, who continues to make my ideas come alive in his chapter-opener and cover art.

I am grateful for the support of my colleagues Allan Nishimura, David Marten, Stephen Contakes, Kristi Lazar, Carrie Hill, and Heidi Henes-Vanbergen, who have supported me in my department while I worked on this book. I am particularly grateful to Allan, who started me on this whole chemistry business—he remains an inspiration in my life. I am also grateful to Brittany Hammer, who helped me with manuscript preparation, and to Andrew Schwemmer, Ryan Fields, and Michael Tro who helped to review end-of-chapter problems for accuracy.

I am grateful to those who have given so much to me personally while writing this book. First on that list is my wife, Ann. Her patience and love for me are beyond description. I also thank my children, Michael, Ali, Kyle, and Kaden, whose smiling faces and love of life always inspire me. I come from a large Cuban family, whose closeness and support most people would envy. Thanks to my parents, Nivaldo and Sara; my siblings, Sarita, Mary, and Jorge; my siblings-in-law, Jeff, Nachy, Karen, and John; my nephews and nieces, Germain, Danny, Lisette, Sara, and Kenny. These are the people with whom I celebrate life.

Lastly, I am indebted to the many reviewers, listed next, whose ideas are scattered throughout this book. They have corrected me, inspired me, and sharpened my thinking on how best to teach this subject we call chemistry. I deeply appreciate their commitment to this project.

Reviewers of the 4th Edition

Jeffrey Allison,
Austin Community College

Mikhail V. Barybin,
The University of Kansas

Lara Baxley,
California Polytechnic State University

Kelly Beefus,
Annoka-Ramsey Community College

Joseph Bergman,
Illinois Central College

Simon Bott,
University of Houston

Maria Cecilia D. de Mesa,
Baylor University

Guy Dadson,
Fullerton College

Brian G. Dixon,
Massachusetts Maritime Academy

Timothy Dudley,
Villanova University

Jeannine Eddleton,
Virginia Tech

Ron Erickson,
University of Iowa

Donna Friedman,
St. Louis Community College—Florissant Valley

Luther D. Giddings,
Salt Lake Community College

Marcus Giotto,
Quinsigamond Community College

Melodie Graber,
Oakton Community College

Maru Grant,
Ohlone College

Jerod Gross,
Roanoke Benson High School

Tammy S. Gummersheimer,
Schenectady County Community College

Tamara E. Hanna,
Texas Tech

Michael A. Hauser,
St. Louis Community College

Bruce E. Hodson,
Baylor University

Donald R. Jones,
Lincoln Land Community College

Martha R. Kellner,
Westminster College

Farkhondeh Khalili,
Massachusetts Bay Community College

Margaret Kiminsky,
Monroe Community College

Rebecca Krystyniak,
Saint Cloud State

Chuck Laland,
Black Hawk College

Richard Lavallee,
Santa Monica College

Laurie Leblanc,
Cuyamaca College

Vicki MacMurdo,
Anoka Ramsey Community College

Carmela Magliocchi Brynes,
MiraCosta College

Jack F. McKenna,
St. Cloud State University

Virginia Miller,
Montgomery College

Meg Osterby,
Western Technical College

John Petty,
University of South Alabama

Jason Serin,
Glendale Community College

Youngju Sohn,
Florida Institute of Technology

Clarissa Sorenson-Unruh,
Central New Mexico Community College

Vidyullata C. Waghulde,
St. Louis Community—Meramec

4TH EDITION ACCURACY REVIEWERS

Steven Socol,
McHenry Community College

David Vanderlinden,
Des Moines Area Community College

FOCUS GROUP PARTICIPANTS

Carmela Byrnes,
Miracosta College

Tammy Gummersheimer,
Schenectady County Community College

Tamara Hanna,
Texas Tech University

Nancy Lee,
MiraCosta College

Geoff Mitchell,
Washington International School

Jie Song,
Univeristy of Michigan, Flint

Reviewers of the 3rd Edition

Benjamin Arrowood
Ohio University

Joe Bergman
Illinois Central College

Timothy Dudley
Villanova University

Sharlene J. Dzugan
University of Cumberlands

Thomas Dzugan
University of Cumberlands

Donna G. Friedman
St. Louis Community College

Erick Fuoco
Daley College

Melodie A. Graber
Oakton Community College

Michael A. Hauser
St Louis Community College, Meramec Campus

Martha R. Joseph
Westminster College

Timothy Kreider
University of Medicine & Dentistry of New Jersey

Laurie Leblanc
Grossmont College

Carol A. Martinez
Central New Mexico Community College

Kresimir Rupnik
Louisiana State University

Kathleen Thrush Shaginaw
Particular Solutions, Inc.

Pong (David) Shieh
Wharton College

Mary Sohn
Florida Tech

Kurt Allen Teets
Okaloosa-Walton College

John Thurston
University of Iowa

Anthony P. Toste
Missouri State University

Carrie Woodcock
Eastern Michigan University

Reviewers of the 2nd Edition

David S. Ballantine, Jr.
Northern Illinois University

Colin Bateman
Brevard Community College

Michele Berkey
San Juan College

Steven R. Boone
Central Missouri State University

Morris Bramlett
University of Arkansas—Monticello

Bryan E. Breyfogle
Southwest Missouri State University

Frank Carey
Wharton County Junior College

Robbey C. Culp
Fresno City College

Michelle Driessen
University of Minnesota—Minneapolis

Donna G. Friedman
St. Louis Community College—Florissant Valley

Crystal Gambino
Manatee Community College

Steve Gunther
Albuquerque Technical Vocational Institute

Michael Hauser
St. Louis Community College—Meramec

Newton P. Hillard, Jr.
Eastern New Mexico University

Carl A. Hoeger
University of California—San Diego

Donna K. Howell
Angelo State University

Nichole Jackson
Odessa College

T. G. Jackson
University of South Alabama

Donald R. Jones
Lincoln Land Community College

Kirk Kawagoe
Fresno City College

Roy Kennedy
Massachusetts Bay Community College

Blake Key
Northwestern Michigan College

Rebecca A. Krystyniak
St. Cloud State University

Laurie LeBlanc
Cuyamaca College

Ronald C. Marks
Warner Southern College

Carol A. Martinez
Albuquerque Technical Vocational Institute

Charles Michael McCallum
University of the Pacific

Robin McCann
Shippensburg University

Victor Ryzhov
Northern Illinois University

Theodore Sakano
Rockland Community College

Deborah G. Simon
Santa Fe Community College

Mary Sohn
Florida Institute of Technology

Peter-John Stanskas
San Bernardino Valley College

James G. Tarter
College of Southern Idaho

Ruth M. Topich
Virginia Commonwealth University

Eric L. Trump
Emporia State University

Mary Urban
College of Lake County

Richard Watt
University of New Mexico

Lynne Zeman
Kirkwood Community College

Reviewers of the 1st Edition

Lori Allen
University of Wisconsin—Parkside

Laura Andersson
Big Bend Community College

Danny R. Bedgood
Arizona State University

Christine V. Bilicki
Pasadena City College

Warren Bosch
Elgin Community College

Bryan E. Breyfogle
Southwest Missouri State University

Carl J. Carrano
Southwest Texas State University

Donald C. Davis
College of Lake County

Donna G. Friedman
St. Louis Community College at Florissant Valley

Leslie Wo-Mei Fung
Loyola University of Chicago

Dwayne Gergens
San Diego Mesa College

George Goth
Skyline College

Jan Gryko
Jacksonville State University

Roy Kennedy
Massachusetts Bay Community College

C. Michael McCallum
University of the Pacific

Kathy Mitchell
St. Petersburg Junior College

Bill Nickels
Schoolcraft College

Bob Perkins
Kwantlen University College

Mark Porter
Texas Tech University

Caryn Prudenté
University of Southern Maine

Connie M. Roberts
Henderson State University

Rill Ann Reuter
Winona State University

Jeffery A. Schneider
SUNY—Oswego

Kim D. Summerhays
University of San Francisco

Ronald H. Takata
Honolulu Community College

Calvin D. Tormanen
Central Michigan University

Eric L. Trump
Emporia State University

Drawing from Professor Tro's experience in the classroom with his own students, *Introductory Chemistry*, **Fourth Edition** brings chemistry out of the laboratory and into the world—helping you learn chemistry by showing you how it is manifested in our daily lives. Clear, specific examples are woven throughout this text to tell the story of chemistry. The **Fourth Edition** is also available with MasteringChemistry®, the premier online homework and assessment tool.

NEW! A CONSISTENT STRATEGY FOR SOLVING PROBLEMS

helps you develop the skills you need to succeed in your chemistry course. A new student-friendly, step-by-step problem-solving approach adds four steps to many of the worked examples (Sort, Strategize, Solve, and Check).

Solution Maps
Many of the Examples use a unique visual approach in the Strategize Step, where you'll be shown how to draw a solution map for a problem.

EXAMPLE 2.13 Solving Multistep Conversion Problems Involving Units Raised to a Power

The average annual per person crude oil consumption in the United States is 15,615 dm³. What is this value in cubic inches?

SORT
You are given a volume in cubic decimeters and asked to convert it to cubic inches.

GIVEN: 15,615 dm³
FIND: in.³

STRATEGIZE
Build a solution map beginning with dm³ and ending with in.³ Each of the conversion factors must be cubed, since the quantities involve cubic units.

SOLUTION MAP

$$ dm^3 \rightarrow m^3 \rightarrow cm^3 \rightarrow in^3 $$

$$ \frac{(0.1\ m)^3}{(1\ dm)^3} \qquad \frac{(1\ cm)^3}{(0.01\ m)^3} \qquad \frac{(1\ in.)^3}{(2.54\ cm)^3} $$

RELATIONSHIPS USED
1 dm = 0.1 m (from Table 2.2)
1 cm = 0.01 m (from Table 2.2)
2.54 cm = 1 in. (from Table 2.3)

SOLVE
Follow the solution map to solve the problem. Begin with the given value in dm³ and multiply by the string of conversion factors to arrive at in.³ Make sure to cube each conversion factor as you carry out the calculation.

Round the answer to five significant figures to reflect the five significant figures in the least precisely known quantity (15,615 dm³). The conversion factors are all exact and therefore do not limit the number of significant figures.

SOLUTION

$$ 15{,}615\ dm^3 \times \frac{(0.1\ m)^3}{(1\ dm)^3} \times \frac{(1\ cm)^3}{(0.01\ m)^3} \times \frac{(1\ in.)^3}{(2.54\ cm)^3} $$

$$ = 9.5289 \times 10^5\ in.^3 $$

CHECK
Check your answer. Are the units correct? Does the answer make physical sense?

The units of the answer are correct and the magnitude makes sense. A cubic inch is smaller than a cubic decimeter, so the value in cubic inches should be larger than the value in cubic decimeters.

▶**SKILLBUILDER 2.13 | Solving Multistep Problems Involving Units Raised to a Power**
How many cubic inches are there in 3.25 yd³?

▶**FOR MORE PRACTICE** Problems 93, 94.

Two-Column Examples

All but the simplest examples are presented in a unique two-column format.

- The left column explains the purpose of each step, while the right column shows how the step is executed.

- This format will help you think about the reason for each step in the solution and to fit the steps together.

Writing Formulas for Ionic Compounds	**EXAMPLE 5.5** Write a formula for the ionic compound that forms from aluminum and oxygen.	**EXAMPLE 5.6** Write a formula for the ionic compound that forms from magnesium and oxygen.
1. Write the symbol for the metal and its charge followed by the symbol of the nonmetal and its charge. For many elements, you can determine these charges from their group number in the periodic table (refer to Figure 4.14).	**SOLUTION** $Al^{3+}\ \ O^{2-}$	**SOLUTION** $Mg^{2+}\ \ O^{2-}$
2. Make the magnitude of the charge on each ion (without the sign) become the subscript for the other ion.	$Al_2^{3+}\ O_3^{2-}$ Al_2O_3	$Mg_2^{2+}\ O_2^{2-}$ Mg_2O_2
3. If possible, reduce the subscripts to give a ratio with the smallest whole numbers.	In this case, the numbers cannot be reduced any further; the correct formula is Al_2O_3.	To reduce the subscripts, divide both subscripts by 2. $Mg_2O_2 \div 2 = MgO$
4. Check to make sure that the sum of the charges of the cations exactly cancels the sum of the charges of the anions.	Cations: 2(3+) = 6+ Anions: 3(2−) = 6− The charges cancel.	Cations: 2+ Anions: 2− The charges cancel.
	▶**SKILLBUILDER 5.5** Write a formula for the compound formed from strontium and chlorine.	▶**SKILLBUILDER 5.6** Write a formula for the compound formed from aluminum and nitrogen. ▶**FOR MORE PRACTICE** Example 5.21; Problems 53, 54, 55, 56.

Skillbuilder Exercises
Every worked example is followed by at least one similar (but un-worked) Skillbuilder exercise.

For More Practice
These follow every worked example, linking you to in-chapter examples and end-of-chapter problems that give you a chance to practice the skills in each worked example.

Three-Column Examples

Procedures for solving particular types of problems are presented in a unique three-column format.

- The first column outlines the problem-solving procedure and explains the reasoning that underlies each step.

- The second and third columns show two similar but slightly different examples to solve this problem.

- Seeing the method applied to solve two related problems helps you understand the general procedure in a way that no single example could convey.

CONCEPTUAL UNDERSTANDING completes the picture. In every chemistry course you take, success requires more than problem-solving skills. Real understanding of concepts will help you see why these skills are important to you and to your world.

✓ **CONCEPTUAL CHECKPOINT 7.1**

These images portray molecular views of one substance before and after a change. Determine whether a chemical reaction has occurred in each case.

Conceptual Checkpoints

These conceptual questions enhance understanding of chemical principles, encourage you to stop and think about the ideas just presented, and provide a tool to assess your own progress. Answers and explanations are given at the end of each chapter. There are approximately 25% more Conceptual Checkpoints in the Fourth Edition.

Chapter Review

Consistent review material at the end of each chapter helps reinforce what you've learned.

CHEMICAL PRINCIPLES

Formula Mass: The formula mass of a compound is the sum of the atomic masses of all the atoms in the chemical formula for the compound. Like atomic mass for elements, formula mass characterizes the average mass of a molecule or formula unit.

RELEVANCE

Formula Mass: Besides being the characteristic mass of a molecule or formula unit, formula mass is important in many calculations involving the composition of compounds and quantities in chemical reactions.

CHEMICAL SKILLS

Constant Composition of Compounds (Section 5.2)

The law of constant composition states that all samples of a given compound should have the same ratio of their constituent elements.

To determine whether experimental data are consistent with the law of constant composition, compute the ratios of the masses of each element in all samples. When computing these ratios, it is most convenient to put the larger number in the numerator (top) and the smaller one in the denominator (bottom); that way, the ratio is greater than 1. If the ratios are the same, then the data are consistent with the law of constant composition.

EXAMPLES

EXAMPLE 5.16 Constant Composition of Compounds

Two samples said to be carbon disulfide (CS_2) are decomposed into their constituent elements. One sample produced 8.08 g S and 1.51 g C, while the other produced 31.3 g S and 3.85 g C. Are these results consistent with the law of constant composition?

SOLUTION
Sample 1

$$\frac{\text{Mass S}}{\text{Mass C}} = \frac{8.08 \text{ g}}{1.51 \text{ g}} = 5.35$$

Sample 2

$$\frac{\text{Mass S}}{\text{Mass C}} = \frac{31.3 \text{ g}}{3.85 \text{ g}} = 8.13$$

These results are not consistent with the law of constant composition, so the information that the two samples are the same substance must therefore be in error.

Chemical Principles
The left column summarizes the key principles that you should take away from the chapter, and the right column tells why each topic is important for you to understand.

Chemical Skills
The left column describes the key skills you should gain after reading the chapter, and the right column contains a worked example illustrating that skill.

HIGHLIGHT PROBLEMS

99. Consider the molecular view of water shown here. Pick a molecule in the interior and draw a line to each of its direct neighbors. Pick a molecule near the edge (analogous to a molecule on the surface in three dimensions) and do the same. Which molecule has the most neighbors? Which molecule is more likely to evaporate?

100. Water does not easily remove grease from dirty hands because grease is nonpolar and water is polar; therefore they are immiscible. The addition of soap, however, results in the removal of the grease. Examine the structure of soap shown here and explain how soap works.

Nonpolar tail   Polar head
Sodium stearate
a soap

Highlight Problems
These are set within a context that will be of particular interest to students because of its timeliness, familiarity, or relevance to an important issue.

Additional End-of-Chapter Features
- Key Terms
- Review Questions
- Problems by Topic
- Cumulative Problems
- Conceptual Problems

VISUALIZING CHEMISTRY
CREATES DEEPER UNDERSTANDING

BY CONNECTING the macroscopic and microscopic worlds, visualizing concepts brings chemistry to life and creates a deeper understanding that will serve you throughout the course.

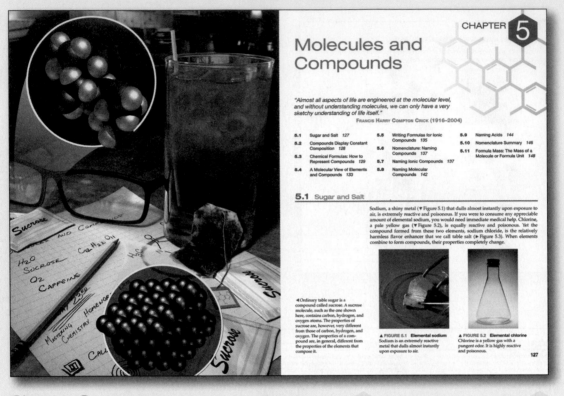

Chapter Openers

Every chapter opens by describing an everyday situation or practical application that demonstrates the importance of the material covered in that chapter. Each chapter-opening image combines macroscopic and molecular views that bring the content to life.

Interest Boxes

Four different types of interest boxes apply chemistry to everyday events and topics.

- **Chemistry in the Environment** boxes discuss environmental issues that are closely tied to chemistry, such as the reactions involved in ozone depletion.

- **Everyday Chemistry** boxes demonstrate the importance of chemistry in everyday situations, such as bleaching your hair.

- **Chemistry in the Media** boxes discuss chemical topics that have been in the news recently, such as the controversy over oxygenated fuels.

- **Chemistry and Health** boxes focus on personal health and fitness topics, as well as biomedical topics.

Microscopic image

Macroscopic image

$$2\,Na(s)\ +\ Cl_2(g)\ \longrightarrow\ 2\,NaCl(s)$$

Symbolic representation

▲ In a synthesis reaction, two simpler substances combine to make a more complex substance. In this series of photographs we see sodium metal and chlorine gas. When they combine, a chemical reaction occurs that forms sodium chloride.

Macroscopic to Microscopic Art

Many illustrations have three parts:

- **a macroscopic image**
  (what you can see with your eyes)

- **a microscopic image**
  (what the molecules are doing)

- **a symbolic representation**
  (how chemists represent the
  process with symbols and equations)

The goal is for you to connect what you see and experience with the molecules responsible, and with the way chemists represent those molecules.

▲ FIGURE 12.3 **Solids have a definite shape** In a solid such as ice, the molecules are fixed in place. However, they vibrate about fixed points.

▶ FIGURE 9.2 **Components of white light** Light is separated into its constituent colors—red, orange, yellow, green, blue, indigo, and violet—when it is passed through a prism.

Red

Prism

Slit

White light contains the entire spectrum of visible colors.

Light source

Violet

Labeling Figures

NEW! to the **Fourth Edition** is a consistent hierarchy for labeling figures. Up to three types of labels appear on the art: the most important in white shadow boxes; the second most important, in tinted boxes with no border; and the third, unboxed. This new feature helps you to efficiently navigate the figure, and to grasp its main points.

MASTERINGCHEMISTRY®
EXTENDS LEARNING BEYOND THE CLASSROOM

MasteringChemistry® is the most effective and widely used online tutorial, homework and assessment system for chemistry. It helps instructors maximize class time with customizable, easy-to-assign, and automatically graded assessments that motivate students to learn outside of class and arrive prepared for lecture. These assessments can easily be customized and personalized by instructors to suit their individual teaching style. To learn more visit: www.masteringchemistry.com

Student Tutorials

MasteringChemistry is the only system to provide instantaneous feedback specific to the most-common wrong answers. Students can submit an answer and receive immediate, error-specific feedback. Simpler sub-problems—"hints"—are provided upon request.

Gradebook

Every assignment is automatically graded. Shades of red highlight vulnerable students and challenging assignments.

Gradebook Diagnostic

This screen provides your favorite diagnostics. With a single click, charts summarize the most difficult problems, vulnerable students, grade distribution, and even score improvement over the course.

Pearson eText

Pearson eText provides access to the text when and wherever students have access to the Internet. eText pages look exactly like the printed text, offering powerful new functionality. Users can create notes, highlight text in different colors, create bookmarks, zoom, click hyperlinked words and phrases to view definitions, view as single or two-pages. eText also links to associated media files, enabling students to view an animation as they read the text. Pearson eText offers a full-text search and the ability to save and export notes.

NEW! Visualizations

These new tutorials, including PhET simulations, enable students to make connections between real-life phenomena and the underlying chemistry that explains such phenomena. The tutorials increase students' understanding of chemistry and clearly illustrate cause-and-effect relationships.

Reading Quizzes

Chapter-specific quizzes and activities focus on important, hard-to-grasp chemistry concepts.

CHAPTER 1

The Chemical World

"Imagination is more important than knowledge."

ALBERT EINSTEIN (1879–1955)

1.1 Soda Pop Fizz

Carbon dioxide molecule

Oxygen atom   Carbon atom   Oxygen atom

Water molecule

Hydrogen atom   Oxygen atom   Hydrogen atom

◄ Soda pop is a mixture of carbon dioxide and water and a few other substances that contribute flavor and color. When soda pop is poured into a glass, some of the carbon dioxide molecules come out of the mixture, producing the familiar fizz.

Open a can of soda pop and you hear the familiar "chchchch" of pressure release. Take a sip and you feel the carbon dioxide bubbles on your tongue. Shake the can before you open it and you will be sprayed with the bubbly liquid. A can of soda pop, like most familiar items in our daily lives, is a chemical mixture. Soda pop consists primarily of sugar, water, and carbon dioxide. It is the unique combination of these substances that gives soda pop its properties. Have you ever wondered why soda pop tastes sweet? To understand why, you need to understand sugar and solutions of sugar with water. We will learn about solutions in Chapter 13. Have you ever wondered why soda fizzes when you open it? To understand the reason, you need to understand gases and their ability to dissolve in liquids and how that ability changes with changing pressure. We will learn about gases in Chapter 11. And if you want to know how drinking too much soda pop makes you gain weight, you need to understand energy and the production of energy by chemical reactions. We will discuss energy in Chapter 3 and chemical reactions in Chapter 7. You need not go any farther than your own home and your own everyday experiences to encounter chemical questions. Chemicals compose virtually everything in our world: the soda; this book; your pencil; indeed, even your own body.

Chemists are particularly interested in the connections between the properties of substances and the properties of the particles that compose them. For example, why does soda pop fizz? Like all common substances, soda pop is ultimately composed of tiny particles called *atoms*. Atoms are so small that a single drop of soda pop contains about one billion trillion of them. In soda pop, as in many substances, these atoms are bound together to form several different types of *molecules*. The molecules important to fizzing are carbon dioxide and water. Carbon dioxide molecules consist of three atoms—one carbon and two oxygen atoms—held together in a straight line by chemical bonds. Water molecules also consist of three atoms—one oxygen and two hydrogen atoms—bonded together, but rather than being straight like carbon dioxide, the water molecule is bent.

▶ Virtually everything around you is composed of chemicals.

We will explore the nature of atoms, molecules, and chemical bonds more fully in later chapters. For now, think of atoms and molecules as tiny particles that compose all common matter, and chemical bonds as the attachments that hold atoms together.

The details of how atoms bond together to form a molecule—straight, bent, or some other shape—as well as the type of atoms in the molecule, determine *everything* about the substance that the molecule composes. The characteristics of water molecules make water a liquid at room temperature. The characteristics of carbon dioxide molecules make carbon dioxide a gas at room temperature. The characteristics of sugar molecules allow them to interact with our taste buds to produce the sensation of sweetness.

The makers of soda pop use *pressure* (the result of collisions between gaseous molecules and the surfaces around them) to force gaseous carbon dioxide molecules to mix with liquid water molecules. As long as the can of soda is sealed, the carbon dioxide molecules remain mixed with the water molecules, held there by pressure. When the can is opened, the pressure is released and carbon dioxide molecules escape out of the soda mixture (▼ Figure 1.1). As they do, they create bubbles—the familiar fizz of soda pop.

▶ FIGURE 1.1 **Where the fizz comes from** Bubbles in soda pop are pockets of carbon dioxide gas molecules escaping out of the liquid water.

▶ The public often has a very narrow view of chemicals, thinking of them only as dangerous poisons or pollutants.

1.2   Chemicals Compose Ordinary Things

▲ Chemists are interested in knowing why ordinary things, such as water, are the way they are. When a chemist sees a pitcher of water, she thinks of the molecules that compose the liquid and how those molecules determine its properties.

Is soda pop composed of chemicals? Yes. In fact, there is nothing you can hold or touch that is *not* made of chemicals. When most people think of chemicals, however, they envision a can of paint thinner in their garage, or they recall a headline about a river polluted by industrial waste. But chemicals compose more than just these things—they compose ordinary things, too. Chemicals compose the air we breathe and the water we drink. They compose toothpaste, Tylenol, and toilet paper. Chemicals make up virtually everything we come into contact with. Chemistry explains the properties and behavior of chemicals, in the broadest sense, by helping us understand the molecules that compose them.

As you experience the world around you, molecules are interacting to create your experience. Imagine watching a sunset. Molecules are involved in every step. Molecules in air interact with light from the sun, scattering away the blue and green light and leaving the red and orange light to create the color you see. Molecules in your eyes absorb that light and as a result are altered in a way that sends a signal to your brain. Molecules in your brain then interpret the signal to produce images and emotions. This whole process—mediated by molecules—creates the experience of seeing a sunset.

Chemists are interested in why ordinary substances are the way they are. Why is water a liquid at room temperature? Why is salt a solid? Why does soda fizz? Why is a sunset red? Throughout this book, you will learn the answers to these questions and many others. *You will learn the connections between the behavior of matter and the behavior of the particles that compose it.*

1.3   All Things Are Made of Atoms and Molecules

▲ Richard Feynman (1918–1988), Nobel Prize–winning physicist and popular professor at California Institute of Technology.

Professor Richard Feynman, in a lecture to first-year physics students at the California Institute of Technology, said that the most important idea in all human knowledge is that *all things are made of atoms*. Since atoms are usually bound together to form molecules, however, a chemist might add the concept of *molecules* to Feynman's bold assertion. This simple idea—that all things are made of atoms and molecules—explains much about our world and our experience of it. Atoms and molecules determine how matter behaves—if they were different, matter would be different. The nature of water molecules, for example, determines how water behaves. The nature of sugar molecules determines how sugar behaves, and the molecules that compose humans determine much about how our bodies behave.

There is a direct connection between the world of atoms and molecules and the world you and I experience every day. Chemists explore this connection. They seek to understand it. A good, simple definition of **chemistry** is *the science that tries to understand how matter behaves by studying how atoms and molecules behave.*

Chemistry—The science that seeks to understand what matter does by studying what atoms and molecules do.

1.4 The Scientific Method: How Chemists Think

Chemists use the **scientific method**—a way of learning that emphasizes observation and experimentation—to understand the world. The scientific method stands in contrast to ancient Greek philosophies that emphasized *reason* as the way to understand the world. Although the scientific method is not a rigid procedure that automatically leads to a definitive answer, it does have key characteristics that distinguish it from other ways of acquiring knowledge. These key characteristics include observation, the formulation of hypotheses, the testing of hypotheses by experiment, and the formulation of laws and theories.

The first step in acquiring scientific knowledge (▼ Figure 1.2) is often the **observation** or measurement of some aspect of nature. Some observations are simple, requiring nothing more than the naked eye. Other observations rely on the use of sensitive instrumentation. Occasionally, an important observation happens entirely by chance. Alexander Fleming, for example, discovered penicillin when he observed a bacteria-free circle around a certain mold that had accidentally grown on his culture plate. Regardless of how the observation occurs, it usually involves the measurement or description of some aspect of the physical world. For example, Antoine Lavoisier (1743–1794), a French chemist who studied combustion, burned substances in closed containers. He carefully measured the mass of each container and its contents before and after burning the substance inside, noting that there was no change in the mass during combustion. Lavoisier made an *observation* about the physical world.

> Combustion means burning. The mass of an object is a measure of the quantity of matter within it.

Observations often lead scientists to formulate a **hypothesis**, a tentative interpretation or explanation of the observations. Lavoisier explained his observations on combustion by hypothesizing that the process involved the combination of a substance with a component of air. A good hypothesis is *falsifiable*, which means that further testing has the potential to prove it wrong. Hypotheses are tested by **experiments**, highly controlled observations designed to validate or invalidate hypotheses. The results of an experiment may confirm a hypothesis or show it to be mistaken in some way. In the latter case, the hypothesis may have to be modified, or even discarded and replaced by an alternative. Either way, the new or revised hypothesis must also be tested through further experimentation.

Sometimes a number of similar observations lead to the development of a **scientific law**, a brief statement that synthesizes past observations and predicts future ones. For example, based on his observations of combustion, Lavoisier developed the **law of conservation of mass**, which states, "In a chemical reaction matter is neither created nor destroyed." This statement grew out of Lavoisier's observations, and it predicted the outcome of similar experiments on *any* chemical reaction. Laws are also subject to experiments, which can prove them wrong or validate them.

> Scientific theories are also called *models*.

One or more well-established hypotheses may form the basis for a scientific **theory**. Theories provide a broader and deeper explanation for observations and laws. They are models of the way nature is, and they often predict behavior that

▼ FIGURE 1.2 The scientific method.

▲ (Left) Painting of the French chemist Antoine Lavoisier and his wife, Marie, who helped him in his work by illustrating his experiments, recording results, and translating scientific articles from English. (*Source:* Jacques Louis David (French, 1748–1825). "Antoine-Laurent Lavoisier (1743–1794) and His Wife (Marie-Anne-Pierrette Paulze, 1758–1836)," 1788, oil on canvas, H. 102-1/4 in. W. 76-5/8 in. (259.7 × 194.6 cm). The Metropolitan Museum of Art, Purchase, Mr. and Mrs. Charles Wrightsman Gift, in honor of Everett Fahy, 1977. (1977.10) Image copyright © The Metropolitan Museum of Art.) (Right) John Dalton, the English chemist who formulated the atomic theory.

▲ **FIGURE 1.3 Are atoms real?** The atomic theory has 200 years of experimental evidence to support it, including recent images, such as this one, of atoms themselves. This image shows the Kanji characters for "atom" written with individual iron atoms on top of a copper surface.

extends well beyond the observations and laws on which they are founded. A good example of a theory is the **atomic theory** of John Dalton (1766–1844). Dalton explained the law of conservation of mass, as well as other laws and observations, by proposing that all matter was composed of small, indestructible particles called atoms. Dalton's theory was a model of the physical world—it went beyond the laws and observations of the time to explain these laws and observations.

Theories are also tested and validated by experiments. Notice that the scientific method begins with observation, and then laws, hypotheses, and theories are developed based on those observations. Experiments, which are carefully controlled observations, are then used to determine the validity of laws, hypotheses, or theories. If a law, hypothesis, or theory is inconsistent with the findings of an experiment, it must be revised and new experiments must be conducted to test the revisions. Over time, poor theories are eliminated and good theories—those consistent with experiments—remain. Established theories with strong experimental support are the most powerful pieces of scientific knowledge. People unfamiliar with science sometimes say, "That is just a theory," as if theories were mere speculations. However, well-tested theories are as close to truth as we get in science. For example, the idea that all matter is made of atoms is "just a theory," but it is a theory with 200 years of experimental evidence to support it, including the recent imaging of atoms themselves (◀ Figure 1.3). Established theories should not be taken lightly—they are the pinnacle of scientific understanding.

CONCEPTUAL CHECKPOINT 1.1

Which statement most resembles a scientific theory?

(a) When the pressure on a sample of oxygen gas is increased 10%, the volume of the gas decreases by 10%.

(b) The volume of a gas is inversely proportional to its pressure.

(c) A gas is composed of small particles in constant motion.

(d) A gas sample has a mass of 15.8 grams and a volume of 10.5 liters.

Note: The answers to all Conceptual Checkpoints appear at the end of the chapter.

EVERYDAY CHEMISTRY

Combustion and the Scientific Method

Early chemical theories attempted to explain common phenomena such as combustion. Why did things burn? What was happening to a substance when it burned? Could something that was burned be unburned? Early chemists burned different substances and made observations to try to answer these questions. They observed that substances would stop burning if placed in a closed container. They found that many metals would burn to form a white powder that they called a *calx* (now we know that these white powders are oxides of the metal) and that the metal could be recovered from the calx, or unburned, by combining the calx with charcoal and heating it.

Chemists in the first part of the eighteenth century formed a theory about combustion to explain these observations. In this theory, combustion involved a fundamental substance that they called *phlogiston*. This substance was present in anything that burned and was released during combustion. Flammable objects were flammable because they contained phlogiston. When things burned in a closed container, they didn't burn for very long because the space within the container became saturated with phlogiston. When things burned in the open, they continued to burn until all of the phlogiston within them was gone. This theory also explained how metals that had burned could be unburned. Charcoal was a phlogiston-rich material—they knew this because it burned so well—and when it was combined with a calx, which was a metal that had been emptied of its phlogiston, it transferred some of its phlogiston into the calx, converting it back into the unburned form of the metal. The phlogiston theory was consistent with all of the observations of the time and was widely accepted as valid.

Like any theory, the phlogiston theory was tested continually by experiment. One set of experiments, conducted in the mid-eighteenth century by Louis-Bernard Guyton de Morveau (1737–1816), consisted of weighing metals before and after burning them. In every case the metals *gained* weight when they were burned. This observation was inconsistent with the phlogiston theory, which predicted that they should *lose* weight because phlogiston was supposed to be lost during combustion. Clearly, the phlogiston theory needed modification.

The first modification proposed that phlogiston was a very light substance so that it actually "buoyed up" the materials that contained it. Thus when phlogiston was released, the material became heavier. Such a modification seemed to fit the observations but also seemed far-fetched. Antoine Lavoisier developed a more likely explanation by devising a completely new theory of combustion. Lavoisier proposed that, when a substance burned, it actually took something *out* of the air, and when it unburned, it released something back into the air. Lavoisier said that burning objects *fixed* (attached or bonded) the air and that the *fixed* air was released during unburning. In a confirming experiment (▼ Figure 1.4), Lavoisier roasted a mixture of calx and charcoal with the aid of sunlight focused by a giant burning lens, and found that a huge volume of "fixed air" was released in the process. The scientific method had worked. The phlogiston theory was proven wrong, and a new theory of combustion took its place—a theory that, with a few refinements, is still valid today.

CAN YOU ANSWER THIS? *What is the difference between a law and a theory? How does the example of the phlogiston theory demonstrate this difference?*

▶ **FIGURE 1.4 Focusing on combustion** The great burning lens belonging to the Academy of Sciences. Lavoisier used a similar lens in 1777 to show that a mixture of *calx* (metal oxide) and charcoal released a large volume of *fixed air* when heated.

1.5 A Beginning Chemist: How to Succeed

▲ To succeed as a scientist, you must have the curiosity of a child.

You are a beginning chemist. This may be your first chemistry course, but it is probably not your last. To succeed as a beginning chemist, keep the following ideas in mind. First, chemistry requires curiosity and imagination. If you are content knowing that the sky is blue, but don't care *why* it is blue, then you may have to rediscover your curiosity. I say "rediscover" because even children—or better said, *especially* children—have this kind of curiosity. To succeed as a chemist, you must have the curiosity and imagination of a child—*you must want to know the why of things.*

Second, chemistry requires calculation. Throughout this course, you will be asked to calculate answers and quantify information. *Quantification* involves measurement as part of observation—it is one of the most important tools in science. Quantification allows you to go beyond merely saying that this object is hot and that one is cold or that this one is large and that one is small. It allows you to specify the difference precisely. For example, two samples of water may feel equally hot to your hand, but when you measure their temperatures, you may find that one is 40 °C and the other is 44 °C. Even small differences can be important in a calculation or experiment, so assigning numbers to observations and manipulating those numbers become very important in chemistry.

Lastly, chemistry requires commitment. To succeed in this course, you must commit yourself to learning chemistry. Roald Hoffman, winner of the 1981 Nobel Prize for chemistry, said,

> I like the idea that human beings can do anything they want to. They need to be trained sometimes. They need a teacher to awaken the intelligence within them. But to be a chemist requires no special talent, I'm glad to say. Anyone can do it, with hard work.

Professor Hoffman is right. The key to success in this course is hard work—that requires commitment. You must do your work regularly and carefully. If you do, you will succeed, and you will be rewarded by seeing a whole new world—the world of molecules and atoms. This world exists beneath the surface of nearly everything you encounter. I welcome you to this world and consider it a privilege, together with your professor, to be your guide.

CHAPTER IN REVIEW

CHEMICAL PRINCIPLES

Matter and Molecules: Chemists are interested in all matter, even ordinary matter such as water or air. You don't need to go to a chemical storeroom to find chemical questions. Chemicals are all around you. Chemistry is the science that tries to understand what matter does by understanding what molecules do.

RELEVANCE

Matter and Molecules: Chemists want to understand matter for several reasons. First, chemists are simply curious—they want to know why. Why are some substances reactive and others not? Why are some substances gases, some liquids, and others solids? Chemists are also practical; they want to understand matter so that they can control it and produce substances that are useful to society and to humankind.

The Scientific Method: Chemists employ the scientific method, which makes use of observations, hypotheses, laws, theories, and experiments. Observations involve measuring or observing some aspect of nature. Hypotheses are tentative interpretations of the observations. Laws summarize the results of a large number of observations, and theories are models that explain and give the underlying causes for observations and laws. Hypotheses, laws, and theories must be tested and validated by experiment. If they are not confirmed, they are revised and tested through further experimentation.

The Scientific Method: The scientific method is a way to understand the world. Since the inception of the scientific method, knowledge about the natural world has grown rapidly. The application of the scientific method has produced technologies that have raised living standards throughout the world with advances such as increased food production, rapid transportation, unparalleled access to information, and longer life spans.

Success as a Beginning Chemist: To succeed as a beginning chemist, you must be curious and imaginative, be willing to do calculations, and be committed to learning the material.

Success as a Beginning Chemist: Understanding chemistry will give you a deeper appreciation for the world in which you live, and if you choose science as a career, it will be a foundation upon which you will continue to build.

KEY TERMS

atomic theory [1.4]
chemistry [1.3]
experiment [1.4]

hypothesis [1.4]
law of conservation
    of mass [1.4]

observation [1.4]
scientific law [1.4]
scientific method [1.4]

theory [1.4]

EXERCISES

QUESTIONS

Answers to all questions numbered in blue appear in the Answers section at the back of the book.

1. Why does soda fizz?
2. What are chemicals? Give some examples.
3. What do chemists try to do? How do they understand the natural world?
4. What is meant by the statement, "Matter does what molecules do"? Give an example.
5. Define *chemistry*.
6. How is chemistry connected to everyday life? How is chemistry relevant outside the chemistry laboratory?
7. Explain the scientific method.
8. Give an example from this chapter of the scientific method at work.

9. What is the difference between a law and a theory?
10. What is the difference between a hypothesis and a theory?
11. What is wrong with the statement, "It is just a theory"?
12. What is the law of conservation of mass, and who discovered it?
13. What is the atomic theory, and who formulated it?
14. What are three things you need to do to succeed in this course?

PROBLEMS

Note: The exercises in the Problems section are paired, and the answers to the odd-numbered exercises (numbered in blue) appear in the Answers section at the back of the book.

15. Examine the opening figure of this chapter. Use the information in Section 1.1 to identify the two molecules sitting next to the cola glass and identify each of the atoms within each molecule.

16. Examine Figure 1.1 and, from a molecular point of view, explain why soda pop fizzes. What molecules are inside the bubbles in a glass of soda pop?

17. Classify each statement as an observation, a law, or a theory.

    (a) When a metal is burned in a closed container, the sum of the masses of the container and its contents do not change.

    (b) Matter is made of atoms.

    (c) Matter is conserved in chemical reactions.

    (d) When wood is burned in a closed container, its mass does not change.

18. Classify each statement as an observation, a law, or a theory.

    (a) The star closest to Earth is moving away from Earth at high speed.

    (b) A body in motion stays in motion unless acted upon by a force.

    (c) The universe began as a cosmic explosion called the Big Bang.

    (d) A stone dropped from an altitude of 450 m falls to the ground in 9.6 s.

19. A student prepares several samples of the same gas and measures their mass and volume. The results are tabulated as follows. Formulate a tentative law from the measurements.

Mass of Gas (in grams)	Volume of Gas (in L)
22.5	1.60
35.8	2.55
70.2	5.00
98.5	7.01

20. A student measures the volume of a gas sample at several different temperatures. The results are tabulated as follows. Formulate a tentative law from the measurements.

Temperature of Gas (in Kelvin)	Volume of Gas (in L)
298	4.55
315	4.81
325	4.96
335	5.11

21. A chemist in an imaginary universe does an experiment that attempts to correlate the size of an atom with its chemical reactivity. The results are tabulated as follows.

Size of Atom	Chemical Reactivity
small	low
medium	intermediate
large	high

    (a) Formulate a law from this data.

    (b) Formulate a theory to explain the law.

22. A chemist decomposes several samples of water into hydrogen and oxygen and weighs (or more correctly "measures the mass of") the hydrogen and the oxygen obtained. The results are tabulated as follows.

Sample Number	Grams of Hydrogen	Grams of Oxygen
1	1.5	12
2	2	16
3	2.5	20

    (a) Summarize these observations in a short statement.

    Next, the chemist decomposes several samples of carbon dioxide into carbon and oxygen. The results are tabulated as follows.

Sample Number	Grams of Carbon	Grams of Oxygen
1	0.5	1.3
2	1.0	2.7
3	1.5	4.0

    (b) Summarize these observations in a short statement.

    (c) Formulate a law from the observations in (a) and (b).

    (d) Formulate a theory that might explain your law in (c).

▶ANSWERS TO CONCEPTUAL CHECKPOINTS

1.1 (c) Answers (a) and (d) are observations. Answer (b) is a scientific law. Answer (c) is the only answer that proposes a *model* for what a gas is like.

CHAPTER 2

Measurement and Problem Solving

"The important thing in science is not so much to obtain new facts as to discover new ways of thinking about them."

SIR WILLIAM LAWRENCE BRAGG (1890–1971)

2.1 Measuring Global Temperatures

Global warming has become a household term. Average global temperatures affect things from agriculture to weather and ocean levels. The media report that global temperatures are increasing. These reports are based on the work of scientists who—after analyzing records from thousands of temperature-measuring stations around the world—concluded that average global temperatures have risen by 0.6 °C in the last century.

Notice how the scientists reported their results. What if they had reported a temperature increase of simply 0.6 without any *units*? The result would be unclear. Units are extremely important in reporting and working with scientific measurements, and they must always be included. Suppose that the scientists had included additional zeros in their results—for example, 0.60 °C or 0.600 °C—or that they had reported the number their computer displayed after averaging many measurements, something like 0.58759824 °C. Would these convey the same information? Not really. Scientists agree to a standard way of reporting measured quantities in which the number of reported digits reflects the precision in the measurement—more digits, more precision; fewer digits, less precision. Numbers are usually written so that the uncertainty is indicated by the last reported digit. For example, by reporting a temperature increase of 0.6 °C, the scientists mean 0.6 ± 0.1 °C (± means plus or minus). The temperature rise could be as much as 0.7 °C or as little as 0.5 °C, but it is not 1.0 °C. The degree of certainty in this particular measurement is critical, influencing political decisions that directly affect people's lives.

▌ A unit is a standard, agreed-on quantity by which other quantities are measured.

◀ The graph in this image displays average global temperatures (relative to the mean) over the past 100 years.

2.2 Scientific Notation: Writing Large and Small Numbers

▲ Lasers such as this one can measure time periods as short as 1×10^{-15} s.

Science has constantly pushed the boundaries of the very large and the very small. We can, for example, now measure time periods as short as 0.000000000000001 seconds and distances as great as 14,000,000,000 light-years. Because the many zeros in these numbers are cumbersome to write, scientists use **scientific notation** to write them more compactly. In scientific notation, 0.000000000000001 is 1×10^{-15}, and 14,000,000,000 is 1.4×10^{10}. A number written in scientific notation consists of a **decimal part,** a number that is usually between 1 and 10, and an **exponential part,** 10 raised to an **exponent, *n.***

$$1.2 \times 10^{-10} \leftarrow \text{exponent } (n)$$

decimal part   exponential part

A positive exponent means 1 multiplied by 10 *n* times.

$$10^0 = 1$$
$$10^1 = 1 \times 10 = 10$$
$$10^2 = 1 \times 10 \times 10 = 100$$
$$10^3 = 1 \times 10 \times 10 \times 10 = 1000$$

A negative exponent ($-n$) means 1 divided by 10 *n* times.

$$10^{-1} = \frac{1}{10} = 0.1$$

$$10^{-2} = \frac{1}{10 \times 10} = 0.01$$

$$10^{-3} = \frac{1}{10 \times 10 \times 10} = 0.001$$

To convert a number to scientific notation, move the decimal point (either to the left or to the right, as needed) to obtain a number between 1 and 10 and then multiply that number (the decimal part) by 10 raised to the power that reflects the movement of the decimal point. For example, to write 5983 in scientific notation, move the decimal point to the left three places to get 5.983 (a number between 1 and 10) and then multiply the decimal part by 1000 to compensate for moving the decimal point.

$$5983 = 5.983 \times 1000$$

1000 is 10^3

$$= 5.983 \times 10^3$$

You can do this in one step by counting how many places you move the decimal point to obtain a number between 1 and 10 and then writing the decimal part multiplied by 10 raised to the number of places you moved the decimal point.

$$5983 = 5.983 \times 10^3$$

321

If the decimal point is moved to the left, as in the previous example, the exponent is positive. If the decimal is moved to the right, the exponent is negative.

$$0.00034 = 3.4 \times 10^{-4}$$

To express a number in scientific notation:

1. Move the decimal point to obtain a number between 1 and 10.
2. Write the result from Step 1 multiplied by 10 raised to the number of places you moved the decimal point.
   - The exponent is positive if you moved the decimal point to the left.
   - The exponent is negative if you moved the decimal point to the right.

EXAMPLE 2.1   Scientific Notation

The 2010 U.S. population was estimated to be 308,255,000 people. Express this number in scientific notation.

To obtain a number between 1 and 10, move the decimal point to the left 8 decimal places; the exponent is 8. Since you move the decimal point to the left, the sign of the exponent is positive.

SOLUTION

$$308{,}255{,}000 \text{ people} = 3.08255 \times 10^{8} \text{ people}$$

▶**SKILLBUILDER 2.1 | Scientific Notation**

The total U.S national debt in 2010 was approximately $12,102,000,000,000. Express this number in scientific notation.

Note: The answers to all Skillbuilders appear at the end of the chapter.

▶**FOR MORE PRACTICE**   Example 2.18; Problems 31, 32.

EXAMPLE 2.2   Scientific Notation

The radius of a carbon atom is approximately 0.000000000070 m. Express this number in scientific notation.

To obtain a number between 1 and 10, move the decimal point to the right 11 decimal places; therefore, the exponent is 11. Since the decimal point was moved to the right, the sign of the exponent is negative.

SOLUTION

$$0.000000000070 \text{ m} = 7.0 \times 10^{-11} \text{ m}$$

▶**SKILLBUILDER 2.2 | Scientific Notation**

Express the number 0.000038 in scientific notation.

▶**FOR MORE PRACTICE**   Problems 33, 34.

✓ CONCEPTUAL CHECKPOINT 2.1

The radius of a dust speck is 4.5×10^{-3} mm. What is the correct value of this number in decimal notation (i.e., express the number without using scientific notation)?

(a) 4500 mm

(b) 0.045 mm

(c) 0.0045 mm

(d) 0.00045 mm

Note: The answers to all Conceptual Checkpoints appear at the end of the chapter.

2.3 Significant Figures: Writing Numbers to Reflect Precision

▲ Since pennies come in whole numbers, 7 pennies means 7.00000… pennies. This is an exact number and therefore never limits significant figures in calculations.

If we tell someone we have seven pennies, our meaning is clear. Pennies come in whole numbers, and seven pennies means seven whole pennies—it is unlikely that we would have 7.4 pennies. However, if we tell someone that we have a 10-g gold bar, the meaning is *unclear*. Our knowledge of the actual amount of gold in the bar depends on how precisely it was measured, which in turn depends on the scale or balance used to make the measurement. As we just learned, measured quantities are written to reflect the uncertainty in the measurement. If the gold measurement was rough, we could describe the bar as containing "10 g of gold." If a more precise balance was used, we could write the gold content as "10.0 g." We would report an even more precise measurement as "10.00 g."

Scientific numbers are reported so that every digit is certain except the last, which is estimated. For example, suppose a reported measurement is:

$$\underline{45.87}2$$

↑   ↑
certain   *estimated*

The first four digits are certain; the last digit is estimated.

Suppose that we weigh an object on a balance with marks at every 1 g, and the pointer is between the 1-g mark and the 2-g mark (▼ Figure 2.1) but much closer to the 1-g mark. To record the measurement, we mentally divide the space between the 1- and 2-g marks into 10 equal spaces and estimate the position of the pointer. In this case, the pointer indicates about 1.2 g. We then write the measurement as 1.2 g, indicating that we are sure of the "1" but have estimated the ".2."

▲ Our knowledge of the amount of gold in a 10-g gold bar depends on how precisely it was measured.

If we measure the same object using a balance with marks every tenth of a gram, we need to write the result with more digits. For example, suppose that on this more precise balance the pointer is between the 1.2-g mark and the 1.3-g mark (▼ Figure 2.2). We again divide the space between the two marks into 10 equal spaces and estimate the third digit. In the case of the nut shown in Figure 2.2, we report 1.26 g. Digital balances usually have readouts that report the mass to the correct number of digits.

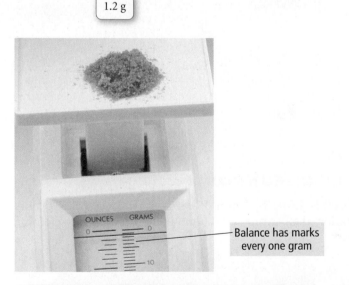

1.2 g

Balance has marks every one gram

▲ **FIGURE 2.1 Estimating tenths of a gram** This balance has markings every 1 g, so we estimate to the tenths place. To estimate between markings, mentally divide the space into 10 equal spaces and estimate the last digit. This reading is 1.2 g.

1.26 g

Balance has marks every tenth of a gram

▲ **FIGURE 2.2 Estimating hundredths of a gram** Since this scale has markings every 0.1 g, we estimate to the hundredths place. The correct reading is 1.26 g.

EXAMPLE 2.3   Reporting the Right Number of Digits

The bathroom scale in ▼ Figure 2.3 has markings at every 1 lb. Report the reading to the correct number of digits.

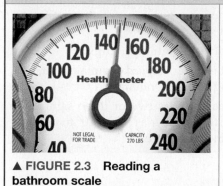

▲ **FIGURE 2.3   Reading a bathroom scale**

SOLUTION

Since the pointer is between the 147- and 148-lb markings, mentally divide the space between the markings into 10 equal spaces and estimate the next digit. In this case, the result should be reported as:

147.7 lb

What if you estimated a little differently and wrote 147.6 lb? In general, one unit of difference in the last digit is acceptable because the last digit is estimated and different people might estimate it slightly differently. However, if you wrote 147.2 lb, you would clearly be wrong.

▶**SKILLBUILDER 2.3 | Reporting the Right Number of Digits**

A thermometer is used to measure the temperature of a backyard hot tub, and the reading is shown in ▼ Figure 2.4. Write the temperature reading to the correct number of digits.

◀ **FIGURE 2.4   Reading a thermometer**

▶**FOR MORE PRACTICE**   Example 2.19; Problems 41, 42.

COUNTING SIGNIFICANT FIGURES

The non–place-holding digits in a measurement are **significant figures** (or **significant digits**) and, as we have seen, represent the precision of a measured quantity. The greater the number of significant figures, the greater the precision of the measurement. We can determine the number of significant figures in a written number fairly easily; however, if the number contains zeros, we must distinguish between the zeros that are significant and those that simply mark the decimal place. In the number 0.002, for example, the leading zeros simply mark the decimal place; they *do not* add to the precision of the measurement. In the number 0.00200, the trailing zeros *do* add to the precision of the measurement.

To determine the number of significant figures in a number, follow these rules:

1. All nonzero digits are significant.

1.05   0.0110

2. Interior zeros (zeros between two numbers) are significant.

4.0208   50.1

3. Trailing zeros (zeros to the right of a nonzero number) that fall after a decimal point are significant.

5.10   3.00

4. Trailing zeros that fall before a decimal point are significant.

$$\underline{50.00} \qquad 17\underline{00}.24$$

When a number is expressed in scientific notation, all trailing zeros are significant.

5. Leading zeros (zeros to the left of the first nonzero number) are not significant. They only serve to locate the decimal point.

   For example, the number 0.0005 has only one significant digit.

6. **Trailing** zeros at the end of a number, but before an *implied* decimal point, are ambiguous and should be avoided by using scientific notation.

Some books put a decimal point after one or more trailing zeros if the zeros are to be considered significant. We avoid that practice in this book, but you should be aware of it.

   For example, it is unclear if the number 350 has two or three significant figures. We can avoid confusion by writing the number as 3.5×10^2 to indicate two significant figures or as 3.50×10^2 to indicate three.

EXACT NUMBERS

Exact numbers have an unlimited number of significant figures. Exact numbers originate from three sources:

- Exact counting of discrete objects. For example, 3 atoms means 3.00000...atoms.
- *Defined quantities*, such as the number of centimeters in 1 m. Because 100 cm is defined as 1 m,

$$100\,\text{cm} = 1\,\text{m means } 100.00000\ldots\text{cm} = 1.0000000\ldots\text{m}$$

  Note that some conversion factors are defined quantities whereas others are not.

- Integral numbers that are part of an equation. For example, in the equation, $radius = \dfrac{diameter}{2}$, the number 2 is exact and therefore has an unlimited number of significant figures.

EXAMPLE 2.4   Determining the Number of Significant Figures in a Number

How many significant figures are in each number?

(a) 0.0035

(b) 1.080

(c) 2371

(d) 2.97×10^5

(e) 1 dozen = 12

(f) 100.00

(g) 100,000

	SOLUTION
The 3 and the 5 are significant (rule 1). The leading zeros only mark the decimal place and are not significant (rule 5).	**(a)** 0.0035 two significant figures
The interior zero is significant (rule 2), and the trailing zero is significant (rule 3). The 1 and the 8 are also significant (rule 1).	**(b)** 1.080 four significant figures
All digits are significant (rule 1).	**(c)** 2371 four significant figures
All digits in the decimal part are significant (rule 1).	**(d)** 2.97×10^5 three significant figures
Defined numbers are exact and therefore have an unlimited number of significant figures.	**(e)** 1 dozen = 12 unlimited significant figures
The 1 is significant (rule 1), and the trailing zeros before the decimal point are significant (rule 4). The trailing zeros after the decimal point are also significant (rule 3).	**(f)** 100.00 five significant figures
This number is ambiguous. Write as 1×10^5 to indicate one significant figure or as 1.00000×10^5 to indicate six significant figures.	**(g)** 100,000 ambiguous

▶**SKILLBUILDER 2.4 | Determining the Number of Significant Figures in a Number**

How many significant figures are in each number?

(a) 58.31
(b) 0.00250
(c) 2.7×10^3
(d) $1\,cm = 0.01\,m$
(e) 0.500
(f) 2100

▶**FOR MORE PRACTICE**   Example 2.20; Problems 43, 44, 45, 46, 47, 48.

  CONCEPTUAL CHECKPOINT 2.2

A researcher reports that the Spirit rover on the surface of Mars recently measured the temperature to be $-25.49\ °F$. What is the actual temperature?

(a) between $-25.490\ °F$ and $-25.499\ °F$
(b) between $-25.48\ °F$ and $-25.50\ °F$
(c) between $-25.4\ °F$ and $-25.5\ °F$
(d) exactly $-25.49\ °F$

2.4 Significant Figures in Calculations

When we use measured quantities in calculations, the results of the calculation must reflect the precision of the measured quantities. We should not lose or gain precision during mathematical operations.

MULTIPLICATION AND DIVISION

In multiplication or division, the result carries the same number of significant figures as the factor with the fewest significant figures.

For example:

$$5.02 \times 89.665 \times 0.10 = 45.0118 = 45$$

(3 sig. figures) (5 sig. figures) (2 sig. figures)   (2 sig. figures)

The intermediate result (in blue) is rounded to two significant figures to reflect the least precisely known factor (0.10), which has two significant figures.

In division, we follow the same rule.

$$5.892 \div 6.10 = 0.96590 = 0.966$$

(4 sig. figures) (3 sig. figures)   (3 sig. figures)

The intermediate result (in blue) is rounded to three significant figures to reflect the least precisely known factor (6.10), which has three significant figures.

ROUNDING

When we round to the correct number of significant figures:

we round down if the last (or leftmost) digit dropped is 4 or less;
we round up if the last (or leftmost) digit dropped is 5 or more.

CHEMISTRY IN THE MEDIA

The COBE Satellite and Very Precise Measurements That Illuminate Our Cosmic Past

Since the earliest times, humans have wondered about the origins of our planet. Science has slowly probed this question and has developed theories for how the universe and the Earth began. The most accepted theory today about the origin of the universe is the Big Bang theory. According to the Big Bang theory, the universe began in a tremendous expansion about 13.7 billion years ago and has been expanding ever since. A measurable prediction of this theory is the presence of a remnant "background radiation" from the expansion of the universe. That remnant is characteristic of the current temperature of the universe. When the Big Bang occurred, the temperature of the universe was very hot and the associated radiation very bright. Today, 13.7 billion years later, the temperature of the universe is very cold and the background radiation very faint.

In the early 1960s, Robert H. Dicke, P. J. E. Peebles, and their coworkers at Princeton University began to build a device to measure this background radiation and thus take a direct look into the cosmological past and provide evidence for the Big Bang theory. At about the same time, quite by accident, Arno Penzias and Robert Wilson of Bell Telephone Laboratories measured excess radio noise on one of their communications satellites. As it turned out, this noise was the background radiation that the Princeton scientists were looking for. The two groups published papers together in 1965 reporting their findings along with the corresponding current temperature of the universe, about 3 degrees above absolute zero, or 3 K. We will define temperature measurement scales in Chapter 3. For now, know that 3 K is an extremely low temperature (460 degrees below zero on the Fahrenheit scale).

In 1989, the Cosmic Background Explorer (COBE) satellite was developed by NASA's Goddard Space Flight Center to measure the background radiation more precisely. The COBE satellite determined that the background radiation corresponded to a universe with a temperature of 2.735 K. (Notice the difference in significant figures from the previous measurement.) It went on to measure tiny fluctuations in the background radiation that amount to temperature differences of 1 part in 100,000. These fluctuations, though small, are an important prediction of the Big Bang theory. Scientists announced that the COBE satellite had produced the strongest evidence yet for the Big Bang theory of the creation of the universe. This is the way that science works. Measurement, and precision in measurement, are important to understanding the world—so important that we dedicate most of this chapter just to the concept of measurement.

CAN YOU ANSWER THIS? *How many significant figures are there in each of the preceding temperature measurements (3 K, 2.735 K)?*

▲ The COBE Satellite, launched in 1989 to measure background radiation. Background radiation is a remnant of the Big Bang—the expansion that is believed to have formed the universe.

Consider rounding each of these numbers to two significant figures.

2.33 rounds to 2.3
2.37 rounds to 2.4
2.34 rounds to 2.3
2.35 rounds to 2.4

We use only the *last (or leftmost) digit being dropped* to decide in which direction to round—we ignore all digits to the right of it. For example, to round 2.349 to two significant figures, only the 4 in the hundredths place (2.349) determines which direction to round—the 9 is irrelevant.

2.349 rounds to 2.3

For calculations involving multiple steps, we round only the final answer—we do not round off between steps. This prevents small rounding errors from affecting the final answer.

EXAMPLE 2.5   Significant Figures in Multiplication and Division

Perform each calculation to the correct number of significant figures.

(a) $1.01 \times 0.12 \times 53.51 \div 96$
(b) $56.55 \times 0.920 \div 34.2585$

Round the intermediate result (in blue) to two significant figures to reflect the two significant figures in the least precisely known quantities (0.12 and 96).	**SOLUTION** **(a)** $1.01 \times 0.12 \times 53.51 \div 96 = 0.067556 = 0.068$
Round the intermediate result (in blue) to three significant figures to reflect the three significant figures in the least precisely known quantity (0.920).	**(b)** $56.55 \times 0.920 \div 34.2585 = 1.51863 = 1.52$

▶**SKILLBUILDER 2.5 | Significant Figures in Multiplication and Division**

Perform each calculation to the correct number of significant figures.

   (a) $1.10 \times 0.512 \times 1.301 \times 0.005 \div 3.4$

   (b) $4.562 \times 3.99870 \div 89.5$

▶**FOR MORE PRACTICE**   Examples 2.21, 2.22; Problems 57, 58, 59, 60.

ADDITION AND SUBTRACTION

In addition or subtraction, the result carries the same number of decimal places as the quantity carrying the fewest decimal places.

For example:

$$
\begin{array}{r}
5.74 \\
0.823 \\
+\,2.651 \\
\hline
9.214 = 9.21
\end{array}
$$

It is sometimes helpful to draw a vertical line directly to the right of the number with the fewest decimal places. The line shows the number of decimal places that should be in the answer.

We round the intermediate answer (in blue) to two decimal places because the quantity with the fewest decimal places (5.74) has two decimal places.

For subtraction, we follow the same rule. For example:

$$
\begin{array}{r}
4.8 \\
-\,3.965 \\
\hline
0.835 = 0.8
\end{array}
$$

We round the intermediate answer (in blue) to one decimal place because the quantity with the fewest decimal places (4.8) has one decimal place. Remember: *For multiplication and division, the quantity with the fewest **significant figures** determines the number of significant figures in the answer. For addition and subtraction, the quantity with the fewest **decimal places** determines the number of decimal places in* *the answer.* In multiplication and division we focus on significant figures, but in addition and subtraction we focus on decimal places. When a problem involves addition and subtraction, the answer may have a different number of significant figures than the initial quantities. For example:

$$28.02 - 27.99 = 0.03$$

Initial quantities each have four significant figures

Answer has only one significant figure

The answer has only one significant figure, even though the initial quantities each had four significant figures.

EXAMPLE 2.6   Significant Figures in Addition and Subtraction

Perform the calculations to the correct number of significant figures.

(a)
```
     0.987
  +125.1
    −1.22
```

(b)
```
    0.765
   −3.449
   −5.98
```

	SOLUTION

Round the intermediate answer (in blue) to one decimal place to reflect the quantity with the fewest decimal places (125.1). Notice that 125.1 is not the quantity with the fewest significant figures—it has four while the other quantities only have three—but because it has the fewest decimal places, it determines the number of decimal places in the answer.	**(a)** 0.987
	  +125.1
	    −1.22
	  124.867 = 124.9

Round the intermediate answer (in blue) to two decimal places to reflect the quantity with the fewest decimal places (5.98).	**(b)** 0.765
	  −3.449
	  −5.98
	  −8.664 = −8.66

▶ **SKILLBUILDER 2.6 | Significant Figures in Addition and Subtraction**

Perform the calculations to the correct number of significant figures.

(a)
```
     2.18
   +5.621
   +1.5870
    −1.8
```

(b)
```
    7.876
   −0.56
 +123.792
```

▶ **FOR MORE PRACTICE**   Example 2.23; Problems 61, 62, 63, 64.

CALCULATIONS INVOLVING BOTH MULTIPLICATION/DIVISION AND ADDITION/SUBTRACTION

In calculations involving both multiplication/division and addition/subtraction, we do the steps in parentheses first; determine the correct number of significant figures in the intermediate answer; then do the remaining steps.

For example:

$$3.489 \times (5.67 - 2.3)$$

We complete the subtraction step first.

$$5.67 - 2.3 = 3.37$$

We use the subtraction rule to determine that the intermediate answer (3.37) has only one significant decimal place. To avoid small errors, it is best not to round at this point; instead, we underline the least significant figure as a reminder.

$$= 3.489 \times 3.\underline{3}7$$

We then do the multiplication step.

$$3.489 \times 3.\underline{3}7 = 11.758 = 12$$

We use the multiplication rule to determine that the intermediate answer (11.758) rounds to two significant figures (12) because it is limited by the two significant figures in 3.$\underline{3}$7.

EXAMPLE 2.7   Significant Figures in Calculations Involving Both Multiplication/Division and Addition/Subtraction

Perform the calculations to the correct number of significant figures.

(a) $6.78 \times 5.903 \times (5.489 - 5.01)$
(b) $19.667 - (5.4 \times 0.916)$

Do the step in parentheses first. Use the subtraction rule to mark 0.479 to two decimal places since 5.01, the number in the parentheses with the least number of decimal places, has two.	**SOLUTION**
	(a) $6.78 \times 5.903 \times (5.489 - 5.01)$
	$= 6.78 \times 5.903 \times (0.479)$
Then perform the multiplication and round the answer to two significant figures since the number with the least number of significant figures has two.	$= 6.78 \times 5.903 \times 0.4\underline{7}9$
	$6.78 \times 5.903 \times 0.4\underline{7}90 = 19.1707$
	$= 19$
Do the step in parentheses first. The number with the least number of significant figures within the parentheses (5.4) has two, so mark the answer to two significant figures.	**(b)** $19.667 - (5.4 \times 0.916)$
	$= 19.667 - (4.9464)$
	$= 19.667 - 4.9\underline{4}64$
Then perform the subtraction and round the answer to one decimal place since the number with the least number of decimal places has one.	$19.667 - 4.9\underline{4}64 = 14.7206$
	$= 14.7$

▶ **SKILLBUILDER 2.7 | Significant Figures in Calculations Involving Both Multiplication/Division and Addition/Subtraction**

Perform each calculation to the correct number of significant figures.

(a) $3.897 \times (782.3 - 451.88)$

(b) $(4.58 \div 1.239) - 0.578$

▶ **FOR MORE PRACTICE**   Example 2.24; Problems 65, 66, 67, 68.

✓ CONCEPTUAL CHECKPOINT 2.3

Which calculation would have its result reported to the *greater* number of significant figures?

(a) 3 + (15/12)

(b) (3 + 15)/12

2.5 The Basic Units of Measurement

By themselves, numbers have limited meaning. Read this sentence: When my son was 7 he walked 3, and when he was 4 he threw his baseball 8 and said his school was 5 away. The sentence is confusing because we don't know what the numbers mean—the **units** are missing. The meaning becomes clear, however, when we add the missing units to the numbers: When my son was 7 *months old* he walked 3 *steps*, and when he was 4 *years old* he threw his baseball 8 *feet* and said his school was 5 *minutes* away. Units make all the difference. In chemistry, units are critical. Never write a number by itself; always use its associated units—otherwise your work will be as confusing as the initial sentence.

The two most common unit systems are the **English system,** used in the United States, and the **metric system,** used in most of the rest of the world. The English system uses units such as inches, yards, and pounds, while the metric system uses centimeters, meters, and kilograms. The most convenient system for science measurements is based on the metric system and is called the **International System** of units or **SI units.** SI units are a set of standard units agreed on by scientists throughout the world.

| The abbreviation *SI* comes from the French *le Système International.*

THE STANDARD UNITS

TABLE 2.1 Important SI Standard Units

Quantity	Unit	Symbol
Length	meter	m
Mass	kilogram	kg
Time	second	s
Temperature*	kelvin	K

*Temperature units are discussed in Chapter 3.

Table 2.1 lists the standard units in the SI system. They include the **meter (m)** as the standard unit of length; the **kilogram (kg)** as the standard unit of mass; and the **second (s)** as the standard unit of time. Each of these standard units is precisely defined. The meter is defined as the distance light travels in a certain period of

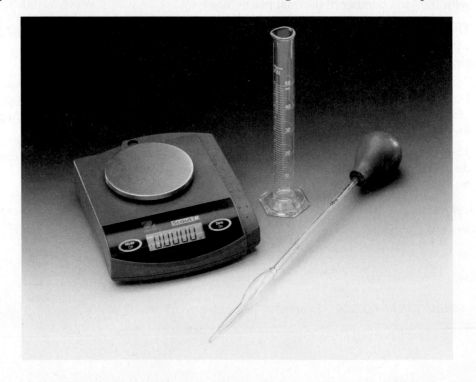

► Science uses instruments to make measurements. Every instrument is calibrated in a particular unit without which the measurements would be meaningless.

▲ **FIGURE 2.5   The standard of length**   The definition of a meter, established by international agreement in 1983, is the distance that light travels in vacuum in 1/299,792,458 s.   **Question:** Why is such a precise standard necessary?

▲ **FIGURE 2.6   The standard of mass**   A duplicate of the international standard kilogram, called kilogram 20, is kept at the National Institute of Standards and Technology near Washington, DC.

▲ **FIGURE 2.7   The standard of time**   The second is defined, using an atomic clock, as the duration of 9,192,631,770 periods of the radiation emitted from a certain transition in a cesium-133 atom.

A nickel (5 cents) has a mass of about 5 grams.

time: 1/299,792,458 s (▲ Figure 2.5). (The speed of light is 3.0×10^8 m/s.) The kilogram is defined as the mass of a block of metal kept at the International Bureau of Weights and Measures at Sèvres, France (▲ Figure 2.6). The second is defined using an atomic standard (▲ Figure 2.7).

Most people are familiar with the SI standard unit of time, the second. However, if you live in the United States, you may be less familiar with the meter and the kilogram. The meter is slightly longer than a yard (a yard is 36 in. while a meter is 39.37 in.). A 100-yd football field measures only 91.4 m.

The kilogram is a measure of mass, which is different from weight. The **mass** of an object is a measure of the quantity of matter within it, whereas the weight of an object is a measure of the gravitational pull on that matter. Consequently, weight depends on gravity while mass does not. If you were to weigh yourself on Mars, for example, the lower gravity would pull you toward the scale less than Earth's gravity would, resulting in a lower weight. A 150-lb person on Earth weighs only 57 lb on Mars. However, the person's mass, the quantity of matter in his or her body, remains the same. A kilogram of mass is the equivalent of 2.205 lb of weight on Earth, so if we express mass in kilograms, a 150-lb person on Earth has a mass of approximately 68 kg. A second common unit of mass is the gram (g), defined as follows:

$$1000\,g = 10^3\,g = 1\,kg$$

PREFIX MULTIPLIERS

The SI system employs **prefix multipliers** (Table 2.2) with the standard units. These multipliers change the value of the unit by powers of 10. For example, the kilometer (km) has the prefix *kilo-*, meaning 1000 or 10^3. Therefore:

$$1\,km = 1000\,m = 10^3\,m$$

Similarly, the millisecond (ms) has the prefix *milli-*, meaning 0.001 or 10^{-3}.

$$1\,ms = 0.001\,s = 10^{-3}\,s$$

▲ The diameter of a quarter is about 2.4 cm.   **Question:** Why would you *not* use meters to make this measurement?

TABLE 2.2   SI Prefix Multipliers

Prefix	Symbol	Multiplier	
tera-	T	1,000,000,000,000	(10^{12})
giga-	G	1,000,000,000	(10^{9})
mega-	M	1,000,000	(10^{6})
kilo-	k	1,000	(10^{3})
deci-	d	0.1	(10^{-1})
centi-	c	0.01	(10^{-2})
milli-	m	0.001	(10^{-3})
micro-	μ	0.000001	(10^{-6})
nano-	n	0.000000001	(10^{-9})
pico-	p	0.000000000001	(10^{-12})
femto-	f	0.000000000000001	(10^{-15})

The prefix multipliers allow us to express a wide range of measurements in units that are similar in size to the quantity we are measuring. You should choose the prefix multiplier that is most convenient for a particular measurement. For example, to measure the diameter of a quarter, use centimeters because a quarter has a diameter of about 2.4 cm. A centimeter is a common metric unit and is about equivalent to the width of a pinky finger (2.54 cm = 1 in.). The millimeter could also work to express the diameter of the quarter; then the quarter would measure 24 mm. The kilometer, however, would not work as well since, in that unit, the quarter's diameter is 0.000024 km. Pick a unit similar in size to (or smaller than) the quantity you are measuring. Consider expressing the length of a short chemical bond, about 1.2×10^{-10} m. Which prefix multiplier should you use? The most convenient one is probably the picometer (pico = 10^{-12}). Chemical bonds measure about 120 pm.

TABLE 2.3   Some Common Units and Their Equivalents

Length

1 kilometer (km) = 0.6214 mile (mi)
1 meter (m) = 39.37 inches (in.)
= 1.094 yards (yd)
1 foot (ft) = 30.48 centimeters (cm)
1 inch (in.) = 2.54 centimeters (cm)
(exact)

Mass

1 kilogram (kg) = 2.205 pounds (lb)
1 pound (lb) = 453.59 grams (g)
1 ounce (oz) = 28.35 grams (g)

Volume

1 liter (L) = 1000 milliliters (mL)
= 1000 cubic centimeters (cm³)
1 liter (L) = 1.057 quarts (qt)
1 U.S. gallon (gal) = 3.785 liters (L)

✔ **CONCEPTUAL CHECKPOINT 2.4**

What would be the most convenient unit to express the dimensions of a polio virus, which is about 2.8×10^{-8} m in diameter?
(a) Mm
(b) mm
(c) μm
(d) nm

DERIVED UNITS

A derived unit is formed from other units. For example, many units of **volume,** a measure of space, are derived units. Any unit of length, when cubed (raised to the third power), becomes a unit of volume. Therefore, cubic meters (m^3), cubic centimeters (cm^3), and cubic millimeters (mm^3) are all units of volume. In these units, a three-bedroom house has a volume of about 630 m^3, a can of soda pop has a volume of about 350 cm^3, and a rice grain has a volume of about 3 mm^3. We also use the **liter (L)** and milliliter (mL) to express volume (although these are not derived units). A gallon is equal to 3.785 L. A milliliter is equivalent to 1 cm^3. Table 2.3 lists some common units and their equivalents.

2.6   Problem Solving and Unit Conversions

Problem solving is one of the most important skills you will acquire in this course. Not only will this skill help you succeed in chemistry, but it will help you to learn how to think critically, which is important in every area of knowledge. My daughter, a freshman in high school, recently came to me for help on an algebra problem. The statement of the problem went something like this:

> Sam and Sara live 11 miles apart. Sam leaves his house traveling at 6 miles per hour toward Sara's house. Sara leaves her house traveling at 3 miles per hour toward Sam's house. How much time until Sam and Sara meet?

Solving the problem requires setting up the equation $11 - 6t = 3t$. Although my daughter could solve this equation for t quite easily, getting to the equation from the problem statement was another matter—that process requires *critical thinking*. You can't succeed in chemistry—or in life, really—without developing critical thinking skills. Learning how to solve chemical problems will help you develop these kinds of skills.

Although no simple formula applies to every problem, you can learn problem-solving strategies and begin to develop some chemical intuition. Many of the problems you will solve in this course can be thought of as *unit conversion problems*, where you are given one or more quantities and asked to convert them into different units. Other problems require the use of *specific equations* to get to the information you are trying to find. In the sections that follow, we examine strategies to help you solve both of these types of problems. Of course, many problems contain both conversions and equations, requiring the combination of these strategies, and some problems may require an altogether different approach but the basic tools you learn here can be applied to those problems as well.

CONVERTING BETWEEN UNITS

Units are critical in calculations. Knowing how to work with and manipulate units in calculations is a very important part of problem solving. In calculations, units help determine correctness. Units should always be included in calculations, and we can think of many calculations as converting from one unit to another. Units are multiplied, divided, and canceled like any other algebraic quantity.

> Using units as a guide to solving problems is called dimensional analysis.

Remember:

1. Always write every number with its associated unit. Never ignore units; they are critical.

2. Always include units in your calculations, dividing them and multiplying them as if they were algebraic quantities. Do not let units magically appear or disappear in calculations. Units must flow logically from beginning to end.

Consider converting 17.6 in. to centimeters. We know from Table 2.3 that 1 in. = 2.54 cm. To determine how many centimeters are in 17.6 in., we perform the conversion:

$$17.6 \; \cancel{\text{in.}} \times \frac{2.54 \, \text{cm}}{1 \, \cancel{\text{in.}}} = 44.7 \, \text{cm}$$

The unit *in.* cancels and we are left with *cm* as our final unit. The quantity $\dfrac{2.54 \, \text{cm}}{1 \, \text{in.}}$ is a **conversion factor** between *in.* and *cm*—it is a quotient with *cm* on top and *in.* on bottom.

For most conversion problems, we are given a quantity in some unit and asked to convert the quantity to another unit. These calculations take the form:

$$\text{information given} \times \text{conversion factor(s)} = \text{information sought}$$

$$\cancel{\text{given unit}} \times \frac{\text{desired unit}}{\cancel{\text{given unit}}} = \text{desired unit}$$

Conversion factors are constructed from any two quantities known to be equivalent. In our example, 2.54 cm = 1 in., so we construct the conversion factor by dividing both sides of the equality by 1 in. and canceling the units.

$$2.54 \text{ cm} = 1 \text{ in.}$$

$$\frac{2.54 \text{ cm}}{1 \text{ in.}} = \frac{1 \cancel{\text{in.}}}{1 \cancel{\text{in.}}}$$

$$\frac{2.54 \text{ cm}}{1 \text{ in.}} = 1$$

The quantity $\dfrac{2.54 \text{ cm}}{1 \text{ in.}}$ is equal to 1 and can be used to convert between inches and centimeters.

What if we want to perform the conversion the other way, from centimeters to inches? If we try to use the same conversion factor, the units do not cancel correctly.

$$44.7 \text{ cm} \times \frac{2.54 \text{ cm}}{1 \text{ in.}} = \frac{114 \text{ cm}^2}{\text{in.}}$$

The units in the answer, as well as the value of the answer, are incorrect. The unit $\text{cm}^2/\text{in.}$ is not correct, and, based on our knowledge that centimeters are smaller than inches, we know that 44.7 cm cannot be equivalent to 114 in. In solving problems, always check if the final units are correct, and consider whether or not the magnitude of the answer makes sense. In this case, our mistake was in how we used the conversion factor. We must invert it.

$$44.7 \cancel{\text{cm}} \times \frac{1 \text{ in.}}{2.54 \cancel{\text{cm}}} = 17.6 \text{ in.}$$

Conversion factors can be inverted because they are equal to 1 and the inverse of 1 is 1.

$$\frac{1}{1} = 1$$

Therefore,

$$\frac{2.54 \text{ cm}}{1 \text{ in.}} = 1 = \frac{1 \text{ in.}}{2.54 \text{ cm}}$$

We can diagram conversions using a **solution map.** A solution map is a visual outline that shows the strategic route required to solve a problem. For unit conversion, the solution map focuses on units and how to convert from one unit to another. The solution map for converting from inches to centimeters is:

$$\frac{2.54 \text{ cm}}{1 \text{ in.}}$$

The solution map for converting from centimeters to inches is:

$$\frac{1 \text{ in.}}{2.54 \text{ cm}}$$

Each arrow in a solution map for a unit conversion has an associated conversion factor with the units of the previous step in the denominator and the units of the following step in the numerator. For one-step problems such as these, the solution map is only moderately helpful, but for multistep problems, it becomes a powerful way to develop a problem-solving strategy. In the section that follows, you will learn how to incorporate solution maps into an overall problem-solving strategy.

GENERAL PROBLEM-SOLVING STRATEGY

In this book, we use a standard problem-solving procedure that can be adapted to many of the problems encountered in chemistry and beyond. Solving any problem essentially requires you to assess the information given in the problem and devise a way to get to the information asked for. In other words, you need to

- Identify the starting point (the *given* information).
- Identify the end point (what you must *find*).
- Devise a way to get from the starting point to the end point using what is given as well as what you already know or can look up. You can use a *solution map* to diagram the steps required to get from the starting point to the end point.

In graphic form, we can represent this progression as

Given $\longrightarrow$ Solution Map $\longrightarrow$ Find

One of the main difficulties beginning students have when trying to solve problems in general chemistry is not knowing where to start. Although no problem-solving procedure is applicable to all problems, the following four-step procedure can be helpful in working through many of the numerical problems you will encounter in this book.

1. **Sort.** Begin by sorting the information in the problem. *Given* information is the basic data provided by the problem—often one or more numbers with their associated units. The given information is the starting point for the problem. *Find* indicates what the problem is asking you to find (the end point of the problem).

2. **Strategize.** This is usually the hardest part of solving a problem. In this step, you must create a solution map—the series of steps that will get you from the given information to the information you are trying to find. You have already seen solution maps for simple unit conversion problems. Each arrow in a solution map represents a computational step. On the left side of the arrow is the quantity (or quantities) you had before the step; on the right side of the arrow is the quantity (or quantities) you will have after the step; and below the arrow is the information you need to get from one to the other—the relationship between the quantities.

   Often such relationships will take the form of conversion factors or equations. These may be given in the problem, in which case you will have written them down under "Given" in Step 1. Usually, however, you will need other information—which may include physical constants, formulas, or conversion factors—to help get you from what you are given to what you must find. You may recall this information from what you have learned or you can look it up in the chapters or tables within the book.

   In some cases, you may get stuck at the strategize step. If you cannot figure out how to get from the given information to the information you are asked to find, you might try working backwards. For example, you may want to look at the units of the quantity you are trying to find and look for conversion factors to get to the units of the given quantity. You may even try a combination of strategies; work forward, backward, or some of both. If you persist, you will develop a strategy to solve the problem.

3. **Solve.** This is the easiest part of solving a problem. Once you set up the problem properly and devise a solution map, you follow the map to solve the problem. Carry out mathematical operations (paying attention to the rules for significant figures in calculations) and cancel units as needed.

4. **Check.** This is the step most often overlooked by beginning students. Experienced problem solvers always ask, Does this answer make physical sense? Are the units correct? Is the number of significant figures correct? When solving multistep problems, errors easily creep into the solution. You can catch most of these errors by simply checking the answer. For example, suppose you are calculating the number of atoms in a gold coin and end up with an answer of 1.1×10^{-6} atoms. Could the gold coin really be composed of one-millionth of one atom?

In Examples 2.8 and 2.9, you will find this problem-solving procedure applied to unit conversion problems. The procedure is summarized in the left column, and two examples of applying the procedure are shown in the middle and right columns. This three-column format is used in selected examples throughout this text. It allows you to see how a particular procedure can be applied to two different problems. Work through one problem first (from top to bottom) and then examine how the same procedure is applied to the other problem. Recognizing the commonalities and differences between problems is a key part of problem solving.

PROBLEM-SOLVING PROCEDURE	**EXAMPLE 2.8** Unit Conversion Convert 7.8 km to miles.	**EXAMPLE 2.9** Unit Conversion Convert 0.825 m to millimeters.
SORT Begin by sorting the information in the problem into *given* and *find*.	**GIVEN:** 7.8 km **FIND:** mi	**GIVEN:** 0.825 m **FIND:** mm
STRATEGIZE Draw a *solution map* for the problem. Begin with the *given* quantity and symbolize each step with an arrow. Below the arrow, write the conversion factor for that step. The solution map ends at the *find* quantity. (In these examples, the relationships used in the conversions are below the solution map.)	**SOLUTION MAP** km → mi $\dfrac{0.6214 \text{ mi}}{1 \text{ km}}$ **RELATIONSHIPS USED** $1 \text{ km} = 0.6214 \text{ mi}$ (This conversion factor is from Table 2.3.)	**SOLUTION MAP** m → mm $\dfrac{1 \text{ mm}}{10^{-3} \text{ m}}$ **RELATIONSHIPS USED** $1 \text{ mm} = 10^{-3} \text{ m}$ (This conversion factor is from Table 2.2.)
SOLVE Follow the *solution map* to solve the problem. Begin with the *given* quantity and its units. Multiply by the appropriate conversion factor, canceling units to arrive at the *find* quantity. Round the answer to the correct number of significant figures. (If possible, obtain conversion factors to enough significant figures so that they do not limit the number of significant figures in the answer.)	**SOLUTION** $7.8 \text{ k}\cancel{m} \times \dfrac{0.6214 \text{ mi}}{1 \text{ k}\cancel{m}} = 4.84692 \text{ mi}$ $4.84692 \text{ mi} = 4.8 \text{ mi}$ Round the answer to two significant figures, since the quantity given has two significant figures.	**SOLUTION** $0.825 \cancel{m} \times \dfrac{1 \text{ mm}}{10^{-3} \cancel{m}} = 825 \text{ mm}$ $825 \text{ mm} = 825 \text{ mm}$ Leave the answer with three significant figures, since the quantity given has three significant figures and the conversion factor is a definition and therefore does not limit the number of significant figures in the answer.

CHECK

Check your answer. Are the units correct? Does the answer make physical sense?

The units, mi, are correct. The magnitude of the answer is reasonable. A mile is longer than a kilometer, so the value in miles should be smaller than the value in kilometers.

The units, mm, are correct and the magnitude is reasonable. A millimeter is shorter than a meter, so the value in millimeters should be larger than the value in meters.

▶**SKILLBUILDER 2.8**
Unit Conversion

Convert 56.0 cm to inches.

▶**SKILLBUILDER 2.9**
Unit Conversion

Convert 5678 m to kilometers.

▶**FOR MORE PRACTICE**   Example 2.25; Problems 73, 74, 75, 76.

▶**FOR MORE PRACTICE**   Problems 69, 70, 71, 72.

✓ CONCEPTUAL CHECKPOINT 2.5

Which conversion factor would you use to convert a distance in meters to kilometers?

(a) $\dfrac{1\,\text{m}}{10^3\,\text{km}}$

(b) $\dfrac{10^3\,\text{m}}{1\,\text{km}}$

(c) $\dfrac{1\,\text{km}}{10^3\,\text{m}}$

(d) $\dfrac{10^3\,\text{km}}{1\,\text{m}}$

2.7 Solving Multistep Unit Conversion Problems

When solving multistep unit conversion problems, we follow the preceding procedure, but we add more steps to the solution map. Each step in the solution map should have a conversion factor with the units of the previous step in the denominator and the units of the following step in the numerator. For example, suppose we want to convert 194 cm to feet. The solution map begins with cm, and we use the relationship 2.54 cm = 1 in to convert to in. We then use the relationship 12 in. = 1 ft to convert to ft.

SOLUTION MAP

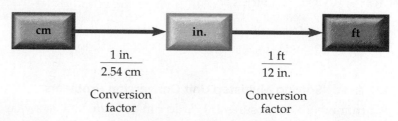

Once the solution map is complete, we follow it to solve the problem.

SOLUTION

$$194 \ \cancel{\text{cm}} \times \frac{1 \ \cancel{\text{in.}}}{2.54 \ \cancel{\text{cm}}} \times \frac{1 \ \text{ft}}{12 \ \cancel{\text{in.}}} = 6.3648 \ \text{ft}$$

| | Since 1 foot is defined as 12 in., it does not limit significant figures.

We then round to the correct number of significant figures—in this case, three (from 194 cm, which has three significant figures).

$$6.3648 \text{ ft} = 6.36 \text{ ft}$$

Finally, we check the answer. The units of the answer, feet, are the correct ones, and the magnitude seems about right. Since a foot is larger than a centimeter, it is reasonable that the value in feet is smaller than the value in centimeters.

EXAMPLE 2.10   Solving Multistep Unit Conversion Problems

A recipe for making creamy pasta sauce calls for 0.75 L of cream. Your measuring cup measures only in cups. How many cups of cream should you use? (4 cups = 1 quart)

SORT
Begin by sorting the information in the problem into given and find.

GIVEN: 0.75 L
FIND: cups

STRATEGIZE
Draw a solution map for the problem. Begin with the *given* quantity and symbolize each step with an arrow. Below the arrow, write the conversion factor for that step. The solution map ends at the find quantity.

SOLUTION MAP

L → qt → cups

$$\frac{1.057 \text{ qt}}{1 \text{ L}} \qquad \frac{4 \text{ cups}}{1 \text{ qt}}$$

RELATIONSHIPS USED
1.057 qt = 1 L (from Table 2.3.)
4 cups = 1 qt (given in problem statement)

SOLVE
Follow the solution map to solve the problem. Begin with 0.75 L and multiply by the appropriate conversion factor, canceling units to arrive at qt. Then, use the second conversion factor to arrive at cups.

Round the answer to the correct number of significant figures. In this case, round the answer to two significant figures, since the quantity given has two significant figures.

SOLUTION

$$0.75 \text{ L} \times \frac{1.057 \text{ qt}}{1 \text{ L}} \times \frac{4 \text{ cups}}{1 \text{ qt}} = 3.171 \text{ cups}$$

$$3.171 \text{ cups} = 3.2 \text{ cups}$$

CHECK
Check your answer. Are the units correct? Does the answer make physical sense?

The answer has the right units (cups) and seems reasonable. A cup is smaller than a liter, so the value in cups should be larger than the value in liters.

▶**SKILLBUILDER 2.10 | Solving Multistep Unit Conversion Problems**
A recipe calls for 1.2 cups of oil. How many liters of oil is this?

▶**FOR MORE PRACTICE**   Problems 85, 86.

EXAMPLE 2.11   Solving Multistep Unit Conversion Problems

One lap of a running track measures 255 m. To run 10.0 km, how many laps should you run?

SORT
Begin by sorting the information in the problem into given and find. You are given a distance in km and asked to find the distance in laps. You are also given the quantity 255 m per lap, which is a conversion factor between m and laps.

GIVEN: 10.0 km
255 m = 1 lap
FIND: number of laps

STRATEGIZE Build the solution map beginning with km and ending at laps. Focus on the units.	**SOLUTION MAP** $$\frac{10^3 \text{ m}}{1 \text{ km}} \qquad \frac{1 \text{ lap}}{255 \text{ m}}$$ **RELATIONSHIPS USED** $1\,\text{km} = 10^3\,\text{m}$ (from Table 2.2) $1\,\text{lap} = 255\,\text{m}$ (given in problem)
SOLVE Follow the solution map to solve the problem. Begin with 10.0 km and multiply by the appropriate conversion factor, canceling units to arrive at m. Then, use the second conversion factor to arrive at laps. Round the intermediate answer (in blue) to three significant figures because it is limited by the three significant figures in the given quantity, 10.0 km.	**SOLUTION** $$10.0 \cancel{\text{ km}} \times \frac{10^3 \cancel{\text{ m}}}{1 \cancel{\text{ km}}} \times \frac{1\,\text{lap}}{255 \cancel{\text{ m}}} = 39.216 \text{ laps} = 39.2 \text{ laps}$$
CHECK Check your answer. Are the units correct? Does the answer make physical sense?	The units of the answer are correct, and the value of the answer makes sense: If a lap is 255 m, there are about 4 laps to each km (1000 m), so it seems reasonable that you would have to run about 40 laps to cover 10 km.

▶**SKILLBUILDER 2.11 | Solving Multistep Unit Conversion Problems**

A running track measures 1056 ft per lap. To run 15.0 km, how many laps should you run? (1 mi = 5280 ft)

▶**SKILLBUILDER PLUS**

An island is 5.72 nautical mi from the coast. How far is the island in meters? (1 nautical mi = 1.151 mi)

▶**FOR MORE PRACTICE**   Problems 83, 84.

2.8   Units Raised to a Power

▌The unit cm³ is often abbreviated as cc.

When converting quantities with units raised to a power, such as cubic centimeters (cm³), the conversion factor must also be raised to that power. For example, suppose we want to convert the size of a motorcycle engine reported as 1255 cm³ to cubic inches. We know that

$$2.54\,\text{cm} = 1\,\text{in}.$$

Most tables of conversion factors do not include conversions between cubic units, but we can derive them from the conversion factors for the basic units. We cube both sides of the preceding equality to obtain the proper conversion factor.

▌2.54 cm = 1 in. is an exact conversion factor. After cubing, we retain five significant figures so that the conversion factor does not limit the four significant figures of our original quantity (1255 cm³).

$$(2.54 \text{ cm})^3 = (1 \text{ in.})^3$$
$$(2.54)^3 \text{ cm}^3 = 1^3 \text{ in.}^3$$
$$16.387 \text{ cm}^3 = 1 \text{ in.}^3$$

We can do the same thing in fractional form.

$$\frac{1\,\text{in.}}{2.54\,\text{cm}} = \frac{(1\,\text{in.})^3}{(2.54\,\text{cm})^3} = \frac{1\,\text{in.}^3}{16.387\,\text{cm}^3}$$

We then proceed with the conversion in the usual manner.

CHEMISTRY AND HEALTH
Drug Dosage

The unit of choice in specifying drug dosage is the milligram (mg). Pick up a bottle of aspirin, Tylenol, or any other common drug, and the label tells you the number of milligrams of the active ingredient contained in each tablet, as well as the number of tablets to take per dose. The following table shows the mass of the active ingredient per pill in several common pain relievers, all reported in milligrams. The remainder of each tablet is composed of inactive ingredients such as cellulose (or fiber) and starch.

The recommended adult dose for many of these pain relievers is one or two tablets every 4 to 8 hours (depending on the specific pain reliever). Notice that the extra-strength version of each pain reliever just contains a higher dose of the same compound found in the regular-strength version. For the pain relievers listed, three regular-strength tablets are the equivalent of two extra-strength tablets (and probably cost less).

The dosages given in the table are fairly standard for each drug, regardless of the brand. When you look on your drugstore shelf, you will find many different brands of regular-strength ibuprofen, some sold under the generic name and others sold under their brand names (such as Advil). However, if you look closely at the labels, you will find that they all contain the same thing: 200 mg of the compound ibuprofen. There is no difference in the compound or in the amount of the compound. Yet these pain relievers will most likely all have different prices. Choose the least expensive. Why pay more for the same thing?

CAN YOU ANSWER THIS? *Convert each of the doses in the table to ounces. Why are drug dosages not listed in ounces?*

Drug Mass per Pill for Common Pain Relievers

Pain Reliever	Mass of Active Ingredient per Pill
Aspirin	325 mg
Aspirin, extra strength	500 mg
Ibuprofen (Advil)	200 mg
Ibuprofen, extra strength	300 mg
Acetaminophen (Tylenol)	325 mg
Acetaminophen, extra strength	500 mg

SOLUTION MAP

$$cm^3 \longrightarrow in.^3$$

$$\frac{1 \text{ in.}^3}{16.387 \text{ cm}^3}$$

SOLUTION

$$1255 \text{ cm}^3 \times \frac{1 \text{ in.}^3}{16.387 \text{ cm}^3} = 76.5851 \text{ in.}^3 = 76.59 \text{ in.}^3$$

EXAMPLE 2.12 Converting Quantities Involving Units Raised to a Power

A circle has an area of 2659 cm². What is its area in square meters?

SORT You are given an area in square centimeters and asked to convert the area to square meters.	**GIVEN:** 2659 cm^2 **FIND:** m^2
STRATEGIZE Build a solution map beginning with cm² and ending with m². Remember that you must square the conversion factor.	**SOLUTION MAP** $$\frac{(0.01 \text{ m})^2}{(1 \text{ cm})^2}$$ **RELATIONSHIPS USED** $1 \text{ cm} = 0.01 \text{ m}$ (from Table 2.2)

SOLVE	**SOLUTION**
Follow the solution map to solve the problem. Square the conversion factor (both the units and the number) as you carry out the calculation.	$$2659\,\text{cm}^2 \times \frac{(0.01\,\text{m})^2}{(1\,\text{cm})^2}$$
Round the answer to four significant figures to reflect the four significant figures in the given quantity. The conversion factor is exact and therefore does not limit the number of significant figures.	$$= 2659\,\text{cm}^2 \times \frac{10^{-4}\,\text{m}^2}{1\,\text{cm}^2}$$ $$= 0.265900\,\text{m}^2$$ $$= 0.2659\,\text{m}^2$$
CHECK	
Check your answer. Are the units correct? Does the answer make physical sense?	The units of the answer are correct, and the magnitude makes physical sense. A square meter is much larger than a square centimeter, so the value in square meters should be much smaller than the value in square centimeters.

▶**SKILLBUILDER 2.12 | Converting Quantities Involving Units Raised to a Power**

An automobile engine has a displacement (a measure of the size of the engine) of 289.7 in.^{3} What is its displacement in cubic centimeters?

▶**FOR MORE PRACTICE**   Example 2.26; Problems 87, 88, 89, 90, 91, 92.

EXAMPLE 2.13   Solving Multistep Conversion Problems Involving Units Raised to a Power

The average annual per person crude oil consumption in the United States is 15,615 dm^3. What is this value in cubic inches?

SORT	**GIVEN:** 15,615 dm^3
You are given a volume in cubic decimeters and asked to convert it to cubic inches.	**FIND:** in.^{3}
STRATEGIZE	**SOLUTION MAP**
Build a solution map beginning with dm^3 and ending with in.^{3} Each of the conversion factors must be cubed, since the quantities involve cubic units.	$$\boxed{\text{dm}^3} \longrightarrow \boxed{\text{m}^3} \longrightarrow \boxed{\text{cm}^3} \longrightarrow \boxed{\text{in.}^3}$$ $$\frac{(0.1\,\text{m})^3}{(1\,\text{dm})^3} \qquad \frac{(1\,\text{cm})^3}{(0.01\,\text{m})^3} \qquad \frac{(1\,\text{in.})^3}{(2.54\,\text{cm})^3}$$
	RELATIONSHIPS USED
	1 dm = 0.1 m (from Table 2.2)
	1 cm = 0.01 m (from Table 2.2)
	2.54 cm = 1 in. (from Table 2.3)
SOLVE	**SOLUTION**
Follow the solution map to solve the problem. Begin with the given value in dm^3 and multiply by the string of conversion factors to arrive at in.^{3} Make sure to cube each conversion factor as you carry out the calculation.	$$15,615\,\text{dm}^3 \times \frac{(0.1\,\text{m})^3}{(1\,\text{dm})^3} \times \frac{(1\,\text{cm})^3}{(0.01\,\text{m})^3} \times \frac{(1\,\text{in.})^3}{(2.54\,\text{cm})^3}$$
Round the answer to five significant figures to reflect the five significant figures in the least precisely known quantity (15,615 dm^3). The conversion factors are all exact and therefore do not limit the number of significant figures.	$$= 9.5289 \times 10^5\,\text{in.}^3$$
CHECK	
Check your answer. Are the units correct? Does the answer make physical sense?	The units of the answer are correct and the magnitude makes sense. A cubic inch is smaller than a cubic decimeter, so the value in cubic inches should be larger than the value in cubic decimeters.

▶**SKILLBUILDER 2.13 | Solving Multistep Problems Involving Units Raised to a Power**

How many cubic inches are there in 3.25 yd^3?

▶**FOR MORE PRACTICE**   Problems 93, 94.

CONCEPTUAL CHECKPOINT 2.6

You know that there are 3 ft in a yard. How many cubic feet are there in a cubic yard?

(a) 3

(b) 6

(c) 9

(d) 27

2.9 Density

▲ Top-end bicycle frames are made of titanium because of titanium's low density and high relative strength. Titanium has a density of 4.50 g/cm³, while iron, for example, has a density of 7.86 g/cm³.

TABLE 2.4 Densities of Some Common Substances

Substance	Density (g/cm³)
Charcoal, oak	0.57
Ethanol	0.789
Ice	0.92
Water	1.0
Glass	2.6
Aluminum	2.7
Titanium	4.50
Iron	7.86
Copper	8.96
Lead	11.4
Gold	19.3
Platinum	21.4

Remember that cubic centimeters and milliliters are equivalent units.

Why do some people pay more than $3000 for a bicycle made of titanium? A steel frame would be just as strong for a fraction of the cost. The difference between the two bikes is their mass—the titanium bike is lighter. For a given volume of metal, titanium has less mass than steel. We describe this property by saying that titanium is *less dense* than steel. The **density** of a substance is the ratio of its mass to its volume.

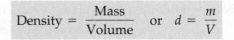

$$\text{Density} = \frac{\text{Mass}}{\text{Volume}} \quad \text{or} \quad d = \frac{m}{V}$$

Density is a fundamental property of substances that differs from one substance to another. The units of density are those of mass divided by those of volume, most conveniently expressed in grams per cubic centimeter (g/cm³) or grams per milliliter (g/mL). See Table 2.4 for a list of the densities of some common substances. Aluminum is among the least dense structural metals with a density of 2.70 g/cm³, while platinum is among the densest with a density of 21.4 g/cm³. Titanium has a density of 4.50 g/cm³.

CALCULATING DENSITY

We calculate the density of a substance by dividing the mass of a given amount of the substance by its volume. For example, a sample of liquid has a volume of 22.5 mL and a mass of 27.2 g. To find its density, we use the equation $d = m/V$.

$$d = \frac{m}{V} = \frac{27.2 \text{ g}}{22.5 \text{ mL}} = 1.21 \text{ g/mL}$$

We can use a solution map for solving problems involving equations, but the solution map will take a slightly different form than for pure conversion problems. In a problem involving an equation, the solution map shows how the *equation* takes you from the *given* quantities to the *find* quantity. The solution map for this problem is:

$$d = \frac{m}{V}$$

The solution map illustrates how the values of m and V, when substituted into the equation $d = \frac{m}{V}$ give the desired result, d.

EXAMPLE 2.14 Calculating Density

A jeweler offers to sell a ring to a woman and tells her that it is made of platinum. Noting that the ring felt a little light, the woman decides to perform a test to determine the ring's density. She places the ring on a balance and finds that it has a mass of 5.84 g. She also finds that the ring *displaces* 0.556 cm³ of water. Is the ring made of platinum? The density of platinum is 21.4 g/cm³. (The displacement of water is a common way to measure the volume of irregularly shaped objects. To say that an object *displaces* 0.556 cm³ of water means that when the object is submerged in a container of water filled to the brim, 0.556 cm³ of water overflows. Therefore, the volume of the object is 0.556 cm³.)

SORT You are given the mass and volume of the ring and asked to find the density.	**GIVEN:** $m = 5.84$ g $V = 0.556$ cm³ **FIND:** density in g/cm³
STRATEGIZE If the ring is platinum, its density should match that of platinum. Build a solution map that represents how you get from the given quantities (mass and volume) to the find quantity (density). Unlike in conversion problems, where you write a conversion factor beneath the arrow, here you write the equation for density beneath the arrow.	**SOLUTION MAP** $$d = \frac{m}{V}$$ **RELATIONSHIPS USED** $d = \dfrac{m}{V}$ (equation for density)
SOLVE Follow the solution map. Substitute the given values into the density equation and compute the density. Round the answer to three significant figures to reflect the three significant figures in the given quantities.	**SOLUTION** $$d = \frac{m}{V} = \frac{5.84 \text{ g}}{0.556 \text{ cm}^3} = 10.5 \text{ g/cm}^3$$ The density of the ring is much too low to be platinum; therefore the ring is a fake.
CHECK Check your answer. Are the units correct? Does the answer make physical sense?	The units of the answer are correct, and the magnitude seems reasonable to be an actual density. As you can see from Table 2.4, the densities of liquids and solids range from below 1 g/cm³ to just over 20 g/cm³.

▶**SKILLBUILDER 2.14 | Calculating Density**

The woman takes the ring back to the jewelry shop, where she is met with endless apologies. They accidentally had made the ring out of silver rather than platinum. They give her a new ring that they promise is platinum. This time when she checks the density, she finds the mass of the ring to be 9.67 g and its volume to be 0.452 cm³. Is this ring genuine?

▶**FOR MORE PRACTICE** Example 2.27; Problems 95, 96, 97, 98, 99, 100.

DENSITY AS A CONVERSION FACTOR

We can use the density of a substance as a conversion factor between the mass of the substance and its volume. For example, suppose we need 68.4 g of a liquid with a density of 1.32 g/cm³ and want to measure the correct amount with a graduated cylinder (a piece of laboratory glassware used to measure volume). How much volume should we measure?

We start with the mass of the liquid and use the density as a conversion factor to convert mass to volume. However, we must use the inverted density expression 1 cm³/1.32 g because we want g, the unit we are converting from, to be on the bottom (in the denominator) and cm³, the unit we are converting to, on the top (in the numerator). Our solution map takes this form:

CHEMISTRY AND HEALTH

Density, Cholesterol, and Heart Disease

Cholesterol is fatty substance found in animal-derived foods such as beef, eggs, fish, poultry, and milk products. Cholesterol is used by the body for several purposes. However, excessive amounts in the blood—which can be caused by both genetic factors and diet—may result in the deposition of cholesterol in arterial walls, leading to a condition called atherosclerosis, or blocking of the arteries. These blockages are dangerous because they inhibit blood flow to important organs, causing heart attacks and strokes. The risk of stroke and heart attack increases with increasing blood cholesterol levels (Table 2.5). Cholesterol is carried in the bloodstream by a class of substances known as lipoproteins. Lipoproteins are often separated and classified according to their density.

The main carriers of blood cholesterol are low-density lipoproteins (LDLs). LDLs, also called bad cholesterol, have a density of 1.04 g/cm^3. They are bad because they tend to deposit cholesterol on arterial walls, increasing the risk of stroke and heart attack. Cholesterol is also carried by high-density lipoproteins (HDLs). HDLs, also called good cholesterol, have a density of 1.13 g/cm^3. HDLs transport cholesterol to the liver for processing and excretion and therefore have a tendency to reduce cholesterol on arterial walls. Too low a level of HDLs (below 35 mg/100 mL) is considered a risk factor for heart disease. Exercise, along with a diet low in saturated fats, is believed to raise HDL levels in the blood while lowering LDL levels.

CAN YOU ANSWER THIS? *What mass of low-density lipoprotein is contained in a cylinder that is 1.25 cm long and 0.50 cm in diameter? (The volume of a cylinder, V, is given by $V = \pi r^2 \ell$, where r is the radius of the cylinder and ℓ is its length.)*

Clogged artery

▲ Too many low-density lipoproteins in the blood can lead to the blocking of arteries.

TABLE 2.5 Risk of Stroke and Heart Attack vs. Blood Cholesterol Level

Risk Level	Total Blood Cholesterol (mg/100 mL)	LDL (mg/100 mL)
low	< 200	< 130
borderline	200–239	130–159
high	240+	160+

SOLUTION MAP

g → cm^3 → mL

$$\frac{1\ cm^3}{1.32\ g} \qquad \frac{1\ mL}{1\ cm^3}$$

SOLUTION

$$68.4\ \cancel{g} \times \frac{1\ \cancel{cm^3}}{1.32\ \cancel{g}} \times \frac{1\ mL}{1\ \cancel{cm^3}} = 51.8\ mL$$

We must measure 51.8 mL to obtain 68.4 g of the liquid.

EXAMPLE 2.15   Density as a Conversion Factor

The gasoline in an automobile gas tank has a mass of 60.0 kg and a density of 0.752 g/cm^3. What is its volume in cm^3?

SORT You are given the mass in kilograms and asked to find the volume in cubic centimeters. Density is the conversion factor between mass and volume.	**GIVEN:** 60.0 kg Density = 0.752 g/cm^3 **FIND:** volume in cm^3

STRATEGIZE
Build the solution map starting with kg and ending with cm^3. Use the density (inverted) to convert from g to cm^3.
	SOLUTION MAP
	
	$\boxed{\text{kg}} \longrightarrow \boxed{\text{g}} \longrightarrow \boxed{\text{cm}^3}$
	
	$\dfrac{1000 \text{ g}}{1 \text{ kg}}$ $\dfrac{1 \text{ cm}^3}{0.752 \text{ g}}$
	
	RELATIONSHIPS USED
	
	   0.752 g/cm^3 (given in problem)
	
	   1000 g = 1 kg (from Table 2.2)

SOLVE Follow the solution map to solve the problem. Round the answer to three significant figures to reflect the three significant figures in the given quantities.	**SOLUTION** $60.0 \text{ kg} \times \dfrac{1000 \text{ g}}{1 \text{ kg}} \times \dfrac{1 \text{ cm}^3}{0.752 \text{ g}} = 7.98 \times 10^4 \text{ cm}^3$
CHECK Check your answer. Are the units correct? Does the answer make physical sense?	The units of the answer are those of volume, so they are correct. The magnitude seems reasonable because the density is somewhat less than 1 g/cm^3; therefore the volume of 60.0 kg should be somewhat more than $60.0 \times 10^3 \text{ cm}^3$.

▶**SKILLBUILDER 2.15 | Density as a Conversion Factor**

A drop of acetone (nail polish remover) has a mass of 35 mg and a density of 0.788 g/cm^3. What is its volume in cubic centimeters?

▶**SKILLBUILDER PLUS**

A steel cylinder has a volume of 246 cm^3 and a density of 7.93 g/cm^3. What is its mass in kilograms?

▶**FOR MORE PRACTICE**   Example 2.28; Problems 101, 102.

2.10   Numerical Problem-Solving Overview

In this chapter, you have seen a few examples of how to solve numerical problems. In Section 2.6, we developed a procedure to solve simple unit conversion problems. We then learned how to modify that procedure to work with multistep unit conversion problems and problems involving an equation. We will now summarize and generalize these procedures and apply them to two additional examples. As we did in Section 2.6, we provide the general procedure for solving numerical problems in the left column and the application of the procedure to two examples in the center and right columns.

	EXAMPLE 2.16	**EXAMPLE 2.17**
SOLVING NUMERICAL PROBLEMS	**Unit Conversion**	**Unit Conversion with Equation**
	A 23.5-kg sample of ethanol is needed for a large-scale reaction. What volume in liters of ethanol should be used? The density of ethanol is 0.789 g/cm³.	A 55.9-kg person displaces 57.2 L of water when submerged in a water tank. What is the density of the person in grams per cubic centimeter?

SORT	
• Scan the problem for one or more numbers and their associated units. This number (or numbers) is (are) the starting point(s) of the calculation. Write them down as given. • Scan the problem to determine what you are asked to find. Sometimes the units of this quantity are implied; other times they are specified. Write down the quantity and/or units you are asked to find.	**GIVEN:** 23.5 kg ethanol density $= 0.789\,\text{g/cm}^3$ **FIND:** volume in L

GIVEN: $m = 55.9\,\text{kg}$

$V = 57.2\,\text{L}$

FIND: density in g/cm^3

STRATEGIZE

• For problems involving only conversions, focus on units. The solution map shows how to get from the units in the given quantity to the units in the quantity you are asked to find.

• For problems involving equations, focus on the equation. The solution map shows how the equation takes you from the given quantity (or quantities) to the quantity you are asked to find.

• Some problems may involve both unit conversions and equations, in which case the solution map employs both of the above points.

SOLUTION MAP

$$\boxed{\text{kg}} \rightarrow \boxed{\text{g}} \rightarrow \boxed{\text{cm}^3} \rightarrow \boxed{\text{mL}} \rightarrow \boxed{\text{L}}$$

$$\frac{1000\,\text{g}}{1\,\text{kg}} \quad \frac{1\,\text{cm}^3}{0.789\,\text{g}} \quad \frac{1\,\text{mL}}{1\,\text{cm}^3} \quad \frac{1\,\text{L}}{1000\,\text{mL}}$$

RELATIONSHIPS USED

$0.789\,\text{g/cm}^3$ (given in problem)

$1000\,\text{g} = 1\,\text{kg}$ (Table 2.2)

$1000\,\text{mL} = 1\,\text{L}$ (Table 2.2)

$1\,\text{mL} = 1\,\text{cm}^3$ (Table 2.3)

SOLUTION MAP

$$\boxed{m, V} \rightarrow \boxed{d}$$

$$d = \frac{m}{V}$$

RELATIONSHIPS USED

$d = \dfrac{m}{V}$ (definition of density)

SOLVE

• For problems involving only conversions, begin with the given quantity and its units. Multiply by the appropriate conversion factor(s), canceling units, to arrive at the quantity you are asked to find.

SOLUTION

$$23.5\,\text{kg} \times \frac{1000\,\text{g}}{1\,\text{kg}} \times \frac{1\,\text{cm}^3}{0.789\,\text{g}} \times$$

$$\frac{1\,\text{mL}}{1\,\text{cm}^3} \times \frac{1\,\text{L}}{1000\,\text{mL}} = 29.7845\,\text{L}$$

$$29.7845\,\text{L} = 29.8\,\text{L}$$

The equation is already solved for the find quantity. Convert mass from kilograms to grams.

$$m = 55.9\,\text{kg} \times \frac{1000\,\text{g}}{1\,\text{kg}}$$

$$= 5.59 \times 10^4\,\text{g}$$

- For problems involving equations, solve the equation to arrive at the quantity you are asked to find. (Use algebra to rearrange the equation so that the quantity you are asked to find is isolated on one side.) Gather each of the quantities that must go into the equation in the correct units. (Convert to the correct units using additional solution maps if necessary.) Finally, substitute the numerical values and their units into the equation and compute the answer.
- Round the answer to the correct number of significant figures. Use the significant-figure rules from Sections 2.3 and 2.4.

Convert volume from liters to cubic centimeters.

$$V = 57.2 \, \cancel{L} \times \frac{1000 \, \cancel{mL}}{1 \, \cancel{L}} \times \frac{1 \, cm^3}{1 \, \cancel{mL}}$$

$$= 57.2 \times 10^3 \, cm^3$$

Compute density.

$$d = \frac{m}{V} = \frac{55.9 \times 10^3}{57.2 \times 10^3 \, cm^3}$$

$$= 0.9772727 \, \frac{g}{cm^3}$$

$$= 0.977 \, \frac{g}{cm^3}$$

CHECK

- Does the magnitude of the answer make physical sense? Are the units correct?

The units are correct (L) and the magnitude is reasonable. Since the density is less than 1 g/cm³, the computed volume (29.8 L) should be greater than the mass (23.5 kg).

The units are correct. Since the mass in kilograms and the volume in liters were very close to each other in magnitude, it makes sense that the density is close to 1 g/cm³.

▶**SKILLBUILDER 2.16**

Unit Conversion

A pure gold metal bar displaces 0.82 L of water. What is its mass in kilograms? (The density of gold is 19.3 g/cm³.)

▶**SKILLBUILDER 2.17**

Unit Conversion with Equation

A gold-colored pebble is found in a stream. Its mass is 23.2 mg, and its volume is 1.20 mm³. What is its density in grams per cubic centimeter? Is it gold? (The density of gold − 19.3 g/cm³.)

▶**FOR MORE PRACTICE** Problems 103, 109, 110, 111, 112.

▶**FOR MORE PRACTICE** Problems 104, 105, 106.

CHAPTER IN REVIEW

CHEMICAL PRINCIPLES

Uncertainty: Scientists report measured quantities so that the number of digits reflects the certainty in the measurement. Write measured quantities so that every digit is certain except the last, which is estimated.

RELEVANCE

Uncertainty: Measurement is a hallmark of science, and the precision of a measurement must be communicated with the measurement so that others know how reliable the measurement is. When you write or manipulate measured quantities, you must show and retain the precision with which the original measurement was made.

Units: Measured quantities usually have units associated with them. The SI unit for length is the meter; for mass, the kilogram; and for time, the second. Prefix multipliers such as *kilo-* or *milli-* are often used in combination with these basic units. The SI units of volume are units of length raised to the third power; liters or milliliters are often used as well.

Units: The units in a measured quantity communicate what the quantity actually is. Without an agreed-on system of units, scientists could not communicate their measurements. Units are also important in calculations, and the tracking of units throughout a calculation is essential.

Density: The density of a substance is its mass divided by its volume, $d = m/V$, and is usually reported in units of grams per cubic centimeter or grams per milliliter. Density is a fundamental property of all substances and generally differs from one substance to another.

Density: The density of substances is an important consideration in choosing materials from which to make things. Airplanes, for example, are made of low-density materials, while bridges are made of higher-density materials. Density is important as a conversion factor between mass and volume and vice versa.

CHEMICAL SKILLS

EXAMPLES

Scientific Notation (Section 2.2)

To express a number in scientific notation:

- Move the decimal point to obtain a number between 1 and 10.

- Write the decimal part multiplied by 10 raised to the number of places you moved the decimal point.

- The exponent is positive if you moved the decimal point to the left and negative if you moved the decimal point to the right.

EXAMPLE 2.18   Scientific Notation

Express the number 45,000,000 in scientific notation.

$$45{,}000{,}000$$
$$\overset{\frown}{7\ 6\ 5\ 4\ 3\ 2\ 1}$$

$$4.5 \times 10^7$$

Reporting Measured Quantities to the Right Number of Digits (Section 2.3)

Report measured quantities so that every digit is certain except the last, which is estimated.

EXAMPLE 2.19   Reporting Measured Quantities to the Right Number of Digits

Record the volume of liquid in the graduated cylinder to the correct number of digits. Laboratory glassware is calibrated (and should therefore be read) from the bottom of the meniscus (see figure).

Since the graduated cylinder has markings every 0.1 mL, the measurement should be recorded to the nearest 0.01 mL. In this case, that is 4.57 mL.

Counting Significant Digits (Section 2.3)

The following digits should always be counted as significant:

- nonzero digits
- interior zeros
- trailing zeros after a decimal point
- trailing zeros before a decimal point but after a nonzero number

The following digits should never be counted as significant:

- zeros to the left of the first nonzero number

The following digits are ambiguous and should be avoided by using scientific notation:

- zeros at the end of a number, but before a decimal point

Rounding (Section 2.4)

When rounding numbers to the correct number of significant figures, round down if the last digit dropped is 4 or less; round up if the last digit dropped is 5 or more.

Significant Figures in Multiplication and Division (Section 2.4)

The result of a multiplication or division should carry the same number of significant figures as the factor with the least number of significant figures.

Significant Figures in Addition and Subtraction (Section 2.4)

The result of an addition or subtraction should carry the same number of decimal places as the quantity carrying the least number of decimal places.

EXAMPLE 2.20 Counting Significant Digits

How many significant figures are in the following numbers?

1.0050	five significant figures
0.00870	three significant figures
100.085	six significant digits
5400	It is not possible to tell in its current form.

In order for us to know, the number needs to be written as 5.4×10^3, 5.40×10^3, or 5.400×10^3, depending on the number of significant figures intended.

EXAMPLE 2.21 Rounding

Round 6.442 and 6.456 to two significant figures each.

6.442 rounds to 6.4
6.456 rounds to 6.5

EXAMPLE 2.22 Significant Figures in Multiplication and Division

Perform the following calculation and report the answer to the correct number of significant figures.

$$8.54 \times 3.589 \div 4.2$$
$$- 7.2976$$
$$= 7.3$$

Round the final result to two significant figures to reflect the two significant figures in the factor with the least number of significant figures (4.2).

EXAMPLE 2.23 Significant Figures in Addition and Subtraction

Perform the following operation and report the answer to the correct number of significant figures.

3.098
0.67
−0.9452
2.8228 = 2.82

Round the final result to two decimal places to reflect the two decimal places in the quantity with the least number of decimal places (0.67).

Significant Figures in Calculations Involving Both Addition/Subtraction and Multiplication/Division (Section 2.4)

In calculations involving both addition/subtraction and multiplication/division, do the steps in parentheses first, keeping track of how many significant figures are in the answer by underlining the least significant figure, then proceeding with the remaining steps. Do not round off until the very end.

EXAMPLE 2.24   Significant Figures in Calculations Involving Both Addition/Subtraction and Multiplication/Division

Perform the following operation and report the answer to the correct number of significant figures.

$$8.16 \times (5.4323 - 5.411)$$
$$= 8.16 \times 0.021\underline{3}$$
$$= 0.173\underline{8} = 0.17$$

Unit Conversion (Sections 2.6, 2.7)

Solve unit conversion problems by following these steps.

1. **Sort** Write down the given quantity and its units and the quantity you are asked to find and its units.

2. **Strategize** Draw a solution map showing how to get from the given quantity to the quantity you are asked to find.

3. **Solve** Follow the solution map. Starting with the given quantity and its units, multiply by the appropriate conversion factor(s), canceling units, to arrive at the quantity to find in the desired units. Round the final answer to the correct number of significant figures.

4. **Check** Are the units correct? Does the answer make physical sense?

EXAMPLE 2.25   Unit Conversion

Convert 108 ft to meters.

GIVEN: 108 ft

FIND: m

SOLUTION MAP

$$\frac{12 \text{ in.}}{1 \text{ ft}} \qquad \frac{1 \text{ m}}{39.37 \text{ in.}}$$

RELATIONSHIPS USED

$$1 \text{ m} = 39.37 \text{ in.} \quad \text{(Table 2.3)}$$
$$1 \text{ ft} = 12 \text{ in.} \quad \text{(by definition)}$$

SOLUTION

$$108 \text{ ft} \times \frac{12 \text{ in.}}{1 \text{ ft}} \times \frac{1 \text{ m}}{39.37 \text{ in.}}$$
$$= 32.918 \text{ m}$$
$$= 32.9 \text{ m}$$

The answer has the right units (meters), and it makes sense; since a meter is longer than a foot, the number of meters should be less than the number of feet.

Unit Conversion Involving Units Raised to a Power (Section 2.8)

When working problems involving units raised to a power, raise the conversion factors to the same power.

1. **Sort** Write down the given quantity and its units and the quantity you are asked to find and its units.

2. **Strategize** Draw a solution map showing how to get from the given quantity to the quantity you are asked to find. Since the units are squared, you must square the conversion factor.

EXAMPLE 2.26   Unit Conversion Involving Units Raised to a Power

How many square meters are in 1.0 km²?

GIVEN: 1.0 km^2

FIND: m^2

SOLUTION MAP

$$\frac{(1000 \text{ m})^2}{(1 \text{ km})^2}$$

RELATIONSHIPS USED

$$1 \text{ km} = 1000 \text{ m} \quad \text{(Table 2.2)}$$

3. **Solve** Follow the solution map. Starting with the given quantity and its units, multiply by the appropriate conversion factor(s), canceling units, to arrive at the quantity you are asked to find in the desired units. Don't forget to square the conversion factor for squared units.

4. **Check** Are the units correct? Does the answer make physical sense?

SOLUTION

$$1.0 \, km^2 \times \frac{(1000 \, m)^2}{(1 \, km)^2}$$

$$= 1.0 \, km^2 \times \frac{1 \times 10^6 \, m^2}{1 \, km^2}$$

$$= 1.0 \times 10^6 \, m^2$$

The units are correct. The answer makes physical sense; a square meter is much smaller than a square kilometer, so the number of square meters should be much larger than the number of square kilometers.

Calculating Density (Section 2.10)

The density of an object or substance is its mass divided by its volume.

$$d = \frac{m}{V}$$

1. **Sort** Write down the given quantity and its units and the quantity you are asked to find and its units.

2. **Strategize** Draw a solution map showing how to get from the given quantity to the quantity you are asked to find. Use the definition of density as the equation that takes you from the mass and the volume to the density.

3. **Solve** Substitute the correct values into the equation for density.

4. **Check** Are the units correct? Does the answer make physical sense?

EXAMPLE 2.27 Calculating Density

An object has a mass of 23.4 g and displaces 5.7 mL of water. Determine its density in grams per milliliter.

GIVEN:

$$m = 23.4 \, g$$
$$V = 5.7 \, mL$$

FIND: density in g/mL

SOLUTION MAP

$$d = \frac{m}{V}$$

RELATIONSHIPS USED

$$d = \frac{m}{V} \text{ (definition of density)}$$

SOLUTION

$$d = \frac{m}{V}$$

$$= \frac{23.4 \, g}{5.7 \, mL}$$

$$= 4.11 \, g/mL$$

$$= 4.1 \, g/mL$$

The units (g/mL) are units of density. The answer is in the range of values for the densities of liquids and solids (see Table 2.4).

Density as a Conversion Factor (Section 2.10)

Density can be used as a conversion factor from mass to volume or from volume to mass. To convert between volume and mass, use density directly. To convert between mass and volume, invert the density.

1. **Sort** Write down the given quantity and its units and the quantity you are asked to find and its units.

2. **Strategize** Draw a solution map showing how to get from the given quantity to the quantity you are asked to find. Use the inverse of the density to convert from g to mL.

3. **Solve** Begin with given quantity and multiply by the appropriate conversion factors to arrive at the quantity you are asked to find. Round to the correct number of significant figures.

4. **Check** Are the units correct? Does the answer make physical sense?

EXAMPLE 2.28 Density as a Conversion Factor

What is the volume in liters of 321 g of a liquid with a density of 0.84 g/mL?

GIVEN: 321 g

FIND: volume in L

SOLUTION MAP

$$\frac{1\ mL}{0.84\ g} \qquad \frac{1\ L}{1000\ mL}$$

RELATIONSHIPS USED

0.84 g/mL (given in the problem)
1 L = 1000 mL (Table 2.2)

SOLUTION

$$321\ g \times \frac{1\ \cancel{mL}}{0.84\ \cancel{g}} \times \frac{1\ L}{1000\ \cancel{mL}}$$
$$= 0.382\ L = 0.38\ L$$

The answer is in the correct units. The magnitude seems right because the density is slightly less than 1; therefore the volume (382 mL) should be slightly greater than the mass (321 g).

KEY TERMS

conversion factor **[2.6]**
decimal part **[2.2]**
density **[2.9]**
English system **[2.5]**
exponent **[2.2]**
exponential part **[2.2]**

International System **[2.5]**
kilogram (kg) **[2.5]**
liter (L) **[2.5]**
mass **[2.5]**
meter (m) **[2.5]**
metric system **[2.5]**

prefix multipliers **[2.5]**
scientific notation **[2.2]**
second (s) **[2.5]**
SI units **[2.5]**
significant figures
(digits) **[2.3]**

solution map **[2.6]**
units **[2.5]**
volume **[2.5]**

EXERCISES

QUESTIONS

Answers to all questions numbered in blue appear in the Answers section at the back of the book.

1. Why is it important to report units with scientific measurements?

2. Why are the number of digits reported in scientific measurements important?

3. Why is scientific notation useful?

4. If a measured quantity is written correctly, which digits are certain? Which are uncertain?

5. Explain when zeros count as significant digits and when they do not.

6. How many significant digits are there in exact numbers? What kinds of numbers are exact?

7. What limits the number of significant digits in a calculation involving only multiplication and division?

8. What limits the number of significant digits in a calculation involving only addition and subtraction?

9. How are significant figures determined in calculations involving both addition/subtraction and multiplication/division?

10. What are the rules for rounding numbers?

11. What are the basic SI units of length, mass, and time?

12. List the common units of volume.

13. Suppose you are trying to measure the diameter of a Frisbee. What unit and prefix multiplier should you use?

14. What is the difference between mass and weight?

15. Obtain a metric ruler and measure these objects to the correct number of significant figures.
    (a) quarter (diameter)
    (b) dime (diameter)
    (c) notebook paper (width)
    (d) this book (width)
16. Obtain a stopwatch and measure each time to the correct number of significant figures.
    (a) time between your heartbeats
    (b) time it takes you to do the next problem
    (c) time between your breaths
17. Explain why units are important in calculations.
18. How are units treated in a calculation?
19. What is a conversion factor?
20. Why is the fundamental value of a quantity not changed when the quantity is multiplied by a conversion factor?
21. Write the conversion factor that converts a measurement in inches to feet. How would the conversion factor change for converting a measurement in feet to inches?
22. Write conversion factors for each:
    (a) miles to kilometers
    (b) kilometers to miles
    (c) gallons to liters
    (d) liters to gallons
23. This book outlines a four-step problem-solving strategy. Describe each step and its significance.
    (a) Sort
    (b) Strategize
    (c) Solve
    (d) Check
24. Experienced problem solvers always consider both the value and units of their answer to a problem. Why?
25. Draw a solution map to convert a measurement in grams to pounds.
26. Draw a solution map to convert a measurement in milliliters to gallons.
27. Draw a solution map to convert a measurement in meters to feet.
28. Draw a solution map to convert a measurement in ounces to grams. (1 lb = 16 oz)
29. What is density? Explain why density can work as a conversion factor. Between what quantities does it convert?
30. Explain how you would calculate the density of a substance. Include a solution map in your explanation.

PROBLEMS

Note: The exercises in the Problems section are paired, and the answers to the odd-numbered exercises (numbered in blue) appear in the Answers section at the back of the book.

SCIENTIFIC NOTATION

31. Express each number in scientific notation.
    (a) 36,756,000 (population of California)
    (b) 1,288,000 (population of Hawaii)
    (c) 19,490,000 (population of New York)
    (d) 532,000 (population of Wyoming)
32. Express each number in scientific notation.
    (a) 6,796,000,000 (population of the world)
    (b) 1,338,000,000 (population of China)
    (c) 11,451,000 (population of Cuba)
    (d) 4,203,000 (population of Ireland)

33. Express each number in scientific notation.
    (a) 0.00000000007461 m (length of a hydrogen–hydrogen chemical bond)
    (b) 0.0000158 mi (number of miles in an inch)
    (c) 0.000000632 m (wavelength of red light)
    (d) 0.000015 m (diameter of a human hair)
34. Express each number in scientific notation.
    (a) 0.000000001 s (time it takes light to travel 1 ft)
    (b) 0.143 s (time it takes light to travel around the world)
    (c) 0.000000000001 s (time it takes a chemical bond to undergo one vibration)
    (d) 0.000001 m (approximate size of a dust particle)

35. Express each number in decimal notation (i.e., express the number without using scientific notation).
    (a) 6.022×10^{23} (number of carbon atoms in 12.01 g of carbon)
    (b) 1.6×10^{-19} C (charge of a proton in coulombs)
    (c) 2.99×10^{8} m/s (speed of light)
    (d) 3.44×10^{2} m/s (speed of sound)
36. Express each number in decimal notation (i.e., express the number without using scientific notation).
    (a) 450×10^{-9} m (wavelength of blue light)
    (b) 13.7×10^{9} years (approximate age of the universe)
    (c) 5×10^{9} years (approximate age of Earth)
    (d) 4.7×10^{1} years (approximate age of this author)

37. Express each number in decimal notation (i.e., express the number without using scientific notation).
    (a) 3.22×10^{7}
    (b) 7.2×10^{-3}
    (c) 1.18×10^{11}
    (d) 9.43×10^{-6}
38. Express each number in decimal notation. (i.e., express the number without using scientific notation)
    (a) 1.30×10^{6}
    (b) 1.1×10^{-4}
    (c) 1.9×10^{2}
    (d) 7.41×10^{-10}

39. Complete the table.

Decimal Notation	Scientific Notation
2,000,000,000	_____
_____	1.211×10^9
0.000874	_____
_____	3.2×10^{11}

40. Complete the table.

Decimal Notation	Scientific Notation
_____	4.2×10^{-3}
315,171,000	_____
_____	1.8×10^{-11}
1,232,000	_____

SIGNIFICANT FIGURES

41. Read each instrument to the correct number of significant figures. Laboratory glassware should always be read from the bottom of the *meniscus* (the curved surface at the top of the liquid column).

(a)

(b) Celsius

(c) Celsius

(d)

42. Read each instrument to the correct number of significant figures. Laboratory glassware should always be read from the bottom of the meniscus (the curved surface at the top of the liquid column).

Note: A burette reads from the top down.

(a)

(b)

Note: A pipette reads from the top down.

(c)

Note: Digital balances normally display mass to the correct number of significant figures for that **(d)** particular balance.

---

43. For each measured quantity, underline the zeros that are significant and draw an X through the zeros that are not.
  (a) 0.005050 m
  (b) 0.0000000000000060 s
  (c) 220,103 kg
  (d) 0.00108 in.

44. For each measured quantity, underline the zeros that are significant and draw an X through the zeros that are not.
  (a) 0.00010320 s
  (b) 1,322,600,324 kg
  (c) 0.0001240 in.
  (d) 0.02061 m

45. How many significant figures are in each measured quantity?
  (a) 0.001125 m
  (b) 0.1125 m
  (c) 1.12500 × 10⁴ m
  (d) 11205 m

46. How many significant figures are in each measured quantity?
  (a) 13001 kg
  (b) 13111 kg
  (c) 1.30 × 10⁴ kg
  (d) 0.00013 kg

47. Determine whether each of the entries in the table is correct. Correct the entries that are wrong.

Quantity	Significant Figures
(a) 895675 m	6
(b) 0.000869 kg	6
(c) 0.5672100 s	5
(d) 6.022 × 10²³ atoms	4

48. Determine whether each of the entries in the table is correct. Correct the entries that are wrong.

Quantity	Significant Figures
(a) 24 days	2
(b) 5.6 × 10⁻¹² s	3
(c) 3.14 m	3
(d) 0.00383 g	5

ROUNDING

49. Round each number to four significant figures.
  (a) 255.98612
  (b) 0.0004893222
  (c) 2.900856 × 10⁻⁴
  (d) 2,231,479

50. Round each number to three significant figures.
  (a) 10,776.522
  (b) 4.999902 × 10⁶
  (c) 1.3499999995
  (d) 0.0000344988

51. Round each number to two significant figures.
  (a) 2.34
  (b) 2.35
  (c) 2.349
  (d) 2.359

52. Round each number to three significant figures.
  (a) 65.74
  (b) 65.749
  (c) 65.75
  (d) 65.750

53. Each number was supposed to be rounded to three significant figures. Find the ones that were incorrectly rounded and correct them.
  (a) 42.3492 to 42.4
  (b) 56.9971 to 57.0
  (c) 231.904 to 232
  (d) 0.04555 to 0.046

54. Each number was supposed to be rounded to two significant figures. Find the ones that were incorrectly rounded and correct them.
  (a) 1.249 × 10³ to 1.3 × 10³
  (b) 3.999 × 10² to 40
  (c) 56.21 to 56.2
  (d) 0.009964 to 0.010

55. Round the number on the left to the number of significant figures indicated as shown by the example in the first row. (Use scientific notation as needed to avoid ambiguity.)

Number	Rounded to 4 Significant Figures	Rounded to 2 Significant Figures	Rounded to 1 Significant Figure
1.45815	1.458	1.5	1
8.32466			
84.57225			
132.5512			

56. Round the number on the left to the number of significant figures indicated as shown by the example in the first row. (Use scientific notation as needed to avoid ambiguity.)

Number	Rounded to 4 Significant Figures	Rounded to 2 Significant Figures	Rounded to 1 Significant Figure
94.52118	94.52	95	9 × 10¹
105.4545			
0.455981			
0.009999991			

SIGNIFICANT FIGURES IN CALCULATIONS

57. Perform each calculation to the correct number of significant figures.
  (a) 4.5 × 0.03060 × 0.391
  (b) 5.55 ÷ 8.97
  (c) (7.890 × 10¹²) ÷ (6.7 × 10⁴)
  (d) 67.8 × 9.8 ÷ 100.04

58. Perform each calculation to the correct number of significant figures.
  (a) 89.3 × 77.0 × 0.08
  (b) (5.01 × 10⁵) ÷ (7.8 × 10²)
  (c) 4.005 × 74 × 0.007
  (d) 453 ÷ 2.031

59. Determine whether the answer to each calculation has the correct number of significant figures. If not, correct it.
(a) $34.00 \times 567 \div 4.564 = 4.2239 \times 10^3$
(b) $79.3 \div 0.004 \times 35.4 = 7 \times 10^5$
(c) $89.763 \div 22.4581 = 3.997$
(d) $(4.32 \times 10^{12}) \div (3.1 \times 10^{-4}) = 1.4 \times 10^{16}$

60. Determine whether the answer to each calculation has the correct number of significant figures. If not, correct it.
(a) $45.3254 \times 89.00205 = 4034.05$
(b) $0.00740 \times 45.0901 = 0.334$
(c) $49857 \div 904875 = 0.05510$
(d) $0.009090 \times 6007.2 = 54.605$

61. Perform each calculation to the correct number of significant figures.
(a) $87.6 + 9.888 + 2.3 + 10.77$
(b) $43.7 - 2.341$
(c) $89.6 + 98.33 - 4.674$
(d) $6.99 - 5.772$

62. Perform each calculation to the correct number of significant figures.
(a) $1459.3 + 9.77 + 4.32$
(b) $0.004 + 0.09879$
(c) $432 + 7.3 - 28.523$
(d) $2.4 + 1.777$

63. Determine whether the answer to each calculation has the correct number of significant figures. If not, correct it.
(a) $(3.8 \times 10^5) - (8.45 \times 10^5) = -4.7 \times 10^5$
(b) $0.00456 + 1.0936 = 1.10$
(c) $8475.45 - 34.899 = 8440.55$
(d) $908.87 - 905.34095 = 3.5291$

64. Determine whether the answer to each calculation has the correct number of significant figures. If not, correct it.
(a) $78.9 + 890.43 - 23 = 9.5 \times 10^2$
(b) $9354 - 3489.56 + 34.3 = 5898.74$
(c) $0.00407 + 0.0943 = 0.0984$
(d) $0.00896 - 0.007 = 0.00196$

65. Perform each calculation to the correct number of significant figures.
(a) $(78.4 - 44.889) \div 0.0087$
(b) $(34.6784 \times 5.38) + 445.56$
(c) $(78.7 \times 10^5 \div 88.529) + 356.99$
(d) $(892 \div 986.7) + 5.44$

66. Perform each calculation to the correct number of significant figures.
(a) $(1.7 \times 10^6 \div 2.63 \times 10^5) + 7.33$
(b) $(568.99 - 232.1) \div 5.3$
(c) $(9443 + 45 - 9.9) \times 8.1 \times 10^6$
(d) $(3.14 \times 2.4367) - 2.34$

67. Determine whether the answer to each calculation has the correct number of significant figures. If not, correct it.
(a) $(78.56 - 9.44) \times 45.6 = 3152$
(b) $(8.9 \times 10^5 \div 2.348 \times 10^2) + 121 = 3.9 \times 10^3$
(c) $(45.8 \div 3.2) - 12.3 = 2$
(d) $(4.5 \times 10^3 - 1.53 \times 10^3) \div 34.5 = 86$

68. Determine whether the answer to each calculation has the correct number of significant figures. If not, correct it.
(a) $(908.4 - 3.4) \div 3.52 \times 10^4 = 0.026$
(b) $(1206.7 - 0.904) \times 89 = 1.07 \times 10^5$
(c) $(876.90 + 98.1) \div 56.998 = 17.11$
(d) $(455 \div 407859) + 1.00098 = 1.00210$

UNIT CONVERSION

69. Perform each conversion within the metric system.
(a) 3.55 kg to grams
(b) 8944 mm to meters
(c) 4598 mg to kilograms
(d) 0.0187 L to milliliters

70. Perform each conversion within the metric system.
(a) 155.5 cm to meters
(b) 2491.6 g to kilograms
(c) 248 cm to millimeters
(d) 6781 mL to liters

71. Perform each conversion within the metric system.
(a) 5.88 dL to liters
(b) 3.41×10^{-5} g to micrograms
(c) 1.01×10^{-8} s to nanoseconds
(d) 2.19 pm to meters

72. Perform each conversion within the metric system.
(a) 1.08 Mm to kilometers
(b) 4.88 fs to picoseconds
(c) 7.39×10^{11} m to gigameters
(d) 1.15×10^{-10} m to picometers

73. Perform each conversion between the English and metric systems.
(a) 22.5 in. to centimeters
(b) 126 ft to meters
(c) 825 yd to kilometers
(d) 2.4 in. to millimeters

74. Perform each conversion between the English and metric systems.
(a) 78.3 in. to centimeters
(b) 445 yd to meters
(c) 336 ft to centimeters
(d) 45.3 in. to millimeters

75. Perform each conversion between the metric and English systems.
  (a) 40.0 cm to inches
  (b) 27.8 m to feet
  (c) 10.0 km to miles
  (d) 3845 kg to pounds

76. Perform each conversion between the metric and English systems.
  (a) 254 cm to inches
  (b) 89 mm to inches
  (c) 7.5 L to quarts
  (d) 122 kg to pounds

77. Complete the table:

m	km	Mm	Gm	Tm
5.08×10^8 m	_____	508 Mm	_____	_____
_____	_____	27,976 Mm	_____	
_____	_____	_____	_____	1.77 Tm
_____	1.5×10^5 km	_____	_____	_____
_____	_____	_____	423 Gm	_____

78. Complete the table:

s	ms	μs	ns	ps
1.31×10^{-4} s	_____	131 μs	_____	_____
_____	_____	_____	_____	12.6 ps
_____	_____	_____	155 ns	_____
_____	1.99×10^{-3} ms	_____	_____	_____
_____	_____	8.66×10^{-5} μs	_____	_____

79. Convert 2.255×10^{10} g to each unit:
  (a) kg
  (b) Mg
  (c) mg
  (d) metric tons (1 metric ton = 1000 kg)

80. Convert 1.88×10^{-6} g to each unit.
  (a) mg
  (b) cg
  (c) ng
  (d) μg

81. A student loses 3.3 lb in one month. How many grams did he lose?

82. A student gains 1.9 lb in two weeks. How many grams did he gain?

83. A runner wants to run 10.0 km. She knows that her running pace is 7.5 mi/h. How many minutes must she run? *Hint:* Use 7.5 mi/h as a conversion factor between distance and time.

84. A cyclist rides at an average speed of 24 mi/h. If she wants to bike 195 km, how long (in hours) must she ride?

85. A recipe calls for 5.0 qt of milk. What is this quantity in cubic centimeters?

86. A gas can holds 2.0 gal of gasoline. What is this quantity in cubic centimeters?

UNITS RAISED TO A POWER

87. Fill in the blanks.
  (a) $1.0 \text{ km}^2 =$ _____ m^2
  (b) $1.0 \text{ cm}^3 =$ _____ m^3
  (c) $1.0 \text{ mm}^3 =$ _____ m^3

88. Fill in the blanks.
  (a) $1.0 \text{ ft}^2 =$ _____ in.^2
  (b) $1.0 \text{ yd}^2 =$ _____ ft^2
  (c) $1.0 \text{ m}^2 =$ _____ yd^2

89. The hydrogen atom has a volume of approximately 6.2×10^{-31} m^3. What is this volume in each unit?
  (a) cubic picometers
  (b) cubic nanometers
  (c) cubic angstroms (1 angstrom = 10^{-10} m)

90. Earth has a surface area of 197 million square miles. What is its area in each unit?
  (a) square kilometers
  (b) square megameters
  (c) square decimeters

91. A modest-sized house has an area of 215 m^2. What is its area in each unit?
  (a) km^2
  (b) dm^2
  (c) cm^2

92. A classroom has a volume of 285 m^3. What is its volume in each unit?
  (a) km^3
  (b) dm^3
  (c) cm^3

93. Total U.S. farmland occupies 954 million acres. How many square miles is this?

(1 acre = 43,560 ft^2; 1 mi = 5280 ft)

94. The average U.S. farm occupies 435 acres. How many square miles is this?

(1 acre = 43,560 ft^2; 1 mi = 5280 ft)

DENSITY

95. A sample of an unknown metal has a mass of 35.4 g and a volume of 3.11 cm^3. Calculate its density and identify the metal by comparison to Table 2.4.

96. A new penny has a mass of 2.49 g and a volume of 0.349 cm^3. Is the penny pure copper?

97. Glycerol is a syrupy liquid often used in cosmetics and soaps. A 2.50-L sample of pure glycerol has a mass of 3.15×10^3 g. What is the density of glycerol in grams per cubic centimeter?

98. An aluminum engine block has a volume of 4.77 L and a mass of 12.88 kg. What is the density of the aluminum in grams per cubic centimeter?

99. A supposedly gold tooth crown is tested to determine its density. It displaces 10.7 mL of water and has a mass of 206 g. Could the crown be made of gold?

100. A vase is said to be solid platinum. It displaces 18.65 mL of water and has a mass of 157 g. Could the vase be solid platinum?

101. Ethylene glycol (antifreeze) has a density of 1.11 g/cm^3.
  (a) What is the mass in grams of 387 mL of this liquid?
  (b) What is the volume in liters of 3.46 kg of this liquid?

102. Acetone (fingernail-polish remover) has a density of 0.7857 g/cm^3.
  (a) What is the mass in grams of 17.56 mL of acetone?
  (b) What is the volume in milliliters of 7.22 g of acetone?

CUMULATIVE PROBLEMS

103. A thief uses a bag of sand to replace a gold statue that sits on a weight-sensitive, alarmed pedestal. The bag of sand and the statue have exactly the same volume, 1.75 L. (Assume that the mass of the bag is negligible.)
  (a) Calculate the mass of each object. (density of gold = 19.3 g/cm^3; density of sand = 3.00 g/cm^3)
  (b) Did the thief set off the alarm? Explain.

104. One of the particles that composes an atom is the proton. A proton has a radius of approximately 1.0×10^{-13} cm and a mass of 1.7×10^{-24} g. Determine the density of a proton.

$$\left(\text{volume of a sphere} = -\frac{4}{3}\,\pi r^3; \pi = 3.14 \right)$$

105. A block of metal has a volume of 13.4 in.^{3} and weighs 5.14 lb. What is its density in grams per cubic centimeter?

106. A log is either oak or pine. It displaces 2.7 gal of water and weighs 19.8 lb. Is the log oak or pine? (density of oak = 0.9 g/cm^3; density of pine = 0.4 g/cm^3)

107. The density of aluminum is 2.7 g/cm^3. What is its density in kilograms per cubic meter?

108. The density of platinum is 21.4 g/cm^3. What is its density in pounds per cubic inch?

109. A typical backyard swimming pool holds 150 yd^3 of water. What is the mass in pounds of the water?

110. An iceberg has a volume of 8975 ft^3. What is the mass in kilograms of the iceberg?

111. The mass of fuel in an airplane must be carefully accounted for before takeoff. If a 747 contains 155,211 L of fuel, what is the mass of the fuel in kilograms? Assume the density of the fuel to be 0.768 g/cm^3.

112. A backpacker carries 2.5 L of white gas as fuel for her stove. How many pounds does the fuel add to her load? Assume the density of white gas to be 0.79 g/cm^3.

113. Honda produces a hybrid electric car called the Honda Insight. The Insight has both a gasoline-powered engine and an electric motor and has an EPA gas mileage rating of 43 miles per gallon on the highway. What is the Insight's rating in kilometers per liter?

114. You rent a car in Germany with a gas mileage rating of 12.8 km/L. What is its rating in miles per gallon?

115. A car has a mileage rating of 38 miles per gallon of gasoline. How many miles can the car travel on 76.5 liters of gasoline?

116. A hybrid SUV consumes fuel at a rate of 12.8 km/L. How many miles can the car travel on 22.5 gallons of gasoline?

117. Block A of an unknown metal has a volume of 125 cm³. Block B of a different metal has a volume of 145 cm³. If block A has a greater mass than block B, what can be said of the relative densities of the two metals? (Assume that both blocks are solid.)

118. Block A of an unknown metal has a volume of 125 cm³. Block B of a different metal has a volume of 105 cm³. If block A has a greater mass than block B, what can be said of the relative densities of the two metals? (Assume that both blocks are solid.)

119. The masses and volumes of two cylinders are measured. The mass of cylinder 1 is 1.35 times the mass of cylinder 2. The volume of cylinder 1 is 0.792 times the volume of cylinder 2. If the density of cylinder 1 is 3.85 g/cm³, what is the density of cylinder 2?

120. A bag contains a mixture of copper and lead BBs. The average density of the BBs is 9.87 g/cm³. Assuming that the copper and lead are pure, determine the relative amounts of each kind of BB.

HIGHLIGHT PROBLEMS

121. In 1999, NASA lost a $94 million orbiter because one group of engineers used metric units in their calculations while another group used English units. Consequently, the orbiter descended too far into the Martian atmosphere and burned up. Suppose that the orbiter was to have established orbit at 155 km and that one group of engineers specified this distance as 1.55×10^5 m. Suppose further that a second group of engineers programmed the orbiter to go to 1.55×10^5 ft. What was the difference in kilometers between the two altitudes? How low did the probe go?

122. A NASA satellite showed that in 2009 the ozone hole over Antarctica had a maximum surface area of 24.1 million km². The largest ozone hole on record occurred in 2006 and had a surface area of 29.6 million km². Calculate the difference in diameter (in meters) between the ozone hole in 2009 and in 2006.

▲ A layer of ozone gas (a form of oxygen) in the upper atmosphere protects Earth from harmful ultraviolet radiation in sunlight. Human-made chemicals react with the ozone and deplete it, especially over the Antarctic at certain times of the year (the so-called ozone hole). The region of low ozone concentration in 2006 (represented here by the dark purple color) was the largest on record.

▲ The $94 million Mars Climate Orbiter was lost in the Martian atmosphere in 1999 because two groups of engineers failed to communicate to each other the units that they used in their calculations.

123. In 1999, scientists discovered a new class of black holes with masses 100 to 10,000 times the mass of our sun, but occupying less space than our moon. Suppose that one of these black holes has a mass of 1×10^3 suns and a radius equal to one-half the radius of our moon. What is its density in grams per cubic centimeter? The mass of the sun is 2.0×10^{30} kg, and the radius of the moon is

2.16×10^3 mi. $\left(\text{Volume of a sphere} = \dfrac{4}{3}\pi r^3.\right)$

124. A titanium bicycle frame contains the same amount of titanium as a titanium cube measuring 6.8 cm on a side. Use the density of titanium to calculate the mass in kilograms of titanium in the frame. What would be the mass of a similar frame composed of iron?

Titanium

6.8 cm

▲ A titanium bicycle frame contains the same amount of titanium as a titanium cube measuring 6.8 cm on a side.

►ANSWERS TO SKILLBUILDER EXERCISES

Skillbuilder 2.1 $\$1.2102 \times 10^{13}$

Skillbuilder 2.2 3.8×10^{-5}

Skillbuilder 2.3 ·103.4 °F

Skillbuilder 2.4
- (a) four significant figures
- (b) three significant figures
- (c) two significant figures
- (d) unlimited significant figures
- (e) three significant figures
- (f) ambiguous

Skillbuilder 2.5
- (a) 0.001 or 1×10^{-3}
- (b) 0.204

Skillbuilder 2.6
- (a) 7.6
- (b) 131.11

Skillbuilder 2.7
- (a) 1288
- (b) 3.12

Skillbuilder 2.8 22.0 in.

Skillbuilder 2.9 5.678 km

Skillbuilder 2.10 0.28 L

Skillbuilder 2.11 46.6 laps

Skillbuilder Plus, p. 31 1.06×10^4 m

Skillbuilder 2.12 4747 cm^3

Skillbuilder 2.13 1.52×10^5 in.^{3}

Skillbuilder 2.14 Yes, the density is 21.4 g/cm^3 and matches that of platinum.

Skillbuilder 2.15 4.4×10^{-2} cm^3

Skillbuilder Plus, p. 37 1.95 kg

Skillbuilder 2.16 16 kg

Skillbuilder 2.17 $d = 19.3$ g/cm^3; yes, the density is consistent with that of gold.

►ANSWERS TO CONCEPTUAL CHECKPOINTS

2.1 (c) Multiplying by 10^{-3} is equivalent to moving the decimal point three places to the left.

2.2 (b) The last digit is considered to be uncertain by ±1.

2.3 (b) The result of the calculation in (a) would be reported as 4; the result of the calculation in (b) would be reported as 1.5.

2.4 (d) The diameter would be expressed as 28 nm.

2.5 (c) Kilometers must appear in the numerator and meters in the denominator, and the conversion factor in (d) is incorrect (10^3 km $\neq$ 1 m).

2.6 (d) (3 ft) × (3 ft) × (3 ft) = 27 ft^3

Matter and Energy

"Thus, the task is, not so much to see what no one has yet seen; but to think what nobody has yet thought, about that which everybody sees."

ERWIN SCHRÖDINGER (1887–1961)

3.1 In Your Room

◄ Everything that you can see in this room is made of matter. As students of chemistry, we are interested in how the differences between different kinds of matter are related to the differences between the molecules and atoms that compose the matter. The molecular structures shown here are water molecules on the left and carbon atoms in graphite on the right.

Look around the room you are in—what do you see? You might see your desk, your bed, or a glass of water. Maybe you have a window and can see trees, grass, or mountains. You can certainly see this book and possibly the table it sits on. What are these things made of? They are all made of *matter*, which we will define more carefully shortly. For now, know that all you see is matter—your desk, your bed, the glass of water, the trees, the mountains, and this book. Some of what you don't see is matter as well. For example, you are constantly breathing air, which is also matter, into and out of your lungs. You feel the matter in air when you feel wind on your skin. Virtually everything is made of matter.

Think about the differences between different kinds of matter. Air is different from water, and water is different from wood. One of our first tasks as we learn about matter is to identify the similarities and differences among different kinds of matter. How are sugar and salt similar? How are air and water different? Why are they different? Why is a mixture of sugar and water similar to a mixture of salt and water but different from a mixture of sand and water? As students of chemistry, we are particularly interested in the similarities and differences between various kinds of matter and how these reflect the similarities and differences between their component atoms and molecules. We strive to understand the connection between the macroscopic world and the molecular one.

3.2 What Is Matter?

Matter is defined as anything that occupies space and has mass. Some types of matter—such as steel, water, wood, and plastic—are easily visible to our eyes. Other types of matter—such as air or microscopic dust—are impossible to see without magnification. Matter may sometimes appear smooth and continuous, but actually it is not. Matter is ultimately composed of **atoms**, submicroscopic particles that are the fundamental building blocks of matter (▼ Figure 3.1a). In many cases, these atoms are bonded together to form **molecules**, two or more atoms joined to one another in specific geometric arrangements (Figure 3.1b). Recent advances in microscopy have allowed us to image the atoms (▼ Figure 3.2) and molecules (▼ Figure 3.3) that compose matter, sometimes with stunning clarity.

Aluminum atoms

Aluminum can

Isopropyl alcohol molecules

Isopropyl alcohol (Rubbing alcohol)

(a) (b)

▲ **FIGURE 3.1 Atoms and molecules** All matter is ultimately composed of atoms. **(a)** In some substances, such as aluminum, the atoms exist as independent particles. **(b)** In other substances, such as rubbing alcohol, several atoms bond together in well-defined structures called molecules.

▲ **FIGURE 3.2 Scanning tunneling microscope image of nickel atoms** A scanning tunneling microscope (STM) creates an image by scanning a surface with a tip of atomic dimensions. It can distinguish individual atoms, seen as blue bumps, in this image. (*Source:* Reprint Courtesy of International Business Machines Corporation, copyright © International Business Machines Corporation.)

▲ **FIGURE 3.3 Scanning tunneling microscope image of a DNA molecule** DNA is the hereditary material that encodes the operating instructions for most cells in living organisms. In this image, the DNA molecule is yellow, and the double-stranded structure of DNA is discernible.

3.3 Classifying Matter According to Its State: Solid, Liquid, and Gas

The common **states of matter** are **solid**, **liquid**, and **gas** (▼ Figure 3.4). In solid matter, atoms or molecules pack close to each other in fixed locations. Although neighboring atoms or molecules in a solid may vibrate or oscillate, they do not move around each other, giving solids their familiar fixed volume and rigid shape.

▶ **FIGURE 3.4 Three states of matter** Water exists as ice (solid), water (liquid), and steam (gas). In ice, the water molecules are closely spaced and, although they vibrate about a fixed point, they do not generally move relative to one another. In liquid water, the water molecules are also closely spaced but are free to move around and past each other. In steam, water molecules are separated by large distances and do not interact significantly with one another.

Ice, diamond, quartz, and iron are examples of solid matter. Solid matter may be **crystalline**, in which case its atoms or molecules arrange in geometric patterns with long-range, repeating order (◀ Figure 3.5a), or it may be **amorphous**, in which case its atoms or molecules do not have long-range order (Figure 3.5b). Examples of *crystalline* solids include salt (▼ Figure 3.6) and diamond; the well-ordered, geometric shapes of salt and diamond crystals reflect the well-ordered geometric arrangement of their atoms. Examples of *amorphous* solids include glass, rubber, and plastic.

(a) Crystalline solid

(b) Amorphous solid

▲ **FIGURE 3.5 Types of solid matter** **(a)** In a crystalline solid, atoms or molecules occupy specific positions to create a well-ordered, three-dimensional structure. **(b)** In an amorphous solid, atoms do not have any long-range order.

▲ **FIGURE 3.6 Salt: a crystalline solid** Sodium chloride is an example of a crystalline solid. The well-ordered, cubic shape of salt crystals is due to the well-ordered, cubic arrangement of its atoms.

Solid—not compressible

Gas—compressible

In liquid matter, atoms or molecules are close to each other (about as close as molecules in a solid) but are free to move around and by each other. Like solids, liquids have a fixed volume because their atoms or molecules are in close contact. Unlike solids, however, liquids assume the shape of their container because the atoms or molecules are free to move relative to one another. Water, gasoline, alcohol, and mercury are all examples of liquid matter.

In gaseous matter, atoms or molecules are separated by large distances and are free to move relative to one another. Since the atoms or molecules that compose gases are not in contact with one another, gases are **compressible** (◄ Figure 3.7). When you inflate a bicycle tire, for example, you push more atoms and molecules into the same space, compressing them and making the tire harder. Gases always assume the shape and volume of their containers. Oxygen, helium, and carbon dioxide are all good examples of gases. Table 3.1 summarizes the properties of solids, liquids, and gases.

TABLE 3.1   Properties of Liquids, Solids, and Gases

State	Atomic/Molecular Motion	Atomic/Molecular Spacing	Shape	Volume	Compressibility
Solid	Oscillation/ vibration about fixed point	Close together	Definite	Definite	Incompressible
Liquid	Free to move relative to one another	Close together	Indefinite	Definite	Incompressible
Gas	Free to move relative to one another	Far apart	Indefinite	Indefinite	Compressible

◄ FIGURE 3.7   **Gases are compressible**   Since the atoms or molecules that compose gases are not in contact with one another, gases can be compressed.

3.4   Classifying Matter According to Its Composition: Elements, Compounds, and Mixtures

In addition to classifying matter according to its state, we can classify it according to its composition (▶ Figure 3.8). Matter may be either a **pure substance**, composed of only one type of atom or molecule, or a **mixture**, composed of two or more different types of atoms or molecules combined in variable proportions.

Pure substances are composed of only one type of atom or molecule. Helium and water are both pure substances. The atoms that compose helium are all helium atoms, and the molecules that compose water are all water molecules—no other atoms or molecules are mixed in.

Pure substances can themselves be divided into two types: elements and compounds. Copper is an example of an **element**, a substance that cannot be broken down into simpler substances. The graphite in pencils is also an element—carbon. No chemical transformation can decompose graphite into simpler substances; it is pure carbon. All known elements are listed in the periodic table in the inside front cover of this book and in alphabetical order on the inside back cover of this book.

A pure substance can also be a **compound**, a substance composed of two or more elements in fixed definite proportions. Compounds are more common than pure elements because most elements are chemically reactive and combine with other elements to form compounds. Water, table salt, and sugar are examples of compounds; they can all be decomposed into simpler substances. If you heat sugar on a pan over a flame, you decompose it into several substances including carbon

A compound is composed of different atoms that are chemically united (bonded). A mixture is composed of different substances that are not chemically united, but simply mixed together.

Pure substance

Helium atoms

▲ Helium is a pure substance composed only of helium atoms.

Pure substance

Water molecules

▲ Water is a pure substance composed only of water molecules.

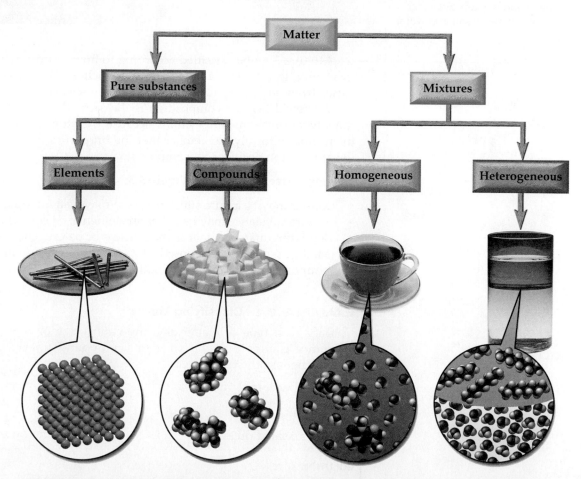

▲ **FIGURE 3.8   Classification of matter**   Matter may be a pure substance or a mixture. A pure substance may be either an element (such as copper) or a compound (such as sugar), and a mixture may be either homogeneous (such as sweetened tea) or heterogeneous (such as hydrocarbon and water).

(an element) and gaseous water (a different compound). The black substance left on your pan after burning contains the carbon; the water escapes into the air as steam.

The majority of matter that we encounter is in the form of mixtures. Apple juice, a flame, salad dressing, and soil are all examples of mixtures; they each contain several substances mixed together in proportions that vary from one sample to another. Other common mixtures include air, seawater, and brass. Air is a mixture composed primarily of nitrogen and oxygen gas, seawater is a mixture composed primarily of salt and water, and brass is a mixture composed of copper and zinc. Each of these mixtures can have different proportions of its constituent components. For example, metallurgists vary the relative amounts of copper and zinc in brass to tailor the metal's properties to its intended use—the higher the zinc content relative to the copper content, the more brittle the brass.

Air and seawater are mixtures

Oxygen

Nitrogen

Water

Salt

▶ Air and seawater are examples of mixtures. Air contains primarily nitrogen and oxygen. Seawater contains primarily salt and water.

Mixtures can be classified according to how uniformly the substances within them mix. In a **heterogeneous mixture**, such as oil and water, the composition varies from one region to another. In a **homogeneous mixture**, such as salt water or sweetened tea, the composition is the same throughout. Homogeneous mixtures have uniform compositions because the atoms or molecules that compose them mix uniformly. Remember that the properties of matter are determined by the atoms or molecules that compose it.

To summarize, as shown in Figure 3.8:

- Matter may be a pure substance, or it may be a mixture.
- A pure substance may be either an element or a compound.
- A mixture may be either homogeneous or heterogeneous.
- Mixtures may be composed of two or more elements, two or more compounds, or a combination of both.

EXAMPLE 3.1 Classifying Matter

Classify each type of matter as a pure substance or a mixture. If it is a pure substance, classify it as an element or a compound; if it is a mixture, classify it as homogeneous or heterogeneous.

(a) a lead weight
(b) seawater
(c) distilled water
(d) Italian salad dressing

SOLUTION
Begin by examining the alphabetical listing of pure elements inside the back cover of this text. If the substance appears in that table, it is a pure substance and an element. If it is not in the table but is a pure substance, then it is a compound.

If the substance is not a pure substance, then it is a mixture. Refer to your everyday experience with each mixture to determine if it is homogeneous or heterogeneous.

(a) Lead is listed in the table of elements. It is a pure substance and an element.

(b) Seawater is composed of several substances, including salt and water; it is a mixture. It has a uniform composition, so it is a homogeneous mixture.

(c) Distilled water is not listed in the table of elements, but it is a pure substance (water); therefore, it is a compound.

(d) Italian salad dressing contains a number of substances and is therefore a mixture. It usually separates into at least two distinct regions with different composition and is therefore a heterogeneous mixture.

▶**SKILLBUILDER 3.1 | Classifying Matter**

Classify each type of matter as a pure substance or a mixture. If it is a pure substance, classify it as an element or a compound. If it is a mixture, classify it as homogeneous or heterogeneous.

(a) mercury in a thermometer

(b) exhaled air

(c) minestrone soup

(d) sugar

▶**FOR MORE PRACTICE**   Example 3.12; Problems 31, 32, 33, 34, 35, 36.

Note: The answers to all Skillbuilders appear at the end of the chapter.

3.5   How We Tell Different Kinds of Matter Apart: Physical and Chemical Properties

The characteristics that distinguish one substance from another are called **properties**. Different substances have unique properties that characterize them and distinguish them from other substances. For example, we can distinguish water from alcohol based on their different smells, or we can distinguish gold from silver based on their different colors.

In chemistry, we categorize properties into two different types: physical and chemical. A **physical property** is one that a substance displays without changing its composition. A **chemical property** is one that a substance displays only through changing its composition. For example, the characteristic odor of gasoline is a physical property—gasoline does not change its composition when it exhibits its odor. On the other hand, the flammability of gasoline is a chemical property—gasoline does change its composition when it burns.

The atomic or molecular composition of a substance does not change when the substance displays its physical properties. For example, the boiling point of water—a physical property—is 100 °C. When water boils, it changes from a liquid to a gas, but the gas is still water (◀ Figure 3.9).

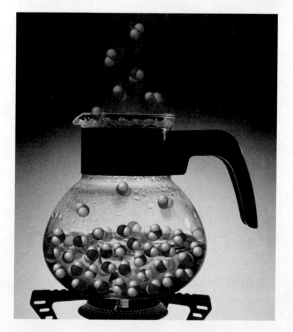

◀ **FIGURE 3.9   A physical property**   The boiling point of water is a physical property, and boiling is a physical change. When water boils, it turns into a gas, but the water molecules are the same in both the liquid water and the gaseous steam.

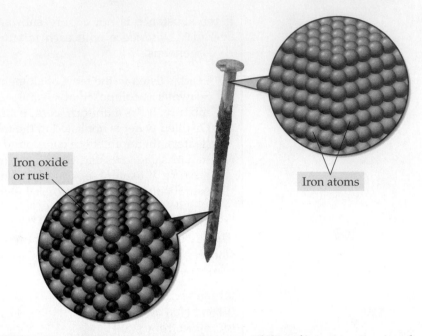

▲ **FIGURE 3.10 A chemical property** The susceptibility of iron to rusting is a chemical property, and rusting is a chemical change. When iron rusts, it turns from iron to iron oxide.

On the other hand, the susceptibility of iron to rust is a chemical property—iron must change into iron oxide to display this property (▲ Figure 3.10). Physical properties include odor, taste, color, appearance, melting point, boiling point, and density. Chemical properties include corrosiveness, flammability, acidity, and toxicity.

EXAMPLE 3.2 Physical and Chemical Properties

Determine whether each property is physical or chemical.

(a) the tendency of copper to turn green when exposed to air
(b) the tendency of automobile paint to dull over time
(c) the tendency of gasoline to evaporate quickly when spilled
(d) the low mass (for a given volume) of aluminum relative to other metals

SOLUTION

(a) Copper turns green because it reacts with gases in air to form compounds; this is a chemical property.
(b) Automobile paint dulls over time because it can fade (decompose) due to sunlight or it can react with oxygen in air. In either case, this is a chemical property.
(c) Gasoline evaporates quickly because it has a low boiling point; this is a physical property.
(d) Aluminum's low mass (for a given volume) relative to other metals is due to its low density; this is a physical property.

▶**SKILLBUILDER 3.2 | Physical and Chemical Properties**

Determine whether each property is physical or chemical.

(a) the explosiveness of hydrogen gas
(b) the bronze color of copper
(c) the shiny appearance of silver
(d) the ability of dry ice to sublime (change from solid directly to vapor)

▶**FOR MORE PRACTICE** Example 3.13; Problems 37, 38, 39, 40.

3.6   How Matter Changes: Physical and Chemical Changes

Every day, we witness changes in matter: Ice melts, iron rusts, and fruit ripens, for example. What happens to the atoms and molecules that make up these substances during the change? The answer depends on the kind of change. In a **physical change**, matter changes its appearance but not its composition. For example, when ice melts, it looks different—water looks different from ice—but its composition is the same. Solid ice and liquid water are both composed of water molecules, so melting is a physical change. Similarly, when glass shatters, it looks different, but its composition remains the same—it is still glass. Again, this is a physical change. On the other hand, in a **chemical change**, matter *does* change its composition. For example, copper turns green upon continued exposure to air because it reacts with gases in air to form new compounds. This is a chemical change. Matter undergoes a chemical change when it undergoes a **chemical reaction**. In a chemical reaction, the substances present before the chemical change are called **reactants**, and the substances present after the change are called **products**:

$$\text{Reactants} \xrightarrow[\text{Chemical Change}]{} \text{Products}$$

We cover chemical reactions in much more detail in Chapter 7.

The differences between physical and chemical changes are not always apparent. Only chemical examination of the substances before and after the change can verify whether the change is physical or chemical. For many cases, however, we can identify chemical and physical changes based on what we know about the changes. Changes in state, such as melting or boiling, or changes that involve merely appearance, such as those produced by cutting or crushing, are always physical changes. Changes involving chemical reactions—often evidenced by heat exchange or color changes—are always chemical changes.

The main difference between chemical and physical changes is related to the changes at the molecular and atomic level. In physical changes, the atoms that compose the matter *do not* change their fundamental associations, even though the matter may change its appearance. In chemical changes, atoms do change their fundamental associations, resulting in matter with a new identity. *A physical change results in a different form of the same substance, while a chemical change results in a completely new substance.*

Consider physical and chemical changes in liquid butane, the substance used to fuel butane lighters. In many lighters, you can see the liquid butane through the plastic case of the lighter. If you push the fuel button on the lighter without turning the flint, some of the liquid butane *vaporizes* (changes from liquid to gas). If you listen carefully you can usually hear hissing as the gaseous butane leaks out (◀ Figure 3.11). Since the liquid butane and the gaseous butane are both composed of butane molecules, the change is physical. On the other hand, if you push the button *and* turn the flint to create a spark, a chemical change occurs. The butane molecules react with oxygen molecules in air to form new molecules, carbon dioxide and water (◀ Figure 3.12). The change is chemical because the molecular composition changes upon burning.

> State changes—transformations from one state of matter (such as solid or liquid) to another—are always physical changes.

▼ FIGURE 3.11   Vaporization: a physical change If you push the button on a lighter without turning the flint, some of the liquid butane vaporizes to gaseous butane. Since the liquid butane and the gaseous butane are both composed of butane molecules, this is a physical change.

Physical change

Gaseous butane

Liquid butane

Chemical change

Carbon dioxide and water molecules

Liquid butane

◀ FIGURE 3.12   Burning: a chemical change If you push the button *and* turn the flint to create a spark, you produce a flame. The butane molecules react with oxygen molecules in air to form new molecules, carbon dioxide and water. This is a chemical change.

EXAMPLE 3.3   Physical and Chemical Changes

Determine whether each change is physical or chemical.

(a) the rusting of iron
(b) the evaporation of fingernail-polish remover (acetone) from the skin
(c) the burning of coal
(d) the fading of a carpet upon repeated exposure to sunlight

SOLUTION

(a) Iron rusts because it reacts with oxygen in air to form iron oxide; therefore, this is a chemical change.
(b) When fingernail-polish remover (acetone) evaporates, it changes from liquid to gas, but it remains acetone; therefore, this is a physical change.
(c) Coal burns because it reacts with oxygen in air to form carbon dioxide; this is a chemical change.
(d) A carpet fades on repeated exposure to sunlight because the molecules that give the carpet its color are decomposed by sunlight; this is a chemical change.

▶SKILLBUILDER 3.3 | Physical and Chemical Changes

Determine whether each change is physical or chemical.

(a) copper metal forming a blue solution when it is dropped into colorless nitric acid
(b) a train flattening a penny placed on a railroad track
(c) ice melting into liquid water
(d) a match igniting a firework

▶FOR MORE PRACTICE   Example 3.14; Problems 41, 42, 43, 44.

✔ CONCEPTUAL CHECKPOINT 3.1

In this figure liquid water is being vaporized into steam.

$H_2O(l)$

Which diagram best represents the molecules in the steam?

(a)   (b)   (c)

Note: The answers to all Conceptual Checkpoints appear at the end of the chapter.

▶ **FIGURE 3.13   Separating a mixture of two liquids by distillation** The liquid with the lower boiling point vaporizes first. The vapors are collected and cooled (with cold water) until they condense back into liquid form.

Distillation

Condenser

Mixture of liquids

Boiling flask

Cold water out

Cold water in

Clamp

Receiving flask

Burner

More volatile liquid

Filtration

Stirring rod

Mixture of liquid and solid

Funnel

Filter paper traps solid

Liquid component of mixture

▲ **FIGURE 3.14   Separating a solid from a liquid by filtration**

SEPARATING MIXTURES THROUGH PHYSICAL CHANGES

Chemists often want to separate mixtures into their components. Such separations can be easy or difficult, depending on the components in the mixture. In general, mixtures are separable because the different components have different properties. Various techniques that exploit these differences can be used to achieve separation. For example, oil and water are immiscible (do not mix) and have different densities. For this reason, oil floats on top of water and can be separated from water by **decanting**—carefully pouring off—the oil into another container. Mixtures of miscible liquids can usually be separated by **distillation**, a process in which the mixture is heated to boil off the more **volatile**—the more easily vaporizable—liquid. The volatile liquid is then recondensed in a condenser and collected in a separate flask (▲ Figure 3.13). If a mixture is composed of a solid and a liquid, the two can be separated by **filtration**, in which the mixture is poured through filter paper usually held in a funnel (◀ Figure 3.14).

3.7   Conservation of Mass: There Is No New Matter

As we have seen, our planet, our air, and even our own bodies are composed of matter. Physical and chemical changes do not destroy matter, nor do they create new matter. Recall from Chapter 1 that Antoine Lavoisier, by studying combustion, established the law of conservation of mass, which states:

> Matter is neither created nor destroyed in a chemical reaction.

This law is a slight oversimplification. In nuclear reactions, covered in Chapter 17, significant changes in mass can occur. In chemical reactions, however, the changes are so minute that they can be ignored.

During physical and chemical changes, the total amount of matter remains constant even though it may not initially appear that it has. When we burn butane in a lighter, for example, the butane slowly disappears. Where does it go? It combines with oxygen to form carbon dioxide and water that travel into the surrounding air. The mass of the carbon dioxide and water that form, however, exactly equals the mass of the butane and oxygen that combined.

We examine the quantitative relationships in chemical reactions in Chapter 8.

Suppose that we burn 58 g of butane in a lighter. It will react with 208 g of oxygen to form 176 g of carbon dioxide and 90 g of water.

$$\text{Butane} + \text{Oxygen} \longrightarrow \text{Carbon Dioxide} + \text{Water}$$
$$\underbrace{58\,\text{g} + 208\,\text{g}}_{266\,\text{g}} \qquad \underbrace{176\,\text{g} + 90\,\text{g}}_{266\,\text{g}}$$

The sum of the masses of the butane and oxygen, 266 g, is equal to the sum of the masses of the carbon dioxide and water, which is also 266 g. In this chemical reaction, as in all chemical reactions, matter is conserved.

EXAMPLE 3.4 Conservation of Mass

A chemist forms 16.6 g of potassium iodide by combining 3.9 g of potassium with 12.7 g of iodine. Show that these results are consistent with the law of conservation of mass.

SOLUTION

The sum of the masses of the potassium and iodine is:

$$3.9\,\text{g} + 12.7\,\text{g} = 16.6\,\text{g}$$

The sum of the masses of potassium and iodine equals the mass of the product, potassium iodide. The results are consistent with the law of conservation of mass.

▶SKILLBUILDER 3.4 | Conservation of Mass

Suppose 12 g of natural gas combines with 48 g of oxygen in a flame. The chemical change produces 33 g of carbon dioxide. How many grams of water form?

▶FOR MORE PRACTICE Example 3.15; Problems 45, 46, 47, 48, 49, 50.

CONCEPTUAL CHECKPOINT 3.2

Consider a drop of water that is put into a flask, sealed with a cap, and heated until the droplet vaporizes. Is the mass of the container and water different after heating?

3.8 Energy

▲ Water behind a dam contains potential energy.

Matter is one of the two major components of our universe. The other major component is **energy**, *the capacity to do work*. **Work** is defined as the result of a force acting on a distance. For example, if you push this book across your desk, you have done work. You may at first think that chemistry is concerned only with matter, but the behavior of matter is driven in large part by energy, so understanding energy is critical to understanding chemistry. Like matter, energy is conserved. The **law of conservation of energy** states that *energy is neither created nor destroyed*. The total amount of energy is constant; energy can be changed from one form to another or transferred from one object to another, but it cannot be created out of nothing, and it does not vanish into nothing.

Virtually all samples of matter have energy. The total energy of a sample of matter is the sum of its **kinetic energy**, the energy associated with its motion, and its **potential energy**, the energy associated with its position or composition. For example, a moving billiard ball contains *kinetic energy* because it is *moving* at some speed across the billiard table. Water behind a dam contains *potential energy* because it is held at a high *position* in the Earth's gravitational field by the dam.

CHEMISTRY IN THE ENVIRONMENT
Getting Energy out of Nothing?

The law of conservation of energy has significant implications for energy use. The best we can do with energy is break even (and even that is not really possible); we can't continually draw energy from a device without putting energy into it. A device that supposedly produces energy without the need for energy input is sometimes called a *perpetual motion machine* (▼ Figure 3.15) and, according to the law of conservation of energy, cannot exist. Occasionally, the media report or speculate on the discovery of a system that appears to produce more energy than it consumes. For example, I once heard a radio talk show on the subject of energy and gasoline costs. The reporter suggested that we simply design an electric car that recharges itself while being driven. The battery in the electric car would charge during operation in the same way that the battery in a conventional car recharges, except the electric car would run with energy from the battery. Although people have dreamed of machines such as this for decades, such ideas violate the law of conservation of energy because they produce energy without any energy input. In the case of the perpetually moving electric car, the fault lies in the idea that driving the electric car can recharge the battery—it can't.

The battery in a conventional car recharges because energy from gasoline combustion is converted into electrical energy that then charges the battery. The electric car needs energy to move forward, and the battery will eventually discharge as it provides that energy. Hybrid cars (electric and gasoline-powered) such as the Toyota Prius can capture some limited energy from braking and use that energy to recharge the battery. However, they could never run indefinitely without the addition of fuel. Our society has a continual need for energy, and as our current energy resources dwindle, new energy sources will be required. Unfortunately, those sources must also follow the law of conservation of energy—energy must be conserved.

◀ **FIGURE 3.15**
A proposed perpetual motion machine The rolling balls supposedly keep the wheel perpetually spinning. **Question:** Can you explain why this would not work?

CAN YOU ANSWER THIS? *A friend asks you to invest in a new flashlight he invented that never needs batteries. What questions should you ask before writing a check?*

When the water flows through the dam from a higher position to a lower position, it can turn a turbine and produce electrical energy. **Electrical energy** is the energy associated with the flow of electrical charge. **Thermal energy** is the energy associated with the random motions of atoms and molecules in matter. The hotter an object, the more thermal energy it contains.

Chemical systems contain **chemical energy**, a form of potential energy associated with the positions of the particles that compose the chemical system. For example, the molecules that compose gasoline contain a substantial amount of chemical energy. They are a bit like the water behind a dam. Burning the gasoline is analogous to releasing the water from the dam. The chemical energy present in the gasoline is released upon burning. When we drive a car, we use that chemical energy to move the car forward. When we heat a home, we use chemical energy stored in natural gas to produce heat and warm the air in the house.

UNITS OF ENERGY

Several different energy units are in common use. The SI unit of energy is the joule (J), named after the English scientist James Joule (1818–1889), who demonstrated that energy could be converted from one type to another as long as the total energy was conserved. A second unit of energy is the **calorie (cal)**, the amount of energy required to raise the temperature of 1 g of water by 1 °C. A calorie is a larger unit than a joule: 1 cal = 4.184 J. A related energy unit is the nutritional or *capital C* **Calorie (Cal)**, equivalent to 1000 *little c* calories. Electricity bills usually come in yet another energy unit, the **kilowatt-hour (kWh)**. The average cost of residential electricity in the United States is about $0.12 per kilowatt-hour. Table 3.2 lists various energy units and their conversion factors. Table 3.3 shows the amount of energy required for various processes in each of these units.

TABLE 3.2   Energy Conversion Factors

1 calorie (cal)	=	4.184 joules (J)
1 Calorie (Cal)	=	1000 calories (cal)
1 kilowatt-hour (kWh)	=	3.60×10^6 joules (J)

TABLE 3.3 Energy Use in Various Units

Unit	Energy Required to Raise Temperature of 1 g of Water by 1 °C	Energy Required to Light 100-W Bulb for 1 Hour	Total Energy Used by Average U.S. Citizen in 1 Day
joule (J)	4.18	3.6×10^5	9.0×10^8
calorie (cal)	1.00	8.60×10^4	2.2×10^8
Calorie (Cal)	0.00100	86.0	2.2×10^5
kilowatt-hour (kWh)	1.16×10^{-6}	0.100	2.50×10^2

EXAMPLE 3.5 Conversion of Energy Units

A candy bar contains 225 Cal of nutritional energy. How many joules does it contain?

SORT Begin by sorting the information in the problem. Here you are *given* energy in Calories and asked to *find* energy in joules.	**GIVEN:** 225 Cal **FIND:** J

STRATEGIZE
Draw a solution map. Begin with Cal, convert to cal, and then convert to J.

SOLUTION MAP

$$\frac{1000 \text{ cal}}{1 \text{ Cal}} \qquad \frac{4.184 \text{ J}}{1 \text{ cal}}$$

RELATIONSHIPS USED

1000 calories = 1 Cal (Table 3.2)
4.184 J = 1 cal (Table 3.2)

SOLVE
Follow the solution map to solve the problem. Begin with 225 Cal and multiply by the appropriate conversion factors to arrive at J. Round the answer to the correct number of significant figures (in this case, three because of the three significant figures in 225 Cal).

SOLUTION

$$225 \text{ Cal} \times \frac{1000 \text{ cal}}{1 \text{ Cal}} \times \frac{4.184 \text{ J}}{1 \text{ cal}} = 9.41 \times 10^5 \text{ J}$$

CHECK
Check your answer. Are the units correct? Does the answer make physical sense?

The units of the answer (J) are the desired units. The magnitude of the answer makes sense because the J is a smaller unit than the Cal; therefore, the quantity of energy in J should be greater than the quantity in Cal.

▶**SKILLBUILDER 3.5 | Conversion of Energy Units**

The complete combustion of a small wooden match produces approximately 512 cal of heat. How many kilojoules are produced?

▶**SKILLBUILDER PLUS**

Convert 2.75×10^4 kJ to calories.

▶**FOR MORE PRACTICE**   Example 3.16; Problems 51, 52, 53, 54, 55, 56, 57, 58.

✔ **CONCEPTUAL CHECKPOINT 3.3**

Suppose a salesperson wants to make an appliance seem as efficient as possible. In which units would the yearly energy consumption of the appliance have the lowest numerical value and therefore seem most efficient?

(a) J

(b) cal

(c) Cal

(d) kWh

3.9 Energy and Chemical and Physical Change

When discussing energy transfer, we often define the object of our study (such as a flask in which a chemical reaction is occurring) as the *system.* The system then exchanges energy with its *surroundings.* In other words, we view energy changes as an exchange of energy between the system and the surroundings.

The physical and chemical changes that we discussed in Section 3.6 are usually accompanied by energy changes. For example, when water evaporates from your skin (a physical change), the water molecules absorb energy, cooling your skin. When you burn natural gas on the stove (a chemical change), energy is released, heating the food you are cooking.

The release of energy during a chemical reaction is analogous to the release of energy that occurs when you drop a weight to the ground. When you lift a weight, you raise its potential energy; when you drop it, the potential energy is released. *Systems with high potential energy—like the raised weight—have a tendency to change in a way that lowers their potential energy.* For this reason, objects or systems with high potential energy tend to be *unstable.* A weight lifted several meters from the ground is unstable because it contains a significant amount of localized potential energy. Unless restrained, the weight will fall, lowering its potential energy.

Some chemical substances are like the raised weight just described. For example, the molecules that compose TNT (trinitrotoluene) have a relatively high potential energy—energy is concentrated in them just as energy is concentrated in the raised weight. TNT molecules therefore tend to undergo rapid chemical changes that lower their potential energy, which is why TNT is explosive. Chemical reactions that *release* energy, like the explosion of TNT, are said to be **exothermic**.

Some chemical reactions behave in just the opposite way—they *absorb* energy from their surroundings as they occur. Such reactions are said to be **endothermic**. The use of a chemical cold pack is a good example of an endothermic reaction. When a barrier separating the reactants in a chemical cold pack is broken, the substances mix, react, and absorb heat from the surroundings. The surroundings—possibly including your bruised ankle—get colder.

We can represent the energy changes that occur during a chemical reaction with an energy diagram, as shown in ◄ Figure 3.16. In an exothermic reaction (Figure 3.16a), the reactants have greater energy than the products, and energy is released as the reaction occurs. In an endothermic reaction (Figure 3.16b), the products have more energy than the reactants, and energy is absorbed as the reaction occurs.

If a particular reaction or process is exothermic, then the reverse process must be endothermic. For example, the evaporation of water from your skin is endothermic (and therefore cools you off), but the condensation of water onto your skin is exothermic (which is why steam burns can be so painful and dangerous).

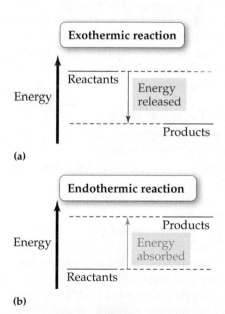

▲ **FIGURE 3.16 Exothermic and endothermic reactions** **(a)** In an exothermic reaction, energy is released. **(b)** In an endothermic reaction, energy is absorbed.

EXAMPLE 3.6 Exothermic and Endothermic Processes

Identify each change as exothermic or endothermic.

(a) wood burning in a fire
(b) ice melting

SOLUTION

(a) When wood burns, it emits heat into the surroundings. Therefore, the process is exothermic.
(b) When ice melts, it absorbs heat from the surroundings. For example, when ice melts in a glass of water, it cools the water as the melting ice absorbs heat from the water. Therefore, the process is endothermic.

▶**SKILLBUILDER 3.6 | Exothermic and Endothermic Processes**

Identify each change as exothermic or endothermic.

(a) water freezing into ice
(b) natural gas burning

▶**FOR MORE PRACTICE** Problems 61, 62, 63, 64.

3.10 Temperature: Random Motion of Molecules and Atoms

The atoms and molecules that compose matter are in constant random motion—they contain *thermal energy*. The **temperature** of a substance is a measure of its thermal energy. The hotter an object, the greater the random motion of the atoms and molecules that compose it, and the higher its temperature. We must be careful to not confuse *temperature* with *heat*. **Heat**, which has units of energy, is the *transfer* or *exchange* of thermal energy caused by a temperature difference. For example, when a cold ice cube is dropped into a warm cup of water, heat is transferred from the water to the ice, resulting in the cooling of the water. Temperature, by contrast, is a *measure* of the thermal energy of matter (not the exchange of thermal energy).

Three different temperature scales are in common use. The most familiar in the United States is the **Fahrenheit (°F) scale**. On the Fahrenheit scale, water freezes at 32 °F and boils at 212 °F. Room temperature is approximately 72 °F. The Fahrenheit scale was initially set up by assigning 0 °F to the freezing point of a concentrated saltwater solution and 96 °F to normal body temperature (although body temperature is now known to be 98.6 °F).

The scale used by scientists is the **Celsius (°C) scale**. On this scale, water freezes at 0 °C and boils at 100 °C. Room temperature is approximately 22 °C.

The Fahrenheit and Celsius scales differ in both the size of their respective degrees and the temperature each calls "zero" (▶ Figure 3.17). Both the Fahrenheit and Celsius scales contain negative temperatures. A third temperature scale, called the **Kelvin (K) scale**, avoids negative temperatures by assigning 0 K to the coldest temperature possible, absolute zero. Absolute zero (−273.15 °C or −459.7 °F) is the temperature at which molecular motion virtually stops. There is no lower temperature. The kelvin degree, or kelvin (K), is the same size as the Celsius degree—the only difference is the temperature that each scale designates as zero.

We can convert between these temperature scales using the following formulas.

The degree symbol is used with the Celsius and Fahrenheit scales, but not with the Kelvin scale.

$$K = °C + 273.15$$

$$°C = \frac{(°F - 32)}{1.8}$$

▶ FIGURE 3.17 **Comparison of the Fahrenheit, Celsius, and Kelvin temperature scales** The Fahrenheit degree is five-ninths the size of a Celsius degree. The Celsius degree and the kelvin degree are the same size.

For example, suppose we want to convert 212 K to Celsius. Following the procedure for solving numerical problems (Section 2.6), we first sort the information in the problem statement:

GIVEN: 212 K

FIND: °C

In a solution map involving a formula, the formula establishes the relationship between the variables. However, the formula under the arrow is not necessarily solved for the correct variable until later, as is the case here.

SOLUTION MAP
We then strategize by building a solution map.

$$K = °C + 273.15$$

RELATIONSHIPS USED
$K = °C + 273.15$ (This equation relates the *given* quantity (K) to the *find* quantity (°C) and is given in this section.)

SOLUTION
Finally, we follow the solution map to solve the problem. The equation below the arrow shows the relationship between K and °C, but it is not solved for the correct variable. Before using the equation, we must solve it for °C.

$$K = °C + 273.15$$

$$°C = K - 273.15$$

We can now substitute the given value for K and compute the answer to the correct number of significant figures.

$$°C = 212 - 273.15$$

$$= -61 °C$$

EXAMPLE 3.7 Converting between Celsius and Kelvin Temperature Scales

Convert −25 °C to kelvins.

SORT You are given a temperature in degrees Celsius and asked to find the value of the temperature in kelvins.	GIVEN: −25 °C FIND: K
STRATEGIZE Draw a solution map. Use the equation that relates the temperature in kelvins to the temperature in Celsius to convert from the given quantity to the quantity you are asked to find.	SOLUTION MAP K = °C + 273.15 RELATIONSHIPS USED K = °C + 273.15 (presented in this section)
SOLVE Follow the solution map to solve the problem by substituting the correct value for °C and calculating the answer to the correct number of significant figures.	SOLUTION K = °C + 273.15 K = −25 °C + 273.15 = 248 K
CHECK Check your answer. Are the units correct? Does the answer make physical sense?	The units (K) are correct. The answer makes sense because the value in kelvins should be a more positive number than the value in degrees Celsius.

▶ SKILLBUILDER 3.7 | Converting between Celsius and Kelvin Temperature Scales

Convert 358 K to Celsius.

▶ FOR MORE PRACTICE Example 3.17; Problems 65c, 66d.

EXAMPLE 3.8 Converting between Fahrenheit and Celsius Temperature Scales

Convert 55 °F to Celsius.

SORT You are given a temperature in degrees Fahrenheit and asked to find the value of the temperature in degrees Celsius.	GIVEN: 55 °F FIND: °C
STRATEGIZE Draw the solution map. Use the equation that shows the relationship between the given quantity (°F) and the find quantity (°C).	SOLUTION MAP $°C = \dfrac{(°F - 32)}{1.8}$ RELATIONSHIPS USED $°C = \dfrac{(°F - 32)}{1.8}$ (presented in this section)
SOLVE Substitute the given value into the equation and calculate the answer to the correct number of significant figures.	SOLUTION $°C = \dfrac{(°F - 32)}{1.8}$ $°C = \dfrac{(55 - 32)}{1.8} = 12.778 °C = 13 °C$

CHECK

Check your answer. Are the units correct? Does the answer make physical sense?

The units (°C) are correct. The value of the answer (13 °C) is smaller than the value in degrees Fahrenheit. For positive temperatures, the value of a temperature in degrees Celsius will always be smaller than the value in degrees Fahrenheit because the Fahrenheit degree is smaller than the Celsius degree and the Fahrenheit scale is offset by 32 degrees (see Figure 3.17).

▶**SKILLBUILDER 3.8 | Converting between Fahrenheit and Celsius Temperature Scales**

Convert 139 °C to Fahrenheit.

▶**FOR MORE PRACTICE**   Example 3.18; Problems 65a, 66a, c.

EXAMPLE 3.9   Converting between Fahrenheit and Kelvin Temperature Scales

Convert 310 K to Fahrenheit.

SORT

You are given a temperature in kelvins and asked to find the value of the temperature in degrees Fahrenheit.

GIVEN: 310 K

FIND: °F

STRATEGIZE

Build the solution map, which requires two steps: one to convert kelvins to degrees Celsius and one to convert degrees Celsius to degrees Fahrenheit.

SOLUTION MAP

$$\boxed{K} \longrightarrow \boxed{°C} \longrightarrow \boxed{°F}$$

$$K = °C + 273.15 \qquad °C = \frac{(°F - 32)}{1.8}$$

RELATIONSHIPS USED

$K = °C + 273.15$ (presented in this section)

$°C = \dfrac{(°F - 32)}{1.8}$ (presented in this section)

SOLVE

Solve the first equation for °C and substitute the given quantity in K to convert it to °C.

Solve the second equation for °F. Substitute the value of the temperature in °C (from the previous step) to convert it to °F and round the answer to the correct number of significant figures.

SOLUTION

$K = °C + 273.15$
$°C = K - 273.15$
$°C = 310 - 273.15 = 37 \,°C$

$°C = \dfrac{(°F - 32)}{1.8}$

$1.8 \,°C = (°F - 32)$
$°F = 1.8 \,°C + 32$
$°F = 1.8(37) + 32 = 98.6 \,°F = 99 \,°F$

CHECK

Check your answer. Are the units correct? Does the answer make physical sense?

The units (°F) are correct. The magnitude of the answer is a bit trickier to judge. In this temperature range, a temperature in Fahrenheit should indeed be smaller than the temperature in kelvins. However, because the Fahrenheit degree is smaller, temperatures in Fahrenheit become larger than temperatures in kelvins above 575 °F.

▶**SKILLBUILDER 3.9 | Converting between Fahrenheit and Kelvin Temperature Scales**

Convert −321 °F to kelvins.

▶**FOR MORE PRACTICE**   Problems 65b, d, 66b.

CONCEPTUAL CHECKPOINT 3.4

Which temperature is identical on both the Celsius and the Fahrenheit scale?

(a) 100°

(b) 32°

(c) 0°

(d) −40°

3.11 Temperature Changes: Heat Capacity

TABLE 3.4 Specific Heat Capacities of Some Common Substances

Substance	Specific Heat Capacity (J/g °C)
Lead	0.128
Gold	0.128
Silver	0.235
Copper	0.385
Iron	0.449
Aluminum	0.903
Ethanol	2.42
Water	4.184

▲ San Francisco enjoys cool weather even in summer months because of the high heat capacity of the surrounding ocean.

All substances change temperature when they are heated, but how much they change for a given amount of heat varies significantly from one substance to another. For example, if you put a steel skillet on a flame, its temperature rises rapidly. However, if you put some water in the skillet, the temperature increases more slowly. Why? One reason is that when you add water, the same amount of heat energy must warm more matter, so the temperature rise is slower. The second and more interesting reason is that water is more resistant to temperature change than steel because water has a higher *heat capacity*. The **heat capacity** of a substance is the quantity of heat (usually in joules) required to change the temperature of a given amount of the substance by 1 °C. When the amount of the substance is expressed in grams, the heat capacity is called the **specific heat capacity** (or simply the **specific heat**) and has units of joules per gram per degree Celsius (J/g °C). Table 3.4 lists the values of the specific heat capacity for several substances.

Notice that water has the highest specific heat capacity on the list—changing its temperature requires a lot of heat. If you have traveled from an inland geographical region to a coastal one and have felt the drop in temperature, you have experienced the effects of water's high specific heat capacity. On a summer day in California, for example, the temperature difference between Sacramento (an inland city) and San Francisco (a coastal city) can be 30 °F; San Francisco enjoys a cool 68 °F, while Sacramento bakes at near 100 °F. Yet the intensity of sunlight falling on these two cities is the same. Why the large temperature difference? The difference between the two locations is due to the presence of the Pacific Ocean, which practically surrounds San Francisco. Water, with its high heat capacity, absorbs much of the sun's heat without undergoing a large increase in temperature, keeping San Francisco cool. The land surrounding Sacramento, on the other hand, with its low heat capacity, cannot absorb a lot of heat without a large increase in temperature—it has a lower *capacity* to absorb heat without a large temperature increase.

Similarly, only two U.S. states have never recorded a temperature above 100 °F. One of them is obvious: Alaska. It is too far north to get that hot. The other one, however, may come as a surprise. It is Hawaii. The water that surrounds America's only island state moderates the temperature, preventing Hawaii from ever getting too hot.

CONCEPTUAL CHECKPOINT 3.5

If you want to heat a metal plate to as high a temperature as possible for a given energy input, you should make the plate out of:

(a) copper

(b) iron

(c) aluminum

(d) it would make no difference

EVERYDAY CHEMISTRY
Coolers, Camping, and the Heat Capacity of Water

Have you ever loaded a cooler with ice and then added room-temperature drinks? If you have, you know that the ice quickly melts. In contrast, if you load your cooler with chilled drinks, the ice lasts for hours. Why the difference? The answer is related to the high heat capacity of the water within the drinks. As we just learned, water must absorb a lot of heat to raise its temperature, and it must also release a lot of heat to lower its temperature. When the warm drinks are placed into the ice, they release heat, which then melts the ice. The chilled drinks, on the other hand, are already cold, so they do not release much heat. It is always better to load your cooler with chilled drinks—that way, the ice will last the rest of the day.

CAN YOU ANSWER THIS? *Suppose you are cold-weather camping and decide to heat some objects to bring into your sleeping bag for added warmth. You place a large water jug and a rock of equal mass close to the fire. Over time, both the rock and the water jug warm to about 38 °C (100 °F). If you could bring only one into your sleeping bag, which one should you bring to keep you the warmest? Why?*

▲ The ice in a cooler loaded with cold drinks lasts much longer than the ice in a cooler loaded with warm drinks.
Question: Can you explain why?

3.12 Energy and Heat Capacity Calculations

When a substance absorbs heat (which we represent with the symbol q), its temperature change (which we represent as ΔT) is in direct proportion to the amount of heat absorbed.

$$\xrightarrow{\quad q \quad} \boxed{\text{System}}$$
$$\Delta T$$

In other words, the more heat absorbed, the greater the temperature change. The specific heat capacity of the substance can be used to *quantify* the relationship between the amount of heat added to a given amount of the substance and the corresponding temperature increase. The equation that relates these quantities is:

$$\text{Heat} = \text{Mass} \times \text{Specific Heat Capacity} \times \text{Temperature Change}$$
$$q = m \times C \times \Delta T$$

where q is the amount of heat in joules, m is the mass of the substance in grams, C is the specific heat capacity in joules per gram per degree Celsius, and ΔT is the temperature change in Celsius. The symbol Δ means *the change in*, so ΔT means *the change in temperature*. For example, suppose you are making a cup of tea and

ΔT in °C is equal to ΔT in K but is not equal to ΔT in °F.

want to know how much heat energy will warm 235 g of water (about 8 oz) from 25 °C to 100.0 °C (boiling). We begin by sorting the information in the problem.

GIVEN: 235 g water (m)
25 °C initial temperature (T_i)
100.0 °C final temperature (T_f)

FIND: amount of heat needed (q)

SOLUTION MAP

Then we strategize by building a solution map.

$$q = m \cdot C \cdot \Delta T$$

In addition to m and ΔT, the equation requires C, the specific heat capacity of water. The next step is to gather all of the required quantities for the equation (C, m, and ΔT) in the correct units. These are:

$$C = 4.18 \, J/g \, °C$$

$$m = 235 \, g$$

The other required quantity is ΔT. The change in temperature is the difference between the final temperature (T_f) and the initial temperature (T_i).

$$\Delta T = T_f - T_i$$
$$= 100.0 \, °C - 25 \, °C = 75 \, °C$$

SOLUTION

Finally, we solve the problem. Substitute the correct values into the equation and calculate the answer to the correct number of significant figures.

$$q = m \cdot C \cdot \Delta T$$

$$= 235 \, \cancel{g} \times 4.18 \times \frac{J}{\cancel{g} \, \cancel{°C}} \times 75 \, \cancel{°C}$$

$$= 7.367 \times 10^4 \, J = 7.4 \times 10^4 \, J$$

It is critical that you substitute each of the correct variables into the equation in the correct units and cancel units as you compute the answer. If, during this process, you learn that one of your variables is not in the correct units, convert it to the correct units using the skills you learned in Chapter 2. Notice that the sign of q is positive (+) if the substance is increasing in temperature (heat entering the substance) and negative (−) if the substance is decreasing in temperature (heat leaving the substance).

EXAMPLE 3.10 Relating Heat Energy to Temperature Changes

Gallium is a solid metal at room temperature but melts at 29.9 °C. If you hold gallium in your hand, it melts from your body heat. How much heat must 2.5 g of gallium absorb from your hand to raise its temperature from 25.0 °C to 29.9 °C? The specific heat capacity of gallium is 0.372 J/g °C.

SORT You are given the mass of gallium, its initial and final temperatures, and its specific heat capacity, and are asked to find the amount of heat absorbed.	**GIVEN:** 2.5 g gallium (m) $T_i = 25.0$ °C $T_f = 29.9$ °C $C = 0.372\,\text{J/g °C}$ **FIND:** q
STRATEGIZE The equation that relates the *given* and *find* quantities is the specific heat capacity equation. The solution map indicates that this equation takes you from the *given* quantities to the quantity you are asked to *find*.	**SOLUTION MAP** $\boxed{C, m, \Delta T} \longrightarrow \boxed{q}$ $q = m \cdot C \cdot \Delta T$ **RELATIONSHIPS USED** $q = m \cdot C \cdot \Delta T$ (presented in this section)
SOLVE Before solving the problem, you must gather the necessary quantities—C, m, and ΔT—in the correct units. Substitute C, m, and ΔT into the equation, canceling units, and calculate the answer to the correct number of significant figures.	**SOLUTION** $C = 0.372\,\text{J/g °C}$ $m = 2.5$ g $\Delta T = T_f - T_i$ $\quad\quad = 29.9\,\text{°C} - 25.0\,\text{°C}$ $\quad\quad = 4.9\,\text{°C}$ $q = m \cdot C \cdot \Delta T$ $\quad = 2.5\ \cancel{g} \times 0.372\ \dfrac{\text{J}}{\cancel{g}\ \cancel{\text{°C}}} \times 4.9\ \cancel{\text{°C}} = 4.557\,\text{J} = 4.6\,\text{J}^{*}$
CHECK Check your answer. Are the units correct? Does the answer make physical sense?	The units (J) are correct. The magnitude of the answer makes sense because it takes almost 1 J to heat the 2.5 g sample of the metal by 1 °C; therefore, it should take about 5 J to heat the sample by 5 °C.

▶**SKILLBUILDER 3.10 | Relating Heat Energy to Temperature Changes**

You find a copper penny (pre-1982) in the snow and pick it up. How much heat is absorbed by the penny as it warms from the temperature of the snow, −5.0 °C, to the temperature of your body, 37.0 °C? Assume the penny is pure copper and has a mass of 3.10 g. You can find the heat capacity of copper in Table 3.4 (p. 74).

▶**SKILLBUILDER PLUS |**

The temperature of a lead fishing weight rises from 26 °C to 38 °C as it absorbs 11.3 J of heat. What is the mass of the fishing weight in grams?

▶**FOR MORE PRACTICE** Example 3.19; Problems 75, 76, 77, 78.

---

* This is the amount of heat required to raise the temperature to the melting point. Actually melting the gallium requires additional heat.

EXAMPLE 3.11 Relating Heat Capacity to Temperature Changes

A chemistry student finds a shiny rock that she suspects is gold. She weighs the rock on a balance and obtains the mass, 14.3 g. She then finds that the temperature of the rock rises from 25 °C to 52 °C upon absorption of 174 J of heat. Find the heat capacity of the rock and determine whether the value is consistent with the heat capacity of gold.

SORT	**GIVEN:** 14.3 g
You are given the mass of the "gold" rock, the amount of heat absorbed, and the initial and final temperature. You are asked to find the heat capacity.	174 J of heat absorbed T_i = 25 °C T_f = 52 °C **FIND:** C

STRATEGIZE	**SOLUTION MAP**
The solution map shows how the heat capacity equation relates the given and find quantities.	$\boxed{m, q, \Delta T} \longrightarrow \boxed{C}$ $q = m \cdot C \cdot \Delta T$ **RELATIONSHIPS USED** $q = m \cdot C \cdot \Delta T$ (presented in this section)

SOLVE	**SOLUTION**
First, gather the necessary quantities—m, q, and ΔT—in the correct units.	$m = 14.3$ g $q = 174$ J $\Delta T = 52\ °C - 25\ °C = 27\ °C$
Then solve the equation for C and substitute the correct variables into the equation. Finally, calculate the answer to the right number of significant figures.	$q = m \cdot C \cdot \Delta T$ $C = \dfrac{q}{m \cdot \Delta T}$ $C = \dfrac{174\ \text{J}}{14.3\ \text{g} \times 27\ °C}$ $= 0.4507\ \dfrac{\text{J}}{\text{g}\,°C} = 0.45\ \dfrac{\text{J}}{\text{g}\,°C}$
	By comparing the calculated value of the specific heat capacity (0.45 J/g °C) with the specific heat capacity of gold from Table 3.4 (0.128 J/g °C), we conclude that the rock could not be pure gold.

CHECK	
Check your answer. Are the units correct? Does the answer make physical sense?	The units of the answer are those of specific heat capacity, so they are correct. The magnitude of the answer falls in the range of specific heat capacities given in Table 3.4. A value of heat capacity that falls far outside this range would immediately be suspect.

▶**SKILLBUILDER 3.11 | Relating Heat Capacity to Temperature Changes**

A 328-g sample of water absorbs 5.78×10^3 J of heat. Calculate the change in temperature for the water. If the water is initially at 25.0 °C, what is its final temperature?

▶**FOR MORE PRACTICE** Problems 85, 86, 87, 88.

 CONCEPTUAL CHECKPOINT 3.6

The heat capacity of substance A is twice that of substance B. If samples of equal mass of both substances absorb the same amount of heat, which substance undergoes the largest change in temperature?

CHAPTER IN REVIEW

CHEMICAL PRINCIPLES

Matter: Matter is anything that occupies space and has mass. It is composed of atoms, which are often bonded together as molecules. Matter can exist as a solid, a liquid, or a gas. Solid matter can be either amorphous or crystalline.

Classification of Matter: Matter can be classified according to its composition. Pure matter is composed of only one type of substance; that substance may be an element (a substance that cannot be decomposed into simpler substances), or it may be a compound (a substance composed of two or more elements in fixed definite proportions). Mixtures are composed of two or more different substances the proportions of which may vary from one sample to the next. Mixtures can be either homogeneous, having the same composition throughout, or heterogeneous, having a composition that varies from region to region.

Properties and Changes of Matter: The properties of matter can be divided into two types: physical and chemical. The physical properties of matter do not involve a change in composition. The chemical properties of matter involve a change in composition. The changes in matter can be divided into physical and chemical. In a physical change, the appearance of matter may change, but its composition does not. In a chemical change, the composition of matter changes.

Conservation of Mass: Whether the changes in matter are chemical or physical, matter is always conserved. In a chemical change, the masses of the matter undergoing the chemical change must equal the sum of the masses of matter resulting from the chemical change.

Energy: Besides matter, energy is the other major component of our universe. Like matter, energy is conserved—it can be neither created nor destroyed. Energy exists in various different types, and these can be converted from one to another. Some common units of energy are the joule (J), the calorie (cal), the nutritional Calorie (Cal), and the kilowatt-hour (kWh). Chemical reactions that emit energy are said to be exothermic; those that absorb energy are said to be endothermic.

Temperature: The temperature of matter is related to the random motions of the molecules and atoms that compose it—the greater the motion, the higher the temperature. Temperature is commonly measured on three scales: Fahrenheit (°F), Celsius (°C), and Kelvin (K).

RELEVANCE

Matter: Everything is made of matter—you, me, the chair you sit on, and the air we breathe. The physical universe basically contains only two things: matter and energy. We begin our study of chemistry by defining and classifying these two building blocks of the universe.

Classification of Matter: Since ancient times, humans have tried to understand matter and harness it for their purposes. The earliest humans shaped matter into tools and used the transformation of matter—especially fire—to keep warm and to cook food. To manipulate matter, we must understand it. Fundamental to this understanding is the connection between the properties of matter and the molecules and atoms that compose it.

Properties and Changes of Matter: The physical and chemical properties of matter make the world around us the way it is. For example, a physical property of water is its boiling point at sea level—100 °C. The physical properties of water—and all matter—are determined by the atoms and molecules that compose it. If water molecules were different—even slightly different—water would boil at a different temperature. Imagine a world where water boiled at room temperature.

Conservation of Mass: The conservation of matter is relevant to, for example, pollution. We often think that humans create pollution, but, actually, we are powerless to create anything. Matter cannot be created. So, pollution is simply misplaced matter—matter that has been put into places where it does not belong.

Energy: Our society's energy sources will not last forever because as we burn fossil fuels—our primary energy source—we convert chemical energy, stored in molecules, to kinetic and thermal energy. The kinetic and thermal energy is not readily available to be used again. Consequently, our energy resources are dwindling, and the conservation of energy implies that we will not be able simply to create new energy—it must come from somewhere. All of the chemical reactions that we use for energy are exothermic.

Temperature: The temperature of matter and its measurement are relevant to many everyday phenomena. Humans are understandibly interested in the weather, and air temperature is a fundamental part of weather. We use body temperature as one measure of human health and global temperature as one measure of the planet's health.

Heat Capacity: The temperature change that a sample of matter undergoes upon absorption of a given amount of heat is related to the heat capacity of the substance composing the matter. Water has one of the highest heat capacities, meaning that it is most resistant to rapid temperature changes.

Heat Capacity: The heat capacity of water explains why it is cooler in coastal areas, which are near large bodies of high-heat-capacity water, than in inland areas, which are surrounded by low-heat-capacity land. It also explains why it takes longer to cool a refrigerator filled with liquids than an empty one.

CHEMICAL SKILLS

EXAMPLES

Classifying Matter (Sections 3.3, 3.4)
Begin by examining the alphabetical listing of elements in the back of this book. If the substance is listed in that table, it is a pure substance and an element.

If the substance is not listed in that table, refer to your everyday experience with the substance to determine whether it is a pure substance. If it is a pure substance not listed in the table, then it is a compound.

If it is not a pure substance, then it is a mixture. Refer to your everyday experience with the mixture to determine whether it has uniform composition throughout (homogeneous) or nonuniform composition (heterogeneous).

EXAMPLE 3.12 Classifying Matter

Classify each type of matter as a pure substance or a mixture. If it is a pure substance, classify it as an element or compound. If it is a mixture, classify it as homogeneous or heterogeneous.

(a) pure silver
(b) swimming-pool water
(c) dry ice (solid carbon dioxide)
(d) blueberry muffin

SOLUTION

(a) Pure element; silver appears in the element table.
(b) Homogeneous mixture; pool water contains at least water and chlorine, and it is uniform throughout.
(c) Compound; dry ice is a pure substance (carbon dioxide), but it is not listed in the table.
(d) Heterogeneous mixture; a blueberry muffin is a mixture of several things and has nonuniform composition.

Physical and Chemical Properties (Section 3.5)
To distinguish between physical and chemical properties, consider whether the substance changes composition while displaying the property. If it *does not* change composition, the property is physical; if it *does*, the property is chemical.

EXAMPLE 3.13 Physical and Chemical Properties

Determine whether each property is physical or chemical.

(a) the tendency for platinum jewelry to scratch easily
(b) the ability of sulfuric acid to burn the skin
(c) the ability of hydrogen peroxide to bleach hair
(d) the density of lead relative to other metals

SOLUTION

(a) Physical; scratched platinum is still platinum.
(b) Chemical; the acid chemically reacts with the skin to produce the burn.
(c) Chemical; the hydrogen peroxide chemically reacts with hair to bleach it.
(d) Physical; the heaviness can be felt without changing the lead into anything else.

Physical and Chemical Changes (Section 3.6)
To distinguish between physical and chemical changes, consider whether the substance changes composition during the change. If it *does not* change composition, the change is physical; if it *does*, the change is chemical.

EXAMPLE 3.14 Physical and Chemical Changes

Determine whether each change is physical or chemical.

(a) the explosion of gunpowder in the barrel of a gun
(b) the melting of gold in a furnace
(c) the bubbling that occurs upon mixing baking soda and vinegar
(d) the bubbling that occurs when water boils

SOLUTION

(a) Chemical; the gunpowder reacts with oxygen during the explosion.
(b) Physical; the liquid gold is still gold.
(c) Chemical; the bubbling is a result of a chemical reaction between the two substances to form new substances, one of which is carbon dioxide released as bubbles.
(d) Physical; the bubbling is due to liquid water turning into gaseous water, but it is still water.

Conservation of Mass (Section 3.7)

The sum of the masses of the substances involved in a chemical change must be the same before and after the change.

EXAMPLE 3.15 Conservation of Mass

An automobile runs for 10 minutes and burns 47 g of gasoline. The gasoline combined with oxygen from air and formed 132 g of carbon dioxide and 34 g of water. How much oxygen was consumed in the process?

SOLUTION

The total mass after the chemical change is:

$$132 \text{ g} + 34 \text{ g} = 166 \text{ g}$$

The total mass before the change must also be 166 g.

$$47 \text{ g} + \text{oxygen} = 166 \text{ g}$$

So, the mass of oxygen consumed is the total mass (166 g) minus the mass of gasoline (47 g).

$$\text{grams of oxygen} = 166\text{ g} - 47\text{ g} = 119 \text{ g}$$

Conversion of Energy Units (Section 3.8)

Solve unit conversion problems using the problem-solving strategies outlined in Section 2.6.

EXAMPLE 3.16 Conversion of Energy Units

Convert 1.7×10^3 kWh (the amount of energy used by the average U.S. citizen in one week) into calories.

SORT
You are given an amount of energy in kilowatt-hours and asked to find the amount in calories.

GIVEN: 1.7×10^3 kWh

FIND: cal

STRATEGIZE
Draw a solution map. Begin with kilowatt-hours and determine the conversion factors to get to calories.

SOLUTION MAP

kWh → J → cal

$$\frac{3.60 \times 10^6 \text{ J}}{1 \text{ kWh}} \qquad \frac{1 \text{ cal}}{4.18 \text{ J}}$$

RELATIONSHIPS USED

$$1\,\text{kWh} = 3.60 \times 10^6 \text{ J (Table 3.2)}$$
$$1\,\text{cal} = 4.18 \text{ J (Table 3.2)}$$

SOLVE
Follow the solution map to solve the problem. Begin with the given quantity and multiply by the conversion factors to arrive at calories. Round the answer to the correct number of significant figures.

SOLUTION

$$1.7 \times 10^3 \text{ kWh} \times \frac{3.60 \times 10^6 \text{ J}}{1 \text{ kWh}} \times \frac{1 \text{ cal}}{4.18 \text{ J}}$$
$$= 1.464 \times 10^9 \text{ cal}$$
$$1.464 \times 10^9 \text{ cal} = 1.5 \times 10^9 \text{ cal}$$

CHECK
Are the units correct? Does the answer make physical sense?

The unit of the answer, cal, is correct. The magnitude of the answer makes sense since cal is a smaller unit than kWh; therefore, the value in cal should be larger than the value in kWh.

Converting between Celsius and Kelvin Temperature Scales (Section 3.10)

Solve temperature conversion problems using the problem-solving procedure in Sections 2.6 and 2.10. Take the steps appropriate for equations.

SORT

You are given the temperature in kelvins and asked to convert it to degrees Celsius.

STRATEGIZE

Draw a solution map. Use the equation that relates the *given* quantity to the *find* quantity.

SOLVE

Solve the equation for the *find* quantity (°C) and substitute the temperature in K into the equation. Calculate the answer to the correct number of significant figures.

CHECK

Are the units correct? Does the answer make physical sense?

EXAMPLE 3.17 Converting between Celsius and Kelvin Temperature Scales

Convert 257 K to Celsius.

GIVEN: 257 K

FIND: °C

SOLUTION MAP

$$K = °C + 273.15$$

RELATIONSHIPS USED

$$K = °C + 273.15 \text{ (Section 3.10)}$$

SOLUTION

$$K = °C + 273.15$$
$$°C = K - 273.15$$
$$°C = 257 - 273.15 = -16 °C$$

The answer has the correct unit, and its magnitude seems correct (see Figure 3.17).

Converting between Fahrenheit and Celsius Temperature Scales (Section 3.10)

Solve temperature conversion problems using the problem-solving procedure in Sections 2.6 and 2.10. Take the steps appropriate for equations.

SORT

You are given the temperature in degrees Celsius and asked to convert it to degrees Fahrenheit.

STRATEGIZE

Draw a solution map. Use the equation that relates the *given* quantity to the *find* quantity.

SOLVE

Solve the equation for the *find* quantity (°F) and substitute the temperature in °C into the equation. Calculate the answer to the correct number of significant figures.

CHECK

Are the units correct? Does the answer make physical sense?

EXAMPLE 3.18 Converting between Fahrenheit and Celsius Temperature Scales

Convert 62.0 °C to Fahrenheit.

GIVEN: 62.0 °C

FIND: °F

SOLUTION MAP

$$°C = \frac{(°F - 32)}{1.8}$$

RELATIONSHIPS USED

$$°C = \frac{(°F - 32)}{1.8}$$

SOLUTION

$$°C = \frac{(°F - 32)}{1.8}$$
$$1.8\,(°C) = °F - 32$$
$$°F = 1.8\,(°C) + 32$$
$$°F = 1.8\,(62.0) + 32 = 143.60 \; °F = 144 \; °F$$

The answer is in the correct units, and its magnitude seems correct (see Figure 3.17).

Energy, Temperature Change, and Heat Capacity Calculations (Sections 3.11, 3.12)

Solve heat capacity problems using the problem-solving procedure in Sections 2.6 and 2.10. Take the steps appropriate for equations.

SORT

You are given the volume of water and the amount of heat absorbed. You are asked to find the change in temperature.

STRATEGIZE

The solution map shows how the heat capacity equation relates the *given* and *find* quantities.

SOLVE

First, gather the necessary quantities—m, q, and ΔT—in the correct units. The value for q must be converted from kJ to J.

You must convert the value for m from milliliters to grams; use the density of water, 1.0 g/mL, to convert milliliters to grams.

Look up the heat capacity for water in Table 3.4.

Then solve the equation for ΔT and substitute the correct variables into the equation. Finally, calculate the answer to the right number of significant figures.

CHECK

Check your answer. Are the units correct? Does the answer make physical sense?

EXAMPLE 3.19 Energy, Temperature Change, and Heat Capacity Calculations

What is the temperature change in 355 mL of water upon absorption of 34 kJ of heat?

GIVEN: 355 mL water
34 kJ of heat

FIND: ΔT

SOLUTION MAP

$$q = m \cdot C \cdot \Delta T$$

RELATIONSHIPS USED

$$q = m \cdot C \cdot \Delta T$$

SOLUTION

$$q = 34 \text{ kj} \times \frac{1000 \text{ J}}{1 \text{ kj}} = 3.4 \times 10^4 \text{ J}$$

$$m = 355 \text{ mL} \times \frac{1.0 \text{ g}}{1 \text{ mL}} = 355 \text{ g}$$

$$C = 4.18 \text{ J/g} \,°\text{C}$$

$$q = m \cdot C \cdot \Delta T$$

$$\Delta T = \frac{q}{mC}$$

$$\Delta T = \frac{3.4 \times 10^4 \text{ J}}{355 \text{ g} \times 4.18 \text{ J/g} \,°\text{C}}$$

$$= 22.91 \,°\text{C} = 23 \,°\text{C}$$

The answer has the correct units, and the magnitude seems correct. If the magnitude of the answer were a huge number—3×10^6, for example—we would go back and look for a mistake. Above 100 °C, water boils, so such a large answer would be unlikely.

KEY TERMS

molecule [3.2]
physical change [3.6]
physical property [3.5]
potential energy [3.8]

product [3.7]
property [3.5]
pure substance [3.4]
reactant [3.7]

solid [3.3]
specific heat capacity
  (specific heat) [3.11]
state of matter [3.3]

temperature [3.10]
thermal energy [3.8]
volatile [3.6]
work [3.8]

EXERCISES

QUESTIONS

Answers to all odd-numbered questions (numbered in blue) appear in the Answers section at the back of the book.

1. Define matter and list some examples.
2. What is matter composed of?
3. What are the three states of matter?
4. What are the properties of a solid?
5. What is the difference between a crystalline solid and an amorphous solid?
6. What are the properties of a liquid?
7. What are the properties of a gas?
8. Why are gases compressible?
9. What is a mixture?
10. What is the difference between a homogeneous mixture and a heterogeneous mixture?
11. What is a pure substance?
12. What is an element? A compound?
13. What is the difference between a mixture and a compound?
14. What is the definition of a physical property? What is the definition of a chemical property?
15. What is the difference between a physical change and a chemical change?
16. What is the law of conservation of mass?
17. What is the definition of energy?
18. What is the law of conservation of energy?
19. Expain the difference between kinetic energy and potential energy.
20. What is chemical energy? List some examples of common substances that contain chemical energy.

21. What are three common units for energy?
22. What is an exothermic reaction? Which has greater energy in an exothermic reaction, the reactants or the products?
23. What is an endothermic reaction? Which has greater energy in an endothermic reaction, the reactants or the products?
24. List three common units for measuring temperature.
25. Explain the difference between heat and temperature.
26. How do the three temperature scales differ?
27. What is heat capacity?
28. Why are coastal geographic regions normally cooler in the summer than inland geographic regions?
29. The following equation can be used to convert Fahrenheit temperature to Celsius temperature.

$$°C = \frac{(°F - 32)}{1.8}$$

Use algebra to change the equation to convert Celsius temperature to Fahrenheit temperature.
30. The following equation can be used to convert Celsius temperature to Kelvin temperature.

$$K = °C + 273$$

Use algebra to change the equation to convert Kelvin temperature to Celsius temperature.

PROBLEMS

Note: The exercises in the Problems section are paired, and the answers to the odd-numbered exercises (numbered in blue) appear in the Answers section at the back of the book.

CLASSIFYING MATTER

31. Classify each pure substance as an element or a compound.
    (a) aluminum
    (b) sulfur
    (c) methane
    (d) acetone

32. Classify each pure substance as an element or a compound.
    (a) carbon
    (b) baking soda (sodium bicarbonate)
    (c) nickel
    (d) gold

33. Classify each mixture as homogeneous or heterogeneous.
    (a) coffee
    (b) chocolate sundae
    (c) apple juice
    (d) gasoline

34. Classify each mixture as homogeneous or heterogeneous.
    (a) baby oil
    (b) chocolate chip cookie
    (c) water and gasoline
    (d) wine

35. Classify each substance as a pure substance or a mixture. If it is a pure substance, classify it as an element or a compound. If it is a mixture, classify it as homogeneous or heterogeneous.
    (a) helium gas
    (b) clean air
    (c) rocky road ice cream
    (d) concrete

36. Classify each substance as a pure substance or a mixture. If it is a pure substance, classify it as an element or a compound. If it is a mixture, classify it as homogeneous or heterogeneous.
    (a) urine
    (b) pure water
    (c) Snickers™ bar
    (d) soil

PHYSICAL AND CHEMICAL PROPERTIES AND PHYSICAL AND CHEMICAL CHANGES

37. Classify each property as physical or chemical.
    (a) the tendency of silver to tarnish
    (b) the shine of chrome
    (c) the color of gold
    (d) the flammability of propane gas

38. Classify each property as physical or chemical.
    (a) the boiling point of ethyl alcohol
    (b) the temperature at which dry ice sublimes
    (c) the flammability of ethyl alcohol
    (d) the smell of perfume

39. The following list contains several properties of ethylene (a ripening agent for bananas). Which are physical properties, and which are chemical?
    • colorless
    • odorless
    • flammable
    • gas at room temperature
    • 1 L has a mass of 1.260 g under standard conditions
    • mixes with acetone
    • polymerizes to form polyethylene

40. The following list contains several properties of ozone (a pollutant in the lower atmosphere but part of a protective shield against UV light in the upper atmosphere). Which are physical, and which are chemical?
    • bluish color
    • pungent odor
    • very reactive
    • decomposes on exposure to ultraviolet light
    • gas at room temperature

41. Determine whether each change is physical or chemical.
    (a) A balloon filled with hydrogen gas explodes upon contact with a spark.
    (b) The liquid propane in a barbecue evaporates away because the user left the valve open.
    (c) The liquid propane in a barbecue ignites upon contact with a spark.
    (d) Copper metal turns green on exposure to air and water.

42. Determine whether each change is physical or chemical.
    (a) Sugar dissolves in hot water.
    (b) Sugar burns in a pot.
    (c) A metal surface becomes dull because of continued abrasion.
    (d) A metal surface becomes dull on exposure to air.

43. A block of aluminum is (a) ground into aluminum powder and then (b) ignited. It then emits flames and smoke. Classify (a) and (b) as chemical or physical changes.

44. Several pieces of graphite from a mechanical pencil are (a) broken into tiny pieces. Then the pile of graphite is (b) ignited with a hot flame. Classify (a) and (b) as chemical or physical changes.

THE CONSERVATION OF MASS

45. An automobile gasoline tank holds 42 kg of gasoline. When the gasoline burns, 168 kg of oxygen are consumed and carbon dioxide and water are produced. What is the total combined mass of carbon dioxide and water that is produced?

46. In the explosion of a hydrogen-filled balloon, 0.50 g of hydrogen reacted with 4.0 g of oxygen. How many grams of water vapor are formed? (Water vapor is the only product.)

47. Are these data sets on chemical changes consistent with the law of conservation of mass?

    (a) A 7.5-g sample of hydrogen gas completely reacts with 60.0 g of oxygen gas to form 67.5 g of water.

    (b) A 60.5-g sample of gasoline completely reacts with 243 g of oxygen to form 206 g of carbon dioxide and 88 g of water.

48. Are these data sets on chemical changes consistent with the law of conservation of mass?

    (a) A 12.8-g sample of sodium completely reacts with 19.6 g of chlorine to form 32.4 g of sodium chloride.

    (b) An 8-g sample of natural gas completely reacts with 32 g of oxygen gas to form 17 g of carbon dioxide and 16 g of water.

49. In a butane lighter, 9.7 g of butane combine with 34.7 g of oxygen to form 29.3 g carbon dioxide and how many grams of water?

50. A 56-g sample of iron reacts with 24 g of oxygen to form how many grams of iron oxide?

CONVERSION OF ENERGY UNITS

51. Perform each conversion.

    (a) 588 cal to joules

    (b) 17.4 J to Calories

    (c) 134 kJ to Calories

    (d) 56.2 Cal to joules

52. Perform each conversion.

    (a) 45.6 J to calories

    (b) 355 cal to joules

    (c) 43.8 kJ to calories

    (d) 215 cal to kilojoules

53. Perform each conversion.

    (a) 25 kWh to joules

    (b) 249 cal to Calories

    (c) 113 cal to kilowatt-hours

    (d) 44 kJ to calories

54. Perform each conversion.

    (a) 345 Cal to kilowatt-hours

    (b) 23 J to calories

    (c) 5.7×10^3 J to kilojoules

    (d) 326 kJ to joules

55. Complete the table:

J	cal	Cal	kWh
225 J	_____	5.38×10^{-2} Cal	_____
_____	8.21×10^5 cal	_____	_____
_____	_____	_____	295 kWh
_____	_____	155 Cal	_____

56. Complete the table:

J	cal	Cal	kWh
7.88×10^6 J	1.88×10^6 cal	_____	_____
_____	_____	1154 Cal	_____
_____	88.4 cal	_____	_____
_____	_____	_____	125 kWh

57. An energy bill indicates that a customer used 1027 kWh in July. How many joules did the customer use?

58. A television uses 32 kWh of energy per year. How many joules does it use?

59. An adult eats food whose nutritional energy totals approximately 2.2×10^3 Cal per day. The adult burns 2.0×10^3 Cal per day. How much excess nutritional energy, in kilojoules, does the adult consume per day? If 1 lb of fat is stored by the body for each 14.6×10^3 kJ of excess nutritional energy consumed, how long will it take this person to gain 1 lb?

60. How many joules of nutritional energy are in a bag of chips whose label reads 245 Cal? If 1 lb of fat is stored by the body for each 14.6×10^3 kJ of excess nutritional energy consumed, how many bags of chips contain enough nutritional energy to result in 1 lb of body fat?

ENERGY AND CHEMICAL AND PHYSICAL CHANGE

61. A common type of handwarmer contains iron powder that reacts with oxygen to form an oxide of iron. As soon as the handwarmer is exposed to air, the reaction begins and heat is emitted. Is the reaction between the iron and oxygen exothermic or endothermic? Draw an energy diagram showing the relative energies of the reactants and products in the reaction.

62. In a chemical cold pack, two substances are kept separate by a divider. When the divider is broken, the substances mix and absorb heat from the surroundings. The chemical cold pack feels cold. Is the reaction exothermic or endothermic? Draw an energy diagram showing the relative energies of the reactants and products in the reaction.

63. Determine whether each process is exothermic or endothermic.

(a) gasoline burning in a car

(b) isopropyl alcohol evaporating from skin

(c) water condensing as dew during the night

64. Determine whether each process is exothermic or endothermic.

(a) dry ice subliming (changing from a solid directly to a gas)

(b) the wax in a candle burning

(c) a match burning

CONVERTING BETWEEN TEMPERATURE SCALES

65. Perform each temperature conversion.

(a) 212 °F to Celsius (temperature of boiling water)

(b) 77 K to Fahrenheit (temperature of liquid nitrogen)

(c) 25 °C to Kelvin (room temperature)

(d) 98.6 °F to Kelvin (body temperature)

66. Perform each temperature conversion.

(a) 102 °F to Celsius

(b) 0 K to Fahrenheit

(c) −48 °C to Fahrenheit

(d) 273 K to Celsius

67. The coldest temperature ever measured in the United States was −80 °F on January 23, 1971, in Prospect Creek, Alaska. Convert that temperature to degrees Celsius and Kelvin. (Assume that −80 °F is accurate to two significant figures.)

68. The warmest temperature ever measured in the United States was 134 °F on July 10, 1913, in Death Valley, California. Convert that temperature to degrees Celsius and Kelvin.

69. Vodka will not freeze in the freezer because it contains a high percentage of ethanol. The freezing point of pure ethanol is −114 °C. Convert that temperature to degrees Fahrenheit and Kelvin.

70. Liquid helium boils at 4.2 K. Convert this temperature to degrees Fahrenheit and Celsius.

71. The temperature in the South Pole during the Antarctic winter is so cold that planes cannot land or take off, effectively leaving the inhabitants of the South Pole isolated for the winter. The average daily temperature at the South Pole in July is −59.7 °C. Convert this temperature to degrees Fahrenheit.

72. The coldest temperature ever recorded in Iowa was −47 °F on February 3, 1998. Convert this temperature to Kelvin and degrees Celsius.

73. Complete the table.

Kelvin	Fahrenheit	Celsius
0.0 K	_____	−273.0 °C
_____	82.5 °F	_____
_____		8.5 °C

74. Complete the table.

Kelvin	Fahrenheit	Celsius
273.0 K	_____	0.0 °C
_____	−40.0 °F	_____
385 K		_____

ENERGY, HEAT CAPACITY, AND TEMPERATURE CHANGES

75. Calculate the amount of heat required to raise the temperature of a 65-g sample of water from 32 °C to 65 °C.

76. Calculate the amount of heat required to raise the temperature of a 22-g sample of water from 7 °C to 18 °C.

77. Calculate the amount of heat required to heat a 45-kg sample of ethanol from 11.0 °C to 19.0 °C.

78. Calculate the amount of heat required to heat a 3.5-kg gold bar from 21 °C to 67 °C.

79. If 89 J of heat are added to a pure gold coin with a mass of 12 g, what is its temperature change?

80. If 57 J of heat are added to an aluminum can with a mass of 17.1 g, what is its temperature change?

81. An iron nail with a mass of 12 g absorbs 15 J of heat. If the nail was initially at 28 °C, what is its final temperature?

82. A 45-kg sample of water absorbs 345 kJ of heat. If the water was initially at 22.1 °C, what is its final temperature?

83. Calculate the temperature change that occurs when 248 cal of heat are added to 24 g of water.

84. A lead fishing weight with a mass of 57 g absorbs 146 cal of heat. If its initial temperature is 47 °C, what is its final temperature?

85. An unknown metal with a mass of 28 g absorbs 58 J of heat. Its temperature rises from 31.1 °C to 39.9 °C. Calculate the heat capacity of the metal and identify it referring to Table 3.4.

86. An unknown metal is suspected to be gold. When 2.8 J of heat are added to 5.6 g of the metal, its temperature rises by 3.9 °C. Are these data consistent with the metal being gold?

87. When 56 J of heat are added to 11 g of a liquid, its temperature rises from 10.4 °C to 12.7 °C. What is the heat capacity of the liquid?

88. When 47.5 J of heat are added to 13.2 g of a liquid, its temperature rises by 1.72 °C. What is the heat capacity of the liquid?

89. Two identical coolers are packed for a picnic. Each cooler is packed with eighteen 12-oz soft drinks and 3 lb of ice. However, the drinks that went into cooler A were refrigerated for several hours before they were packed in the cooler, while the drinks that went into cooler B were at room temperature. When the two coolers are opened three hours later, most of the ice in cooler A is still ice, while nearly all of the ice in cooler B has melted. Explain.

90. A 100-g block of iron metal and 100 g of water are each warmed to 75 °C and placed into two identical insulated containers. Two hours later, the two containers are opened and the temperature of each substance is measured. The iron metal has cooled to 38 °C while the water has cooled only to 69 °C. Explain.

CUMULATIVE PROBLEMS

91. Calculate the final temperature of 245 mL of water initially at 32 °C upon absorption of 17 kJ of heat.

92. Calculate the final temperature of 32 mL of ethanol initially at 11 °C upon absorption of 562 J of heat. (density of ethanol = 0.789 g/mL)

93. A pure gold ring with a volume of 1.57 cm^3 is initially at 11.4 °C. When it is put on, it warms to 29.5 °C. How much heat did the ring absorb? (density of gold = 19.3 g/cm^3)

94. A block of aluminum with a volume of 98.5 cm^3 absorbs 67.4 J of heat. If its initial temperature was 32.5 °C, what is its final temperature? (density of aluminum = 2.70 g/cm^3)

95. How much heat in kilojoules is required to heat 56 L of water from 85 °F to 212 °F?

96. How much heat in joules is required to heat a 43-g sample of aluminum from 72 °F to 145 °F?

97. What is the temperature change in Celsius when 29.5 L of water absorbs 2.3 kWh of heat?

98. If 1.45 L of water is initially at 25.0 °C, what will its temperature be after absorption of 9.4 × 10^{-2} kWh of heat?

99. A water heater contains 55 gal of water. How many kilowatt-hours of energy are necessary to heat the water in the water heater by 25 °C?

100. A room contains 48 kg of air. How many kilowatt-hours of energy are necessary to heat the air in the house from 7 °C to 28 °C? The heat capacity of air is 1.03 J/g °C.

101. A backpacker wants to carry enough fuel to heat 2.5 kg of water from 25 °C to 100.0 °C. If the fuel he carries produces 36 kJ of heat per gram when it burns, how much fuel should he carry? (For the sake of simplicity, assume that the transfer of heat is 100% efficient.)

102. A cook wants to heat 1.35 kg of water from 32.0 °C to 100.0 °C. If he uses the combustion of natural gas (which is exothermic) to heat the water, how much natural gas will he need to burn? Natural gas produces 49.3 kJ of heat per gram. (For the sake of simplicity, assume that the transfer of heat is 100% efficient.)

103. Evaporating sweat cools the body because evaporation is endothermic and absorbs 2.44 kJ per gram of water evaporated. Estimate the mass of water that must evaporate from the skin to cool a body by 0.50 °C, if the mass of the body is 95 kg and its heat capacity is 4.0 J/g °C. (Assume that the heat transfer is 100% efficient.)

104. When ice melts, it absorbs 0.33 kJ per gram. How much ice is required to cool a 12.0-oz drink from 75 °F to 35 °F, if the heat capacity of the drink is 4.18 J/g °C (Assume that the heat transfer is 100% efficient.)

105. A 15.7-g aluminum block is warmed to 53.2 °C and plunged into an insulated beaker containing 32.5 g of water initially at 24.5 °C. The aluminum and the water are allowed to come to thermal equilibrium. Assuming that no heat is lost, what is the final temperature of the water and aluminum?

106. 25.0 mL of ethanol (density = 0.789 g/mL) initially at 7.0 °C is mixed with 35.0 mL of water (density = 1.0 g/mL) initially at 25.3 °C in an insulated beaker. Assuming that no heat is lost, what is the final temperature of the mixture?

107. The wattage of an appliance indicates the average power consumption in watts (W), where 1 W = 1 J/s. What is the difference in the number of kJ of energy consumed per month between a refrigeration unit that consumes 625 W and one that consumes 855 W? If electricity costs $0.15 per kWh, what is the monthly cost difference to operate the two refrigerators? (Assume 30.0 days in one month and 24.0 hours per day.)

108. A portable electric water heater transfers 255 watts (W) of power to 5.5 L of water, where 1 W = 1 J/s. How much time (in minutes) will it take for the water heater to heat the 5.5 L of water from 25 °C to 42 °C? (Assume that the water has a density of 1.0 g/mL.)

109. What temperature is the same whether it is expressed in the Celsius or Fahrenheit scale?

110. What temperature on the Celsius scale is equal to twice its value when expressed on the Fahrenheit scale?

HIGHLIGHT PROBLEMS

111. Classify each molecular picture as a pure substance or a mixture.

(a)

(b)

(c)

(d)

112. Classify each molecular picture as a pure substance or a mixture. If it is a pure substance, classify it as an element or a compound. If it is a mixture, classify it as homogeneous or heterogeneous.

(a)

(b)

(c)

(d)

113. This molecular drawing shows images of acetone molecules before and after a change. Was the change chemical or physical?

114. This molecular drawing shows images of methane molecules and oxygen molecules before and after a change. Was the change chemical or physical?

115. A major event affecting global climate is the El Niño/La Niña cycle. In this cycle, equatorial Pacific Ocean waters warm by several degrees Celsius above normal (El Niño) and then cool by several degrees Celsius below normal (La Niña). This cycle affects weather not only in North and South America, but also in places as far away as Africa. Why does a seemingly small change in ocean temperature have such a large impact on weather?

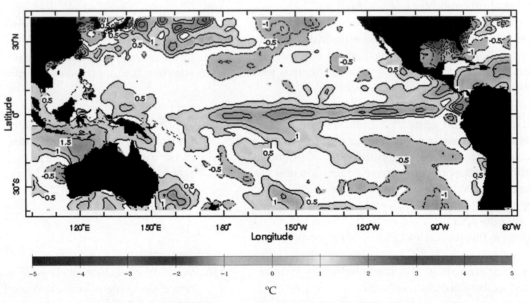

▲ Temperature anomaly plot of the world's oceans for January 17–23, 2010. The large red-orange section in the middle of the map indicates the El Niño effect, a warming of the Pacific Ocean along the equator.

116. Global warming refers to the rise in average global temperature due to the increased concentration of certain gases, called greenhouse gases, in our atmosphere. Earth's oceans, because of their high heat capacity, absorb heat and therefore act to slow down global warming. How much heat would be required to warm Earth's oceans by 1.0 °C? Assume that the volume of water in Earth's oceans is $137 \times 10^7 \text{ km}^3$ and that the density of seawater is 1.03 g/cm^3. Also assume that the heat capacity of seawater is the same as that of water.

◀ Earth's oceans moderate temperatures by absorbing heat during warm periods.

117. Examine the data for the maximum and minimum average temperatures of San Francisco and Sacramento in the summer and in the winter.

San Francisco (Coastal City)

January		August	
High	Low	High	Low
57.4 °F	43.8 °F	64.4 °F	54.5 °F

Sacramento (Inland City)

January		August	
High	Low	High	Low
53.2 °F	37.7 °F	91.5 °F	57.7 °F

(a) Notice the difference between the August high in San Francisco and Sacramento. Why is it much hotter in the summer in Sacramento?

(b) Notice the difference between the January low in San Francisco and Sacramento. How might the heat capacity of the ocean contribute to this difference?

►ANSWERS TO SKILLBUILDER EXERCISES

Skillbuilder 3.1
    (a) pure substance, element
    (b) mixture, homogeneous
    (c) mixture, heterogeneous
    (d) pure substance, compound

Skillbuilder 3.2
    (a) chemical
    (b) physical
    (c) physical
    (d) physical

Skillbuilder 3.3
    (a) chemical
    (b) physical
    (c) physical
    (d) chemical

Skillbuilder 3.4   27 g

Skillbuilder 3.5   2.14 kJ

Skillbuilder Plus, p. 68 6.57×10^6 cal

Skillbuilder 3.6
    (a) exothermic
    (b) exothermic

Skillbuilder 3.7   85 °C

Skillbuilder 3.8   282 °F

Skillbuilder 3.9   77 K

Skillbuilder 3.10   50.1 J

Skillbuilder Plus, p. 77   7.4 g

Skillbuilder 3.11 $\Delta T = 4.21\ °C;\ T_f = 29.2\ °C$

►ANSWERS TO CONCEPTUAL CHECKPOINTS

3.1 (a) Vaporization is a physical change, so the water molecules are the same before and after the boiling.

3.2 No In the vaporization, the liquid water becomes gaseous, but its mass does not change. Like chemical changes, physical changes also follow the law of conservation of mass.

3.3 (d) kWh is the largest of the four units listed, so the numerical value of the yearly energy consumption is lowest if expressed in kWh.

3.4 (d) You can confirm this by substituting each of the Fahrenheit temperatures into the equation in Section 3.10 and solving for the Celsius temperature.

3.5 (a) Because copper has the lowest specific heat capacity of the three metals, it experiences the greatest temperature change for a given energy input.

3.6 Substance B will undergo a greater change in temperature because it has the lower heat capacity. A substance with a lower heat capacity is less resistant to temperature changes.

CHAPTER 4

Atoms and Elements

"Nothing exists except atoms and empty space; everything else is opinion."

DEMOCRITUS (460–370 B.C.)

4.1 Experiencing Atoms at Tiburon

As we learned in Chapter 3, many atoms exist not as free particles but as groups of atoms bound together to form molecules. Nevertheless, all matter is ultimately made of atoms.

◀ Seaside rocks are typically composed of silicates, compounds of silicon and oxygen atoms. Seaside air, like all air, contains nitrogen and oxygen molecules, and it may also contain substances called amines. The amine shown here is triethylamine, which is emitted by decaying fish. Triethylamine is one of the compounds responsible for the fishy smell of the seaside.

My wife and I recently enjoyed a visit to the northern California seaside town of Tiburon. Tiburon sits next to San Francisco Bay with views of the water, the city of San Francisco, and the surrounding mountains. As we walked along a waterside path, I could feel the wind as it blew over the bay. I could hear the water splashing on the shore, and I could smell the sea air. What was the cause of these sensations? The answer is simple—atoms.

Since all matter is made of atoms, atoms are at the foundation of our sensations. The atom is the fundamental building block of everything you hear, feel, see, and experience. When you feel wind on your skin, you are feeling atoms. When you hear sounds, you are in a sense hearing atoms. When you touch a shoreside rock, you are touching atoms, and when you smell sea air, you are smelling atoms. You eat atoms, you breathe atoms, and you excrete atoms. Atoms are the building blocks of matter; they are the basic units from which nature builds. They are all around us and compose everything, including our own bodies.

Atoms are incredibly small. A single pebble from the shoreline contains more atoms than you could ever count. The number of atoms in a single pebble far exceeds the number of pebbles on the bottom of San Francisco Bay. To get an idea of how small atoms are, imagine this: If every atom within a small pebble were the size of the pebble itself, the pebble would be larger than Mount Everest (▶ Figure 4.1). Atoms are small—yet they compose everything.

The key to connecting the microscopic world with the macroscopic world is the atom. Atoms compose matter; their properties determine matter's properties. An **atom** is the smallest identifiable unit of an element. Recall from Section 3.4 that an *element* is a substance that cannot be broken down into simpler substances.

▲ **FIGURE 4.1   The size of the atom**   If every atom within a pebble were the size of the pebble itself, then the pebble would be larger than Mount Everest.

The exact number of naturally occurring elements is controversial because some elements previously considered only synthetic may actually occur in nature in very small quantities.

There are about 91 different elements in nature, and consequently about 91 different kinds of atoms. In addition, scientists have succeeded in making about 20 synthetic elements (not found in nature). In this chapter, we examine atoms: what they are made of, how they differ from one another, and how they are structured. We also examine the elements that atoms compose and some of the properties of those elements.

4.2   Indivisible: The Atomic Theory

▲ Diogenes and Democritus, as imagined by a medieval artist. Democritus is the first person on record to have postulated that matter was composed of atoms.

If we simply look at matter, even under a microscope, it is not obvious that matter is composed of tiny particles. In fact, it appears to be just the opposite. If we divide a sample of matter into smaller and smaller pieces, it seems that we could divide it forever. From our perspective, matter seems continuous. The first people recorded as thinking otherwise were Leucippus (fifth century B.C., exact dates unknown) and Democritus (460–370 B.C.). These Greek philosophers theorized that matter was ultimately composed of small, indivisible particles. Democritus suggested that if you divided matter into smaller and smaller pieces, you would eventually end up with tiny, indestructible particles called *atomos*, or "atoms," meaning "indivisible."

The ideas of Leucippus and Democritus were not widely accepted, and it was not until 1808—over 2000 years later—that John Dalton formalized a theory of atoms that gained broad acceptance. Dalton's atomic theory has three parts:

1. Each element is composed of tiny indestructible particles called atoms.
2. All atoms of a given element have the same mass and other properties that distinguish them from the atoms of other elements.
3. Atoms combine in simple, whole-number ratios to form compounds.

Today, the evidence for the atomic theory is overwhelming. Recent advances in microscopy have allowed scientists not only to image individual atoms but also to pick them up and move them (▶ Figure 4.2). Matter is indeed composed of atoms.

EVERYDAY CHEMISTRY

Atoms and Humans

All matter is composed of atoms. What does that mean? What does it imply? It means that everything before you is composed of tiny particles too small to see. It means that even you and I are composed of these same particles. We acquired those particles from the food we have eaten over the years. The average carbon atom in our own bodies has been used by 20 other living organisms before we get to it and will be used by other organisms when we are done with it. In fact, it is likely that at this moment, your body contains over 1 trillion carbon atoms that were at one time part of your chemistry professor.*

The idea that all matter is composed of atoms has far-reaching implications. It implies that our bodies, our hearts, and even our brains are composed of atoms acting according to the laws of chemistry and physics. Some people view this as a devaluation of human life. We have always wanted to distinguish ourselves from everything else, and the idea that we are made of the same basic particles as all other matter takes something away from that distinction . . . or does it?

CAN YOU ANSWER THIS? *Do you find the idea that you are made of atoms disturbing? Why or why not?*

This calculation assumes that all of the carbon atoms metabolized by your professor over the last 40 years have been uniformly distributed into atmospheric carbon dioxide and subsequently incorporated into plants that you eat.

Xenon atoms

▶ **FIGURE 4.2 Writing with atoms**
Scientists at IBM used a special microscope, called a scanning tunneling microscope (STM), to move xenon atoms to form the letters I, B, and M. The cone shape of these atoms is due to the peculiarities of the instrumentation. Atoms are, in general, spherical in shape.

4.3 The Nuclear Atom

Electric charge is more fully defined in Section 4.4. For now, think of it as an inherent property of electrons that causes them to interact with other charged particles.

By the end of the nineteenth century, scientists were convinced that matter was composed of atoms, the permanent, indestructible building blocks from which all substances are constructed. However, an English physicist named J. J. Thomson (1856–1940) complicated the picture by discovering an even smaller and more fundamental particle called the electron. Thomson discovered that electrons are negatively charged, that they are much smaller and lighter than atoms, and that they are uniformly present in many different kinds of substances. The indestructible building block called the atom could apparently be "chipped."

The discovery of negatively charged particles within atoms raised the question of a balancing positive charge. Atoms were known to be charge-neutral, so it was believed that they must contain positive charge that balanced the negative charge of electrons. But how did the positive and negative charges within the atom fit together? Were atoms just a jumble of even more fundamental particles? Were they solid spheres, or did they have some internal structure? Thomson proposed that the negatively charged electrons were small particles held within a positively charged sphere. This model, the most popular of the time, became

Electron

Sphere of positive charge

▲ **FIGURE 4.3 Plum pudding model of the atom** In the model suggested by J. J. Thomson, negatively charged electrons (yellow) were held in a sphere of positive charge (red).

known as the plum pudding model (plum pudding is an English dessert) (◄ Figure 4.3). The picture suggested by Thomson was—to those of us not familiar with plum pudding—like a blueberry muffin, where the blueberries are the electrons and the muffin is the positively charged sphere.

In 1909, Ernest Rutherford (1871–1937), who had worked under Thomson and adhered to his plum pudding model, performed an experiment in an attempt to confirm it. His experiment instead proved it wrong. In his experiment, Rutherford directed tiny, positively charged particles—called alpha-particles—at an ultrathin sheet of gold foil (▼ Figure 4.4). Alpha-particles are about 7000 times more massive than electrons and carry a positive charge. These particles were to act as probes of the gold atoms' structure. If the gold atoms were indeed like blueberry muffins or plum pudding—with their mass and charge spread throughout the entire volume of the atom—these speeding probes should pass right through the gold foil with minimum deflection. Rutherford's results were not as he expected. A majority of the particles did pass directly through the foil, but some particles were deflected, and some (1 in 20,000) even bounced back. The results puzzled Rutherford, who found them "about as credible as if you had fired a 15-inch shell at a piece of tissue paper and it came back and hit you." What must the structure of the atom be in order to explain this odd behavior?

Rutherford created a new model to explain his results (▶ Figure 4.5). He concluded that matter must not be as uniform as it appears. It must contain large regions of empty space dotted with small regions of very dense matter. In order to explain the deflections he observed, the mass and positive charge of an atom must all be concentrated in a space much smaller than the size of the atom itself. Based on this idea, he developed the **nuclear theory of the atom**, which has three basic parts:

1. Most of the atom's mass and all of its positive charge are contained in a small core called the *nucleus*.
2. Most of the volume of the atom is empty space through which the tiny, negatively charged electrons are dispersed.
3. There are as many negatively charged electrons outside the nucleus as there are positively charged particles (*protons*) inside the nucleus, so that the atom is electrically neutral.

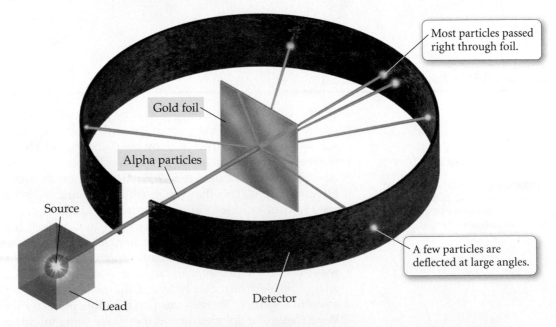

Most particles passed right through foil.

Gold foil

Alpha particles

Source

Lead

Detector

A few particles are deflected at large angles.

▲ **FIGURE 4.4 Rutherford's gold foil experiment** Tiny particles called alpha-particles were directed at a thin sheet of gold foil. Most of the particles passed directly through the foil. A few, however, were deflected—some of them at sharp angles.

(a) Rutherford's expected result

(b) Rutherford's actual result

◀ **FIGURE 4.5 Discovery of the atomic nucleus** **(a)** Expected result of Rutherford's gold foil experiment. If the plum pudding model were correct, the alpha-particles would pass right through the gold foil with minimal deflection. **(b)** Actual result of Rutherford's gold foil experiment. A small number of alpha-particles were deflected or bounced back. The only way to explain the deflections was to suggest that most of the mass and all of the positive charge of an atom must be concentrated in a space much smaller than the size of the atom itself—the nucleus. The nucleus itself is composed of positively charged particles (protons) and neutral particles (neutrons).

Later work by Rutherford and others demonstrated that the atom's **nucleus** contains both positively charged **protons** and neutral particles called **neutrons**. The dense nucleus makes up more than 99.9% of the mass of the atom, but occupies only a small fraction of its volume. The electrons are distributed through a much larger region, but don't have much mass (▼ Figure 4.6). For now, you can think of these electrons like the water droplets that make up a cloud—they are dispersed throughout a large volume but weigh almost nothing.

Rutherford's nuclear theory was a success and is still valid today. The revolutionary part of this theory is the idea that matter—at its core—is much less uniform than it appears. If the nucleus of the atom were the size of this dot ·, the average electron would be about 10 m away. Yet the dot would contain almost the entire mass of the atom. Imagine what matter would be like if atomic structure broke down. What if matter were composed of atomic nuclei piled on top of each other like marbles? Such matter would be incredibly dense; a single grain of sand composed of solid atomic nuclei would have a mass of 5 million kg (or a weight of about 10 million lb). Astronomers believe that black holes and neutron stars are composed of this kind of incredibly dense matter.

▶ **FIGURE 4.6 The nuclear atom** In this model, 99.9% of the atom's mass is concentrated in a small, dense nucleus that contains protons and neutrons. The rest of the volume of the atom is mostly empty space occupied by negatively charged electrons. The number of electrons outside the nucleus is equal to the number of protons inside the nucleus. In this image, the nucleus is greatly enlarged and the electrons are portrayed as particles.

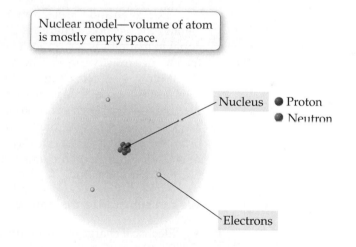

Nuclear model—volume of atom is mostly empty space.

4.4 The Properties of Protons, Neutrons, and Electrons

Protons and neutrons have very similar masses. In SI units, the mass of the proton is 1.67262×10^{-27} kg, and the mass of the neutron is a close 1.67493×10^{-27} kg. A more common unit to express these masses, however, is the **atomic mass unit (amu)**, defined as one-twelfth of the mass of a carbon atom containing six protons and six neutrons. In this unit, a proton has a mass of 1.0073 amu and a neutron has a mass of 1.0087 amu. Electrons, by contrast, have an almost negligible mass of 0.00091×10^{-27} kg, or approximately 0.00055 amu.

◀ If a proton had the mass of a baseball, an electron would have the mass of a rice grain. The proton is nearly 2000 times as massive as an electron.

EVERYDAY CHEMISTRY

Solid Matter?

If matter really is mostly empty space as Rutherford suggested, then why does it appear so solid? Why can I tap my knuckles on the table and feel a solid thump? Matter appears solid because the variation in the density is on such a small scale that our eyes can't see it. Imagine a jungle gym 100 stories high and the size of a football field. It is mostly empty space. Yet if you viewed it from an airplane, it would appear as a solid mass. Matter is similar. When you tap your knuckles on the table, it is much like one giant jungle gym (your finger) crashing into another (the table). Even though they are both primarily empty space, one does not fall into the other.

CAN YOU ANSWER THIS? *Use the jungle gym analogy to explain why most of Rutherford's alpha-particles went right through the gold foil and why a few bounced back. Remember that his gold foil was extremely thin.*

▲ Matter appears solid and uniform because the variation in density is on a scale too small for our eyes to see. Just as this scaffolding appears solid at a distance, so matter appears solid to us.

The proton and the electron both have electrical **charge**. The proton's charge is 1+ and the electron's charge is 1−. The charges of the proton and the electron are equal in magnitude but opposite in sign, so that when the two particles are paired, the charges exactly cancel. The neutron has no charge.

What is electrical charge? Electrical charge is a fundamental property of protons and electrons, just as mass is a fundamental property of matter. Most matter is charge-neutral because protons and electrons occur together and their charges cancel. However, you may have experienced excess electrical charge when brushing your hair on a dry day. The brushing action results in the accumulation of electrical charge on the hair strands, which then repel each other, causing your hair to stand on end.

We can summarize the nature of electrical charge as follows (◀ Figure 4.7)

- Electrical charge is a fundamental property of protons and electrons.
- Positive and negative electrical charges attract each other.
- Positive–positive and negative–negative charges repel each other.
- Positive and negative charges cancel each other so that a proton and an electron, when paired, are charge-neutral.

Note that matter is usually charge-neutral due to the canceling effect of protons and electrons. When matter does acquire charge imbalances, these imbalances usually equalize quickly, often in dramatic ways. For example, the shock you receive when touching a doorknob during dry weather is the equalization of a charge imbalance that developed as you walked across the carpet. Lightning is an equalization of charge imbalances that develop during electrical storms.

If you had a sample of matter—even a tiny sample, such as a sand grain—that was composed of only protons or only electrons, the forces around that matter would be extraordinary, and the matter would be unstable. Fortunately, matter is not that way—protons and electrons exist together, canceling each other's charge and making matter charge-neutral. Table 4.1 summarizes the properties of protons, neutrons, and electrons.

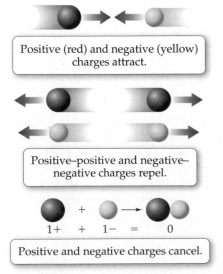

Positive (red) and negative (yellow) charges attract.

Positive–positive and negative–negative charges repel.

$$1+ \ + \ 1- \ = \ 0$$

Positive and negative charges cancel.

▲ **FIGURE 4.7** The properties of electrical charge

Negative charge buildup

Charge equalization

Positive charge buildup

▶ Matter is normally charge-neutral, having equal numbers of positive and negative charges that exactly cancel. When the charge balance of matter is disturbed, as in an electrical storm, it quickly rebalances, often in dramatic ways such as lightning.

TABLE 4.1 Subatomic Particles

	Mass (kg)	Mass (amu)	Charge
proton	1.67262×10^{-27}	1.0073	1+
neutron	1.67493×10^{-27}	1.0087	0
electron	0.00091×10^{-27}	0.00055	1−

✔ **CONCEPTUAL CHECKPOINT 4.1**

An atom composed of which of these particles would have a mass of approximately 12 amu and be charge-neutral?

(a) 6 protons and 6 electrons

(b) 3 protons, 3 neutrons, and 6 electrons

(c) 6 protons, 6 neutrons, and 6 electrons

(d) 12 neutrons and 12 electrons

4.5 Elements: Defined by Their Numbers of Protons

▼ **FIGURE 4.8 The number of protons in the nucleus defines the element**

We have seen that atoms are composed of protons, neutrons, and electrons. However, it is the number of protons in the nucleus of an atom that identifies it as a particular element. For example, atoms with 2 protons in their nucleus are helium atoms, atoms with 13 protons in their nucleus are aluminum atoms, and atoms with 92 protons in their nucleus are uranium atoms. The number of protons in an atom's nucleus defines the element (▼ Figure 4.8). Every aluminum atom has

Helium nucleus

 2 protons

Aluminum nucleus

● 13 protons

Atomic number (Z)
Chemical symbol

4
Be
9.01
beryllium

Atomic mass (defined in section 4.9)

Name

1A 1	2A 2	3B 3	4B 4	5B 5	6B 6	7B 7	8B 8	8B 9	8B 10	1B 11	2B 12	3A 13	4A 14	5A 15	6A 16	7A 17	8A 18
1 **H** 1.01 hydrogen																	2 **He** 4.00 helium
3 **Li** 6.94 lithium	4 **Be** 9.01 beryllium											5 **B** 10.81 boron	6 **C** 12.01 carbon	7 **N** 14.01 nitrogen	8 **O** 16.00 oxygen	9 **F** 19.00 fluorine	10 **Ne** 20.18 neon
11 **Na** 22.99 sodium	12 **Mg** 24.31 magnesium											13 **Al** 26.98 aluminum	14 **Si** 28.09 silicon	15 **P** 30.97 phosphorus	16 **S** 32.07 sulfur	17 **Cl** 35.45 chlorine	18 **Ar** 39.95 argon
19 **K** 39.10 potassium	20 **Ca** 40.08 calcium	21 **Sc** 44.96 scandium	22 **Ti** 47.88 titanium	23 **V** 50.94 vanadium	24 **Cr** 52.00 chromium	25 **Mn** 54.94 manganese	26 **Fe** 55.85 iron	27 **Co** 58.93 cobalt	28 **Ni** 58.69 nickel	29 **Cu** 63.55 copper	30 **Zn** 65.39 zinc	31 **Ga** 69.72 gallium	32 **Ge** 72.61 germanium	33 **As** 74.92 arsenic	34 **Se** 78.96 selenium	35 **Br** 79.90 bromine	36 **Kr** 83.80 krypton
37 **Rb** 85.47 rubidium	38 **Sr** 87.62 strontium	39 **Y** 88.91 yttrium	40 **Zr** 91.22 zirconium	41 **Nb** 92.91 niobium	42 **Mo** 95.94 molybdenum	43 **Tc** (99) technetium	44 **Ru** 101.07 ruthenium	45 **Rh** 102.91 rhodium	46 **Pd** 106.42 palladium	47 **Ag** 107.87 silver	48 **Cd** 112.41 cadmium	49 **In** 114.82 indium	50 **Sn** 118.71 tin	51 **Sb** 121.75 antimony	52 **Te** 127.60 tellurium	53 **I** 126.91 iodine	54 **Xe** 131.29 xenon
55 **Cs** 132.91 cesium	56 **Ba** 137.33 barium	57 **La** 138.91 lanthanum	72 **Hf** 178.49 hafnium	73 **Ta** 180.95 tantalum	74 **W** 183.85 tungsten	75 **Re** 186.21 rhenium	76 **Os** 190.2 osmium	77 **Ir** 192.22 iridium	78 **Pt** 195.08 platinum	79 **Au** 196.97 gold	80 **Hg** 200.59 mercury	81 **Tl** 204.38 thallium	82 **Pb** 207.2 lead	83 **Bi** 208.98 bismuth	84 **Po** (209) polonium	85 **At** (210) astatine	86 **Rn** (222) radon
87 **Fr** (223) francium	88 **Ra** (226) radium	89 **Ac** (227) actinium	104 **Rf** (261) rutherfordium	105 **Db** (262) dubnium	106 **Sg** (263) seaborgium	107 **Bh** (262) bohrium	108 **Hs** (265) hassium	109 **Mt** (266) meitnerium	110 **Ds** (281) darmstadtium	111 **Rg** (280) roentgenium	112 **Cn** (285)	113 — (284)	114 — (289)	115 — (288)	116 — (292)	117 ** (292)	118 — (294)

Periods

Lanthanides

58 **Ce** 140.12 cerium	59 **Pr** 140.91 praseodymium	60 **Nd** 144.24 neodymium	61 **Pm** (147) promethium	62 **Sm** 150.36 samarium	63 **Eu** 151.97 europium	64 **Gd** 157.25 gadolinium	65 **Tb** 158.93 terbium	66 **Dy** 162.50 dysprosium	67 **Ho** 164.93 holmium	68 **Er** 167.26 erbium	69 **Tm** 168.93 thulium	70 **Yb** 173.04 ytterbium	71 **Lu** 174.97 lutetium

Actinides

90 **Th** (232) thorium	91 **Pa** (231) protactinium	92 **U** (238) uranium	93 **Np** (237) neptunium	94 **Pu** (244) plutonium	95 **Am** (243) americium	96 **Cm** (247) curium	97 **Bk** (247) berkelium	98 **Cf** (251) californium	99 **Es** (252) einsteinium	100 **Fm** (257) fermium	101 **Md** (258) mendelevium	102 **No** (259) nobelium	103 **Lr** (260) lawrencium

▲ **FIGURE 4.9** **The periodic table of the elements**

**Element 117 is currently under review by IUPAC.

13 protons in its nucleus; if it had a different number of protons, it would be a different element. The number of protons in the nucleus of an atom is its **atomic number** and is given the symbol **Z**.

The periodic table of the elements (▲ Figure 4.9) lists all known elements according to their atomic numbers. Each element is represented by a unique **chemical symbol**, a one- or two-letter abbreviation for the element that appears directly below its atomic number on the periodic table. The chemical symbol for helium is He; for aluminum, Al; and for uranium, U. The chemical symbol and the atomic number always go together. If the atomic number is 13, the chemical symbol must be Al. If the atomic number is 92, the chemical symbol must be U. This is just another way of saying that the number of protons defines the element.

Most chemical symbols are based on the English name of the element. For example, the symbol for carbon is C; for silicon, Si; and for bromine, Br. Some elements, however, have symbols based on their Latin names. For example, the symbol for potassium is K, from the Latin *kalium*, and the symbol for sodium is Na, from the Latin *natrium*. Additional elements with symbols based on their Greek or Latin names include the following:

lead	Pb	*plumbum*
mercury	Hg	*hydrargyrum*
iron	Fe	*ferrum*
silver	Ag	*argentum*
tin	Sn	*stannum*
copper	Cu	*cuprum*

Early scientists often gave newly discovered elements names that reflected their properties. For example, *argon* originates from the Greek word *argos*, meaning "inactive," referring to argon's chemical inertness (it does not react with other elements). *Bromine* originates from the Greek word *bromos*, meaning "stench," referring to bromine's strong odor. Other elements were named after countries.

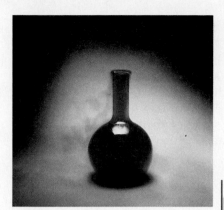

▲ The name *bromine* originates from the Greek word *bromos*, meaning "stench." Bromine vapor, seen as the red-brown gas in this photograph, has a strong odor.

Curium
96
Cm
(247)

▲ Curium is named after Marie Curie, a chemist who helped discover radioactivity and also discovered two new elements. Curie won two Nobel Prizes for her work.

For example, polonium was named after Poland, francium after France, and americium after the United States of America. Still other elements were named after scientists. Curium was named after Marie Curie, and mendelevium after Dmitri Mendeleev. Every element's name, symbol, and atomic number are included in the periodic table (inside front cover) and in an alphabetical listing (inside back cover) in this book.

EXAMPLE 4.1 Atomic Number, Atomic Symbol, and Element Name

Find the atomic symbol and atomic number for each element.

(a) silicon
(b) potassium
(c) gold
(d) antimony

SOLUTION

As you become familiar with the periodic table, you will be able to quickly locate elements on it. For now, it might be easier to find them in the alphabetical listing on the inside back cover of this book, but you should also find their position in the periodic table.

Element	Symbol	Atomic Number
silicon	Si	14
potassium	K	19
gold	Au	79
antimony	Sb	51

▶**SKILLBUILDER 4.1 | Atomic Number, Atomic Symbol, and Element Name**

Find the name and atomic number for each element.

(a) Na
(b) Ni
(c) P
(d) Ta

▶**FOR MORE PRACTICE** Problems 41, 42, 45, 46, 47, 48, 49, 50.

4.6 Looking for Patterns: The Periodic Law and the Periodic Table

▲ Dmitri Mendeleev, a Russian chemistry professor who proposed the periodic law and arranged early versions of the periodic table, shown on a Russian postage stamp.

The organization of the periodic table has its origins in the work of Dmitri Mendeleev (1834–1907), a nineteenth-century Russian chemistry professor. In his time, about 65 different elements had been discovered. Through the work of a number of chemists, much was known about each of these elements, including their relative masses, chemical activity, and some of their physical properties. However, there was no systematic way of organizing them.

1	2	3	4	5	6	7	8	9	10	11	12	13	14	15	16	17	18	19	20
H	He	Li	Be	B	C	N	O	F	Ne	Na	Mg	Al	Si	P	S	Cl	Ar	K	Ca

▲ **FIGURE 4.10 Recurring properties** The elements shown are listed in order of increasing atomic number (Mendeleev used relative mass, which is similar). The color of each element represents its properties. Notice that the properties (colors) of these elements form a repeating pattern.

In 1869, Mendeleev noticed that certain groups of elements had similar properties. He found that if he listed the elements in order of increasing relative mass, those similar properties recurred in a regular pattern (▲ Figure 4.10). Mendeleev summarized these observations in the **periodic law**:

Periodic means "recurring regularly." The properties of the elements, when listed in order of increasing relative mass, formed a *repeating pattern*.

> When the elements are arranged in order of increasing relative mass, certain sets of properties recur periodically.

Mendeleev organized all the known elements in a table in which relative mass increased from left to right and elements with similar properties were aligned in the same vertical columns (◄ Figure 4.11). Since many elements had not yet been discovered, Mendeleev's table contained some gaps, which allowed him to predict the existence of yet-undiscovered elements. For example, Mendeleev predicted the existence of an element he called *eka-silicon*, which fell below silicon on the table and between gallium and arsenic. In 1886, eka-silicon was discovered by German chemist Clemens Winkler (1838–1904) and was found to have almost exactly the properties that Mendeleev had anticipated. Winkler named the element germanium, after his home country.

▲ **FIGURE 4.11 Making a periodic table** If we place the elements from Figure 4.10 in a table, we can arrange them in rows so that similar properties align in the same vertical columns. This is similar to Mendeleev's first periodic table.

Mendeleev's original listing has evolved into the modern **periodic table**. In the modern table, elements are listed in order of increasing atomic number rather than increasing relative mass. The modern periodic table also contains more elements than Mendeleev's original table because many more have been discovered since his time.

Mendeleev's periodic law was based on observation. Like all scientific laws, the periodic law summarized many observations but did not give the underlying reason for the observation—only theories do that. For now, we accept the periodic law as it is, but in Chapter 9 we will examine a powerful theory that explains the law and gives the underlying reasons for it.

The elements in the periodic table can be broadly classified as metals, nonmetals, and metalloids (▶ Figure 4.12). **Metals** occupy the left side of the periodic table and have similar properties: They are good conductors of heat and electricity; they can be pounded into flat sheets (malleability); they can be drawn into wires (ductility); they are often shiny; and they tend to lose electrons when they undergo chemical changes. Good examples of metals are iron, magnesium, chromium, and sodium.

Nonmetals occupy the upper right side of the periodic table. The dividing line between metals and nonmetals is the zigzag diagonal line running from boron to astatine (see Figure 4.12). Nonmetals have more varied properties—some are solids at room temperature, others are gases—but as a whole they tend to be poor conductors of heat and electricity, and they all tend to gain electrons when they undergo chemical changes. Good examples of nonmetals are oxygen, nitrogen, chlorine, and iodine.

1A 1												3A 13	4A 14	5A 15	6A 16	7A 17	8A 18
1 **H**	2A 2																2 **He**
3 **Li**	4 **Be**	3B 3	4B 4	5B 5	6B 6	7B 7	─8B─ 8	9	10	1B 11	2B 12	5 **B**	6 **C**	7 **N**	8 **O**	9 **F**	10 **Ne**
11 **Na**	12 **Mg**											13 **Al**	14 **Si**	15 **P**	16 **S**	17 **Cl**	18 **Ar**
19 **K**	20 **Ca**	21 **Sc**	22 **Ti**	23 **V**	24 **Cr**	25 **Mn**	26 **Fe**	27 **Co**	28 **Ni**	29 **Cu**	30 **Zn**	31 **Ga**	32 **Ge**	33 **As**	34 **Se**	35 **Br**	36 **Kr**
37 **Rb**	38 **Sr**	39 **Y**	40 **Zr**	41 **Nb**	42 **Mo**	43 **Tc**	44 **Ru**	45 **Rh**	46 **Pd**	47 **Ag**	48 **Cd**	49 **In**	50 **Sn**	51 **Sb**	52 **Te**	53 **I**	54 **Xe**
55 **Cs**	56 **Ba**	57 **La**	72 **Hf**	73 **Ta**	74 **W**	75 **Re**	76 **Os**	77 **Ir**	78 **Pt**	79 **Au**	80 **Hg**	81 **Tl**	82 **Pb**	83 **Bi**	84 **Po**	85 **At**	86 **Rn**
87 **Fr**	88 **Ra**	89 **Ac**	104 **Rf**	105 **Db**	106 **Sg**	107 **Bh**	108 **Hs**	109 **Mt**	110 **Ds**	111 **Rg**	112 **Cn**	113	114	115	116	117 **	118

Metals
Nonmetals
Metalloids

Lanthanides

58 **Ce**	59 **Pr**	60 **Nd**	61 **Pm**	62 **Sm**	63 **Eu**	64 **Gd**	65 **Tb**	66 **Dy**	67 **Ho**	68 **Er**	69 **Tm**	70 **Yb**	71 **Lu**
90 **Th**	91 **Pa**	92 **U**	93 **Np**	94 **Pu**	95 **Am**	96 **Cm**	97 **Bk**	98 **Cf**	99 **Es**	100 **Fm**	101 **Md**	102 **No**	103 **Lr**

Actinides

▲ **FIGURE 4.12   Metals, nonmetals, and metalloids**
The elements in the periodic table can be broadly classified as metals, nonmetals, and metalloids.

▲ Silicon is a metalloid used extensively in the computer and electronics industries.

Most of the elements that lie along the zigzag diagonal line dividing metals and nonmetals are called **metalloids**, or semimetals, and display mixed properties. Metalloids are also called **semiconductors** because of their intermediate electrical conductivity, which can be changed and controlled. This property makes semiconductors useful in the manufacture of the electronic devices that are central to computers, cell phones, and many other modern gadgets. Silicon, arsenic, and germanium are good examples of metalloids.

EXAMPLE 4.2   Classifying Elements as Metals, Nonmetals, or Metalloids

Classify each element as a metal, nonmetal, or metalloid.

(a) Ba
(b) I
(c) O
(d) Te

SOLUTION

(a) Barium is on the left side of the periodic table; it is a metal.
(b) Iodine is on the right side of the periodic table; it is a nonmetal.
(c) Oxygen is on the right side of the periodic table; it is a nonmetal.
(d) Tellurium is in the middle-right section of the periodic table, along the line that divides the metals from the nonmetals; it is a metalloid.

▶**SKILLBUILDER 4.2 | Classifying Elements as Metals, Nonmetals, or Metalloids**

Classify each element as a metal, nonmetal, or metalloid.

(a) S
(b) Cl
(c) Ti
(d) Sb

▶**FOR MORE PRACTICE**   Problems: 51, 52, 53, 54.

▶ **FIGURE 4.13   Main-group and transition elements** The periodic table can be broadly divided into main-group elements, whose properties can generally be predicted based on their position, and transition elements, whose properties tend to be less predictable based on their position.

The periodic table can also be broadly divided into **main-group elements**, whose properties tend to be more predictable based on their position in the periodic table, and **transition elements** or **transition metals**, whose properties are less easily predictable based simply on their position in the periodic table (▲ Figure 4.13). Main-group elements are in columns labeled with a number and the letter A. Transition elements are in columns labeled with a number and the letter B. A competing numbering system does not use letters, but only the numbers 1–18. Both numbering systems are shown in the periodic table in the inside front cover of this book.

Each column within the periodic table is called a **family** or **group** of elements. The elements within a family of main-group elements usually have similar properties, and some have a group name. For example, the Group 8A elements, called the **noble gases**, are chemically inert gases. The most familiar noble gas is probably helium, used to fill balloons. Helium, like the other noble gases, is chemically stable—it won't combine with other elements to form compounds—and is therefore safe to put into balloons. Other noble gases include neon, often used in neon signs; argon, which makes up a small percentage of our atmosphere; krypton; and xenon. The Group 1A elements, called the **alkali metals**, are all very reactive metals. A marble-sized piece of sodium can explode when dropped into water. Other alkali metals include lithium, potassium, and rubidium. The Group 2A elements, called the **alkaline earth metals**, are also fairly reactive, although not quite as reactive as the alkali metals. Calcium, for example, reacts fairly vigorously when

The noble gases are inert (or unreactive) compared to other elements. However, some noble gases, especially the heavier ones, will form a limited number of compounds with other elements under special conditions.

Noble gases

▲ The noble gases include helium (used in balloons), neon (used in neon signs), argon, krypton, and xenon.

▲ The periodic table with Groups 1A, 2A, 7A, and 8A highlighted.

Alkali metals

▲ The alkali metals include lithium (shown in the first photo), sodium (shown in the second photo reacting with water), potassium, rubidium, and cesium.

dropped into water but will not explode as readily as sodium. Other alkaline earth metals are magnesium, a common low-density structural metal; strontium; and barium. The Group 7A elements, called the **halogens**, are very reactive nonmetals. Chlorine, a greenish-yellow gas with a pungent odor is probably the most familiar halogen. Because of its reactivity, chlorine is often used as a sterilizing and disinfecting agent (because it reacts with and kills bacteria and other microscopic organisms). Other halogens include bromine, a red-brown liquid that easily evaporates into a gas; iodine, a purple solid; and fluorine, a pale yellow gas.

EXAMPLE 4.3 Groups and Families of Elements

To which group or family of elements does each element belong?

(a) Mg
(b) N
(c) K
(d) Br

SOLUTION

(a) Mg is in Group 2A; it is an alkaline earth metal.
(b) N is in Group 5A.
(c) K is in Group 1A; it is an alkali metal.
(d) Br is in Group 7A; it is a halogen.

▶**SKILLBUILDER 4.3 | Groups and Families of Elements**

To which group or family of elements does each element belong?

(a) Li
(b) B
(c) I
(d) Ar

▶**FOR MORE PRACTICE**   Problems 57, 58, 59, 60, 61, 62, 63, 64.

Alkaline earth metals

◀ The alkaline earth metals include beryllium, magnesium (shown burning in the first photo), calcium (shown reacting with water in the second photo), strontium, and barium.

Halogens

▶ The halogens include fluorine, chlorine (shown in the first photo), bromine, iodine (shown in the second photo), and astatine.

> ✓ **CONCEPTUAL CHECKPOINT 4.2**
>
> Which statement can NEVER be true?
>
> **(a)** An element can be both a transition element and a metal.
>
> **(b)** An element can be both a transition element and a metalloid.
>
> **(c)** An element can be both a metalloid and a halogen.
>
> **(d)** An element can be both a main-group element and a halogen.

4.7 Ions: Losing and Gaining Electrons

| The charge of an ion is shown in the upper right corner of the symbol.

In chemical reactions, atoms often lose or gain electrons to form charged particles called **ions**. For example, neutral lithium (Li) atoms contain 3 protons and 3 electrons; however, in reactions, lithium atoms lose one electron (e^-) to form Li^+ ions.

$$Li \longrightarrow Li^+ + e^-$$

The Li^+ *ion* contains 3 protons but only 2 electrons, resulting in a net charge of 1+. Ion charges are usually written with the magnitude of the charge first followed by the sign of the charge. For example, a positive two charge is written as 2+, and a negative two charge is written as 2−. The charge of an ion depends on how many electrons were gained or lost and is given by the formula:

$$\text{Ion charge} = \text{number of protons} - \text{number of electrons}$$

$$= \#p^+ - \#e^-$$

where p^+ stands for *proton* and e^- stands for *electron*.

For the Li^+ ion with 3 protons and 2 electrons the charge is:

$$\text{Ion Charge} = 3 - 2 = 1+$$

Neutral fluorine (F) atoms contain 9 protons and 9 electrons; however, in chemical reactions fluorine atoms gain 1 electron to form F^- ions:

$$F + e^- \longrightarrow F^-$$

The F^- *ion* contains 9 protons and 10 electrons, resulting in a 1− charge.

$$\text{Ion charge} = 9 - 10$$

$$= 1-$$

Positively charged ions, such as Li^+, are **cations**, and negatively charged ions, such as F^-, are **anions**. Ions behave very differently than the atoms from which they are formed. Neutral sodium atoms, for example, are extremely reactive, interacting violently with most things they contact. Sodium cations (Na^+), on the other hand, are relatively inert—we eat them all the time in sodium chloride (table salt). In nature, cations and anions always occur together so that, again, matter is charge-neutral. For example, in table salt, the sodium cation occurs together with the chloride anion (Cl^-).

EXAMPLE 4.4   Determining Ion Charge from Numbers of Protons and Electrons

Determine the charge of each ion.

(a) a magnesium ion with 10 electrons
(b) a sulfur ion with 18 electrons
(c) an iron ion with 23 electrons

SOLUTION

To determine the charge of each ion, use the ion charge equation.

$$\text{Ion charge} = \#p - \#e^-$$

The number of electrons is given in the problem. The number of protons is obtained from the element's atomic number in the periodic table.

(a) magnesium with atomic number 12

$$\text{Ion charge} = 12 - 10 = 2+ (Mg^{2+})$$

(b) sulfur with atomic number 16

$$\text{Ion charge} = 16 - 18 = 2- (S^{2-})$$

(c) iron with atomic number 26

$$\text{Ion charge} = 26 - 23 = 3+ (Fe^{3+})$$

▶**SKILLBUILDER 4.4** | **Determining Ion Charge from Numbers of Protons and Electrons**

Determine the charge of each ion.

(a) a nickel ion with 26 electrons
(b) a bromine ion with 36 electrons
(c) a phosphorus ion with 18 electrons

▶**FOR MORE PRACTICE**   Example 4.10; Problems 73, 74.

EXAMPLE 4.5   **Determining the Number of Protons and Electrons in an Ion**

Find the number of protons and electrons in the Ca^{2+} ion.

From the periodic table, find that the atomic number for calcium is 20, so calcium has 20 protons. The number of electrons can be found using the ion charge equation.	**SOLUTION** $$\text{Ion charge} = \#p - \#e^-$$ $$2+ = 20 - \#e^-$$ $$\#e^- = 20 - 2 = 18$$ Therefore the number of electrons is 18. The Ca^{2+} ion has 20 protons and 18 electrons.

▶**SKILLBUILDER 4.5** | **Determining the Number of Protons and Electrons in an Ion**

Find the number of protons and electrons in the S^{2-} ion.

▶**FOR MORE PRACTICE**   Example 4.11; Problems 75, 76.

IONS AND THE PERIODIC TABLE

For many main-group elements, we can use the periodic table to predict how many electrons tend to be lost or gained when an atom of that particular element ionizes. The number associated with the letter A above each *main-group* column in the periodic table—1 through 8—gives the number of *valence electrons* for the elements in that column. We will discuss the concept of valence electrons more fully in Chapter 9, but for now, think of valence electrons as the outermost electrons in an atom. Since oxygen is in column 6A, we can deduce that it has 6 valence

▲ FIGURE 4.14 **Elements that form predictable ions**

electrons; since magnesium is in column 2A, it has 2 valence electrons, and so on. An important exception to this rule is helium—it is in column 8A, but has only 2 valence electrons. Valence electrons are particularly important because, as we shall see in Chapter 10, it is these electrons that take part in chemical bonding.

The key to predicting the charge acquired by a particular element when it ionizes is its position in the periodic table relative to the noble gases:

> Main-group elements tend to form ions that have the same number of valence electrons as the nearest noble gas.

For example, the closest noble gas to oxygen is neon. When oxygen ionizes, it *acquires* two additional electrons for a total of 8 valence electrons—the same number as neon. When determining the closest noble gas, we can move either forward or backward on the periodic table. For example, the closest noble gas to magnesium is also neon, even though neon (atomic number 10) falls before magnesium (atomic number 12) in the periodic table. Therefore magnesium *loses* its 2 valence electrons to attain the same number of valence electrons as neon.

In accordance with this principle, the alkali metals (Group 1A) tend to lose 1 electron and therefore form 1+ ions, while the alkaline earth metals (Group 2A) tend to lose 2 electrons and therefore form 2+ ions. The halogens (Group 7A) tend to gain 1 electron and therefore form 1− ions. The groups in the periodic table that form predictable ions are shown in ▲ Figure 4.14. Be familiar with these groups and the ions they form. In Chapter 9, we will examine a theory that more fully explains why these groups form ions as they do.

EXAMPLE 4.6 Charge of Ions from Position in Periodic Table

Based on their position in the periodic table, what ions do barium and iodine tend to form?

SOLUTION

Since barium is in Group 2A, it tends to form a cation with a 2+ charge (Ba^{2+}). Since iodine is in Group 7A, it tends to form an anion with a 1− charge (I^-).

▶SKILLBUILDER 4.6 | **Charge of Ions from Position in Periodic Table**

Based on their position in the periodic table, what ions do potassium and selenium tend to form?

▶**FOR MORE PRACTICE** Problems 79, 80.

✔ CONCEPTUAL CHECKPOINT 4.3

Which of these pairs of ions have the same total number of electrons?

(a) Na^+ and Mg^{2+}

(b) F^- and Cl^-

(c) O^- and O^{2-}

(d) Ga^{3+} and Fe^{3+}

4.8   Isotopes: When the Number of Neutrons Varies

All atoms of a given element have the same number of protons; however, they do not necessarily have the same number of neutrons. Since neutrons and protons have nearly the same mass (approximately 1 amu), and since the number of neutrons in the atoms of a given element can vary, all atoms of a given element *do not* have the same mass (contrary to what John Dalton originally proposed in his atomic theory). For example, all neon atoms in nature contain 10 protons, but they may have 10, 11, or 12 neutrons (▼ Figure 4.15). All three types of neon atoms exist, and each has a slightly different mass. Atoms with the same number of protons but different numbers of neutrons are called **isotopes**. Some elements, such as beryllium (Be) and aluminum (Al), have only one naturally occurring isotope, while other elements, such as neon (Ne) and chlorine (Cl), have two or more.

For a given element, the relative amounts of each different isotope in a naturally occurring sample of that element is always the same. For example, in any natural sample of neon atoms, 90.48% of them are the isotope with 10 neutrons, 0.27% are the isotope with 11 neutrons, and 9.25% are the isotope with 12 neutrons as summarized in Table 4.2. This means that out of 10,000 neon atoms, 9048 have

> There are a few exceptions to this rule, such as boron, but they are beyond our scope in this text.

TABLE 4.2   Neon Isotopes

Symbol	Number of Protons	Number of Neutrons	A (Mass Number)	Percent Natural Abundance
Ne-20 or $^{20}_{10}$Ne	10	10	20	90.48%
Ne-21 or $^{21}_{10}$Ne	10	11	21	0.27%
Ne-22 or $^{22}_{10}$Ne	10	12	22	9.25%

▶ **FIGURE 4.15**
Isotopes of neon
Naturally occurring neon contains three different isotopes, Ne-20 (with 10 neutrons), Ne-21 (with 11 neutrons), and Ne-22 (with 12 neutrons).

● Ne-21
● Ne-22
● Ne-20

Percent means "per hundred." 90.48% means that 90.48 atoms out of 100 are the isotope with 10 neutrons.

10 neutrons, 27 have 11 neutrons, and 925 have 12 neutrons. These percentages are referred to as the **percent natural abundance** of the isotopes. The preceding numbers are for neon only; all elements have their own unique percent natural abundance of isotopes.

The sum of the number of neutrons and protons in an atom is its **mass number** and is given the symbol **A**.

$$A = \text{Number of protons} + \text{Number of neutrons}$$

For neon, which has 10 protons, the mass numbers of the three different naturally occurring isotopes are 20, 21, and 22, corresponding to 10, 11, and 12 neutrons, respectively.

Isotopes are often symbolized in the following way:

where X is the chemical symbol, A is the mass number, and Z is the atomic number.

For example, the symbols for the neon isotopes are:

$$^{20}_{10}\text{Ne} \quad ^{21}_{10}\text{Ne} \quad ^{22}_{10}\text{Ne}$$

Notice that the chemical symbol, Ne, and the atomic number, 10, are redundant: If the atomic number is 10, the symbol must be Ne, and vice versa. The mass numbers, however, are different, reflecting the different number of neutrons in each isotope.

A second common notation for isotopes is the chemical symbol (or chemical name) followed by a hyphen and the mass number of the isotope.

$$X - A$$

Chemical symbol or name   Mass number

In this notation, the neon isotopes are:

Ne-20	neon-20
Ne-21	neon-21
Ne-22	neon-22

Notice that all isotopes of a given element have the same number of protons (otherwise they would be a different element). Notice also that the mass number is the *sum* of the number of protons and the number of neutrons. The number of neutrons in an isotope is the difference between the mass number and the atomic number.

In general, mass number increases with increasing atomic number.

EXAMPLE 4.7 Atomic Numbers, Mass Numbers, and Isotope Symbols

What are the atomic number (Z), mass number (A), and symbols of the carbon isotope with 7 neutrons?

SOLUTION

Find that the atomic number (Z) of carbon is 6 (from the periodic table). This tells you that carbon atoms have 6 protons. The mass number (A) for the isotope with 7 neutrons is the sum of the number of protons and the number of neutrons.

$$A = 6 + 7 = 13$$

So, Z = 6, A = 13, and the symbols for the isotope are C-13 and $^{13}_{6}\text{C}$.

▶SKILLBUILDER 4.7 | Atomic Numbers, Mass Numbers, and Isotope Symbols

What are the atomic number, mass number, and symbols for the chlorine isotope with 18 neutrons?

▶FOR MORE PRACTICE   Example 4.12; Problems: 85, 87, 89, 90.

EXAMPLE 4.8   Numbers of Protons and Neutrons from Isotope Symbols

How many protons and neutrons are in the chromium isotope $^{52}_{24}\text{Cr}$?

The number of protons is equal to Z (lower left number).	**SOLUTION** $\#p^+ = Z = 24$
The number of neutrons is equal to A (upper left number) $-$Z (lower left number).	$\#n = A - Z$ $= 52 - 24$ $= 28$

▶SKILLBUILDER 4.8 | Numbers of Protons and Neutrons from Isotope Symbols

How many protons and neutrons are in the potassium isotope $^{39}_{19}\text{K}$?

▶FOR MORE PRACTICE   Example 4.13; Problems 91, 92.

 CONCEPTUAL CHECKPOINT 4.4

If an atom with a mass number of 27 has 14 neutrons, it is an isotope of which element?

(a) silicon

(b) aluminum

(c) cobalt

(d) niobium

 CONCEPTUAL CHECKPOINT 4.5

Throughout this book, we represent atoms as spheres. For example, a carbon atom is represented by a black sphere as shown here. In light of the nuclear theory of the atom, would C-12 and C-13 look different in this representation of atoms? Why or why not?

Carbon

4.9   Atomic Mass: The Average Mass of an Element's Atoms

An important part of Dalton's atomic theory was that all atoms of a given element have the same mass. But as we just learned, the atoms of a given element may have different masses (because of isotopes). So Dalton was not completely correct. We can, however, calculate an average mass—called the **atomic mass**—for each

CHEMISTRY IN THE ENVIRONMENT
Radioactive Isotopes at Hanford, Washington

Nuclei of the isotopes of a given element are not all equally stable. For example, naturally occurring lead is composed primarily of Pb-206, Pb-207, and Pb-208. Other isotopes of lead also exist, but their nuclei are unstable. Scientists can make some of these other isotopes, such as Pb-185, in the laboratory. However, within seconds Pb-185 atoms emit a few energetic subatomic particles from their nuclei and change into different isotopes of different elements (which are themselves unstable). These emitted subatomic particles are called **nuclear radiation**, and the isotopes that emit them are termed **radioactive**. Nuclear radiation, always associated with unstable nuclei, can be harmful to humans and other living organisms because the energetic particles interact with and damage biological molecules. Some isotopes, such as Pb-185, emit significant amounts of radiation only for a very short time. Others, however, remain radioactive for a long time—in some cases millions or even billions of years.

The nuclear power and nuclear weapons industries produce by-products containing unstable isotopes of several different elements. Many of these isotopes emit nuclear radiation for a long time, and their disposal is an environmental problem. For example, in Hanford, Washington, which for 50 years produced fuel for nuclear weapons, 177 underground storage tanks contain 55 million gallons of highly radioactive nuclear waste. Certain radioactive isotopes within that waste will produce nuclear radiation for the foreseeable future. Unfortunately, some of the underground storage tanks are aging, and leaks have allowed some of the waste to seep into the environment. While the danger from short-term external exposure to this waste is minimal, ingestion of the waste through contamination of drinking water or food supplies would pose significant health risks. Consequently, Hanford is now the site of the largest environmental cleanup project in U.S. history. The U.S. government expects the project to take more than 20 years and cost about $10 billion.

Radioactive isotopes are not always harmful, however, and many have beneficial uses. For example, technetium-99 (Tc-99) is often given to patients to diagnose disease. The radiation emitted by Tc-99 helps doctors image internal organs or detect infection.

CAN YOU ANSWER THIS? *Give the number of neutrons in each of the following isotopes: Pb-206, Pb-207, Pb-208, Pb-185, Tc-99.*

◀ Storage tanks at Hanford, Washington, contain 55 million gallons of high-level nuclear waste. Each tank pictured here holds 1 million gallons.

element. The atomic mass of each element is listed in the periodic table directly beneath the element's symbol; it represents the average mass of the atoms that compose that element. For example, the periodic table lists the atomic mass of chlorine as 35.45 amu. Naturally occurring chlorine consists of 75.77% chlorine-35 (mass 34.97 amu) and 24.23% chlorine-37 (mass 36.97 amu). Its atomic mass is:

Some books call this *average atomic mass* or atomic weight instead of simply *atomic mass*.

$$\text{Atomic mass} = (0.7577 \times 34.97 \text{ amu}) + (0.2423 \times 36.97 \text{ amu})$$
$$= 35.45 \text{ amu}$$

Notice that the atomic mass of chlorine is closer to 35 than 37 because naturally occurring chlorine contains more chlorine-35 atoms than chlorine-37 atoms. Notice also that when percentages are used in these calculations, they must always be converted to their decimal value. To convert a percentage to its decimal value, divide by 100. For example:

$$75.77\% = 75.77/100 = 0.7577$$
$$24.23\% = 24.23/100 = 0.2423$$

In general, atomic mass is calculated according to the following equation:

Atomic mass = (Fraction of isotope 1 × Mass of isotope 1) +
(Fraction of isotope 2 × Mass of isotope 2) +
(Fraction of isotope 3 × Mass of isotope 3) + ...

where the fractions of each isotope are the percent natural abundances converted to their decimal values. Atomic mass is useful because it allows us to assign a characteristic mass to each element and, as we will see in Chapter 6, it allows us to quantify the number of atoms in a sample of that element.

EXAMPLE 4.9   Calculating Atomic Mass

Gallium has two naturally occurring isotopes: Ga-69 with mass 68.9256 amu and a natural abundance of 60.11%, and Ga-71 with mass 70.9247 amu and a natural abundance of 39.89%. Calculate the atomic mass of gallium.

Convert the percent natural abundances into decimal form by dividing by 100.	**SOLUTION** Fraction Ga-69 $= \dfrac{60.11}{100} = 0.6011$ Fraction Ga-71 $= \dfrac{39.89}{100} = 0.3989$
Use the fractional abundances and the atomic masses of the isotopes to compute the atomic mass according to the atomic mass definition given earlier.	Atomic mass = $(0.6011 \times 68.9256 \text{ amu}) + (0.3989 \times 70.9247 \text{ amu})$ $= 41.4321 \text{ amu} + 28.2919 \text{ amu}$ $= 69.7231 = 69.72 \text{ amu}$

▶**SKILLBUILDER 4.9 | Calculating Atomic Mass**

Magnesium has three naturally occurring isotopes with masses of 23.99, 24.99, and 25.98 amu and natural abundances of 78.99%, 10.00%, and 11.01%. Calculate the atomic mass of magnesium.

▶**FOR MORE PRACTICE**   Example 4.14; Problems 95, 96.

✔ CONCEPTUAL CHECKPOINT 4.6

A fictitious element is composed of isotopes A and B with masses of 61.9887 and 64.9846 amu, respectively. The atomic mass of the element is 64.52. What can you conclude about the natural abundances of the two isotopes?

(a) The natural abundance of isotope A must be greater than the natural abundance of isotope B.

(b) The natural abundance of isotope B must be greater than the natural abundance of isotope A.

(c) The natural abundances of both isotopes must be about equal.

(d) Nothing can be concluded about the natural abundances of the two isotopes from the given information.

CHAPTER IN REVIEW

CHEMICAL PRINCIPLES

The Atomic Theory: Democritus and Leucippus, ancient Greek philosophers, were the first to assert that matter is ultimately composed of small, indestructible particles. It was not until 2000 years later, however, that John Dalton introduced a formal atomic theory stating that matter is composed of atoms; atoms of a given element have unique properties that distinguish them from atoms of other elements; and atoms combine in simple, whole-number ratios to form compounds.

Discovery of the Atom's Nucleus: Rutherford's gold foil experiment probed atomic structure, and his results led to the nuclear model of the atom, which, with minor modifications to accommodate neutrons, is still valid today. In this model, the atom is composed of protons and neutrons—which compose most of the atom's mass and are grouped together in a dense nucleus—and electrons, which compose most of the atom's volume. Protons and neutrons have similar masses (1 amu), while electrons have a much smaller mass (0.00055 amu).

Charge: Protons and electrons both have electrical charge; the charge of the proton is +1 and the charge of the electron is −1. The neutron has no charge. When protons and electrons combine in atoms, their charges cancel.

The Periodic Table: The periodic table tabulates all known elements in order of increasing atomic number. The periodic table is arranged so that similar elements are grouped in columns. Columns of elements in the periodic table have similar properties and are called groups or families. Elements on the left side of the periodic table are metals and tend to lose electrons in their chemical changes. Elements on the upper right side of the periodic table are nonmetals and tend to gain electrons in their chemical changes. Elements between the two are called metalloids.

Atomic Number: The characteristic that defines an element is the number of protons in the nuclei of its atoms; this number is called the atomic number (Z).

Ions: When an atom gains or loses electrons, it becomes an ion. Positively charged ions are called cations, and negatively charged ions are called anions. Cations and anions occur together so that matter is ordinarily charge-neutral.

RELEVANCE

The Atomic Theory: The concept of atoms is important because it explains the physical world. You and everything you see are made of atoms. To understand the physical world, we must begin by understanding atoms. Atoms are the key concept—they determine the properties of matter.

Discovery of the Atom's Nucleus: We can understand why this is relevant by asking, what if it were otherwise? What if matter were *not* mostly empty space? While we cannot know for certain, it seems probable that such matter would not form the diversity of substances required for life—and then, of course, we would not be around to ask the question.

Charge: Electrical charge is relevant to much of our modern world. Many of the machines and computers we depend on are powered by electricity, which is the movement of electrical charge.

The Periodic Table: The periodic table helps us organize the elements in ways that allow us to predict their properties. Helium, for example, is not toxic in small amounts because it is an inert gas—it does not react with anything. The gases in the column below it on the periodic table are also inert gases and form a family or group of elements called the noble gases. By tabulating the elements and grouping similar ones together, we begin to understand their properties.

Atomic Number: Elements are the fundamental building blocks from which all compounds are made.

Ions: Ions occur in many compounds, such as sodium chloride.

Isotopes: While all atoms of a given element have the same number of protons, they do not necessarily have the same number of neutrons. Atoms of the same element with different numbers of neutrons are called isotopes. Isotopes are characterized by their mass number (A), the sum of the number of protons and the number of neutrons in their nucleus.

Each naturally occurring sample of an element has the same percent natural abundance of each isotope. These percentages, together with the mass of each isotope, are used to compute the atomic mass of the element, a weighted average of the masses of the individual isotopes.

Isotopes: Isotopes are relevant because they influence tabulated atomic masses. To understand these masses, we must understand the presence and abundance of isotopes. In nuclear processes—processes in which the nuclei of atoms actually change—the presence of different isotopes becomes even more important.

Some isotopes are not stable—they lose subatomic particles and are transformed into other elements. The emission of subatomic particles by unstable nuclei is called radioactive decay. In many situations, such as in diagnosing and treating certain diseases, nuclear radiation is extremely useful. In other situations, such as in the disposal of radioactive waste, it can pose environmental problems.

CHEMICAL SKILLS

EXAMPLES

Determining Ion Charge from Numbers of Protons and Electrons (Section 4.7)

- From the periodic table or from the alphabetical list of elements, find the atomic number of the element; this number is equal to the number of protons.

- Use the ion charge equation to compute charge.

  Ion charge = $\#p^+ - \#e^-$

EXAMPLE 4.10 Determining Ion Charge from Numbers of Protons and Electrons

Determine the charge of a selenium ion with 36 electrons.

SOLUTION

Selenium is atomic number 34; therefore, it has 34 protons.

$$\text{Ion charge} = 34 - 36 = 2-$$

Determining the Number of Protons and Electrons in an Ion (Section 4.7)

- From the periodic table or from the alphabetical list of elements, find the atomic number of the element; this number is equal to the number of protons.

- Use the ion charge equation and substitute in the known values.

  Ion charge = $\#p^+ - \#e^-$

- Solve the equation for the number of electrons.

EXAMPLE 4.11 Determining the Number of Protons and Electrons in an Ion

Find the number of protons and electrons in the O^{2-} ion.

SOLUTION

The atomic number of O is 8; therefore, it has 8 protons.

$$\text{Ion charge} = \#p^+ - \#e^-$$
$$2- = 8 - \#e^-$$
$$\#e^- = 8 + 2 = 10$$

The ion has 8 protons and 10 electrons.

Determining Atomic Numbers, Mass Numbers, and Isotope Symbols for an Isotope (Section 4.8)

- From the periodic table or from the alphabetical list of elements, find the atomic number of the element.

- The mass number (A) is equal to the atomic number plus the number of neutrons.

- Write the symbol for the isotope by writing the symbol for the element with the mass number in the upper left corner and the atomic number in the lower left corner.

- The other symbol for the isotope is simply the chemical symbol followed by a hyphen and the mass number.

EXAMPLE 4.12 Determining Atomic Numbers, Mass Numbers, and Isotope Symbols for an Isotope

What are the atomic number (Z), mass number (A), and symbols for the iron isotope with 30 neutrons?

SOLUTION

The atomic number of iron is 26.

$$A = 26 + 30 = 56$$

The mass number is 56.

$$^{56}_{26}\text{Fe}$$

Fe-56

Number of Protons and Neutrons from Isotope Symbols (Section 4.8)

- The number of protons is equal to Z (lower left number).

- The number of neutrons is equal to

    A (upper left number) − Z (lower left number.)

EXAMPLE 4.13 Number of Protons and Neutrons from Isotope Symbols

How many protons and neutrons are in $^{62}_{28}\text{Ni}$?

SOLUTION

28 protons

$$\#n = 62 - 28 = 34 \text{ neutrons}$$

Calculating Atomic Mass from Percent Natural Abundances and Isotopic Masses (Section 4.9)

- Convert the natural abundances from percent to decimal values by dividing by 100.

- Find the atomic mass by multiplying the fractions of each isotope by their respective masses and adding.

- Round to the correct number of significant figures.

- Check your work.

EXAMPLE 4.14 Calculating Atomic Mass from Percent Natural Abundances and Isotopic Masses

Copper has two naturally occurring isotopes: Cu-63 with mass 62.9395 amu and a natural abundance of 69.17%, and Cu-65 with mass 64.9278 amu and a natural abundance of 30.83%. Calculate the atomic mass of copper.

SOLUTION

$$\text{Fraction Cu-63} = \frac{69.17}{100} = 0.6917$$

$$\text{Fraction Cu-65} = \frac{30.83}{100} = 0.3083$$

$$\begin{aligned}
\text{Atomic mass} &= (0.6917 \times 62.9395 \text{ amu}) \\
&= 43.5353 \text{ amu} + 20.0107 \text{ amu} \\
&= 63.5460 \text{ amu} \\
&= 63.55 \text{ amu}
\end{aligned}$$

KEY TERMS

alkali metals [4.6]
alkaline earth metals [4.6]
anions [4.7]
atom [4.1]
atomic mass [4.9]
atomic mass unit (amu) [4.4]
atomic number (Z) [4.5]
cation [4.7]
charge [4.4]

chemical symbol [4.5]
electron [4.3]
family (of elements) [4.6]
group (of elements) [4.6]
halogens [4.6]
ion [4.7]
isotope [4.8]
main-group elements [4.6]
mass number (A) [4.8]
metalloids [4.6]

metals [4.6]
neutron [4.3]
noble gases [4.6]
nonmetals [4.6]
nuclear radiation [4.9]
nuclear theory of the atom [4.3]
nucleus [4.3]
percent natural abundance [4.8]

periodic law [4.6]
periodic table [4.6]
proton [4.3]
radioactive [4.9]
semiconductor [4.6]
transition elements [4.6]
transition metals [4.6]

EXERCISES

QUESTIONS

1. What did Democritus contribute to our modern understanding of matter?
2. What are three main ideas in Dalton's atomic theory?
3. Describe Rutherford's gold foil experiment and the results of that experiment. How did these results contradict the plum pudding model of the atom?
4. What are the main ideas in the nuclear theory of the atom?

5. List the three subatomic particles and their properties.
6. What is electrical charge?
7. Is matter usually charge-neutral? How would matter be different if it were not charge-neutral?
8. What does the atomic number of an element specify?
9. What is a chemical symbol?
10. List some examples of how elements got their names.

11. What was Dmitri Mendeleev's main contribution to our modern understanding of chemistry?
12. What is the main idea in the periodic law?
13. How is the periodic table organized?
14. What are the properties of metals? Where are metals found on the periodic table?
15. What are the properties of nonmetals? Where are nonmetals found on the periodic table?
16. Where on the periodic table are metalloids found?
17. What is a family or group of elements?
18. Locate each group of elements on the periodic table and list its group number.
    (a) alkali metals
    (b) alkaline earth metals
    (c) halogens
    (d) noble gases

19. What is an ion?
20. What is an anion? What is a cation?
21. Locate each group on the periodic table and list the charge of the ions it tends to form.
    (a) Group 1A
    (b) Group 2A
    (c) Group 3A
    (d) Group 6A
    (e) Group 7A
22. What are isotopes?
23. What is the percent natural abundance of isotopes?
24. What is the mass number of an isotope?
25. What notations are commonly used to specify isotopes? What do each of the numbers in these symbols mean?
26. What is the atomic mass of an element?

PROBLEMS

ATOMIC AND NUCLEAR THEORY

27. Which statements are *inconsistent* with Dalton's atomic theory as it was originally stated? Why?
    (a) All carbon atoms are identical.
    (b) Helium atoms can be split into two hydrogen atoms.
    (c) An oxygen atom combines with 1.5 hydrogen atoms to form water molecules.
    (d) Two oxygen atoms combine with a carbon atom to form carbon dioxide molecules.

28. Which statements are *consistent* with Dalton's atomic theory as it was originally stated? Why?
    (a) Calcium and titanium atoms have the same mass.
    (b) Neon and argon atoms are the same.
    (c) All cobalt atoms are identical.
    (d) Sodium and chlorine atoms combine in a 1:1 ratio to form sodium chloride.

29. Which statements are *inconsistent* with Rutherford's nuclear theory as it was originally stated? Why?
    (a) Helium atoms have two protons in the nucleus and two electrons outside the nucleus.
    (b) Most of the volume of hydrogen atoms is due to the nucleus.
    (c) Aluminum atoms have 13 protons in the nucleus and 22 electrons outside the nucleus.
    (d) The majority of the mass of nitrogen atoms is due to their 7 electrons.

30. Which statements are *consistent* with Rutherford's nuclear theory as it was originally stated? Why?
    (a) Atomic nuclei are small compared to the size of atoms.
    (b) The volume of an atom is mostly empty space.
    (c) Neutral potassium atoms contain more protons than electrons.
    (d) Neutral potassium atoms contain more neutrons than protons.

31. If atoms are mostly empty space, and atoms compose all ordinary matter, then why does solid matter seem to have no space within it?

32. Rutherford's experiment suggested that matter was not as uniform as it appears. What part of his experimental results implied this idea? Explain.

PROTONS, NEUTRONS, AND ELECTRONS

33. Which statements about electrons are true?
    (a) Electrons repel each other.
    (b) Electrons are attracted to protons.
    (c) Some electrons have a charge of 1− and some have no charge.
    (d) Electrons are much lighter than neutrons.

34. Which statements about electrons are false?
    (a) Most atoms have more electrons than protons.
    (b) Electrons have a charge of 1−.
    (c) If an atom has an equal number of protons and electrons, it will be charge-neutral.
    (d) Electrons experience an attraction to protons.

35. Which statements about protons are true?

(a) Protons have twice the mass of neutrons.

(b) Protons have the same magnitude of charge as electrons but are opposite in sign.

(c) Most atoms have more protons than electrons.

(d) Protons have a charge of 1+.

36. Which statements about protons are false?

(a) Protons have about the same mass as neutrons.

(b) Protons have about the same mass as electrons.

(c) Some atoms don't have any protons.

(d) Protons have the same magnitude of charge as neutrons, but are opposite in sign.

37. How many electrons would it take to equal the mass of a proton?

38. A helium nucleus has two protons and two neutrons. How many electrons would it take to equal the mass of a helium nucleus?

39. What mass of electrons would be required to just neutralize the charge of 1.0 g of protons?

40. What mass of protons would be required to just neutralize the charge of 1.0 g of electrons?

ELEMENTS, SYMBOLS, AND NAMES

41. Find the atomic number (Z) for each element.

(a) Fr

(b) Kr

(c) Pa

(d) Ge

(e) Al

42. Find the atomic number (Z) for each element.

(a) Si

(b) W

(c) Ni

(d) Rn

(e) Sr

43. How many protons are in the nucleus of an atom of each element?

(a) Ar

(b) Sn

(c) Xe

(d) O

(e) Tl

44. How many protons are in the nucleus of an atom of each element?

(a) Ti

(b) Li

(c) U

(d) Br

(e) F

45. List the symbol and atomic number corresponding to each element.

(a) carbon

(b) nitrogen

(c) sodium

(d) potassium

(e) copper

46. List the symbol and atomic number corresponding to each element.

(a) boron

(b) neon

(c) silver

(d) mercury

(e) curium

47. List the name and the atomic number corresponding to the symbol for each element.

(a) Mn

(b) Ag

(c) Au

(d) Pb

(e) S

48. List the name and the atomic number corresponding to the symbol for each element.

(a) Y

(b) N

(c) Ne

(d) K

(e) Mo

49. Fill in the blanks to complete the table.

Element Name	Element Symbol	Atomic Number
___	Au	79
Tin	___	___
___	As	___
Copper	___	29
___	Fe	___
___	___	80

50. Fill in the blanks to complete the table.

Element Name	Element Symbol	Atomic Number
___	Al	13
Iodine	___	___
___	Sb	___
Sodium	___	___
___	Rn	86
___	___	82

THE PERIODIC TABLE

51. Classify each element as a metal, nonmetal, or metalloid.

(a) Sr

(b) Mg

(c) F

(d) N

(e) As

52. Classify each element as a metal, nonmetal, or metalloid.

(a) Na

(b) Ge

(c) Si

(d) Br

(e) Ag

53. Which elements would you expect to lose electrons in chemical changes?

(a) potassium

(b) sulfur

(c) fluorine

(d) barium

(e) copper

54. Which elements would you expect to gain electrons in chemical changes?

(a) nitrogen

(b) iodine

(c) tungsten

(d) strontium

(e) gold

55. Which elements are main-group elements?

(a) Te

(b) K

(c) V

(d) Re

(e) Ag

56. Which elements are *not* main-group elements?

(a) Al

(b) Br

(c) Mo

(d) Cs

(e) Pb

57. Which elements are alkaline earth metals?

(a) sodium

(b) aluminum

(c) calcium

(d) barium

(e) lithium

58. Which elements are alkaline earth metals?

(a) rubidium

(b) tungsten

(c) magnesium

(d) cesium

(e) beryllium

59. Which elements are alkali metals?

(a) barium

(b) sodium

(c) gold

(d) tin

(e) rubidium

60. Which elements are alkali metals?

(a) scandium

(b) iron

(c) potassium

(d) lithium

(e) cobalt

61. Classify each element as a halogen, a noble gas, or neither.
    (a) Cl
    (b) Kr
    (c) F
    (d) Ga
    (e) He

62. Classify each element as a halogen, a noble gas, or neither.
    (a) Ne
    (b) Br
    (c) S
    (d) Xe
    (e) I

63. To what group number does each element belong?
    (a) oxygen
    (b) aluminum
    (c) silicon
    (d) tin
    (e) phosphorus

64. To what group number does each element belong?
    (a) germanium
    (b) nitrogen
    (c) sulfur
    (d) carbon
    (e) boron

65. Which element do you expect to be most like sulfur? Why?
    (a) nitrogen
    (b) oxygen
    (c) fluorine
    (d) lithium
    (e) potassium

66. Which element do you expect to be most like magnesium? Why?
    (a) potassium
    (b) silver
    (c) bromine
    (d) calcium
    (e) lead

67. Which pair of elements do you expect to be most similar? Why?
    (a) Si and P
    (b) Cl and F
    (c) Na and Mg
    (d) Mo and Sn
    (e) N and Ni

68. Which pair of elements do you expect to be most similar? Why?
    (a) Ti and Ga
    (b) N and O
    (c) Li and Na
    (d) Ar and Br
    (e) Ge and Ga

69. Fill in the blanks to complete the table.

Chemical Symbol	Group Number	Group Name	Metal or Nonmetal
K	____	____	metal
Br	____	halogens	____
Sr	____	____	____
He	8A	____	____
Ar	____	____	____

70. Fill in the blanks to complete the table.

Chemical Symbol	Group Number	Group Name	Metal or Nonmetal
Cl	7A	____	____
Ca	____	____	metal
Xe	____	____	nonmetal
Na	____	alkali metal	____
F	____	____	____

IONS

71. Complete each ionization equation.
    (a) $Na \longrightarrow Na^+ + \underline{\quad}$
    (b) $O + 2e^- \longrightarrow \underline{\quad}$
    (c) $Ca \longrightarrow Ca^{2+} + \underline{\quad}$
    (d) $Cl + e^- \longrightarrow \underline{\quad}$

72. Complete each ionization equation.
    (a) $Mg \longrightarrow \underline{\quad} + 2e^-$
    (b) $Ba \longrightarrow Ba^{2+} + \underline{\quad}$
    (c) $I + e^- \longrightarrow \underline{\quad}$
    (d) $Al \longrightarrow \underline{\quad} + 3e^-$

73. Determine the charge of each ion.

(a) oxygen ion with 10 electrons

(b) aluminum ion with 10 electrons

(c) titanium ion with 18 electrons

(d) iodine ion with 54 electrons

74. Determine the charge of each ion.

(a) tungsten ion with 68 electrons

(b) tellurium ion with 54 electrons

(c) nitrogen ion with 10 electrons

(d) barium ion with 54 electrons

75. Determine the number of protons and electrons in each ion.

(a) Na^+

(b) Ba^{2+}

(c) O^{2+}

(d) Co^{3+}

76. Determine the number of protons and electrons in each ion.

(a) Al^{3+}

(b) S^{2-}

(c) I^-

(d) Ag^+

77. Determine whether each statement is true or false. If false, correct it.

(a) The Ti^{2+} ion contains 22 protons and 24 electrons.

(b) The I^- ion contains 53 protons and 54 electrons.

(c) The Mg^{2+} ion contains 14 protons and 12 electrons.

(d) The O^{2-} ion contains 8 protons and 10 electrons.

78. Determine whether each statement is true or false. If false, correct it.

(a) The Fe^+ ion contains 29 protons and 26 electrons.

(b) The Cs^+ ion contains 55 protons and 56 electrons.

(c) The Se^{2-} ion contains 32 protons and 34 electrons.

(d) The Li^+ ion contains 3 protons and 2 electrons.

79. Predict the ion formed by each element:

(a) Rb

(b) K

(c) Al

(d) O

80. Predict the ion formed by each element:

(a) F

(b) N

(c) Mg

(d) Na

81. Predict how many electrons will most likely be gained or lost by each element:

(a) Ga

(b) Li

(c) Br

(d) S

82. Predict how many electrons will most likely be gained or lost by each element:

(a) I

(b) Ba

(c) Cs

(d) Se

83. Fill in the blanks to complete the table.

Symbol	Ion Commonly Formed	Number of Electrons in Ion	Number of Protons in Ion
Te	___	54	___
In	___	___	49
Sr	Sr^{2+}	___	___
___	Mg^{2+}	___	12
Cl	___	___	___

84. Fill in the blanks to complete the table.

Symbol	Ion Commonly Formed	Number of Electrons in Ion	Number of Protons in Ion
F	___	___	9
___	Be^{2+}	2	___
Br	___	36	___
Al	___	___	13
O	___	___	___

ISOTOPES

85. What are the atomic number and mass number for each isotope?

(a) the hydrogen isotope with 2 neutrons

(b) the chromium isotope with 28 neutrons

(c) the calcium isotope with 22 neutrons

(d) the tantalum isotope with 109 neutrons

86. How many neutrons are in an atom with each set of atomic numbers and mass numbers?

(a) Z = 28, A = 59

(b) Z = 92, A = 235

(c) Z = 21, A = 46

(d) Z = 18, A = 42

87. Write isotopic symbols of the form $^A_Z X$ for each isotope.

(a) the oxygen isotope with 8 neutrons

(b) the fluorine isotope with 10 neutrons

(c) the sodium isotope with 12 neutrons

(d) the aluminum isotope with 14 neutrons

88. Write isotopic symbols of the form X-A (for example, C-13) for each isotope.

(a) the iodine isotope with 74 neutrons

(b) the phosphorus isotope with 16 neutrons

(c) the uranium isotope with 234 neutrons

(d) the argon isotope with 22 neutrons

89. Write the symbol for each isotope in the form $^A_Z X$.

(a) cobalt-60

(b) neon-22

(c) iodine-131

(d) plutonium-244

90. Write the symbol for each isotope in the form $^A_Z X$.

(a) U-235

(b) V-52

(c) P-32

(d) Xe-144

91. Determine the number of protons and neutrons in each isotope:

(a) $^{23}_{11}Na$

(b) $^{266}_{88}Ra$

(c) $^{208}_{82}Pb$

(d) $^{14}_{7}N$

92. Determine the number of protons and neutrons in each isotope:

(a) $^{33}_{15}P$

(b) $^{40}_{19}K$

(c) $^{222}_{86}Rn$

(d) $^{99}_{43}Tc$

93. Carbon-14, present within living organisms and substances derived from living organisms, is often used to establish the age of fossils and artifacts. Determine the number of protons and neutrons in a carbon-14 isotope and write its symbol in the form $^A_Z X$.

94. Plutonium-239 is used in nuclear bombs. Determine the number of protons and neutrons in plutonium-239 and write its symbol in the form $^A_Z X$.

ATOMIC MASS

95. Rubidium has two naturally occurring isotopes: Rb-85 with mass 84.9118 amu and a natural abundance of 72.17%, and Rb-87 with mass 86.9092 amu and a natural abundance of 27.83%. Calculate the atomic mass of rubidium.

96. Silicon has three naturally occurring isotopes: Si-28 with mass 27.9769 amu and a natural abundance of 92.21%, Si-29 with mass 28.9765 amu and a natural abundance of 4.69%, and Si-30 with mass 29.9737 amu and a natural abundance of 3.10%. Calculate the atomic mass of silicon.

97. Bromine has two naturally occurring isotopes (Br-79 and Br-81) and an atomic mass of 79.904 amu.

(a) If the natural abundance of Br-79 is 50.69%, what is the natural abundance of Br-81?

(b) If the mass of Br-81 is 80.9163 amu, what is the mass of Br-79?

98. Silver has two naturally occurring isotopes (Ag-107 and Ag-109).

(a) Use the periodic table to find the atomic mass of silver.

(b) If the natural abundance of Ag-107 is 51.84%, what is the natural abundance of Ag-109?

(c) If the mass of Ag-107 is 106.905 amu, what is the mass of Ag-109?

99. An element has two naturally occurring isotopes. Isotope 1 has a mass of 120.9038 amu and a relative abundance of 57.4%, and isotope 2 has a mass of 122.9042 amu and a relative abundance of 42.6%. Find the atomic mass of this element and, referring to the periodic table, identify it.

100. Copper has two naturally occurring isotopes. Cu-63 has a mass of 62.939 amu and relative abundance of 69.17%. Use the atomic weight of copper to determine the mass of the other copper isotope.

CUMULATIVE PROBLEMS

101. Electrical charge is sometimes reported in coulombs (C). On this scale, 1 electron has a charge of -1.6×10^{-19} C. Suppose your body acquires -125 mC (millicoulombs) of charge on a dry day. How many excess electrons has it acquired? (*Hint:* Use the charge of an electron in coulombs as a conversion factor between charge and electrons.)

102. How many excess protons are in a positively charged object with a charge of $+398$ mC (millicoulombs)? The charge of 1 proton is $+1.6 \times 10^{-19}$ C.

103. The hydrogen atom contains 1 proton and 1 electron. The radius of the proton is approximately 1.0 fm (femtometers), and the radius of the hydrogen atom is approximately 53 pm (picometers). Calculate the volume of the nucleus and the volume of the atom for hydrogen. What percentage of the hydrogen atom's volume is occupied by the nucleus?

104. Carbon-12 contains 6 protons and 6 neutrons. The radius of the nucleus is approximately 2.7 fm, and the radius of the atom is approximately 70 pm. Calculate the volume of the nucleus and the volume of the atom. What percentage of the carbon atom's volume is occupied by the nucleus?

105. Prepare a table such as Table 4.2 for the four different isotopes of Sr that have the following natural abundances and masses.

Sr-84	0.56%	83.9134 amu
Sr-86	9.86%	85.9093 amu
Sr-87	7.00%	86.9089 amu
Sr-88	82.58%	87.9056 amu

Use your table and the preceding atomic masses to calculate the atomic mass of strontium.

106. Determine the number of protons and neutrons in each isotope of chromium and use the following natural abundances and masses to calculate its atomic mass.

Cr-50	4.345%	49.9460 amu
Cr-52	83.79%	51.9405 amu
Cr-53	9.50%	52.9407 amu
Cr-54	2.365%	53.9389 amu

107. Fill in the blanks to complete the table.

Symbol	Z	A	Number of Protons	Number of Electrons	Number of Neutrons	Charge
Zn$^+$	___	___	___	___	34	1+
___	25	55	___	22	___	___
___	___	___	15	15	16	___
O^{2-}	___	16	___	___	___	2−
___	___	___	16	18	18	___

108. Fill in the blanks to complete the table.

Symbol	Z	A	Number of Protons	Number of Electrons	Number of Neutrons	Charge
Mg^{2+}	___	25	___	___	13	2+
___	22	48	___	18	___	___
___	16	___	___	___	16	2−
Ga^{3+}	___	71	___	___	___	___
___	___	___	82	80	125	___

109. Europium has two naturally occurring isotopes: Eu-151 with a mass of 150.9198 amu and a natural abundance of 47.8%, and Eu-153. Use the atomic mass of europium to find the mass and natural abundance of Eu-153.

110. Rhenium has two naturally occurring isotopes: Re-185 with a natural abundance of 37.40%, and Re-187 with a natural abundance of 62.60%. The sum of the masses of the two isotopes is 371.9087 amu. Find the masses of the individual isotopes.

111. Chapter 1 describes the difference between observations, laws, and theories. Provide two examples of theories from this chapter and explain why they are theories.

112. Chapter 1 describes the difference between observations, laws, and theories. Provide one example of a law from this chapter and explain why it is a law.

113. The atomic mass of fluorine is 19.00 amu, and all fluorine atoms in a naturally occurring sample of fluorine have this mass. The atomic mass of chlorine is 35.45 amu, but no chlorine atoms in a naturally occurring sample of chlorine have this mass. Explain the difference.

114. The atomic mass of germanium is 72.61 amu. Is it likely that any individual germanium atoms have a mass of 72.61 amu?

115. Copper has only two naturally occurring isotopes, Cu-63 and Cu-65. The mass of Cu-63 is 62.9396 amu, and the mass of Cu-65 is 64.9278 amu. Use the atomic mass of copper to determine the relative abundance of each isotope in a naturally occurring sample.

116. Gallium has only two naturally occurring isotopes, Ga-69 and Ga-71. The mass of Ga-69 is 68.9256 amu, and the mass of Ga-71 is 70.9247 amu. Use the atomic mass of gallium to determine the relative abundance of each isotope in a naturally occurring sample.

HIGHLIGHT PROBLEMS

117. The figure is a representation of 50 atoms of a fictitious element with the symbol Nt and atomic number 120. Nt has three isotopes represented by the following colors: Nt-304 (red), Nt-305 (blue), and Nt-306 (green).

(a) Assuming that the figure is statistically representative of naturally occurring Nt, what is the percent natural abundance of each Nt isotope?

(b) Use the following masses of each isotope to calculate the atomic mass of Nt. Then draw a box for the element similar to the boxes for each element shown in the periodic table in the inside front cover of this book. Make sure your box includes the atomic number, symbol, and atomic mass. (Assume that the percentages from part (a) are correct to four significant figures.)

Nt-304	303.956 amu
Nt-305	304.962 amu
Nt-306	305.978 amu

118. Neutron stars are believed to be composed of solid nuclear matter, primarily neutrons.

(a) If the radius of a neutron is 1.0×10^{-13} cm, calculate its density in g/cm^3.

(volume of a sphere $= \frac{4}{3}\pi r^3$)

(b) Assuming that a neutron star has the same density as a neutron, calculate the mass in kilograms of a small piece of a neutron star the size of a spherical pebble with a radius of 0.10 mm.

►ANSWERS TO SKILLBUILDER EXERCISES

Skillbuilder 4.1
(a) sodium, 11
(b) nickel, 28
(c) phosphorus, 15
(d) tantalum, 73

Skillbuilder 4.2
(a) nonmetal
(b) nonmetal
(c) metal
(d) metalloid

Skillbuilder 4.3
(a) alkali metal, group 1A
(b) group 3A
(c) halogen, group 7A
(d) noble gas, group 8A

Skillbuilder 4.4
(a) 2+
(b) 1−
(c) 3−

Skillbuilder 4.5 16 protons, 18 electrons

Skillbuilder 4.6 K^+ and Se^{2-}

Skillbuilder 4.7 Z = 17, A = 35, Cl-35, and $^{35}_{17}Cl$

Skillbuilder 4.8 19 protons, 20 neutrons

Skillbuilder 4.9 24.31 amu

►ANSWERS TO CONCEPTUAL CHECKPOINTS

4.1 (c) The mass in amu is approximately equal to the number of protons plus the number of neutrons. In order to be charge-neutral, the number of protons must equal the number of electrons.

4.2 (b) All of the metalloids are main-group elements (see Figures 4.12 and 4.13).

4.3 (a) Both of these ions have 10 electrons.

4.4 (b) This atom must have (27 − 14) = 13 protons; the element with an atomic number of 13 is Al.

4.5 The isotopes C-12 and C-13 would not look different in this representation of atoms because the only difference between the two isotopes is that C-13 has an extra neutron in the nucleus. The illustration represents the whole atom and does not attempt to illustrate its nucleus. Since the nucleus of an atom is miniscule compared to the size of the atom itself, the extra neutron would not affect the size of the atom.

4.6 (b) The natural abundance of isotope B must be greater than the natural abundance of isotope A because the atomic mass is closer to the mass of isotope B than to the mass of isotope A.

Sucrose

Molecules and Compounds

H_2O

Sucrose $C_{12}H_{22}O_{11}$

O_2

Caffeine

Tuesday 23rd

Mastering Homework Due
Chemistry

Cal

Sucrose

Sucrose

CHAPTER 5

Molecules and Compounds

"Almost all aspects of life are engineered at the molecular level, and without understanding molecules, we can only have a very sketchy understanding of life itself."

FRANCIS HARRY COMPTON CRICK (1916–2004)

5.1 Sugar and Salt

Sodium, a shiny metal (▼ Figure 5.1) that dulls almost instantly upon exposure to air, is extremely reactive and poisonous. If you were to consume any appreciable amount of elemental sodium, you would need immediate medical help. Chlorine, a pale yellow gas (▼ Figure 5.2), is equally reactive and poisonous. Yet the compound formed from these two elements, sodium chloride, is the relatively harmless flavor enhancer that we call table salt (▶ Figure 5.3). When elements combine to form compounds, their properties completely change.

◀ Ordinary table sugar is a compound called sucrose. A sucrose molecule, such as the one shown here, contains carbon, hydrogen, and oxygen atoms. The properties of sucrose are, however, very different from those of carbon (also shown in the form of graphite), hydrogen, and oxygen. The properties of a compound are, in general, different from the properties of the elements that compose it.

▲ FIGURE 5.1 **Elemental sodium**
Sodium is an extremely reactive metal that dulls almost instantly upon exposure to air.

▲ FIGURE 5.2 **Elemental chlorine**
Chlorine is a yellow gas with a pungent odor. It is highly reactive and poisonous.

▲ FIGURE 5.3 Sodium chloride
The compound formed by sodium and chlorine is table salt.

Consider ordinary sugar. Sugar is a compound composed of carbon, hydrogen, and oxygen. Each of these elements has its own unique properties. Carbon is most familiar to us as the graphite found in pencils or as the diamonds in jewelry. Hydrogen is an extremely flammable gas used as a fuel for the space shuttle, and oxygen is one of the gases that compose air. When these three elements combine to form sugar, however, a sweet, white, crystalline solid results.

In Chapter 4, we learned how protons, neutrons, and electrons combine to form different elements, each with its own properties and its own chemistry, each different from the other. In this chapter, we learn how these elements combine with each other to form different compounds, each with its own properties and its own chemistry, each different from all the others and different from the elements that compose it. This is the great wonder of nature: how from such simplicity—protons, neutrons, and electrons—we get such great complexity. It is exactly this complexity that makes life possible. Life could not exist with just 91 different elements if they did not combine to form compounds. It takes compounds in all of their diversity to make living organisms.

5.2 Compounds Display Constant Composition

Although some of the substances we encounter in everyday life are elements, most are not—they are compounds. Free atoms are rare in nature. As we learned in Chapter 3, a compound is different from a mixture of elements. In a compound, the elements combine in fixed, definite proportions, whereas in a mixture, they can have any proportions whatsoever. Consider the difference between a mixture of hydrogen and oxygen gas (▼ Figure 5.4) and the compound water (▼ Figure 5.5). A mixture of hydrogen and oxygen gas can contain any propor-

The ratio of hydrogen to oxygen in water is fixed.

The ratio of hydrogen to oxygen in a mixture is variable.

Hydrogen molecule

Oxygen molecule

Water molecule

2 H atoms (◗) to every
1 O atom (●)

▲ FIGURE 5.4 A mixture This balloon is filled with a mixture of hydrogen and oxygen gas. The relative amounts of hydrogen and oxygen are variable. We could easily add either more hydrogen or more oxygen to the balloon.

▲ FIGURE 5.5 A chemical compound This balloon is filled with water, composed of molecules that have a fixed ratio of hydrogen to oxygen. (*Source:* JoLynn E. Funk.)

tions of hydrogen and oxygen. Water, on the other hand, is composed of water molecules that consist of two hydrogen atoms bonded to one oxygen atom. Consequently, water has a definite proportion of hydrogen to oxygen.

The first chemist to formally state the idea that elements combine in fixed proportions to form compounds was Joseph Proust (1754–1826) in the **law of constant composition**, which states:

> All samples of a given compound have the same proportions of their constituent elements.

For example, if we decompose an 18.0 g sample of water, we would get 16.0 g of oxygen and 2.0 g of hydrogen, or an oxygen-to-hydrogen mass ratio of

$$\text{Mass ratio} = \frac{16.0 \text{ g O}}{2.0 \text{ g H}} = 8.0 \quad \text{or} \quad 8.0{:}1$$

This is true of any sample of pure water, no matter what its origin. The law of constant composition applies not only to water, but to every compound. If we decomposed a 17.0 g sample of ammonia, a compound composed of nitrogen and hydrogen, we would get 14.0 g of nitrogen and 3.0 g of hydrogen, or a nitrogen-to–hydrogen mass ratio of

Even though atoms combine in whole-number ratios, their mass ratios are not necessarily whole numbers.

$$\text{Mass ratio} = \frac{14.0 \text{ g N}}{3.0 \text{ g H}} = 4.7 \quad \text{or} \quad 4.7{:}1$$

Again, this ratio is the same for every sample of ammonia—the composition of each compound is constant.

EXAMPLE 5.1 Constant Composition of Compounds

Two samples of carbon dioxide, obtained from different sources, are decomposed into their constituent elements. One sample produces 4.8 g of oxygen and 1.8 g of carbon, and the other sample produces 17.1 g of oxygen and 6.4 g of carbon. Show that these results are consistent with the law of constant composition.

Compute the mass ratio of one element to the other by dividing the larger mass by the smaller one. For the first sample:	**SOLUTION** $$\frac{\text{Mass oxygen}}{\text{Mass carbon}} = \frac{4.8 \text{ g}}{1.8 \text{ g}} = 2.7$$
For the second sample:	$$\frac{\text{Mass oxygen}}{\text{Mass carbon}} = \frac{17.1 \text{ g}}{6.4 \text{ g}} = 2.7$$

Since the ratios are the same for the two samples, these results are consistent with the law of constant composition.

▶**SKILLBUILDER 5.1 | Constant Composition of Compounds**

Two samples of carbon monoxide, obtained from different sources, are decomposed into their constituent elements. One sample produces 4.3 g of oxygen and 3.2 g of carbon, and the other sample produces 7.5 g of oxygen and 5.6 g of carbon. Are these results consistent with the law of constant composition?

▶**FOR MORE PRACTICE** Example 5.16; Problems 25, 26.

5.3 Chemical Formulas: How to Represent Compounds

Compounds have constant composition with respect to mass (as we learned in the previous section) because they are composed of atoms in fixed ratios.

We represent a compound with a **chemical formula**, which indicates the elements present in the compound and the relative number of atoms of each. For example, H_2O is the chemical formula for water; it indicates that water consists of hydrogen and oxygen atoms in a 2:1 ratio. The formula contains the symbol for each element, accompanied by a subscript indicating the number of atoms of that element. By convention, a subscript of 1 is omitted.

Other common chemical formulas include NaCl for table salt, indicating sodium and chlorine atoms in a 1:1 ratio; CO_2 for carbon dioxide, indicating carbon and oxygen atoms in a 1:2 ratio; and $C_{12}H_{22}O_{11}$ for table sugar (sucrose), indicating carbon, hydrogen, and oxygen atoms in a 12:22:11 ratio. The subscripts in a chemical formula are part of the compound's definition—if they change, the formula no longer specifies the same compound. For example, CO is the chemical formula for carbon monoxide, an air pollutant with adverse health effects on humans. When inhaled, carbon monoxide interferes with the blood's ability to carry oxygen, which can be fatal. CO is the primary substance responsible for the deaths of people who inhale too much automobile exhaust. If you change the subscript of the O in CO from 1 to 2, however, you get the formula for a totally different compound. CO_2 is the chemical formula for carbon dioxide, the relatively harmless product of combustion and human respiration. We breathe small amounts of CO_2 all the time with no harmful effects. So, remember that:

CO CO_2

> The subscripts in a chemical formula represent the relative numbers of each type of atom in a chemical compound; they never change for a given compound.

Chemical formulas normally list the most metallic elements first. Therefore, the formula for table salt is NaCl, not ClNa. In compounds that do not include a metal, the more metal-like element is listed first. Recall from Chapter 4 that metals are found on the left side of the periodic table and nonmetals on the upper right side. Among nonmetals, those to the left in the periodic table are more metal-like than those to the right and are normally listed first. Therefore, we write CO_2 and NO, not O_2C and ON. Within a single column in the periodic table, elements toward the bottom are more metal-like than elements toward the top. So, we write SO_2, not O_2S. The specific order for listing nonmetal elements in a chemical formula is shown in Table 5.1.

There are a few historical exceptions to the practice in which the most metallic element is listed first, such as the hydroxide ion, which is written as OH^-.

TABLE 5.1 Order of Listing Nonmetal Elements in a Chemical Formula

C	P	N	H	S	I	Br	Cl	O	F

Elements on the left are generally listed before elements on the right.

EXAMPLE 5.2 Writing Chemical Formulas

Write a chemical formula for each compound.

(a) the compound containing two aluminum atoms to every three oxygen atoms
(b) the compound containing three oxygen atoms to every sulfur atom
(c) the compound containing four chlorine atoms to every carbon atom

	SOLUTION
Since aluminum is the metal, it is listed first.	(a) Al_2O_3
Since sulfur is below oxygen on the periodic table and since it occurs before oxygen in Table 5.1, it is listed first.	(b) SO_3
Since carbon is to the left of chlorine on the periodic table and since it occurs before chlorine in Table 5.1, it is listed first.	(c) CCl_4

▶**SKILLBUILDER 5.2** | **Writing Chemical Formulas**

Write a chemical formula for each compound.

(a) the compound containing two silver atoms to every sulfur atom
(b) the compound containing two nitrogen atoms to every oxygen atom
(c) the compound containing two oxygen atoms to every titanium atom

▶**FOR MORE PRACTICE**   Example 5.17; Problems 31, 32, 33, 34.

Some chemical formulas contain groups of atoms that act as a unit. When several groups of the same kind are present, their formula is set off in parentheses with a subscript to indicate the number of that group. Many of these groups of atoms have a charge associated with them and are called **polyatomic ions**. For example, NO_3^- is a polyatomic ion with a 1− charge. Polyatomic ions are described in more detail in Section 5.7.

To determine the total number of each type of atom in a compound containing a group within parentheses, multiply the subscript outside the parentheses by the subscript for each atom inside the parentheses. For example, $Mg(NO_3)_2$ indicates a compound containing one magnesium atom (present as the Mg^{2+} ion) and two NO_3^- groups.

Therefore, the preceding formula has the following numbers of each type of atom.

Mg:	1 Mg	
N:	$1 \times 2 = 2\,N$	(implied 1 inside parentheses times 2 outside parentheses)
O:	$3 \times 2 = 6\,O$	(3 inside parentheses times 2 outside parentheses)

EXAMPLE 5.3   Total Number of Each Type of Atom in a Chemical Formula

Determine the number of each type of atom in $Mg_3(PO_4)_2$.

SOLUTION

Mg: There are three Mg atoms (present as Mg^{2+} ions), as indicated by the subscript 3.

P: There are two P atoms. We determine this by multiplying the subscript outside the parentheses (2) by the subscript for P inside the parentheses, which is 1 (implied).

O: There are eight O atoms. We determine this by multiplying the subscript outside the parentheses (2) by the subscript for O inside the parentheses (4).

▶**SKILLBUILDER 5.3** | **Total Number of Each Type of Atom in a Chemical Formula**

Determine the number of each type of atom in K_2SO_4.

▶**SKILLBUILDER PLUS**

Determine the number of each type of atom in $Al_2(SO_4)_3$.

▶**FOR MORE PRACTICE**   Example 5.18; Problems 35, 36, 37, 38.

✓ CONCEPTUAL CHECKPOINT 5.1

Which formula represents the greatest total number of atoms?

(a) $Al(C_2H_3O_2)_3$

(b) $Al_2(Cr_2O_7)_3$

(c) $Pb(HSO_4)_4$

(d) $Pb_3(PO_4)_4$

(e) $(NH_4)_3PO_4$

TYPES OF CHEMICAL FORMULAS

Hydrogen

Carbon

Nitrogen

Oxygen

Fluorine

Phosphorus

Sulfur

Chlorine

We can categorize chemical formulas as three different types: empirical, molecular, and structural. An **empirical formula** gives the simplest whole-number ratio of atoms of each element in a compound. A **molecular formula** gives the *actual* number of atoms of each element in a molecule of the compound. For example, the molecular formula for hydrogen peroxide is H_2O_2, and its empirical formula is HO. The molecular formula is always a whole number multiple of the empirical formula. For many compounds, the molecular and empirical formula are the same. For example, the empirical and molecular formula for water is H_2O because water molecules contain two hydrogen atoms and one oxygen atom; no simpler whole number ratio can express the relative number of hydrogen atoms to oxygen atoms.

A **structural** formula uses lines to represent chemical bonds and shows how the atoms in a molecule are connected to each other. The structural formula for hydrogen peroxide is H—O—O—H. In addition to formulas, we also use **molecular models**—three-dimensional representations of molecules—to represent compounds. In this book, we use two types of molecular models: ball-and-stick and space-filling. In **ball-and-stick models**, we represent atoms as balls and chemical bonds as sticks. The balls and sticks are connected to represent the molecule's shape. The balls are color coded, and each element is assigned a color as shown in the margin.

In **space-filing models**, atoms fill the space between each other to more closely represent our best idea for how a molecule might appear if we could scale it to a visible size. Consider the following ways to represent a molecule of methane, the main component of natural gas:

CH_4

H
|
H—C—H
|
H

Molecular formula   Structural formula   Ball-and-stick model   Space-filling model

The molecular formula of methane indicates that methane has one carbon atom and four hydrogen atoms. The structural formula shows how the atoms are connected: Each hydrogen atom is bonded to the central carbon atom. The ball-and-stick model and the space-filling model illustrate the *geometry* of the molecule: how the atoms are arranged in three dimensions.

Throughout this book, you have seen and will continue to see images that show the connection between the *macroscopic world* (what we see), the *atomic and molecular world* (the particles that compose matter), and the *symbolic way* that chemists represent the atomic and molecular world. For example, at left is a representation of water using this kind of image.

The main goal of these images is to help you visualize the main theme of this book: *the connection between the world around us and the world of atoms and molecules.*

Macroscopic   Molecular

H_2O

Symbolic

5.4   A Molecular View of Elements and Compounds

In Chapter 3, we learned that pure substances could be categorized as either elements or compounds. We can further subcategorize elements and compounds according to the basic units that compose them (▼ Figure 5.6). Pure substances may be elements, or they may be compounds. Elements may be either atomic or molecular. Compounds may be either molecular or ionic.

▶ **FIGURE 5.6   A molecular view of elements and compounds**

Example: Ne   Example: O₂   Example: H₂O   Example: NaCl

Mercury atoms

Hg

▲ **FIGURE 5.7   An atomic element** The basic units that compose mercury, an atomic element and a metal, are single mercury atoms.

A few molecular elements, such as S₈ and P₄, are composed of molecules containing several atoms.

ATOMIC ELEMENTS

Atomic elements are those that exist in nature with single atoms as their basic units. Most elements fall into this category. For example, helium is composed of helium atoms, copper is composed of copper atoms, and mercury of mercury atoms (◀ Figure 5.7).

Chlorine molecule

Cl₂

MOLECULAR ELEMENTS

Molecular elements do not normally exist in nature with single atoms as their basic units. Instead, these elements exist as *diatomic molecules*—two atoms of that element bonded together—as their basic units. For example, hydrogen is composed of H₂ molecules, oxygen is composed of O₂ molecules, and chlorine of Cl₂ molecules (◀ Figure 5.8). Elements that exist as diatomic molecules are shown in Table 5.2 and ▶ Figure 5.9.

◀ **FIGURE 5.8   A molecular element** The basic units that compose chlorine, a molecular element, are diatomic chlorine molecules, each composed of two chlorine atoms.

TABLE 5.2   Elements That Occur as Diatomic Molecules

Name of Element	Formula of Basic Unit
hydrogen	H_2
nitrogen	N_2
oxygen	O_2
fluorine	F_2
chlorine	Cl_2
bromine	Br_2
iodine	I_2

▲ **FIGURE 5.9   Elements that form diatomic molecules**   Elements that normally exist as diatomic molecules are highlighted in yellow on this periodic table. Note that they are all nonmetals, and include four of the halogens.

Carbon dioxide molecule

CO_2

▲ **FIGURE 5.10   A molecular compound**   The basic units that compose dry ice, a molecular compound, are CO_2 molecules.

MOLECULAR COMPOUNDS

Molecular compounds are compounds formed from two or more nonmetals. The basic units of molecular compounds are molecules composed of the constituent atoms. For example, water is composed of H_2O molecules, dry ice is composed of CO_2 molecules (◄ Figure 5.10), and acetone (finger nail–polish remover) of C_3H_6O molecules.

IONIC COMPOUNDS

Ionic compounds contain one or more cations paired with one or more anions. In most cases, the cations are metals and the anions are nonmetals. When a metal, which has a tendency to lose electrons (see Section 4.6), combines with a nonmetal, which has a tendency to gain electrons, one or more electrons transfer from the metal to the nonmetal, creating positive and negative ions that are then attracted to each other. You can assume that a compound composed of a metal and a nonmetal is ionic. The basic unit of ionic compounds is the **formula unit**, the smallest electrically neutral collection of ions. Formula units are different from molecules in that they do not exist as discrete entities, but rather as part of a larger lattice. For example, salt (NaCl) is composed of Na^+ and Cl^- ions in a 1:1 ratio. In table salt, Na^+ and Cl^- ions exist in an alternating three-dimensional array (► Figure 5.11). However, any one Na^+ ion does not pair with one specific Cl^- ion. Sometimes chemists refer to formula units as molecules, but this is not strictly correct since ionic compounds do not contain distinct molecules.

Sodium chloride formula unit

NaCl

▲ **FIGURE 5.11   An ionic compound**   The basic units that compose table salt, an ionic compound, are NaCl formula units. Unlike molecular compounds, ionic compounds do not contain individual molecules but rather sodium and chloride ions in an alternating three-dimensional array.

EXAMPLE 5.4   Classifying Substances as Atomic Elements, Molecular Elements, Molecular Compounds, or Ionic Compounds

Classify each substance as an atomic element, molecular element, molecular compound, or ionic compound.

(a) krypton
(b) $CoCl_2$
(c) nitrogen
(d) SO_2
(e) KNO_3

SOLUTION

(a) Krypton is an element that is not listed as diatomic in Table 5.2; therefore, it is an atomic element.
(b) $CoCl_2$ is a compound composed of a metal (left side of periodic table) and nonmetal (right side of the periodic table); therefore, it is an ionic compound.
(c) Nitrogen is an element that is listed as diatomic in Table 5.2; therefore, it is a molecular element.
(d) SO_2 is a compound composed of two nonmetals; therefore, it is a molecular compound.
(e) KNO_3 is a compound composed of a metal and two nonmetals; therefore, it is an ionic compound.

▶**SKILLBUILDER 5.4 | Classifying Substances as Atomic Elements, Molecular Elements, Molecular Compounds, or Ionic Compounds**

Classify each substance as an atomic element, molecular element, molecular compound, or ionic compound.

(a) chlorine
(b) NO
(c) Au
(d) Na_2O
(e) $CrCl_3$

▶**FOR MORE PRACTICE**   Example 5.19, Example 5.20; Problems 43, 44, 45, 46.

(a)

(b)

✓ CONCEPTUAL CHECKPOINT 5.2

Which of the figures at left (in the margin) represents a molecular compound?

5.5   Writing Formulas for Ionic Compounds

Revisit Section 4.7 and Figure 4.14 to review the elements that form ions with a predictable charge.

Since ionic compounds must be charge-neutral, and since many elements form only one type of ion with a predictable charge, we can determine the formulas for many ionic compounds based on their constituent elements. For example, the formula for the ionic compound composed of sodium and chlorine must be NaCl and not anything else because in compounds Na always forms 1+ cations and Cl always forms 1− anions. In order for the compound to be charge-neutral, it must contain one Na^+ cation to every Cl^- anion. The formula for the ionic compound composed of magnesium and chlorine, however, must be $MgCl_2$, because Mg always forms 2+ cations and Cl always forms 1− anions. In order for the compound to be charge-neutral, it must contain one Mg^{2+} cation to every two Cl^- anions. In general:

- Ionic compounds always contain positive and negative ions.
- In the chemical formula, the sum of the charges of the positive ions (cations) must always equal the sum of the charges of the negative ions (anions).

To write the formula for an ionic compound, follow the procedure in the left column of the following table. Two examples of how to apply the procedure are provided in the center and right columns.

Writing Formulas for Ionic Compounds	EXAMPLE 5.5 Write a formula for the ionic compound that forms from aluminum and oxygen.	EXAMPLE 5.6 Write a formula for the ionic compound that forms from magnesium and oxygen.
1. Write the symbol for the metal and its charge followed by the symbol of the nonmetal and its charge. For many elements, you can determine these charges from their group number in the periodic table (refer to Figure 4.14).	SOLUTION Al^{3+} O^{2-}	SOLUTION Mg^{2+} O^{2-}
2. Make the magnitude of the charge on each ion (without the sign) become the subscript for the other ion.	Al^{3+} O^{2-} Al_2O_3	Mg^{2+} O^{2-} Mg_2O_2
3. If possible, reduce the subscripts to give a ratio with the smallest whole numbers.	In this case, the numbers cannot be reduced any further; the correct formula is Al_2O_3.	To reduce the subscripts, divide both subscripts by 2. $Mg_2O_2 \div 2 = MgO$
4. Check to make sure that the sum of the charges of the cations exactly cancels the sum of the charges of the anions.	Cations: 2(3+) = 6+ Anions: 3(2−) = 6− The charges cancel.	Cations: 2+ Anions: 2− The charges cancel.

▶SKILLBUILDER 5.5

Write a formula for the compound formed from strontium and chlorine.

▶SKILLBUILDER 5.6

Write a formula for the compound formed from aluminum and nitrogen.

▶FOR MORE PRACTICE Example 5.21; Problems 53, 54, 55, 56.

EXAMPLE 5.7   Writing Formulas for Ionic Compounds

Write a formula for the compound composed of potassium and oxygen.

SOLUTION

First write the symbol for each ion along with its appropriate charge from its group number in the periodic table.

$$K^+ \quad O^{2-}$$

Then make the magnitude of each ion's charge become the subscript for the other ion.

$$K^+ \quad O^{2-} \text{ becomes } K_2O$$

No reduction of subscripts is necessary in this case. Finally, check to see that the sum of the charges of the cations $[2(1+) = 2+]$ exactly cancels the sum of the charges of the anion $(2-)$. The correct formula is K_2O.

▶**SKILLBUILDER 5.7** | **Writing Formulas for Ionic Compounds**

Write a formula for the compound that forms from calcium and bromine.

▶**FOR MORE PRACTICE** Problems 57, 58.

5.6 Nomenclature: Naming Compounds

Since there are so many different compounds, chemists have developed systematic ways to name them. If you learn these naming rules, you can examine a compound's formula and determine its name or vice versa. Many compounds also have a common name. For example, H_2O has the common name *water* and the systematic name *dihydrogen monoxide*. A common name is like a nickname for a compound, used by those who are familiar with it. Since water is such a familiar compound, everyone uses its common name and not its systematic name. In the sections that follow, you will learn how to systematically name simple ionic and molecular compounds. Keep in mind, however, that some compounds also have common names that are often used instead of the systematic name. Common names can be learned only through familiarity.

5.7 Naming Ionic Compounds

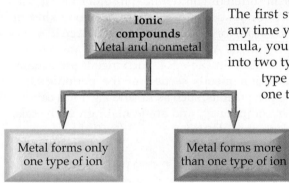

▲ FIGURE 5.12 **Classification of ionic compounds** Ionic compounds can be categorized into two types, depending on the metal in the compound.

The first step in naming an ionic compound is identifying it as one. Remember, any time you have a metal and one or more nonmetals together in a chemical formula, you can assume the compound is ionic. Ionic compounds are categorized into two types (◀ Figure 5.12) depending on the metal in the compound. The first type (sometimes called Type I) contains a metal with an invariant charge—one that does not vary from one compound to another. Sodium, for instance, has a 1+ charge in all of its compounds. Table 5.3 lists more examples of metals whose charge is invariant from one compound to another. The charge of most of these metals can be inferred from their group number in the periodic table (see Figure 4.14).

TABLE 5.3 **Metals Whose Charge Is Invariant from One Compound to Another**

Metal	Ion	Name	Group Number
Li	Li^+	lithium	1A
Na	Na^+	sodium	1A
K	K^+	potassium	1A
Rb	Rb^+	rubidium	1A
Cs	Cs^+	cesium	1A
Mg	Mg^{2+}	magnesium	2A
Ca	Ca^{2+}	calcium	2A
Sr	Sr^{2+}	strontium	2A
Ba	Ba^{2+}	barium	2A
Al	Al^{3+}	aluminum	3A
Zn	Zn^{2+}	zinc	*
Ag	Ag^+	silver	*

*The charge of these metals cannot be inferred from their group number.

TABLE 5.4 Some Metals That Form More Than One Type of Ion and Their Common Charges

Metal	Symbol Ion	Name	Older Name*
chromium	Cr^{2+}	chromium(II)	chromous
	Cr^{3+}	chromium(III)	chromic
iron	Fe^{2+}	Iron(II)	ferrous
	Fe^{3+}	iron(III)	ferric
cobalt	Co^{2+}	cobalt(II)	cobaltous
	Co^{3+}	cobalt(III)	cobaltic
copper	Cu^{+}	copper(I)	cuprous
	Cu^{2+}	copper(II)	cupric
tin	Sn^{2+}	tin(II)	stannous
	Sn^{4+}	tin(IV)	stannic
mercury	Hg_2^{2+}	mercury(I)	mercurous
	Hg^{2+}	mercury(II)	mercuric
lead	Pb^{2+}	lead(II)	plumbous
	Pb^{4+}	lead(IV)	plumbic

*An older naming system substitutes the names found in this column for the name of the metal and its charge. Under this system, chromium(II) oxide is named chromous oxide. We do *not* use this older system in this text.

▲ **FIGURE 5.13 The transition metals** The metals that form more than one type of ion are usually (but not always) transition metals.

The second type of ionic compound (sometimes called Type II) contains a metal with a charge that can differ in different compounds. In other words, the metal in this second type of ionic compound can form more than one kind of cation (depending on the compound). Iron, for instance, has a 2+ charge in some of its compounds and a 3+ charge in others. Additional examples of metals that form more than one type of cation are listed in Table 5.4. Such metals are usually (but not always) found in the **transition metals** section of the periodic table (◄ Figure 5.13). However, some transition metals, such as Zn and Ag, form cations with the same charge in all of their compounds, and some main group metals, such as Pb and Sn, form more than one type of cation.

NAMING BINARY IONIC COMPOUNDS CONTAINING A METAL THAT FORMS ONLY ONE TYPE OF CATION

Binary compounds are those that contain only two different elements. The names for binary ionic compounds containing a metal that forms only one type of ion have the form:

Since the charge of the metal is always the same for these types of compounds, it need not be specified in the compound's name. For example, the name for NaCl consists of the name of the cation, *sodium*, followed by the base name of the anion, *chlor*, with the ending *-ide*. The full name is *sodium chloride*.

The name of the cation in ionic compounds is the same as the name of the metal.

NaCl   sodium chloride

The name for MgO consists of the name of the cation, *magnesium*, followed by the base name of the anion, *ox*, with the ending *-ide*. The full name is *magnesium oxide*.

MgO   magnesium oxide

Table 5.5 contains the base names for various nonmetals and their most common charges in ionic compounds.

TABLE 5.5   Some Common Anions

Nonmetal	Symbol for Ion	Base Name	Anion Name
fluorine	F^-	fluor-	fluoride
chlorine	Cl^-	chlor-	chloride
bromine	Br^-	brom-	bromide
iodine	I^-	iod-	iodide
oxygen	O^{2-}	ox-	oxide
sulfur	S^{2-}	sulf-	sulfide
nitrogen	N^{3-}	nitr-	nitride

EXAMPLE 5.8   Naming Ionic Compounds Containing a Metal That Forms Only One Type of Cation

Name the compound MgF_2.

SOLUTION

The cation is magnesium. The anion is fluorine, which becomes *fluoride*. Its correct name is *magnesium fluoride*.

▶**SKILLBUILDER 5.8 | Naming Ionic Compounds Containing a Metal That Forms Only One Type of Ion**

Name the compound KBr.

▶**SKILLBUILDER PLUS**

Name the compound Zn_3N_2.

▶**FOR MORE PRACTICE**   Example 5.22; Problems 59, 60.

NAMING BINARY IONIC COMPOUNDS CONTAINING A METAL THAT FORMS MORE THAN ONE TYPE OF CATION

Since the charge of the metal cation in these types of compounds is not always the same, the charge must be specified in the metal's name. We specify the charge with a roman numeral (in parentheses) following the name of the metal. For example, we distinguish between Cu^+ and Cu^{2+} by writing a (I) to indicate the 1+ ion or a (II) to indicate the 2+ ion:

$$Cu^+ \quad Copper(I)$$

$$Cu^{2+} \quad Copper(II)$$

The full name for these types of compounds have the form:

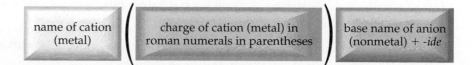

name of cation (metal) (charge of cation (metal) in roman numerals in parentheses) base name of anion (nonmetal) + -ide

We can determine the charge of the metal from the chemical formula of the compound—remember that the sum of all the charges must be zero. For example, the charge of iron in $FeCl_3$ must be 3+ in order for the compound to be charge neutral with the three Cl^- anions. The name for $FeCl_3$ is therefore the name of the cation, *iron*, followed by the charge of the cation in parentheses *(III)*, followed by the base name of the anion, *chlor*, with the ending *-ide*. The full name is *iron(III) chloride*.

$FeCl_3$ iron(III) chloride

Likewise, the name for CrO consists of the name of the cation, *chromium*, followed by the charge of the cation in parentheses *(II)*, followed by the base name of the anion, *ox-*, with the ending *-ide*. The full name is *chromium(II) oxide*.

CrO   chromium(II) oxide

The charge of chromium must be 2+ in order for the compound to be charge-neutral with one O^{2-} anion.

EXAMPLE 5.9 Naming Ionic Compounds Containing a Metal That Forms More Than One Type of Cation

Name the compound $PbCl_4$.

SOLUTION

The name for $PbCl_4$ consists of the name of the cation, *lead*, followed by the charge of the cation in parentheses *(IV)*, followed by the base name of the anion, *chlor-*, with the ending *-ide*. The full name is *lead(IV) chloride*. We know the charge on Pb is 4+ because the charge on Cl is 1−. Since there are 4 Cl^- anions, the Pb cation must be Pb^{4+}.

$PbCl_4$ lead(IV) chloride

▶ **SKILLBUILDER 5.9 | Naming Ionic Compounds Containing a Metal That Forms More Than One Type of Cation**

Name the compound PbO.

▶ **FOR MORE PRACTICE**   Example 5.23; Problems 61, 62.

 CONCEPTUAL CHECKPOINT 5.3

Explain why CaO is NOT named calcium(II) oxide.

NAMING IONIC COMPOUNDS CONTAINING A POLYATOMIC ION

As we saw previously, some ionic compounds contain polyatomic ions (ions that are themselves composed of a group of atoms with an overall charge). The most common polyatomic ions are listed in Table 5.6. We name ionic compounds containing polyatomic ions using the same procedure we apply to other ionic compounds, except that we use the name of the polyatomic ion whenever it occurs. For example, KNO_3 is named using its cation, K^+, *potassium*, and its polyatomic anion, NO_3^-, *nitrate*. The full name is *potassium nitrate*.

KNO_3 potassium nitrate

$Fe(OH)_2$ is named according to its cation, *iron*, its charge *(II)*, and its polyatomic ion, *hydroxide*. Its full name is *iron(II) hydroxide*.

$Fe(OH)_2$ iron(II) hydroxide

If the compound contains both a polyatomic cation and a polyatomic anion, use the names of both polyatomic ions. For example, NH_4NO_3 is *ammonium nitrate*.

NH_4NO_3 ammonium nitrate

EVERYDAY CHEMISTRY

Polyatomic Ions

Aglance at the labels of household products reveals the importance of polyatomic ions in everyday compounds. For example, the active ingredient in household bleach is sodium hypochlorite, which acts to decompose color-causing molecules in clothes (bleaching action) and to kill bacteria (disinfection). A box of baking soda contains sodium bicarbonate (sodium hydrogen carbonate), which acts as an antacid when consumed in small quantities and as a source of carbon dioxide gas in baking. The pockets of carbon dioxide gas make baked goods fluffy rather than flat.

Calcium carbonate is the active ingredient in many antacids such as Tums™ and Alka-Mints™. It neutralizes stomach acids, relieving the symptoms of indigestion and heartburn. Too much calcium carbonate, however, can cause constipation, so Tums should not be overused. Sodium nitrite is a common food additive used to preserve packaged meats such as ham, hot dogs, and bologna. Sodium nitrite inhibits the growth of bacteria, especially those that cause botulism, an often fatal type of food poisoning.

◀ Compounds containing polyatomic ions are present in many consumer products.

▶ The active ingredient in bleach is sodium hypochlorite.

CAN YOU ANSWER THIS? *Write a formula for each of these compounds that contain polyatomic ions: sodium hypochlorite, sodium bicarbonate, calcium carbonate, sodium nitrite.*

TABLE 5.6 Some Common Polyatomic Ions

Name	Formula	Name	Formula
acetate	$C_2H_3O_2^-$	hypochlorite	ClO^-
carbonate	CO_3^{2-}	chlorite	ClO_2^-
hydrogen carbonate (or bicarbonate)	HCO_3^-	chlorate	ClO_3^-
hydroxide	OH^-	perchlorate	ClO_4^-
nitrate	NO_3^-	permanganate	MnO_4^-
nitrite	NO_2^-	sulfate	SO_4^{2-}
chromate	CrO_4^{2-}	sulfite	SO_3^{2-}
dichromate	$Cr_2O_7^{2-}$	hydrogen sulfite (or bisulfite)	HSO_3^-
phosphate	PO_4^{3-}	hydrogen sulfate (or bisulfate)	HSO_4^-
hydrogen phosphate	HPO_4^{2-}	peroxide	O_2^{2-}
ammonium	NH_4^+	cyanide	CN^-

You will need to be able to recognize polyatomic ions in a chemical formula, so become familiar with Table 5.6. Most polyatomic ions are **oxyanions**, anions containing oxygen. Notice that when a series of oxyanions contain different numbers of oxygen atoms, they are named systematically according to the number of oxygen atoms in the ion. If there are two ions in the series, the one with more oxygen atoms is given the ending -*ate* and the one with fewer is given the ending -*ite*. For example, NO_3^- is called *nitrate* and NO_2^- is called *nitrite*.

NO_3^- nitrate
NO_2^- nitrite

If there are more than two ions in the series, then the prefixes *hypo-*, meaning "less than," and *per-*, meaning "more than," are used. So ClO^- is called *hypochlorite*, meaning "less oxygen than chlorite," and ClO_4^- is called *perchlorate*, meaning "more oxygen than chlorate."

ClO^- hypochlorite
ClO_2^- chlorite
ClO_3^- chlorate
ClO_4^- perchlorate

EXAMPLE 5.10   Naming Ionic Compounds Containing a Polyatomic Ion

Name the compound K_2CrO_4.

SOLUTION

The name for K_2CrO_4 consists of the name of the cation, *potassium*, followed by the name of the polyatomic ion, *chromate*.

K_2CrO_4 potassium chromate

▶**SKILLBUILDER 5.10 | Naming Ionic Compounds Containing a Polyatomic Ion**

Name the compound $Mn(NO_3)_2$.

▶**FOR MORE PRACTICE**   Example 5.24; Problems 65, 66.

✓ CONCEPTUAL CHECKPOINT 5.4

You have just learned that the anion ClO_3^- is named chlorate. What is the name of the anion IO_3^-?

5.8   Naming Molecular Compounds

The first step in naming a molecular compound is identifying it as one. Remember, nearly all molecular compounds form from two or more nonmetals. In this section, we learn how to name binary (two-element) molecular compounds. Their names have the form:

prefix | name of 1st element | prefix | base name of 2nd element + -*ide*

When writing the name of a molecular compound, as when writing the formula, the first element is the more metal-like one (see Table 5.1). The prefixes given to each element indicate the number of atoms present.

mono- 1	*hexa-* 6
di- 2	*hepta-* 7
tri- 3	*octa-* 8
tetra- 4	*nona-* 9
penta- 5	*deca-* 10

If there is only one atom of the *first element* in the formula, the prefix *mono-* is normally omitted. For example, CO_2 is named according to the first element, *carbon*, with no prefix because *mono-* is omitted for the first element, followed by the prefix *di-*, to indicate two oxygen atoms, followed by the base name of the second element, *ox*, with the ending *-ide*.

carbon di- ox -ide

The full name is *carbon dioxide*.

CO_2 carbon dioxide

The compound N_2O, also called laughing gas, is named according to the first element, *nitrogen*, with the prefix *di-*, to indicate that there are two of them, followed by the base name of the second element, *ox*, prefixed by *mono-*, to indicate one, and the suffix *-ide*. Since *mono-* ends with a vowel and *oxide* begins with one, an *o* is dropped and the two are combined as *monoxide*. The entire name is *dinitrogen monoxide*.

N_2O dinitrogen monoxide

> When the prefix ends with a vowel and the base name starts with a vowel, the first vowel is sometimes dropped, especially in the case of mono oxide, which becomes monoxide.

EXAMPLE 5.11   Naming Molecular Compounds

Name each compound.

(a) CCl_4
(b) BCl_3
(c) SF_6

SOLUTION

(a) The name of the compound is the name of the first element, *carbon*, followed by the base name of the second element, *chlor*, prefixed by *tetra-* to indicate four, and the suffix *–ide*.

CCl_4 carbon tetrachloride

(b) The name of the compound is the name of the first element, *boron*, followed by the base name of the second element, *chlor*, prefixed by *tri-* to indicate three, and the suffix *-ide*.

BCl_3 boron trichloride

(c) The name of the compound is the name of the first element, *sulfur*, followed by the base name of the second element, *fluor*, prefixed by *hexa-* to indicate six, and the suffix *-ide*. The entire name is *sulfur hexafluoride*.

SF_6 sulfur hexafluoride

▶**SKILLBUILDER 5.11 | Naming Molecular Compounds**
Name the compound N_2O_4.

▶**FOR MORE PRACTICE**   Example 5.25; Problems 71, 72.

5.9 Naming Acids

Acids are molecular compounds that produce H^+ ions when dissolved in water. They are composed of hydrogen, usually written first in their formula, and one or more nonmetals, written second. Acids are characterized by their sour taste and their ability to dissolve some metals. For example, HCl(*aq*) is an acid—the (*aq*) indicates that the compound is "aqueous" or "dissolved in water". HCl(*aq*) has a characteristically sour taste. Since HCl(*aq*) is present in stomach fluids, its sour taste becomes painfully obvious during vomiting. HCl(*aq*) also dissolves some metals. If you drop a strip of zinc into a beaker of HCl(*aq*), it will slowly disappear as the acid converts the zinc metal into dissolved Zn^{2+} cations.

Acids are present in many foods, such as lemons and limes, and they are used in some household products such as toilet bowl cleaner and Lime-A-Way. In this section, we simply learn how to name them, but in Chapter 14 we will learn more about their properties. We can categorize acids into two groups: **binary acids**, those containing only hydrogen and a nonmetal, and **oxyacids**, those containing hydrogen, a nonmetal, and oxygen (◄ Figure 5.14).

| HCl(*g*) refers to HCl molecules in the gas phase.

Acids formula has H as first element

Binary contain only two elements

Oxyacids contain oxygen

◄ **FIGURE 5.14 Classification of acids** Acids are classified into two types, depending on the number of elements in the acid. If the acid contains only two elements, it is a binary acid. If it contains oxygen, it is an oxyacid.

NAMING BINARY ACIDS

Binary acids are composed of hydrogen and a nonmetal. The names for binary acids have the following form:

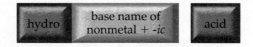

hydro | base name of nonmetal + *-ic* | acid

For example, HCl(*aq*) is hydro*chlor*ic acid and HBr(*aq*) is hydro*brom*ic acid.

HCl(*aq*)   hydrochloric acid   HBr(*aq*)   hydrobromic acid

EXAMPLE 5.12   Naming Binary Acids

Give the name of H_2S(*aq*).

| The base name of S is *sulfur*, so the name is *hydrosulfuric acid* | **SOLUTION** H_2S(*aq*)   hydrosulfuric acid |

▶**SKILLBUILDER 5.12 | Naming Binary Acids**
Name HF(*aq*).

▶**FOR MORE PRACTICE**   Example 5.26; Problems 77b, 78d.

NAMING OXYACIDS

Oxyacids are acids that contain oxyanions, which can be found in the table of polyatomic ions (Table 5.6). For example, HNO_3 (*aq*) contains the nitrate (NO_3^-) ion, H_2SO_3(*aq*) contains the sulfite (SO_3^{2-}) ion, and H_2SO_4 (*aq*) contains the sulfate (SO_4^{2-}) ion. All of these acids are a combination of one or more H^+ ions with an oxyanion. The number of H^+ ions depends on the charge of the oxyanion, so that the formula is always charge-neutral. The names of oxyacids depend on the ending of the oxyanion (▶ Figure 5.15).

► **FIGURE 5.15 Classification of oxyacids** Oxyacids are classified into two types, depending on the endings of the oxyanions that they contain.

The names of acids containing oxyanions ending with *-ite* take this form:

The saying, "*ic* I *ate* an acid" is sometimes used to help remember the association of -ic with -*ate*.

The names of acids containing oxyanions ending with *-ate* take this form:

So H_2SO_3 is *sulfurous acid* (oxyanion is sulfite), and HNO_3 is *nitric acid* (oxyanion is nitrate).

$H_2SO_3(aq)$ sulfurous acid $HNO_3(aq)$ nitric acid

Table 5.7 lists the names of some common oxyacids and their oxyanions.

EXAMPLE 5.13   Naming Oxyacids

Name $HC_2H_3O_2(aq)$.

The oxyanion is acetate, which ends in -*ate*; therefore, the name of the acid is *acetic acid*.	**SOLUTION** $HC_2H_3O_2(aq)$ acetic acid

►**SKILLBUILDER 5.13 | Naming Oxyacids**

Name $HNO_2(aq)$.

►**FOR MORE PRACTICE**   Examples 5.27, 5.28; Problems 77acd, 78abc.

TABLE 5.7   Names of Some Common Oxyacids and Their Oxyanions

Acid Formula	Acid Name	Oxyanion Name	Oxyanion Formula
HNO_2	nitrous acid	nitrite	NO_2^-
HNO_3	nitric acid	nitrate	NO_3^-
H_2SO_3	sulfurous acid	sulfite	SO_3^{2-}
H_2SO_4	sulfuric acid	sulfate	SO_4^{2-}
$HClO_2$	chlorous acid	chorite	ClO_2^-
$HClO_3$	chloric acid	chlorate	ClO_3^-
$HC_2H_3O_2$	acetic acid	acetate	$C_2H_3O_2^-$
H_2CO_3	carbonic acid	carbonate	CO_3^{2-}

CHEMISTRY IN THE ENVIRONMENT

Acid Rain

Acid rain occurs when rainwater mixes with air pollutants—such as NO, NO_2, and SO_2—that form acids. NO and NO_2, primarily from vehicular emission, combine with water to form HNO_3 (*aq*). SO_2, primarily from coal-powered electricity generation, combines with water and oxygen in air to form H_2SO_4 (*aq*). HNO_3 (*aq*) and H_2SO_4 (*aq*) both cause rainwater to become acidic. The problem is greatest in the northeastern United States, where pollutants from midwestern electrical power plants combine with rainwater to produce rain with acid levels that are up to 10 times higher than normal.

When acid rain falls or flows into lakes and streams, it makes them more acidic. Some species of aquatic animals—such as trout, bass, snails, salamanders, and clams—cannot tolerate the increased acidity and die. This then disturbs the ecosystem of the lake, resulting in imbalances that may lead to the death of other aquatic species. Acid rain also weakens trees by dissolving nutrients in the soil and by damaging their leaves. Appalachian red spruce trees have been the hardest hit, with many forests showing significant acid rain damage.

Acid rain also damages building materials. Acids dissolve $CaCO_3$ (limestone), a main component of marble and concrete, and iron, the main component of steel. Consequently, many statues, buildings, and bridges in the northeastern United States show significant deterioration, and some historical gravestones made of limestone are barely legible due to acid rain damage.

Although acid rain has been a problem for many years, innovative legislation has offered hope for change. In 1990, Congress passed several amendments to the Clean Air Act that included provisions requiring electrical utilities to reduce SO_2 emissions. Since then, SO_2 emissions have decreased, and rain in the northeastern United States has become somewhat less acidic. For example, in the early 1990s, scientists categorized 30% of the lakes in the Northeast as being of *acute concern*; today, the percentage of lakes in that category has been reduced to 18%. With time, and continued enforcement of the acid rain program, lakes, streams, and forests damaged by acid rain should recover. However, acid rain continues to worsen in countries such as China, where industrial growth is outpacing environmental controls. International cooperation is essential to solving environmental problems such as acid rain.

CAN YOU ANSWER THIS? *Name each compound, given here as formulas:*

NO, NO_2, SO_2, HNO_3 (*aq*), $CaCO_3$

▲ Acid rain harms many materials, including the limestone often used for tombstones, buildings, and statues.

▲ A forest damaged by acid rain.

5.10 Nomenclature Summary

Acids are technically a subclass of molecular compounds; that is, they are molecular compounds that form H^+ ions when dissolved in water.

Naming compounds requires several steps. The flowchart in ▶ Figure 5.16 summarizes the different categories of compounds that we have covered in the chapter and how to identify and name them. The first step is to decide whether the compound is ionic, molecular, or an acid. You can recognize ionic compounds by the presence of a metal and a nonmetal, molecular compounds by two or more nonmetals, and acids by the presence of hydrogen (written first) and one or more nonmetals.

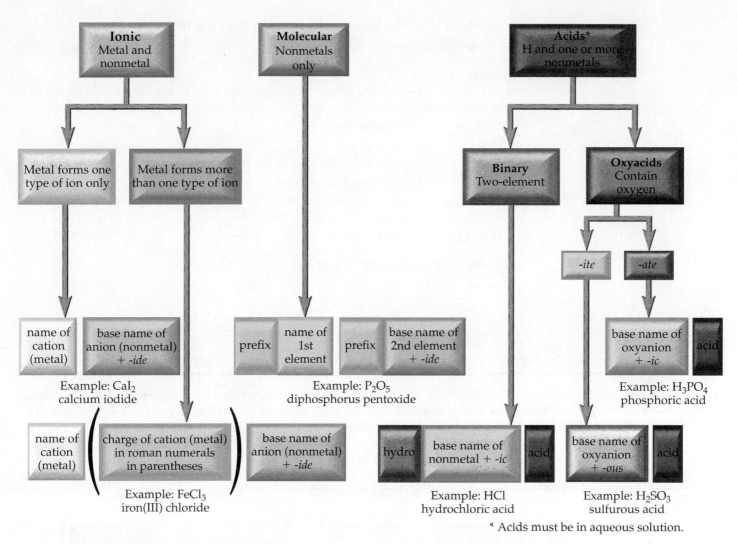

▲ **FIGURE 5.16   Nomenclature flowchart**

IONIC COMPOUNDS

For an ionic compound, you must next decide whether the metal forms only one type of ion or more than one type of ion. Group 1A (alkali) metals, Group 2A (alkaline earth) metals, and aluminum will always form only one type of ion (Figure 4.14). Most of the transition metals will form more than one type of ion. Once you have identified the type of ionic compound, name it according to the scheme in the chart. If the ionic compound contains a polyatomic ion—something you must learn to recognize by familiarity—insert the name of the polyatomic ion in place of the metal (positive polyatomic ion) or the nonmetal (negative polyatomic ion).

Zinc (Zn^{2+}), scandium (Sc^{3+}), and silver (Ag^+) also form only one type of ion.

MOLECULAR COMPOUNDS

We have learned how to name only one type of molecular compound, the binary (two-element) compound. If you identify a compound as molecular, name it according to the scheme in Figure 5.16.

ACIDS

To name an acid, you must first decide whether it is a binary (two-element) acid or an oxyacid (an acid containing oxygen). Binary acids are named according to the scheme in Figure 5.16. Oxyacids must be further subdivided based on the name of their corresponding oxyanion. If the oxyanion ends in *-ite*, use one scheme; if it ends with *-ate*, use the other.

EXAMPLE 5.14 Nomenclature Using Figure 5.16

Name each compound: CO, CaF$_2$, HF(*aq*), Fe(NO$_3$)$_3$, HClO$_4$ (*aq*), H$_2$SO$_3$ (*aq*).

SOLUTION

For each compound, the following table shows how to use Figure 5.16 to arrive at a name for the compound.

Formula	Flowchart Path	Name
CO	molecular	carbon monoxide
CaF$_2$	ionic $\longrightarrow$ one type of ion $\longrightarrow$	calcium fluoride
HF(*aq*)	acid $\longrightarrow$ binary $\longrightarrow$	hydrofluoric acid
Fe(NO$_3$)$_3$	ionic $\longrightarrow$ more than one type of ion $\longrightarrow$	iron(III) nitrate
HClO$_4$ (*aq*)	acid $\longrightarrow$ oxyacid $\longrightarrow$ -*ate* $\longrightarrow$	perchloric acid
H$_2$SO$_3$ (*aq*)	acid $\longrightarrow$ oxyacid $\longrightarrow$ -*ite* $\longrightarrow$	sulfurous acid

▶**FOR MORE PRACTICE**   Problems 93, 94.

5.11 Formula Mass: The Mass of a Molecule or Formula Unit

The terms *molecular mass* and *molecular weight*, which are also commonly used, have the same meaning as formula mass.

In Chapter 4, we discussed atoms and elements and designated the average mass of the atoms that compose an element as the atomic mass for that element. Similarly, in this chapter, which introduces molecules and compounds, we designate the average mass of the molecules (or formula units) that compose a compound as the **formula mass**.

For any compound, the formula mass is the sum of the atomic masses of all the atoms in its chemical formula:

$$\text{Formula mass} = \left(\begin{array}{c} \text{\# atoms of 1st} \\ \text{element in} \\ \text{chemical formula} \end{array} \times \begin{array}{c} \text{Atomic mass} \\ \text{of} \\ \text{1st element} \end{array} \right) + \left(\begin{array}{c} \text{\# atoms of 2nd} \\ \text{element in} \\ \text{chemical formula} \end{array} \times \begin{array}{c} \text{Atomic mass} \\ \text{of} \\ \text{2nd element} \end{array} \right) + \dots$$

Like atomic mass for atoms, formula mass characterizes the average mass of a molecule or formula unit. For example, the formula mass of water, H$_2$O, is:

$$\text{Formula mass} = 2(1.01 \text{ amu}) + 16.00 \text{ amu}$$
$$= 18.02 \text{ amu}$$

and that of sodium chloride, NaCl, is:

$$\text{Formula mass} = 22.99 \text{ amu} + 35.45 \text{ amu}$$
$$= 58.44 \text{ amu}$$

In addition to giving a characteristic mass to the molecules or formula units of a compound, formula mass—as we will learn in Chapter 6—allows us to quantify the number of molecules or formula units in a sample of a given mass.

EXAMPLE 5.15 Calculating Formula Mass

Calculate the formula mass of carbon tetrachloride, CCl$_4$.

SOLUTION

To find the formula mass, sum the atomic masses of each atom in the chemical formula.

$$\text{Formula mass} = 1 \times (\text{Atomic mass C}) + 4 \times (\text{Atomic mass Cl})$$
$$= 12.01 \text{ amu} + 4(35.45 \text{ amu})$$
$$= 12.01 \text{ amu} + 141.\underline{8}0 \text{ amu}$$
$$= 153.8 \text{ amu}$$

▶**SKILLBUILDER 5.15 | Calculating Formula Masses**

Calculate the formula mass of dinitrogen monoxide, N_2O, also called laughing gas.

▶**FOR MORE PRACTICE**   Example 5.29; Problems 83, 84.

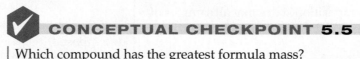

CONCEPTUAL CHECKPOINT 5.5

Which compound has the greatest formula mass?

(a) O_2 **(b)** O_3 **(c)** H_2O **(d)** H_2O_2

CHAPTER IN REVIEW

CHEMICAL PRINCIPLES

Compounds: Matter is ultimately composed of atoms, but those atoms are often combined in compounds. The most important characteristic of a compound is its constant composition. The elements that make up a particular compound are in fixed, definite proportions in all samples of the compound.

Chemical Formulas: Compounds are represented by chemical formulas, which indicate the elements present in the compound and the relative number of atoms of each. These formulas represent the basic units that make up a compound. Pure substances can be categorized according to the basic units that compose them. Elements can be composed of atoms or molecules. Compounds can be molecular, in which case their basic units are molecules, or ionic, in which case their basic units are ions. The formulas for many ionic compounds can be written simply by knowing the elements in the compound.

Chemical Nomenclature: The names of simple ionic compounds, molecular compounds, and acids can all be written by examining their chemical formula. The nomenclature flowchart (Figure 5.16) shows the basic procedure for determining these names.

RELEVANCE

Compounds: Most of the matter we encounter is in the form of compounds. Water, salt, and carbon dioxide are all examples of common simple compounds. More complex compounds include caffeine, aspirin, acetone, and testosterone.

Chemical Formulas: To understand compounds, we must understand their composition, which is represented by a chemical formula. The connection between the microscopic world and the macroscopic world hinges on the particles that compose matter. Since most matter is in the form of compounds, the properties of most matter depend on the molecules or ions that compose it. Molecular matter does what its molecules do; ionic matter does what its ions do. The world we see and experience is governed by what these particles are doing.

Chemical Nomenclature: Since there are so many compounds, we need a systematic way to name them. By learning these few simple rules, you will be able to name thousands of different compounds. The next time you look at the label on a consumer product, try to identify as many of the compounds as you can by examining their names.

Formula Mass: The formula mass of a compound is the sum of the atomic masses of all the atoms in the chemical formula for the compound. Like atomic mass for elements, formula mass characterizes the average mass of a molecule or formula unit.

Formula Mass: Besides being the characteristic mass of a molecule or formula unit, formula mass is important in many calculations involving the composition of compounds and quantities in chemical reactions.

CHEMICAL SKILLS

EXAMPLES

Constant Composition of Compounds (Section 5.2)

The law of constant composition states that all samples of a given compound should have the same ratio of their constituent elements.

To determine whether experimental data are consistent with the law of constant composition, compute the ratios of the masses of each element in all samples. When computing these ratios, it is most convenient to put the larger number in the numerator (top) and the smaller one in the denominator (bottom); that way, the ratio is greater than 1. If the ratios are the same, then the data are consistent with the law of constant composition.

EXAMPLE 5.16 Constant Composition of Compounds

Two samples said to be carbon disulfide (CS_2) are decomposed into their constituent elements. One sample produced 8.08 g S and 1.51 g C, while the other produced 31.3 g S and 3.85 g C. Are these results consistent with the law of constant composition?

SOLUTION
Sample 1

$$\frac{\text{Mass S}}{\text{Mass C}} = \frac{8.08 \text{ g}}{1.51 \text{ g}} = 5.35$$

Sample 2

$$\frac{\text{Mass S}}{\text{Mass C}} = \frac{31.3 \text{ g}}{3.85 \text{ g}} = 8.13$$

These results are not consistent with the law of constant composition, so the information that the two samples are the same substance must therefore be in error.

Writing Chemical Formulas (Section 5.3)

Chemical formulas indicate the elements present in a compound and the relative number of atoms of each. When writing formulas, put the more metallic element first.

EXAMPLE 5.17 Writing Chemical Formulas

Write a chemical formula for the compound containing one nitrogen atom for every two oxygen atoms.

SOLUTION

NO_2

Total Number of Each Type of Atom in a Chemical Formula (Section 5.3)

The numbers of atoms not enclosed in parentheses are given directly by their subscript.

Find the numbers of atoms within parentheses by multiplying their subscript within the parentheses by their subscript outside the parentheses.

EXAMPLE 5.18 Total Number of Each Type of Atom in a Chemical Formula

Determine the number of each type of atom in $Pb(ClO_3)_2$.

SOLUTION

One Pb atom

Two Cl atoms

Six O atoms

Classifying Elements as Atomic or Molecular (Section 5.4)

Most elements exist as atomic elements, their basic units in nature being individual atoms. However, several elements (H_2, N_2, O_2, F_2, Cl_2, Br_2, and I_2) exist as molecular elements, their basic units in nature being diatomic molecules.

EXAMPLE 5.19 Classifying Elements as Atomic or Molecular

Classify each element as atomic or molecular: sodium, iodine, and nitrogen.

SOLUTION

sodium: atomic

iodine: molecular (I_2)

nitrogen: molecular (N_2)

Classifying Compounds as Ionic or Molecular (Section 5.4)

Compounds containing a metal and a nonmetal are ionic. If the metal is a transition metal, it will likely form more than one type of ion (see exceptions in Tables 5.3 and 5.4). If the metal is not a transition metal, it will likely form only one type of ion (see exceptions in Tables 5.3 and 5.4). Compounds composed of nonmetals are molecular.

EXAMPLE 5.20 Classifying Compounds as Ionic or Molecular

Classify each compound as ionic or molecular. If they are ionic, determine whether the metal forms only one type of ion or more than one type of ion.

$$FeCl_3, K_2SO_4, CCl_4$$

SOLUTION

$FeCl_3$: ionic, metal forms more than one type of ion

K_2SO_4: ionic, metal forms only one type of ion

CCl_4: molecular

Writing Formulas for Ionic Compounds (Section 5.5)

1. Write the symbol for the metal ion followed by the symbol for the nonmetal ion (or polyatomic ion) and their charges. These charges can be deduced from the group numbers in the periodic table. (In the case of polyatomic ions, the charges come from Table 5.6.)

2. Make the magnitude of the charge on each ion become the subscript for the other ion.

3. Check to see if the subscripts can be reduced to simpler whole numbers. Subscripts of 1 can be dropped, since they are normally implied.

4. Check that the sum of the charges of the cations exactly cancels the sum of the charges of the anions.

EXAMPLE 5.21 Writing Formulas for Ionic Compounds

Write a formula for the compound that forms from lithium and sulfate ions.

SOLUTION

$$Li^+ \quad SO_4^{2-}$$

$$Li_2(SO_4)$$

In this case, the subscripts cannot be further reduced.

$$Li_2SO_4$$

Cations	Anions
$2(1+) = 2+$	$2-$

Naming Binary Ionic Compounds Containing a Metal That Forms Only One Type of Ion (Section 5.7)

The name of the metal is unchanged. The name of the nonmetal is its base name with the ending *-ide*.

EXAMPLE 5.22 Naming Binary Ionic Compounds Containing a Metal That Forms Only One Type of Ion

Name the compound Al_2O_3.

SOLUTION

aluminum oxide

Naming Binary Ionic Compounds Containing a Metal That Forms More than One Type of Ion (Section 5.7)

Since the names of these compounds include the charge of the metal ion, you must first determine that charge. To do this, calculate the total charge of the nonmetal ions.

The total charge of the metal ions must equal the total charge of the nonmetal ions, but have the opposite sign.

The name of the compound is the name of the metal ion, followed by the charge of the metal ion, followed by the base name of the nonmetal + -ide.

EXAMPLE 5.23 Naming Binary Ionic Compounds Containing a Metal that Forms More than One Type of Ion

Name the compound Fe_2S_3.

SOLUTION

$$3 \text{ sulfide ions} \times (2-) = 6-$$
$$2 \text{ iron ions} \times (ion\ charge) = 6+$$
$$ion\ charge = 3+$$
$$\text{Charge of each iron ion} = 3+$$

iron (III) sulfide

Naming Compounds Containing a Polyatomic Ion (Section 5.7)

Name ionic compounds containing a polyatomic ion in the normal way, except substitute the name of the polyatomic ion (from Table 5.6) in place of the nonmetal.

Since the metal in this example forms more than one type of ion, you need to determine the charge on the metal ion. The charge of the metal ion must be equal in magnitude to the sum of the charges of the polyatomic ions but opposite in sign.

The name of the compound is the name of the metal ion, followed by the charge of the metal ion, followed by the name of the polyatomic ion.

EXAMPLE 5.24 Naming Compounds Containing a Polyatomic Ion

Name the compound $Co(ClO_4)_2$.

SOLUTION

$$2 \text{ perchlorate ions} \times (1-) = 2-$$
$$\text{Charge of cobalt ion} = 2+$$

cobalt(II) perchlorate

Naming Molecular Compounds (Section 5.8)

The name consists of a prefix indicating the number of atoms of the first element, followed by the name of the first element, and a prefix for the number of atoms of the second element followed by the base name of the second element plus the suffix -ide. When *mono-* occurs on the first element, it is normally dropped.

EXAMPLE 5.25 Naming Molecular Compounds

Name the compound NO_2.

SOLUTION

nitrogen dioxide

Naming Binary Acids (Section 5.9)

The name begins with *hydro-*, followed by the base name of the nonmetal, plus the suffix -ic and then the word *acid*.

EXAMPLE 5.26 Naming Binary Acids

Name the acid $HI(aq)$.

SOLUTION

hydroiodic acid

Naming Oxyacids with an Oxyanion Ending in -ate (Section 5.9)

The name is the base name of the oxyanion + -ic, followed by the word *acid* (sulfate violates the rule somewhat, since in strict term, the base name would be *sulf*).

EXAMPLE 5.27 Naming Oxyacids with an Oxyanion Ending in -ate

Name the acid $H_2SO_4\ (aq)$.

SOLUTION

The oxyanion is sulfate. The name of the acid is *sulfuric acid*.

Naming Oxyacids with an Oxyanion Ending in -ite (Section 5.9)

The name is the base name of the oxyanion + -ous, followed by the word *acid*.

EXAMPLE 5.28 Naming Oxyacids with an Oxyanion Ending in -ite

Name the acid $HClO_2(aq)$.

SOLUTION

The oxyanion is chlorite. The name of the acid is *chlorous acid*.

Calculating Formula Mass (Section 5.11)

The formula mass is the sum of the atomic masses of all the atoms in the chemical formula. In determining the number of each type of atom, don't forget to multiply subscripts inside parentheses by subscripts outside parentheses.

EXAMPLE 5.29 Calculating Formula Mass

Calculate the formula mass of $Mg(NO_3)_2$.

SOLUTION

$$Formula\ mass = 24.31 + 2(14.01) + 6(16.00)$$
$$= 148.33\ amu$$

KEY TERMS

acid [5.9]	empirical formula [5.3]	molecular compound [5.4]	oxyanion [5.7]
atomic element [5.4]	formula mass [5.11]	molecular element [5.4]	polyatomic ion [5.7]
ball-and-stick model [5.3]	formula unit [5.4]	molecular formula [5.3]	space-filling model [5.3]
binary acid [5.9]	ionic compound [5.4]	molecular model [5.3]	transition metals [5.7]
binary compound [5.7]	law of constant	oxyacid [5.9]	
chemical formula [5.3]	composition [5.2]		

EXERCISES

QUESTIONS

1. Do the properties of an element change when it combines with another element to form a compound? Explain.
2. How might the world be different if elements did not combine to form compounds?
3. What is the law of constant composition? Who discovered it?
4. What is a chemical formula? List some examples.
5. In a chemical formula, which element is listed first?
6. In a chemical formula, how do you calculate the number of atoms of an element within parentheses? Give an example.
7. Explain the difference between a molecular formula and an empirical formula.
8. What is structural formula? What is the difference between a structural formula and a molecular model?
9. What is the difference between a molecular element and an atomic element? List the elements that occur as diatomic molecules.
10. What is the difference between an ionic compound and a molecular compound?

11. What is the difference between a common name for a compound and a systematic name?
12. List the metals that form only one type of ion (that is, metals whose charge is invariant from one compound to another). What are the group numbers of these metals?
13. Find the block in the periodic table of metals that tend to form more than one type of ion. What is the name of this block?
14. What is the basic form for the names of ionic compounds containing a metal that forms only one type of ion?
15. What is the basic form for the names of ionic compounds containing a metal that forms more than one type of ion?
16. Why are numbers needed in the names of ionic compounds containing a metal that forms more than one type of ion?
17. How are compounds containing a polyatomic ion named?
18. What polyatomic ions have a 2− charge? What polyatomic ions have a 3− charge?

19. What is the basic form for the names of molecular compounds?

20. How many atoms does each prefix specify? *mono-, di-, tri-, tetra-, penta-, hexa-.*

21. What is the basic form for the names of binary acids?

22. What is the basic form for the name of oxyacids whose oxyanions end with -*ate*?

23. What is the basic form for the name of oxyacids whose oxyanions end with -*ite*?

24. What is the formula mass of a compound?

PROBLEMS

CONSTANT COMPOSITION OF COMPOUNDS

25. Two samples of sodium chloride were decomposed into their constituent elements. One sample produced 4.65 g of sodium and 7.16 g of chlorine, and the other sample produced 7.45 g of sodium and 11.5 g of chlorine. Are these results consistent with the law of constant composition? Explain your answer.

26. Two samples of carbon tetrachloride were decomposed into their constituent elements. One sample produced 32.4 g of carbon and 373 g of chlorine, and the other sample produced 12.3 g of carbon and 112 g of chlorine. Are these results consistent with the law of constant composition? Explain your answer.

27. Upon decomposition, one sample of magnesium fluoride produced 1.65 kg of magnesium and 2.57 kg of fluorine. A second sample produced 1.32 kg of magnesium. How much fluorine (in grams) did the second sample produce?

28. The mass ratio of sodium to fluorine in sodium fluoride is 1.21:1. A sample of sodium fluoride produced 34.5 g of sodium upon decomposition. How much fluorine (in grams) was formed?

29. Use the law of constant composition to complete the table summarizing the amounts of nitrogen and oxygen produced upon the decomposition of several samples of dinitrogen monoxide.

	Mass N_2O	Mass N	Mass O
Sample A	2.85 g	1.82 g	1.03 g
Sample B	4.55 g	_____	_____
Sample C	_____	_____	1.35 g
Sample D	_____	1.11 g	_____

30. Use the law of constant composition to complete the table summarizing the amounts of iron and chlorine produced upon the decomposition of several samples of iron(III) chloride.

	Mass $FeCl_3$	Mass Fe	Mass Cl
Sample A	3.785 g	1.302 g	2.483 g
Sample B	2.175 g	_____	_____
Sample C	_____	2.012 g	_____
Sample D	_____	_____	2.329 g

CHEMICAL FORMULAS

31. Write a chemical formula for the compound containing one nitrogen atom for every three iodine atoms.

32. Write a chemical formula for the compound containing one carbon atom for every four bromine atoms.

33. Write chemical formulas for compounds containing:

   (a) three iron atoms for every four oxygen atoms

   (b) one phosphorus atom for every three chlorine atoms

   (c) one phosphorus atom for every five chlorine atoms

   (d) two silver atoms for every oxygen atom

34. Write chemical formulas for compounds containing:

   (a) one calcium atom for every two iodine atoms

   (b) two nitrogen atoms for every four oxygen atoms

   (c) one silicon atom for every two oxygen atoms

   (d) one zinc atom for every two chlorine atoms

35. How many oxygen atoms are in each chemical formula?

   (a) H_3PO_4

   (b) Na_2HPO_4

   (c) $Ca(HCO_3)_2$

   (d) $Ba(C_2H_3O_2)_2$

36. How many hydrogen atoms are in each of the formulas in Problem 35?

37. Determine the number of each type of atom in each formula.
- **(a)** $MgCl_2$
- **(b)** $NaNO_3$
- **(c)** $Ca(NO_2)_2$
- **(d)** $Sr(OH)_2$

38. Determine the number of each type of atom in each formula.
- **(a)** NH_4Cl
- **(b)** $Mg_3(PO_4)_2$
- **(c)** $NaCN$
- **(d)** $Ba(HCO_3)_2$

39. Complete the table.

Formula	Number of $C_2H_3O_2^-$ Units	Number of Carbon Atoms	Number of Hydrogen Atoms	Number of Oxygen Atoms	Number of Metal Atoms
$Mg(C_2H_3O_2)_2$	_____	_____	_____	_____	_____
$NaC_2H_3O_2$	_____	_____	_____	_____	_____
$Cr_2(C_2H_3O_2)_4$	_____	_____	_____	_____	_____

40. Complete the table.

Formula	Number of SO_4^{2-} Units	Number of Sulfur Atoms	Number of Oxygen Atoms	Number of Metal Atoms
$CaSO_4$	_____	_____	_____	_____
$Al_2(SO_4)_3$	_____	_____	_____	_____
K_2SO_4	_____	_____	_____	_____

41. Give the empirical formula that corresponds to each molecular formula.
- **(a)** C_2H_6
- **(b)** N_2O_4
- **(c)** $C_4H_6O_2$
- **(d)** NH_3

42. Give the empirical formula that corresponds to each molecular formula.
- **(a)** C_2H_2
- **(b)** CO_2
- **(c)** $C_6H_{12}O_6$
- **(d)** B_2H_6

MOLECULAR VIEW OF ELEMENTS AND COMPOUNDS

43. Classify each element as atomic or molecular.
- **(a)** chlorine
- **(b)** argon
- **(c)** cobalt
- **(d)** hydrogen

44. Which elements have molecules as their basic units?
- **(a)** helium
- **(b)** oxygen
- **(c)** iron
- **(d)** bromine

45. Classify each compound as ionic or molecular.
- **(a)** CS_2
- **(b)** CuO
- **(c)** KI
- **(d)** PCl_3

46. Classify each compound as ionic or molecular.
- **(a)** PtO_2
- **(b)** CF_2Cl_2
- **(c)** CO
- **(d)** SO_3

47. Match the substances on the left with the basic units that compose them on the right.

helium	molecules
CCl_4	formula units
K_2SO_4	diatomic molecules
bromine	single atoms

48. Match the substances on the left with the basic units that compose them on the right.

NI_3	molecules
copper metal	single atoms
$SrCl_2$	diatomic molecules
nitrogen	formula units

49. What are the basic units—single atoms, molecules, or formula units—that compose each substance?
- **(a)** $BaBr_2$
- **(b)** Ne
- **(c)** I_2
- **(d)** CO

50. What are the basic units—single atoms, molecules, or formula units—that compose each substance?
- **(a)** Rb_2O
- **(b)** N_2
- **(c)** $Fe(NO_3)_2$
- **(d)** N_2F_4

51. Classify each compound as ionic or molecular. If it is ionic, determine whether the metal forms only one type of ion or more than one type of ion.
    (a) KCl
    (b) CBr_4
    (c) NO_2
    (d) $Sn(SO_4)_2$

52. Classify each compound as ionic or molecular. If it is ionic, determine whether the metal forms only one type of ion or more than one type of ion.
    (a) $CoCl_2$
    (b) CF_4
    (c) $BaSO_4$
    (d) NO

WRITING FORMULAS FOR IONIC COMPOUNDS

53. Write a formula for the ionic compound that forms from each pair of elements.
    (a) sodium and sulfur
    (b) strontium and oxygen
    (c) aluminum and sulfur
    (d) magnesium and chlorine

54. Write a formula for the ionic compound that forms from each pair of elements.
    (a) aluminum and oxygen
    (b) beryllium and iodine
    (c) calcium and sulfur
    (d) calcium and iodine

55. Write a formula for the compound that forms from potassium and
    (a) acetate
    (b) chromate
    (c) phosphate
    (d) cyanide

56. Write a formula for the compound that forms from calcium and
    (a) hydroxide
    (b) carbonate
    (c) phosphate
    (d) hydrogen phosphate

57. Write formulas for the compounds formed from the element on the left and each element on the right.
    (a) Li   N, O, F
    (b) Ba   N, O, F
    (c) Al   N, O, F

58. Write formulas for the compounds formed from the element on the left and each polyatomic ion on the right.
    (a) Rb NO_3^-, SO_4^{2-}, PO_4^{3-}
    (b) Sr NO_3^-, SO_4^{2-}, PO_4^{3-}
    (c) In NO_3^-, SO_4^{2-}, PO_4^{3-}
    (Assume In charge is 3+.)

NAMING IONIC COMPOUNDS

59. Name each ionic compound. In each of these compounds, the metal forms only one type of ion.
    (a) CsCl
    (b) $SrBr_2$
    (c) K_2O
    (d) LiF

60. Name each ionic compound. In each of these compounds, the metal forms only one type of ion.
    (a) LiI
    (b) MgS
    (c) BaF_2
    (d) NaF

61. Name each ionic compound. In each of these compounds, the metal forms more than one type of ion.
    (a) $CrCl_2$
    (b) $CrCl_3$
    (c) SnO_2
    (d) PbI_2

62. Name each ionic compound. In each of these compounds, the metal forms more than one type of ion.
    (a) $HgBr_2$
    (b) Fe_2O_3
    (c) CuI_2
    (d) $SnCl_4$

63. Determine whether the metal in each ionic compound forms only one type of ion or more than one type of ion and name the compound accordingly.

   (a) Cr_2O_3

   (b) NaI

   (c) $CaBr_2$

   (d) SnO

64. Determine whether the metal in each ionic compound forms only one type of ion or more than one type of ion and name the compound accordingly.

   (a) FeI_3

   (b) $PbCl_4$

   (c) SrI_2

   (d) BaO

65. Name each ionic compound containing a polyatomic ion.

   (a) $Ba(NO_3)_2$

   (b) $Pb(C_2H_3O_2)_2$

   (c) NH_4I

   (d) $KClO_3$

   (e) $CoSO_4$

   (f) $NaClO_4$

66. Name each ionic compound containing a polyatomic ion.

   (a) $Ba(OH)_2$

   (b) $Fe(OH)_3$

   (c) $Cu(NO_2)_2$

   (d) $PbSO_4$

   (e) KClO

   (f) $Mg(C_2H_3O_2)_2$

67. Name each polyatomic ion.

   (a) BrO^-

   (b) BrO_2^-

   (c) BrO_3^-

   (d) BrO_4^-

68. Name each polyatomic ion.

   (a) IO^-

   (b) IO_2^-

   (c) IO_3^-

   (d) IO_4^-

69. Write a formula for each ionic compound.

   (a) copper(II) bromide

   (b) silver nitrate

   (c) potassium hydroxide

   (d) sodium sulfate

   (e) potassium hydrogen sulfate

   (f) sodium hydrogen carbonate

70. Write a formula for each ionic compound.

   (a) copper(I) chlorate

   (b) potassium permanganate

   (c) lead(II) chromate

   (d) calcium fluoride

   (e) iron(II) phosphate

   (f) lithium hydrogen sulfite

NAMING MOLECULAR COMPOUNDS

71. Name each molecular compound.

   (a) SO_2

   (b) NI_3

   (c) BrF_5

   (d) NO

   (e) N_4Se_4

72. Name each molecular compound.

   (a) XeF_4

   (b) PI_3

   (c) SO_3

   (d) $SiCl_4$

   (e) I_2O_5

73. Write a formula for each molecular compound.

   (a) carbon monoxide

   (b) disulfur tetrafluoride

   (c) dichlorine monoxide

   (d) phosphorus pentafluoride

   (e) boron tribromide

   (f) diphosphorus pentasulfide

74. Write a formula for each molecular compound.

   (a) chlorine monoxide

   (b) xenon tetroxide

   (c) xenon hexafluoride

   (d) carbon tetrabromide

   (e) diboron tetrachloride

   (f) tetraphosphorus triselenide

75. Determine whether the name shown for each molecular compound is correct. If not, provide the compound's correct name.

(a) PBr$_5$ phosphorus(V) pentabromide

(b) P$_2$O$_3$ phosphorus trioxide

(c) SF$_4$ monosulfur hexafluoride

(d) NF$_3$ nitrogen trifluoride

76. Determine whether the name shown for each molecular compound is correct. If not, provide the compound's correct name.

(a) NCl$_3$ nitrogen chloride

(b) CI$_4$ carbon(IV) iodide

(c) CO   carbon oxide

(d) SCl$_4$ sulfur tetrachloride

NAMING ACIDS

77. Determine whether each acid is a binary acid or an oxyacid and name each acid. If the acid is an oxyacid, also provide the name of the oxyanion.

(a) HNO$_2$ (aq)

(b) HI (aq)

(c) H$_2$SO$_4$ (aq)

(d) HNO$_3$ (aq)

78. Determine whether each acid is a binary acid or an oxyacid and name each acid. If the acid is an oxyacid, also provide the name of the oxyanion.

(a) H$_2$CO$_3$ (aq)

(b) HC$_2$H$_3$O$_2$ (aq)

(c) H$_3$PO$_4$ (aq)

(d) HCl (aq)

79. Name each acid.

(a) HClO

(b) HClO$_2$

(c) HClO$_3$

(d) HClO$_4$

80. Name each acid. (*Hint:* The names of the oxyanions are analogous to the names of the oxyanions of chlorine.)

(a) HBrO$_3$

(b) HIO$_3$

81. Write a formula for each acid.

(a) phosphoric acid

(b) hydrobromic acid

(c) sulfurous acid

82. Write a formula for each acid.

(a) hydrofluoric acid

(b) hydrocyanic acid

(c) chlorous acid

FORMULA MASS

83. Calculate the formula mass for each compound.

(a) HNO$_3$

(b) CaBr$_2$

(c) CCl$_4$

(d) Sr(NO$_3$)$_2$

84. Calculate the formula mass for each compound.

(a) CS$_2$

(b) C$_6$H$_{12}$O$_6$

(c) Fe(NO$_3$)$_3$

(d) C$_7$H$_{16}$

85. Arrange the compounds in order of decreasing formula mass.

Ag$_2$O, PtO$_2$, Al(NO$_3$)$_3$, PBr$_3$

86. Arrange the compounds in order of decreasing formula mass.

WO$_2$, Rb$_2$SO$_4$, Pb(C$_2$H$_3$O$_2$)$_2$, RbI

CUMULATIVE PROBLEMS

87. Write a molecular formula for each molecular model. (White = hydrogen; red = oxygen; black = carbon; blue = nitrogen; yellow = sulfur)

(a)

(b)

(c)

88. Write a molecular formula for each molecular model. (White = hydrogen; red = oxygen; black = carbon; blue = nitrogen; yellow = sulfur)

(a)

(b)

(c)

89. How many chlorine atoms are in each set?
    (a) three carbon tetrachloride molecules
    (b) two calcium chloride formula units
    (c) four phosphorus trichloride molecules
    (d) seven sodium chloride formula units

90. How many oxygen atoms are in each set?
    (a) four dinitrogen monoxide molecules
    (b) two calcium carbonate formula units
    (c) three sulfur dioxide molecules
    (d) five perchlorate ions

91. Specify the number of hydrogen atoms (white) represented in each set of molecular models:

(a)   (b)   (c)

92. Specify the number of oxygen atoms (red) represented in each set of molecular models:

(a)   (b)   (c)

93. Complete the table:

Formula	Type of Compound (Ionic, Molecular, Acid)	Name
N_2H_4	molecular	_____
_____	_____	potassium chloride
H_2CrO_4 (aq)	_____	_____
_____	_____	cobalt(III) cyanide

94. Complete the table:

Formula	Type of Compound (Ionic, Molecular, Acid)	Name
$K_2Cr_2O_7$	ionic	_____
HBr (aq)	_____	hydrobromic acid
_____	_____	dinitrogen pentoxide
PbO_2	_____	_____

95. Determine whether each name is correct for the given formula. If not, provide the correct name.
    (a) $Ca(NO_2)_2$ calcium nitrate
    (b) K_2O dipotassium monoxide
    (c) PCl_3 phosphorus chloride
    (d) $PbCO_3$ lead(II) carbonate
    (e) KIO_2 potassium hypoiodite

96. Determine whether each name is correct for the given formula. If not, provide the correct name.
    (a) HNO_3 (aq)   hydrogen nitrate
    (b) $NaClO$ sodium hypochlorite
    (c) CaI_2 calcium diiodide
    (d) $SnCrO_4$ tin chromate
    (e) $NaBrO_3$ sodium bromite

97. For each compound, list the correct formula and calculate the formula mass.
    (a) tin(IV) sulfate
    (b) nitrous acid
    (c) sodium bicarbonate
    (d) phosphorus pentafluoride

98. For each compound, list the correct formula and calculate the formula mass.
    (a) barium bromide
    (b) dinitrogen trioxide
    (c) copper(I) sulfate
    (d) hydrobromic acid

99. Name each compound and calculate its formula mass.
    (a) PtO_2
    (b) N_2O_5
    (c) $Al(ClO_3)_3$
    (d) PBr_5

100. Name each compound and calculate its formula mass.
    (a) $Al_2(SO_4)_3$
    (b) P_2O_3
    (c) $HClO$ (aq)
    (d) $Cr(C_2H_3O_2)_3$

101. A compound contains only carbon and hydrogen and has a formula mass of 28.06 amu. What is its molecular formula?

102. A compound contains only nitrogen and oxygen and has a formula mass of 44.02 amu. What is its molecular formula?

103. Carbon has two naturally occurring isotopes: carbon-12 (mass = 12.00 amu) and carbon-13 (mass = 13.00 amu). Chlorine also has two naturally occurring isotopes: chlorine-35 (mass = 34.97 amu) and chlorine-37 (mass = 36.97 amu). How many CCl_4 molecules of different masses can exist? Determine the mass (in amu) of each of them.

104. Nitrogen has two naturally occurring isotopes: nitrogen-14 (mass = 14.00 amu) and nitrogen-15 (mass = 15.00 amu). Bromine also has two naturally occurring isotopes: bromine-79 (mass = 78.92 amu) and bromine-81 (mass = 80.92 amu). How many NBr_3 molecules of different masses can exist? Determine the mass (in amu) of each of them.

HIGHLIGHT PROBLEMS

105. Examine each substance and the corresponding molecular view and classify it as an atomic element, a molecular element, a molecular compound, or an ionic compound.

(a)

(b)

(c)

(d)

106. Molecules can be as small as two atoms or as large as thousands of atoms. In 1962, Max F. Perutz and John C. Kendrew were awarded the Nobel Prize for their discovery of the structure of hemoglobin, a very large molecule that transports oxygen from the lungs to cells through the bloodstream. The chemical formula of hemoglobin is $C_{2952}H_{4664}O_{832}N_{812}S_8Fe_4$. Calculate the formula mass of hemoglobin.

▶ Max Perutz and John C. Kendrew won a Nobel Prize in 1962 for determining the structure of hemoglobin by X-ray diffraction.

◀ Computer generated model of hemoglobin.

107. Examine each consumer product label. Write chemical formulas for as many of the compounds as possible based on what you have learned in this chapter.

(a)

Active Ingredient:
Sodium Hypochlorite 6.0%
Other Ingredients: 94.0%
Total: 100.0%
(Yields 5.7% available chlorine)

KEEP OUT OF REACH OF CHILDREN

DANGER: CORROSIVE.

FIRST AID: IF IN EYES: Hold eye open and rinse slowly and gently with water for 15–20 minutes. Remove contact lenses, if present, after the first 5 minutes, then continue rinsing eye. IF ON SKIN OR CLOTHING: Take off contaminated clothing. Rinse skin immediately with plenty of water for 15–20 minutes. IN EITHER CASE, CALL A POISON CONTROL CENTER OR DOCTOR IMMEDIATELY FOR TREATMENT ADVICE. See back panel for additional precautionary labeling.

Kills Methicillin Resistant Staphylococcus aureus (MRSA) *Staphylococcus aureus, Streptococcus pyogenes, Salmonella enterica, Escherichia coli O157:H7 and Influenza A2

3 QT (96 FL OZ) 2.83 L

(b)

NOTICE! PROTECTIVE INNER SEAL BENEATH CAP. IF MISSING OR DAMAGED, DO NOT USE CONTENTS.

Drug Facts

Active ingredients (in each tablet)	Purpose
Calcium carbonate 1000 mg	Antacid
Simethicone 60 mg	Antigas

Uses for the relief of
• acid indigestion
• heartburn
• sour stomach
• upset stomach associated with these symptoms
• bloating and pressure commonly referred to as gas

Warning
Do not take more than 8 tablets in a 24-hour period or use the maximum dosage for more than 2 weeks except under the advice and supervision of a physician

Ask a doctor before use if you have
• kidney stones   • a calcium-restricted diet

Ask a doctor or pharmacist before use if your are presently taking a prescription drug. Antacids may interact with certain prescription drugs.

When using this product
• at maximum dose, constipation may occur

Stop use and ask a doctor if
• symptoms last more than two weeks

Keep out of reach of children.

Directions

(c)

Drug Facts

Active ingredients (in each 5 mL teaspoon)	Purposes
Aluminum hydroxide (equivalent to dried gel, USP) 400 mg	Antacid
Magnesium hydroxide 400 mg	Antacid
Simethicone 40 mg	Antigas

Use relieves: ■ heartburn ■ acid indigestion ■ sour stomach ■ upset stomach due to these symptoms ■ pressure and bloating commonly referred to as gas

Warnings
Ask a doctor before use if you have
■ kidney disease ■ a magnesium-restricted diet

Ask a doctor or pharmacist if you are taking a prescription drug. Antacids may interact with certain prescription drugs.

Stop use and ask a doctor if symptoms last more than 2 weeks.

Keep out of reach of children.

Directions ■ shake well ■ adults/children 12 years and older: take 2-4 teaspoonfuls between meals, at bedtime, or as directed by a doctor ■ do not take more than 12 teaspoonfuls in a 24-hour period, or use the maximum dosage for more than 2 weeks ■ children under 12 years: ask a doctor

Other information ■ each teaspoon contains: magnesium 171 mg ■ do not use if breakaway band on plastic cap is broken or missing ■ does not meet USP requirements for preservative effectiveness ■ do not freeze

Inactive ingredients butylparaben, carboxymethylcellulose sodium, flavors, hypromellose, microcrystalline cellulose, propylparaben, purified water, sodium saccharin, sorbitol

Questions or comments?
1-800-469-5268 (English) or
1-888-466-8746 (Spanish)

Johnson-Johnson • MERCK
Consumer Pharmaceuticals Co.
FORT WASHINGTON, PA 19034 USA
© 2009 JJMCPC
Patent Pending

(d)

Nutrition Facts

Serving Size 1/8 tsp (0.6g)
Servings Per Container about 472

Amount Per Serving

Calories 0

	% **Daily Value***
Total Fat 0g	0%
Sodium 65mg	3%
Total Carb. 0g	0%
Protein 0g	
Calcium 2%	

Not a significant source of calories from fat, saturated fat, trans fat, cholesterol, dietary fiber, sugars, vitamin A, vitamin C and iron.

*Percent Daily Values are based on a 2,000 calorie diet.

Ingredients: Cornstarch, Sodium Bicarbonate, Sodium Aluminum Sulfate, Monocalcium Phosphate.

**CLABBER GIRL CORPORATION
TERRE HAUTE, IN 47808**

davisbakingpowder.com

MADE IN USA

▶ANSWERS TO SKILLBUILDER EXERCISES

Skillbuilder 5.1   Yes, because in both cases

$$\frac{\text{Mass O}}{\text{Mass C}} = 1.3$$

Skillbuilder 5.2   (a) Ag_2S (b) N_2O (c) TiO_2

Skillbuilder 5.3   two K atoms, one S atom, four O atoms

Skillbuilder Plus, p. 131   two Al atoms, three S atoms, twelve O atoms

Skillbuilder 5.4

    (a) molecular element
    (b) molecular compound
    (c) atomic element
    (d) ionic compound
    (e) ionic compound

Skillbuilder 5.5 $SrCl_2$

Skillbuilder 5.6   AlN

Skillbuilder 5.7 $CaBr_2$

Skillbuilder 5.8   potassium bromide

Skillbuilder Plus, p. 139   zinc nitride

Skillbuilder 5.9   lead(II) oxide

Skillbuilder 5.10   manganese(II) nitrate

Skillbuilder 5.11   dinitrogen tetroxide

Skillbuilder 5.12   hydrofluoric acid

Skillbuilder 5.13   nitrous acid

Skillbuilder 5.15   44.02 amu

▶ANSWERS TO CONCEPTUAL CHECKPOINTS

5.1 (b)   This formula represents 2 Al atoms + 3 (2 Cr atoms + 7 O atoms) = 29 atoms.

5.2 (b)   The figure represents a molecular compound because the compound exists as individual molecules. Figure (a) represents an ionic compound with formula units in a lattice structure.

5.3   Because calcium forms only one type of ion (Ca^{2+}); therefore, the charge of the ion is not included in the name (because it is always the same, 2+).

5.4   Iodate

5.5 (b)

Moonlight Diner
Specials

Breakfast Special
Two eggs, Three strips of bacon, Sausage link, Toast ,NaCl $5.95

Omelettes
Served with side of NaCl
Denver Omelette
BP Delight
Atherosclerosis
Sodium Chloride
Hypertension

On the Healthy Side
120/80 mmHg

Lunch Special

1500 milligrams a day

142/93 mmHg

Symptoms:

Chest pain
Confusion
Ear noise or buzzing
Irregular heartbeat
Nosebleed
Tiredness

CHAPTER 6

Chemical Composition

"In science, you don't ask why, you ask how much."

ERWIN CHARGAFF (1905–2002)

6.1 How Much Sodium?

Sodium is an important dietary mineral that we eat in our food, primarily as sodium chloride (table salt). Sodium is involved in the regulation of body fluids, and eating too much of it can lead to high blood pressure. High blood pressure, in turn, increases the risk of stroke and heart attack. Consequently, people with high blood pressure should limit their sodium intake. The FDA recommends that a person consume less than 2.4 g (2400 mg) of sodium per day. However, sodium is usually consumed as sodium chloride, so the mass of sodium that we eat is not the same as the mass of sodium chloride that we eat. How many grams of sodium chloride can we consume and still stay below the FDA recommendation for sodium?

To answer this question, we need to know the *chemical composition* of sodium chloride. From Chapter 5, we are familiar with its formula, NaCl, so we know that there is one sodium ion to every chloride ion. However, since the masses of sodium and chlorine are different, the relationship between the mass of sodium and the mass of sodium chloride is not clear from the chemical formula alone. In this chapter, we learn how to use the information in a chemical formula, together with atomic and formula masses, to calculate the amount of a constituent element in a given amount of a compound (or vice versa).

Chemical composition is important not just for assessing dietary sodium intake, but for addressing many other issues as well. A company that mines iron, for example, wants to know how much iron it can extract from a given amount of iron ore; a company interested in developing hydrogen as a potential fuel would want to know how much hydrogen it can extract from a given amount of water. Many environmental issues also require knowledge of chemical composition. An estimate of the threat of ozone depletion requires knowing how much chlorine is in a given amount of a particular chlorofluorocarbon such as freon-12. To determine

◄ Ordinary table salt is a compound called sodium chloride. The sodium within sodium chloride is linked to high blood pressure. In this chapter, we learn how to determine how much sodium is in a given amount of sodium chloride.

▲ The mining of iron requires knowing how much iron is in a given amount of iron ore.

Freon-12

▲ Estimating the threat of ozone depletion requires knowing the amount of chlorine in a given amount of a chlorofluorocarbon.

these kinds of quantities, we must understand the relationships inherent in a chemical formula and the relationship between numbers of atoms or molecules and their masses. In this chapter, we examine these relationships.

6.2 Counting Nails by the Pound

3.4 lbs nails

8.25 grams carbon

▲ Asking how many nails are in a given weight of nails is similar to asking how many atoms are in a given mass of an element. In both cases, we count the objects by weighing them.

Some hardware stores sell nails by the pound, which is easier than selling them by the nail because customers often need hundreds of nails and counting them takes too long. However, a customer may still want to know the number of nails contained in a given weight of nails. This problem is similar to asking how many atoms are in a given mass of an element. With atoms, however, we *must* use their mass as a way to count them because atoms are too small and too numerous to count individually. Even if you could see atoms and counted them 24 hours a day for as long as you lived, you would barely begin to count the number of atoms in something as small as a grain of sand. However, just as the hardware store customer wants to know the number of nails in a given weight, we want to know the number of atoms in a given mass. How do we do that?

Suppose the hardware store customer buys 2.60 lb of medium-sized nails and a dozen nails weigh 0.150 lb. How many nails did the customer buy? This calculation requires two conversions: one between pounds and dozens and another between dozens and number of nails. The conversion factor for the first part is the weight per dozen nails.

$$0.150 \text{ lb nails} = 1 \text{ doz nails}$$

The conversion factor for the second part is the number of nails in one dozen.

$$1 \text{ doz nails} = 12 \text{ nails}$$

The solution map for the problem is:

$$\frac{1 \text{ doz nails}}{0.150 \text{ lb nails}} \qquad \frac{12 \text{ nails}}{1 \text{ doz nails}}$$

Beginning with 2.60 lb and using the solution map as a guide, we convert from lb to number of nails.

$$2.60 \text{ lb nails} \times \frac{1 \text{ doz nails}}{0.150 \text{ lb nails}} \times \frac{12 \text{ nails}}{1 \text{ doz nails}} = 208 \text{ nails}$$

The customer who bought 2.60 lb of nails has 208 nails. He counted the nails by weighing them. If the customer purchased a different size of nail, the first conversion factor—relating pounds to dozens—would change, but the second conversion factor would not. One dozen corresponds to 12 nails, regardless of their size.

6.3 Counting Atoms by the Gram

1 mole of copper atoms

▲ Twenty-two *copper* pennies contain approximately 1 mol of copper atoms. Pennies were mostly copper until 1982, at which point the U.S. Mint started making them out of zinc with only a copper coating (because copper became too valuable).

| The value of the mole is actually an empirically measured quantity.

1 mole of helium atoms

▲ Two large helium balloons contain approximately 1 mol of helium atoms.

Determining the number of atoms in a sample with a certain mass is similar to determining the number of nails in a sample with a certain weight. With nails, we used a dozen as a convenient number in our conversions, but a dozen is too small to use with atoms. We need a larger number because atoms are so small. The chemist's "dozen" is called the **mole (mol)** and has a value of 6.022×10^{23}.

$$1 \text{ mol} = 6.022 \times 10^{23}$$

This number is also called **Avogadro's number**, named after Amadeo Avogadro (1776–1856).

The first thing to understand about the mole is that it can specify Avogadro's number of anything. *One mole of anything is 6.022×10^{23} units of that thing.* For example, one mole of marbles corresponds to 6.022×10^{23} marbles, and one mole of sand grains corresponds to 6.022×10^{23} sand grains. One mole of atoms, ions, or molecules generally makes up objects of reasonable size. For example, 22 *copper* pennies (pennies were mostly copper until 1982) contain approximately 1 mol of copper (Cu) atoms, and a couple of large helium balloons contain approximately 1 mol of helium (He) atoms.

The second thing to understand about the mole is how it gets its specific value. *The numerical value of the mole is defined as being equal to the number of atoms in exactly 12 g of pure carbon-12.*

This definition of the mole establishes a relationship between mass (grams of carbon) and number of atoms (Avogadro's number). This relationship, as we will see shortly, allows us to count atoms by weighing them.

CONVERTING BETWEEN MOLES AND NUMBER OF ATOMS

Converting between moles and number of atoms is similar to converting between dozens and number of nails. To convert between moles of atoms and number of atoms, we use the conversion factors:

$$\frac{1 \text{ mol}}{6.022 \times 10^{23} \text{ atoms}} \quad \text{or} \quad \frac{6.022 \times 10^{23} \text{ atoms}}{1 \text{ mol}}$$

For example, suppose we want to convert 3.5 mol of helium to a number of helium atoms. We set up the problem in the standard way.

GIVEN: 3.5 mol He

FIND: He atoms

RELATIONSHIPS USED 1 mol He $= 6.022 \times 10^{23}$ He atoms

SOLUTION MAP We draw a solution map showing the conversion from moles of He to He atoms.

$$\frac{6.022 \times 10^{23} \text{ He atoms}}{1 \text{ mol He}}$$

SOLUTION

Beginning with 3.5 mol He, we use the conversion factor to get to He atoms.

$$3.5 \ \cancel{\text{mol He}} \times \frac{6.022 \times 10^{23} \text{ He atoms}}{1 \ \cancel{\text{mol He}}} = 2.1 \times 10^{24} \text{ He atoms}$$

EXAMPLE 6.1   Converting between Moles and Number of Atoms

A silver ring contains 1.1×10^{22} silver atoms. How many moles of silver are in the ring?

SORT You are given the number of silver atoms and asked to find the number of moles.	**GIVEN:** 1.1×10^{22} Ag atoms **FIND:** mol Ag
STRATEGIZE Draw a solution map, beginning with silver atoms and ending at moles. The conversion factor is Avogadro's number.	**SOLUTION MAP** $$\frac{1 \text{ mol Ag}}{6.022 \times 10^{23} \text{ Ag atoms}}$$ **RELATIONSHIPS USED** $1 \text{ mol Ag} = 6.022 \times 10^{23}$ Ag atoms (Avogadro's number)
SOLVE Follow the solution map to solve the problem. Beginning with 1.1×10^{22} Ag atoms, use the conversion factor to get to moles of Ag.	**SOLUTION** $$1.1 \times 10^{22} \ \cancel{\text{Ag atoms}} \times \frac{1 \text{ mol Ag}}{6.022 \times 10^{23} \ \cancel{\text{Ag atoms}}} = 1.8 \times 10^{-2} \text{ mol Ag}$$
CHECK Are the units correct? Does the answer make physical sense?	The units, mol Ag, are the desired units. The magnitude of the answer is orders of magnitude smaller than the given quantity because it takes many atoms to make a mole, so we expect the answer to be orders of magnitude smaller than the given quantity.

▶**SKILLBUILDER 6.1 | Converting between Moles and Number of Atoms**

How many gold atoms are in a pure gold ring containing 8.83×10^{-2} mol Au?

▶**FOR MORE PRACTICE**   Example 6.13, Problems 17, 18, 19, 20.

CONVERTING BETWEEN GRAMS AND MOLES OF AN ELEMENT

We just learned how to convert between moles and number of atoms, which is like converting between dozens and number of nails. We need one more conversion factor to convert from the mass of a sample to the number of atoms in the sample. For nails, we used the weight of one dozen nails; for atoms, we use the mass of one mole of atoms.

The mass of 1 mol of atoms of an element is its **molar mass**. The value of an element's molar mass in grams per mole is numerically equal to the element's atomic mass in atomic mass units.

Recall that Avogadro's number, the number of atoms in a mole, is defined as the number of atoms in exactly 12 g of carbon-12. Since the atomic mass unit is defined as one-twelfth of the mass of a carbon-12 atom, it follows that the molar mass of any element—the mass of 1 mol of atoms in grams of that element—is equal to the atomic mass of that element expressed in atomic mass units. For example, copper has an atomic mass of 63.55 amu; therefore, 1 mol of copper atoms has a mass of 63.55 g, and the molar mass of copper is 63.55 g/mol. Just as the weight of 1 doz nails changes for different nails, so the mass of 1 mol of atoms changes for different elements: 1 mol of sulfur atoms (sulfur atoms are lighter than copper atoms) has a mass of 32.07 g; 1 mol of carbon atoms (lighter than sulfur) has a mass of 12.01 g; and 1 mol of lithium atoms (lighter yet) has a mass of 6.94 g.

$$32.07 \text{ g sulfur} = 1 \text{ mol sulfur} = 6.022 \times 10^{23} \text{ S atoms}$$

$$12.01 \text{ g carbon} = 1 \text{ mol carbon} = 6.022 \times 10^{23} \text{ C atoms}$$

$$6.94 \text{ g lithium} = 1 \text{ mol lithium} = 6.022 \times 10^{23} \text{ Li atoms}$$

The lighter the atom, the less mass in one mole of that atom (▼ Figure 6.1).

Therefore, the molar mass of any element becomes a conversion factor between grams of that element and moles of that element. For carbon:

$$12.01 \text{ g C} = 1 \text{ mol C} \quad \text{or} \quad \frac{12.01 \text{ g C}}{1 \text{ mol C}} \quad \text{or} \quad \frac{1 \text{ mol C}}{12.01 \text{ g C}}$$

1 dozen large nails   1 dozen small nails

(a)

1 mole S (32.07 g)   1 mole C (12.01 g)

▶ **FIGURE 6.1   The mass of 1 mol**
(a) Each of these pictures shows the same number of nails: 12. As you can see, 12 large nails have more weight and occupy more space than 12 small nails. The same is true for atoms.
(b) Each of these samples has the same number of atoms: 6.022×10^{23}. Since sulfur atoms are more massive and larger than carbon atoms, 1 mol of S atoms is heavier and occupies more space than 1 mol of C atoms.

(b)

A 0.58-g diamond would be about a three-carat diamond.

Suppose we want to calculate the number of moles of carbon in a 0.58-g diamond (pure carbon).

We first sort the information in the problem.

GIVEN: 0.58 g C

FIND: mol C

SOLUTION MAP We then strategize by drawing a solution map showing the conversion from grams of C to moles of C. The conversion factor is the molar mass of carbon.

$$\frac{1 \text{ mol}}{12.01 \text{ g}}$$

RELATIONSHIPS USED

12.01 g C = 1 mol C (molar mass of carbon, from periodic table)

SOLUTION

Finally, we solve the problem by following the solution map.

$$0.58 \text{ g C} \times \frac{1 \text{ mol C}}{12.01 \text{ g C}} = 4.8 \times 10^{-2} \text{ mol C}$$

EXAMPLE 6.2   The Mole Concept—Converting between Grams and Moles

Calculate the number of moles of sulfur in 57.8 g of sulfur.

SORT Begin by sorting the information in the problem. You are given the mass of sulfur and asked to find the number of moles.	**GIVEN:** 57.8 g S **FIND:** mol S
STRATEGIZE Draw a solution map showing the conversion from g S to mol S. The conversion factor is the molar mass of sulfur.	**SOLUTION MAP** g S → mol S $$\frac{1 \text{ mol S}}{32.07 \text{ g S}}$$ **RELATIONSHIPS USED** 32.07 g S = 1 mol S (molar mass of sulfur, from periodic table)
SOLVE Follow the solution map to solve the problem. Begin with 57.8 g S and use the conversion factor to get to mol S.	**SOLUTION** $$57.8 \text{ g S} \times \frac{1 \text{ mol S}}{32.07 \text{ g S}} = 1.80 \text{ mol S}$$
CHECK Check your answer. Are the units correct? Does the answer make physical sense?	The units (mol S) are correct. The magnitude of the answer makes sense because 1 mole of S has a mass of 32.07 g; therefore, 57.8 g of S should be close to 2 moles.

▶SKILLBUILDER 6.2 | The Mole Concept—Converting between Grams and Moles

Calculate the number of grams of sulfur in 2.78 mol of sulfur.

▶FOR MORE PRACTICE   Example 6.14; Problems 25, 26, 27, 28, 29, 30.

CONVERTING BETWEEN GRAMS OF AN ELEMENT AND NUMBER OF ATOMS

Now, suppose we want to know the number of carbon *atoms* in the 0.58-g diamond. We first convert from grams to moles and then from moles to number of atoms. The solution map is:

$$\frac{1\text{ mol C}}{12.01\text{ g C}} \qquad \frac{6.022 \times 10^{23}\text{ C atoms}}{1\text{ mol C}}$$

Notice the similarity between this solution map and the one we used for nails:

$$\frac{1\text{ doz nails}}{0.150\text{ lb nails}} \qquad \frac{12\text{ nails}}{1\text{ doz nails}}$$

Beginning with 0.58 g carbon and using the solution map as a guide, we convert to the number of carbon atoms.

$$0.58 \text{ g C} \times \frac{1\text{ mol C}}{12.01\text{ g C}} \times \frac{6.022 \times 10^{23}\text{ C atoms}}{1\text{ mol C}} = 2.9 \times 10^{22}\text{ C atoms}$$

EXAMPLE 6.3   The Mole Concept—Converting between Grams and Number of Atoms

How many aluminum atoms are in an aluminum can with a mass of 16.2 g?

SORT You are given the mass of aluminum and asked to find the number of aluminum atoms.	**GIVEN:** 16.2 g Al **FIND:** Al atoms

STRATEGIZE
The solution map has two steps. In the first step, convert from g Al to mol Al. In the second step, convert from mol Al to the number of Al atoms. The required conversion factors are the molar mass of aluminum and the number of atoms in a mole.
	SOLUTION MAP
	
	$$\frac{1\text{ mol Al}}{26.98\text{ g Al}} \qquad \frac{6.022 \times 10^{23}\text{ Al atoms}}{1\text{ mol Al}}$$
	
	RELATIONSHIPS USED
	    26.98 g Al = 1 mol Al (molar mass of aluminum, from periodic table)
	
	    6.022×10^{23} = 1 mol (Avogadro's number)

SOLVE Follow the solution map to solve the problem, beginning with 16.2 g Al and multiplying by the appropriate conversion factors to arrive at Al atoms.	**SOLUTION** $$16.2 \text{ g Al} \times \frac{1\text{ mol Al}}{26.98\text{ g Al}} \times \frac{6.022 \times 10^{23}\text{ Al atoms}}{1\text{ mol Al}} = 3.62 \times 10^{23}\text{ Al atoms}$$
CHECK Are the units correct? Does the answer make physical sense?	The units, Al atoms, are correct. The answer makes sense because the number of atoms in any macroscopic-sized sample of matter should be very large.

▶**SKILLBUILDER 6.3 | The Mole Concept—Converting between Grams and Number of Atoms**

Calculate the mass of 1.23×10^{24} helium atoms.

▶**FOR MORE PRACTICE**   Example 6.15; Problems 35, 36, 37, 38, 39, 40, 41, 42.

Before we move on, notice that numbers with large exponents, such as 6.022×10^{23}, are almost unimaginably large. Twenty-two copper pennies contain 6.022×10^{23} or 1 mol of copper atoms, but 6.022×10^{23} pennies would cover Earth's entire surface to a depth of 300 m. Even objects that are small by everyday standards occupy a huge space when we have a mole of them. For example, one crystal of granulated sugar has a mass of less than 1 mg and a diameter of less than 0.1 mm, yet 1 mol of sugar crystals would cover the state of Texas to a depth of several feet. For every increase of 1 in the exponent of a number, the number increases by 10. So a number with an exponent of 23 is incredibly large. A mole has to be a large number because atoms are so small.

CONCEPTUAL CHECKPOINT 6.1

Which statement is *always* true for samples of atomic elements, regardless of the type of element present in the samples?

(a) If two samples of different elements contain the same number of atoms, they contain the same number of moles.

(b) If two samples of different elements have the same mass, they contain the same number of moles.

(c) If two samples of different elements have the same mass, they contain the same number of atoms.

CONCEPTUAL CHECKPOINT 6.2

Without doing any calculations, determine which sample contains the most atoms.

(a) one gram of cobalt

(b) one gram of carbon

(c) one gram of lead

6.4 Counting Molecules by the Gram

Remember, ionic compounds do not contain individual molecules. In loose language, the smallest electrically neutral collection of ions is sometimes called a molecule but is more correctly called a formula unit.

The calculations we just performed for atoms can also be applied to molecules for covalent compounds or formula units for ionic compounds. We first convert between the mass of a compound and moles of the compound, and then we calculate the number of molecules (or formula units) from moles.

CONVERTING BETWEEN GRAMS AND MOLES OF A COMPOUND

Remember, the formula mass for a compound is the sum of the atomic masses of all of the atoms in a chemical formula.

For elements, the molar mass is the mass of 1 mol of atoms of that element. For compounds, the molar mass is the mass of 1 mol of molecules or formula units of that compound. The molar mass of a compound in grams per mole is numerically equal to the formula mass of the compound in atomic mass units. For example, the formula mass of CO_2 is:

$$\text{Formula mass} = 1(\text{Atomic mass of C}) + 2(\text{Atomic mass of O})$$

$$= 1(12.01 \text{ amu}) + 2(16.00 \text{ amu})$$

$$= 44.01 \text{ amu}$$

The molar mass of CO_2 is therefore:

$$\text{Molar mass} = 44.01 \text{ g/mol}$$

Just as the molar mass of an element serves as a conversion factor between grams and moles of that element, the molar mass of a compound serves as a conversion factor between grams and moles of that compound. For example, suppose we want to find the number of moles in a 22.5-g sample of dry ice (solid CO_2). We set up the problem in the normal way.

We begin by sorting the information.

GIVEN: 22.5 g CO_2

FIND: mol CO_2

SOLUTION MAP

We then strategize by drawing a solution map which shows how the molar mass converts grams of the compound to moles of the compound.

$$\frac{1 \text{ mol } CO_2}{44.01 \text{ g } CO_2}$$

RELATIONSHIPS USED

44.01 g CO_2 = 1 mol CO_2 (molar mass of CO_2)

SOLUTION

Finally, we solve the problem.

$$22.5 \; \cancel{g} \times \frac{1 \text{ mol } CO_2}{44.01 \; \cancel{g}} = 0.511 \text{ mol } CO_2$$

EXAMPLE 6.4   The Mole Concept—Converting between Grams and Moles for Compounds

Calculate the mass (in grams) of 1.75 mol of water.

SORT You are given moles of water and asked to find the mass.	**GIVEN:** 1.75 mol H_2O **FIND:** g H_2O
STRATEGIZE Draw a solution map showing the conversion from mol H_2O to g H_2O. The conversion factor is the molar mass of water, which you can determine by summing the atomic masses of all the atoms in the chemical formula.	**SOLUTION MAP** mol H_2O → g H_2O $$\frac{18.02 \text{ g } H_2O}{1 \text{ mol } H_2O}$$ **RELATIONSHIPS USED** H_2O molar mass = 2(Atomic mass H) + 1(Atomic mass O) = 2(1.01) + 1(16.00) = 18.02 g/mol
SOLVE Follow the solution map to solve the problem. Begin with 1.75 mol of water and use the molar mass to convert to grams of water.	**SOLUTION** $$1.75 \; \cancel{\text{mol } H_2O} \times \frac{18.02 \text{ g } H_2O}{\cancel{\text{mol } H_2O}} = 31.5 \text{ g } H_2O$$
CHECK Check your answer. Are the units correct? Does the answer make physical sense?	The units (g H_2O) are the desired units. The magnitude of the answer makes sense because 1 mole of water has a mass of 18.02 g; therefore, 1.75 moles should have a mass that is slightly less than 36 g.

▶SKILLBUILDER 6.4 | The Mole Concept—Converting between Grams and Moles

Calculate the number of moles of NO_2 in 1.18 g of NO_2.

▶FOR MORE PRACTICE Problems 47, 48, 49, 50.

CONVERTING BETWEEN GRAMS OF A COMPOUND AND NUMBER OF MOLECULES

Suppose that we want to find the *number of CO_2 molecules* in a sample of dry ice (solid CO_2) with a mass of 22.5 g.

The solution map for the problem is:

$$\frac{1 \text{ mol } CO_2}{44.01 \text{ g } CO_2} \quad \frac{6.022 \times 10^{23} \text{ CO}_2 \text{ molecules}}{1 \text{ mol } CO_2}$$

Notice that the first part of the solution map is identical to computing the number of moles of CO_2 in 22.5 g of dry ice. The second part of the solution map shows the conversion from moles to number of molecules. Following the solution map, we calculate:

$$22.5 \text{ g } \cancel{CO_2} \times \frac{1 \cancel{\text{ mol } CO_2}}{44.01 \text{ g } \cancel{CO_2}} \times \frac{6.022 \times 10^{23} \text{ CO}_2 \text{ molecules}}{\cancel{\text{mol } CO_2}}$$

$$= 3.08 \times 10^{23} \text{ CO}_2 \text{ molecules}$$

EXAMPLE 6.5 The Mole Concept—Converting between Mass of a Compound and Number of Molecules

What is the mass of 4.78×10^{24} NO_2 molecules?

SORT You are given the number of NO_2 molecules and asked to find the mass.	**GIVEN:** 4.78×10^{24} NO_2 molecules **FIND:** g NO_2
STRATEGIZE The solution map has two steps. In the first step, convert from molecules of NO_2 to moles of NO_2. In the second step, convert from moles of NO_2 to mass of NO_2. The required conversion factors are the molar mass of NO_2 and the number of molecules in a mole.	**SOLUTION MAP** $\boxed{NO_2 \text{ molecules}} \longrightarrow \boxed{\text{mol } NO_2} \longrightarrow \boxed{\text{g } NO_2}$ $\dfrac{1 \text{ mol } NO_2}{6.022 \times 10^{23} \text{ } NO_2 \text{ molecules}} \quad \dfrac{46.01 \text{ g } NO_2}{1 \text{ mol } NO_2}$ **RELATIONSHIPS USED** 6.022×10^{23} molecules = 1 mol (Avogadro's number) NO_2 molar mass = 1(Atomic mass N) + 2(Atomic mass O) $\quad = 14.01 + 2(16.00)$ $\quad = 46.01 \text{ g/mol}$
SOLVE Using the solution map as a guide, begin with molecules of NO_2 and multiply by the appropriate conversion factors to arrive at g NO_2.	**SOLUTION** $4.78 \times 10^{24} \cancel{NO_2 \text{ molecules}} \times \dfrac{1 \cancel{\text{ mol } NO_2}}{6.022 \times 10^{23} \cancel{NO_2 \text{ molecules}}}$ $\times \dfrac{46.01 \text{ g } NO_2}{1 \cancel{\text{ mol } NO_2}} = 365 \text{ g } NO_2$
CHECK Check your answer. Are the units correct? Does the answer make physical sense?	The units, g NO_2, are correct. Since the number of NO_2 molecules is more than one mole, the answer should be more than one molar mass (more than 46.01 g), which it is; therefore, the magnitude of the answer is reasonable.

▶SKILLBUILDER 6.5 | The Mole Concept—Converting between Mass and Number of Molecules

How many H_2O molecules are in a sample of water with a mass of 3.64 g?

▶FOR MORE PRACTICE   Problems 51, 52, 53, 54.

 CONCEPTUAL CHECKPOINT 6.3

Compound A has a molar mass of 100 g/mol and Compound B has a molar mass of 200 g/mol. If you have samples of equal mass of both compounds, which sample contains the greatest number of molecules?

6.5 Chemical Formulas as Conversion Factors

3 leaves : 1 clover

▲ From our knowledge of clovers, we know that each clover has three leaves. We can express that as a ratio: 3 leaves : 1 clover.

We are almost ready to address the sodium problem in our opening example. To determine how much of a particular element (such as sodium) is in a given amount of a particular compound (such as sodium chloride), we must understand the numerical relationships inherent in a chemical formula. We can understand these relationships with a simple analogy: Asking how much sodium is in a given amount of sodium chloride is similar to asking how many leaves are on a given number of clovers. For example, suppose we want to know the number of leaves on 14 clovers. We need a conversion factor between leaves and clovers. For clovers, the conversion factor comes from our knowledge about them—we know that each clover has 3 leaves. We can express that relationship as a ratio between clovers and leaves.

3 leaves : 1 clover

Like other conversion factors, this ratio gives the relationship between leaves and clovers. With this ratio, we can write a conversion factor to determine the number of leaves in 14 clovers. The solution map is:

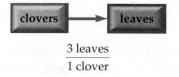

$$\frac{3 \text{ leaves}}{1 \text{ clover}}$$

We solve the problem by beginning with clovers and converting to leaves.

$$14 \ \cancel{\text{clovers}} \times \frac{3 \text{ leaves}}{1 \ \cancel{\text{clover}}} = 42 \text{ leaves}$$

Similarly, a chemical formula gives us ratios between elements and molecules for a particular compound. For example, the formula for carbon dioxide (CO_2) means there are two O atoms per CO_2 molecule. We write this as:

2 O atoms : 1 CO_2 molecule

Just as 3 leaves : 1 clover can also be written as 3 dozen leaves : 1 dozen clovers, for molecules we can write:

2 doz O atoms : 1 doz CO_2 molecules

However, for atoms and molecules, we normally work in moles.

$$2 \text{ mol O} : 1 \text{ mol CO}_2$$

| Chemical formulas are discussed in Chapter 5.

With conversion factors such as these—which come directly from the chemical formula—we can determine the amounts of the constituent elements present in a given amount of a compound.

▶ Each of these shows a ratio.

8 legs : 1 spider   4 legs : 1 chair   2 H atoms : 1 H₂O molecule

CONVERTING BETWEEN MOLES OF A COMPOUND AND MOLES OF A CONSTITUENT ELEMENT

Suppose we want to know the number of moles of O in 18 mol of CO_2. Our solution map is:

mol CO₂ → mol O

$$\frac{2 \text{ mol O}}{1 \text{ mol CO}_2}$$

We can then calculate the moles of O.

$$18 \ \cancel{\text{mol CO}_2} \times \frac{2 \text{ mol O}}{1 \ \cancel{\text{mol CO}_2}} = 36 \text{ mol O}$$

EXAMPLE 6.6   Chemical Formulas as Conversion Factors—Converting between Moles of a Compound and Moles of a Constituent Element

Determine the number of moles of O in 1.7 mol of $CaCO_3$.

SORT You are given the number of moles of $CaCO_3$ and asked to find the number of moles of O.	**GIVEN:** 1.7 mol CaCO₃ **FIND:** mol O
STRATEGIZE The solution map begins with moles of calcium carbonate and ends with moles of oxygen. Determine the conversion factor from the chemical formula, which indicates three O atoms for every $CaCO_3$ unit.	**SOLUTION MAP** mol CaCO₃ → mol O $\dfrac{3 \text{ mol O}}{1 \text{ mol CaCO}_3}$ **RELATIONSHIPS USED** 3 mol O : 1 mol CaCO₃ (from chemical formula)
SOLVE Follow the solution map to solve the problem. The subscripts in a chemical formula are exact, so they never limit significant figures.	**SOLUTION** $1.7 \ \cancel{\text{mol CaCO}_3} \times \dfrac{3 \text{ mol O}}{1 \ \cancel{\text{mol CaCO}_3}} = 5.1 \text{ mol O}$

CHECK

Check your answer. Are the units correct? Does the answer make physical sense?

The units (mol O) are correct. The magnitude is reasonable as the number of moles of oxygen should be larger than the number of moles of $CaCO_3$ (because each $CaCO_3$ unit contains 3 O atoms).

▶**SKILLBUILDER 6.6 | Chemical Formulas as Conversion Factors—Converting between Moles of a Compound and Moles of a Constituent Element**

Determine the number of moles of O in 1.4 mol of H_2SO_4.

▶**FOR MORE PRACTICE**   Example 6.16; Problems 63, 64.

CONVERTING BETWEEN GRAMS OF A COMPOUND AND GRAMS OF A CONSTITUENT ELEMENT

Now, we have the tools we need to solve our sodium problem. Suppose we want to know the mass of sodium in 15 g of NaCl. The chemical formula gives us the relationship between moles of Na and moles of NaCl:

$$1 \text{ mol Na} : 1 \text{ mol NaCl}$$

To use this relationship, we need *mol* NaCl, but we have *g* NaCl. We can, however, use the *molar mass* of NaCl to convert from g NaCl to mol NaCl. Then we use the conversion factor from the chemical formula to convert to mol Na. Finally, we use the molar mass of Na to convert to g Na. The solution map is:

g NaCl → mol NaCl → mol Na → g Na

$$\frac{1 \text{ mol NaCl}}{58.44 \text{ g NaCl}} \qquad \frac{1 \text{ mol Na}}{1 \text{ mol NaCl}} \qquad \frac{22.99 \text{ g Na}}{1 \text{ mol Na}}$$

Notice that we must convert from g NaCl to mol NaCl *before* we can use the chemical formula as a conversion factor.

> The chemical formula gives us a relationship between moles of substances, not between grams.

We follow the solution map to solve the problem.

$$15 \text{ g NaCl} \times \frac{1 \text{ mol NaCl}}{58.44 \text{ g NaCl}} \times \frac{1 \text{ mol Na}}{1 \text{ mol NaCl}} \times \frac{22.99 \text{ g Na}}{1 \text{ mol Na}} = 5.9 \text{ g Na}$$

The general form for solving problems where you are asked to find the mass of an element present in a given mass of a compound is:

Mass compound ⟶ **Moles** compound ⟶ **Moles** element ⟶ **Mass** element

Use the atomic or molar mass to convert between mass and moles, and use the relationships inherent in the chemical formula to convert between moles and moles (▼ Figure 6.2).

▶ **FIGURE 6.2   Mole relationships from a chemical formula**   The relationships inherent in a chemical formula allow us to convert between moles of the compound and moles of a constituent element (or vice versa).

$1 \text{ mol } CCl_4 : 4 \text{ mol Cl}$

EXAMPLE 6.7 Chemical Formulas as Conversion Factors—Converting between Grams of a Compound and Grams of a Constituent Element

Carvone ($C_{10}H_{14}O$) is the main component of spearmint oil. It has a pleasant aroma and mint flavor. Carvone is often added to chewing gum, liqueurs, soaps, and perfumes. Calculate the mass of carbon in 55.4 g of carvone.

SORT You are given the mass of carvone and asked to find the mass of one of its constituent elements.	**GIVEN:** 55.4 g $C_{10}H_{14}O$ **FIND:** g C
STRATEGIZE Base the solution map on Grams $\longrightarrow$ Mole $\longrightarrow$ Mole $\longrightarrow$ Grams	**SOLUTION MAP** $\boxed{\text{g } C_{10}H_{14}O} \longrightarrow \boxed{\text{mol } C_{10}H_{14}O} \longrightarrow \boxed{\text{mol C}} \longrightarrow \boxed{\text{g C}}$ $\dfrac{1 \text{ mol } C_{10}H_{14}O}{150.2 \text{ g } C_{10}H_{14}O} \qquad \dfrac{10 \text{ mol C}}{1 \text{ mol } C_{10}H_{14}O} \qquad \dfrac{12.01 \text{ g C}}{1 \text{ mol C}}$
You need three conversion factors. The first is the molar mass of carvone.	**RELATIONSHIPS USED** $\text{Molar mass carvone} = 10(12.01) + 14(1.01) + 1(16.00)$ $\qquad\qquad = 120.1 + 14.14 + 16.00$ $\qquad\qquad = 150.2 \text{ g/mol}$
The second conversion factor is the relationship between moles of carbon and moles of carvone from the molecular formula.	$10 \text{ mol C} : 1 \text{ mol } C_{10}H_{14}O$ (from chemical formula)
The third conversion factor is the molar mass of carbon.	$1 \text{ mol C} = 12.01 \text{ g C}$ (molar mass C, from periodic table)
SOLVE Follow the solution map to solve the problem, beginning with g $C_{10}H_{14}O$ and multiplying by the appropriate conversion factors to arrive at g C.	**SOLUTION** $55.4 \text{ g } \cancel{C_{10}H_{14}O} \times \dfrac{1 \text{ mol } \cancel{C_{10}H_{14}O}}{150.2 \text{ g } \cancel{C_{10}H_{14}O}} \times$ $\qquad\qquad \dfrac{10 \text{ mol } \cancel{C}}{1 \text{ mol } \cancel{C_{10}H_{14}O}} \times \dfrac{12.01 \text{ g C}}{1 \text{ mol } \cancel{C}} = 44.3 \text{ g C}$
CHECK Check your answer. Are the units correct? Does the answer make physical sense?	The units, g C, are correct. The magnitude of the answer is reasonable since the mass of carbon with the compound must be less than the mass of the compound itself. If you had arrived at a mass of carbon that was greater than the mass of the compound, you would immediately know that you had made a mistake; the mass of a constituent element can never be greater than the mass of the compound itself.

▶**SKILLBUILDER 6.7** | **Chemical Formulas as Conversion Factors—Converting between Grams of a Compound and Grams of a Constituent Element**

Determine the mass of oxygen in a 5.8-g sample of sodium bicarbonate ($NaHCO_3$).

▶**SKILLBUILDER PLUS**

Determine the mass of oxygen in a 7.20-g sample of $Al_2(SO_4)_3$.

▶**FOR MORE PRACTICE** Example 6.17; Problems 67, 68, 69, 70.

CHEMISTRY IN THE ENVIRONMENT

Chlorine in Chlorofluorocarbons

About 30 years ago, scientists began to suspect that synthetic compounds known as chlorofluorocarbons (CFCs) were destroying a vital compound called ozone (O_3) in Earth's upper atmosphere. Upper atmospheric ozone is important because it acts as a shield to protect life on Earth from harmful ultraviolet light (▼ Figure 6.3). CFCs are chemically inert molecules (they do not readily react with other substances) used primarily as refrigerants and industrial solvents. Their inertness has allowed them to leak into the atmosphere and stay there for many years. In the upper atmosphere, however, sunlight eventually breaks bonds within CFCs, resulting in the release of chlorine atoms. The chlorine atoms then react with ozone and destroy it by converting it from O_3 into O_2.

In 1985, scientists discovered a large hole in the ozone layer over Antarctica that has since been attributed to CFCs. The amount of ozone over Antarctica had depleted by a startling 50%. The ozone hole is transient, existing only in the Antarctic spring, from late August to November. Examination of data from previous years showed that this gradually expanding ozone hole has formed each spring since 1977 (▼ Figure 6.4), and it continues to form today.

A similar hole has been observed during some years over the North Pole, and a smaller, but still significant, drop in ozone has been observed over more populated areas such as the northern United States and Canada. The thinning of ozone over these areas is dangerous because ultraviolet light can harm living things and induce skin cancer in humans. Based on this evidence, most developed nations banned the production of CFCs on January 1, 1996. However, CFCs still lurk in most older refrigerators and air conditioning units and can leak into the atmosphere and destroy ozone.

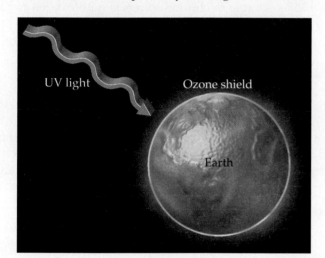

▲ **FIGURE 6.3** **The ozone shield** Atmospheric ozone shields life on Earth from harmful ultraviolet light.

CAN YOU ANSWER THIS? *Suppose a car air conditioner contains 2.5 kg of freon-12 (CCl_2F_2), a CFC. How many kilograms of Cl are contained within the freon?*

▶ **FIGURE 6.4 Growth of the ozone hole** Antarctic ozone levels in three Septembers from 1979 to 2000. The darkest blue colors indicate the lowest ozone levels.

Sep 1979

Sep 1991

Sep 2000

CONCEPTUAL CHECKPOINT 6.4

Without doing any detailed calculations, determine which sample contains the most fluorine atoms.

(a) 25 g of HF

(b) 1.5 mol of CH_3F

(c) 1.0 mol of F_2

6.6   Mass Percent Composition of Compounds

Another way to express how much of an element is in a given compound is to use the element's mass percent composition for that compound. The **mass percent composition** or simply **mass percent** of an element is the element's percentage of the total mass of the compound. For example, the mass percent composition of sodium in sodium chloride is 39%. This information tells us that a 100-g sample of sodium chloride contains 39 g of sodium. The mass percent composition for a compound can be determined from experimental data using the formula:

$$\text{Mass percent of element } X = \frac{\text{Mass of } X \text{ in a sample of the compound}}{\text{Mass of the sample of the compound}} \times 100\%$$

For example, suppose a 0.358-g sample of chromium reacts with oxygen to form 0.523 g of the metal oxide. Then the mass percent of chromium is:

$$\text{Mass percent Cr} = \frac{\text{Mass Cr}}{\text{Mass metal oxide}} \times 100\%$$

$$= \frac{0.358 \text{ g}}{0.523 \text{ g}} \times 100\% = 68.5\%$$

We can use mass percent composition as a conversion factor between grams of a constituent element and grams of the compound. For example, we just saw that the mass percent composition of sodium in sodium chloride is 39%. This can be written as:

$$39 \text{ g sodium} : 100 \text{ g sodium chloride}$$

or in fractional form:

$$\frac{39 \text{ g Na}}{100 \text{ g NaCl}} \quad \text{or} \quad \frac{100 \text{ g NaCl}}{39 \text{ g Na}}$$

These fractions are conversion factors between g Na and g NaCl, as shown in Example 6.8.

EXAMPLE 6.8   Using Mass Percent Composition as a Conversion Factor

The FDA recommends that adults consume less than 2.4 g of sodium per day. How many grams of sodium chloride can you consume and still be within the FDA guidelines? Sodium chloride is 39% sodium by mass.

SORT You are given the mass of sodium and the mass percent of sodium in sodium chloride. When mass percent is given, write it as a fraction. *Percent* means *per hundred*, so 39% sodium indicates that there are 39 g Na per 100 g NaCl. You are asked to find the mass of sodium chloride that contains the given mass of sodium.	**GIVEN:** 2.4 g Na $\dfrac{39 \text{ g Na}}{100 \text{ g NaCl}}$ **FIND:** g NaCl
STRATEGIZE Draw a solution map that starts with the mass of sodium and uses the mass percent as a conversion factor to get to the mass of sodium chloride.	**SOLUTION MAP** $\dfrac{100 \text{ g NaCl}}{39 \text{ g Na}}$ **RELATIONSHIPS USED** 39 g Na : 100 g NaCl (given in the problem)

SOLVE

Follow the solution map to solve the problem, beginning with grams Na and ending with grams of NaCl. The amount of salt you can consume and still be within the FDA guideline is 6.2 g NaCl.

SOLUTION

$$2.4 \ \text{g Na} \times \frac{100 \ \text{g NaCl}}{39 \ \text{g Na}} = 6.2 \ \text{g NaCl}$$

CHECK

Check your answer. Are the units correct? Does the answer make physical sense?

The units, g NaCl, are correct. The answer makes physical sense because the mass of NaCl should be *larger* than the mass of Na. The mass of a compound containing a given mass of a particular element is always larger than the mass of the element itself.

▲ Twelve and a half salt packets contain 6.2 g NaCl.

▶**SKILLBUILDER 6.8** | **Using Mass Percent Composition as a Conversion Factor**

If a woman consumes 22 g of sodium chloride, how much sodium does she consume? Sodium chloride is 39% sodium by mass.

▶**FOR MORE PRACTICE**   Example 6.19; Problems 75, 76, 77, 78.

6.7   Mass Percent Composition from a Chemical Formula

In the previous section, we learned how to calculate mass percent composition from experimental data and how to use mass percent composition as a conversion factor. We can also calculate the mass percent of any element in a compound from the chemical formula for the compound. Based on the chemical formula, the mass percent of element X in a compound is:

$$\text{Mass percent of element } X = \frac{\text{Mass of element } X \text{ in 1 mol of compound}}{\text{Mass of 1 mol of compound}} \times 100\%$$

Suppose, for example, that we want to calculate the mass percent composition of Cl in the chlorofluorocarbon CCl_2F_2. The mass percent of Cl is given by:

$$\text{Mass percent Cl} = \frac{2 \times \text{Molar mass Cl}}{\text{Molar mass CCl}_2\text{F}_2} \times 100\%$$

We must multiply the molar mass of Cl by 2 because the chemical formula has a subscript of 2 for Cl, meaning that 1 mol of CCl_2F_2 contains 2 mol of Cl atoms. We calculate the molar mass of CCl_2F_2 as follows:

$$\text{Molar mass} = 1(12.01) + 2(35.45) + 2(19.00) = 120.91 \ \text{g/mol}$$

So the mass percent of Cl in CCl_2F_2 is

$$\text{Mass percent Cl} = \frac{2 \times \text{Molar mass Cl}}{\text{Molar mass CCl}_2\text{F}_2} \times 100\% = \frac{2 \times 35.45 \ \text{g}}{120.91 \ \text{g}} \times 100\%$$

$$= 58.64\%$$

EXAMPLE 6.9 Mass Percent Composition

Calculate the mass percent of Cl in freon-114 ($C_2Cl_4F_2$).

SORT	GIVEN: $C_2Cl_4F_2$
You are given the molecular formula of freon-114 and asked to find the mass percent of Cl.	FIND: Mass % Cl

STRATEGIZE	SOLUTION MAP
The solution map shows how you can use the information in the chemical formula to substitute into the mass percent equation and obtain the mass percent Cl.	Chemical formula → mass % Cl $$\text{Mass \% Cl} = \frac{4 \times \text{Molar mass Cl}}{\text{Molar mass } C_2Cl_4F_2} \times 100\%$$

RELATIONSHIPS USED

Mass percent of element $X =$

$$\frac{\text{Mass of element } X \text{ in 1 mol of compound}}{\text{Mass of 1 mol of compound}} \times 100\%$$

(mass percent equation, introduced in this section)

SOLVE	SOLUTION
Calculate the molar mass of freon-114 and substitute the values into the equation to find mass percent Cl.	$4 \times$ Molar mass Cl = 4(35.45 g) = 141.8 g

$$\text{Molar mass } C_2Cl_4F_2 = 2(12.01) + 4(35.45) + 2(19)$$
$$= 24.02 + 141.8 + 38.00$$
$$= \frac{203.8 \text{ g}}{\text{mol}}$$

$$\text{Mass \% Cl} = \frac{4 \times \text{Molar mass Cl}}{\text{Molar mass } C_2Cl_4F_2} \times 100\%$$

$$= \frac{141.8 \text{ g}}{203.8 \text{ g}} \times 100\%$$

$$= 69.58\%$$

CHECK	
Check your answer. Are the units correct? Does the answer make physical sense?	The units (%) are correct. The answer makes physical sense. Mass percent composition should never exceed 100%. If your answer is greater than 100%, you have made an error.

▶SKILLBUILDER 6.9 | Mass Percent Composition

Acetic acid ($HC_2H_3O_2$) is the active ingredient in vinegar. Calculate the mass percent composition of O in acetic acid.

▶FOR MORE PRACTICE Example 6.20; Problems 79, 80, 81, 82, 83, 84.

 CONCEPTUAL CHECKPOINT 6.5

Which compound has the highest mass percent of O? (You should not have to perform any detailed calculations to answer this question.)

(a) CrO

(b) CrO_2

(c) Cr_2O_3

CHEMISTRY AND HEALTH

Fluoridation of Drinking Water

In the early 1900s, scientists discovered that people whose drinking water naturally contained fluoride (F⁻) ions had fewer cavities than people whose water did not. At the proper levels, fluoride strengthens tooth enamel, which prevents tooth decay. In an effort to improve public health, fluoride has been artificially added to drinking water supplies since 1945. In the United States today, about 62% of the population drinks artificially fluoridated drinking water. The American Dental Association and public health agencies estimate that water fluoridation reduces tooth decay by 40 to 65%.

The fluoridation of public drinking water, however, is often controversial. Some opponents argue that fluoride is available from other sources—such as toothpaste, mouthwash, drops, and pills—and therefore should not be added to drinking water. Anyone who wants fluoride can get it from these optional sources, they argue, and the government should not impose fluoride on the general population. Other opponents argue that the risks associated with fluoridation are too great. Indeed, too much fluoride can cause teeth to become brown and spotted, a condition known as dental fluorosis. Extremely high levels can lead to skeletal fluorosis, a condition in which the bones become brittle and arthritic.

The scientific consensus is that, like many minerals, fluoride shows some health benefits at certain levels—about 1–4 mg/day for adults—but can have detrimental effects at higher levels. Consequently, most major cities fluoridate their drinking water at a level of about 1 mg/L. Since adults drink between 1 and 2 L of water per day, they should receive the beneficial amounts of fluoride from the water. Bottled water does not normally contain fluoride, and therefore does not have the benefit of fluoride to teeth. Fluoridated bottled water can sometimes be found in the infant section of supermarkets.

CAN YOU ANSWER THIS? *Fluoride is often added to water as sodium fluoride (NaF). What is the mass percent composition of F⁻ in NaF? How many grams of NaF should be added to 1500 L of water to fluoridate it at a level of 1.0 mg F⁻/L?*

6.8 Calculating Empirical Formulas for Compounds

In Section 6.7, we learned how to calculate mass percent composition from a chemical formula. But can we go the other way? Can we calculate a chemical formula from mass percent composition? This is important because laboratory analyses of compounds do not often give chemical formulas directly; rather, they give the relative masses of each element present in a compound. For example, if we decompose water into hydrogen and oxygen in the laboratory, we could measure the masses of hydrogen and oxygen produced. Can we determine a chemical formula for water from this kind of data?

▶ We just learned how to go from the chemical formula of a compound to its mass percent composition. Can we also go the other way?

The answer is a qualified yes. We can determine a chemical formula, but it is the **empirical formula**, not the molecular formula. As we saw in Section 5.3, an empirical formula only gives the smallest whole-number ratio of each type of atom in a compound, not the specific number of each type of atom in a molecule. Recall that the **molecular formula** is always a whole-number multiple of the empirical formula:

Molecular formula = Empirical formula × n, where n = 1, 2, 3 . . .

For example, the molecular formula for hydrogen peroxide is H_2O_2, and its empirical formula is HO.

A chemical formula represents a ratio of atoms or moles of atoms, not a ratio of masses.

$$HO \times 2 \longrightarrow H_2O_2$$

CALCULATING AN EMPIRICAL FORMULA FROM EXPERIMENTAL DATA

Suppose we decompose a sample of water in the laboratory and find that it produces 3.0 g of hydrogen and 24 g of oxygen. How do we determine an empirical formula from these data?

oxygen gas   hydrogen gas

▲ Water can be decomposed by an electric current into hydrogen and oxygen. How can we find the empirical formula for water from the masses of its component elements?

We know that an empirical formula represents a ratio of atoms or a ratio of moles of atoms, but it *does not* represent a ratio of masses. So the first thing we must do is convert our data from grams to moles. How many moles of each element formed during the decomposition? To convert to moles, we divide each mass by the molar mass of that element.

$$\text{Moles H} = 3.0 \ \cancel{\text{g H}} \times \frac{1 \ \text{mol H}}{1.01 \ \cancel{\text{g H}}} = 3.0 \ \text{mol H}$$

$$\text{Moles O} = 24 \ \cancel{\text{g O}} \times \frac{1 \ \text{mol O}}{16.00 \ \cancel{\text{g O}}} = 1.5 \ \text{mol O}$$

From these data, we know there are 3 mol of H for every 1.5 mol of O. We can now write a pseudoformula for water:

$$H_3O_{1.5}$$

To get whole-number subscripts in our formula, we divide all the subscripts by the smallest one, in this case 1.5.

$$H_{\frac{3}{1.5}}O_{\frac{1.5}{1.5}} = H_2O$$

Our empirical formula for water, which in this case also happens to be the molecular formula, is H_2O. The following procedure can be used to obtain the empirical formula of any compound from experimental data. The left column outlines the procedure, and the center and right columns show two examples of how to apply the procedure.

Obtaining an Empirical Formula from Experimental Data	**EXAMPLE 6.10**	**EXAMPLE 6.11**
	A compound containing nitrogen and oxygen is decomposed in the laboratory and produces 24.5 g of nitrogen and 70.0 g of oxygen. Calculate the empirical formula of the compound.	A laboratory analysis of aspirin determines the following mass percent composition: C 60.00% H 4.48% O 35.53% Find the empirical formula.
1. Write down (or calculate) as given the masses of each element present in a sample of the compound. If you are given mass percent composition, assume a 100-g sample and calculate the masses of each element from the given percentages.	**GIVEN:** 24.5 g N 70.0 g O **FIND:** empirical formula	**GIVEN:** In a 100-g sample: 60.00 g C 4.48 g H 35.53 g O **FIND:** empirical formula
2. Convert each of the masses in Step 1 to moles by using the appropriate molar mass for each element as a conversion factor.	**SOLUTION** $24.5 \text{ g N} \times \dfrac{1 \text{ mol N}}{14.01 \text{ g N}}$ $= 1.75 \text{ mol N}$ $70.0 \text{ g O} \times \dfrac{1 \text{ mol O}}{16.00 \text{ g O}}$ $= 4.38 \text{ mol O}$	**SOLUTION** $60.00 \text{ g C} \times \dfrac{1 \text{ mol C}}{12.01 \text{ g C}}$ $= 4.996 \text{ mol C}$ $4.48 \text{ g H} \times \dfrac{1 \text{ mol H}}{1.01 \text{ g H}}$ $= 4.44 \text{ mol H}$ $35.53 \text{ g O} \times \dfrac{1 \text{ mol O}}{16.00 \text{ g O}}$ $= 2.221 \text{ mol O}$
3. Write down a pseudoformula for the compound, using the moles of each element (from Step 2) as subscripts.	$N_{1.75}O_{4.38}$	$C_{4.996}H_{4.44}O_{2.221}$
4. Divide all the subscripts in the formula by the smallest subscript.	$N_{\frac{1.75}{1.75}}O_{\frac{4.38}{1.75}} \longrightarrow N_1O_{2.5}$	$C_{\frac{4.996}{2.221}}H_{\frac{4.44}{2.221}}O_{\frac{2.221}{2.221}} \longrightarrow C_{2.25}H_2O_1$
5. If the subscripts are not whole numbers, multiply all the subscripts by a small whole number (see the following table) to arrive at whole-number subscripts.	$N_1O_{2.5} \times 2 \longrightarrow N_2O_5$ The correct empirical formula is N_2O_5.	$C_{2.25}H_2O_1 \times 4 \longrightarrow C_9H_8O_4$ The correct empirical formula is $C_9H_8O_4$.

Fractional Subscript	Multiply by This Number to Get Whole-Number Subscripts
_.10	10
_.20	5
_.25	4
_.33	3
_.50	2
_.66	3
_.75	4

▶**SKILLBUILDER 6.10**

A sample of a compound is decomposed in the laboratory and produces 165 g of carbon, 27.8 g of hydrogen, and 220.2 g O. Calculate the empirical formula of the compound.

▶**FOR MORE PRACTICE**
Problems 85, 86, 87, 88.

▶**SKILLBUILDER 6.11**

Ibuprofen, an aspirin substitute, has the mass percent composition: C 75.69%; H 8.80%; O 15.51%. Calculate the empirical formula of the ibuprofen.

▶**FOR MORE PRACTICE**
Example 6.21; Problems 89, 90, 91, 92.

EXAMPLE 6.12   Calculating an Empirical Formula from Reaction Data

A 3.24-g sample of titanium reacts with oxygen to form 5.40 g of the metal oxide. What is the empirical formula of the metal oxide?

You are given the mass of titanium and the mass of the metal oxide that forms. You are asked to find the empirical formula. You must recognize this problem as one requiring a special procedure and apply that procedure, which is outlined below.	**GIVEN:** 3.24 g Ti 5.40 g metal oxide **FIND:** empirical formula

1. Write down (or calculate) the masses of each element present in a sample of the compound.
In this case, you are given the mass of the initial Ti sample and the mass of its oxide after the sample reacts with oxygen. The mass of oxygen is the difference between the mass of the oxide and the mass of titanium.
	SOLUTION
	  3.24 g Ti
	  Mass O = Mass oxide − Mass titanium
	        = 5.40 g − 3.24 g
	        = 2.16 g O

2. Convert each of the masses in Step 1 to moles by using the appropriate molar mass for each element as a conversion factor.	$3.24 \ \text{g Ti} \times \dfrac{1 \ \text{mol Ti}}{47.88 \ \text{g Ti}} = 0.0677 \ \text{mol Ti}$ $2.16 \ \text{g O} \times \dfrac{1 \ \text{mol O}}{16.00 \ \text{g O}} = 0.135 \ \text{mol O}$
3. Write down a pseudoformula for the compound, using the moles of each element obtained in Step 2 as subscripts.	$Ti_{0.0677}O_{0.135}$
4. Divide all the subscripts in the formula by the smallest subscript.	$Ti_{\frac{0.0677}{0.0677}} O_{\frac{0.135}{0.0677}} \longrightarrow TiO_2$
5. If the subscripts are not whole numbers, multiply all the subscripts by a small whole number to arrive at whole-number subscripts.	Since the subscripts are already whole numbers, this last step is unnecessary. The correct empirical formula is TiO_2.

▶**SKILLBUILDER 6.12 | Calculating an Empirical Formula from Reaction Data**

A 1.56-g sample of copper reacts with oxygen to form 1.95 g of the metal oxide. What is the formula of the metal oxide?

▶**FOR MORE PRACTICE**   Problems 93, 94, 95, 96.

6.9 Calculating Molecular Formulas for Compounds

▲ Fructose, a sugar found in fruit.

You can determine the *molecular* formula of a compound from the empirical formula if you also know the molar mass of the compound. Recall from Section 6.8 that the molecular formula is always a whole-number multiple of the empirical formula.

$$\text{Molecular formula} = \text{Empirical formula} \times n, \text{ where } n = 1, 2, 3 \ldots$$

Suppose we want to find the molecular formula for fructose (a sugar found in fruit) from its empirical formula, CH_2O, and its molar mass, 180.2 g/mol. We know that the molecular formula is a whole-number multiple of CH_2O.

$$\text{Molecular formula} = CH_2O \times n$$

We also know that the molar mass is a whole-number multiple of the **empirical formula molar mass**, the sum of the masses of all the atoms in the empirical formula.

$$\text{Molar mass} = \text{Empirical formula molar mass} \times n$$

For a particular compound, the value of n in both cases is the same. Therefore, we can find n by calculating the ratio of the molar mass to the empirical formula molar mass.

$$n = \frac{\text{Molar mass}}{\text{Empirical formula molar mass}}$$

For fructose, the empirical formula molar mass is:

$$\text{Empirical formula molar mass} = 1(12.01) + 2(1.01) + 16.00 = 30.03 \text{ g/mol}$$

Therefore, n is:

$$n = \frac{180.2 \text{ g/mol}}{30.03 \text{ g/mol}} = 6$$

We can then use this value of n to find the molecular formula.

$$\text{Molecular formula} = CH_2O \times 6 = C_6H_{12}O_6$$

EXAMPLE 6.13   Calculating Molecular Formula from Empirical Formula and Molar Mass

Naphthalene is a compound containing carbon and hydrogen that is often used in mothballs. Its empirical formula is C_5H_4 and its molar mass is 128.16 g/mol. What is its molecular formula?

SORT You are given the empirical formula and the molar mass of a compound and asked to find its molecular formula.	**GIVEN:** empirical formula = C_5H_4 molar mass = 128.16 g/mol **FIND:** molecular formula
STRATEGIZE In the first step, use the molar mass (which is given) and the empirical formula molar mass (which you can calculate based on the empirical formula) to determine n (the integer by which you must multiply the empirical formula to get the molecular formula). In the second step, multiply the subscripts in the empirical formula by n to arrive at the molecular formula.	**SOLUTION MAP** molar mass, empirical formula molar mass $\rightarrow$ n $n = \dfrac{\text{Molar mass}}{\text{Empirical formula molar mass}}$ n, empirical formula $\rightarrow$ molecular formula Molecular formula = Empirical formula $\times$ n
SOLVE First find the empirical formula molar mass. Now follow the solution map. Find n by dividing the molar mass by the empirical formula molar mass (which you just calculated). Multiply the empirical formula by n to get the molecular formula.	**SOLUTION** $\begin{aligned} \text{Empirical formula molar mass} &= 5(12.01) + 4(1.01) \\ &= 64.09 \text{ g/mol} \end{aligned}$ $n = \dfrac{\text{Molar mass}}{\text{Empirical formula mass}} = \dfrac{128.16 \text{ g/mol}}{64.09 \text{ g/mol}} = 2$ Molecular formula = $C_5H_4 \times 2 = C_{10}H_8$
CHECK Check your answer. Does the answer make physical sense?	The answer makes physical sense because it is a whole-number multiple of the empirical formula. Any answer containing fractional subscripts would be an error.

▶**SKILLBUILDER 6.13 | Calculating Molecular Formula from Empirical Formula and Molar Mass**

Butane is a compound containing carbon and hydrogen that is used as a fuel in butane lighters. Its empirical formula is C_2H_5, and its molar mass is 58.12 g/mol. Find its molecular formula.

▶**SKILLBUILDER PLUS**

A compound with the following mass percent composition has a molar mass of 60.10 g/mol. Find its molecular formula.

C	39.97%	H	13.41%	N	46.62%

▶**FOR MORE PRACTICE**   Example 6.22; Problems 97, 98, 99, 100.

CHAPTER IN REVIEW

CHEMICAL PRINCIPLES

The Mole Concept: The mole is a specific number (6.022×10^{23}) that allows us to easily count atoms or molecules by weighing them. One mole of any element has a mass equivalent to its atomic mass in grams, and a mole of any compound has a mass equivalent to its formula mass in grams. The mass of 1 mol of an element or compound is its molar mass.

Chemical Formulas and Chemical Composition: Chemical formulas indicate the relative number of each kind of element in a compound. These numbers are based on atoms or moles. By using molar masses, we can use the information in a chemical formula to determine the relative masses of each kind of element in a compound. We can then relate the mass of a sample of a compound to the masses of the elements contained in the compound.

Empirical and Molecular Formulas from Laboratory Data: We can refer to the relative masses of the elements within a compound to determine the empirical formula of the compound. If the chemist also knows the molar mass of the compound, he or she can also determine its molecular formula.

RELEVANCE

The Mole Concept: The mole concept allows us to determine the number of atoms or molecules in a sample from its mass. Just as a hardware store customer wants to know the number of nails in a certain weight of nails, so we want to know the number of atoms in a certain mass of atoms. Since atoms are too small to count, we use their mass.

Chemical Formulas and Chemical Composition: The chemical composition of compounds is important because it lets us determine how much of a particular element is contained within a particular compound. For example, an assessment of the threat to the Earth's ozone layer from chlorofluorocarbons (CFCs) requires knowing how much chlorine is in a particular CFC.

Empirical and Molecular Formulas from Laboratory Data: The first thing a chemist wants to know about an unknown compound is its chemical formula, because the formula reveals the compound's composition. Chemists often arrive at formulas by analyzing compounds in the laboratory—either by decomposing them or by synthesizing them—to determine the relative masses of the elements they contain.

CHEMICAL SKILLS

Converting between Moles and Number of Atoms (Section 6.3)

SORT
You are given moles of copper and asked to find the number of copper atoms.

STRATEGIZE
To convert between moles and number of atoms, use Avogadro's number, 6.022×10^{23} atoms = 1 mol, as a conversion factor.

EXAMPLES

EXAMPLE 6.13 Converting between Moles and Number of Atoms

Calculate the number of atoms in 4.8 mol of copper.

GIVEN: 4.8 mol Cu

FIND: Cu atoms

SOLUTION MAP

$$\frac{6.022 \times 10^{23} \text{ Cu atoms}}{1 \text{ mol Cu}}$$

RELATIONSHIPS USED
1 mol Cu = 6.022×10^{23} Cu atoms (Avogadro's number, from inside back cover)

SOLVE

Follow the solution map to solve the problem.

CHECK

Check your answer. Are the units correct? Does the answer make physical sense?

SOLUTION

$$4.8 \ \cancel{\text{mol Cu}} \times \frac{6.022 \times 10^{23} \text{ Cu atoms}}{1 \ \cancel{\text{mol Cu}}} =$$

$$2.9 \times 10^{24} \text{ Cu atoms}$$

The units, Cu atoms, are correct. The answer makes physical sense because the number is very large, as you would expect for nearly 5 moles of atoms.

Converting between Grams and Moles (Section 6.3)

SORT

You are given the number of moles of aluminum and asked to find the mass of aluminum in grams.

STRATEGIZE

Use the molar mass of aluminum to convert between moles and grams.

SOLVE

Follow the solution map to solve the problem.

CHECK

Check your answer. Are the units correct? Does the answer make physical sense?

EXAMPLE 6.14   Converting between Grams and Moles

Calculate the mass of aluminum (in grams) of 6.73 moles of aluminum.

GIVEN: 6.73 mol Al

FIND: g Al

SOLUTION MAP

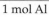

$$\frac{26.98 \text{ g Al}}{1 \text{ mol Al}}$$

RELATIONSHIPS USED

26.98 g Al = 1 mol Al (molar mass of Al from periodic table)

SOLUTION

$$6.73 \ \cancel{\text{mol Al}} \times \frac{26.98 \text{ g Al}}{1 \ \cancel{\text{mol Al}}} = 182 \text{ g Al}$$

The units, g Al, are correct. The answer makes physical sense because each mole has a mass of about 27 g; therefore, nearly 7 moles should have a mass of nearly 190 g.

Converting between Grams and Number of Atoms or Molecules (Section 6.3)

SORT
You are given the mass of a zinc sample and asked to find the number of Zn atoms that it contains.

STRATEGIZE
First use the molar mass of the element to convert from grams to moles, and then use Avogadro's number to convert moles to number of atoms.

SOLVE
Follow the solution map to solve the problem.

CHECK
Check your answer. Are the units correct? Does the answer make physical sense?

EXAMPLE 6.15   Converting between Grams and Number of Atoms or Molecules

Determine the number of atoms in a 48.3-g sample of zinc.

GIVEN: 48.3 g Zn

FIND: Zn atoms

SOLUTION MAP

$$\frac{1 \text{ mol Zn}}{65.39 \text{ g Zn}} \qquad \frac{6.022 \times 10^{23} \text{ Zn atoms}}{1 \text{ mol Zn}}$$

RELATIONSHIPS USED

65.39 g Zn = 1 mol Zn (molar mass of Zn from periodic table)

1 mol = 6.022×10^{23} atoms (Avogadro's number, from inside back cover)

SOLUTION

$$48.3 \text{ g Zn} \times \frac{1 \text{ mol Zn}}{65.39 \text{ g Zn}} \times \frac{6.022 \times 10^{23} \text{ Zn atoms}}{1 \text{ mol Zn}}$$

$$= 4.45 \times 10^{23} \text{ Zn atoms}$$

The units, Zn atoms, are correct. The answer makes physical sense because the number of atoms in any macroscopic-sized sample should be very large.

Converting between Moles of a Compound and Moles of a Constituent Element (Section 6.5)

SORT
You are given the number of moles of sulfuric acid and asked to find the number of moles of oxygen.

STRATEGIZE
To convert between moles of a compound and moles of a constituent element, use the chemical formula of the compound to determine a ratio between the moles of the element and the moles of the compound.

SOLVE
Follow the solution map to solve the problem.

CHECK
Check your answer. Are the units correct? Does the answer make physical sense?

EXAMPLE 6.16   Converting between Moles of a Compound and Moles of a Constituent Element

Determine the number of moles of oxygen in 7.20 mol of H_2SO_4.

GIVEN: 7.20 mol H_2SO_4

FIND: mol O

SOLUTION MAP

$$\frac{4 \text{ mol O}}{1 \text{ mol H}_2\text{SO}_4}$$

RELATIONSHIPS USED

4 mol O : 1 mol H_2SO_4

SOLUTION

$$7.20 \text{ mol H}_2\text{SO}_4 \times \frac{4 \text{ mol O}}{1 \text{ mol H}_2\text{SO}_4} = 28.8 \text{ mol O}$$

The units, mol O, are correct. The answer makes physical sense because the number of moles of an element in a compound is equal to or greater than the number of moles of the compound itself.

Converting between Grams of a Compound and Grams of a Constituent Element (Section 6.5)

SORT

You are given the mass of iron (III) oxide and asked to find the mass of iron contained within it.

STRATEGIZE

Use the molar mass of the compound to convert from grams of the compound to moles of the compound. Then use the chemical formula to obtain a conversion factor to convert from moles of the compound to moles of the constituent element. Finally, use the molar mass of the constituent element to convert from moles of the element to grams of the element.

SOLVE

Follow the solution map to solve the problem.

CHECK

Check your answer. Are the units correct? Does the answer make physical sense?

EXAMPLE 6.17 Converting between Grams of a Compound and Grams of a Constituent Element

Find the grams of iron in 79.2 g of Fe_2O_3.

GIVEN: 79.2 g Fe_2O_3

FIND: g Fe

SOLUTION MAP

$$\frac{1\ mol\ Fe_2O_3}{159.70\ g\ Fe_2O_3} \qquad \frac{2\ mol\ Fe}{1\ mol\ Fe_2O_3} \qquad \frac{55.85\ g\ Fe}{1\ mol\ Fe}$$

RELATIONSHIPS USED

Molar mass Fe_2O_3

$$= 2(55.85) + 3(16.00)$$

$$= 159.70\ g/mol$$

2 mol Fe : 1 mol Fe_2O_3 (from given chemical formula)

SOLUTION

$$79.2\ \cancel{g\ Fe_2O_3} \times \frac{1\ \cancel{mol\ Fe_2O_3}}{159.70\ \cancel{g\ Fe_2O_3}} \times \frac{2\ \cancel{mol\ Fe}}{1\ \cancel{mol\ Fe_2O_3}} \times$$

$$\frac{55.85\ g\ Fe}{1\ \cancel{mol\ Fe}} = 55.4\ g\ Fe$$

The units, g Fe, are correct. The answer makes physical sense because the mass of a constituent element within a compound should be less than the mass of the compound itself.

Using Mass Percent Composition as a Conversion Factor (Section 6.6)

SORT

You are given the mass of titanium(IV) oxide and the mass percent titanium in the oxide. You are asked to find the mass of titanium in the sample.

STRATEGIZE

Use the percent composition as a conversion factor between grams of titanium(IV) oxide and grams of titanium.

SOLVE

Follow the solution map to solve the problem.

CHECK

Check your answer. Are the units correct? Does the answer make physical sense?

EXAMPLE 6.18 Using Mass Percent Composition as a Conversion Factor

Determine the mass of titanium in 57.2 g of titanium(IV) oxide. The mass percent of titanium in titanium(IV) oxide is 59.9%.

GIVEN: 57.2 g TiO_2

$$\frac{59.9 \text{ g Ti}}{100 \text{ g } TiO_2}$$

FIND: g Ti

SOLUTION MAP

$$\frac{59.9 \text{ g Ti}}{100 \text{ g } TiO_2}$$

RELATIONSHIPS USED

59.9 g Ti : 100 g TiO_2

SOLUTION

$$57.2 \text{ g } \cancel{TiO_2} \times \frac{59.9 \text{ g Ti}}{100 \text{ g } \cancel{TiO_2}} = 34.3 \text{ g Ti}$$

The units, g Ti, are correct. The answer makes physical sense because the mass of an element within a compound should be less than the mass of the compound itself.

Determining Mass Percent Composition from a Chemical Formula (Section 6.7)

SORT

You are given the formula of potassium oxide and asked to determine the mass percent of potassium within it.

STRATEGIZE

The solution map shows how the information derived from the chemical formula can be substituted into the mass percent equation to yield the mass percent of the element.

EXAMPLE 6.19 Determining Mass Percent Composition from a Chemical Formula

Calculate the mass percent composition of potassium in potassium oxide (K_2O).

GIVEN: K_2O

FIND: Mass % K

SOLUTION MAP

```
Chemical  →  Mass % K
formula
```

$$\text{Mass \% K} = \frac{2 \times \text{Molar mass K}}{\text{Molar mass } K_2O} \times 100\%$$

RELATIONSHIPS USED

Mass percent of element X

$$= \frac{\text{Mass of element } X \text{ in 1 mol of compound}}{\text{Mass of 1 mol of compound}} \times 100\%$$

(mass percent equation, from Section 6.6)

SOLVE

Calculate the molar mass of potassium oxide and then follow the solution map to solve the problem.

SOLUTION

Molar mass K_2O = 2(39.10) + 16.00

$$= 94.20 \text{ g/mol}$$

$$\text{Mass \% K} = \frac{2(39.10 \text{ g K})}{94.20 \text{ g K}_2\text{O}} \times 100\% = 83.01\% \text{ K}$$

CHECK

Check your answer. Are the units correct? Does the answer make physical sense?

The units, % K, are correct. The answer makes physical sense because it should be below 100%.

Determining an Empirical Formula from Experimental Data (Section 6.8)

You must recognize this problem as one requiring a special procedure. Follow these steps to solve the problem.

EXAMPLE 6.20 Determining an Empirical Formula from Experimental Data

A laboratory analysis of vanillin, the flavoring agent in vanilla, determined the mass percent composition: C, 63.15%; H, 5.30%; O, 31.55%. Determine the empirical formula of vanillin.

GIVEN: 63.15 % C, 5.30 % H, and 31.55 % O.

FIND: empirical formula

1. Write down (or calculate) the masses of each element present in a sample of the compound. If you are given mass percent composition, assume a 100-g sample and calculate the masses of each element from the given percentages.

SOLUTION

In a 100 g sample:

63.15 g C

5.30 g H

31.55 g O

2. Convert each of the masses in Step 1 to moles by using the appropriate molar mass for each element as a conversion factor.

$$63.15 \text{ g C} \times \frac{1 \text{ mol C}}{12.01 \text{ g C}} = 5.258 \text{ mol C}$$

$$5.30 \text{ g H} \times \frac{1 \text{ mol H}}{1.01 \text{ g H}} = 5.25 \text{ mol H}$$

$$31.55 \text{ g O} \times \frac{1 \text{ mol O}}{16.00 \text{ g O}} = 1.972 \text{ mol O}$$

3. Write down a pseudoformula for the compound using the moles of each element (from Step 2) as subscripts.

$$C_{5.258}H_{5.25}O_{1.972}$$

4. Divide all the subscripts in the formula by the smallest subscript.

$$C_{\frac{5.258}{1.972}} H_{\frac{5.25}{1.972}} O_{\frac{1.972}{1.972}} \longrightarrow C_{2.67}H_{2.66}O_1$$

5. If the subscripts are not whole numbers, multiply all the subscripts by a small whole number to arrive at whole-number subscripts.

$$C_{2.67}H_{2.66}O_1 \times 3 \longrightarrow C_8H_8O_3$$

The correct empirical formula is $C_8H_8O_3$.

Calculating a Molecular Formula from an Empirical Formula and Molar Mass (Section 6.9)

EXAMPLE 6.21   Calculating a Molecular Formula from an Empirical Formula and Molar Mass

Acetylene, a gas often used in welding torches, has the empirical formula CH and a molar mass of 26.04 g/mol. Find its molecular formula.

SORT

You are given the empirical formula and molar mass of acetylene and asked to find the molecular formula.

GIVEN: empirical formula = CH

molar mass = 26.04 g/mol

FIND: molecular formula

STRATEGIZE

In the first step, use the molar mass (which is given) and the empirical formula molar mass (which you can calculate based on the empirical formula) to determine n (the integer by which you must multiply the empirical formula to arrive at the molecular formula).

In the second step, multiply the coefficients in the empirical formula by n to arrive at the molecular formula.

SOLUTION MAP

$$n = \frac{\text{Molar mass}}{\text{Empirical formula molar mass}}$$

Molecular formula = Empirical formula × n

SOLVE

Follow the solution map to solve the problem. Calculate the empirical formula molar mass, which is the sum of the masses of all the atoms in the empirical formula.

SOLUTION

Empirical formula molar mass

$= 12.01 + 1.01$

$= 13.02$ g/mol

Next, find n, the ratio of the molar mass to empirical mass.

$$n = \frac{\text{Molar mass}}{\text{Empirical formula molar mass}}$$

$$= \frac{26.04 \text{ g/mol}}{13.02 \text{ g/mol}} = 2$$

Finally, multiply the empirical formula by n to get the molecular formula.

Molecular formula = CH × 2 ⟶ C_2H_2

CHECK

Check your answer. Does the answer make physical sense?

The answer makes physical sense because the formula subscripts are all integers. Any answer with fractional integers would be suspect.

KEY TERMS

EXERCISES

QUESTIONS

1. Why is chemical composition important?
2. How can you determine the number of atoms in a sample of an element? Why is counting them not an option?
3. How many atoms are in 1 mol of atoms?
4. How many molecules are in 1 mol of molecules?
5. What is the mass of 1 mol of atoms for an element?
6. What is the mass of 1 mol of molecules for a compound?
7. What is the mass of 1 mol of atoms of each element?

   (a) P

   (b) Pt

   (c) C

   (d) Cr

8. What is the mass of 1 mol of molecules of each compound?

   (a) CO_2

   (b) CH_2Cl_2

   (c) $C_{12}H_{22}O_{11}$

   (d) SO_2

9. The subscripts in a chemical formula give relationships between moles of the constituent elements and moles of the compound. Explain why these subscripts *do not* give relationships between grams of the constituent elements and grams of the compound.

10. Write the conversion factors between moles of each constituent element and moles of the compound for $C_{12}H_{22}O_{11}$.
11. Mass percent composition can be used as a conversion factor between grams of a constituent element and grams of the compound. Write the conversion factor (including units) inherent in each mass percent composition.

    (a) Water is 11.19% hydrogen by mass.

    (b) Fructose, also known as fruit sugar, is 53.29% oxygen by mass.

    (c) Octane, a component of gasoline, is 84.12% carbon by mass.

    (d) Ethanol, the alcohol in alcoholic beverages, is 52.14% carbon by mass.

12. What is the mathematical formula for calculating mass percent composition from a chemical formula?
13. How are the empirical formula and the molecular formula of a compound related?
14. Why is it important to be able to calculate an empirical formula from experimental data?
15. What is the empirical formula mass of a compound?
16. How are the molar mass and empirical formula mass for a compound related?

PROBLEMS

THE MOLE CONCEPT

17. How many mercury atoms are in 5.8 mol of mercury?

18. How many moles of gold atoms do 3.45×10^{24} gold atoms constitute?

19. How many atoms are in each elemental sample?

    (a) 3.4 mol Cu

    (b) 9.7×10^{-3} mol C

    (c) 22.9 mol Hg

    (d) 0.215 mol Na

20. How many moles of atoms are in each elemental sample?

    (a) 4.6×10^{24} Pb atoms

    (b) 2.87×10^{22} He atoms

    (c) 7.91×10^{23} K atoms

    (d) 4.41×10^{21} Ca atoms

21. Complete the table:

Element	Moles	Number of Atoms
Ne	0.552	_____
Ar	_____	3.25×10^{24}
Xe	1.78	_____
He	_____	1.08×10^{20}

22. Complete the table:

Element	Moles	Number of Atoms
Cr	_____	9.61×10^{23}
Fe	1.52×10^{-5}	_____
Ti	0.0365	_____
Hg	_____	1.09×10^{23}

23. Consider these definitions.

    1 doz = 12

    1 gross = 144

    1 ream = 500

    1 mol = 6.022×10^{23}

    Suppose you have 872 sheets of paper. How many _____ of paper do you have?

    (a) dozens
    (b) gross
    (c) reams
    (d) moles

24. A pure copper penny contains approximately 3.0×10^{22} copper atoms. Use the definitions in the previous problem to determine how many _____ of copper atoms are in a penny.

    (a) dozens
    (b) gross
    (c) reams
    (d) moles

25. How many moles of tin atoms are in a pure tin cup with a mass of 38.1 g?

26. A lead fishing weight contains 0.12 mol of lead atoms. What is its mass?

27. A pure gold coin contains 0.145 mol of gold. What is its mass?

28. A helium balloon contains 0.46 g of helium. How many moles of helium does it contain?

29. How many moles of atoms are in each elemental sample?

    (a) 1.34 g Zn
    (b) 24.9 g Ar
    (c) 72.5 g Ta
    (d) 0.0223 g Li

30. What is the mass in grams of each elemental sample?

    (a) 6.64 mol W
    (b) 0.581 mol Ba
    (c) 68.1 mol Xe
    (d) 1.57 mol S

31. Complete the table:

Element	Moles	Mass
Ne	_____	22.5 g
Ar	0.117	_____
Xe	_____	1.00 kg
He	1.44×10^{-4}	_____

32. Complete the table:

Element	Moles	Mass
Cr	0.00442	_____
Fe	_____	73.5 mg
Ti	1.009×10^{-3}	_____
Hg	_____	1.78 kg

33. A pure silver ring contains 0.0134 mmol (millimol) Ag. How many silver atoms does it contain?

34. A pure gold ring contains 0.0102 mmol (millimol) Au. How many gold atoms does it contain?

35. How many aluminum atoms are in 3.78 g of aluminum?

36. What is the mass of 4.91×10^{21} platinum atoms?

37. How many atoms are in each elemental sample?

    (a) 16.9 g Sr
    (b) 26.1 g Fe
    (c) 8.55 g Bi
    (d) 38.2 g P

38. Calculate the mass in grams of each elemental sample:

    (a) 1.32×10^{20} uranium atoms
    (b) 2.55×10^{22} zinc atoms
    (c) 4.11×10^{23} lead atoms
    (d) 6.59×10^{24} silicon atoms

39. How many carbon atoms are in a diamond (pure carbon) with a mass of 38 mg?

40. How many helium atoms are in a helium blimp containing 495 kg of helium?

41. How many titanium atoms are in a pure titanium bicycle frame with a mass of 1.28 kg?

42. How many copper atoms are in a pure copper statue with a mass of 133 kg?

43. Complete the table:

Element	Mass	Moles	Number of Atoms
Na	38.5 mg	——	——
C	——	1.12	——
V	——	——	214
Hg	1.44 kg	——	——

44. Complete the table:

Element	Mass	Moles	Number of Atoms
Pt	——	0.0449	——
Fe	——	——	1.14×10^{25}
Ti	23.8 mg	——	——
Hg	——	2.05	——

45. Which sample contains the greatest number of atoms?

(a) 27.2 g Cr

(b) 55.1 g Ti

(c) 205 g Pb

46. Which sample contains the greatest number of atoms?

(a) 10.0 g He

(b) 25.0 g Ne

(c) 115 g Xe

47. Determine the number of moles of molecules (or formula units) in each sample.

(a) 38.2 g sodium chloride

(b) 36.5 g nitrogen monoxide

(c) 4.25 kg carbon dioxide

(d) 2.71 mg carbon tetrachloride

48. Determine the mass of each sample.

(a) 1.32 mol carbon tetrafluoride

(b) 0.555 mol magnesium fluoride

(c) 1.29 mmol carbon disulfide

(d) 1.89 kmol sulfur trioxide

49. Complete the table:

Compound	Mass	Moles	Number of molecules
H_2O	112 kg	——	——
N_2O	6.33 g	——	——
SO_2	——	2.44	——
CH_2Cl_2	——	0.0643	——

50. Complete the table:

Compound	Mass	Moles	Number of molecules
CO_2	——	0.0153	——
CO	——	0.0150	——
BrI	23.8 mg	——	——
CF_2Cl_2	1.02 kg	——	——

51. A mothball, composed of naphthalene ($C_{10}H_8$), has a mass of 1.32 g. How many naphthalene molecules does it contain?

52. Calculate the mass in grams of a single water molecule.

53. How many molecules are in each sample?

(a) 3.5 g H_2O

(b) 56.1 g N_2

(c) 89 g CCl_4

(d) 19 g $C_6H_{12}O_6$

54. Calculate the mass in grams of each sample.

(a) 5.94×10^{20} H_2O_2 molecules

(b) 2.8×10^{22} SO_2 molecules

(c) 4.5×10^{25} O_3 molecules

(d) 9.85×10^{19} CH_4 molecules

55. A sugar crystal contains approximately 1.8×10^{17} sucrose ($C_{12}H_{22}O_{11}$) molecules. What is its mass in milligrams?

56. A salt crystal has a mass of 0.12 mg. How many NaCl formula units does it contain?

57. How much money, in dollars, does one mole of pennies represent? If this amount of money were evenly distributed among the entire world's population (about 6.6 billion people), how much would each person get? Would each person be a millionaire? Billionaire? Trillionaire?

58. A typical dust particle has a diameter of about 10.0 μm. If 1.0 mol of dust particles were laid end to end along the equator, how many times would they encircle the planet? The circumference of the Earth at the equator is 40,076 km.

CHEMICAL FORMULAS AS CONVERSION FACTORS

59. Determine the number of moles of Cl in 2.7 mol CaCl$_2$.

60. How many moles of O are in 12.4 mol Fe(NO$_3$)$_3$?

61. Which sample contains the greatest number of moles of O?
    (a) 2.3 mol H$_2$O
    (b) 1.2 mol H$_2$O$_2$
    (c) 0.9 mol NaNO$_3$
    (d) 0.5 mol Ca(NO$_3$)$_2$

62. Which sample contains the greatest number of moles of Cl?
    (a) 3.8 mol HCl
    (b) 1.7 mol CH$_2$Cl$_2$
    (c) 4.2 mol NaClO$_3$
    (d) 2.2 mol Mg(ClO$_4$)$_2$

63. Determine the number of moles of C in each sample.
    (a) 2.5 mol CH$_4$
    (b) 0.115 mol C$_2$H$_6$
    (c) 5.67 mol C$_4$H$_{10}$
    (d) 25.1 mol C$_8$H$_{18}$

64. Determine the number of moles of H in each sample.
    (a) 4.67 mol H$_2$O
    (b) 8.39 mol NH$_3$
    (c) 0.117 mol N$_2$H$_4$
    (d) 35.8 mol C$_{10}$H$_{22}$

65. For each set of molecular models, write a relationship between moles of hydrogen and moles of molecules. Then determine the total number of hydrogen atoms present. (H—white; O—red; C—black; N—blue)

(a)   (b)   (c)

66. For each set of molecular models, write a relationship between moles of oxygen and moles of molecules. Then determine the total number of oxygen atoms present. (H—white; O—red; C—black; S—yellow)

(a)   (b)   (c)

67. How many grams of Cl are in 38.0 g of each sample of chlorofluorocarbons (CFCs)?
    (a) CF$_2$Cl$_2$
    (b) CFCl$_3$
    (c) C$_2$F$_3$Cl$_3$
    (d) CF$_3$Cl

68. Calculate the number of grams of sodium in 1.00 g of each sodium-containing food additive.
    (a) NaCl (table salt)
    (b) Na$_3$PO$_4$ (sodium phosphate)
    (c) NaC$_7$H$_5$O$_2$ (sodium benzoate)
    (d) Na$_2$C$_6$H$_6$O$_7$ (sodium hydrogen citrate)

69. Iron is found in Earth's crust as several different iron compounds. Calculate the mass (in kg) of each compound that contains 1.0×10^3 kg of iron.
    (a) Fe$_2$O$_3$ (hematite)
    (b) Fe$_3$O$_4$ (magnetite)
    (c) FeCO$_3$ (siderite)

70. Lead is often found in Earth's crust as several lead compounds. Calculate the mass (in kg) of each compound that contains 1.0×10^3 kg of lead.
    (a) PbS (galena)
    (b) PbCO$_3$ (cerussite)
    (c) PbSO$_4$ (anglesite)

MASS PERCENT COMPOSITION

71. A 2.45-g sample of strontium completely reacts with oxygen to form 2.89 g of strontium oxide. Use this data to calculate the mass percent composition of strontium in strontium oxide.

72. A 4.78-g sample of aluminum completely reacts with oxygen to form 6.67 g of aluminum oxide. Use this data to calculate the mass percent composition of aluminum in aluminum oxide.

73. A 1.912-g sample of calcium chloride is decomposed into its constituent elements and found to contain 0.690 g Ca and 1.222 g Cl. Calculate the mass percent composition of Ca and Cl in calcium chloride.

74. A 0.45-g sample of aspirin is decomposed into its constituent elements and found to contain 0.27 g C, 0.020 g H, and 0.16 g O. Calculate the mass percent composition of C, H, and O in aspirin.

75. Copper(II) fluoride contains 37.42% F by mass. Use this percentage to calculate the mass of fluorine in grams contained in 28.5 g of copper(II) fluoride.

76. Silver chloride, often used in silver plating, contains 75.27% Ag. Calculate the mass of silver chloride in grams required to make 4.8 g of silver plating.

77. In small amounts, the fluoride ion (often consumed as NaF) prevents tooth decay. According to the American Dental Association, an adult female should consume 3.0 mg of fluorine per day. Calculate the amount of sodium fluoride (45.24% F) that a woman should consume to get the recommended amount of fluorine.

78. The iodide ion, usually consumed as potassium iodide, is a dietary mineral essential to good nutrition. In countries where potassium iodide is added to salt, iodine deficiency or goiter has been almost completely eliminated. The recommended daily allowance (RDA) for iodine is 150 μg/day. How much potassium iodide (76.45% I) should you consume to meet the RDA?

MASS PERCENT COMPOSITION FROM CHEMICAL FORMULA

79. Calculate the mass percent composition of nitrogen in each compound.
- **(a)** N_2O
- **(b)** NO
- **(c)** NO_2
- **(d)** N_2O_5

80. Calculate the mass percent composition of carbon in each compound.
- **(a)** C_2H_2
- **(b)** C_3H_6
- **(c)** C_2H_6
- **(d)** C_2H_6O

81. Calculate the mass percent composition of each element in each compound.
- **(a)** $C_2H_4O_2$
- **(b)** CH_2O_2
- **(c)** C_3H_9N
- **(d)** $C_4H_{12}N_2$

82. Calculate the mass percent composition of each element in each compound.
- **(a)** $FeCl_3$
- **(b)** TiO_2
- **(c)** H_3PO_4
- **(d)** HNO_3

83. Iron ores have different amounts of iron per kilogram of ore. Calculate the mass percent composition of iron for each iron ore: Fe_2O_3 (hematite), Fe_3O_4 (magnetite), $FeCO_3$ (siderite). Which ore has the highest iron content?

84. Plants need nitrogen to grow; so many fertilizers consist of nitrogen-containing compounds. Calculate the mass percent composition of nitrogen in each fertilizer: NH_3, $CO(NH_2)_2$, NH_4NO_3, $(NH_4)_2SO_4$. Which fertilizer has the highest nitrogen content?

CALCULATING EMPIRICAL FORMULAS

85. A compound containing nitrogen and oxygen is decomposed in the laboratory and produces 1.78 g of nitrogen and 4.05 g of oxygen. Calculate the empirical formula of the compound.

86. A compound containing selenium and fluorine is decomposed in the laboratory and produces 2.231 g of selenium and 3.221 g of fluorine. Calculate the empirical formula of the compound.

87. Samples of several compounds were decomposed, and the masses of their constituent elements were measured. Calculate the empirical formula for each compound.
    (a) 1.245 g Ni, 5.381 g I
    (b) 1.443 g Se, 5.841 g Br
    (c) 2.128 g Be, 7.557 g S, 15.107 g O

88. Samples of several compounds were decomposed, and the masses of their constituent elements were measured. Calculate the empirical formula for each compound.
    (a) 2.677 g Ba, 3.115 g Br
    (b) 1.651 g Ag, 0.1224 g O
    (c) 0.672 g Co, 0.569 g As, 0.486 g O

89. The rotten smell of a decaying animal carcass is partially due to a nitrogen-containing compound called putrescine. Elemental analysis of putrescine shows that it consists of 54.50% C, 13.73% H, and 31.77% N. Calculate the empirical formula of putrescine.

90. Citric acid, the compound responsible for the sour taste of lemons, has the elemental composition: C, 37.51%; H, 4.20%; O, 58.29%. Calculate the empirical formula of citric acid.

91. The compounds listed here are often found in many natural flavors and scents. Calculate the empirical formula for each compound.
    (a) ethyl butyrate (pineapple oil): C, 62.04%; H, 10.41%; O, 27.55%
    (b) methyl butyrate (apple flavor): C, 58.80%; H, 9.87%; O, 31.33%
    (c) benzyl acetate (oil of jasmine): C, 71.98%; H, 6.71%; O, 21.31%

92. The compounds listed here are all over-the-counter pain relievers. Calculate the empirical formula for each compound.
    (a) acetaminophen (Tylenol): C, 63.56%; H, 6.00%; N, 9.27%; O, 21.17%
    (b) naproxen (Aleve): C, 73.03%; H, 6.13%; O, 20.84%

93. A 1.45-g sample of phosphorus burns in air and forms 2.57 g of a phosphorus oxide. Calculate the empirical formula of the oxide.

94. A 2.241-g sample of nickel reacts with oxygen to form 2.852 g of the metal oxide. Calculate the empirical formula of the oxide.

95. A 0.77-mg sample of nitrogen reacts with chlorine to form 6.61 mg of the chloride. What is the empirical formula of the nitrogen chloride?

96. A 45.2-mg sample of phosphorus reacts with selenium to form 131.6 mg of the selenide. What is the empirical formula of the phosphorus selenide?

CALCULATING MOLECULAR FORMULAS

97. A compound containing carbon and hydrogen has a molar mass of 56.11 g/mol and an empirical formula of CH_2. Determine its molecular formula.

98. A compound containing phosphorus and oxygen has a molar mass of 219.9 g/mol and an empirical formula of P_2O_3. Determine its molecular formula.

99. The molar masses and empirical formulas of several compounds containing carbon and chlorine are as follows. Find the molecular formula of each compound.
    (a) 284.77 g/mol, CCl
    (b) 131.39 g/mol, C_2HCl_3
    (c) 181.44 g/mol, C_2HCl

100. The molar masses and empirical formulas of several compounds containing carbon and nitrogen are as follows. Find the molecular formula of each compound.
    (a) 163.26 g/mol, $C_{11}H_{17}N$
    (b) 186.24 g/mol, C_6H_7N
    (c) 312.29 g/mol, C_3H_2N

CUMULATIVE PROBLEMS

101. A pure copper cube has an edge length of 1.42 cm. How many copper atoms does it contain? (volume of a cube = (edge length)^3; density of copper = 8.96 g/cm^3)

102. A pure silver sphere has a radius of 0.886 cm. How many silver atoms does it contain? (volume of a sphere = $\frac{4}{3}\pi r^3$; density of silver = 10.5 g/cm^3)

103. A drop of water has a volume of approximately 0.05 mL. How many water molecules does it contain? (density of water = 1.0 g/cm^3)

104. Fingernail-polish remover is primarily acetone (C_3H_6O). How many acetone molecules are in a bottle of acetone with a volume of 325 mL? (density of acetone = 0.788 g/cm^3)

105. Complete the table:

Substance	Mass	Moles	Number of Particles (atoms or molecules)
Ar	——	4.5×10^{-4}	——
NO_2	——	——	1.09×10^{20}
K	22.4 mg	——	——
C_8H_{18}	3.76 kg	——	——

106. Complete the table:

Substance	Mass	Moles	Number of Particles (atoms or molecules)
$C_6H_{12}O_6$	15.8 g	——	——
Pb	——	——	9.04×10^{21}
CF_4	22.5 kg	——	——
C	——	0.0388	——

107. Determine the chemical formula of each compound and then refer to it to calculate the mass percent composition of each constituent element.
  (a) copper(II) iodide
  (b) sodium nitrate
  (c) lead(II) sulfate
  (d) calcium fluoride

108. Determine the chemical formula of each compound and then refer to it to calculate the mass percent composition of each constituent element.
  (a) nitrogen triiodide
  (b) xenon tetrafluoride
  (c) phosphorus trichloride
  (d) carbon monoxide

109. The rock in a particular iron ore deposit contains 78% Fe_2O_3 by mass. How many kilograms of the rock must be processed to obtain 1.0×10^3 kg of iron?

110. The rock in a lead ore deposit contains 84% PbS by mass. How many kilograms of the rock must be processed to obtain 1.0 kg of Pb?

111. A leak in the air conditioning system of an office building releases 12 kg of CHF_2Cl per month. If the leak continues, how many kilograms of Cl will be emitted into the atmosphere each year?

112. A leak in the air conditioning system of an older car releases 55 g of CF_2Cl_2 per month. How much Cl is emitted into the atmosphere each year by this car?

113. Hydrogen is a possible future fuel. However, elemental hydrogen is rare, so it must be obtained from a hydrogen-containing compound such as water. If hydrogen were obtained from water, how much hydrogen in grams could be obtained from 1.0 L of water? (density of water = 1.0 g/cm^3)

114. Hydrogen, a possible future fuel mentioned in Problem 113, can also be obtained from other compounds such as ethanol. Ethanol can be made from the fermentation of crops such as corn. How much hydrogen in grams can be obtained from 1.0 kg of ethanol (C_2H_5OH)?

115. Complete the table of compounds that contain only carbon and hydrogen.

Formula	Molar Mass	% C (by mass)	% H (by mass)
C_2H_4	——	——	——
——	58.12	82.66%	——
C_4H_8	——	——	——
——	44.09	——	18.29%

116. Complete the table of compounds that contain only chromium and oxygen.

Formula	Name	Molar Mass	% Cr (by mass)	% O (by mass)
——	Chromium (III) oxide	——	——	——
——	——	84.00	61.90%	——
——	——	100.00	——	48.00%

117. Butanedione, a component of butter and body odor, has a cheesy smell. Elemental analysis of butanedione gave the mass percent composition: C, 55.80%; H, 7.03%; O, 37.17%. The molar mass of butanedione is 86.09 g/mol. Determine the molecular formula of butanedione.

118. Caffeine, a stimulant found in coffee and soda, has the mass percent composition: C, 49.48%; H, 5.19%; N, 28.85%; O, 16.48%. The molar mass of caffeine is 194.19 g/mol. Find the molecular formula of caffeine.

119. Nicotine, a stimulant found in tobacco, has the mass percent composition: C, 74.03%; H, 8.70%; N, 17.27%. The molar mass of nicotine is 162.23 g/mol. Find the molecular formula of nicotine.

120. Estradiol is a female sexual hormone that causes maturation and maintenance of the female reproductive system. Elemental analysis of estradiol gave the mass percent composition: C, 79.37%; H, 8.88%; O, 11.75%. The molar mass of estradiol is 272.37 g/mol. Find the molecular formula of estradiol.

121. A sample contains both KBr and KI in unknown quantities. If the sample has a total mass of 5.00 g and contains 1.51 g K, what are the percentages of KBr and KI in the sample by mass?

122. A sample contains both CO_2 and Ne in unknown quantities. If the sample contains a combined total of 1.75 mol and has a total mass of 65.3 g, what are the percentages of CO_2 and Ne in the sample by mole?

123. Ethanethiol (C_2H_6S) is a compound with a disagreeable odor that can be used to impart an odor to natural gas. When ethanethiol is burned, the sulfur reacts with oxygen to form SO_2. What mass of SO_2 forms upon the complete combustion of 28.7 g of ethanethiol?

124. Methanethiol (CH_4S) has a disagreeable odor and is often a component of bad breath. When methanethiol is burned, the sulfur reacts with oxygen to form SO_2. What mass of SO_2 forms upon the complete combustion of 1.89 g of methanethiol?

125. An iron ore contains 38% Fe_2O_3 by mass. What is the maximum mass of iron that can be recovered from 10.0 kg of this ore?

126. Seawater contains approximately 3.5% NaCl by mass and has a density of 1.02 g/mL. What volume of seawater contains 1.0 g of sodium?

HIGHLIGHT PROBLEMS

127. You can use the concepts in this chapter to obtain an estimate of the number of atoms in the universe. These steps will guide you through this calculation.

   (a) Begin by calculating the number of atoms in the sun. Assume that the sun is pure hydrogen with a density of 1.4 g/cm³. The radius of the sun is 7×10^8 m, and the volume of a sphere is $V = \frac{4}{3}\pi r^3$.

   (b) Since the sun is an average-sized star, and since stars are believed to compose most of the mass of the visible universe (planets are so small they can be ignored), we can estimate the number of atoms in a galaxy by assuming that every star in the galaxy has the same number of atoms as our sun. The Milky Way galaxy is believed to contain 1×10^{11} stars. Use your answer from part (a) to calculate the number of atoms in the Milky Way galaxy.

   (c) The universe is estimated to contain approximately 1×10^{11} galaxies. If each of these galaxies contains the same number of atoms as the Milky Way galaxy, what is the total number of atoms in the universe?

128. Because of increasing evidence of damage to the ozone layer, chlorofluorocarbon (CFC) production was banned in 1996. However, there are about 100 million auto air conditioners that still use CFC-12 (CF_2Cl_2). These air conditioners are recharged from stockpiled supplies of CFC-12. If each of the 100 million automobiles contains 1.1 kg of CFC-12 and leaks 25% of its CFC-12 into the atmosphere per year, how much Cl in kilograms is added to the atmosphere each year by auto air conditioners? (Assume two significant figures in your calculations.)

▲ The ozone hole over Antarctica on September 24, 2009. The dark blue and purple areas over the South Pole represent very depressed ozone concentrations.

▲ Our sun is one of the 100 billion stars in the Milky Way galaxy. The universe is estimated to contain about 100 billion galaxies.

129. In 1996, the media reported that possible evidence of life on Mars was found on a meteorite called Allan Hills 84001 (AH 84001). The meteorite was discovered in Antarctica in 1984 and is believed to have originated on Mars. Elemental analysis of substances within its crevices revealed carbon-containing compounds that normally derive only from living organisms. Suppose that one of those compounds had a molar mass of 202.23 g/mol and the mass percent composition: C, 95.02%; H, 4.98%. What is the molecular formula for the carbon-containing compound?

▲ The Allan Hills 84001 meteorite. Elemental analysis of the substances within the crevices of this meteorite revealed carbon-containing compounds that normally originate from living organisms.

▶ANSWERS TO SKILLBUILDER EXERCISES

Skillbuilder 6.1	5.32×10^{22} Au atoms	**Skillbuilder 6.8**	8.6 g Na
Skillbuilder 6.2	89.2 g S	**Skillbuilder 6.9**	53.28% O
Skillbuilder 6.3	8.17 g He	**Skillbuilder 6.10**	CH_2O
Skillbuilder 6.4	2.56×10^{-2} mol NO_2	**Skillbuilder 6.11**	$C_{13}H_{18}O_2$
Skillbuilder 6.5	1.22×10^{23} H_2O molecules	**Skillbuilder 6.12**	CuO
Skillbuilder 6.6	5.6 mol O	**Skillbuilder 6.13**	C_4H_{10}
Skillbuilder 6.7	3.3 g O	**Skillbuilder Plus, p. 187**	$C_2H_8N_2$
Skillbuilder Plus, p. 178	4.04 g O		

▶ANSWERS TO CONCEPTUAL CHECKPOINTS

6.1 (a) The mole is a counting unit; it represents a definite number (Avogadro's number, 6.022×10^{23}). Therefore, a given number of atoms always represents a precise number of moles, regardless of what atom is involved. Atoms of different elements have different masses, so if samples of different elements have the same mass, they *cannot* contain the same number of atoms or moles.

6.2 (b) Since carbon has lower molar mass than cobalt or lead, a one-gram sample of carbon contains more atoms than one gram of cobalt or lead.

6.3 (a) Sample A would have the greatest number of molecules. Since sample A has a lower molar mass than sample B, a given mass of sample A has more moles and therefore more molecules than the same mass of sample B.

6.4 (c) 1.0 mole of F_2 contains 2.0 mol of F atoms. The other two options each contain less than two moles of F atoms.

6.5 (b) This compound has the highest ratio of oxygen atoms to chromium atoms and so must have the greatest mass percent of oxygen.

CHAPTER 7

Chemical Reactions

"Chemistry . . . is one of the broadest branches of science if for no other reason than, when we think about it, everything is chemistry."

LUCIANO CAGLIOTI (1933–)

7.1 Kindergarten Volcanoes, Automobiles, and Laundry Detergents

Did you ever make a clay volcano in kindergarten that erupted when filled with vinegar, baking soda, and red food coloring for effect? Have you pushed the gas pedal of a car and felt the acceleration as the car moved forward? Have you wondered why laundry detergents work better than normal soap to clean your clothes? Each of these processes involves a *chemical reaction*—the transformation of one or more substances into different substances.

In the classic kindergarten volcano, the baking soda (which is sodium bicarbonate) reacts with acetic acid in the vinegar to form carbon dioxide gas, water, and sodium acetate. The newly formed carbon dioxide bubbles out of the mixture, causing the eruption. Reactions that occur in liquids and form a gas are *gas evolution reactions*. A similar reaction causes the fizzing of antacids such as Alka-Seltzer™.

When you drive a car, hydrocarbons such as octane (in gasoline) react with oxygen from the air to form carbon dioxide gas and water (▶ Figure 7.1). This reaction produces heat, which expands the gases in the car's cylinders, accelerating it forward. Reactions such as this one—in which a substance reacts with oxygen, emitting heat and forming one or more oxygen-containing compounds—are *combustion reactions*. Combustion reactions are a subcategory of *oxidation–reduction reactions*, in which electrons are transferred from one substance to another. The formation of rust and the dulling of automobile paint are other examples of oxidation–reduction reactions.

Laundry detergent works better than soap to wash clothes because it contains substances that soften hard water. Hard water contains dissolved calcium (Ca^{2+}) and magnesium (Mg^{2+}) ions. These ions interfere with the action of soap by reacting

Hydrocarbons are covered in detail in Chapter 18.

◀ In the space shuttle's main engines, hydrogen molecules, H_2 (white), and oxygen molecules, O_2 (red), which are stored in the central fuel tank, react violently to form water molecules, H_2O. The reaction emits the energy that helps propel the shuttle into space.

Octane
(a component of gasoline)

Oxygen

Auto engine

Carbon dioxide

Water

▲ FIGURE 7.1 **A combustion reaction** In an automobile engine, hydrocarbons such as octane (C_8H_{18}) from gasoline combine with oxygen from the air and react to form carbon dioxide and water.

Soap in pure water

Soap in hard water

▲ FIGURE 7.2 **Soap and water** Soap forms suds with pure water (left), but reacts with the ions in hard water (right) to form a gray residue that adheres to clothes.

with it to form a gray, slimy substance called *curd* or *soap scum* (◄ Figure 7.2). If you have ever washed your clothes in ordinary soap, you may have noticed gray soap scum residue on your clothes.

Laundry detergents inhibit curd formation because they contain substances such as sodium carbonate (Na_2CO_3) that remove calcium and magnesium ions from the water. When sodium carbonate dissolves in water, it *dissociates*, or separates into sodium ions (Na^+) and carbonate ions (CO_3^{2-}). The dissolved carbonate ions react with calcium and magnesium ions in the hard water to form solid calcium carbonate ($CaCO_3$) and solid magnesium carbonate ($MgCO_3$). These solids simply settle to the bottom of the laundry mixture, resulting in the removal of the ions from the water. In other words, laundry detergents contain substances that react with the ions in hard water to immobilize them. Reactions such as these—that form solid substances in water—are *precipitation reactions*. Precipitation reactions are also used to remove dissolved toxic metals in industrial wastes.

Chemical reactions take place all around us and even inside us. They are involved in many of the products we use daily and in many of our experiences. Chemical reactions can be relatively simple, like the combination of hydrogen and oxygen to form water, or they can be complex, like the synthesis of a protein molecule from thousands of simpler molecules. In some cases, such as the neutralization reaction that occurs in a swimming pool when acid is added to adjust the water's acidity level, chemical reactions are not noticeable to the naked eye. In other cases, such as the combustion reaction that produces a pillar of smoke and fire under the space shuttle during liftoff, chemical reactions are very obvious. In all cases, however, chemical reactions produce changes in the arrangements of the molecules and atoms that compose matter. Often, these molecular changes cause macroscopic changes that we can directly experience.

7.2 Evidence of a Chemical Reaction

If we could see the atoms and molecules that compose matter, we could easily identify a chemical reaction. Do atoms combine with other atoms to form compounds? Do new molecules form? Do the original molecules decompose? Do atoms in one molecule change places with atoms in another? If the answer to one or more of these questions is yes, a chemical reaction has occurred. Of course, we are not normally able to see atoms and molecules, so we need other ways to identify a chemical reaction.

Although we can't see atoms, many chemical reactions do produce easily detectable changes as they occur. For example, when the color-causing molecules in a brightly colored shirt decompose with repeated exposure to sunlight, the color of the shirt fades. Similarly, when the molecules embedded in the plastic of a child's temperature-sensitive spoon transform upon warming, the color of the spoon changes. These *color changes* are evidence that a chemical reaction has occurred.

Solid formation

Gas formation

▲ **FIGURE 7.3   A precipitation reaction**   The formation of a solid in a previously clear solution is evidence of a chemical reaction.

▲ **FIGURE 7.4   A gas evolution reaction**   The formation of a gas is evidence of a chemical reaction.

Other changes that identify chemical reactions include the *formation of a solid* (▶ Figure 7.3) or *the formation of a gas* (▶ Figure 7.4). Dropping Alka-Seltzer tablets into water or combining baking soda and vinegar (as in our opening example of the kindergarten volcano) are both good examples of chemical reactions that produce a gas—the gas is visible as bubbles in the liquid.

Heat absorption or *emission*, as well as *light emission*, are also evidence of reactions. For example, a natural gas flame produces heat and light. A chemical cold pack becomes cold when the plastic barrier separating two substances is broken. Both of these changes suggest that a chemical reaction is occurring.

Recall from Section 3.9 that a reaction that emits heat is an *exothermic* reaction and one that absorbs heat is an *endothermic* reaction.

Color change

Heat absorption

▲ A child's temperature-sensitive spoon changes color upon warming due to a reaction induced by the higher temperature.

▲ A change in temperature due to absorption or emission of heat is evidence of a chemical reaction. This chemical cold pack becomes cold when the barrier separating two substances is broken.

▲ **FIGURE 7.5 Boiling: a physical change** When water boils, bubbles are formed and a gas is evolved. However, no chemical change has occurred because the gas, like the liquid water, is also composed of water molecules.

While these changes provide evidence of a chemical reaction, they are not *definitive* evidence. Only chemical analysis showing that the initial substances have changed into other substances conclusively proves that a chemical reaction has occurred. We can be fooled. For example, when water boils, bubbles form, but no chemical reaction has occurred. Boiling water forms gaseous steam, but both water and steam are composed of water molecules—no chemical change has occurred (◀ Figure 7.5). On the other hand, chemical reactions may occur without any obvious signs, yet chemical analysis may show that a reaction has indeed occurred. The changes occurring at the atomic and molecular level determine whether a chemical reaction has occurred.

In summary, each of the following provides *evidence of a chemical reaction*.

- a *color change*

- the *formation of a solid* in a previously clear solution

- the *formation of a gas* when you add a substance to a solution

- the *emission of light*

- the *emission* or *absorption of heat*

EXAMPLE 7.1   Evidence of a Chemical Reaction

Which changes involve a chemical reaction? Explain your answers.

(a) ice melting upon warming
(b) an electric current passing through water, resulting in the formation of hydrogen and oxygen gas that appear as bubbles rising in the water
(c) iron rusting
(d) bubbles forming when a soda can is opened

SOLUTION

(a) not a chemical reaction; melting ice forms water, but both the ice and water are composed of water molecules.
(b) chemical reaction; water decomposes into hydrogen and oxygen, as evidenced by the bubbling.
(c) chemical reaction; iron changes into iron oxide, changing color in the process.
(d) not a chemical reaction; even though there is bubbling, it is just carbon dioxide coming out of the liquid.

▶**SKILLBUILDER 7.1 | Evidence of a Chemical Reaction**

Which changes involve a chemical reaction? Explain your answers.

(a) butane burning in a butane lighter
(b) butane evaporating out of a butane lighter
(c) wood burning
(d) dry ice subliming

▶**FOR MORE PRACTICE**   Example 7.16; Problems 25, 26, 27, 28, 29, 30.

✔ **CONCEPTUAL CHECKPOINT 7.1**

These images portray molecular views of one substance before and after a change. Determine whether a chemical reaction has occurred in each case.

(a)

(b)

(c)

7.3 The Chemical Equation

As we saw in Section 3.6, we represent chemical reactions with *chemical equations*. For example, the reaction occurring in a natural-gas flame, such as the flame on a kitchen stove, is methane (CH_4) reacting with oxygen (O_2) to form carbon dioxide (CO_2) and water (H_2O). We represent this reaction with the equation:

$$CH_4 + O_2 \longrightarrow CO_2 + H_2O$$
$$\text{reactants} \qquad\qquad \text{products}$$

The substances on the left side of the equation are the *reactants*, and the substances on the right side are the *products*. We often specify the state of each reactant or product in parentheses next to the formula. If we add states to our equation, it becomes:

$$CH_4(g) + O_2(g) \longrightarrow CO_2(g) + H_2O(g)$$

The (g) indicates that these substances are gases in the reaction. Table 7.1 summarizes the common states of reactants and products and their symbols used in chemical reactions.

Let's look more closely at the equation for the burning of natural gas. How many oxygen atoms are on each side of the equation?

The left side of the equation has two oxygen atoms, and the right side has three. Since chemical equations represent real chemical reactions, atoms cannot simply appear or disappear in chemical equations because, as we know, atoms don't simply appear or disappear in nature. We must account for the atoms on both sides of the equation. Notice also that the left side of the equation has four hydrogen atoms and the right side only two.

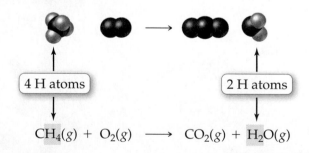

To correct these problems, we must create a **balanced equation**, one in which the numbers of each type of atom on both sides of the equation are equal. To balance an equation, we insert coefficients—not subscripts—in front of the chemical formulas as needed to make the number of each type of atom in the reactants equal to the number of each type of atom in the products. New atoms do not form during a reaction, nor do atoms vanish—matter must be conserved.

When we balance chemical equations by inserting coefficients as needed in front of the formulas of the reactants and products, it changes the number of molecules in the equation, but it does not change the *kinds* of molecules. To balance the

---

TABLE 7.1 Abbreviations Indicating the States of Reactants and Products in Chemical Equations

Abbreviation	State
(g)	gas
(l)	liquid
(s)	solid
(aq)	aqueous (water solution)*

*The (aq) designation stands for *aqueous*, which indicates that a substance is dissolved in water. When a substance dissolves in water, the mixture is called a *solution* (see Section 7.5).

In chemical equations, atoms cannot change from one type to another—hydrogen atoms cannot change into oxygen atoms, for example. Nor can atoms disappear (recall the law of conservation of mass from Section 3.7).

preceding equation, for example, we put the coefficient 2 before O_2 in the reactants, and the coefficient 2 before H_2O in the products.

$$CH_4(g) + 2\,O_2(g) \longrightarrow CO_2(g) + 2\,H_2O$$

The equation is now balanced because the numbers of each type of atom on both sides of the equation are equal. We can verify this by summing the number of each type of atom.

> The number of a particular type of atom within a chemical formula embedded in an equation is obtained by multiplying the subscript for the atom by the coefficient for the chemical formula.

If there is no coefficient or subscript, a 1 is implied. So, the balanced equation for the combustion of natural gas is:

$$CH_4(g) + 2\,O_2(g) \longrightarrow CO_2(g) + 2\,H_2O(g)$$

Reactants	Products
1 C atom (1 × <u>C</u>H₄)	1 C atom (1 × <u>C</u>O₂)
4 H atoms (1 × C<u>H</u>₄)	4 H atoms (2 × <u>H</u>₂O)
4 O atoms (2 × <u>O</u>₂)	4 O atoms (1 × C<u>O</u>₂ + 2 × H₂<u>O</u>)

The numbers of each type of atom on both sides of the equation are equal—the equation is balanced.

$$CH_4(g) + 2\,O_2(g) \longrightarrow CO_2(g) + 2\,H_2O(g)$$

▶ A balanced chemical equation represents a chemical reaction. In this image, methane molecules combine with oxygen to form carbon dioxide and water.

✔ CONCEPTUAL CHECKPOINT 7.2

In photosynthesis, plants make the sugar glucose, $C_6H_{12}O_6$, from carbon dioxide and water. The equation for the reaction is

$$6\,CO_2 + 6\,H_2O \longrightarrow C_6H_{12}O_6 + x\,O_2$$

In order for this equation to be balanced, the coefficient x must be

(a) 3   **(b)** 6   **(c)** 9   **(d)** 12

7.4 How to Write Balanced Chemical Equations

The following procedure box details the steps for writing balanced chemical equations. As in other procedures, we show the steps in the left column and examples of applying each step in the center and right columns. Remember, change only the *coefficients* to balance a chemical equation; *never change the subscripts*.

Writing Balanced Chemical Equations	EXAMPLE 7.2 Write a balanced equation for the reaction between solid silicon dioxide and solid carbon to produce solid silicon carbide and carbon monoxide gas.	EXAMPLE 7.3 Write a balanced equation for the combustion of liquid octane (C_8H_{18}), a component of gasoline, in which it combines with gaseous oxygen to form gaseous carbon dioxide and gaseous water.
1. Write a skeletal equation by writing chemical formulas for each of the reactants and products. Review Chapter 5 for nomenclature rules. (If a skeletal equation is provided, skip this step and go to Step 2.)	**SOLUTION** $$SiO_2(s) + C(s) \longrightarrow SiC(s) + CO(g)$$	**SOLUTION** $$C_8H_{18}(l) + O_2(g) \longrightarrow CO_2(g) + H_2O(g)$$
2. If an element occurs in only one compound on both sides of the equation, balance it first. If there is more than one such element, balance metals before nonmetals.	**Begin with Si** $$SiO_2(s) + C(s) \longrightarrow SiC(s) + CO(g)$$ **1 Si atom $\longrightarrow$ 1 Si atom** Si is already balanced. **Balance O next** $$SiO_2(s) + C(s) \longrightarrow SiC(s) + CO(g)$$ **2 O atoms $\longrightarrow$ 1 O atom** To balance O, put a 2 before CO(g). $$SiO_2(s) + C(s) \longrightarrow SiC(s) + 2\,CO(g)$$ **2 O atoms $\longrightarrow$ 2 O atoms**	**Begin with C** $$C_8H_{18}(l) + O_2(g) \longrightarrow CO_2(g) + H_2O(g)$$ **8 C atoms $\longrightarrow$ 1 C atom** To balance C, put an 8 before $CO_2(g)$. $$C_8H_{18}(l) + O_2(g) \longrightarrow 8\,CO_2(g) + H_2O(g)$$ **8 C atoms $\longrightarrow$ 8 C atoms** **Balance H next** $$C_8H_{18}(l) + O_2(g) \longrightarrow 8\,CO_2(g) + H_2O(g)$$ **18 H atoms $\longrightarrow$ 2 H atoms** To balance H, put a 9 before $H_2O(g)$. $$C_8H_{18}(l) + O_2(g) \longrightarrow 8\,CO_2(g) + 9\,H_2O(g)$$ **18 H atoms $\longrightarrow$ 18 H atoms**
3. If an element occurs as a free element on either side of the chemical equation, balance it last. Always balance free elements by adjusting the coefficient *on the free element*.	**Balance C** $$SiO_2(s) + C(s) \longrightarrow SiC(s) + 2\,CO(g)$$ **1 C atom $\longrightarrow$ 1 C + 2 C = 3 C atoms** To balance C, put a 3 before C(s). $$SiO_2(s) + 3\,C(s) \longrightarrow SiC(s) + 2\,CO(g)$$ **3 C atoms $\longrightarrow$ 1 C + 2 C = 3 C atoms**	**Balance O** $$C_8H_{18}(l) + O_2(g) \longrightarrow 8\,CO_2(g) + 9\,H_2O(g)$$ **2 O atoms $\longrightarrow$ 16 O + 9 O = 25 O atoms** To balance O, put a $\frac{25}{2}$ before $O_2(g)$. $$C_8H_{18}(l) + \tfrac{25}{2}\,O_2(g) \longrightarrow 8\,CO_2(g) + 9\,H_2O(g)$$ **25 O atoms $\longrightarrow$ 16 O + 9 O = 25 O atoms**
4. If the balanced equation contains coefficient fractions, change these into whole numbers by multiplying the entire equation by the appropriate factor.	This step is not necessary in this example. Proceed to Step 5.	$$[C_8H_{18}(l) + \tfrac{25}{2}\,O_2(g) \longrightarrow 8\,CO_2(g) + 9\,H_2O(g)] \times 2$$ $$2\,C_8H_{18}(l) + 25\,O_2(g) \longrightarrow 16\,CO_2(g) + 18\,H_2O(g)$$

5. Check to make certain the equation is balanced by summing the total number of each type of atom on both sides of the equation.

$$SiO_2(s) + 3\,C(s) \longrightarrow SiC(s) + 2\,CO(g)$$

Reactants		Products
1 Si atom	$\longrightarrow$	1 Si atom
2 O atoms	$\longrightarrow$	2 O atoms
3 C atoms	$\longrightarrow$	3 C atoms

The equation is balanced.

▶SKILLBUILDER 7.2

Write a balanced equation for the reaction between solid chromium(III) oxide and solid carbon to produce solid chromium and carbon dioxide gas.

$$2\,C_8H_{18}(l) + 25\,O_2(g) \longrightarrow 16\,CO_2(g) + 18\,H_2O(g)$$

Reactants		Products
16 C atoms	$\longrightarrow$	16 C atoms
36 H atoms	$\longrightarrow$	36 H atoms
50 O atoms	$\longrightarrow$	50 O atoms

The equation is balanced.

▶SKILLBUILDER 7.3

Write a balanced equation for the combustion of gaseous C_4H_{10} in which it combines with gaseous oxygen to form gaseous carbon dioxide and gaseous water.

▶FOR MORE PRACTICE   Example 7.17; Problems 33, 34, 35, 36, 37, 38.

EXAMPLE 7.4   Balancing Chemical Equations

Write a balanced equation for the reaction of solid aluminum with aqueous sulfuric acid to form aqueous aluminum sulfate and hydrogen gas.

Use your knowledge of chemical nomenclature from Chapter 5 to write a skeletal equation containing formulas for each of the reactants and products. The formulas for each compound MUST BE CORRECT before you begin to balance the equation.	**SOLUTION** $Al(s) + H_2SO_4(aq) \longrightarrow Al_2(SO_4)_3(aq) + H_2(g)$
Since both aluminum and hydrogen occur as pure elements, balance those last. Sulfur and oxygen occur in only one compound on each side of the equation, so balance these first. Sulfur and oxygen are also part of a polyatomic ion that stays intact on both sides of the equation. *Balance polyatomic ions such as these as a unit.* There are $3\,SO_4^{2-}$ ions on the right side of the equation, so put a 3 in front of H_2SO_4.	$Al(s) + \mathbf{3}\,H_2SO_4(aq) \longrightarrow Al_2(SO_4)_3(aq) + H_2(g)$
Balance Al next. Since there are 2 Al atoms on the right side of the equation, place a 2 in front of Al on the left side of the equation.	$\mathbf{2}\,Al(s) + 3\,H_2SO_4(aq) \longrightarrow Al_2(SO_4)_3(aq) + H_2(g)$
Balance H next. Since there are 6 H atoms on the left side, place a 3 in front of $H_2(g)$ on the right side.	$2\,Al(s) + 3\,H_2SO_4(aq) \longrightarrow Al_2(SO_4)_3(aq) + \mathbf{3}\,H_2(g)$
Finally, sum the number of atoms on each side to make sure that the equation is balanced.	$2\,Al(s) + 3\,H_2SO_4(aq) \longrightarrow Al_2(SO_4)_3(aq) + 3\,H_2(g)$

Reactants		Products
2 Al atoms	$\longrightarrow$	2 Al atoms
6 H atoms	$\longrightarrow$	6 H atoms
3 S atoms	$\longrightarrow$	3 S atoms
12 O atoms	$\longrightarrow$	12 O atoms

▶SKILLBUILDER 7.4 | Balancing Chemical Equations

Write a balanced equation for the reaction of aqueous lead(II) acetate with aqueous potassium iodide to form solid lead(II) iodide and aqueous potassium acetate.

▶FOR MORE PRACTICE   Problems 39, 40, 41, 42, 43, 44.

EXAMPLE 7.5   Balancing Chemical Equations

Balance this chemical equation.

$$Fe(s) + HCl(aq) \longrightarrow FeCl_3(aq) + H_2(g)$$

	SOLUTION
Since Cl occurs in only one compound on each side of the equation, balance it first. There is 1 Cl atom on the left side of the equation and 3 Cl atoms on the right side. To balance Cl, place a 3 in front of HCl.	$Fe(s) + \mathbf{3} HCl(aq) \longrightarrow FeCl_3(aq) + H_2(g)$
Since H and Fe occur as free elements, balance them last. There is 1 Fe atom on the left side of the equation and 1 Fe atom on the right, so Fe is balanced. There are 3 H atoms on the left and 2 H atoms on the right. Balance H by placing a $\frac{3}{2}$ in front of H_2. (That way you don't alter other elements that are already balanced.)	$Fe(s) + 3 HCl(aq) \longrightarrow FeCl_3(aq) + \frac{3}{2} H_2(g)$
Since the equation now contains a coefficient fraction, clear it by multiplying the entire equation (both sides) by 2.	$[Fe(s) + 3 HCl(aq) \longrightarrow FeCl_3(aq) + \frac{3}{2} H_2(g)] \times 2$ $2 Fe(s) + 6 HCl(aq) \longrightarrow 2 FeCl_3(aq) + 3 H_2(g)$
Finally, sum the number of atoms on each side to check that the equation is balanced.	$2 Fe(s) + 6 HCl(aq) \longrightarrow 2 FeCl_3(aq) + 3 H_2(g)$

Reactants		Products
2 Fe atoms	$\longrightarrow$	2 Fe atoms
6 Cl atoms	$\longrightarrow$	6 Cl atoms
6 H atoms	$\longrightarrow$	6 H atoms

▶**SKILLBUILDER 7.5 | Balancing Chemical Equations**

Balance this chemical equation.

$$HCl(g) + O_2(g) \longrightarrow H_2O(l) + Cl_2(g)$$

▶**FOR MORE PRACTICE**   Problems 45, 46, 47, 48, 49, 50.

 CONCEPTUAL CHECKPOINT 7.3

Which quantity must always be the same on both sides of a balanced chemical equation?

(a) the number of each type of atom

(b) the number of each type of molecule

(c) the sum of all of the coefficients

7.5 Aqueous Solutions and Solubility: Compounds Dissolved in Water

Reactions occurring in aqueous solution are among the most common and important. An **aqueous solution** is a homogeneous mixture of a substance with water. For example, a sodium chloride (NaCl) solution, also called a saline solution, is composed of sodium chloride dissolved in water. Sodium chloride

A sodium chloride solution contains independent **Na$^+$** and **Cl$^-$** ions.

A silver nitrate solution contains independent **Ag$^+$** and **NO$_3$$^-$** ions.

When silver chloride is added to water, it remains as solid AgCl—it does not dissolve into independent ions.

solutions are common both in the oceans and in living cells. You can form a sodium chloride solution yourself by adding table salt to water. As you stir the salt into the water, it seems to disappear. However, you know the salt is still there because if you taste the water, it has a salty flavor. How does sodium chloride dissolve in water?

When ionic compounds such as NaCl dissolve in water, they usually dissociate into their component ions. A sodium chloride solution, represented as NaCl(*aq*), does not contain any NaCl units; only dissolved Na$^+$ ions and Cl$^-$ ions are present.

We know that NaCl is present as independent sodium and chloride ions in solution because sodium chloride solutions conduct electricity, which requires the presence of freely moving charged particles. Substances (such as NaCl) that completely dissociate into ions in solution are called *strong electrolytes* and the resultant solutions are called **strong electrolyte solutions** (▼ Figure 7.6). Similarly, a silver nitrate solution, represented as AgNO$_3$(*aq*), does not contain any AgNO$_3$ units, but only dissolved Ag$^+$ ions and NO$_3$$^-$ ions. It, too, is a strong electrolyte solution. When compounds containing polyatomic ions such as NO$_3$$^-$ dissolve, the polyatomic ions dissolve as intact units.

Not all ionic compounds, however, dissolve in water. AgCl, for example, does not. If we add AgCl to water, it remains as solid AgCl and appears as a white solid at the bottom of the beaker.

▲ **FIGURE 7.6   Ions as conductors**   **(a)** Pure water does not conduct electricity. **(b)** Ions in a sodium chloride solution conduct electricity, causing the bulb to light. Solutions such as NaCl are called strong electrolyte solutions.

SOLUBILITY

A compound is **soluble** in a particular liquid if it dissolves in that liquid; a compound is **insoluble** if it does not dissolve in the liquid. NaCl, for example, is soluble in water. If we mix solid sodium chloride into water, it dissolves and forms a strong electrolyte solution. AgCl, on the other hand, is insoluble in water. If we mix solid silver chloride into water, it remains as a solid within the liquid water.

There is no easy way to tell whether a particular compound will be soluble or insoluble in water. For ionic compounds, however, empirical rules have been deduced from observations of many compounds. These **solubility rules** are summarized in Table 7.2 and ▼ Figure 7.7. For example, the solubility rules indicate that compounds containing the lithium ion are *soluble*. That means that compounds such as LiBr, $LiNO_3$, Li_2SO_4, LiOH, and Li_2CO_3 all dissolve in water to form strong electrolyte solutions. If a compound contains Li^+, it is soluble. Similarly, the solubility rules state that compounds containing the NO_3^- ion are soluble. Therefore, compounds such as $AgNO_3$, $Pb(NO_3)_2$, $NaNO_3$, $Ca(NO_3)_2$ and $Sr(NO_3)_2$ all dissolve in water to form strong electrolyte solutions.

The solubility rules also state that, with some exceptions, compounds containing the CO_3^{2-} ion are *insoluble*. Therefore, compounds such as $CuCO_3$, $CaCO_3$, $SrCO_3$, and $FeCO_3$ do not dissolve in water. Note that the solubility rules contain many exceptions. For example, compounds containing CO_3^{2-} are *soluble when paired with* Li^+, Na^+, K^+, or NH_4^+. Thus Li_2CO_3, Na_2CO_3, K_2CO_3, and $(NH_4)_2CO_3$ are all soluble.

The solubility rules apply only to the solubility of the compounds in water.

▼ **FIGURE 7.7** Solubility rules flowchart.

Compounds that are mostly soluble

Compounds containing Li^+, Na^+, K^+, NH_4^+, NO_3^-, or $C_2H_3O_2^-$ → soluble

Compounds containing Cl^-, Br^-, or I^- → Does the compound also contain Ag^+, Hg_2^{2+}, or Pb^{2+}? — no → soluble / yes → insoluble

Compounds containing SO_4^{2-} → Does the compound also contain Sr^{2+}, Ba^{2+}, Pb^{2+}, or Ca^{2+}? — no → soluble / yes → insoluble

Compounds that are mostly insoluble

Compounds containing OH^- → Does the compound also contain Li^+, Na^+, K^+, or NH_4^+? — no → Does the compound also contain Ca^{2+}, Sr^{2+}, or Ba^{2+}? — no → insoluble / yes → slightly soluble — yes → soluble

Compounds containing S^{2-} → Does the compound also contain Li^+, Na^+, K^+, NH_4^+, Ca^{2+}, Sr^{2+}, or Ba^{2+}? — no → insoluble / yes → soluble

Compounds containing CO_3^{2-} or PO_4^{3-} → Does the compound also contain Li^+, Na^+, K^+, or NH_4^+? — no → insoluble / yes → soluble

TABLE 7.2 Solubility Rules

Compounds Containing the Following Ions Are Mostly Soluble	Exceptions
Li^+, Na^+, K^+, NH_4^+	None
NO_3^-, $C_2H_3O_2^-$	None
Cl^-, Br^-, I^-	When any of these ions pairs with Ag^+, Hg_2^{2+}, or Pb^{2+}, the compound is insoluble.
SO_4^{2-}	When SO_4^{2-} pairs with Sr^{2+}, Ba^{2+}, Pb^{2+}, or Ca^{2+}, the compound is insoluble.

Compounds Containing the Following Ions Are Mostly Insoluble	Exceptions
OH^-, S^{2-}	When either of these ions pairs with Li^+, Na^+, K^+, or NH_4^+ the compound is soluble. When S^{2-} pairs with Ca^{2+}, Sr^{2+}, or Ba^{2+}, the compound is soluble. When OH^- pairs with Ca^{2+}, Sr^{2+}, or Ba^{2+}, the compound is slightly soluble.*
CO_3^{2-}, PO_4^{3-}	When either of these ions pairs with Li^+, Na^+, K^+, or NH_4^+, the compound is soluble.

*For many purposes these can be considered insoluble.

EXAMPLE 7.6 Determining Whether a Compound Is Soluble

Determine whether each compound is soluble or insoluble.

(a) AgBr   **(b)** $CaCl_2$ **(c)** $Pb(NO_3)_2$ **(d)** $PbSO_4$

SOLUTION

(a) Insoluble; compounds containing Br^- are normally soluble, but Ag^+ is an exception.
(b) Soluble; compounds containing Cl^- are normally soluble, and Ca^{2+} is not an exception.
(c) Soluble; compounds containing NO_3^- are always soluble.
(d) Insoluble; compounds containing SO_4^{2-} are normally soluble, but Pb^{2+} is an exception.

▶SKILLBUILDER 7.6 | **Determining Whether a Compound Is Soluble**

Determine whether each compound is soluble or insoluble.
(a) CuS   **(b)** $FeSO_4$ **(c)** $PbCO_3$ **(d)** NH_4Cl

▶**FOR MORE PRACTICE**   Example 7.18; Problems 57, 58, 59, 60, 61, 62.

✔ **CONCEPTUAL CHECKPOINT 7.4**

Which image best depicts a mixture of $BaCl_2$ and water?

(a)   (b)   (c)

7.6 Precipitation Reactions: Reactions in Aqueous Solution That Form a Solid

Recall from Section 7.1 that sodium carbonate in laundry detergent reacts with dissolved Mg^{2+} and Ca^{2+} ions to form solids that precipitate (or come out of) solution. These reactions are examples of **precipitation reactions**—reactions that form a solid, called a **precipitate**, upon mixing two aqueous solutions.

Precipitation reactions are common in chemistry. Potassium iodide and lead nitrate, for example, both form colorless, strong electrolyte solutions when dissolved in water (see the solubility rules). When the two solutions are combined, however, a brilliant yellow precipitate forms (▼ Figure 7.8). We can describe this precipitation reaction with the chemical equation:

$$2 KI(aq) + Pb(NO_3)_2(aq) \longrightarrow PbI_2(s) + 2 KNO_3(aq)$$

Precipitation reactions do not always occur when mixing two aqueous solutions. For example, when solutions of $KI(aq)$ and $NaCl(aq)$ are combined, nothing happens (▼ Figure 7.9).

$$KI(aq) + NaCl(aq) \longrightarrow NO\ REACTION$$

$$2 KI(aq) + Pb(NO_3)_2(aq) \longrightarrow PbI_2(s) + 2 KNO_3(aq)$$

▲ **FIGURE 7.8** **Precipitation** When a potassium iodide solution is mixed with a lead(II) nitrate solution, a brilliant yellow precipitate of $PbI_2(s)$ forms.

$$KI(aq) + NaCl(aq) \longrightarrow NO\ REACTION$$

▲ **FIGURE 7.9** **No reaction** When a potassium iodide solution is mixed with a sodium chloride solution, no reaction occurs.

PREDICTING PRECIPITATION REACTIONS

The key to predicting precipitation reactions is understanding that *only insoluble compounds form precipitates.* In a precipitation reaction, two solutions containing soluble compounds combine and an insoluble compound precipitates. Consider the precipitation reaction from Figure 7.8.

$$2 KI(aq) + Pb(NO_3)_2(aq) \longrightarrow PbI_2(s) + 2 KNO_3(aq)$$
soluble   soluble   insoluble   soluble

KI and $Pb(NO_3)_2$ are both soluble, but the precipitate, PbI_2, is *insoluble*. Before mixing, $KI(aq)$ and $Pb(NO_3)_2(aq)$ are both dissociated in their respective solutions.

The instant that the solutions are mixed, all four ions are present.

However, new compounds—potentially insoluble ones—are now possible. Specifically, the cation from one compound can now pair with the anion from the other compound to form new (and potentially insoluble) products.

If the *potentially insoluble* products are both *soluble*, then no reaction occurs. If, on the other hand, one or both of the potentially insoluble products are *indeed insoluble*, a precipitation reaction occurs. In this case, KNO_3 is soluble, but PbI_2 is insoluble. Consequently, PbI_2 precipitates.

To predict whether a precipitation reaction will occur when two solutions are mixed and to write an equation for the reaction, follow the steps in the procedure box. As usual, the steps are shown in the left column, and two examples of applying the procedure are shown in the center and right columns.

Writing Equations for Precipitation Reactions	**EXAMPLE 7.7** Write an equation for the precipitation reaction that occurs (if any) when solutions of sodium carbonate and copper(II) chloride are mixed.	**EXAMPLE 7.8** Write an equation for the precipitation reaction that occurs (if any) when solutions of lithium nitrate and sodium sulfate are mixed.
1. Write the formulas of the two compounds being mixed as reactants in a chemical equation.	**SOLUTION** $Na_2CO_3(aq) + CuCl_2(aq) \longrightarrow$	**SOLUTION** $LiNO_3(aq) + Na_2SO_4(aq) \longrightarrow$
2. Below the equation, write the formulas of the potentially insoluble products that could form from the reactants. Obtain these by combining the cation from one reactant with the anion from the other. Make sure to write correct (charge neutral) formulas for these ionic compounds as described in Section 5.5.	$Na_2CO_3(aq) + CuCl_2(aq) \longrightarrow$ **Potentially Insoluble Products** NaCl $CuCO_3$	$LiNO_3(aq) + Na_2SO_4(aq) \longrightarrow$ **Potentially Insoluble Products** $NaNO_3$ Li_2SO_4
3. Use the solubility rules to determine whether any of the potential new products are indeed insoluble.	NaCl is *soluble* (compounds containing Cl^- are usually soluble, and Na^+ is not an exception). $CuCO_3$ is *insoluble* (compounds containing CO_3^{2-} are usually insoluble, and Cu^{2+} is not an exception).	$NaNO_3$ is *soluble* (compounds containing NO_3^- are soluble and Na^+ is not an exception). Li_2SO_4 is *soluble* (compounds containing SO_4^{2-} are soluble and Li^+ is not an exception).
4. If all of the potentially insoluble products are soluble, there will be no precipitate. Write NO REACTION next to the arrow.	Since this example has an insoluble product, proceed to the next step.	$LiNO_3(aq) + Na_2SO_4(aq) \longrightarrow$ NO REACTION
5. If one or both of the potentially insoluble products are insoluble, write their formula(s) as the product(s) of the reaction, using (s) to indicate solid. Write any soluble products with (aq) to indicate aqueous.	$Na_2CO_3(aq) + CuCl_2(aq) \longrightarrow$ $CuCO_3(s) + NaCl(aq)$	
6. Balance the equation. Remember to adjust only coefficients, not subscripts.	$Na_2CO_3(aq) + CuCl_2(aq) \longrightarrow$ $CuCO_3(s) + \mathbf{2}\,NaCl(aq)$	
	▶**SKILLBUILDER 7.7** Write an equation for the precipitation reaction that occurs (if any) when solutions of potassium hydroxide and nickel(II) bromide are mixed.	▶**SKILLBUILDER 7.8** Write an equation for the precipitation reaction that occurs (if any) when solutions of ammonium chloride and iron(III) nitrate are mixed.
		▶**FOR MORE PRACTICE** Example 7.19; Problems 63, 64, 65, 66.

EXAMPLE 7.9   Predicting and Writing Equations for Precipitation Reactions

Write an equation for the precipitation reaction (if any) that occurs when solutions of lead(II) acetate and sodium sulfate are mixed. If no reaction occurs, write *NO REACTION*.

1. Write the formulas of the two compounds being mixed as reactants in a chemical equation.	**SOLUTION** $Pb(C_2H_3O_2)_2(aq) + Na_2SO_4(aq) \longrightarrow$
2. Below the equation, write the formulas of the potentially insoluble products that could form from the reactants. These are obtained by combining the cation from one reactant with the anion from the other. Make sure to adjust the subscripts so that all formulas are charge-neutral.	 **Potentially insoluble products** $NaC_2H_3O_2$ $PbSO_4$
3. Use the solubility rules to determine whether any of the potentially insoluble products are indeed insoluble.	$NaC_2H_3O_2$ is *soluble* (compounds containing Na^+ are always soluble). $PbSO_4$ is *insoluble* (compounds containing SO_4^{2-} are normally soluble, but Pb^{2+} is an exception).
4. If all of the potentially insoluble products are soluble, there will be no precipitate. Write *NO REACTION* next to the arrow.	Since this reaction has an insoluble product, proceed to the next step.
5. If one or both of the potentially insoluble products are indeed insoluble, write their formula(s) as the product(s) of the reaction, using (*s*) to indicate solid. Write any soluble products with (*aq*) to indicate aqueous.	$Pb(C_2H_3O_2)_2(aq) + Na_2SO_4(aq) \longrightarrow PbSO_4(s) + NaC_2H_3O_2(aq)$
6. Balance the equation.	$Pb(C_2H_3O_2)_2(aq) + Na_2SO_4(aq) \longrightarrow PbSO_4(s) + \mathbf{2}\,NaC_2H_3O_2(aq)$

▶**SKILLBUILDER 7.9 | Predicting and Writing Equations for Precipitation Reactions**

Write an equation for the precipitation reaction (if any) that occurs when solutions of potassium sulfate and strontium nitrate are mixed. If no reaction occurs, write *NO REACTION*.

▶**FOR MORE PRACTICE**   Problems 67, 68.

✓ **CONCEPTUAL CHECKPOINT 7.5**

Which of these reactions would result in the formation of a precipitate?

(a) $NaNO_3(aq) + CaS(aq)$

(b) $MgSO_4(aq) + CaS(aq)$

(c) $NaNO_3(aq) + MgSO_4(aq)$

7.7 Writing Chemical Equations for Reactions in Solution: Molecular, Complete Ionic, and Net Ionic Equations

Consider the following equation for a precipitation reaction.

$$AgNO_3(aq) + NaCl(aq) \longrightarrow AgCl(s) + NaNO_3(aq)$$

This equation is written as a **molecular equation**, an equation showing the complete neutral formulas for every compound in the reaction. Equations for reactions occurring in aqueous solution may also be written to show that aqueous ionic compounds normally dissociate in solution. For example, the previous equation can be written as:

$$Ag^+(aq) + NO_3^-(aq) + Na^+(aq) + Cl^-(aq) \longrightarrow AgCl(s) + Na^+(aq) + NO_3^-(aq)$$

When writing complete ionic equations, separate only aqueous ionic compounds into their constituent ions. Do NOT separate solid, liquid, or gaseous compounds.

Equations such as this one, showing the reactants and products as they are actually present in solution, are called **complete ionic equations**.

Notice that in the complete ionic equation, some of the ions in solution appear unchanged on both sides of the equation. These ions are called **spectator ions** because they do not participate in the reaction.

Spectator ions

To simplify the equation, and to more clearly show what is happening, spectator ions can be omitted.

$$Ag^+(aq) + Cl^-(aq) \longrightarrow AgCl(s)$$

Species refers to a kind or sort of thing. In this case, the species are all the different molecules and ions that are present during the reaction.

Equations such as this one, which show only the *species* that actually participate in the reaction, are called **net ionic equations**.

As another example, consider the reaction between HCl(*aq*) and NaOH(*aq*).

$$HCl(aq) + NaOH(aq) \longrightarrow H_2O(l) + NaCl(aq)$$

HCl, NaOH, and NaCl exist in solution as independent ions. The *complete ionic equation* for this reaction is:

$$H^+(aq) + Cl^-(aq) + Na^+(aq) + OH^-(aq) \longrightarrow H_2O(l) + Na^+(aq) + Cl^-(aq)$$

To write the *net ionic equation*, we remove the spectator ions, those that are unchanged on both sides of the equation.

Spectator ions

The net ionic equation is $H^+(aq) + OH^-(aq) \longrightarrow H_2O(l)$

To summarize:

- A molecular equation is a chemical equation showing the complete, neutral formulas for every compound in a reaction.
- A complete ionic equation is a chemical equation showing all of the species as they are actually present in solution.
- A net ionic equation is an equation showing only the species that actually participate in the reaction.

EXAMPLE 7.10   Writing Complete Ionic and Net Ionic Equations

Consider this precipitation reaction occurring in aqueous solution.

$$Pb(NO_3)_2(aq) + 2\ LiCl(aq) \longrightarrow PbCl_2(s) + 2\ LiNO_3(aq)$$

Write a complete ionic equation and a net ionic equation for the reaction.

| Write the complete ionic equation by separating aqueous ionic compounds into their constituent ions. The $PbCl_2(s)$ remains as one unit. | **SOLUTION**
Complete ionic equation
$$Pb^{2+}(aq) + 2\,NO_3^-(aq) + 2\,Li^+(aq) + 2\,Cl^-(aq) \longrightarrow$$ $$PbCl_2(s) + 2\,Li^+(aq) + 2\,NO_3^-(aq)$$ |
| Write the net ionic equation by eliminating the spectator ions, those that do not change during the reaction. | **Net ionic equation**
$$Pb^{2+}(aq) + 2\,Cl^-(aq) \longrightarrow PbCl_2(s)$$ |

▶**SKILLBUILDER 7.10 | Writing Complete Ionic and Net Ionic Equations**

Consider this reaction occurring in aqueous solution.

$$2\,HBr(aq) + Ca(OH)_2(aq) \longrightarrow 2\,H_2O(l) + CaBr_2(aq)$$

Write a complete ionic equation and net ionic equation for the reaction.

▶**FOR MORE PRACTICE**   Example 7.20; Problems 69, 70, 71, 72.

7.8   Acid–Base and Gas Evolution Reactions

Basic

▲ Milk of magnesia is basic and tastes bitter.

| Even though coffee itself is acidic overall, it contains some naturally occurring bases (such as caffeine) that give it a bitter taste.

Two other kinds of reactions that occur in solution are **acid–base reactions**—reactions that form water upon mixing of an acid and a base—and **gas evolution reactions**—reactions that evolve a gas. Like precipitation reactions, these reactions occur when the cation of one reactant combines with the anion of another. As we will see in the next section, many gas evolution reactions also happen to be acid–base reactions.

ACID–BASE (NEUTRALIZATION) REACTIONS

As we saw in Chapter 5, an acid is a compound characterized by its sour taste, its ability to dissolve some metals, and its tendency to form H^+ ions in solution. A base is a compound characterized by its bitter taste, its slippery feel, and its tendency to form OH^- ions in solution. Some common acids and bases are listed in Table 7.3. Acids and bases are also found in many everyday substances. Foods such as lemons, limes, and vinegar contain acids. Soap, coffee, and milk of magnesia all contain bases.

When an acid and base are mixed, the $H^+(aq)$ from the acid combines with the $OH^-(aq)$ from the base to form $H_2O(l)$. Consider the reaction between hydrochloric acid and sodium hydroxide mentioned earlier.

$$\underset{\text{Acid}}{HCl(aq)} + \underset{\text{Base}}{NaOH(aq)} \longrightarrow \underset{\text{Water}}{H_2O(l)} + \underset{\text{Salt}}{NaCl(aq)}$$

TABLE 7.3   Some Common Acids and Bases

Acid	Formula	Base	Formula
hydrochloric acid	HCl	sodium hydroxide	NaOH
hydrobromic acid	HBr	lithium hydroxide	LiOH
nitric acid	HNO_3	potassium hydroxide	KOH
sulfuric acid	H_2SO_4	calcium hydroxide	$Ca(OH)_2$
perchloric acid	$HClO_4$	barium hydroxide	$Ba(OH)_2$
acetic acid	$HC_2H_3O_2$		

Acidic

▲ Common foods and everyday substances such as oranges, lemons, vinegar, and vitamin C contain acids.

Acid–base reactions (also called **neutralization reactions**) generally form water and an ionic compound—called a **salt**—that usually remains dissolved in the solution. The net ionic equation for many acid–base reactions is:

$$H^+(aq) + OH^-(aq) \longrightarrow H_2O(l)$$

Another example of an acid–base reaction is the reaction that occurs between sulfuric acid and potassium hydroxide.

$$\underset{\text{acid}}{H_2SO_4(aq)} + \underset{\text{base}}{2\,KOH} \longrightarrow \underset{\text{water}}{2\,H_2O(l)} + \underset{\text{salt}}{K_2SO_4(aq)}$$

Notice the pattern of acid and base reacting to form water and a salt.

$$\text{Acid} + \text{Base} \longrightarrow \text{Water} + \text{Salt} \qquad \text{(acid–base reactions)}$$

When writing equations for acid–base reactions, write the formula of the salt using the procedure for writing formulas of ionic compounds given in Section 5.5.

EXAMPLE 7.11 Writing Equations for Acid–Base Reactions

Write a molecular and net ionic equation for the reaction between aqueous HNO_3 and aqueous $Ca(OH)_2$.

You must recognize these substances as an acid and a base. Write the skeletal reaction following the general pattern of acid plus base goes to water plus salt.	**SOLUTION** $$\underset{\text{acid}}{HNO_3(aq)} + \underset{\text{base}}{Ca(OH)_2(aq)} \longrightarrow \underset{\text{water}}{H_2O(l)} + \underset{\text{salt}}{Ca(NO_3)_2(aq)}$$
Next, balance the equation.	$$2\,HNO_3(aq) + Ca(OH)_2(aq) \longrightarrow 2\,H_2O(l) + Ca(NO_3)_2(aq)$$
Write the net ionic equation by eliminating those ions that remain the same on both sides of the equation.	$$2\,H^+(aq) + 2\,OH^-(aq) \longrightarrow 2\,H_2O(l)$$ or simply $$H^+(aq) + OH^-(aq) \longrightarrow H_2O(l)$$

▶ **SKILLBUILDER 7.11 | Writing Equations for Acid–Base Reactions**

Write a molecular and net ionic equation for the reaction that occurs between aqueous H_2SO_4 and aqueous KOH.

▶ **FOR MORE PRACTICE** Example 7.21; Problems 77, 78, 79, 80.

GAS EVOLUTION REACTIONS

Some aqueous reactions form a gas as a product. These reactions, as we learned in the opening section of this chapter, are called gas evolution reactions. Some gas evolution reactions form a gaseous product directly when the cation of one reactant reacts with the anion of the other. For example, when sulfuric acid reacts with lithium sulfide, dihydrogen sulfide gas is formed.

$$\underset{}{H_2SO_4(aq)} + Li_2S(aq) \longrightarrow \underset{\text{Gas}}{H_2S(g)} + Li_2SO_4(aq)$$

Many gas evolution reactions such as this one are also acid–base reactions. In Chapter 14 we learn how ions such as HCO_3^- act as bases in aqueous solution.

Other gas evolution reactions form an intermediate product that then decomposes into a gas. For example, when aqueous hydrochloric acid is mixed with aqueous sodium bicarbonate, the following reaction occurs.

$$HCl(aq) + NaHCO_3(aq) \longrightarrow H_2CO_3(aq) + NaCl(aq) \longrightarrow H_2O(l) + \underset{\text{Gas}}{CO_2(g)} + NaCl(aq)$$

Gas evolution reaction

▲ In this gas evolution reaction, vinegar (a dilute solution of acetic acid) and baking soda (sodium bicarbonate) produce carbon dioxide.

The intermediate product, H_2CO_3, is not stable and decomposes to form H_2O and gaseous CO_2. This reaction is almost identical to the reaction in the kindergarten volcano of Section 7.1, which involves the mixing of acetic acid and sodium bicarbonate.

$$HC_2H_3O_2(aq) + NaHCO_3(aq) \longrightarrow H_2CO_3(aq) + NaC_2H_3O_2(aq) \longrightarrow$$
$$H_2O(l) + CO_2(g) + NaC_2H_3O_2(aq)$$

The bubbling is caused by the newly formed carbon dioxide gas. Other important gas evolution reactions form either H_2SO_3 or NH_4OH as intermediate products.

$$HCl(aq) + NaHSO_3(aq) \longrightarrow H_2SO_3(aq) + NaCl(aq) \longrightarrow$$
$$H_2O(l) + SO_2(g) + NaCl(aq)$$

$$NH_4Cl(aq) + NaOH(aq) \longrightarrow NH_4OH(aq) + NaCl(aq) \longrightarrow$$
$$H_2O(l) + NH_3(g) + NaCl(aq)$$

The main types of compounds that form gases in aqueous reactions, as well as the gases that they form, are listed in Table 7.4.

TABLE 7.4   Types of Compounds That Undergo Gas Evolution Reactions

Reactant Type	Intermediate Product	Gas Evolved	Example
sulfides	none	H_2S	$2\,HCl(aq) + K_2S(aq) \longrightarrow H_2S(g) + 2\,KCl(aq)$
carbonates and bicarbonates	H_2CO_3	CO_2	$2\,HCl(aq) + K_2CO_3(aq) \longrightarrow H_2O(l) + CO_2(g) + 2\,KCl(aq)$
sulfites and bisulfites	H_2SO_3	SO_2	$2\,HCl(aq) + K_2SO_3(aq) \longrightarrow H_2O(l) + SO_2(g) + 2\,KCl(aq)$
ammonium	NH_4OH	NH_3	$NH_4Cl(aq) + KOH(aq) \longrightarrow H_2O(l) + NH_3(g) + KCl(aq)$

EXAMPLE 7.12   Writing Equations for Gas Evolution Reactions

Write a molecular equation for the gas evolution reaction that occurs when you mix aqueous nitric acid and aqueous sodium carbonate.

Begin by writing a skeletal equation that includes the reactants and products that form when the cation of each reactant combines with the anion of the other.	**SOLUTION** $$HNO_3(aq) + Na_2CO_3(aq) \longrightarrow H_2CO_3(aq) + NaNO_3(aq)$$
You must recognize that $H_2CO_3(aq)$ decomposes into $H_2O(l)$ and $CO_2(g)$ and write the corresponding equation.	$$HNO_3(aq) + Na_2CO_3(aq) \longrightarrow H_2O(l) + CO_2(g) + NaNO_3(aq)$$
Finally, balance the equation.	$$2\,HNO_3(aq) + Na_2CO_3(aq) \longrightarrow H_2O(l) + CO_2(g) + 2\,NaNO_3(aq)$$

▶SKILLBUILDER 7.12 | **Writing Equations for Gas Evolution Reactions**

Write a molecular equation for the gas evolution reaction that occurs when you mix aqueous hydrobromic acid and aqueous potassium sulfite.

▶SKILLBUILDER PLUS

Write a net ionic equation for the previous reaction.

▶FOR MORE PRACTICE   Example 7.22; Problems 81, 82.

CHEMISTRY AND HEALTH

Neutralizing Excess Stomach Acid

Your stomach normally contains acids that are involved in food digestion. Certain foods and stress, however, can increase the acidity of your stomach to uncomfortable levels, causing acid stomach or heartburn. Antacids are over-the-counter medicines that work by reacting with and neutralizing stomach acid. Antacids employ different bases as neutralizing agents. Tums™, for example, contains $CaCO_3$; milk of magnesia contains $Mg(OH)_2$; and Mylanta™ contains $Al(OH)_3$. They all, however, have the same effect of neutralizing stomach acid and relieving heartburn.

CAN YOU ANSWER THIS? *Assume that stomach acid is HCl and write equations showing how each of these antacids neutralizes stomach acid.*

▲ The base in an antacid neutralizes excess stomach acid, relieving heartburn and acid stomach.

▲ Antacids contain bases such $Mg(OH)_2$, $Al(OH)_3$, and $NaHCO_3$.

7.9 Oxidation–Reduction Reactions

We will cover oxidation–reduction reactions in more detail in Chapter 16.

Reactions involving the transfer of electrons are called **oxidation–reduction reactions** or **redox reactions**. Redox reactions are responsible for the rusting of iron, the bleaching of hair, and the production of electricity in batteries. Many redox reactions involve the reaction of a substance with oxygen.

$$2 H_2(g) + O_2(g) \longrightarrow 2 H_2O(g)$$
(reaction that powers the space shuttle)

$$4 Fe(s) + 3 O_2(g) \longrightarrow 2 Fe_2O_3(s)$$
(rusting of iron)

$$CH_4(g) + 2 O_2(g) \longrightarrow CO_2(g) + 2 H_2O(g)$$
(combustion of natural gas)

However, redox reactions need not involve oxygen. Consider, for example, the reaction between sodium and chlorine to form table salt (NaCl).

$$2 Na(s) + Cl_2(g) \longrightarrow 2 NaCl(s)$$

This reaction is similar to the reaction between sodium and oxygen which can form sodium oxide.

The reaction between sodium and oxygen also forms other oxides besides Na_2O.

$$4 Na(s) + O_2(g) \longrightarrow 2 Na_2O(s)$$

Helpful mnemonics: OIL RIG—**O**xidation **I**s **L**oss; **R**eduction **I**s **G**ain. LEO GER—**L**ose **E**lectrons **O**xidation; **G**ain **E**lectrons **R**eduction.

What do these two reactions have in common? In both cases, sodium (a metal with a tendency to lose electrons) reacts with a nonmetal (that has a tendency to gain electrons). In both cases, sodium atoms lose electrons to nonmetal atoms. A fundamental definition of oxidation is *the loss of electrons*, and a fundamental definition of reduction is *the gain of electrons*.

Notice that oxidation and reduction must occur together. If one substance loses electrons (oxidation), then another substance must gain electrons (reduction). For now, you simply need to be able to identify redox reactions. In Chapter 16 we will examine them more thoroughly.

A reaction can be classified as a redox reaction if it meets any one of these requirements.

Redox reactions are those in which:

*   A substance reacts with elemental oxygen.
*   A metal reacts with a nonmetal.
*   More generally, one substance transfers electrons to another substance.

EXAMPLE 7.13   Identifying Redox Reactions

Which of these are redox reactions?

(a) $2 Mg(s) + O_2(g) \longrightarrow 2 MgO(s)$
(b) $2 HBr(aq) + Ca(OH)_2(aq) \longrightarrow 2 H_2O(l) + CaBr_2(aq)$
(c) $Ca(s) + Cl_2(g) \longrightarrow CaCl_2(s)$
(d) $Zn(s) + Fe^{2+}(aq) \longrightarrow Zn^{2+}(aq) + Fe(s)$

SOLUTION

(a) Redox reaction; Mg reacts with elemental oxygen.
(b) Not a redox reaction; it is an acid–base reaction.
(c) Redox reaction; a metal reacts with a nonmetal.
(d) Redox reaction; Zn transfers two electrons to Fe^{2+}.

▶**SKILLBUILDER 7.13** | Identifying Redox Reactions

Which of these are redox reactions?

(a) $2 Li(s) + Cl_2(g) \longrightarrow 2 LiCl(s)$
(b) $2 Al(s) + 3 Sn^{2+}(aq) \longrightarrow 2 Al^{3+}(aq) + 3 Sn(s)$
(c) $Pb(NO_3)_2(aq) + 2 LiCl(aq) \longrightarrow PbCl_2(s) + 2 LiNO_3(aq)$
(d) $C(s) + O_2(g) \longrightarrow CO_2(g)$

▶**FOR MORE PRACTICE**   Example 7.23; Problems 83, 84.

COMBUSTION REACTIONS

Combustion reactions are a type of redox reaction. They are important to us because most of our society's energy is derived from combustion reactions. Combustion reactions are characterized by the reaction of a substance with O_2 to form one or more oxygen-containing compounds, often including water. Combustion reactions are exothermic (they emit heat). For example, as we saw in Section 7.3, natural gas (CH_4) reacts with oxygen to form carbon dioxide and water.

The water formed in combustion reactions may be gaseous (*g*) or liquid (*l*) depending on the reaction conditions.

$$CH_4(g) + 2 O_2(g) \longrightarrow CO_2(g) + 2 H_2O(g)$$

As mentioned in the opening section of this chapter, combustion reactions power automobiles. For example, octane, a component of gasoline, reacts with oxygen to form carbon dioxide and water.

$$2 C_8H_{18}(l) + 25 O_2(g) \longrightarrow 16 CO_2(g) + 18 H_2O(g)$$

▲ Combustion of octane occurs in the cylinders of an automobile engine.

Ethanol, the alcohol in alcoholic beverages, also reacts with oxygen in a combustion reaction to form carbon dioxide and water.

$$C_2H_5OH(l) + 3\,O_2(g) \longrightarrow 2\,CO_2(g) + 3\,H_2O(g)$$

Compounds containing carbon and hydrogen—or carbon, hydrogen, and oxygen—always form carbon dioxide and water upon combustion. Other combustion reactions include the reaction of carbon with oxygen to form carbon dioxide:

$$C(s) + O_2(g) \longrightarrow CO_2(g)$$

and the reaction of hydrogen with oxygen to form water:

$$2\,H_2(g) + O_2(g) \longrightarrow 2\,H_2O(g)$$

EXAMPLE 7.14 Writing Combustion Reactions

Write a balanced equation for the combustion of liquid methyl alcohol (CH_3OH).

Begin by writing a skeletal equation showing the reaction of CH_3OH with O_2 to form CO_2 and H_2O.	**SOLUTION** $$CH_3OH(l) + O_2(g) \longrightarrow CO_2(g) + H_2O(g)$$
Balance the skeletal equation using the rules in Section 7.4.	$$2\,CH_3OH(l) + 3\,O_2(g) \longrightarrow 2\,CO_2(g) + 4\,H_2O(g)$$

▶ SKILLBUILDER 7.14 | Writing Combustion Reactions

Write a balanced equation for the combustion of liquid pentane (C_5H_{12}), a component of gasoline.

▶ SKILLBUILDER PLUS

Write a balanced equation for the combustion of liquid propanol (C_3H_7OH).

▶ FOR MORE PRACTICE Example 7.24; Problems 85, 86.

7.10 Classifying Chemical Reactions

Throughout this chapter, we have examined different types of chemical reactions. We have seen examples of precipitation reactions, acid–base reactions, gas evolution reactions, oxidation–reduction reactions, and combustion reactions. We can organize these different types of reactions with the following flowchart.

*Many gas evolution reactions are also acid–base reactions.

This classification scheme focuses on the type of chemistry or phenomenon that is occurring during the reaction (such as the formation of a precipitate or the transfer of electrons). However, an alternative way to classify chemical reactions is by what atoms or groups of atoms do during the reaction.

CLASSIFYING CHEMICAL REACTIONS BY WHAT ATOMS DO

In this alternative way of classifying reactions, we focus on the pattern of the reaction by classifying it into one of the following four categories. In this classification scheme, the letters (A, B, C, D) represent atoms or groups of atoms.

Type of Reaction	Generic Equation
synthesis or combination	$A + B \longrightarrow AB$
decomposition	$AB \longrightarrow A + B$
displacement	$A + BC \longrightarrow AC + B$
double-displacement	$AB + CD \longrightarrow AD + CB$

SYNTHESIS OR COMBINATION REACTIONS. In a **synthesis** or **combination reaction**, simpler substances combine to form more complex substances. The simpler substances may be elements, such as sodium and chlorine combining to form sodium chloride.

$$2 \, Na(s) + Cl_2(g) \longrightarrow 2 \, NaCl(s)$$

The simpler substances may also be compounds, such as calcium oxide and carbon dioxide combining to form calcium carbonate.

$$CaO(s) + CO_2(g) \longrightarrow CaCO_3(s)$$

In either case, a synthesis reaction follows the general equation:

$$A + B \longrightarrow AB$$

$$2 \, Na(s) + Cl_2(g) \longrightarrow 2 \, NaCl(s)$$

▲ In a synthesis reaction, two simpler substances combine to make a more complex substance. In this series of photographs we see sodium metal and chlorine gas. When they combine, a chemical reaction occurs that forms sodium chloride.

Note that the first two of these reactions are also redox reactions.

Other examples of synthesis reactions include:

$$2 H_2(g) + O_2(g) \longrightarrow 2 H_2O(l)$$
$$2 Mg(s) + O_2(g) \longrightarrow 2 MgO(s)$$
$$SO_3(g) + H_2O(l) \longrightarrow H_2SO_4(aq)$$

DECOMPOSITION REACTIONS. In a **decomposition reaction**, a complex substance decomposes to form simpler substances. The simpler substances may be elements, such as the hydrogen and oxygen gases that form upon the decomposition of water when electrical current passes through it.

$$2 H_2O(l) \xrightarrow{\text{electrical current}} 2 H_2(g) + O_2(g)$$

The simpler substances may also be compounds, such as the calcium oxide and carbon dioxide that form upon heating calcium carbonate.

$$CaCO_3(s) \xrightarrow{\text{heat}} CaO(s) + CO_2(g)$$

In either case, a decomposition reaction follows the general equation:

$$AB \longrightarrow A + B$$

Other examples of decomposition reactions include:

$$2 HgO(s) \xrightarrow{\text{heat}} 2 Hg(l) + O_2(g)$$
$$2 KClO_3(s) \xrightarrow{\text{heat}} 2 KCl(s) + 3 O_2(g)$$
$$CH_3I(g) \xrightarrow{\text{light}} CH_3(g) + I(g)$$

Notice that these decomposition reactions require energy in the form of heat, electrical current, or light to make them happen. This is because compounds are normally stable and energy is required to decompose them. A number of decomposition reactions require *ultraviolet* or *UV light*, which is light in the ultraviolet region of the spectrum. UV light carries more energy than visible light and can therefore initiate the decomposition of many compounds. (We will discuss light in more detail in Chapter 9.)

DISPLACEMENT REACTIONS. In a **displacement** or **single-displacement reaction**, one element displaces another in a compound. For example, when metallic zinc is added to a solution of copper(II) chloride, the zinc replaces the copper.

$$Zn(s) + CuCl_2(aq) \longrightarrow ZnCl_2(aq) + Cu(s)$$

A displacement reaction follows the general equation:

$$A + BC \longrightarrow AC + B$$

Other examples of displacement reactions include:

$$Mg(s) + 2 HCl(aq) \longrightarrow MgCl_2(aq) + H_2(g)$$
$$2 Na(s) + 2 H_2O(l) \longrightarrow 2 NaOH(aq) + H_2(g)$$

The last reaction can be identified more easily as a displacement reaction if we write water as HOH(l).

$$2 Na(s) + 2 HOH(l) \longrightarrow 2 NaOH(aq) + H_2(g)$$

$$2 H_2O(l) \longrightarrow 2 H_2(g) + O_2(g)$$

▲ When electrical current is passed through water, the water undergoes a decomposition reaction to form hydrogen gas and oxygen gas.

▶ In a single-displacement reaction, one element displaces another in a compound. When zinc metal is immersed in a copper(II) chloride solution, the zinc atoms displace the copper ions in solution.

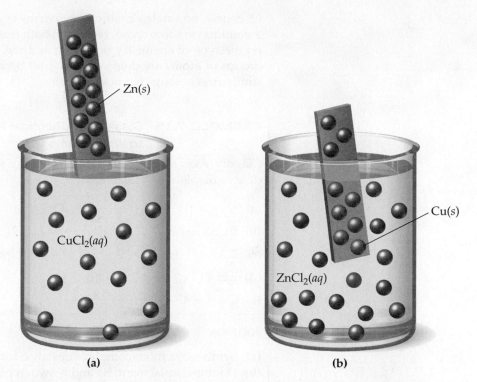

(a)   (b)

DOUBLE-DISPLACEMENT REACTIONS. In a **double-displacement reaction**, two elements or groups of elements in two different compounds exchange places to form two new compounds. For example, in aqueous solution, the silver in silver nitrate changes places with the sodium in sodium chloride to form solid silver chloride and aqueous sodium nitrate.

This double-displacement reaction is also a precipitation reaction.

$$AgNO_3(aq) + NaCl(aq) \longrightarrow AgCl(s) + NaNO_3(aq)$$

A double-displacement reaction follows the general form:

$$AB + CD \longrightarrow AD + CB$$

Other examples of double-displacement reactions include:

These double-displacement reactions are also acid–base reactions.

$$HCl(aq) + NaOH(aq) \longrightarrow H_2O(l) + NaCl(aq)$$
$$2\,HCl(aq) + Na_2CO_3(aq) \longrightarrow H_2CO_3(aq) + 2\,NaCl(aq)$$

As we learned in Section 7.8, $H_2CO_3(aq)$ is not stable and decomposes to form $H_2O(l) + CO_2(g)$, so the overall equation is:

This double-displacement reacton is also a gas evolution reaction and an acid–base reaction.

$$2\,HCl(aq) + Na_2CO_3(aq) \longrightarrow H_2O(l) + CO_2(g) + 2\,NaCl(aq)$$

CLASSIFICATION FLOWCHART

A flowchart for this classification scheme of chemical reactions is as follows:

Of course, no single classification scheme is perfect because all chemical reactions are unique in some sense. However, both classification schemes—one that focuses on the type of chemistry occurring and the other that focuses on what atoms or groups of atoms are doing—are helpful because they help us see differences and similarities among chemical reactions.

EXAMPLE 7.15   Classifying Chemical Reactions According to What Atoms Do

Classify each reaction as a synthesis, decomposition, single-displacement, or double-displacement reaction.

(a) $Na_2O(s) + H_2O(l) \longrightarrow 2\,NaOH(aq)$

(b) $Ba(NO_3)_2(aq) + K_2SO_4(aq) \longrightarrow BaSO_4(s) + 2\,KNO_3(aq)$

(c) $2\,Al(s) + Fe_2O_3(s) \longrightarrow Al_2O_3(s) + 2\,Fe(l)$

(d) $2\,H_2O_2(aq) \longrightarrow 2\,H_2O(l) + O_2(g)$

(e) $Ca(s) + Cl_2(g) \longrightarrow CaCl_2(s)$

SOLUTION

(a) Synthesis; a more complex substance forms from two simpler ones.
(b) Double-displacment; Ba and K switch places to form two new compounds.
(c) Single-displacement; Al displaces Fe in Fe_2O_3.
(d) Decomposition; a complex substance decomposes into simpler ones.
(e) Synthesis; a more complex substance forms from two simpler ones.

▶**SKILLBUILDER 7.15** | Classifying Chemical Reactions According to What Atoms Do

Classify each reaction as a synthesis, decomposition, single-displacement, or double-displacement reaction.

(a) $2\,Al(s) + 2\,H_3PO_4(aq) \longrightarrow 2\,AlPO_4(aq) + 3\,H_2(g)$

(b) $CuSO_4(aq) + 2\,KOH(aq) \longrightarrow Cu(OH)_2(s) + K_2SO_4(aq)$

(c) $2\,K(s) + Br_2(l) \longrightarrow 2\,KBr(s)$

(d) $CuCl_2(aq) \xrightarrow[\text{electrical current}]{} Cu(s) + Cl_2(g)$

▶**FOR MORE PRACTICE**   Example 7.25; Problems 89, 90, 91, 92.

✓ CONCEPTUAL CHECKPOINT 7.6

Precipitation reactions and acid–base reactions can both also be classified as:

(a) synthesis reactions

(b) decomposition reactions

(c) single-displacement reactions

(d) double-displacement reactions

CHEMISTRY IN THE ENVIRONMENT
The Reactions Involved in Ozone Depletion

*C*hemistry in the Environment: Chlorine in Chlorofluorocarbons in Chapter 6 explained that chlorine atoms from chlorofluorocarbons deplete the ozone layer, which normally protects life on Earth from harmful ultraviolet light. Through research, chemists have discovered the reactions by which this depletion occurs.

Ozone normally forms in the upper atmosphere according to this reaction.

(a) $O_2(g) + O(g) \longrightarrow O_3(g)$

When chlorofluorocarbons drift to the upper atmosphere, they are exposed to ultraviolet light and undergo the reaction.

(b) $CF_2Cl_2(g) \xrightarrow{\text{UV light}} CF_2Cl(g) + Cl(g)$

Atomic chlorine then reacts with and depletes ozone according to this cycle of reactions.

(c) $Cl(g) + O_3(g) \longrightarrow ClO(g) + O_2(g)$

(d) $O_3(g) \xrightarrow{\text{UV light}} O_2(g) + O(g)$

(e) $O(g) + ClO(g) \longrightarrow O_2(g) + Cl(g)$

Notice that in the final reaction, atomic chlorine is regenerated and can go through the cycle again to deplete more ozone. Through this cycle of reactions, a single chlorofluorocarbon molecule can deplete thousands of ozone molecules.

CAN YOU ANSWER THIS? *Classify each of these reactions (a–e) as a synthesis, decomposition, single-displacement, or double-displacement reaction.*

CHAPTER IN REVIEW

CHEMICAL PRINCIPLES

RELEVANCE

Chemical Reactions: In a chemical reaction, one or more substances—either elements or compounds—changes into a different substance.

Chemical Reactions: Chemical reactions are central to many processes, including transportation, energy generation, manufacture of household products, vision, and life itself.

Evidence of a Chemical Reaction: The only absolute evidence for a chemical reaction is chemical analysis showing that one or more substances has changed into another substance. However, one or more of the following often accompanies a chemical reaction: a color change; the formation of a solid or precipitate; the formation of a gas; the emission of light; and the emission or absorption of heat.

Evidence of a Chemical Reaction: We can often perceive the changes that accompany chemical reactions. In fact, we often employ chemical reactions for the changes they produce. For example, we use the heat emitted by the combustion of fossil fuels to heat our homes, drive our cars, and generate electricity.

Chemical Equations: Chemical equations represent chemical reactions. They include formulas for the reactants (the substances present before the reaction) and for the products (the new substances formed by the reaction). Chemical equations must be balanced to reflect the conservation of matter in nature; atoms do not spontaneously appear or disappear.

Chemical Equations: Chemical equations allow us to represent and understand chemical reactions. For example, the equations for the combustion reactions of fossil fuels let us see that carbon dioxide, a gas that contributes to global warming, is one of the products of these reactions.

Aqueous Solutions and Solubility: Aqueous solutions are mixtures of a substance dissolved in water. If a substance dissolves in water it is soluble. Otherwise, it is insoluble.

Aqueous Solutions and Solubility: Aqueous solutions are common. Oceans, lakes, and most of the fluids in our bodies are aqueous solutions.

Some Specific Types of Reactions:

    Precipitation reaction: A solid or precipitate forms upon mixing two aqueous solutions.

    Acid–base reaction: Water forms upon mixing an acid and a base.

    Gas evolution reaction: A gas forms upon mixing two aqueous solutions.

    Redox reaction: Electrons are transferred from one substance to another.

    Combustion reaction: A substance reacts with oxygen, emitting heat, and forming an oxygen-containing compound and, in many cases, water.

Some Specific Types of Reactions: Many of the specific types of reactions discussed in this chapter occur in aqueous solutions and are therefore important to living organisms. Acid–base reactions, for example, constantly occur in the blood of living organisms to maintain constant blood acidity levels. In humans, a small change in blood acidity levels would result in death, so the body carries out chemical reactions to prevent this. Combustion reactions are important because they are the main energy source for our society.

Classifying Chemical Reactions: Many chemical reactions can be classified into one of the following four categories according to what atoms or groups of atoms do:

- synthesis: $(A + B \longrightarrow AB)$
- decomposition: $(AB \longrightarrow A + B)$
- single-displacement: $(A + BC \longrightarrow AC + B)$
- double-displacement: $(AB + CD \longrightarrow AD + CB)$

Classifying Chemical Reactions: We classify chemical reactions to better understand them and to recognize similarities and differences among reactions.

CHEMICAL SKILLS

Identifying a Chemical Reaction (Section 7.2)

To identify a chemical reaction, determine whether one or more of the initial substances changed into a different substance. If so, a chemical reaction occurred. One or more of the following often accompanies a chemical reaction: a color change; the formation of a solid or precipitate; the formation of a gas; the emission of light; and the emission or absorption of heat.

EXAMPLES

EXAMPLE 7.16   Identifying a Chemical Reaction

Which of these are chemical reactions?

(a) Copper turns green on exposure to air.
(b) When sodium bicarbonate is combined with hydrochloric acid, bubbling is observed.
(c) Liquid water freezes to form solid ice.
(d) A pure copper penny forms bubbles of a dark brown gas when dropped into nitric acid. The nitric acid solution turns blue.

SOLUTION

(a) Chemical reaction, as evidenced by the color change.
(b) Chemical reaction, as evidenced by the evolution of a gas.
(c) Not a chemical reaction; solid ice is still water.
(d) Chemical reaction, as evidenced by the evolution of a gas and by a color change.

Writing Balanced Chemical Equations (Sections 7.3, 7.4)

To write balanced chemical equations, follow these steps.

1. Write a skeletal equation by writing chemical formulas for each of the reactants and products. (If a skeletal equation is provided, proceed to Step 2.)

2. If an element occurs in only one compound on both sides of the equation, balance that element first. If there is more than one such element, and the equation contains both metals and nonmetals, balance metals before nonmetals.

3. If an element occurs as a free element on either side of the chemical equation, balance that element last.

4. If the balanced equation contains coefficient fractions, clear these by multiplying the entire equation by the appropriate factor.

5. Check to make certain the equation is balanced by summing the total number of each type of atom on both sides of the equation.

Reminders

- Change only the *coefficients* to balance a chemical equation, *never the subscripts*. Changing the subscripts would change the compound itself.

- If the equation contains polyatomic ions that stay intact on both sides of the equation, balance the polyatomic ions as a group.

Determining Whether a Compound Is Soluble (Section 7.5)

To determine whether a compound is soluble, refer to the solubility rules in Table 7.2. It is easiest to begin by looking for those ions that always form soluble compounds (Li^+, Na^+, K^+, NH_4^+, NO_3^-, and $C_2H_3O_2^-$). If a compound contains one of those, it is soluble. If it does not, look at the anion and determine whether it is mostly soluble (Cl^-, Br^-, I^-, or SO_4^{2-}) or mostly insoluble (OH^-, S^{2-}, CO_3^{2-}, or PO_4^{3-}). Look also at the cation to determine whether it is one of the exceptions.

EXAMPLE 7.17 Writing Balanced Chemical Equations

Write a balanced chemical equation for the reaction of solid vanadium(V) oxide with hydrogen gas to form solid vanadium(III) oxide and liquid water.

$$V_2O_5(s) + H_2(g) \longrightarrow V_2O_3(s) + H_2O(l)$$

SOLUTION

Skeletal equation is given. Proceed to Step 2.

Vanadium occurs in only one compound on both sides of the equation. However, it is balanced, so you can proceed and balance oxygen by placing a 2 in front of H_2O on the right side.

$$V_2O_5(s) + H_2(g) \longrightarrow V_2O_3(s) + 2\,H_2O(l)$$

Hydrogen occurs as a free element, so balance it last by placing a 2 in front of H_2 on the left side.

$$V_2O_5(s) + 2\,H_2(g) \longrightarrow V_2O_3(s) + 2\,H_2O(l)$$

Equation does not contain coefficient fractions. Proceed to Step 5.

Check the equation.

$$V_2O_5(s) + 2\,H_2(g) \longrightarrow V_2O_3(s) + 2\,H_2O(l)$$

Reactants		Products
2 V atoms	$\longrightarrow$	2 V atoms
5 O atoms	$\longrightarrow$	5 O atoms
4 H atoms	$\longrightarrow$	4 H atoms

EXAMPLE 7.18 Determining Whether a Compound Is Soluble

Determine whether each compound is soluble.

(a) $CuCO_3$
(b) $BaSO_4$
(c) $Fe(NO_3)_3$

SOLUTION

(a) Insoluble; compounds containing CO_3^{2-} are insoluble, and Cu^{2+} is not an exception.
(b) Insoluble; compounds containing SO_4^{2-} are usually soluble, but Ba^{2+} is an exception.
(c) Soluble; all compounds containing NO_3^- are soluble.

Predicting Precipitation Reactions (Section 7.6)

To predict whether a precipitation reaction occurs when two solutions are mixed and to write an equation for the reaction, follow these steps.

1. Write the formulas of the two compounds being mixed as reactants in a chemical equation.
2. Below the equation, write the formulas of the potentially insoluble products that could form from the reactants. These are obtained by combining the cation from one reactant with the anion from the other. Make sure to adjust the subscripts so that all formulas are charge-neutral.
3. Use the solubility rules to determine whether any of the potentially insoluble products are indeed insoluble.
4. If all of the potentially insoluble products are soluble, there will be no precipitate. Write *NO REACTION* next to the arrow.
5. If one or both of the potentially insoluble products are insoluble, write their formula(s) as the product(s) of the reaction using (*s*) to indicate *solid*. Write any soluble products with (*aq*) to indicate *aqueous*.
6. Balance the equation.

EXAMPLE 7.19 Predicting Precipitation Reactions

Write an equation for the precipitation reaction that occurs, if any, when solutions of sodium phosphate and cobalt(II) chloride are mixed.

SOLUTION

$$Na_3PO_4(aq) + CoCl_2(aq) \longrightarrow$$

Potentially Insoluble Products:

$$NaCl \qquad Co_3(PO_4)_2$$

NaCl is soluble.
$Co_3(PO_4)_2$ is insoluble.
    Reaction contains an insoluble product; proceed to Step 5.

$$Na_3PO_4(aq) + CoCl_2(aq) \longrightarrow Co_3(PO_4)_2(s) + NaCl(aq)$$
$$2\,Na_3PO_4(aq) + 3\,CoCl_2(aq) \longrightarrow$$
$$Co_3(PO_4)_2(s) + 6\,NaCl(aq)$$

Writing Complete Ionic and Net Ionic Equations (Section 7.7)

To write a complete ionic equation from a molecular equation, separate all aqueous ionic compounds into independent ions. Do not separate solid, liquid, or gaseous compounds.

    To write a net ionic equation from a complete ionic equation, eliminate all species that do not change (spectator ions) in the course of the reaction.

EXAMPLE 7.20 Writing Complete Ionic and Net Ionic Equations

Write a complete ionic and net ionic equation for the reaction.

$$2\,NH_4Cl(aq) + Hg_2(NO_3)_2(aq) \longrightarrow$$
$$Hg_2Cl_2(s) + 2\,NH_4NO_3(aq)$$

SOLUTION
Complete ionic equation:

$$2\,NH_4{}^+(aq) + 2\,Cl^-(aq) + Hg_2{}^{2+}(aq) + 2\,NO_3{}^-(aq) \longrightarrow$$
$$Hg_2Cl_2(s) + 2\,NH_4{}^+(aq) + 2\,NO_3{}^-(aq)$$

    Net ionic equation:

$$2\,Cl^-(aq) + Hg_2{}^{2+}(aq) \longrightarrow Hg_2Cl_2(s)$$

Writing Equations for Acid–Base Reactions (Section 7.8)

When you see an acid and a base (see Table 7.3) as reactants in an equation, write a reaction in which the acid and the base react to form water and a salt.

EXAMPLE 7.21 Writing Equations for Acid–Base Reactions

Write an equation for the reaction that occurs when aqueous hydroiodic acid is mixed with aqueous barium hydroxide.

SOLUTION

$$\underset{\text{acid}}{2\,HI(aq)} + \underset{\text{base}}{Ba(OH)_2(aq)} \longrightarrow \underset{\text{water}}{2\,H_2O(l)} + \underset{\text{salt}}{BaI_2(aq)}$$

Writing Equations for Gas Evolution Reactions (Section 7.8)

See Table 7.4 to identify gas evolution reactions.

EXAMPLE 7.22 Writing Equations for Gas Evolution Reactions

Write an equation for the reaction that occurs when aqueous hydrobromic acid is mixed with aqueous potassium bisulfite.

SOLUTION

$$HBr(aq) + KHSO_3(aq) \longrightarrow H_2SO_3(aq) + KBr(aq) \longrightarrow$$
$$H_2O(l) + SO_2(g) + KBr(aq)$$

Identifying Redox Reactions (Section 7.9)

Redox reactions are those in which any of the following occurs:

- a substance reacts with elemental oxygen
- a metal reacts with a nonmetal
- one substance transfers electrons to another substance

EXAMPLE 7.23 Identifying Redox Reactions

Which of these reactions is a redox reaction?

(a) $4\,Fe(s) + 3\,O_2(g) \longrightarrow 2\,Fe_2O_3(s)$
(b) $CaO(s) + CO_2(g) \longrightarrow CaCO_3(s)$
(c) $AgNO_3(aq) + NaCl(aq) \longrightarrow AgCl(s) + NaNO_3(aq)$

SOLUTION
Only **(a)** is a redox reaction.

Writing Equations for Combustion Reactions (Section 7.9)

In a combustion reaction, a substance reacts with O_2 to form one or more oxygen-containing compounds and, in many cases, water.

EXAMPLE 7.24 Writing Equations for Combustion Reactions

Write a balanced equation for the combustion of gaseous ethane (C_2H_6), a minority component of natural gas.

SOLUTION
The skeletal equation is:

$$C_2H_6(g) + O_2(g) \longrightarrow CO_2(g) + H_2O(g)$$

The balanced equation is:

$$2\,C_2H_6(g) + 7\,O_2(g) \longrightarrow 4\,CO_2(g) + 6\,H_2O(g)$$

Classifying Chemical Reactions (Section 7.10)

Chemical reactions can be classified by inspection. The four major categories are:

Synthesis or combination

$$A + B \longrightarrow AB$$

Decomposition

$$AB \longrightarrow A + B$$

Single-displacement

$$A + BC \longrightarrow AC + B$$

Double-displacement

$$AB + CD \longrightarrow AD + CB$$

EXAMPLE 7.25 Classifying Chemical Reactions

Classify each chemical reaction as a synthesis, decomposition, single-displacement, or double-displacement reaction.

(a) $2\,K(s) + Br_2(g) \longrightarrow 2\,KBr(s)$
(b) $Fe(s) + 2\,AgNO_3(aq) \longrightarrow Fe(NO_3)_2(aq) + 2\,Ag(s)$
(c) $CaSO_3(s) \longrightarrow CaO(s) + SO_2(g)$
(d) $CaCl_2(aq) + Li_2SO_4(aq) \longrightarrow CaSO_4(s) + 2\,LiCl(aq)$

SOLUTION

(a) Synthesis; KBr, a more complex substance, is formed from simpler substances.
(b) Single-displacement; Fe displaces Ag in $AgNO_3$.
(c) Decomposition; $CaSO_3$ decomposes into simpler substances.
(d) Double-displacement; Ca and Li switch places to form new compounds.

KEY TERMS

acid–base reaction [7.8]
aqueous solution [7.5]
balanced equation [7.3]
combination reaction [7.10]
combustion reaction [7.9]
complete ionic equation [7.7]
decomposition reaction [7.10]

displacement reaction [7.10]
double-displacement reaction [7.10]
gas evolution reaction [7.8]
insoluble [7.5]
molecular equation [7.7]
net ionic equation [7.7]

neutralization reaction [7.8]
oxidation–reduction (redox) reaction [7.9]
precipitate [7.6]
precipitation reaction [7.6]
salt [7.8]
single-displacement reaction [7.10]

solubility rules [7.5]
soluble [7.5]
spectator ion [7.7]
strong electrolyte solution [7.5]
synthesis reaction [7.10]

EXERCISES

QUESTIONS

1. What is a chemical reaction? List some examples.
2. If you could observe atoms and molecules, what would you look for as conclusive evidence of a chemical reaction?
3. What are the main indications that a chemical reaction has occurred?
4. What is a chemical equation? Provide an example and identify the reactants and products.
5. What does each abbreviation, often used in chemical equations, represent?
   (a) (g) (b) (l) (c) (s) (d) (aq)
6. To balance a chemical equation, adjust the _____ as necessary to make the numbers of each type of atom on both sides of the equation equal. Never adjust the _____ to balance a chemical equation.
7. List the number of each type of atom on both sides of each equation. Are the equations balanced?
   (a) $2\,Ag_2O(s) + C(s) \longrightarrow CO_2(g) + 4\,Ag(s)$
   (b) $Pb(NO_3)_2(aq) + 2\,NaCl(aq) \longrightarrow$
   $PbCl_2(s) + 2\,NaNO_3(aq)$
   (c) $C_3H_8(g) + O_2(g) \longrightarrow 3\,CO_2(g) + 4\,H_2O(g)$
8. What is an aqueous solution? List two examples.
9. What does it mean for a compound to be soluble? Insoluble?
10. Explain what happens to an ionic substance when it dissolves in water.

11. Do polyatomic ions dissociate when they dissolve in water, or do they remain intact?
12. What is a strong electrolyte solution?
13. What are the solubility rules, and how are they useful?
14. What is a precipitation reaction? Provide an example and identify the precipitate.
15. Will the precipitate in a precipitation reaction always be a compound that is soluble or insoluble? Explain.
16. Describe the differences between a molecular equation, a complete ionic equation, and net ionic equation. Give an example of each to illustrate the differences.
17. What is an acid–base reaction? List an example and identify the acid and the base.
18. What are the properties of acids and bases?
19. What is a gas evolution reaction? Give an example.
20. What is a redox reaction? Give an example.
21. What is a combustion reaction? Give an example.
22. What are two different ways to classify chemical reactions presented in Section 7.10? Explain the differences between them.
23. Explain the difference between a synthesis reaction and a decomposition reaction and give an example of each.
24. Explain the difference between a single-displacement reaction and a double-displacement reaction and give an example of each.

PROBLEMS

EVIDENCE OF CHEMICAL REACTIONS

25. Which observation is consistent with a chemical reaction occurring? Why?
    (a) Solid copper deposits on a piece of aluminum foil when the foil is placed in a blue copper nitrate solution. The blue color of the solution fades.
    (b) Liquid ethyl alcohol turns into a solid when placed in a low-temperature freezer.
    (c) A white precipitate forms when solutions of barium nitrate and sodium sulfate are mixed.
    (d) A mixture of sugar and water bubbles when yeasts are added. After several days, the sugar is gone and ethyl alcohol is found in the water.

26. Which observation is consistent with a chemical reaction occurring? Why?
    (a) Propane forms a flame and emits heat as it burns.
    (b) Acetone feels cold as it evaporates from the skin.
    (c) Bubbling is observed when potassium carbonate and hydrochloric acid solutions are mixed.
    (d) Heat is felt when a warm object is placed in your hand.

27. Vinegar forms bubbles when it is poured onto the calcium deposits on a faucet, and some of the calcium dissolves. Has a chemical reaction occurred? Explain your answer.

28. When a chemical drain opener is added to a clogged sink, bubbles form and the water in the sink gets warmer. Has a chemical reaction occurred? Explain your answer.

29. When a commercial hair bleaching mixture is applied to brown hair, the hair turns blond. Has a chemical reaction occurred? Explain your answer.

30. When water is boiled in a pot, it bubbles. Has a chemical reaction occurred? Explain your answer.

WRITING AND BALANCING CHEMICAL EQUATIONS

31. Consider the unbalanced chemical equation:

$$H_2O(l) \xrightarrow{\text{electrical current}} H_2(g) + O_2(g)$$

A chemistry student tries to balance the equation by placing the subscript 2 after the oxygen atom in H_2O. Explain why this is not correct. What is the correct balanced equation?

32. Consider the unbalanced chemical equation:

$$Al(s) + Cl_2(g) \longrightarrow AlCl_3(s)$$

A student tries to balance the equation by changing the subscript 2 on Cl to a 3. Explain why this is not correct. What is the correct balanced equation?

33. Write a balanced chemical equation for each chemical reaction:
    (a) Solid lead(II) sulfide reacts with aqueous hydrochloric acid to form solid lead(II) chloride and dihydrogen sulfide gas.
    (b) Gaseous carbon monoxide reacts with hydrogen gas to form gaseous methane (CH_4) and liquid water.
    (c) Solid iron(III) oxide reacts with hydrogen gas to form solid iron and liquid water.
    (d) Gaseous ammonia (NH_3) reacts with gaseous oxygen to form gaseous nitrogen monoxide and gaseous water.

34. Write a balanced chemical equation for each chemical reaction:
    (a) Solid copper reacts with solid sulfur to form solid copper(I) sulfide.
    (b) Sulfur dioxide gas reacts with oxygen gas to form sulfur trioxide gas.
    (c) Aqueous hydrochloric acid reacts with solid manganese(IV) oxide to form aqueous manganese(II) chloride, liquid water, and chlorine gas.
    (d) Liquid benzene (C_6H_6) reacts with gaseous oxygen to form carbon dioxide and liquid water.

35. Write a balanced chemical equation for each chemical reaction:
    (a) Solid magnesium reacts with aqueous copper(I) nitrate to form aqueous magnesium nitrate and solid copper.
    (b) Gaseous dinitrogen pentoxide decomposes to form nitrogen dioxide and oxygen gas.
    (c) Solid calcium reacts with aqueous nitric acid to form aqueous calcium nitrate and hydrogen gas.
    (d) Liquid methanol (CH_3OH) reacts with oxygen gas to form gaseous carbon dioxide and gaseous water.

36. Write a balanced chemical equation for each chemical reaction:
    (a) Gaseous acetylene (C_2H_2) reacts with oxygen gas to form gaseous carbon dioxide and gaseous water.
    (b) Chlorine gas reacts with aqueous potassium iodide to form solid iodine and aqueous potassium chloride.
    (c) Solid lithium oxide reacts with liquid water to form aqueous lithium hydroxide.
    (d) Gaseous carbon monoxide reacts with oxygen gas to form carbon dioxide gas.

37. Hydrogen has been widely proposed as a potential fuel to replace fossil fuels. Some scientists are trying to anticipate any potential problems that might be associated with a hydrogen-based economy. One group of scientists has calculated that the amount of atmospheric hydrogen could increase by a factor of four due to leaks in hydrogen transport and storage. Upper atmospheric hydrogen gas reacts with oxygen gas to form liquid water. An increase in upper atmospheric water would enhance processes that release atmospheric chlorine atoms. The gaseous chlorine atoms would then react with gaseous ozone (O_3) to form gaseous chlorine monoxide and gaseous oxygen, resulting in the depletion of ozone. Write balanced chemical equations for the two reactions described in this problem.

38. Waste water from certain industrial chemical processes contains aqueous Hg_2^{2+} ions. Since the mercury ion is toxic, it is removed from the waste water by reaction with aqueous sodium sulfide. The products of the reaction are solid mercury(I) sulfide and aqueous sodium ions. Write a balanced equation for this reaction.

39. When solid sodium is added to liquid water, it reacts with the water to produce hydrogen gas and aqueous sodium hydroxide. Write a balanced chemical equation for this reaction.

40. When iron rusts, solid iron reacts with gaseous oxygen to form solid iron(III) oxide. Write a balanced chemical equation for this reaction.

41. Sulfuric acid in acid rain forms when gaseous sulfur dioxide pollutant reacts with gaseous oxygen and liquid water to form aqueous sulfuric acid. Write a balanced chemical equation for this reaction.

42. Nitric acid in acid rain forms when gaseous nitrogen dioxide pollutant reacts with gaseous oxygen and liquid water to form aqueous nitric acid. Write a balanced chemical equation for this reaction.

43. Write a balanced chemical equation for the reaction of solid vanadium(V) oxide with hydrogen gas to form solid vanadium(III) oxide and liquid water.

44. Write a balanced chemical equation for the reaction of gaseous nitrogen dioxide with hydrogen gas to form gaseous ammonia and liquid water.

45. Write a balanced chemical equation for the fermentation of sugar ($C_{12}H_{22}O_{11}$) by yeasts in which the aqueous sugar reacts with water to form aqueous ethyl alcohol (C_2H_5OH) and carbon dioxide gas.

46. Write a balanced chemical equation for the photosynthesis reaction in which gaseous carbon dioxide and liquid water react in the presence of chlorophyll to produce aqueous glucose ($C_6H_{12}O_6$) and oxygen gas.

47. Balance each chemical equation.
 (a) $Na_2S(aq) + Cu(NO_3)_2(aq) \longrightarrow$
 $\qquad\qquad\qquad\quad NaNO_3(aq) + CuS(s)$
 (b) $HCl(aq) + O_2(g) \longrightarrow H_2O(l) + Cl_2(g)$
 (c) $H_2(g) + O_2(g) \longrightarrow H_2O(l)$
 (d) $FeS(s) + HCl(aq) \longrightarrow FeCl_2(aq) + H_2S(g)$

48. Balance each chemical equation.
 (a) $N_2H_4(l) \longrightarrow NH_3(g) + N_2(g)$
 (b) $H_2(g) + N_2(g) \longrightarrow NH_3(g)$
 (c) $Cu_2O(s) + C(s) \longrightarrow Cu(s) + CO(g)$
 (d) $H_2(g) + Cl_2(g) \longrightarrow HCl(g)$

49. Balance each chemical equation.
 (a) $BaO_2(s) + H_2SO_4(aq) \longrightarrow$
 $\qquad\qquad\qquad\quad BaSO_4(s) + H_2O_2(aq)$
 (b) $Co(NO_3)_3(aq) + (NH_4)_2S(aq) \longrightarrow$
 $\qquad\qquad\qquad\quad Co_2S_3(s) + NH_4NO_3(aq)$
 (c) $Li_2O(s) + H_2O(l) \longrightarrow LiOH(aq)$
 (d) $Hg_2(C_2H_3O_2)_2(aq) + KCl(aq) \longrightarrow$
 $\qquad\qquad\qquad\quad Hg_2Cl_2(s) + KC_2H_3O_2(aq)$

50. Balance each chemical equation.
 (a) $MnO_2(s) + HCl(aq) \longrightarrow$
 $\qquad\qquad\qquad\quad Cl_2(g) + MnCl_2(aq) + H_2O(l)$
 (b) $CO_2(g) + CaSiO_3(s) + H_2O(l) \longrightarrow$
 $\qquad\qquad\qquad\quad SiO_2(s) + Ca(HCO_3)_2(aq)$
 (c) $Fe(s) + S(l) \longrightarrow Fe_2S_3(s)$
 (d) $NO_2(g) + H_2O(l) \longrightarrow HNO_3(aq) + NO(g)$

51. Determine whether each chemical equation is correctly balanced. If not, correct it.

(a) $Rb(s) + H_2O(l) \longrightarrow RbOH(aq) + H_2(g)$

(b) $2 N_2H_4(g) + N_2O_4(g) \longrightarrow 3 N_2(g) + 4 H_2O(g)$

(c) $NiS(s) + O_2(g) \longrightarrow NiO(s) + SO_2(g)$

(d) $PbO(s) + 2 NH_3(g) \longrightarrow$
$$Pb(s) + N_2(g) + H_2O(l)$$

52. Determine whether each chemical equation is correctly balanced. If not, correct it.

(a) $SiO_2(s) + 4 HF(aq) \longrightarrow SiF_4(g) + 2 H_2O(l)$

(b) $2 Cr(s) + 3 O_2(g) \longrightarrow Cr_2O_3(s)$

(c) $Al_2S_3(s) + H_2O(l) \longrightarrow 2 Al(OH)_3(s) + 3 H_2S(g)$

(d) $Fe_2O_3(s) + CO(g) \longrightarrow 2 Fe(s) + CO_2(g)$

53. Human cells obtain energy from a reaction called cellular respiration. Balance the skeletal equation for cellular respiration.

$$C_6H_{12}O_6(aq) + O_2(g) \longrightarrow CO_2(g) + H_2O(l)$$

54. Propane camping stoves produce heat by the combustion of gaseous propane (C_3H_8). Balance the skeletal equation for the combustion of propane.

$$C_3H_8(g) + O_2(g) \longrightarrow CO_2(g) + H_2O(g)$$

55. Catalytic converters work to remove nitrogen oxides and carbon monoxide from exhaust. Balance the skeletal equation for one of the reactions that occurs in a catalytic converter.

$$NO(g) + CO(g) \longrightarrow N_2(g) + CO_2(g)$$

56. Billions of pounds of urea are produced annually for use as a fertilizer. Balance the skeletal equation for the synthesis of urea.

$$NH_3(g) + CO_2(g) \longrightarrow CO(NH_2)_2(s) + H_2O(l)$$

SOLUBILITY

57. Determine whether each compound is soluble or insoluble. For the soluble compounds, identify the ions present in solution.

(a) $NaC_2H_3O_2$

(b) $Sn(NO_3)_2$

(c) AgI

(d) $Na_3(PO_4)$

58. Determine whether each compound is soluble or insoluble. For the soluble compounds, identify the ions present in solution.

(a) $(NH_4)_2S$

(b) $CuCO_3$

(c) ZnS

(d) $Pb(C_2H_3O_2)_2$

59. Pair each cation on the left with an anion on the right that will form an *insoluble* compound with it and write a formula for the insoluble compound. Use each anion only once.

Ag^+ SO_4^{2-}
Ba^{2+} Cl^-
Cu^{2+} CO_3^{2-}
Fe^{3+} S^{2-}

60. Pair each cation on the left with an anion on the right that will form a *soluble* compound with it and write a formula for the soluble compound. Use each anion only once.

Na^+ NO_3^-
Sr^{2+} SO_4^{2-}
Co^{2+} S^{2-}
Pb^{2+} CO_3^{2-}

61. Determine whether each compound is in the correct column. Move any misplaced compounds to the correct column.

Soluble	Insoluble
K_2S	K_2SO_4
$PbSO_4$	Hg_2I_2
BaS	$Cu_3(PO_4)_2$
$PbCl_2$	MgS
Hg_2Cl_2	$CaSO_4$
NH_4Cl	SrS
Na_2CO_3	Li_2S

62. Determine whether each compound is in the correct column. Move any misplaced compounds to the correct column.

Soluble	Insoluble
$LiOH$	$CaCl_2$
Na_2CO_3	$Cu(OH)_2$
$AgCl$	$Ca(C_2H_3O_2)_2$
K_3PO_4	$SrSO_4$
CuI_2	Hg_2Br_2
$Pb(NO_3)_2$	$PbBr_2$
$CoCO_3$	PbI_2

PRECIPITATION REACTIONS

63. Complete and balance each equation. If no reaction occurs, write *NO REACTION*.

    (a) $KI(aq) + BaS(aq) \longrightarrow$

    (b) $K_2SO_4(aq) + BaBr_2(aq) \longrightarrow$

    (c) $NaCl(aq) + Hg_2(C_2H_3O_2)_2(aq) \longrightarrow$

    (d) $NaC_2H_3O_2(aq) + Pb(NO_3)_2(aq) \longrightarrow$

64. Complete and balance each equation. If no reaction occurs, write *NO REACTION*.

    (a) $NaOH(aq) + FeBr_3(aq) \longrightarrow$

    (b) $BaCl_2(aq) + AgNO_3(aq) \longrightarrow$

    (c) $Na_2CO_3(aq) + CoCl_2(aq) \longrightarrow$

    (d) $K_2S(aq) + BaCl_2(aq) \longrightarrow$

65. Write a molecular equation for the precipitation reaction that occurs (if any) when each pair of solutions is mixed. If no reaction occurs, write *NO REACTION*.

    (a) sodium carbonate and lead(II) nitrate

    (b) potassium sulfate and lead(II) acetate

    (c) copper(II) nitrate and barium sulfide

    (d) calcium nitrate and sodium iodide

66. Write a molecular equation for the precipitation reaction that occurs (if any) when each pair of solutions is mixed. If no reaction occurs, write *NO REACTION*.

    (a) potassium chloride and lead(II) acetate

    (b) lithium sulfate and strontium chloride

    (c) potassium bromide and calcium sulfide

    (d) chromium(III) nitrate and potassium phosphate

67. Determine whether each equation for a precipitation reaction is correct. Correct any incorrect equations. If no reaction occurs, write *NO REACTION*.

    (a) $Ba(NO_3)_2(aq) + (NH_4)_2SO_4(aq) \longrightarrow$
$$BaSO_4(s) + 2\,NH_4NO_3(aq)$$

    (b) $BaS(aq) + 2\,KCl(aq) \longrightarrow BaCl_2(s) + K_2S(aq)$

    (c) $2\,KI(aq) + Pb(NO_3)_2(aq) \longrightarrow$
$$PbI_2(s) + 2\,KNO_3(aq)$$

    (d) $Pb(NO_3)_2(aq) + 2\,LiCl(aq) \longrightarrow$
$$2\,LiNO_3(s) + PbCl_2(aq)$$

68. Determine whether each equation for a precipitation reaction is correct. Correct any incorrect equations. If no reaction occurs, write *NO REACTION*.

    (a) $AgNO_3(aq) + NaCl(aq) \longrightarrow$
$$NaCl(s) + AgNO_3(aq)$$

    (b) $K_2SO_4(aq) + Co(NO_3)_2(aq) \longrightarrow$
$$CoSO_4(s) + 2\,KNO_3(aq)$$

    (c) $Cu(NO_3)_2(aq) + (NH_4)_2S(aq) \longrightarrow$
$$CuS(s) + 2\,NH_4NO_3(aq)$$

    (d) $Hg_2(NO_3)_2(aq) + 2\,LiCl(aq) \longrightarrow$
$$Hg_2Cl_2(s) + 2\,LiNO_3(aq)$$

IONIC AND NET IONIC EQUATIONS

69. Identify the spectator ions in the complete ionic equation.

$$2\,K^+(aq) + S^{2-}(aq) + Pb^{2+}(aq) + 2\,NO_3{}^-(aq) \longrightarrow$$
$$PbS(s) + 2\,K^+(aq) + 2\,NO_3{}^-(aq)$$

70. Identify the spectator ions in the complete ionic equation.

$$Ba^{2+}(aq) + 2\,I^-(aq) + 2\,Na^+(aq) + SO_4{}^{2-}(aq) \longrightarrow$$
$$BaSO_4(s) + 2\,I^-(aq) + 2\,Na^+(aq)$$

71. Write balanced complete ionic and net ionic equations for each reaction.

    (a) $AgNO_3(aq) + KCl(aq) \longrightarrow$
$$AgCl(s) + KNO_3(aq)$$

    (b) $CaS(aq) + CuCl_2(aq) \longrightarrow CuS(s) + CaCl_2(aq)$

    (c) $NaOH(aq) + HNO_3(aq) \longrightarrow$
$$H_2O(l) + NaNO_3(aq)$$

    (d) $2\,K_3PO_4(aq) + 3\,NiCl_2(aq) \longrightarrow$
$$Ni_3(PO_4)_2(s) + 6\,KCl(aq)$$

72. Write balanced complete ionic and net ionic equations for each reaction.

    (a) $HI(aq) + KOH(aq) \longrightarrow H_2O(l) + KI(aq)$

    (b) $Na_2SO_4(aq) + CaI_2(aq) \longrightarrow$
$$CaSO_4(s) + 2\,NaI(aq)$$

    (c) $2\,HC_2H_3O_2(aq) + Na_2CO_3(aq) \longrightarrow$
$$H_2O(l) + CO_2(g) + 2\,NaC_2H_3O_2(aq)$$

    (d) $NH_4Cl(aq) + NaOH(aq) \longrightarrow$
$$H_2O(l) + NH_3(g) + NaCl(aq)$$

73. Mercury(I) ions ($Hg_2{}^{2+}$) can be removed from solution by precipitation with Cl^-. Suppose a solution contains aqueous $Hg_2(NO_3)_2$. Write complete ionic and net ionic equations to show the reaction of aqueous $Hg_2(NO_3)_2$ with aqueous sodium chloride to form solid Hg_2Cl_2 and aqueous sodium nitrate.

74. Lead ions can be removed from solution by precipitation with sulfate ions. Suppose a solution contains lead(II) nitrate. Write a complete ionic and net ionic equation to show the reaction of aqueous lead(II) nitrate with aqueous potassium sulfate to form solid lead(II) sulfate and aqueous potassium nitrate.

75. Write complete ionic and net ionic equations for each of the reactions in Problem 65.

76. Write complete ionic and net ionic equations for each of the reactions in Problem 66.

ACID–BASE AND GAS EVOLUTION REACTIONS

77. When a hydrochloric acid solution is combined with a potassium hydroxide solution, an acid–base reaction occurs. Write a balanced molecular equation and a net ionic equation for this reaction.

78. A beaker of nitric acid is neutralized with calcium hydroxide. Write a balanced molecular equation and a net ionic equation for this reaction.

79. Complete and balance each acid–base reaction.

(a) $HCl(aq) + Ba(OH)_2(aq) \longrightarrow$

(b) $H_2SO_4(aq) + KOH(aq) \longrightarrow$

(c) $HClO_4(aq) + NaOH(aq) \longrightarrow$

80. Complete and balance each acid–base reaction.

(a) $HC_2H_3O_2(aq) + Ca(OH)_2(aq) \longrightarrow$

(b) $HBr(aq) + LiOH(aq) \longrightarrow$

(c) $H_2SO_4(aq) + Ba(OH)_2(aq) \longrightarrow$

81. Complete and balance each gas evolution reaction.

(a) $HBr(aq) + NaHCO_3(aq) \longrightarrow$

(b) $NH_4I(aq) + KOH(aq) \longrightarrow$

(c) $HNO_3(aq) + K_2SO_3(aq) \longrightarrow$

(d) $HI(aq) + Li_2S(aq) \longrightarrow$

82. Complete and balance each gas evolution reaction.

(a) $HClO_4(aq) + K_2CO_3(aq) \longrightarrow$

(b) $HC_2H_3O_2(aq) + LiHSO_3(aq) \longrightarrow$

(c) $(NH_4)_2SO_4(aq) + Ca(OH)_2(aq) \longrightarrow$

(d) $HCl(aq) + ZnS(s) \longrightarrow$

OXIDATION–REDUCTION AND COMBUSTION

83. Which reactions are redox reactions?

(a) $Ba(NO_3)_2(aq) + K_2SO_4(aq) \longrightarrow$
$BaSO_4(s) + 2\,KNO_3(aq)$

(b) $Ca(s) + Cl_2(g) \longrightarrow CaCl_2(s)$

(c) $HCl(aq) + NaOH(aq) \longrightarrow H_2O(l) + NaCl(aq)$

(d) $Zn(s) + Fe^{2+}(aq) \longrightarrow Zn^{2+}(aq) + Fe(s)$

84. Which reactions are redox reactions?

(a) $Al(s) + 3\,Ag^+(aq) \longrightarrow Al^{3+}(aq) + 3\,Ag(s)$

(b) $4\,K(s) + O_2(g) \longrightarrow 2\,K_2O(s)$

(c) $SO_3(g) + H_2O(l) \longrightarrow H_2SO_4(aq)$

(d) $Mg(s) + Br_2(l) \longrightarrow MgBr_2(s)$

85. Complete and balance each combustion reaction.

(a) $C_2H_6(g) + O_2(g) \longrightarrow$

(b) $Ca(s) + O_2(g) \longrightarrow$

(c) $C_3H_8O(l) + O_2(g) \longrightarrow$

(d) $C_4H_{10}S(l) + O_2(g) \longrightarrow$

86. Complete and balance each combustion reaction.

(a) $S(s) + O_2(g) \longrightarrow$

(b) $C_7H_{16}(l) + O_2(g) \longrightarrow$

(c) $C_4H_{10}O(l) + O_2(g) \longrightarrow$

(d) $CS_2(l) + O_2(g) \longrightarrow$

87. Write a balanced chemical equation for the synthesis reaction of $Br_2(g)$ with each metal:

(a) $Ag(s)$

(b) $K(s)$

(c) $Al(s)$

(d) $Ca(s)$

88. Write a balanced chemical equation for the synthesis reaction of $Cl_2(g)$ with each metal:

(a) $Zn(s)$

(b) $Ga(s)$

(c) $Rb(s)$

(d) $Mg(s)$

CLASSIFYING CHEMICAL REACTIONS BY WHAT ATOMS DO

89. Classify each chemical reaction as a synthesis, decomposition, single-displacement, or double-displacement reaction.

(a) $K_2S(aq) + Co(NO_3)_2(aq) \longrightarrow$
$2\,KNO_3(aq) + CoS(s)$

(b) $3\,H_2(g) + N_2(g) \longrightarrow 2\,NH_3(g)$

(c) $Zn(s) + CoCl_2(aq) \longrightarrow ZnCl_2(aq) + Co(s)$

(d) $CH_3Br(g) \xrightarrow{\text{UV light}} CH_3(g) + Br(g)$

90. Classify each chemical reaction as a synthesis, decomposition, single-displacement, or double-displacement reaction.

(a) $CaSO_4(s) \xrightarrow{\text{heat}} CaO(s) + SO_3(g)$

(b) $2\,Na(s) + O_2(g) \longrightarrow Na_2O_2(s)$

(c) $Pb(s) + 2\,AgNO_3(aq) \longrightarrow$
$Pb(NO_3)_2(aq) + 2\,Ag(s)$

(d) $HI(aq) + NaOH(aq) \longrightarrow H_2O(l) + NaI(aq)$

91. NO is a pollutant emitted by motor vehicles. It is formed by the reaction:

    (a) $N_2(g) + O_2(g) \longrightarrow 2\,NO(g)$

    Once in the atmosphere, NO (through a series of reactions) adds one oxygen atom to form NO_2. NO_2 then interacts with UV light according to the reaction:

    (b) $NO_2(g) \xrightarrow[\text{UV light}]{} NO(g) + O(g)$

    These freshly formed oxygen atoms then react with O_2 in the air to form ozone (O_3), a main component of smog:

    (c) $O(g) + O_2(g) \longrightarrow O_3(g)$

    Classify each of the preceding reactions as a synthesis, decomposition, single-displacement, or double-displacement reaction.

92. A main source of sulfur oxide pollutants are smelters where sulfide ores are converted into metals. The first step in this process is the reaction of the sulfide ore with oxygen in reactions such as:

    (a) $2\,PbS(s) + 3\,O_2(g) \longrightarrow 2\,PbO(s) + 2\,SO_2(g)$

    Sulfur dioxide can then react with oxygen in air to form sulfur trioxide:

    (b) $2\,SO_2(g) + O_2(g) \longrightarrow 2\,SO_3(g)$

    Sulfur trioxide can then react with water from rain to form sulfuric acid that falls as acid rain:

    (c) $SO_3(g) + H_2O(l) \longrightarrow H_2SO_4(aq)$

    Classify each of the preceding reactions as a synthesis, decomposition, single-displacement, or double-displacement reaction.

CUMULATIVE PROBLEMS

93. Predict the products of each reaction and write balanced complete ionic and net ionic equations for each. If no reaction occurs, write *NO REACTION*.

    (a) $NaI(aq) + Hg_2(NO_3)_2(aq) \longrightarrow$

    (b) $HClO_4(aq) + Ba(OH)_2(aq) \longrightarrow$

    (c) $Li_2CO_3(aq) + NaCl(aq) \longrightarrow$

    (d) $HCl(aq) + Li_2CO_3(aq) \longrightarrow$

94. Predict the products of each reaction and write balanced complete ionic and net ionic equations for each. If no reaction occurs, write *NO REACTION*.

    (a) $LiCl(aq) + AgNO_3(aq) \longrightarrow$

    (b) $H_2SO_4(aq) + Li_2SO_3(aq) \longrightarrow$

    (c) $HC_2H_3O_2(aq) + Ca(OH)_2(aq) \longrightarrow$

    (d) $HCl(aq) + KBr(aq) \longrightarrow$

95. Predict the products of each reaction and write balanced complete ionic and net ionic equations for each. If no reaction occurs, write *NO REACTION*.

    (a) $BaS(aq) + NH_4Cl(aq) \longrightarrow$

    (b) $NaC_2H_3O_2(aq) + KCl(aq) \longrightarrow$

    (c) $KHSO_3(aq) + HNO_3(aq) \longrightarrow$

    (d) $MnCl_3(aq) + K_3PO_4(aq) \longrightarrow$

96. Predict the products of each reaction and write balanced complete ionic and net ionic equations for each. If no reaction occurs, write *NO REACTION*.

    (a) $H_2SO_4(aq) + HNO_3(aq) \longrightarrow$

    (b) $NaOH(aq) + LiOH(aq) \longrightarrow$

    (c) $Cr(NO_3)_3(aq) + LiOH(aq) \longrightarrow$

    (d) $HCl(aq) + Hg_2(NO_3)_2(aq) \longrightarrow$

97. Predict the type of reaction (if any) that occurs between each pair of substances. Write balanced molecular equations for each. If no reaction occurs, write *NO REACTION*.

    (a) aqueous potassium hydroxide and aqueous acetic acid

    (b) aqueous hydrobromic acid and aqueous potassium carbonate

    (c) gaseous hydrogen and gaseous oxygen

    (d) aqueous ammonium chloride and aqueous lead(II) nitrate

98. Predict the type of reaction (if any) that occurs between each pair of substances. Write balanced molecular equations for each. If no reaction occurs, write *NO REACTION*.

    (a) aqueous hydrochloric acid and aqueous copper(II) nitrate

    (b) liquid pentanol ($C_5H_{12}O$) and gaseous oxygen

    (c) aqueous ammonium chloride and aqueous calcium hydroxide

    (d) aqueous strontium sulfide and aqueous copper(II) sulfate

99. Classify each reaction in as many ways as possible.

(a) $2\,Al(s) + 3\,Cu(NO_3)_2(aq) \longrightarrow$
$$2\,Al(NO_3)_3(aq) + 3\,Cu(s)$$

(b) $HBr(aq) + KHSO_3(aq) \longrightarrow$
$$H_2O(l) + SO_2(g) + NaBr(aq)$$

(c) $2\,HI(aq) + Na_2S(aq) \longrightarrow H_2S(g) + 2\,NaI(aq)$

(d) $K_2CO_3(aq) + FeBr_2(aq) \longrightarrow$
$$FeCO_3(s) + 2\,KBr(aq)$$

100. Classify each reaction in as many ways as possible.

(a) $NaCl(aq) + AgNO_3(aq) \longrightarrow$
$$AgCl(s) + NaNO_3(aq)$$

(b) $2\,Rb(s) + Br_2(g) \longrightarrow 2\,RbBr(s)$

(c) $Zn(s) + NiBr_2(aq) \longrightarrow Ni(s) + ZnBr_2(aq)$

(d) $Ca(s) + 2\,H_2O(l) \longrightarrow Ca(OH)_2(aq) + H_2(g)$

101. Hard water often contains dissolved Ca^{2+} and Mg^{2+} ions. One way to soften water is to add phosphates. The phosphate ion forms insoluble precipitates with calcium and magnesium ions, removing them from solution. Suppose that a solution contains aqueous calcium chloride and aqueous magnesium nitrate. Write molecular, complete ionic, and net ionic equations showing how the addition of sodium phosphate precipitates the calcium and magnesium ions.

102. Lakes that have been acidified by acid rain (HNO_3 and H_2SO_4) can be neutralized by a process called *liming*, in which limestone ($CaCO_3$) is added to the acidified water. Write ionic and net ionic equations to show how limestone reacts with HNO_3 and H_2SO_4 to neutralize them. How would you be able to tell if the neutralization process was working?

103. What solution can you add to each cation mixture to precipitate one cation while keeping the other cation in solution? Write a net ionic equation for the precipitation reaction that occurs.

(a) $Fe^{2+}(aq)$ and $Pb^{2+}(aq)$

(b) $K^+(aq)$ and $Ca^{2+}(aq)$

(c) $Ag^+(aq)$ and $Ba^{2+}(aq)$

(d) $Cu^{2+}(aq)$ and $Hg_2^{2+}(aq)$

104. What solution can you add to each cation mixture to precipitate one cation while keeping the other cation in solution? Write a net ionic equation for the precipitation reaction that occurs.

(a) $Sr^{2+}(aq)$ and $Hg_2^{2+}(aq)$

(b) $NH_4^+(aq)$ and $Ca^{2+}(aq)$

(c) $Ba^{2+}(aq)$ and $Mg^{2+}(aq)$

(d) $Ag^+(aq)$ and $Zn^{2+}(aq)$

105. A solution contains one or more of the following ions: Ag^+, Ca^{2+}, and Cu^{2+}. When sodium chloride is added to the solution, no precipitate occurs. When sodium sulfate is added to the solution, a white precipitate occurs. The precipitate is filtered off and sodium carbonate is added to the remaining solution, producing a precipitate. Which ions were present in the original solution? Write net ionic equations for the formation of each of the precipitates observed.

106. A solution contains one or more of the following ions: Hg_2^{2+}, Ba^{2+}, and Fe^{2+}. When potassium chloride is added to the solution, a precipitate forms. The precipitate is filtered off and potassium sulfate is added to the remaining solution, producing no precipitate. When potassium carbonate is added to the remaining solution, a precipitate occurs. Which ions were present in the original solution? Write net ionic equations for the formation of each of the precipitates observed.

107. A solution contains an unknown amount of dissolved calcium. Addition of 0.112 mol of K_3PO_4 causes complete precipitation of all of the calcium. How many moles of calcium were dissolved in the solution? What mass of calcium was dissolved in the solution?

108. A solution contains an unknown amount of dissolved magnesium. Addition of 0.0877 mol of Na_2CO_3 causes complete precipitation of all of the magnesium. What mass of magnesium was dissolved in the solution?

109. A solution contains 0.133 g of dissolved lead. How many moles of sodium chloride must be added to the solution to completely precipitate all of the dissolved lead? What mass of sodium chloride must be added?

110. A solution contains 1.77 g of dissolved silver. How many moles of potassium chloride must be added to the solution to completely precipitate all of the silver? What mass of potassium chloride must be added?

HIGHLIGHT PROBLEMS

111. The following are molecular views of two different possible mechanisms by which an automobile air bag might function. One of these mechanisms involves a chemical reaction and the other does not. By looking at the molecular views, can you tell which mechanism operates via a chemical reaction?

▲ When an airbag is detonated, the bag inflates. These figures show two possible ways in which the inflation may happen.

112. Precipitation reactions often produce brilliant colors. Look at the photographs of each precipitation reaction and write molecular, complete ionic, and net ionic equations for each one.

▲ (a) The precipitation reaction that occurs when aqueous iron(III) nitrate is added to aqueous sodium hydroxide. (b) The precipitation reaction that occurs when aqueous cobalt(II) chloride is added to aqueous potassium hydroxide. (c) The precipitation reaction that occurs when aqueous $AgNO_3$ is added to aqueous sodium iodide.

►ANSWERS TO SKILLBUILDER EXERCISES

Skillbuilder 7.1 **(a)** Chemical reaction; heat and light are emitted. **(b)** Not a chemical reaction; gaseous and liquid butane are both butane. **(c)** Chemical reaction; heat and light are emitted. **(d)** Not a chemical reaction; solid dry ice is made of carbon dioxide, which sublimes (evaporates) as carbon dioxide gas.

Skillbuilder 7.2
$$2 Cr_2O_3(s) + 3 C(s) \longrightarrow 4 Cr(s) + 3 CO_2(g)$$

Skillbuilder 7.3
$$2 C_4H_{10}(g) + 13 O_2(g) \longrightarrow 8 CO_2(g) + 10 H_2O(g)$$

Skillbuilder 7.4
$$Pb(C_2H_3O_2)_2(aq) + 2 KI(aq) \longrightarrow PbI_2(s) + 2 KC_2H_3O_2(aq)$$

Skillbuilder 7.5 $\quad 4 HCl(g) + O_2(g) \longrightarrow 2 H_2O(l) + 2 Cl_2(g)$

Skillbuilder 7.6 **(a)** insoluble **(b)** soluble **(c)** insoluble **(d)** soluble

Skillbuilder 7.7
$$2 KOH(aq) + NiBr_2(aq) \longrightarrow Ni(OH)_2(s) + 2 KBr(aq)$$

Skillbuilder 7.8
$$NH_4Cl(aq) + Fe(NO_3)_3(aq) \longrightarrow NO\ REACTION$$

Skillbuilder 7.9
$$K_2SO_4(aq) + Sr(NO_3)_2(aq) \longrightarrow SrSO_4(s) + 2 KNO_3(aq)$$

Skillbuilder 7.10 Complete ionic equation:
$$2 H^+(aq) + 2 Br^-(aq) + Ca^{2+}(aq) + 2 OH^-(aq) \longrightarrow$$
$$2 H_2O(l) + Ca^{2+}(aq) + 2 Br^-(aq)$$

Net ionic equation:

$2 H^+(aq) + 2 OH^-(aq) \longrightarrow 2 H_2O(l)$, or simply

$H^+(aq) + OH^-(aq) \longrightarrow H_2O(l)$

Skillbuilder 7.11   Molecular equation:

$H_2SO_4(aq) + 2\,KOH(aq) \longrightarrow 2\,H_2O(l) + K_2SO_4(aq)$

Net ionic equation:

$2\,H^+(aq) + 2\,OH^-(aq) \longrightarrow 2\,H_2O(l)$

Skillbuilder 7.12

$2\,HBr(aq) + K_2SO_3(aq) \longrightarrow H_2O(l) + SO_2(g) + 2\,KBr(aq)$

Skillbuilder Plus, p. 225

$2\,H^+(aq) + SO_3^{2-}(aq) \longrightarrow H_2O(l) + SO_2(g)$

Skillbuilder 7.13   (a), (b), and (d) are all redox reactions; (c) is a precipitation reaction.

Skillbuilder 7.14

$C_5H_{12}(l) + 8\,O_2(g) \longrightarrow 5\,CO_2(g) + 6\,H_2O(g)$

Skillbuilder Plus, p. 228

$2\,C_3H_7OH(l) + 9\,O_2(g) \longrightarrow 6\,CO_2(g) + 8\,H_2O(g)$

Skillbuilder 7.15   (a) single-displacement (b) double-displacement (c) synthesis (d) decomposition

▶ANSWERS TO CONCEPTUAL CHECKPOINTS

7.1 **(a)** No reaction occurred. The molecules are the same before and after the change.
 (b) A reaction occurred; the molecules have changed.
 (c) A reaction occurred; the molecules have changed.

7.2 **(b)** There are 18 oxygen atoms on the left side of the equation, so the same number is needed on the right: $6 + 6(2) = 18$.

7.3 **(a)** (a) The number of each type of atom must be the same on both sides of a balanced chemical equation. Since molecules change during a chemical reaction, their number is not the same on both sides (b), nor is the sum of all of the coefficients the same (c).

7.4 **(a)** Since chlorides are soluble, and since Ba^{2+} is not an exception, $BaCl_2$ is soluble and will dissolve in water. When it dissolves, it dissociates into its component ions, as shown in (a).

7.5 **(b)** Both of the possible products, MgS and $CaSO_4$, are insoluble. The possible products of the other reactions—Na_2S, $Ca(NO_3)_2$, Na_2SO_4, and $Mg(NO_3)_2$— are all soluble.

7.6 **(d)** In a precipitation reaction, cations and anions "exchange partners" to produce at least one insoluble product. In an acid–base reaction, H^+ and OH^- combine to form water, and their partners pair off to form a salt.

Quantities in Chemical Reactions

"Man masters nature not by force but by understanding. That is why science has succeeded where magic failed: because it has looked for no spell to cast."

JACOB BRONOWSKI (1908–1974)

8.1 Global Warming: Too Much Carbon Dioxide

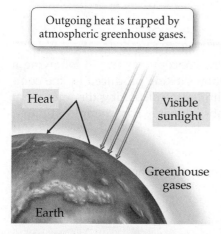

Outgoing heat is trapped by atmospheric greenhouse gases.

Heat

Visible sunlight

Greenhouse gases

Earth

▲ **FIGURE 8.1 The greenhouse effect** Greenhouse gases act like glass in a greenhouse, allowing visible-light energy to enter the atmosphere but preventing heat energy from escaping.

◀ The combustion of fossil fuels such as octane (shown here) produces water and carbon dioxide as products. Carbon dioxide is a greenhouse gas that is believed to be responsible for global warming.

Average global temperatures depend on the balance between incoming sunlight, which warms Earth, and outgoing heat lost to space, which cools it. Certain gases in Earth's atmosphere, called **greenhouse gases**, affect that balance by acting like glass in a greenhouse. They allow sunlight into the atmosphere to warm Earth but prevent heat from escaping (◀ Figure 8.1). Without greenhouse gases, more heat would escape, and Earth's average temperature would be about 60 °F (15 °C) colder. Caribbean tourists would freeze at an icy 21 °F, instead of baking at a tropical 81 °F. On the other hand, if the concentration of greenhouse gases in the atmosphere were to increase, Earth's average temperature would rise.

In recent years scientists have become concerned because the atmospheric concentration of carbon dioxide (CO_2)—Earth's most significant greenhouse gas in terms of its contribution to climate—is rising. This rise in CO_2 concentration enhances the atmosphere's ability to hold heat and may therefore lead to **global warming**, an increase in Earth's average temperature. Since 1860, atmospheric CO_2 levels have risen by 25%, and Earth's average temperature has increased by 0.6 °C (about 1.1 °F) (▶ Figure 8.2 on the next page).

The primary cause of rising atmospheric CO_2 concentration is the burning of fossil fuels. Fossil fuels—natural gas, petroleum, and coal—provide approximately 90% of our society's energy. Combustion of fossil fuels, however, produces CO_2. As an example, consider the combustion of octane (C_8H_{18}), a component of gasoline.

$$2\,C_8H_{18}(l) + 25\,O_2(g) \longrightarrow 16\,CO_2(g) + 18\,H_2O(g)$$

The balanced chemical equation shows that 16 mol of CO_2 are produced for every 2 mol of octane burned. Since we know the world's annual fossil fuel consumption, we can estimate the world's annual CO_2 production. A simple calculation shows that the world's annual CO_2 production—from fossil fuel

▶ FIGURE 8.2 Global warming
Yearly temperature differences from the 120-year average temperature. Earth's average temperature has increased by about 0.6 °C since 1880. (*Source:* NASA GISS Surface Temperature Analysis)

combustion—matches the measured annual atmospheric CO_2 increase, implying that fossil fuel combustion is indeed responsible for increased atmospheric CO_2 levels.

The numerical relationship between chemical quantities in a balanced chemical equation is called reaction **stoichiometry**. Stoichiometry allows us to predict the amounts of products that form in a chemical reaction based on the amounts of reactants. Stoichiometry also allows us to predict how much of the reactants are necessary to form a given amount of product, or how much of one reactant is required to completely react with another reactant. These calculations are central to chemistry, allowing chemists to plan and carry out chemical reactions to obtain products in the desired quantities.

8.2 Making Pancakes: Relationships between Ingredients

The concepts of stoichiometry are similar to the concepts we use in following a cooking recipe. Calculating the amount of carbon dioxide produced by the combustion of a given amount of a fossil fuel is similar to calculating the number of pancakes that can be made from a given number of eggs. For example, suppose you use the following pancake recipe.

For the sake of simplicity, this recipe omits liquid ingredients.

$$1 \text{ cup flour} + 2 \text{ eggs} + \tfrac{1}{2} \text{ tsp baking powder} \longrightarrow 5 \text{ pancakes}$$

| 1 cup flour | 2 eggs | $\frac{1}{2}$ tsp baking powder | 5 pancakes |

▲ A recipe gives numerical relationships between the ingredients and the number of pancakes.

The recipe shows the numerical relationships between the pancake ingredients. It says that if we have 2 eggs—and enough of everything else—we can make 5 pancakes. We can write this relationship as a ratio.

2 eggs : 5 pancakes

What if we have 8 eggs? Assuming that we have enough of everything else, how many pancakes can we make? Using the preceding ratio as a conversion factor, we can determine that 8 eggs are sufficient to make 20 pancakes.

$$8 \text{ eggs} \times \frac{5 \text{ pancakes}}{2 \text{ eggs}} = 20 \text{ pancakes}$$

The pancake recipe contains numerical conversion factors between the pancake ingredients and the number of pancakes. Other conversion factors from this recipe include:

1 cup flour : 5 pancakes

$\frac{1}{2}$ tsp baking powder : 5 pancakes

The recipe also gives us relationships among the ingredients themselves. For example, how much baking powder is required to go with 3 cups of flour? From the recipe:

1 cup flour : $\frac{1}{2}$ tsp baking powder

With this ratio, we can form the conversion factor to calculate the appropriate amount of baking powder.

$$3 \text{ cups flour} \times \frac{\frac{1}{2} \text{ tsp baking powder}}{1 \text{ cup flour}} = \frac{3}{2} \text{ tsp baking powder}$$

8.3   Making Molecules: Mole-to-Mole Conversions

In a balanced chemical equation, we have a "recipe" for how reactants combine to form products. For example, the following equation shows how hydrogen and nitrogen combine to form ammonia (NH_3).

$$3 H_2(g) + N_2(g) \longrightarrow 2 NH_3(g)$$

The balanced equation shows that 3 H_2 molecules react with 1 N_2 molecule to form 2 NH_3 molecules. We can express these relationships as the following ratios.

3 H_2 molecules : 1 N_2 molecule : 2 NH_3 molecules

Since we do not ordinarily deal with individual molecules, we can express the same ratios in moles.

$$3 \text{ mol } H_2 : 1 \text{ mol } N_2 : 2 \text{ mol } NH_3$$

If we have 3 mol of N_2, and more than enough H_2, how much NH_3 can we make? We first sort the information in the problem.

GIVEN: 3 mol N_2

FIND: mol NH_3

SOLUTION MAP

We then strategize by drawing a solution map that begins with mol N_2 and ends with mol NH_3. The conversion factor comes from the balanced chemical equation.

$$\frac{2 \text{ mol } NH_3}{1 \text{ mol } N_2}$$

RELATIONSHIPS USED

1 mol N_2 : 2 mol NH_3 (from balanced equation)

SOLUTION

We can then do the conversion.

$$3 \text{ mol } N_2 \times \frac{2 \text{ mol } NH_3}{1 \text{ mol } N_2} = 6 \text{ mol } NH_3$$

We have enough N_2 to make 6 mol of NH_3.

EXAMPLE 8.1 Mole-to-Mole Conversions

Sodium chloride, NaCl, forms by this reaction between sodium and chlorine.

$$2 \text{ Na}(s) + Cl_2(g) \longrightarrow 2 \text{ NaCl}(s)$$

How many moles of NaCl result from the complete reaction of 3.4 mol of Cl_2? Assume that there is more than enough Na.

SORT You are given the number of moles of a reactant (Cl_2) and asked to find the number of moles of product (NaCl) that will form if the reactant completely reacts.	**GIVEN:** 3.4 mol Cl_2 **FIND:** mol NaCl
STRATEGIZE Draw the solution map beginning with moles of chlorine and using the stoichiometric conversion factor to calculate moles of sodium chloride. The conversion factor comes from the balanced chemical equation.	**SOLUTION MAP** $\dfrac{2 \text{ mol NaCl}}{1 \text{ mol } Cl_2}$ **RELATIONSHIPS USED** 1 mol Cl_2 : 2 mol NaCl (from balanced chemical equation)
SOLVE Follow the solution map to solve the problem. There is enough Cl_2 to produce 6.8 mol of NaCl.	**SOLUTION** $3.4 \text{ mol } Cl_2 \times \dfrac{2 \text{ mol NaCl}}{1 \text{ mol } Cl_2} = 6.8 \text{ mol NaCl}$
CHECK Check your answer. Are the units correct? Does the answer make physical sense?	The answer has the correct units, moles. The answer is reasonable because each mole of Cl_2 makes two moles of NaCl.

Water is formed when hydrogen gas reacts explosively with oxygen gas according to the balanced equation:

$$O_2(g) + 2\,H_2(g) \longrightarrow 2\,H_2O(g)$$

How many moles of H_2O result from the complete reaction of 24.6 mol of O_2? Assume that there is more than enough H_2.

▶**FOR MORE PRACTICE**   Example 8.8; Problems 15, 16, 17, 18.

CONCEPTUAL CHECKPOINT 8.1

Methane (CH_4) undergoes combustion according to the reaction:

$$CH_4(g) + 2\,O_2(g) \longrightarrow CO_2(g) + 2\,H_2O(g)$$

If the figure at left represents the amount of oxygen available to react, which of the following best represents the amount of CH_4 required to completely react with all of the oxygen?

(a)	(b)	(c)

8.4   Making Molecules: Mass-to-Mass Conversions

In Chapter 6, we learned how a chemical *formula* contains conversion factors for converting between moles of a compound and moles of its constituent elements. In this chapter, we have seen how a chemical *equation* contains conversion factors between moles of reactants and moles of products. However, we are often interested in relationships between *mass* of reactants and *mass* of products. For example, we might want to know the mass of carbon dioxide emitted by an automobile per kilogram of gasoline used. Or we might want to know the mass of each reactant required to obtain a certain mass of a product in a synthesis reaction. These calculations are similar to calculations covered in Section 6.5, where we converted between mass of a compound and mass of a constituent element. The general outline for these types of calculations is:

Mass A	Moles A	Moles B	Mass B
Molar mass A	Ratio from balanced equation		Molar mass B

where A and B are two different substances involved in the reaction. We use the molar mass of A to convert from mass of A to moles of A. We use the ratio from the balanced equation to convert from moles of A to moles of B, and we use the molar mass of B to convert moles of B to mass of B. For example, suppose we want to calculate the mass of CO_2 emitted upon the combustion of 5.0×10^2 g of pure octane. The balanced chemical equation for octane combustion is:

$$2\,C_8H_{18}(l) + 25\,O_2(g) \longrightarrow 16\,CO_2(g) + 18\,H_2O(g)$$

CHEMISTRY IN THE MEDIA
The Controversy over Oxygenated Fuels

We have seen that the balanced chemical equation for the combustion of octane, a component of gasoline, is:

$$2\,C_8H_{18}(l) + 25\,O_2(g) \longrightarrow 16\,CO_2(g) + 18\,H_2O(g)$$

We have also learned how balanced chemical equations give numerical relationships between reactants. The preceding equation shows that 25 mol of O_2 are required to completely react with 2 mol of C_8H_{18}. What if there were not enough O_2 in the cylinders of an automobile engine to fully react with the amount of octane flowing into them? For many reactions, a shortage of one reactant simply means that less product forms, something we will learn more about later in this chapter. However, for some reactions, a shortage of one reactant causes other reactions—called *side reactions*—to occur along with the desired reaction. In the case of octane and the other major components of gasoline, those side reactions result in pollutants such as carbon monoxide (CO) and ozone (O_3).

In 1990, the U.S. Congress, in efforts to lower air pollution, passed amendments to the Clean Air Act requiring oil companies to add substances to gasoline that prevent these side reactions. Since these additives have the effect of increasing the amount of oxygen during combustion, the resulting gasoline is called oxygenated fuel. The additive of choice among oil companies used to be a compound called MTBE (methyl tertiary butyl ether). The immediate results were positive. Carbon monoxide and ozone levels in many major cities decreased significantly.

Over time, however, MTBE—a compound that does not readily biodegrade—began to appear in drinking-water supplies across the nation. MTBE made its way into drinking water through gasoline spills at gas stations, from boat motors, and from leaking underground storage tanks. The consequences have been significant. MTBE, even at low levels, imparts a turpentine-like odor and foul taste to drinking water. It is also a suspected carcinogen.

Public response was swift and dramatic. Several multimillion dollar class-action lawsuits were filed and settled against the manufacturers of MTBE, against gas stations suspected of leaking it, and against the oil companies that put MTBE into gasoline. Most states have completely banned MTBE from gasoline. Ethanol, made from the fermentation of grains, has been used as a substitute for MTBE because it has many of the same pollution-reducing effects without the associated health hazards. Oil companies did not use ethanol originally because it was more expensive than MTBE, but now ethanol has become the additive of choice.

CAN YOU ANSWER THIS? *How many moles of oxygen (O_2) are required to completely react with 425 mol of octane (approximate capacity of a 15-gal automobile gasoline tank)?*

▲ The 1990 amendments to the Clean Air Act required oil companies to put additives in gasoline that increased its oxygen content.

▲ MTBE was the additive of choice.

We begin by sorting the information in the problem.

GIVEN: 5.0×10^2 g C_8H_{18}

FIND: g CO_2

Notice that we are given g C_8H_{18} and asked to find g CO_2. The balanced chemical equation, however, gives us a relationship between moles of C_8H_{18} and moles of CO_2. Consequently, before using that relationship, we must convert from grams to moles. The solution map uses the general outline

Mass A $\longrightarrow$ Moles A $\longrightarrow$ Moles B $\longrightarrow$ Mass B

where A is octane and B is carbon dioxide.

SOLUTION MAP

We strategize by drawing the solution map, which begins with mass of octane and ends with mass of carbon dioxide.

$$\dfrac{1 \text{ mol } C_8H_{18}}{114.3 \text{ g } C_8H_{18}} \qquad \dfrac{16 \text{ mol } CO_2}{2 \text{ mol } C_8H_{18}} \qquad \dfrac{44.01 \text{ g } CO_2}{1 \text{ mol } CO_2}$$

RELATIONSHIPS USED

$2 \text{ mol } C_8H_{18} \equiv 16 \text{ mol } CO_2$ (from chemical equation)

Molar mass C_8H_{18} = 114.3 g/mol

Molar mass CO_2 = 44.01 g/mol

SOLUTION

We then follow the solution map to solve the problem, beginning with g C_8H_{18} and canceling units to arrive at g CO_2.

$$5.0 \times 10^2 \text{ g } C_8H_{18} \times \dfrac{1 \text{ mol } C_8H_{18}}{114.3 \text{ g } C_8H_{18}} \times \dfrac{16 \text{ mol } CO_2}{2 \text{ mol } C_8H_{18}} \times \dfrac{44.01 \text{ g } CO_2}{1 \text{ mol } CO_2} = 1.5 \times 10^3 \text{ g } CO_2$$

Upon combustion, 5.0×10^2 g of octane produces 1.5×10^3 g of carbon dioxide.

EXAMPLE 8.2 Mass-to-Mass Conversions

In photosynthesis, plants convert carbon dioxide and water into glucose ($C_6H_{12}O_6$) according to the reaction:

$$6 \, CO_2(g) + 6 \, H_2O(l) \xrightarrow[\text{sunlight}]{} 6 \, O_2(g) + C_6H_{12}O_6(aq)$$

How many grams of glucose can be synthesized from 58.5 g of CO_2? Assume that there is more than enough water present to react with all of the CO_2.

SORT	**GIVEN:** 58.5 g CO_2
You are given the mass of carbon dioxide and asked to find the mass of glucose that can form if the carbon dioxide completely reacts.	**FIND:** g $C_6H_{12}O_6$

STRATEGIZE	**SOLUTION MAP**
The solution map uses the general outline	

Mass A ⟶ Moles A ⟶

    Moles B ⟶ Mass B

where A is carbon dioxide and B is glucose.

The main conversion factor is the stoichiometric relationship between moles of carbon dioxide and moles of glucose. This conversion factor comes from the balanced equation. The other conversion factors are simply the molar masses of carbon dioxide and glucose.

$$\dfrac{1 \text{ mol } CO_2}{44.01 \text{ g } CO_2} \qquad \dfrac{1 \text{ mol } C_6H_{12}O_6}{6 \text{ mol } CO_2} \qquad \dfrac{180.2 \text{ g } C_6H_{12}O_6}{1 \text{ mol } C_6H_{12}O_6}$$

RELATIONSHIPS USED

6 mol CO_2 : 1 mol $C_6H_{12}O_6$ (from balanced chemical equation)

Molar mass CO_2 = 44.01 g/mol

Molar mass $C_6H_{12}O_6$ = 180.2 g/mol

SOLVE	SOLUTION
Follow the solution map to solve the problem. Begin with grams of carbon dioxide and multiply by the appropriate factors to arrive at grams of glucose.	$58.5 \text{ g CO}_2 \times \dfrac{1 \text{ mol CO}_2}{44.01 \text{ g CO}_2} \times \dfrac{1 \text{ mol C}_6\text{H}_{12}\text{O}_6}{6 \text{ mol CO}_2} \times \dfrac{180.2 \text{ g C}_6\text{H}_{12}\text{O}_6}{1 \text{ mol C}_6\text{H}_{12}\text{O}_6} = 39.9 \text{ g C}_6\text{H}_{12}\text{O}_6$
CHECK Are the units correct? Does the answer make physical sense?	The units, g $C_6H_{12}O_6$, are correct. The magnitude of the answer seems reasonable because it is of the same order of magnitude as the given mass of carbon dioxide. An answer that is orders of magnitude different would immediately be suspect.

▶**SKILLBUILDER 8.2 | Mass-to-Mass Conversions**

Magnesium hydroxide, the active ingredient in milk of magnesia, neutralizes stomach acid, primarily HCl, according to the reaction:

$$Mg(OH)_2(aq) + 2\,HCl(aq) \longrightarrow 2\,H_2O(l) + MgCl_2(aq)$$

How much HCl in grams can be neutralized by 5.50 g of $Mg(OH)_2$?

▶**FOR MORE PRACTICE**   Example 8.9; Problems 31, 32, 33, 34.

EXAMPLE 8.3   Mass-to-Mass Conversions

One of the components of acid rain is nitric acid, which forms when NO_2, a pollutant, reacts with oxygen and rainwater according to the following simplified reaction.

$$4\,NO_2(g) + O_2(g) + 2\,H_2O(l) \longrightarrow 4\,HNO_3(aq)$$

Assuming that there is more than enough O_2 and H_2O, how much HNO_3 in kilograms forms from 1.5×10^3 kg of NO_2 pollutant?

SORT You are given the mass of nitrogen dioxide (a reactant) and asked to find the mass of nitric acid that can form if the nitrogen dioxide completely reacts.	**GIVEN:** 1.5×10^3 kg NO_2 **FIND:** kg HNO_3

STRATEGIZE The solution map follows the general format of: Mass $\longrightarrow$ Moles $\longrightarrow$ $\qquad\qquad$ Moles $\longrightarrow$ Mass However, since the original quantity of NO_2 is given in kilograms, you must first convert to grams. Since the final quantity is requested in kilograms, you must convert back to kilograms at the end. The main conversion factor is the stoichiometric relationship between moles of nitrogen dioxide and moles of nitric acid. This conversion factor comes from the balanced equation. The other conversion factors are simply the molar masses of nitrogen dioxide and nitric acid and the relationship between kilograms and grams.	**SOLUTION MAP** **RELATIONSHIPS USED** 4 mol NO_2 : 4 mol HNO_3 (from balanced chemical equation) Molar mass NO_2 = 46.01 g/mol Molar mass HNO_3 = 63.02 g/mol 1 kg = 1000 g

SOLVE	SOLUTION
Follow the solution map to solve the problem. Begin with kilograms of nitrogen dioxide and multiply by the appropriate conversion factors to arrive at kilograms of nitric acid.	$1.5 \times 10^3 \text{ kg NO}_2 \times \dfrac{1000 \text{ g}}{1 \text{ kg}} \times \dfrac{1 \text{ mol NO}_2}{46.01 \text{ g NO}_2} \times \dfrac{4 \text{ mol HNO}_3}{4 \text{ mol NO}_2} \times$ $\dfrac{63.02 \text{ g HNO}_3}{1 \text{ mol HNO}_3} \times \dfrac{1 \text{ kg}}{1000 \text{ g}} = 2.1 \times 10^3 \text{ kg HNO}_3$

CHECK	
Are the units correct? Does the answer make physical sense?	The units, kg HNO_3 are correct. The magnitude of the answer seems reasonable because it is of the same order of magnitude as the given mass of nitrogen dioxide. An answer that is orders of magnitude different would immediately be suspect.

▶ **SKILLBUILDER 8.3 | Mass-to-Mass Conversions**

Another component of acid rain is sulfuric acid, which forms when SO_2, also a pollutant, reacts with oxygen and rainwater according to the following reaction.

$$2 \, SO_2(g) + O_2(g) + 2 \, H_2O(l) \longrightarrow 2 \, H_2SO_4(aq)$$

Assuming that there is more than enough O_2 and H_2O, how much H_2SO_4 in kilograms forms from 2.6×10^3 kg of SO_2?

▶ **FOR MORE PRACTICE**   Problems 35, 36, 37, 38.

8.5   More Pancakes: Limiting Reactant, Theoretical Yield, and Percent Yield

Let's return to our pancake analogy to understand two more concepts important in reaction stoichiometry: limiting reactant and percent yield. Recall our pancake recipe:

$$1 \text{ cup flour} + 2 \text{ eggs} + \tfrac{1}{2} \text{ tsp baking powder} \longrightarrow 5 \text{ pancakes}$$

Suppose we have 3 cups flour, 10 eggs, and 4 tsp baking powder. How many pancakes can we make? We have enough flour to make:

$$3 \text{ cups flour} \times \frac{5 \text{ pancakes}}{1 \text{ cup flour}} = 15 \text{ pancakes}$$

We have enough eggs to make:

$$10 \text{ eggs} \times \frac{5 \text{ pancakes}}{2 \text{ eggs}} = 25 \text{ pancakes}$$

We have enough baking powder to make:

$$4 \text{ tsp baking powder} \times \frac{5 \text{ pancakes}}{\tfrac{1}{2} \text{ tsp baking powder}} = 40 \text{ pancakes}$$

We have enough flour for 15 pancakes, enough eggs for 25 pancakes, and enough baking powder for 40 pancakes. Consequently, unless we get more ingredients, *we can make only 15 pancakes*. The amount of flour we have *limits* the number of pancakes we can make. If this were a chemical reaction, the flour would be the *limiting reactant*, the reactant that limits the amount of product in a chemical reaction. Notice that the **limiting reactant** is simply the reactant that makes *the least amount of product*. If this were a chemical reaction, 15 pancakes would be the **theoretical yield**, the amount of product that can be made in a chemical reaction based on the amount of limiting reactant.

The term *limiting reagent* is sometimes used in place of limiting reactant.

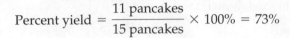

Limiting reactant

Theoretical yield

15 pancakes   25 pancakes   40 pancakes

▲ If this were a chemical reaction, the flour would be the limiting reactant and 15 pancakes would be the theoretical yield.

The actual yield of a chemical reaction, which must be determined experimentally, often depends in various ways on the reaction conditions. We will explore some of the factors involved in Chapter 15.

Let us carry this analogy one step further. Suppose we go on to cook our pancakes. We accidentally burn three of them and one falls on the floor. So even though we had enough flour for 15 pancakes, we finished with only 11 pancakes. If this were a chemical reaction, the 11 pancakes would be our **actual yield**, the amount of product actually produced by a chemical reaction. Finally, our **percent yield**, the percentage of the theoretical yield that was actually attained, is:

$$\text{Percent yield} = \frac{11 \text{ pancakes}}{15 \text{ pancakes}} \times 100\% = 73\%$$

Since four of the pancakes were ruined, we got only 73% of our theoretical yield. In a chemical reaction, the actual yield is almost always less than 100% because at least some of the product does not form or is lost in the process of recovering it (in analogy to some of the pancakes being burned).

To summarize:

- **Limiting reactant (or limiting reagent)**—the reactant that is completely consumed in a chemical reaction.
- **Theoretical yield**—the amount of product that can be made in a chemical reaction based on the amount of limiting reactant.
- **Actual yield**—the amount of product actually produced by a chemical reaction.
- **Percent yield** $= \dfrac{\textbf{Actual yield}}{\textbf{Theoretical yield}} \times \textbf{100\%}$

Consider the reaction.

$$\text{Ti}(s) + 2\,\text{Cl}_2(g) \longrightarrow \text{TiCl}_4(s)$$

If we begin with 1.8 mol of titanium and 3.2 mol of chlorine, what is the limiting reactant and theoretical yield of TiCl_4 in moles? We begin by sorting the information in the problem according to our standard problem-solving procedure.

GIVEN: 1.8 mol Ti

3.2 mol Cl_2

FIND: limiting reactant

theoretical yield

SOLUTION MAP

As in our pancake analogy, we determine the limiting reactant by calculating how much product can be made from each reactant. The reactant that makes the *least amount of product* is the limiting reactant.

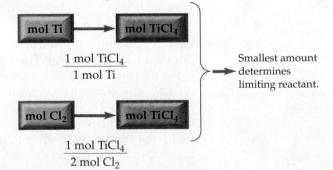

RELATIONSHIPS USED

The conversion factors come from the balanced chemical equation and give the relationships between moles of each of the reactants and moles of product.

$$1 \text{ mol Ti}:1 \text{ mol TiCl}_4$$

$$2 \text{ mol Cl}_2:1 \text{ mol TiCl}_4$$

SOLUTION

$$1.8 \text{ mol Ti} \times \frac{1 \text{ mol TiCl}_4}{1 \text{ mol Ti}} = 1.8 \text{ mol TiCl}_4$$

$$3.2 \text{ mol Cl}_2 \times \frac{1 \text{ mol TiCl}_4}{2 \text{ mol Cl}_2} = 1.6 \text{ mol TiCl}_4$$

Limiting reactant ⤴   ⤴ Least amount of product

> In many industrial applications, the more costly reactant or the reactant that is most difficult to remove from the product mixture is chosen to be the limiting reactant.

Since the 3.2 mol of Cl_2 make the least amount of $TiCl_4$, Cl_2 is the limiting reactant. Notice that we began with more moles of Cl_2 than Ti, but since the reaction requires 2 Cl_2 for each Ti, Cl_2 is still the limiting reactant. The theoretical yield is 1.6 mol of $TiCl_4$.

EXAMPLE 8.4   Limiting Reactant and Theoretical Yield from Initial Moles of Reactants

Consider the reaction:

$$2 \text{ Al}(s) + 3 \text{ Cl}_2(g) \longrightarrow 2 \text{ AlCl}_3(s)$$

If you begin with 0.552 mol of aluminum and 0.887 mol of chlorine, what is the limiting reactant and theoretical yield of $AlCl_3$ in moles?

SORT You are given the number of moles of aluminum and chlorine and asked to find the limiting reactant and theoretical yield of aluminum chloride.	**GIVEN** 0.552 mol Al 0.887 mol Cl_2 **FIND** limiting reactant theoretical yield of $AlCl_3$
STRATEGIZE Draw a solution map that shows how to get from moles of each reactant to moles of $AlCl_3$. The reactant that makes the *least amount of AlCl₃* is the limiting reactant. The conversion factors are the stoichiometric relationships (from the balanced equation).	**SOLUTION MAP** **RELATIONSHIPS USED** 2 mol Al:2 mol $AlCl_3$ (from balanced equation) 3 mol Cl_2:2 mol $AlCl_3$ (from balanced equation)

SOLVE	SOLUTION
Follow the solution map to solve the problem.	$0.552 \ \text{mol Al} \times \dfrac{2 \ \text{mol AlCl}_3}{2 \ \text{mol Al}} = 0.552 \ \text{mol AlCl}_3$ Limiting reactant Least amount of product $0.887 \ \text{mol Cl}_2 \times \dfrac{2 \ \text{mol AlCl}_3}{3 \ \text{mol Cl}_2} = 0.591 \ \text{mol AlCl}_3$ Since the 0.552 mol of Al makes the least amount of AlCl$_3$, Al is the limiting reactant. The theoretical yield is 0.552 mol of AlCl$_3$.
CHECK Are the units correct? Does the answer make physical sense?	The units, mol AlCl$_3$, are correct. The magnitude of the answer seems reasonable because it is of the same order of magnitude as the given number of moles of Al and Cl$_2$. An answer that is orders of magnitude different would immediately be suspect.

▶**SKILLBUILDER 8.4 | Limiting Reactant and Theoretical Yield from Initial Moles of Reactants**

Consider the reaction:

$$2 \, \text{Na}(s) + \text{F}_2(g) \longrightarrow 2 \, \text{NaF}(s)$$

If you begin with 4.8 mol of sodium and 2.6 mol of fluorine, what is the limiting reactant and theoretical yield of NaF in moles?

▶**FOR MORE PRACTICE** Problems 43, 44, 45, 46, 47, 48, 49, 50.

CONCEPTUAL CHECKPOINT 8.2

Consider the reaction:

$$\text{N}_2(g) + 3 \, \text{H}_2(g) \longrightarrow 2 \, \text{NH}_3(g)$$

If the flask at left represents the mixture before the reaction, which flask represents the products after the limiting reactant has completely reacted?

(a)   (b)   (c)

8.6 Limiting Reactant, Theoretical Yield, and Percent Yield from Initial Masses of Reactants

When working in the laboratory, we normally measure the initial amounts of reactants in grams. To find limiting reactants and theoretical yields from initial masses, we must add two steps to our calculations. Consider, for example, the synthesis reaction:

$$2 \, \text{Na}(s) + \text{Cl}_2(g) \longrightarrow 2 \, \text{NaCl}(s)$$

If we have 53.2 g of Na and 65.8 g of Cl_2, what is the limiting reactant and theoretical yield? We begin by sorting the information in the problem.

GIVEN: 53.2 g Na

65.8 g Cl_2

FIND: limiting reactant

theoretical yield

SOLUTION MAP

Again, we find the limiting reactant by calculating how much product can be made from each reactant. Since we are given the initial amounts in grams, we must first convert to moles. After we convert to moles of product, we convert back to grams of product. The reactant that makes the *least amount of product* is the limiting reactant.

RELATIONSHIPS USED

From the balanced chemical equation, we know:

$$2 \text{ mol Na}:2 \text{ mol NaCl}$$
$$1 \text{ mol Cl}_2:2 \text{ mol NaCl}$$

We also use these molar masses:

$$\text{Molar mass Na} = \frac{22.99 \text{ g Na}}{1 \text{ mol Na}}$$

$$\text{Molar mass Cl}_2 = \frac{70.90 \text{ g Cl}_2}{1 \text{ mol Cl}_2}$$

$$\text{Molar mass NaCl} = \frac{58.44 \text{ g NaCl}}{1 \text{ mol NaCl}}$$

SOLUTION

Beginning with the actual amounts of each reactant, we follow the solution map to calculate how much product can be made from each.

The limiting reactant can also be found by calculating the number of moles of NaCl (rather than grams) that can be made from each reactant. However, since theoretical yields are normally calculated in grams, we take the calculation all the way to grams to determine limiting reactant.

$$53.2 \text{ g Na} \times \frac{1 \text{ mol Na}}{22.99 \text{ g Na}} \times \frac{2 \text{ mol NaCl}}{2 \text{ mol Na}} \times \frac{58.44 \text{ g NaCl}}{1 \text{ mol NaCl}} = 135 \text{ g NaCl}$$

$$65.8 \text{ g Cl}_2 \times \frac{1 \text{ mol Cl}_2}{70.90 \text{ g Cl}_2} \times \frac{2 \text{ mol NaCl}}{1 \text{ mol Cl}_2} \times \frac{58.44 \text{ g NaCl}}{1 \text{ mol NaCl}} = 108 \text{ g NaCl}$$

Limiting
reactant

Least amount
of product

The limiting reactant is not necessarily the reactant with the least mass.

Since Cl_2 makes the least amount of product, it is the limiting reactant. Notice that the limiting reactant is not necessarily the reactant with the least mass. In this case, we had fewer grams of Na than Cl_2, yet Cl_2 was the limiting reactant because it made less NaCl. The theoretical yield is therefore 108 g of NaCl, the amount of product possible based on the limiting reactant.

Now suppose that when the synthesis was carried out, the actual yield of NaCl was 86.4 g. What is the percent yield? The percent yield is:

> The actual yield is always less than the theoretical yield because at least a small amount of product is usually lost or does not form during a reaction.

$$\text{Percent yield} = \frac{\text{Actual yield}}{\text{Theoretical yield}} \times 100\% = \frac{86.4 \text{ g}}{108 \text{ g}} \times 100\% = 80.0\%$$

EXAMPLE 8.5 Finding Limiting Reactant and Theoretical Yield

Ammonia, NH_3, can be synthesized by the reaction:

$$2\, NO(g) + 5\, H_2(g) \longrightarrow 2\, NH_3(g) + 2\, H_2O(g)$$

What maximum amount of ammonia in grams can be synthesized from 45.8 g of NO and 12.4 g of H_2?

SORT

You are given the masses of two reactants and asked to find the maximum mass of ammonia that can be formed. Although this problem does not specifically ask for the limiting reactant, it must be found to determine the theoretical yield, which is the maximum amount of ammonia that can be synthesized.

GIVEN: 45.8 g NO, 12.4 g H_2

FIND: maximum amount of NH_3 in g (this is the theoretical yield)

STRATEGIZE

Find the limiting reactant by calculating how much product can be made from each reactant. The reactant that makes the *least amount of product* is the limiting reactant. The mass of ammonia formed by the limiting reactant is the maximum amount of ammonia that can be synthesized.

SOLUTION MAP

$$\boxed{\text{g NO}} \rightarrow \boxed{\text{mol NO}} \rightarrow \boxed{\text{mol NH}_3} \rightarrow \boxed{\text{g NH}_3}$$

$$\frac{1 \text{ mol NO}}{30.01 \text{ g NO}} \qquad \frac{2 \text{ mol NH}_3}{2 \text{ mol NO}} \qquad \frac{17.04 \text{ g NH}_3}{1 \text{ mol NH}_3}$$

$$\boxed{\text{g H}_2} \rightarrow \boxed{\text{mol H}_2} \rightarrow \boxed{\text{mol NH}_3} \rightarrow \boxed{\text{g NH}_3}$$

$$\frac{1 \text{ mol H}_2}{2.02 \text{ g H}_2} \qquad \frac{2 \text{ mol NH}_3}{5 \text{ mol H}_2} \qquad \frac{17.04 \text{ g NH}_3}{1 \text{ mol NH}_3}$$

Smallest amount determines limiting reactant.

The main conversion factors come from the stoichiometric relationship between moles of each reactant and moles of ammonia. The other conversion factors are the molar masses of nitrogen monoxide, hydrogen gas, and ammonia.

RELATIONSHIPS USED

2 mol NO : 2 mol NH_3 5 mol H_2 : 2 mol NH_3

$$\text{Molar mass NO} = \frac{30.01 \text{ g NO}}{1 \text{ mol NO}} \qquad \text{Molar mass H}_2 = \frac{2.02 \text{ g H}_2}{1 \text{ mol H}_2}$$

$$\text{Molar mass NH}_3 = \frac{17.04 \text{ g NH}_3}{1 \text{ mol NH}_3}$$

SOLVE

Follow the solution map, beginning with the actual amount of each reactant given, to calculate the amount of product that can be made from each reactant.

SOLUTION

$$45.8 \text{ g NO} \times \frac{1 \text{ mol NO}}{30.01 \text{ g NO}} \times \frac{2 \text{ mol NH}_3}{2 \text{ mol NO}} \times \frac{17.04 \text{ g NH}_3}{1 \text{ mol NH}_3} = 26.0 \text{ g NH}_3$$

Limiting reactant   Least amount of product

$$12.4 \text{ g H}_2 \times \frac{1 \text{ mol H}_2}{2.02 \text{ g H}_2} \times \frac{2 \text{ mol NH}_3}{5 \text{ mol H}_2} \times \frac{17.04 \text{ g NH}_3}{1 \text{ mol NH}_3} = 41.8 \text{ g NH}_3$$

There is enough NO to make 26.0 g of NH_3 and enough H_2 to make 41.8 g of NH_3. Therefore, NO is the limiting reactant, and the maximum amount of ammonia that can possibly be made is 26.0 g, which is the theoretical yield.

CHECK	The units of the answer, g NH_3, are correct. The magnitude of the answer
Are the units correct? Does the answer make physical sense?	seems reasonable because it is of the same order of magnitude as the given masses of NO and H_2. An answer that is orders of magnitude different would immediately be suspect.

▶**SKILLBUILDER 8.5 | Finding Limiting Reactant and Theoretical Yield**

Ammonia can also be synthesized by the reaction:

$$3 H_2(g) + N_2(g) \longrightarrow 2 NH_3(g)$$

What maximum amount of ammonia in grams can be synthesized from 25.2 g of N_2 and 8.42 g of H_2?

▶**SKILLBUILDER PLUS**

What maximum amount of ammonia in kilograms can be synthesized from 5.22 kg of H_2 and 31.5 kg of N_2?

▶**FOR MORE PRACTICE** Problems 55, 56, 57, 58.

EXAMPLE 8.6 Finding Limiting Reactant, Theoretical Yield, and Percent Yield

Consider the reaction:

$$Cu_2O(s) + C(s) \longrightarrow 2 Cu(s) + CO(g)$$

When 11.5 g of C are allowed to react with 114.5 g of Cu_2O, 87.4 g of Cu are obtained. Find the limiting reactant, theoretical yield, and percent yield.

SORT	GIVEN: 11.5 g C
You are given the mass of the reactants, carbon and copper(I) oxide, as well as the mass of copper formed by the reaction. You are asked to find the limiting reactant, theoretical yield, and percent yield.	114.5 g Cu_2O 87.4 g Cu produced FIND: limiting reactant theoretical yield percent yield

STRATEGIZE	SOLUTION MAP
The solution map shows how to find the mass of Cu formed by the initial masses of Cu_2O and C. The reactant that makes the *least amount of product* is the limiting reactant and determines the theoretical yield.	

	RELATIONSHIPS USED
The main conversion factors are the stoichiometric relationships between moles of each reactant and moles of copper. The other conversion factors are the molar masses of copper(I) oxide, carbon, and copper.	1 mol Cu_2O : 2 mol Cu 1 mol C : 2 mol Cu Molar mass Cu_2O = 143.10 g/mol Molar mass C = 12.01 g/mol Molar mass Cu = 63.55 g/mol

SOLVE

Follow the solution map, beginning with the actual amount of each reactant given, to calculate the amount of product that can be made from each reactant.

Since Cu_2O makes the least amount of product, Cu_2O is the limiting reactant. The theoretical yield is then the amount of product made by the limiting reactant. The percent yield is the actual yield (87.4 g Cu) divided by the theoretical yield (101.7 g Cu) multiplied by 100%.

SOLUTION

$$11.5\text{ g C} \times \frac{1\text{ mol C}}{12.01\text{ g C}} \times \frac{2\text{ mol Cu}}{1\text{ mol C}} \times \frac{63.55\text{ g Cu}}{1\text{ mol Cu}} = 122\text{ g Cu}$$

$$114.5\text{ g }\underbrace{Cu_2O}_{\substack{\text{Limiting}\\\text{reactant}}} \times \frac{1\text{ mol }Cu_2O}{143.10\text{ g }Cu_2O} \times \frac{2\text{ mol Cu}}{1\text{ mol }Cu_2O} \times \frac{63.55\text{ g Cu}}{1\text{ mol Cu}} = \overset{\substack{\text{Least amount}\\\text{of product}}}{101.7\text{ g Cu}}$$

Theoretical yield = 101.7 g Cu

$$\text{Percent yield} = \frac{\text{Actual yield}}{\text{Theoretical yield}} \times 100\%$$

$$= \frac{87.4\text{ g}}{101.7\text{ g}} \times 100\% = 85.9\%$$

CHECK

Are the units correct? Does the answer make physical sense?

The theoretical yield has the right units (g Cu). The magnitude of the theoretical yield seems reasonable because it is of the same order of magnitude as the given masses of C and Cu_2O. The theoretical yield is reasonable because it is less than 100%. Any calculated theoretical yield above 100% would be suspect.

▶**SKILLBUILDER 8.6 | Finding Limiting Reactant, Theoretical Yield, and Percent Yield**

This reaction is used to obtain iron from iron ore:

$$Fe_2O_3(s) + 3\,CO(g) \longrightarrow 2\,Fe(s) + 3\,CO_2(g)$$

The reaction of 185 g of Fe_2O_3 with 95.3 g of CO produces 87.4 g of Fe. Find the limiting reactant, theoretical yield, and percent yield.

▶**FOR MORE PRACTICE**   Example 8.10; Problems 61, 62, 63, 64, 65, 66.

 CONCEPTUAL CHECKPOINT 8.3

Limiting Reactant and Theoretical Yield

Ammonia can by synthesized by the reaction of nitrogen monoxide and hydrogen gas.

$$2\,NO(g) + 5\,H_2(g) \longrightarrow 2\,NH_3(g) + 2\,H_2O(g)$$

A reaction vessel initially contains 4.0 mol of NO and 15.0 mol of H_2. What is in the reaction vessel once the reaction has occurred to the fullest extent possible?

(a) 2 mol NO; 5 mol H_2; 2 mol NH_3; and 2 mol H_2O

(b) 0 mol NO; 0 mol H_2; 6 mol NH_3; and 6 mol H_2O

(c) 2 mol NO; 0 mol H_2; 4 mol NH_3; and 2 mol H_2O

(d) 0 mol NO; 5 mol H_2; 4 mol NH_3; and 4 mol H_2O

8.7   Enthalpy: A Measure of the Heat Evolved or Absorbed in a Reaction

Chapter 3 (see Section 3.9) describes how chemical reactions can be *exothermic* (in which case they *emit* thermal energy when they occur) or *endothermic* (in which case they *absorb* thermal energy when they occur). The *amount* of thermal energy

EVERYDAY CHEMISTRY

Bunsen Burners

In the laboratory, we often use Bunsen burners as heat sources. These burners are normally fueled by methane. The balanced equation for methane (CH_4) combustion is:

$$CH_4(g) + 2 O_2(g) \longrightarrow CO_2(g) + 2 H_2O(g)$$

Most Bunsen burners have a mechanism to adjust the amount of air (and therefore of oxygen) that is mixed with the methane. If you light the burner with the air completely closed off, you get a yellow, smoky flame that is not very hot. As you increase the amount of air going into the burner, the flame becomes bluer, less smoky, and hotter. When you reach the optimum adjustment, the flame has a sharp, inner blue triangle, no smoke, and is hot enough to melt glass easily. Continuing to increase the air beyond this point causes the flame to become cooler again and may actually extinguish it.

CAN YOU ANSWER THIS? *Can you use the concepts from this chapter to explain the changes in the Bunsen burner as the air intake is adjusted?*

(a) No air   (b) Small amount of air   (c) Optimum   (d) Too much air

▲ Bunsen burner at various stages of air intake adjustment.

emitted or absorbed by a chemical reaction, under conditions of constant pressure (which are common for most everyday reactions), can be quantified with a function called **enthalpy**. Specifically, we define a quantity called the **enthalpy of reaction** (ΔH_{rxn}) as the amount of thermal energy (or heat) that flows when a reaction occurs at constant pressure.

SIGN OF ΔH_{rxn}

The *sign* of ΔH_{rxn} (positive or negative) depends on the *direction* in which thermal energy flows when the reaction occurs. If thermal energy flows out of the reaction and into the surroundings (as in an exothermic reaction), then ΔH_{rxn} is negative. For example, we can specify the enthalpy of reaction for the combustion of CH_4, the main component in natural gas, as:

$$CH_4(g) + 2 O_2(g) \longrightarrow CO_2(g) + 2 H_2O(g) \qquad \Delta H_{rxn} = -802.3 \text{ kJ}$$

This reaction is exothermic and therefore has a negative enthalpy of reaction. The magnitude of ΔH_{rxn} tells us that 802.3 kJ of heat are emitted when 1 mol of CH_4 reacts with 2 mol of O_2.

If, by contrast, thermal energy flows into the reaction and out of the surroundings (as in an endothermic reaction), then ΔH_{rxn} is positive. For example, we specify the enthalpy of reaction for the reaction between nitrogen and oxygen gas to form nitrogen monoxide as:

$$N_2(g) + O_2(g) \longrightarrow 2 NO(g) \qquad \Delta H_{rxn} = +182.6 \text{ kJ}$$

▶ **FIGURE 8.3 Exothermic and endothermic reactions** **(a)** In an exothermic reaction, energy is released into the surroundings. **(b)** In an endothermic reaction, energy is absorbed from the surroundings.

This reaction is endothermic and therefore has a positive enthalpy of reaction. When 1 mol of N_2 reacts with 1 mol of O_2, 182.6 kJ of heat are absorbed from the surroundings.

You can think of the energy of a chemical system in the same way that you think about the balance in your checking account. Energy flowing *out* of the chemical system is like a withdrawal and carries a negative sign as shown in ▲ Figure 8.3a. Energy flowing *into* the system is like a deposit and carries a positive sign as shown in Figure 8.3b.

STOICHIOMETRY OF ΔH_{rxn}

The amount of heat emitted or absorbed when a chemical reaction occurs depends on the *amounts* of reactants that actually react. As we have just seen, we usually specify ΔH_{rxn} in combination with the balanced chemical equation for the reaction. The magnitude of ΔH_{rxn} is for the stoichiometric amounts of reactants and products for the reaction *as written*. For example, the balanced equation and ΔH_{rxn} for the combustion of propane (the fuel used in LP gas) is:

$$C_3H_8(g) + 5\,O_2(g) \longrightarrow 3\,CO_2(g) + 4\,H_2O(g) \qquad \Delta H_{rxn} = -2044 \text{ kJ}$$

This means that when 1 mole of C_3H_8 reacts with 5 moles of O_2 to form 3 moles of CO_2 and 4 moles of H_2O, 2044 kJ of heat are emitted. We can write these relationships in the same way that we express stoichiometric relationships: as ratios between two quantities. For example, for the reactants, we write:

$$1 \text{ mol } C_3H_8 : -2044 \text{ kJ} \quad \text{or} \quad 5 \text{ mol } O_2 : -2044 \text{ kJ}$$

The ratios mean that 2044 kJ of thermal energy are evolved when 1 mole of C_3H_8 and 5 moles of O_2 completely react. These ratios can then be used to construct conversion factors between amounts of reactants or products and the quantity of heat emitted (for exothermic reactions) or absorbed (for endothermic reactions). To find out how much heat is emitted upon the combustion of a certain mass in grams of C_3H_8, we can use the following solution map:

We use the molar mass to convert between grams and moles, and the stoichiometric relationship between moles of C_3H_8 and kJ to convert between moles and kJ, as shown in the following example.

EXAMPLE 8.7   Stoichiometry Involving ΔH

An LP gas tank in a home barbecue contains 11.8×10^3 g of propane (C_3H_8). Calculate the heat (in kJ) associated with the complete combustion of all of the propane in the tank.

$$C_3H_8(g) + 5\,O_2(g) \longrightarrow 3\,CO_2(g) + 4\,H_2O(g) \qquad \Delta H_{rxn} = -2044 \text{ kJ}$$

SORT You are given the mass of propane and asked to find the heat evolved (in kJ) in its combustion.	**GIVEN:** 11.8×10^3 g C_3H_8 **FIND:** kJ

STRATEGIZE
Start with the given mass of propane and then use its molar mass to find the number of moles. Next, use the stoichiometric relationship between moles of propane and kilojoules of heat to find the heat evolved.

SOLUTION MAP

$$\boxed{\text{g } C_3H_8} \longrightarrow \boxed{\text{mol } C_3H_8} \longrightarrow \boxed{\text{kJ}}$$

$$\frac{1 \text{ mol } C_3H_8}{44.11 \text{ g } C_3H_8} \qquad \frac{-2044 \text{ kJ}}{1 \text{ mol } C_3H_8}$$

RELATIONSHIPS USED

1 mol C_3H_8 : −2044 kJ (from balanced equation)

Molar mass C_3H_8 = 44.11 g/mol

SOLVE
Follow the solution map to solve the problem. Begin with 11.8×10^3 g C_3H_8 and multiply by the appropriate conversion factors to arrive at kJ.

SOLUTION

$$11.8 \times 10^3 \text{ g } C_3H_8 \times \frac{1 \text{ mol } C_3H_8}{44.11 \text{ g } C_3H_8} \times \frac{-2044 \text{ kJ}}{1 \text{ mol } C_3H_8} = -5.47 \times 10^5 \text{ kJ}$$

CHECK
Check your answer. Are the units correct? Does the answer make physical sense?

The units, kJ, are correct. The answer is negative, as it should be when heat is evolved by a reaction.

▶**SKILLBUILDER 8.7 | Stoichiometry Involving ΔH**

Ammonia reacts with oxygen according to the equation:

$$4 NH_3(g) + 5 O_2(g) \longrightarrow 4 NO(g) + 6 H_2O \qquad \Delta H_{rxn} = -906 \text{ kJ}$$

Calculate the heat (in kJ) associated with the complete reaction of 155 g of NH_3.

▶**SKILLBUILDER PLUS**

What mass of butane in grams is necessary to produce 1.5×10^3 kJ of heat? What mass of CO_2 is produced?

$$C_4H_{10}(g) + \tfrac{13}{2} O_2(g) \longrightarrow 4 CO_2(g) + 5 H_2O(g) \qquad \Delta H_{rxn} = -2658 \text{ kJ}$$

▶**FOR MORE PRACTICE** Example 8.11; Problems 71, 72, 73, 74, 75, 76.

✓ **CONCEPTUAL CONNECTION 8.4**

Stoichiometry Involving ΔH

Consider the generic reaction:

$$2A + 3B \longrightarrow 2C \qquad \Delta H_{rxn} = -100 \text{ kJ}$$

If a reaction mixture initially contains 5 mol of A and 6 mol of B, how much heat (in kJ) will be evolved once the reaction has occurred to the greatest extent possible?

(a) 100 kJ

(b) 150 kJ

(c) 200 kJ

(d) 300 kJ

CHAPTER IN REVIEW

CHEMICAL PRINCIPLES

Stoichiometry: A balanced chemical equation gives quantitative relationships between the amounts of reactants and products. For example, the reaction $2 H_2 + O_2 \longrightarrow 2 H_2O$ says that 2 mol of H_2 reacts with 1 mol of O_2 to form 2 mol of H_2O. These relationships can be used to calculate quantities such as the amount of product possible with a certain amount of reactant, or the amount of one reactant required to completely react with a certain amount of another reactant. The quantitative relationship between reactants and products in a chemical reaction is called reaction stoichiometry.

Limiting Reactant, Theoretical Yield, and Percent Yield: The limiting reactant in a chemical reaction is the reactant that limits the amount of product that can be made. The theoretical yield in a chemical reaction is the amount of product that can be made based on the amount of the limiting reactant. The actual yield in a chemical reaction is the amount of product actually produced. The percent yield in a chemical reaction is the actual yield divided by theoretical yield times 100%.

Enthalpy of Reaction: The amount of heat released or absorbed by a chemical reaction under conditions of constant pressure, is the enthalpy of reaction (ΔH_{rxn}).

RELEVANCE

Stoichiometry: Reaction stoichiometry is important because we often want to know the numerical relationship between the reactants and products in a chemical reaction. For example, we might want to know how much carbon dioxide, a greenhouse gas, is formed when a certain amount of a particular fossil fuel burns.

Limiting Reactant, Theoretical Yield, and Percent Yield: Calculations of limiting reactant, theoretical yield, and percent yield are central to chemistry because they allow for quantitative understanding of chemical reactions. Just as you need to know relationships between ingredients to follow a recipe, so you must know relationships between reactants and products to carry out a chemical reaction. The percent yield in a chemical reaction is often used as a measure of the success of the reaction. Imagine following a recipe and making only 1% of the final product—your cooking would be a failure. Similarly, low percent yields in chemical reactions are usually considered poor, and high percent yields are considered good.

Enthalpy of Reaction: The enthalpy of reaction describes the relationship between the amount of reactant that undergoes reaction and the amount of thermal energy produced. This is important, for example, in determining quantities such as the amount of fuel needed to produce a given amount of energy.

CHEMICAL SKILLS

Mole-to-Mole Conversions (Section 8.3)

SORT
You are given the number of moles of sodium and asked to find the number of moles of sodium oxide formed by the reaction.

EXAMPLES

EXAMPLE 8.8 Mole-to-Mole Conversions

How many moles of sodium oxide can be synthesized from 4.8 mol of sodium? Assume that more than enough oxygen is present. The balanced equation is:

$$4 Na(s) + O_2(g) \longrightarrow 2 Na_2O(s)$$

GIVEN: 4.8 mol Na

FIND: mol Na_2O

STRATEGIZE

Draw a solution map beginning with the number of moles of the given substance and then use the conversion factor from the balanced chemical equation to determine the number of moles of the substance you are trying to find.

SOLUTION MAP

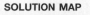

$$\frac{2 \text{ mol Na}_2\text{O}}{4 \text{ mol Na}}$$

RELATIONSHIPS USED

4 mol Na : 2 mol Na$_2$O

SOLVE

Follow the solution map to get to the number of moles of the substance you are trying to find.

SOLUTION

$$4.8 \cancel{\text{ mol Na}} \times \frac{2 \text{ mol Na}_2\text{O}}{4 \cancel{\text{ mol Na}}} = 2.4 \text{ mol Na}_2\text{O}$$

CHECK

Are the units correct? Does the answer make physical sense?

The units of the answer, mol Na$_2$O, are correct. The magnitude of the answer seems reasonable because it is of the same order of magnitude as the given number of moles of Na.

Mass-to-Mass Conversions (Section 8.4)

EXAMPLE 8.9 Mass-to-Mass Conversions

How many grams of sodium oxide can be synthesized from 17.4 g of sodium? Assume that more than enough oxygen is present. The balanced equation is:

$$4 \text{ Na}(s) + \text{O}_2(g) \longrightarrow 2 \text{ Na}_2\text{O}(s)$$

SORT

You are given the mass of sodium and asked to find the mass of sodium oxide that forms upon reaction.

GIVEN: 17.4 g Na

FIND: g Na$_2$O

STRATEGIZE

Draw the solution map by beginning with the mass of the given substance. Convert to moles using the molar mass and then convert to moles of the substance you are trying to find, using the conversion factor obtained from the balanced chemical equation. Finally, convert to mass of the substance you are trying to find, using its molar mass.

SOLUTION MAP

$$\frac{1 \text{ mol Na}}{22.99 \text{ g Na}} \qquad \frac{2 \text{ mol Na}_2\text{O}}{4 \text{ mol Na}}$$

$$\frac{61.98 \text{ g Na}_2\text{O}}{1 \text{ mol Na}_2\text{O}}$$

RELATIONSHIPS USED

4 mol Na : 2 mol Na$_2$O (from balanced equation)

Molar mass Na = 22.99 g/mol

Molar mass Na$_2$O = 61.98 g/mol

SOLVE

Follow the solution map and calculate the answer by beginning with the mass of the given substance and multiplying by the appropriate conversion factors to determine the mass of the substance you are trying to find.

SOLUTION

$$17.4 \cancel{\text{ g Na}} \times \frac{1 \cancel{\text{ mol Na}}}{22.99 \cancel{\text{ g Na}}} \times \frac{2 \cancel{\text{ mol Na}_2\text{O}}}{4 \cancel{\text{ mol Na}}} \times$$

$$\frac{61.98 \text{ g Na}_2\text{O}}{1 \cancel{\text{ mol Na}_2\text{O}}} = 23.5 \text{ g Na}_2\text{O}$$

CHECK

Are the units correct? Does the answer make physical sense?

The units of the answer, g Na$_2$O, are correct. The magnitude of the answer seems reasonable because it is of the same order of magnitude as the given mass of Na.

Limiting Reactant, Theoretical Yield, and Percent Yield (Sections 8.5, 8.6)

EXAMPLE 8.10   Limiting Reactant, Theoretical Yield, and Percent Yield

10.4 g of As reacts with 11.8 g of S to produce 14.2 g of As_2S_3. Find the limiting reactant, theoretical yield, and percent yield for this reaction. The balanced chemical equation is:

$$2 As(s) + 3 S(l) \longrightarrow As_2S_3(s)$$

GIVEN: 10.4 g As
11.8 g S
14.2 g As_2S_3

FIND: limiting reactant
theoretical yield
percent yield

SOLUTION MAP

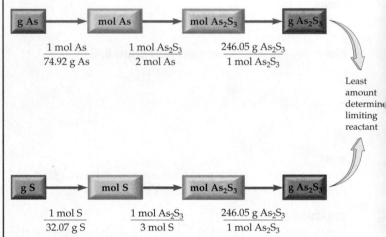

RELATIONSHIPS USED

2 mol As : 1 mol As_2S_3

3 mol S : 1 mol As_2S_3

Molar mass As = 74.92 g/mol

Molar mass S = 32.07 g/mol

Molar mass As_2S_3 = 246.05 g/mol

SORT
You are given the masses of iron and sulfur as well as the mass of iron(III) sulfide formed by the reaction. You are asked to find the limiting reactant, theoretical yield, and percent yield.

STRATEGIZE
The solution map for limiting-reactant problems shows how to convert from the mass of each of the reactants to mass of the product for each reactant. These are mass-to-mass conversions with the basic outline of

$$\text{Mass} \longrightarrow \text{Moles} \longrightarrow \text{Moles} \longrightarrow \text{Mass}$$

The reactant that forms the least amount of product is the limiting reactant.

The conversion factors you need are the stoichiometric relationships between each of the reactants and the product. You also need the molar masses of each reactant and product.

SOLVE
To calculate the amount of product formed by each reactant, begin with the given amount of each reactant and multiply by the appropriate conversion factors, as shown in the solution map, to arrive at the mass of product for each reactant. The reactant that forms the least amount of product is the limiting reactant.

SOLUTION

$$10.4 \text{ g As} \times \frac{1 \text{ mol As}}{74.92 \text{ g As}} \times \frac{1 \text{ mol As}_2\text{S}_3}{2 \text{ mol As}} \times \frac{246.05 \text{ g As}_2\text{S}_3}{1 \text{ mol As}_2\text{S}_3}$$

Limiting reactant

$$= 17.1 \text{ g As}_2\text{S}_3$$

Least amount of product

$$11.8 \text{ g S} \times \frac{1 \text{ mol S}}{32.07 \text{ g S}} \times \frac{1 \text{ mol As}_2\text{S}_3}{3 \text{ mol S}} \times \frac{246.05 \text{ g As}_2\text{S}_3}{1 \text{ mol As}_2\text{S}_3}$$

$$= 30.2 \text{ g As}_2\text{S}_3$$

The limiting reactant is As.

The theoretical yield is the amount of product formed by the limiting reactant.

The percent yield is the actual yield divided by the theoretical yield times 100%.

The theoretical yield is 17.1 g of As_2S_3.

$$Percent\ yield = \frac{Actual\ yield}{Theoretical\ yield} \times 100\%$$

$$= \frac{14.2\ g}{17.1\ g} \times 100\% = 83.0\%$$

The percent yield is 83.0%.

CHECK

Check your answer. Are the units correct? Does the answer make physical sense?

The theoretical yield has the right units (g As_2S_3). The magnitude of the theoretical yield seems reasonable because it is of the same order of magnitude as the given masses of As and S. The theoretical yield is reasonable because it is less than 100%. Any calculated theoretical yield above 100% would be suspect.

Stoichiometry Involving ΔH (Section 8.7)

EXAMPLE 8.11 Stoichiometry Involving ΔH

Calculate the heat evolved (in kJ) upon complete combustion of 25.0 g of methane (CH_4).

$$CH_4(g) + 2\,O_2(g) \longrightarrow CO_2(g) + 2\,H_2O(g)$$

$$\Delta H_{rxn} = -802\ kJ$$

SORT

You are given the mass of methane and asked to find the quantity of heat in kJ emitted upon combustion.

Draw the solution map by beginning with the mass of the given substance. Convert to moles using molar mass and then to kJ using ΔH.

GIVEN: 25 g CH_4

FIND: kJ

SOLUTION MAP

$$\frac{1\ mol\ CH_4}{16.05\ g\ CH_4} \qquad \frac{-802\ kJ}{1\ mol\ CH_4}$$

RELATIONSHIPS USED

1 mol CH_4 : -802 kJ (from balanced equation)

Molar mass CH_4 = 16.05 g/mol

SOLVE

Follow the solution map to solve the problem. Begin with the mass of the given substance and multiply by the appropriate conversion factors to arrive at kJ. A negative answer means that heat is evolved into the surroundings. A positive answer means that heat is absorbed from the surroundings.

SOLUTION

$$25.0\ g\ CH_4 \times \frac{1\ mol\ CH_4}{16.05\ g\ CH_4} \times \frac{-802\ kJ}{1\ mol\ CH_4}$$

$$= -1.25 \times 10^3\ kJ$$

CHECK

Are the units correct? Does the answer make physical sense?

The units, kJ, are correct. The answer is negative, as it should be since heat is evolved by the reaction.

KEY TERMS

EXERCISES

QUESTIONS

1. Why is reaction stoichiometry important? Give some examples.

2. Nitrogen and hydrogen can react to form ammonia.

   $$N_2(g) + 3\,H_2(g) \longrightarrow 2\,NH_3(g)$$

   (a) Write ratios showing the relationships between moles of each of the reactants and products in the reaction.

   (b) How many molecules of H_2 are required to completely react with two molecules of N_2?

   (c) How many moles of H_2 are required to completely react with 2 mol of N_2?

3. Write the conversion factor that you would use to convert from moles of Cl_2 to moles of NaCl in the reaction:

   $$2\,Na(s) + Cl_2(g) \longrightarrow 2\,NaCl$$

4. What is wrong with this statement in reference to the reaction in the previous problem? "Two grams of Na react with 1 g of Cl_2 to form 2 g of NaCl." Correct the statement to make it true.

5. What is the general form of the solution map for problems in which you are given the mass of a reactant in a chemical reaction and asked to find the mass of the product that can be made from the given amount of reactant?

6. Consider the recipe for making tomato and garlic pasta.

   2 cups noodles + 12 tomatoes + 3 cloves garlic
   $$\longrightarrow 4 \text{ servings pasta}$$

   If you have 7 cups of noodles, 27 tomatoes, and 9 cloves of garlic, how many servings of pasta can you make? Which ingredient limits the amount of pasta that it is possible to make?

7. In a chemical reaction, what is the limiting reactant?

8. In a chemical reaction, what is the theoretical yield?

9. In a chemical reaction, what are the actual yield and percent yield?

10. If you are given a chemical equation and specific amounts for each reactant in grams, how would you determine how much product can possibly be made?

11. Consider the generic chemical reaction:

    $$A + 2\,B \longrightarrow C + D$$

    Suppose you have 12 g of A and 24 g of B. Which statement is true?

    (a) A will definitely be the limiting reactant.

    (b) B will definitely be the limiting reactant.

    (c) A will be the limiting reactant if its molar mass is less than B.

    (d) A will be the limiting reactant if its molar mass is greater than B.

12. Consider the generic chemical equation:

    $$A + B \longrightarrow C$$

    Suppose 25 g of A were allowed to react with 8 g of B. Analysis of the final mixture showed that A was completely used up and 4 g of B remained. What was the limiting reactant?

13. What is the enthalpy of reaction (ΔH_{rxn}). Why is this quantity important?

14. Explain the relationship between the sign of ΔH_{rxn} and whether a reaction is exothermic or endothermic.

PROBLEMS

MOLE-TO-MOLE CONVERSIONS

15. Consider the generic chemical reaction:

    $$A + 2\,B \longrightarrow C$$

    How many moles of C are formed upon complete reaction of:

    (a) 2 mol of A

    (b) 2 mol of B

    (c) 3 mol of A

    (d) 3 mol of B

16. Consider the generic chemical reaction:

    $$2\,A + 3\,B \longrightarrow 3\,C$$

    How many moles of B are required to completely react with:

    (a) 6 mol of A

    (b) 2 mol of A

    (c) 7 mol of A

    (d) 11 mol of A

17. For the reaction shown, calculate how many moles of NO_2 form when each amount of reactant completely reacts.

$$2\,N_2O_5(g) \longrightarrow 4\,NO_2(g) + O_2(g)$$

(a) 1.3 mol N_2O_5

(b) 5.8 mol N_2O_5

(c) 4.45×10^3 mol N_2O_5

(d) 1.006×10^{-3} mol N_2O_5

18. For the reaction shown, calculate how many moles of NH_3 form when each amount of reactant completely reacts.

$$3\,N_2H_4(l) \longrightarrow 4\,NH_3(g) + N_2(g)$$

(a) 5.3 mol N_2H_4

(b) 2.28 mol N_2H_4

(c) 5.8×10^{-2} mol N_2H_4

(d) 9.76×10^7 mol N_2H_4

---

19. Dihydrogen monosulfide reacts with sulfur dioxide according to the balanced equation:

$$2\,H_2S(g) + SO_2(g) \longrightarrow 3\,S(s) + 2\,H_2O(g)$$

= SO_2

If the above figure represents the amount of SO_2 available to react, which figure best represents the amount of H_2S required to completely react with all of the SO_2?

20. Chlorine gas reacts with fluorine gas according to the balanced equation:

$$Cl_2(g) + 3\,F_2(g) \longrightarrow 2\,ClF_3(g)$$

= F_2

If the above figure represents the amount of fluorine available to react, and assuming that there is more than enough chlorine, which figure best represents the amount of chlorine trifluoride that would form upon complete reaction of all of the fluorine?

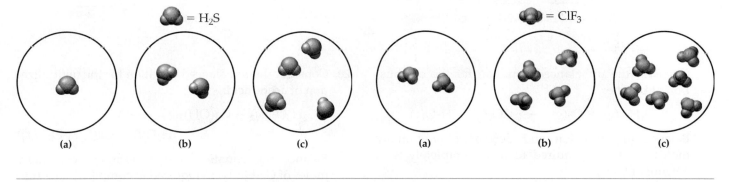

= H_2S = ClF_3

(a)   (b)   (c)   (a)   (b)   (c)

---

21. For each reaction, calculate how many moles of product form when 1.75 mol of the reactant in color completely reacts. Assume there is more than enough of the other reactant.

(a) $H_2(g) + Cl_2(g) \longrightarrow 2\,HCl(g)$

(b) $2\,H_2(g) + O_2(g) \longrightarrow 2\,H_2O(l)$

(c) $2\,Na(s) + O_2(g) \longrightarrow Na_2O_2(s)$

(d) $2\,S(s) + 3\,O_2(g) \longrightarrow 2\,SO_3(g)$

22. For each reaction, calculate how many moles of the product form when 0.112 mol of the reactant in color completely reacts. Assume there is more than enough of the other reactant.

(a) $2\,Ca(s) + O_2(g) \longrightarrow 2\,CaO(s)$

(b) $4\,Fe(s) + 3\,O_2(g) \longrightarrow 2\,Fe_2O_3(s)$

(c) $4\,K(s) + O_2(g) \longrightarrow 2\,K_2O(s)$

(d) $4\,Al(s) + 3\,O_2(g) \longrightarrow 2\,Al_2O_3(s)$

23. For the reaction shown, calculate how many moles of each product form when the given amount of each reactant completely reacts. Assume there is more than enough of the other reactant.

$$2\,PbS(s) + 3\,O_2(g) \longrightarrow 2\,PbO(s) + 2\,SO_2(g)$$

(a) 2.4 mol PbS

(b) 2.4 mol O_2

(c) 5.3 mol PbS

(d) 5.3 mol O_2

24. For the reaction shown, calculate how many moles of each product form when the given amount of each reactant completely reacts. Assume there is more than enough of the other reactant.

$$C_3H_8(g) + 5\,O_2(g) \longrightarrow 3\,CO_2(g) + 4\,H_2O(g)$$

(a) 4.6 mol C_3H_8

(b) 4.6 mol O_2

(c) 0.0558 mol C_3H_8

(d) 0.0558 mol O_2

---

25. Consider the balanced equation:

$$2\,N_2H_4(g) + N_2O_4(g) \longrightarrow 3\,N_2(g) + 4\,H_2O(g)$$

Complete the table showing the appropriate number of moles of reactants and products. If the number of moles of a reactant is provided, fill in the required amount of the other reactant, as well as the moles of each product formed. If the number of moles of a product is provided, fill in the required amount of each reactant to make that amount of product, as well as the amount of the other product that is made.

mol N_2H_4	mol N_2O_4	mol N_2	mol H_2O
___	2	___	___
6	___	___	___
___	___	___	8
___	5.5	___	___
3	___	___	___
___	___	12.4	___

26. Consider the balanced equation:

$$SiO_2(s) + 3\,C(s) \longrightarrow SiC(s) + 2\,CO(g)$$

Complete the table showing the appropriate number of moles of reactants and products. If the number of moles of a reactant is provided, fill in the required amount of the other reactant, as well as the moles of each product formed. If the number of moles of a product is provided, fill in the required amount of each reactant to make that amount of product, as well as the amount of the other product that is made.

mol SiO_2	mol C	mol SiC	mol CO
___	6	___	___
3	___	___	___
___	___	___	10
___	9.5	___	___
3.2	___	___	___

---

27. Consider the unbalanced equation for the combustion of butane:

$$C_4H_{10}(g) + O_2(g) \longrightarrow CO_2(g) + H_2O(g)$$

Balance the equation and determine how many moles of O_2 are required to react completely with 4.9 mol of C_4H_{10}.

28. Consider the unbalanced equation for the neutralization of acetic acid:

$$HC_2H_3O_2(aq) + Ca(OH)_2(aq) \longrightarrow$$
$$H_2O(l) + Ca(C_2H_3O_2)_2(aq)$$

Balance the equation and determine how many moles of $Ca(OH)_2$ are required to completely neutralize 1.07 mol of $HC_2H_3O_2$.

---

29. Consider the unbalanced equation for the reaction of solid lead with silver nitrate:

$$Pb(s) + AgNO_3(aq) \longrightarrow Pb(NO_3)_2(aq) + Ag(s)$$

(a) Balance the equation.

(b) How many moles of silver nitrate are required to completely react with 9.3 mol of lead?

(c) How many moles of Ag are formed by the complete reaction of 28.4 mol of Pb?

30. Consider the unbalanced equation for the reaction of aluminum with sulfuric acid:

$$Al(s) + H_2SO_4(aq) \longrightarrow Al_2(SO_4)_3(aq) + H_2(g)$$

(a) Balance the equation.

(b) How many moles of H_2SO_4 are required to completely react with 8.3 mol of Al?

(c) How many moles of H_2 are formed by the complete reaction of 0.341 mol of Al?

MASS-TO-MASS CONVERSIONS

31. For the reaction shown, calculate how many grams of oxygen form when each quantity of reactant completely reacts.

$$2\,HgO(s) \longrightarrow 2\,Hg(l) + O_2(g)$$

(a) 2.13 g HgO

(b) 6.77 g HgO

(c) 1.55 kg HgO

(d) 3.87 mg HgO

32. For the reaction shown, calculate how many grams of oxygen form when each quantity of reactant completely reacts.

$$2\,KClO_3(s) \longrightarrow 2\,KCl(s) + 3\,O_2(g)$$

(a) 2.72 g KClO$_3$

(b) 0.361g KClO$_3$

(c) 83.6 kg KClO$_3$

(d) 22.4 mg KClO$_3$

33. For each of the reactions shown, calculate how many grams of the product form when 2.4 g of the reactant in color completely reacts. Assume there is more than enough of the other reactant.

(a) $2\,Na(s) + Cl_2(g) \longrightarrow 2\,NaCl(s)$

(b) $CaO(s) + CO_2(g) \longrightarrow CaCO_3(s)$

(c) $2\,Mg(s) + O_2(g) \longrightarrow 2\,MgO(s)$

(d) $Na_2O(s) + H_2O(l) \longrightarrow 2\,NaOH(aq)$

34. For each of the reactions shown, calculate how many grams of the product form when 17.8 g of the reactant in color completely reacts. Assume there is more than enough of the other reactant.

(a) $Ca(s) + Cl_2(g) \longrightarrow CaCl_2(s)$

(b) $2\,K(s) + Br_2(l) \longrightarrow 2\,KBr(s)$

(c) $4\,Cr(s) + 3\,O_2(g) \longrightarrow 2\,Cr_2O_3(s)$

(d) $2\,Sr(s) + O_2(g) \longrightarrow 2\,SrO(s)$

35. For the reaction shown, calculate how many grams of each product form when the given amount of each reactant completely reacts to form products. Assume there is more than enough of the other reactant.

$$2\,Al(s) + Fe_2O_3(s) \longrightarrow Al_2O_3(s) + 2\,Fe(l)$$

(a) 4.7 g Al

(b) 4.7 g Fe$_2$O$_3$

36. For the reaction shown, calculate how many grams of each product form when the given amount of each reactant completely reacts to form products. Assume there is more than enough of the other reactant.

$$2\,HCl(aq) + Na_2CO_3(aq) \longrightarrow$$
$$2\,NaCl(aq) + H_2O(l) + CO_2(g)$$

(a) 10.8 g HCl

(b) 10.8 g Na$_2$CO$_3$

37. Consider the balanced equation for the combustion of methane, a component of natural gas:

$$CH_4(g) + 2\,O_2(g) \longrightarrow CO_2(g) + 2\,H_2O(g)$$

Complete the table with the appropriate masses of reactants and products. If the mass of a reactant is provided, fill in the mass of other reactants required to completely react with the given mass, as well as the mass of each product formed. If the mass of a product is provided, fill in the required masses of each reactant to make that amount of product, as well as the mass of the other product that is formed.

Mass CH$_4$	Mass O$_2$	Mass CO$_2$	Mass H$_2$O
_____	2.57 g	_____	_____
22.32 g	_____	_____	_____
_____	_____	_____	11.32 g
_____	_____	2.94 g	_____
3.18 kg	_____	_____	_____
_____	_____	2.35×10^3 kg	_____

38. Consider the balanced equation for the combustion of butane, a fuel often used in lighters:

$$2\,C_4H_{10}(g) + 13\,O_2(g) \longrightarrow 8\,CO_2(g) + 10\,H_2O(g)$$

Complete the table showing the appropriate masses of reactants and products. If the mass of a reactant is provided, fill in the mass of other reactants required to completely react with the given mass, as well as the mass of each product formed. If the mass of a product is provided, fill in the required masses of each reactant to make that amount of product, as well as the mass of the other product that is formed.

Mass C$_4$H$_{10}$	Mass O$_2$	Mass CO$_2$	Mass H$_2$O
_____	1.11 g	_____	_____
5.22 g	_____	_____	_____
_____	_____	10.12 g	_____
_____	_____	_____	9.04 g
232 mg	_____	_____	_____
_____	_____	118 mg	_____

39. For each acid–base reaction, calculate how many grams of acid are necessary to completely react with and neutralize 2.5 g of the base.

    (a) $HCl(aq) + NaOH(aq) \longrightarrow H_2O(l) + NaCl(aq)$

    (b) $2 HNO_3(aq) + Ca(OH)_2(aq) \longrightarrow$
    $$2 H_2O(l) + Ca(NO_3)_2(aq)$$

    (c) $H_2SO_4(aq) + 2 KOH(aq) \longrightarrow$
    $$2 H_2O(l) + K_2SO_4(aq)$$

40. For each precipitation reaction, calculate how many grams of the first reactant are necessary to completely react with 17.3 g of the second reactant.

    (a) $2 KI(aq) + Pb(NO_3)_2(aq) \longrightarrow$
    $$PbI_2(s) + 2 KNO_3(aq)$$

    (b) $Na_2CO_3(aq) + CuCl_2(aq) \longrightarrow$
    $$CuCO_3(s) + 2 NaCl(aq)$$

    (c) $K_2SO_4(aq) + Sr(NO_3)_2(aq) \longrightarrow$
    $$SrSO_4(s) + 2 KNO_3(aq)$$

41. Sulfuric acid can dissolve aluminum metal according to the reaction:

    $$2 Al(s) + 3 H_2SO_4(aq) \longrightarrow Al_2(SO_4)_3(aq) + 3 H_2(g)$$

    Suppose you wanted to dissolve an aluminum block with a mass of 22.5 g. What minimum amount of H_2SO_4 in grams would you need? How many grams of H_2 gas would be produced by the complete reaction of the aluminum block?

42. Hydrochloric acid can dissolve solid iron according to the reaction:

    $$Fe(s) + 2 HCl(aq) \longrightarrow FeCl_2(aq) + H_2(g)$$

    What minimum mass of HCl in grams would dissolve a 2.8-g iron bar on a padlock? How much H_2 would be produced by the complete reaction of the iron bar?

LIMITING REACTANT, THEORETICAL YIELD, AND PERCENT YIELD

43. Consider the generic chemical equation:
    $$2 A + 4 B \longrightarrow 3 C$$
    What is the limiting reactant when each of the initial quantities of A and B is allowed to react?
    (a) 2 mol A; 5 mol B
    (b) 1.8 mol A; 4 mol B
    (c) 3 mol A; 4 mol B
    (d) 22 mol A; 40 mol B

44. Consider the generic chemical equation:
    $$A + 3 B \longrightarrow C$$
    What is the limiting reactant when each of the initial quantities of A and B is allowed to react?
    (a) 1 mol A; 4 mol B
    (b) 2 mol A; 3 mol B
    (c) 0.5 mol A; 1.6 mol B
    (d) 24 mol A; 75 mol B

45. Determine the theoretical yield of C when each of the initial quantities of A and B is allowed to react in the generic reaction:
    $$A + 2 B \longrightarrow 3 C$$
    (a) 1 mol A; 1 mol B
    (b) 2 mol A; 2 mol B
    (c) 1 mol A; 3 mol B
    (d) 32 mol A; 68 mol B

46. Determine the theoretical yield of C when each of the initial quantities of A and B is allowed to react in the generic reaction:
    $$2 A + 3 B \longrightarrow 2 C$$
    (a) 2 mol A; 4 mol B
    (b) 3 mol A; 3 mol B
    (c) 5 mol A; 6 mol B
    (d) 4 mol A; 5 mol B

47. For the reaction shown, find the limiting reactant for each of the initial quantities of reactants.
    $$2 K(s) + Cl_2(g) \longrightarrow 2 KCl(s)$$
    (a) 1 mol K; 1 mol Cl_2
    (b) 1.8 mol K; 1 mol Cl_2
    (c) 2.2 mol K; 1 mol Cl_2
    (d) 14.6 mol K; 7.8 mol Cl_2

48. For the reaction shown, find the limiting reactant for each of the initial quantities of reactants.
    $$4 Cr(s) + 3 O_2(g) \longrightarrow 2 Cr_2O_3(s)$$
    (a) 1 mol Cr; 1 mol O_2
    (b) 4 mol Cr; 2.5 mol O_2
    (c) 12 mol Cr; 10 mol O_2
    (d) 14.8 mol Cr; 10.3 mol O_2

49. For the reaction shown, calculate the theoretical yield of product in moles for each of the initial quantities of reactants.

$$2 \, Mn(s) + 3 \, O_2(g) \longrightarrow 2 \, MnO_3(s)$$

(a) 2 mol Mn; 2 mol O_2

(b) 4.8 mol Mn; 8.5 mol O_2

(c) 0.114 mol Mn; 0.161 mol O_2

(d) 27.5 mol Mn; 43.8 mol O_2

50. For the reaction shown, calculate the theoretical yield of the product in moles for each of the initial quantities of reactants.

$$Ti(s) + 2 \, Cl_2(g) \longrightarrow TiCl_4(s)$$

(a) 2 mol Ti; 2 mol Cl_2

(b) 5 mol Ti; 9 mol Cl_2

(c) 0.483 mol Ti; 0.911 mol Cl_2

(d) 12.4 mol Ti; 15.8 mol Cl_2

51. Consider the generic reaction between reactants A and B:

$$3 \, A + 4 \, B \longrightarrow 2 \, C$$

If a reaction vessel initially contains 9 mol A and 8 mol B, how many moles of A, B, and C will be in the reaction vessel once the reactants have reacted as much as possible? (Assume 100% actual yield.)

52. Consider the reaction between reactants S and O_2:

$$2 \, S(s) + 3 \, O_2(g) \longrightarrow 2 \, SO_3(g)$$

If a reaction vessel initially contains 5 mol S and 9 mol O_2, how many moles of S, O_2, and SO_3 will be in the reaction vessel once the reactants have reacted as much as possible? (Assume 100% actual yield.)

53. Consider the reaction:

$$4 \, HCl(g) + O_2(g) \longrightarrow 2 \, H_2O(g) + 2 \, Cl_2(g)$$

Each molecular diagram represents an initial mixture of the reactants. How many molecules of Cl_2 would be formed by complete reaction in each case? (Assume 100% actual yield.)

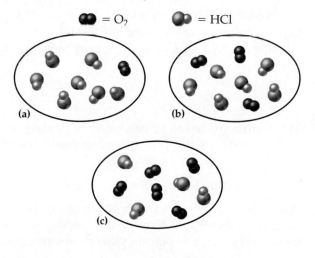

54. Consider the reaction:

$$2 \, CH_3OH(g) + 3 \, O_2(g) \longrightarrow 2 \, CO_2(g) + 4 \, H_2O(g)$$

Each molecular diagram represents an initial mixture of the reactants. How many CO_2 molecules would be formed by complete reaction in each case? (Assume 100% actual yield.)

55. For the reaction shown, find the limiting reactant for each of the initial quantities of reactants.

$$2 \, Li(s) + F_2(g) \longrightarrow 2 \, LiF(s)$$

(a) 1.0 g Li; 1.0 g F_2

(b) 10.5 g Li; 37.2 g F_2

(c) 2.85×10^3 g Li; 6.79×10^3 g F_2

56. For the reaction shown, find the limiting reactant for each of the initial quantities of reactants.

$$4 \, Al(s) + 3 \, O_2(g) \longrightarrow 2 \, Al_2O_3(s)$$

(a) 1.0 g Al; 1.0 g O_2

(b) 2.2 g Al; 1.8 g O_2

(c) 0.353 g Al; 0.482 g O_2

57. For the reaction shown, calculate the theoretical yield of the product in grams for each of initial quantities of reactants.

$$2 Al(s) + 3 Cl_2(g) \longrightarrow 2 AlCl_3(s)$$

(a) 1.0 g Al; 1.0 g Cl_2

(b) 5.5 g Al; 19.8 g Cl_2

(c) 0.439 g Al; 2.29 g Cl_2

58. For the reaction shown, calculate the theoretical yield of the product in grams for each of the initial quantities of reactants.

$$Ti(s) + 2 F_2(g) \longrightarrow TiF_4(s)$$

(a) 1.0 g Ti; 1.0 g F_2

(b) 4.8 g Ti; 3.2 g F_2

(c) 0.388 g Ti; 0.341 g F_2

59. If the theoretical yield of a reaction is 24.8 g and the actual yield is 18.5 g, what is the percent yield?

60. If the theoretical yield of a reaction is 0.118 g and the actual yield is 0.104 g, what is the percent yield?

61. Consider the reaction between calcium oxide and carbon dioxide:

$$CaO(s) + CO_2(g) \longrightarrow CaCO_3(s)$$

A chemist allows 14.4 g of CaO and 13.8 g of CO_2 to react. When the reaction is finished, the chemist collects 19.4 g of $CaCO_3$. Determine the limiting reactant, theoretical yield, and percent yield for the reaction.

62. Consider the reaction between sulfur trioxide and water:

$$SO_3(g) + H_2O(l) \longrightarrow H_2SO_4(aq)$$

A chemist allows 61.5 g of SO_3 and 11.2 g of H_2O to react. When the reaction is finished, the chemist collects 54.9 g of H_2SO_4. Determine the limiting reactant, theoretical yield, and percent yield for the reaction.

63. Consider the reaction between NiS_2 and O_2:

$$2 NiS_2(s) + 5 O_2(g) \longrightarrow 2 NiO(s) + 4 SO_2(g)$$

When 11.2 g of NiS_2 react with 5.43 g of O_2, 4.86 g of NiO are obtained. Determine the limiting reactant, theoretical yield of NiO, and percent yield for the reaction.

64. Consider the reaction between HCl and O_2:

$$4 HCl(g) + O_2(g) \longrightarrow 2 H_2O(l) + 2 Cl_2(g)$$

When 63.1 g of HCl react with 17.2 g of O_2, 49.3 g of Cl_2 are collected. Determine the limiting reactant, theoretical yield of Cl_2, and percent yield for the reaction.

65. Lead ions can be precipitated from solution with NaCl according to the reaction:

$$Pb^{2+}(aq) + 2 NaCl(aq) \longrightarrow PbCl_2(s) + 2 Na^+(aq)$$

When 135.8 g of NaCl are added to a solution containing 195.7 g of Pb^{2+}, a $PbCl_2$ precipitate forms. The precipitate is filtered and dried and found to have a mass of 252.4 g. Determine the limiting reactant, theoretical yield of $PbCl_2$, and percent yield for the reaction.

66. Magnesium oxide can be produced by heating magnesium metal in the presence of oxygen. The balanced equation for the reaction is:

$$2 Mg(s) + O_2(g) \longrightarrow 2 MgO(s)$$

When 10.1 g of Mg react with 10.5 g of O_2, 11.9 g of MgO are collected. Determine the limiting reactant, theoretical yield, and percent yield for the reaction.

67. Consider the reaction between TiO_2 and C:

$$TiO_2(s) + 2 C(s) \longrightarrow Ti(s) + 2 CO(g)$$

A reaction vessel initially contains 10.0 g of each of the reactants. Calculate the masses of TiO_2, C, Ti, and CO that will be in the reaction vessel once the reactants have reacted as much as possible. (Assume 100% yield.)

68. Consider the reaction between N_2H_4 and N_2O_4:

$$2 N_2H_4(g) + N_2O_4(g) \longrightarrow 3 N_2(g) + 4 H_2O(g)$$

A reaction vessel initially contains 27.5 g N_2H_4 and 74.9 g of N_2O_4, Calculate the masses of N_2H_4, N_2O_4, N_2, and H_2O that will be in the reaction vessel once the reactants have reacted as much as possible. (Assume 100% yield.)

ENTHALPY AND STOICHIOMETRY OF ΔH_{rxn}

69. Classify each process as exothermic or endothermic and indicate the sign of ΔH.

(a) butane gas burning in a lighter

(b) the reaction that occurs in the chemical cold packs used to ice athletic injuries

(c) the burning of wax in a candle

70. Classify each process as exothermic or endothermic and indicate the sign of ΔH.

(a) ice melting

(b) a sparkler burning

(c) acetone evaporating from skin

---

71. Consider the generic reaction:

$$A + 2B \longrightarrow C \qquad \Delta H_{rxn} = -55 \text{ kJ}$$

Determine the amount of heat emitted when each amount of reactant completely reacts (assume that there is more than enough of the other reactant):

(a) 1 mol A

(b) 2 mol A

(c) 1 mol B

(d) 2 mol B

72. Consider the generic reaction:

$$2A + 3B \longrightarrow C \qquad \Delta H_{rxn} = -125 \text{ kJ}$$

Determine the amount of heat emitted when each amount of reactant completely reacts (assume that there is more than enough of the other reactant):

(a) 2 mol A

(b) 3 mol A

(c) 3 mol B

(d) 5 mol B

---

73. Consider the equation for the combustion of acetone (C_3H_6O), the main ingredient in nail polish remover.

$$C_3H_6O(l) + 4 O_2(g) \longrightarrow 3 CO_2(g) + 3 H_2O(g)$$
$$\Delta H_{rxn} = -1790 \text{ kJ}$$

If a bottle of nail polish remover contains 155 g of acetone, how much heat is released by its complete combustion?

74. The equation for the combustion of CH_4 (the main component of natural gas) is shown below. How much heat is produced by the complete combustion of 237 g of CH_4?

$$CH_4(g) + 2 O_2(g) \longrightarrow CO_2(g) + 2 H_2O(g)$$
$$\Delta H_{rxn}^{\circ} = -802.3 \text{ kJ}$$

---

75. Octane (C_8H_{18}) is a component of gasoline that burns according to the equation:

$$C_8H_{18}(l) + \tfrac{25}{2} O_2(g) \longrightarrow 8 CO_2(g) + 9 H_2O(g)$$
$$\Delta H_{rxn}^{\circ} = -5074.1 \text{ kJ}$$

What mass of octane (in g) is required to produce 1.55×10^3 kJ of heat?

76. The evaporation of water is endothermic:

$$H_2O(l) \longrightarrow H_2O(g) \qquad \Delta H_{rxn}^{\circ} = +44.01 \text{ kJ}$$

What minimum mass of water (in g) has to evaporate to absorb 175 kJ of heat?

CUMULATIVE PROBLEMS

77. Consider the reaction:

$$2 N_2(g) + 5 O_2(g) + 2 H_2O(g) \longrightarrow 4 HNO_3(g)$$

If a reaction mixture contains 28 g of N_2, 150 g of O_2, and 36 g of H_2O, what is the limiting reactant? (Try to do this problem in your head without any written calculations.)

78. Consider the reaction:

$$2 CO(g) + O_2(g) \longrightarrow 2 CO_2(g)$$

If a reaction mixture contains 28 g of CO and 32 g of O_2, what is the limiting reactant? (Try to do this problem in your head without any written calculations.)

---

79. A solution contains an unknown mass of dissolved barium ions. When sodium sulfate is added to the solution, a white precipitate forms. The precipitate is filtered and dried and found to have a mass of 258 mg. What mass of barium was in the original solution? (Assume that all of the barium was precipitated out of solution by the reaction.)

80. A solution contains an unknown mass of dissolved silver ions. When potassium chloride is added to the solution, a white precipitate forms. The precipitate is filtered and dried and found to have a mass of 212 mg. What mass of silver was in the original solution? (Assume that all of the silver was precipitated out of solution by the reaction.)

81. Sodium bicarbonate is often used as an antacid to neutralize excess hydrochloric acid in an upset stomach. How much hydrochloric acid in grams can be neutralized by 3.5 g of sodium bicarbonate? (*Hint:* Begin by writing a balanced equation for the reaction between aqueous sodium bicarbonate and aqueous hydrochloric acid.)

82. Toilet bowl cleaners often contain hydrochloric acid to dissolve the calcium carbonate deposits that accumulate within a toilet bowl. How much calcium carbonate in grams can be dissolved by 5.8 g of HCl? (*Hint:* Begin by writing a balanced equation for the reaction between hydrochloric acid and calcium carbonate.)

83. The combustion of gasoline produces carbon dioxide and water. Assume gasoline to be pure octane (C_8H_{18}) and calculate how many kilograms of carbon dioxide are added to the atmosphere per 1.0 kg of octane burned. (*Hint:* Begin by writing a balanced equation for the combustion reaction.)

84. Many home barbecues are fueled with propane gas (C_3H_8.) How much carbon dioxide in kilograms is produced upon the complete combustion of 18.9 L of propane (approximate contents of one 5-gal tank)? Assume that the density of the liquid propane in the tank is 0.621 g/mL. (*Hint:* Begin by writing a balanced equation for the combustion reaction.)

85. A hard water solution contains 4.8 g of calcium chloride. How much sodium phosphate in grams should be added to the solution to completely precipitate all of the calcium?

86. Magnesium ions can be precipitated from seawater by the addition of sodium hydroxide. How much sodium hydroxide in grams must be added to a sample of seawater to completely precipitate the 88.4 mg of magnesium present?

87. Hydrogen gas can be prepared in the laboratory by a single-displacement reaction in which solid zinc reacts with hydrochloric acid. How much zinc in grams is required to make 14.5 g of hydrogen gas through this reaction?

88. Sodium peroxide (Na_2O_2) reacts with water to form sodium hydroxide and oxygen gas. Write a balanced equation for the reaction and determine how much oxygen in grams is formed by the complete reaction of 35.23 g of Na_2O_2.

89. Ammonium nitrate reacts explosively upon heating to form nitrogen gas, oxygen gas, and gaseous water. Write a balanced equation for this reaction and determine how much oxygen in grams is produced by the complete reaction of 1.00 kg of ammonium nitrate.

90. Pure oxygen gas can be prepared in the laboratory by the decomposition of solid potassium chlorate to form solid potassium chloride and oxygen gas. How much oxygen gas in grams can be prepared from 45.8 g of potassium chlorate?

91. Aspirin can be made in the laboratory by reacting acetic anhydride ($C_4H_6O_3$) with salicylic acid ($C_7H_6O_3$) to form aspirin ($C_9H_8O_4$) and acetic acid ($C_2H_4O_2$). The balanced equation is:

$$C_4H_6O_3 + C_7H_6O_3 \longrightarrow C_9H_8O_4 + C_2H_4O_2$$

In a laboratory synthesis, a student begins with 5.00 mL of acetic anhydride (density = 1.08 g/mL) and 2.08 g of salicylic acid. Once the reaction is complete, the student collects 2.01 g of aspirin. Determine the limiting reactant, theoretical yield of aspirin, and percent yield for the reaction.

92. The combustion of liquid ethanol (C_2H_5OH) produces carbon dioxide and water. After 3.8 mL of ethanol (density = 0.789 g/mL) is allowed to burn in the presence of 12.5 g of oxygen gas, 3.10 mL of water (density = 1.00 g/mL) is collected. Determine the limiting reactant, theoretical yield of H_2O, and percent yield for the reaction. (*Hint:* Write a balanced equation for the combustion of ethanol.)

93. Urea (CH_4N_2O), a common fertilizer, can be synthesized by the reaction of ammonia (NH_3) with carbon dioxide:

$$2\,NH_3(aq) + CO_2(aq) \longrightarrow CH_4N_2O(aq) + H_2O(l)$$

An industrial synthesis of urea obtains 87.5 kg of urea upon reaction of 68.2 kg of ammonia with 105 kg of carbon dioxide. Determine the limiting reactant, theoretical yield of urea, and percent yield for the reaction.

94. Silicon, which occurs in nature as SiO_2, is the material from which most computer chips are made. If SiO_2 is heated until it melts into a liquid, it reacts with solid carbon to form liquid silicon and carbon monoxide gas. In an industrial preparation of silicon, 52.8 kg of SiO_2 reacts with 25.8 kg of carbon to produce 22.4 kg of silicon. Determine the limiting reactant, theoretical yield, and percent yield for the reaction.

95. The ingestion of lead from food, water, or other environmental sources can cause lead poisoning, a serious condition that affects the central nervous system, causing symptoms such as distractibility, lethargy, and loss of motor function. Lead poisoning is treated with chelating agents, substances that bind to lead and allow it to be eliminated in the urine. A modern chelating agent used for this purpose is succimer ($C_4H_6O_4S_2$). Suppose you are trying to determine the appropriate dose for succimer treatment of lead poisoning. Assume that a patient's blood lead levels are 0.550 mg/L, that total blood volume is 5.0 L, and that 1 mol of succimer binds 1 mol of lead. What minimum mass of succimer in milligrams is needed to bind all of the lead in this patient's bloodstream?

96. An emergency breathing apparatus placed in mines or caves works via the chemical reaction:

$$4\,KO_2(s) + 2\,CO_2(g) \longrightarrow 2\,K_2CO_3(s) + 3\,O_2(g)$$

If the oxygen supply becomes limited or if the air becomes poisoned, a worker can use the apparatus to breathe while exiting the mine. Notice that the reaction produces O_2, which can be breathed, and absorbs CO_2, a product of respiration. What minimum amount of KO_2 is required for the apparatus to produce enough oxygen to allow the user 15 minutes to exit the mine in an emergency? Assume that an adult consumes approximately 4.4 g of oxygen in 15 minutes of normal breathing.

97. The propane fuel (C_3H_8) used in gas barbecues burns according to the equation:

$$C_3H_8(g) + 5\,O_2(g) \longrightarrow 3\,CO_2(g) + 4\,H_2O(g)$$
$$\Delta H^{\circ}_{rxn} = -2044\ kJ$$

If a pork roast must absorb 1.6×10^3 kJ to fully cook, and if only 10% of the heat produced by the barbecue is actually absorbed by the roast, what mass of CO_2 is emitted into the atmosphere during the grilling of the pork roast?

98. Charcoal is primarily carbon. Determine the mass of CO_2 produced by burning enough carbon to produce 5.00×10^2 kJ of heat.

$$C(s) + O_2(g) \longrightarrow CO_2(g) \qquad \Delta H^{\circ}_{rxn} - -393.5\ kJ$$

HIGHLIGHT PROBLEMS

99. A loud classroom demonstration involves igniting a hydrogen-filled balloon. The hydrogen within the balloon reacts explosively with oxygen in the air to form water.

$$2 H_2(g) + O_2(g) \longrightarrow 2 H_2O(g)$$

If the balloon is filled with a mixture of hydrogen and oxygen, the explosion is even louder than if the balloon is filled with only hydrogen; the intensity of the explosion depends on the relative amounts of oxygen and hydrogen within the balloon. Consider the molecular views representing different amounts of hydrogen and oxygen in four different balloons. Based on the balanced chemical equation, which balloon will make the loudest explosion?

(a)

(b)

(c)

(d)

 O$_2$ H$_2$

100. A hydrochloric acid solution will neutralize a sodium hydroxide solution. Consider the molecular views showing one beaker of HCl and four beakers of NaOH. Which NaOH beaker will just neutralize the HCl beaker? Begin by writing a balanced chemical equation for the neutralization reaction.

(a)

(b)

(c)

(d)

101. As we have seen, scientists have grown progressively more worried about the potential for global warming caused by increasing atmospheric carbon dioxide levels. The world burns the fossil fuel equivalent of approximately 9.0×10^{12} kg of petroleum per year. Assume that all of this petroleum is in the form of octane (C_8H_{18}) and calculate how much CO_2 in kilograms is produced by world fossil fuel combustion per year. (*Hint:* Begin by writing a balanced equation for the combustion of octane.) If the atmosphere currently contains approximately 3.0×10^{15} kg of CO_2, how long will it take for the world's fossil fuel combustion to double the amount of atmospheric carbon dioxide?

Atmospheric carbon dioxide (CO_2)

Atmospheric CO_2 levels 1750 to present

102. Lakes that have been acidified by acid rain can be neutralized by the addition of limestone ($CaCO_3$). How much limestone in kilograms would be required to completely neutralize a 5.2×10^9-L lake containing 5.0×10^{-3} g of H_2SO_4 per liter?

▶ANSWERS TO SKILLBUILDER EXERCISES

Skillbuilder 8.1	49.2 mol H_2O	**Skillbuilder Plus, p. 263**	29.4 kg NH_3
Skillbuilder 8.2	6.88 g HCl	**Skillbuilder 8.6**	Limiting reactant is CO; theoretical yield = 127 g Fe; percent yield = 68.8%
Skillbuilder 8.3	4.0×10^3 kg H_2SO_4		
Skillbuilder 8.4	Limiting reactant is Na; theoretical yield is 4.8 mol of NaF	**Skillbuilder 8.7**	-2.06×10^3 kJ
		Skillbuilder Plus, p. 267	33 g C_4H_{10} necessary; 99 g CO_2 produced
Skillbuilder 8.5	30.7 g NH_3		

▶ANSWERS TO CONCEPTUAL CHECKPOINTS

8.1(a) Since the reaction requires 2 O_2 molecules to react with 1 CH_4 molecule, and since there are 4 O_2 molecules available to react, then 2 CH_4 molecules are required for complete reaction.

8.2(c) Hydrogen is the limiting reactant. The reaction mixture contains 3 H_2 molecules; therefore 2 NH_3 molecules will form when the reactants have reacted as completely as possible. Nitrogen is in excess, and there is one leftover nitrogen molecule.

8.3(d) NO is the limiting reagent. The reaction mixture initially contains 4 mol NO; therefore 10 moles of H_2 will be consumed, leaving 5 mol H_2 unreacted. The products will be 4 mol NH_3 and 4 mol H_2O.

8.4(c) B is the limiting reactant. If 4 mol B react, then 200 kJ of heat is produced.

CHAPTER 9

Electrons in Atoms and the Periodic Table

"Anyone who is not shocked by quantum mechanics has not understood it."

NIELS BOHR (1885–1962)

9.1 Blimps, Balloons, and Models of the Atom

▲ The *Hindenburg* was filled with hydrogen, a reactive and flammable gas. **Question:** What makes hydrogen reactive?

◀ Modern blimps are filled with helium, an inert gas. The nucleus of the helium atom (inset) has two protons, so the neutral helium atom has two electrons—a highly stable configuration. In this chapter we learn about models that explain the inertness of helium and the reactivity of other elements.

You have probably seen one of the Goodyear blimps floating in the sky. A Goodyear blimp is often present at championship sporting events such as the Rose Bowl, the Indy 500, or the U.S. Open golf tournament. One was present at the Statue of Liberty's 100th birthday party, and blimps have made appearances in countless movies and television shows. The blimp's inherent stability allows it to provide spectacular views of the world below for television and film.

The Goodyear blimp is similar to a large balloon. Unlike airplanes, which must be moving fast to stay in flight, a blimp or *airship* floats in air because it is filled with a gas that is less dense than air. The Goodyear blimp is filled with helium. Other airships in history, however, have used hydrogen for buoyancy. For example, the *Hindenburg*—the largest airship ever constructed—was filled with hydrogen, which turned out to be a poor choice. Hydrogen is a reactive and flammable gas. On May 6, 1937, while landing in New Jersey on its first transatlantic crossing, the *Hindenburg* burst into flames, destroying the airship and killing 36 of the 97 passengers. Apparently, as the *Hindenburg* was landing, leaking hydrogen gas ignited, resulting in an explosion that destroyed the ship. (The skin of the *Hindenburg*, which was constructed of a flammable material, may have also been partially to blame for its demise.) A similar accident cannot happen to the Goodyear blimp thanks to the inert and therefore nonflammable nature of its helium gas contents. A spark or even a flame would actually be *extinguished* by helium.

Why is helium inert? What is it about helium *atoms* that makes helium *gas* inert? By contrast, why is hydrogen so reactive? Recall from Chapter 5 that elemental hydrogen exists as a diatomic element. Hydrogen atoms are so reactive that they react with each other to form hydrogen molecules. What is it about

The periodic law stated here is a modification of Mendeleev's original formulation. Mendeleev listed elements in order of increasing *mass*; today we list them in order of increasing *atomic number*.

hydrogen atoms that makes them so reactive? What is the difference between hydrogen and helium that accounts for their different reactivities?

When we examine the properties of hydrogen and helium, we make observations about nature. Mendeleev's periodic law, first discussed in Chapter 4, summarizes the results of many similar observations on the properties of elements:

When the elements are arranged in order of increasing atomic number, certain sets of properties recur periodically.

The reactivity exhibited by hydrogen is also seen in other Group I elements, such as lithium and sodium. Likewise, the inertness of helium is seen in neon, argon, and the other noble gases. In this chapter, we consider models and theories that help explain the observed behaviors of groups of elements such as the Group I metals and noble gases. We examine two important models in particular—the **Bohr model** and the **quantum-mechanical model**—that propose explanations for the inertness of helium, the reactivity of hydrogen, and the periodic law. These models explain how electrons exist in atoms and how those electrons affect the chemical and physical properties of elements. You have already learned much about the behavior of elements in this book. You know, for example, that sodium tends to form 1+ ions and that fluorine tends to form 1− ions. You know that some elements are metals and that others are nonmetals. And you know that the noble gases are chemically inert and that the alkali metals are chemically reactive. But you do not know *why*. The models in this chapter explain why.

The Bohr model and the quantum-mechanical model were developed in the early 1900s, and they caused a revolution in the physical sciences, changing our fundamental view of matter at its most basic level. The scientists who devised these models—including Niels Bohr, Erwin Schrödinger, and Albert Einstein—were bewildered by their discoveries. Bohr claimed, "Anyone who is not shocked by quantum mechanics has not understood it." Schrödinger lamented, "I don't like it, and I am sorry I ever had anything to do with it." Einstein disbelieved it, insisting that "God does not play dice with the universe." However, the quantum-mechanical model has such explanatory power that it is rarely questioned today. It forms the basis of the modern periodic table and our understanding of chemical bonding. Its applications include lasers, computers, and semiconductor devices, and it has given us new ways to design drugs that cure disease. The quantum-mechanical model for the atom is, in many ways, the foundation of modern chemistry.

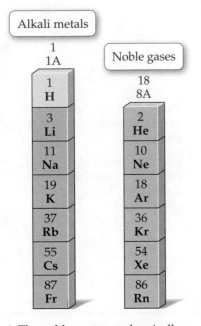

▲ The noble gases are chemically inert, and the alkali metals are chemically reactive. Why? (Hydrogen is a group I element, but it is not considered an alkali metal.)

▲ Niels Bohr (left) and Erwin Schrödinger (right), along with Albert Einstein, played a role in the development of quantum mechanics, yet they were bewildered by their own theory.

9.2 Light: Electromagnetic Radiation

▲ When a water surface is disturbed, waves are created that radiate outward from the site.

| The Greek letter *lambda* (λ) is pronounced "lam-duh."

| Helpful mnemonic: ROY G BIV—Red, Orange, Yellow, Green, Blue, Indigo, Violet

Before we explore models of the atom, we must understand a few things about light, because the interaction of light with atoms helped to shape these models. Light is familiar to all of us—we see the world by it—but what is light? Unlike most of what we have encountered so far in this book, light is not matter—it has no mass. Light is a form of **electromagnetic radiation**, a type of energy that travels through space at a constant speed of 3.0×10^8 m/s (186,000 mi/s). At this speed, a flash of light generated at the equator would travel around the world in one-seventh of a second. This extremely fast speed is part of the reason that you *see* a firework in the sky before you *hear* the sound of its explosion. The light from the exploding firework reaches your eye almost instantaneously. The sound, traveling much more slowly, takes longer.

Before the advent of quantum mechanics, light was described exclusively as a wave of electromagnetic energy traveling through space. You are probably familiar with water waves (think of the waves created by a rock dropped into a still pond), or you may have created a wave on a rope by moving the end of the rope up and down in a quick motion. In either case, the wave carries energy as it moves through the water or along the rope.

Waves are generally characterized by their **wavelength** (λ), the distance between adjacent wave crests (▼ Figure 9.1). For visible light, wavelength determines color. For example, orange light has a longer wavelength than blue light. White light, as produced by the sun or by a lightbulb, contains a spectrum of wavelengths and therefore a spectrum of color. We can see these colors—red, orange, yellow, green, blue, indigo, and violet—in a rainbow or when white light is passed through a prism (▼ Figure 9.2). Red light, with a wavelength of 750 nm (nanometers), has the longest wavelength of visible light. Violet light, with a wavelength of 400 nm, has the shortest (1 nm = 10^{-9} m). The presence of color in

▶ **FIGURE 9.1 Wavelength** The wavelength of light (λ) is defined as the distance between adjacent wave crests.

Wavelength
(λ)

Crest

Direction of travel

Trough

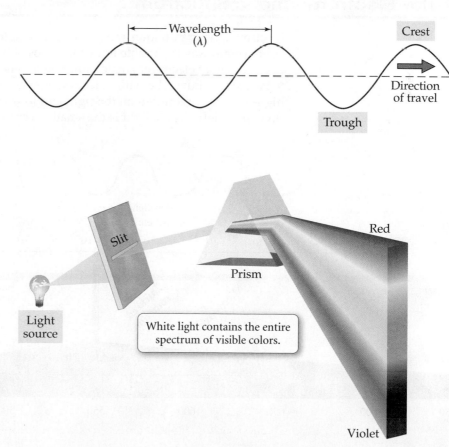

▶ **FIGURE 9.2 Components of white light** Light is separated into its constituent colors—red, orange, yellow, green, blue, indigo, and violet—when it is passed through a prism.

Slit

Prism

Red

Light source

White light contains the entire spectrum of visible colors.

Violet

▲ FIGURE 9.3 Color in objects
A red shirt appears red because it absorbs all colors except red, which it reflects.

▌The Greek letter *nu* (*ν*) is pronounced "noo."

white light is responsible for the colors we see in our everyday vision. For example, a red shirt is red because it reflects red light (◄ Figure 9.3). Our eyes see only the reflected light, making the shirt appear red.

Light waves are also often characterized by their **frequency** (*ν*), the number of cycles or crests that pass through a stationary point in one second. Wavelength and frequency are inversely related—the shorter the wavelength, the higher the frequency. Blue light, for example, has a higher frequency than red light.

In the early twentieth century, scientists such as Albert Einstein discovered that the results of certain experiments could be explained only by describing light, not as waves, but as particles. In this description, the light leaving a flashlight, for example, is viewed as a stream of particles. A particle of light is called a **photon**, and we can think of a photon as a single packet of light energy. The amount of energy carried in the packet depends on the wavelength of the light—the shorter the wavelength, the greater the energy. Just as water waves carry more energy if their crests are closer together—think about surf pounding a beach—light waves carry more energy if their crests are closer together. Therefore, violet light (shorter wavelength) carries more energy per photon than red light (longer wavelength).

To summarize:

- Electromagnetic radiation is a form of energy that travels through space at a constant speed of 3.0×10^8 m/s and can exhibit wavelike or particle-like properties.
- The wavelength of electromagnetic radiation determines the amount of energy carried by one of its photons. The shorter the wavelength, the greater the energy of each photon.
- The frequency and energy of electromagnetic radiation are inversely related to its wavelength.

9.3 The Electromagnetic Spectrum

Electromagnetic radiation ranges in wavelength from 10^{-16} m (gamma rays) to 10^6 m (radio waves). Visible light composes only a tiny portion of that range. The entire range of electromagnetic radiation is called the **electromagnetic spectrum**. ▼ Figure 9.4 shows the entire electromagnetic spectrum, with short-wavelength, high-frequency radiation on the right and long-wavelength, low-frequency radiation on the left. Visible light is the small sliver in the middle.

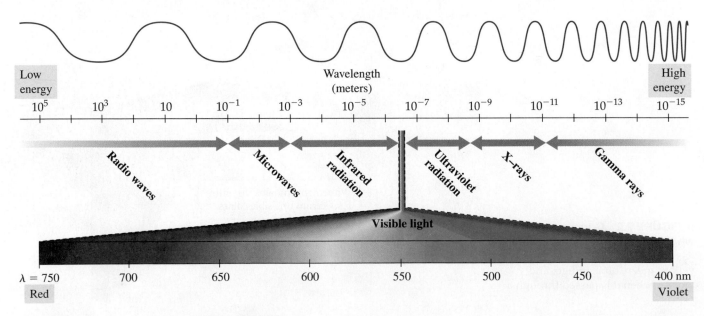

▲ FIGURE 9.4 The electromagnetic spectrum

Remember that the energy carried per photon is greater for short-wavelength light than for long-wavelength light. The shortest wavelength (and therefore most energetic) photons are those of **gamma rays**, shown on the far right of Figure 9.4. Gamma rays are produced by the sun, by stars, and by certain unstable atomic nuclei on Earth. Excessive human exposure to gamma rays is dangerous because the high energy of gamma-ray photons can damage biological molecules.

Next on the electromagnetic spectrum (to the left in Figure 9.4), with longer wavelengths (and lower energy) than gamma rays are **X-rays**, familiar to us from their medical use. X-rays pass through many substances that block visible light and are therefore used to image internal bones and organs. Like gamma-ray photons, X-ray photons carry enough energy to damage biological molecules. While several yearly exposures to X-rays are relatively harmless, excessive exposure to X-rays increases cancer risk.

Sandwiched between X-rays and visible light in the electromagnetic spectrum is **ultraviolet or UV light**, most familiar to us as the component of sunlight that produces a sunburn or suntan. While not as energetic as gamma-ray or X-ray photons, ultraviolet photons still carry enough energy to damage biological molecules. Excessive exposure to ultraviolet light increases the risk of skin cancer and cataracts and causes premature wrinkling of the skin. Next on the spectrum is **visible light**, ranging from violet (shorter wavelength, higher energy) to red (longer wavelength, lower energy). Photons of visible light do not damage biological molecules. They do, however, cause molecules in our eyes to rearrange, which sends a signal to our brains that results in vision.

Infrared light is next, with even longer wavelengths than visible light. The heat you feel when you place your hand near a hot object is infrared light. All warm objects, including human bodies, emit infrared light. While infrared light is invisible to our eyes, infrared sensors can detect it and are often used in night-vision technology to "see" in the dark. In the infrared region of the spectrum, warm objects—such as human bodies—glow, much as a lightbulb glows in the visible region of the spectrum.

Beyond infrared light, at longer wavelengths still, are **microwaves**, used for radar and in microwave ovens. Although microwave light has longer wavelengths—and therefore lower energy per photon—than visible or infrared light, it is efficiently absorbed by water and can therefore heat substances that contain water. For this reason substances that contain water, such as food, are warmed when placed in a microwave oven, but substances that do not contain water, such as a plate, are not.

The longest wavelengths of light are **radio waves**, which are used to transmit the signals used by AM and FM radio, television, and other forms of communication.

Some types of dishes contain substances that absorb microwave radiation, but most do not.

Normal photograph

Infrared photograph

▲ Warm objects, such as human or animal bodies, give off infrared light that is easily detected with an infrared camera. In the infrared photograph, the warmest areas appear as red and the coolest as dark blue. (Note that the photo confirms the familiar idea that healthy dogs have cold noses.) (*Source:* Sierra Pacific Innovations. All rights reserved. SPI CORP, www.x20.org)

CHEMISTRY AND HEALTH

Radiation Treatment for Cancer

X-rays and gamma rays are sometimes called ionizing radiation because the high energy in their photons can ionize atoms and molecules. When ionizing radiation interacts with biological molecules, it can permanently change or even destroy them. Consequently, we normally try to limit our exposure to ionizing radiation. However, doctors can use ionizing radiation to destroy molecules within unwanted cells such as cancer cells.

In radiation therapy (or radiotherapy), doctors aim X-ray or gamma-ray beams at cancerous tumors. The ionizing radiation damages the molecules within the tumor's cells that carry genetic information—information necessary for the cell to grow and divide—and the cell dies or stops dividing. Ionizing radiation also damages molecules within healthy cells; however, cancerous cells divide more quickly than healthy cells, making them more susceptible to genetic damage. Nonetheless, healthy cells often inadvertently sustain damage during treatments, resulting in side effects such as fatigue, skin lesions, and hair loss. Doctors try to minimize the exposure of healthy cells by appropriate shielding and by targeting the tumor from multiple directions, minimizing the exposure of healthy cells while maximizing the exposure of cancerous cells (▶ Figure 9.5).

Another side effect of exposing healthy cells to radiation is that they too may become cancerous. So a treatment for cancer may cause cancer. So why do we continue to use it? Radiation therapy, as most other disease therapies, has associated risks. However, we take risks all the time, many for lesser reasons. For example, every time we drive a car, we risk injury or even death. Why? Because we perceive the benefit—such as getting to the grocery store to buy food—to be worth the risk. The situation is similar in cancer therapy or any other therapy for that matter. The benefit of cancer therapy (possibly curing a cancer that will certainly kill you) is worth the risk (a slight increase in the chance of developing a future cancer).

CAN YOU ANSWER THIS? *Why would visible light not work to destroy cancerous tumors?*

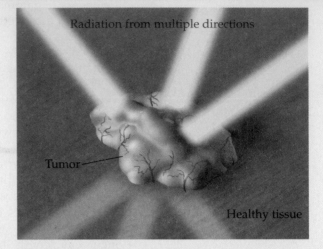

▲ **FIGURE 9.5 Radiation therapy** By targeting the tumor from various different directions, radiologists attempt to limit damage to healthy tissue.

▲ Cancer patient undergoing radiation therapy.

EXAMPLE 9.1 Wavelength, Energy, and Frequency

Arrange the three types of electromagnetic radiation—visible light, X-rays, and microwaves—in order of increasing:

(a) wavelength **(b)** frequency **(c)** energy per photon

	SOLUTION
(a) wavelength Figure 9.4 indicates that X-rays have the shortest wavelength, followed by visible light and then microwaves.	X-rays, visible light, microwaves
(b) frequency Since frequency and wavelength are inversely proportional—the longer the wavelength, the shorter the frequency—the ordering with respect to frequency is exactly the reverse of the ordering with respect to wavelength.	microwaves, visible light, X-rays

(c) energy per photon
Energy per photon decreases with increasing wavelength but increases with increasing frequency; therefore, the ordering with respect to energy per photon is the same as frequency.

microwaves, visible light, X-rays

▶**SKILLBUILDER 9.1 | Wavelength, Energy, and Frequency**
Arrange the colors of visible light—green, red, and blue—in order of increasing:

(a) wavelength   **(b)** frequency   **(c)** energy per photon

▶**FOR MORE PRACTICE**   Example 9.9; Problems 31, 32, 33, 34, 35, 36, 37, 38.

✓ **CONCEPTUAL CHECKPOINT 9.1**

Yellow light has a longer wavelength than violet light. Therefore:

(a) Yellow light has more energy per photon than violet light.

(b) Yellow light has less energy per photon than violet light.

(c) Both yellow light and violet light have the same energy per photon.

9.4 The Bohr Model: Atoms with Orbits

▲ **FIGURE 9.6   A neon sign** Neon atoms inside a glass tube absorb electrical energy and then re-emit the energy as light.

When an atom absorbs energy—in the form of heat, light, or electricity—it often re-emits that energy as light. For example, a neon sign is composed of one or more glass tubes filled with gaseous neon atoms. When an electrical current is passed through the tube, the neon atoms absorb some of the electrical energy and re-emit it as the familiar red light of a neon sign (◀ Figure 9.6). If the atoms in the tube are different, the emitted light is a different color. In other words, atoms of a given element emit light of unique colors (or unique wavelengths). Mercury atoms, for example, emit light that appears blue, hydrogen atoms emit light that appears pink (▼ Figure 9.7), and helium atoms emit light that appears yellow-orange.

▶ **FIGURE 9.7   Light emission by different elements** Light emitted from a mercury lamp (left) appears blue, and light emitted from a hydrogen lamp (right) appears pink.

▲ **FIGURE 9.8 Emission spectra** A white-light spectrum is continuous, with some radiation emitted at every wavelength. The emission spectrum of an individual element, however, includes only certain specific wavelengths. (The different wavelengths appear as *lines* because the light from the source passes through a slit before entering the prism.) Each element produces its own unique and distinctive emission spectrum.

Closer inspection of the light emitted by hydrogen, helium, and neon atoms reveals that the light contains several distinct colors or wavelengths. Just as the white light from a lightbulb can be separated into its constituent wavelengths by passing it through a prism, so the light emitted by glowing hydrogen, helium, or neon can also be separated into its constituent wavelengths (▲ Figure 9.8) by passing it through a prism. The result is called an **emission spectrum**. Notice the differences between a white-light spectrum and the emission spectra of hydrogen, helium, and neon. The white-light spectrum is *continuous*, meaning that the light intensity is uninterrupted or smooth across the entire visible range—there is some radiation at all wavelengths, with no gaps. The emission spectra of hydrogen, helium, and neon, however, are not continuous. They consist of bright spots or lines at specific wavelengths with complete darkness in between. Since the emission of light in atoms is related to the motions of electrons within the atoms, a model for how electrons exist in atoms must account for these spectra.

A major challenge in developing a model for electrons in atoms was the discrete or bright-line nature of the emission spectra. Why did atoms, when excit-

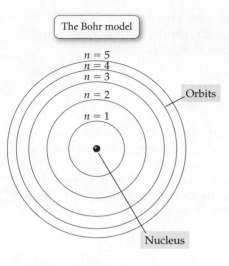

▲ **FIGURE 9.9 Bohr orbits**

▲ **FIGURE 9.10 The Bohr energy ladder** Bohr orbits are like steps on a ladder. It is possible to stand on one step or another, but impossible to stand between steps.

ed with energy, emit light only at particular wavelengths? Why did they *not* emit a continuous spectrum? Niels Bohr developed a simple model to explain these results. In his model, now called the Bohr model, electrons travel around the nucleus in circular orbits that are similar to planetary orbits around the sun. However, unlike planets revolving around the sun—which can theoretically orbit at any distance whatsoever from the sun—electrons in the Bohr model can orbit only at *specific, fixed* distances from the nucleus (◄ Figure 9.9).

The *energy* of each Bohr orbit, specified by a **quantum number** $n = 1, 2, 3 \ldots$, is also fixed, or **quantized**. The energy of each orbit increases with increasing value of n, but the levels become more closely spaced as n increases. Bohr orbits are like steps of a ladder (◄ Figure 9.10), each at a specific distance from the nucleus and each at a specific energy. Just as it is impossible to stand *between steps* on a ladder, so it is impossible for an electron to exist *between orbits* in the Bohr model. An electron in an $n = 3$ orbit, for example, is farther from the nucleus and has more energy than an electron in an $n = 2$ orbit. And an electron cannot exist at an intermediate distance or energy between the two orbits—the orbits are *quantized*. As long as an electron remains in a given orbit, it does not absorb or emit light, and its energy remains fixed and constant.

When an atom absorbs energy, an electron in one of these fixed orbits is *excited* or promoted to an orbit that is farther away from the nucleus (◄ Figure 9.11) and therefore higher in energy (this is analogous to moving up a step on the ladder). However, in this new configuration, the atom is less stable, and the electron quickly falls back or *relaxes* to a lower-energy orbit (this is analogous to moving down a step on the ladder). As it does so, it releases a photon of light containing the precise amount of energy—called a **quantum** of energy—that corresponds to the energy difference between the two orbits.

Since the amount of energy in a photon is directly related to its wavelength, the photon has a specific wavelength. *Consequently, the light emitted by excited atoms consists of specific lines at specific wavelengths, each corresponding to a specific transition between two orbits.* For example, the line at 486 nm in the hydrogen emission spectrum corresponds to an electron relaxing from the $n = 4$ orbit to the $n = 2$ orbit (▼ Figure 9.12). In the same way, the line at 657 nm (longer wavelength and therefore lower energy) corresponds to an electron relaxing from the $n = 3$ orbit to the $n = 2$ orbit. Notice that transitions between orbits that are closer together produce lower-energy (and therefore longer-wavelength) light than transitions between orbits that are farther apart.

The great success of the Bohr model of the atom was that it predicted the lines of the hydrogen emission spectrum. However, it failed to predict the emission

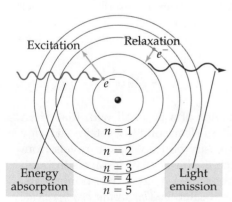

▲ **FIGURE 9.11 Excitation and emission** When a hydrogen atom absorbs energy, an electron is excited to a higher-energy orbit. The electron then relaxes back to a lower-energy orbit, emitting a photon of light.

▲ **FIGURE 9.12 Hydrogen emission lines** The 657-nm line of the hydrogen emission spectrum corresponds to an electron relaxing from the $n = 3$ orbit to the $n = 2$ orbit. The 486-nm line corresponds to an electron relaxing from the $n = 4$ orbit to the $n = 2$ orbit, and the 434-nm line corresponds to an electron relaxing from $n = 5$ to $n = 2$.

The Bohr model is still important because it provides a logical foundation to the quantum-mechanical model and reveals the historical development of scientific understanding.

spectra of other elements that contained more than one electron. For this, and other reasons, the Bohr model was replaced with a more sophisticated model called the quantum-mechanical or wave-mechanical model.

To summarize:

- Electrons exist in quantized orbits at specific, fixed energies and specific, fixed distances from the nucleus.
- When energy is put into an atom, electrons are excited to higher-energy orbits.
- When an electron relaxes (or falls) from a higher-energy orbit to a lower-energy orbit, the atom emits light.
- The energy (and therefore the wavelength) of the emitted light corresponds to the energy difference between the two orbits in the transition. Since these energies are fixed and discrete, the energy (and therefore the wavelength) of the emitted light is fixed and discrete.

✓ CONCEPTUAL CHECKPOINT 9.2

In one transition, an electron in a hydrogen atom falls from the $n = 3$ level to the $n = 2$ level. In a second transition, an electron in a hydrogen atom falls from the $n = 2$ level to the $n = 1$ level. Compared to the radiation emitted by the first of these transitions, the radiation emitted by the second will have:

(a) a lower frequency

(b) a smaller energy per photon

(c) a shorter wavelength

(d) a longer wavelength

9.5 The Quantum-Mechanical Model: Atoms with Orbitals

The quantum-mechanical model of the atom replaced the Bohr model in the early twentieth century. In the quantum-mechanical model *Bohr orbits* are replaced with *quantum-mechanical* **orbitals**. Orbitals are different from orbits in that they represent, not specific paths that electrons follow, but probability maps that show a statistical distribution of where the electron is likely to be found. The idea of an orbital is not easy to visualize. Quantum mechanics revolutionized physics and chemistry because in the quantums-mechanical model, electrons *do not* behave like particles flying through space. We cannot, in general, describe their exact paths. An orbital is a probability map that shows where the electron is *likely* to be found when the atom is probed; it does not represent the exact path that an electron takes as it travels through space.

BASEBALL PATHS AND ELECTRON PROBABILITY MAPS

To understand orbitals, let's contrast the behavior of a baseball with that of an electron. Imagine a baseball thrown from the pitcher's mound to a catcher at home plate (▶ Figure 9.13). The baseball's path can easily be traced as it travels from the pitcher to the catcher. The catcher can watch the baseball as it travels through the air, and he can predict exactly where the baseball will cross over home plate. He can even place his mitt in the correct place to catch it. *This would be impossible for an electron.* Like photons, electrons exhibit a *wave–particle duality*; sometimes they act as particles, and other times as waves. This duality leads to behavior that makes it impossible to trace an electron's path. If an electron were "thrown" from the pitcher's mound to home plate, it would land in a different place every time, *even if it were thrown in exactly the same way.* Baseballs have predictable paths—electrons do not.

▶ FIGURE 9.13   **Baseballs follow predictable paths**   A baseball follows a well-defined path as it travels from the pitcher to the catcher.

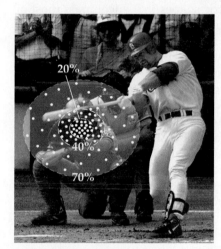

▲ FIGURE 9.14   **Electrons are unpredictable**   To describe the behavior of a "pitched" electron, you would have to construct a probability map of where it would cross home plate.

In the quantum-mechanical world of the electron, the catcher could not know exactly where the electron would cross the plate for any given throw. He would have no way of putting his mitt in the right place to catch it. However, if the catcher kept track of hundreds of electron throws, he could observe a reproducible, statistical pattern of where the electron crosses the plate. He could even draw maps in the strike zone showing the probability of an electron crossing a certain area (◀ Figure 9.14). These maps are called *probability maps*.

FROM ORBITS TO ORBITALS

In the Bohr model, an *orbit* is a circular path—analogous to a baseball's path—that shows the electron's motion around an atomic nucleus. In the quantum-mechanical model, an *orbital* is a probability map, analogous to the probability map drawn by our catcher. It shows the relative likelihood of the electron being found at various locations when the atom is probed. Just as the Bohr model has different orbits with different radii, the quantum-mechanical model has different orbitals with different shapes.

9.6   Quantum-Mechanical Orbitals

In the Bohr model of the atom, a single quantum number (n) specifies each orbit. In the quantum-mechanical model, a number and a letter specify an orbital (or orbitals). For example, the lowest-energy orbital in the quantum-mechanical model—analogous to the $n = 1$ orbit in the Bohr model—is called the *1s orbital*. It is specified by the number 1 and the letter s. The number is called the **principal quantum number** (n)

_____ $n = 4$
_____ $n = 3$

_____ $n = 2$

_____ $n = 1$

Energy

▲ **FIGURE 9.15 Principal quantum numbers** The principal quantum numbers ($n = 1, 2, 3 \ldots$) determine the energy of the hydrogen quantum-mechanical orbitals.

This analogy is purely hypothetical. It is impossible to photograph electrons in this way.

and specifies the **principal shell** of the orbital. The higher the principal quantum number, the higher the energy of the orbital. The possible principal quantum numbers are $n = 1, 2, 3 \ldots$, with energy increasing as n increases (◄ Figure 9.15). Since the 1s orbital has the lowest possible principal quantum number, it is in the lowest-energy shell and has the lowest possible energy.

The letter indicates the **subshell** of the orbital and specifies its shape. The possible letters are s, p, d, and f, each with a different shape. Orbitals within the s subshell have a spherical shape. Unlike the $n = 1$ Bohr orbit, which shows the electron's path, the 1s quantum-mechanical orbital is a three-dimensional probability map. Orbitals are sometimes represented by dots (◄ Figure 9.16), where the dot density is proportional to the probability of finding the electron.

We can understand the dot representation of an orbital better with another analogy. Imagine taking a photograph of an electron in an atom every second for 10 or 15 minutes. One second the electron is very close to the nucleus; the next second it is farther away and so on. Each photo shows a dot representing the electron's position relative to the nucleus at that time. If you took hundreds of photos and superimposed all of them, you would have an image like Figure 9.16—a statistical representation of where the electron is found. Notice that the dot density for the 1s orbital is greatest near the nucleus and decreases farther away from the nucleus. This means that the electron is more likely to be found close to the nucleus than far away from it.

Orbitals can also be represented as geometric shapes that encompass most of the volume where the electron is likely to be found. For example, the 1s orbital can be represented as a sphere (▼ Figure 9.17) that encompasses the volume within which the electron is found 90% of the time. If we superimpose the dot representation of the 1s orbital on the shape representation (▼ Figure 9.18), we can see that most of the dots are within the sphere, meaning that the electron is most likely to be found within the sphere when it is in the 1s orbital.

The single electron of an undisturbed hydrogen atom at room temperature is in the 1s orbital. This is called the **ground state**, or lowest energy state, of the hydrogen atom. However, like the Bohr model, the quantum-mechanical model allows transitions to higher-energy orbitals upon the absorption of energy. What are these higher-energy orbitals? What do they look like?

Dot representation
of 1s orbital

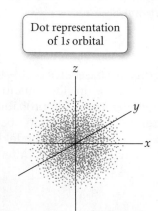

▲ **FIGURE 9.16 1s orbital** The dot density in this plot is proportional to the probability of finding the electron. The greater dot density near the middle represents a higher probability of finding the electron near the nucleus.

Shape representation
of 1s orbital

▲ **FIGURE 9.17 Shape representation of the 1s orbital** Because the distribution of electron density around the nucleus in Figure 9.16 is symmetrical—the same in all directions—we can represent the 1s orbital as a sphere.

Both representations
of 1s superimposed

▲ **FIGURE 9.18 Orbital shape and dot representation for the 1s orbital** The shape representation of the 1s orbital superimposed on the dot density representation. We can see that when the electron is in the 1s orbital, it is most likely to be found within the sphere.

▶ **FIGURE 9.19 Subshells** The number of subshells in a given principal shell is equal to the value of n.

Shell	Number of subshells	Letters specifying subshells			
$n = 4$	4	s	p	d	f
$n = 3$	3	s	p	d	
$n = 2$	2	s	p		
$n = 1$	1	s			

The next orbitals in the quantum-mechanical model are those with principal quantum number $n = 2$. Unlike the $n = 1$ principal shell, which contains only one subshell (specified by s), the $n = 2$ principal shell contains two subshells, specified by s and p.

The number of subshells in a given principal shell is equal to the value of n.

Therefore the $n = 1$ principal shell has one subshell, the $n = 2$ principal shell has two subshells, and so on (▲ Figure 9.19). The s subshell contains the $2s$ orbital, higher in energy than the $1s$ orbital and slightly larger (◀ Figure 9.20), but otherwise similar in shape. The p subshell contains three $2p$ orbitals (▼ Figure 9.21), all with the same dumbbell-like shape but with different orientations.

The next principal shell, $n = 3$, contains three subshells specified by s, p, and d. The s and p subshells contain the $3s$ and $3p$ orbitals, similar in shape to the $2s$ and $2p$ orbitals, but slightly larger and higher in energy. The d subshell contains the five d orbitals shown in ▶ Figure 9.22 on the next page. The next principal shell, $n = 4$, contains four subshells specified by s, p, d, and f. The s, p, and d subshells are similar to those in $n = 3$. The f subshell contains seven orbitals (called the $4f$ orbitals), whose shape we do not consider in this book.

As we have already discussed, hydrogen's single electron is usually in the $1s$ orbital because electrons generally seek out the lowest-energy orbital available. In hydrogen, the rest of the orbitals are normally empty. However, the absorption of energy by a hydrogen atom can cause the electron to jump (or make a transition) from the $1s$ orbital to a higher-energy orbital. When the electron is in a higher-energy orbital, the hydrogen atom is said to be in an **excited state**.

Because of their higher energy, excited states are unstable, and the electron will usually fall (or relax) back to a lower-energy orbital. In the process the electron emits energy, often in the form of light. As in the Bohr model, the energy difference between the two orbitals involved in the transition determines the wavelength of the emitted light (the greater the energy difference, the shorter the wavelength). The quantum-mechanical model predicts the bright-line spectrum of hydrogen as well as the Bohr model. However, it can also predict the bright-line spectra of other elements as well.

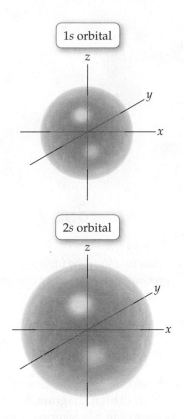

1s orbital

2s orbital

▲ **FIGURE 9.20 The 2s orbital** The $2s$ orbital is similar to the $1s$ orbital, but larger in size.

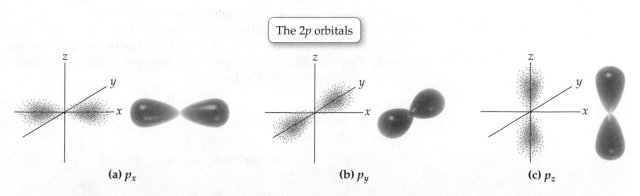

The 2p orbitals

(a) p_x **(b) p_y** **(c) p_z**

▲ **FIGURE 9.21 The 2p orbitals** This figure shows both the dot representation (left) and shape representation (right) for each p-orbital.

The 3d orbitals

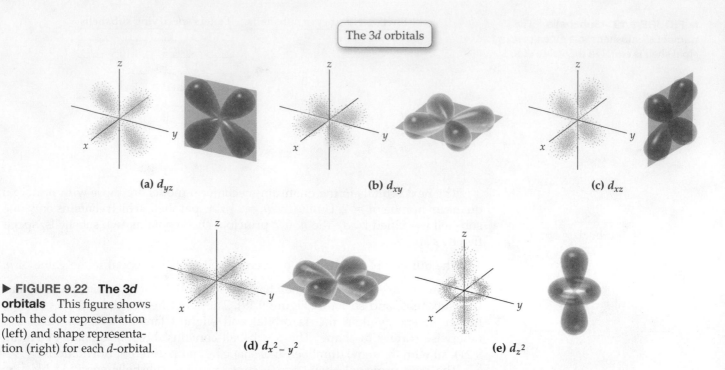

(a) d_{yz}

(b) d_{xy}

(c) d_{xz}

(d) $d_{x^2-y^2}$

(e) d_{z^2}

▶ **FIGURE 9.22   The 3d orbitals**   This figure shows both the dot representation (left) and shape representation (right) for each d-orbital.

ELECTRON CONFIGURATIONS: HOW ELECTRONS OCCUPY ORBITALS

An **electron configuration** shows the occupation of orbitals by electrons for a particular atom. For example, the electron configuration for a ground-state (or lowest energy) hydrogen atom is:

The electron configuration tells us that hydrogen's single electron is in the 1s orbital.

Another way to represent this information is with an **orbital diagram**, which gives similar information but shows the electrons as arrows in a box representing the orbital. The orbital diagram for a ground-state hydrogen atom is:

H   ⌈↑⌉

1s

The box represents the 1s orbital, and the arrow within the box represents the electron in the 1s orbital. In orbital diagrams, the direction of the arrow (pointing up or pointing down) represents **electron spin**, a fundamental property of electrons. All electrons have spin. The **Pauli exclusion principle** states that *orbitals may hold no more than two electrons with opposing spins.* We symbolize this as two arrows pointing in opposite directions

↑↓

A helium atom, for example, has two electrons. The electron configuration and orbital diagram for helium are:

	Electron configuration	**Orbital diagram**
He	$1s^2$	

1s

▶ **FIGURE 9.23 Energy ordering of orbitals for multi-electron atoms** Different subshells within the same principal shell have different energies.

In multi-electron atoms, the subshells within a principal shell do not have the same energy because of electron–electron interactions.

Remember that the number of electrons in an atom is equal to its atomic number.

Since we know that electrons occupy the lowest-energy orbitals available, and since we know that only two electrons (with opposing spins) are allowed in each orbital, we can continue to build ground-state electron configurations for the rest of the elements as long as we know the energy ordering of the orbitals. ▲ Figure 9.23 shows the energy ordering of a number of orbitals for multi-electron atoms.

Notice that, for multi-electron atoms (in contrast to hydrogen which has only one electron), the subshells within a principal shell *do not* have the same energy. In elements other than hydrogen, the energy ordering is not determined by the principal quantum number alone. For example, in multi-electron atoms, the $4s$ subshell is lower in energy than the $3d$ subshell, even though its principal quantum number is higher. Using this relative energy ordering, we can write ground-state electron configurations and orbital diagrams for other elements. For lithium, which has three electrons, the electron configuration and orbital diagram are:

Electron configuration	**Orbital diagram**

Li　　　　$1s^2 2s^1$

For carbon, which has six electrons, the electron configuration and orbital diagram are:

Electron configuration	**Orbital diagram**

C　　　　$1s^2 2s^2 2p^2$

Notice that the $2p$ electrons occupy the p orbitals (of equal energy) singly rather than pairing in one orbital. This is the result of **Hund's rule**, which states that *when filling orbitals of equal energy, electrons fill them singly first, with parallel spins.*

Before we write electron configurations for other elements, let us summarize what we have learned so far:

- Electrons occupy orbitals so as to minimize the energy of the atom; therefore, lower-energy orbitals fill before higher-energy orbitals. Orbitals fill in the following order: $1s\ 2s\ 2p\ 3s\ 3p\ 4s\ 3d\ 4p\ 5s\ 4d\ 5p\ 6s$ (◀ Figure 9.24).
- Orbitals can hold no more than two electrons each. When two electrons occupy the same orbital, they must have opposing spins. This is known as the Pauli exclusion principle.
- When orbitals of identical energy are available, these are first occupied singly with parallel spins rather than in pairs. This is known as Hund's rule.

▲ **FIGURE 9.24 Orbital filling order** The arrows indicate the order in which orbitals fill.

Consider the electron configurations and orbital diagrams for elements with atomic numbers 3 through 10.

Symbol (#e⁻)	Electron configuration	Orbital diagram

Notice how the p orbitals fill. As a result of Hund's rule, the p orbitals fill with single electrons before they fill with paired electrons. The electron configuration of neon represents the complete filling of the $n = 2$ principal shell. When writing electron configurations for elements beyond neon—or beyond any other noble gas—the electron configuration of the previous noble gas is often abbreviated by the symbol for the noble gas in brackets. For example, the electron configuration of sodium is:

$$\text{Na} \qquad 1s^2 2s^2 2p^6 3s^1$$

This can also be written using the noble gas core notation as:

$$\text{Na} \qquad [\text{Ne}]3s^1$$

where [Ne] represents $1s^2 2s^2 2p^6$, the electron configuration for neon.

To write an electron configuration for an element, first find its atomic number from the periodic table—this number equals the number of electrons in the neutral atom. Then use the order of filling from Figure 9.23 or 9.24 to distribute the electrons in the appropriate orbitals. Remember that each orbital can hold a maximum of 2 electrons. Consequently:

- the s subshell has only 1 orbital and therefore can hold only 2 electrons.
- the p subshell has 3 orbitals and therefore can hold 6 electrons.
- the d subshell has 5 orbitals and therefore can hold 10 electrons.
- the f subshell has 7 orbitals and therefore can hold 14 electrons.

EXAMPLE 9.2 Electron Configurations

Write electron configurations for each element.

(a) Mg   **(b)** S   **(c)** Ga

(a) Magnesium has 12 electrons. Distribute two of these into the 1s orbital, two into the 2s orbital, six into the 2p orbitals, and two into the 3s orbital. You can also write the electron configuration more compactly using the noble gas core notation. For magnesium, use [Ne] to represent $1s^2 2s^2 2p^6$.	**SOLUTION** Mg $1s^2 2s^2 2p^6 3s^2$ *or* Mg $[Ne]3s^2$
(b) Sulfur has 16 electrons. Distribute two of these into the 1s orbital, two into the 2s orbital, six into the 2p orbitals, two into the 3s orbital, and four into the 3p orbitals. You can write the electron configuration more compactly by using [Ne] to represent $1s^2 2s^2 2p^6$.	S $1s^2 2s^2 2p^6 3s^2 3p^4$ *or* S $[Ne]3s^2 3p^4$
(c) Gallium has 31 electrons. Distribute two of these into the 1s orbital, two into the 2s orbital, six into the 2p orbitals, two into the 3s orbital, six into the 3p orbitals, two into the 4s orbital, ten into the 3d orbitals, and one into the 4p orbitals. Notice that the d subshell has five orbitals and can therefore hold 10 electrons. You can write the electron configuration more compactly by using [Ar] to represent $1s^2 2s^2 2p^6 3s^2 3p^6$.	Ga $1s^2 2s^2 2p^6 3s^2 3p^6 4s^2 3d^{10} 4p^1$ *or* Ga $[Ar]4s^2 3d^{10} 4p^1$

▶**SKILLBUILDER 9.2 | Electron Configurations**

Write electron configurations for each element.

(a) Al   **(b)** Br   **(c)** Sr

▶**SKILLBUILDER PLUS**

Write electron configurations for each ion. (*Hint:* To determine the number of electrons to include in the electron configuration of an ion, make sure to add or subtract electrons as needed to account for the charge of the ion.)

(a) Al^{3+} **(b)** Cl^- **(c)** O^{2-}

▶**FOR MORE PRACTICE** Problems 49, 50, 53, 54, 55, 56.

EXAMPLE 9.3 Writing Orbital Diagrams

Write an orbital diagram for silicon.

SOLUTION

Since silicon is atomic number 14, it has 14 electrons. Draw a box for each orbital, putting the lowest-energy orbital (1s) on the far left and proceeding to orbitals of higher energy to the right.

Distribute the 14 electrons into the orbitals, allowing a maximum of 2 electrons per orbital and remembering Hund's rule. The complete orbital diagram is:

▶**SKILLBUILDER 9.3 | Writing Orbital Diagrams**

Write an orbital diagram for argon.

▶**FOR MORE PRACTICE** Example 9.10; Problems 51, 52.

> ### ✔ CONCEPTUAL CHECKPOINT 9.3
>
> Which pair of elements has the same *total* number of electrons in p orbitals?
>
> **(a)** Na and K
>
> **(b)** K and Kr
>
> **(c)** P and N
>
> **(d)** Ar and Ca

9.7   Electron Configurations and the Periodic Table

VALENCE ELECTRONS

Valence electrons are the electrons in the outermost principal shell (the principal shell with the highest principal quantum number, n). These electrons are important because, as we will see in the next chapter, they are involved in chemical bonding. Electrons that are *not* in the outermost principal shell are called **core electrons**. For example, silicon, with the electron configuration of $1s^2 2s^2 2p^6 3s^2 3p^2$, has 4 valence electrons (those in the $n = 3$ principal shell) and 10 core electrons.

Si $\qquad 1s^2 2s^2 2p^6 \; 3s^2 3p^2$

Core electrons   Valence electrons

EXAMPLE 9.4   Valence Electrons and Core Electrons

Write an electron configuraton for selenium and identify the valence electrons and the core electrons.

SOLUTION

Write the electron configuration for selenium by determining the total number of electrons from selenium's atomic number (34) and then distributing them into the appropriate orbitals.

$$\text{Se} \qquad 1s^2 2s^2 2p^6 3s^2 3p^6 4s^2 3d^{10} 4p^4$$

The valence electrons are those in the outermost principal shell. For selenium, the outermost principal shell is the $n = 4$ shell, which contains 6 electrons (2 in the $4s$ orbital and 4 in the three $4p$ orbitals). All other electrons, including those in the $3d$ orbitals, are core electrons.

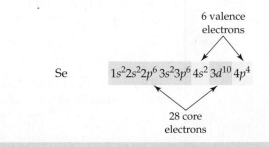

6 valence electrons

Se $\qquad 1s^2 2s^2 2p^6 \; 3s^2 3p^6 \; 4s^2 \; 3d^{10} \; 4p^4$

28 core electrons

▶ **SKILLBUILDER 9.4** | Valence Electrons and Core Electrons

Write an electron configuration for chlorine and identify the valence electrons and core electrons.

▶ **FOR MORE PRACTICE**   Example 9.11; Problems 57, 58, 61, 62.

▶ **FIGURE 9.25 Outer electron configurations of the first 18 elements**

▲ Figure 9.25 shows the first 18 elements in the periodic table with an outer electron configuration listed below each one. As you move across a row, the orbitals are simply filling in the correct order. As you move down a column, the highest principal quantum number increases, but the number of electrons in each subshell remains the same. Consequently, the elements within a column (or family) all have the same number of valence electrons and similar outer electron configurations.

A similar pattern exists for the entire periodic table (▼ Figure 9.26). Notice that, because of the filling order of orbitals, the periodic table can be divided into blocks representing the filling of particular subshells.

- The first two columns on the left side of the periodic table are the s block with outer electron configurations of ns^1 (first column) and ns^2 (second column).
- The six columns on the right side of the periodic table are the p block with outer electron configurations of: ns^2np^1, ns^2np^2, ns^2np^3, ns^2np^4, ns^2np^5 (halogens), and ns^2np^6 (noble gases).
- The transition metals are the d block.
- The lanthanides and actinides (also called the inner transition metals) are the f block.

▲ **FIGURE 9.26 Outer electron configurations of the elements**

Remember that main-group elements are those in the two far left columns (1A, 2A) and the six far right columns (3A–8A) of the periodic table (see Section 4.6).

Notice that, except for helium, the number of valence electrons for any main-group element is equal to the group number of its column. For example, we can tell that chlorine has 7 valence electrons because it is in the column with group number 7A. The row number in the periodic table is equal to the number of the highest principal shell (n value). For example, since chlorine is in row 3, its highest principal shell is the $n = 3$ shell.

The transition metals have electron configurations with trends that differ somewhat from main-group elements. As you move across a row in the d block, the d orbitals are filling (Figure 9.26). However, the principal quantum number of the d orbital being filled across each row in the transition series is equal to the row number minus one (in the fourth row, the $3d$ orbitals fill; in the fifth row, the $4d$ orbitals fill; and so on). For the first transition series, the outer configuration is $4s^2 3d^x$ (x = number of d electrons) with two exceptions: Cr is $4s^1 3d^5$ and Cu is $4s^1 3d^{10}$. These exceptions occur because a half-filled d subshell and a completely filled d subshell are particularly stable. Otherwise, the number of outershell electrons in a transition series does not change as you move across a period. In other words, *the transition series represents the filling of core orbitals, and the number of outershell electrons is mostly constant.*

We can now see that the organization of the periodic table allows us to write the electron configuration for any element based simply on its position in the periodic table. For example, suppose we want to write an electron configuration for P. The inner electrons of P are simply those of the noble gas that precedes P in the periodic table, Ne. So we can represent the inner electrons with [Ne]. We obtain the outer electron configuration by tracing the elements between Ne and P and assigning electrons to the appropriate orbitals (◀ Figure 9.27). Remember that the highest n value is given by the row number (3 for phosphorus). So we begin with [Ne], then add in the two $3s$ electrons as we trace across the s block, followed by three $3p$ electrons as we trace across the p block to P, which is in the third column of the p block. The electron configuration is:

▲ **FIGURE 9.27 Electron configuration of phosphorus** Determining the electron configuration for P from its position in the periodic table.

$$\text{P} \qquad [\text{Ne}]3s^2 3p^3$$

Notice that P is in column 5A and therefore has 5 valence electrons and an outer electron configuration of $ns^2 np^3$.

To summarize writing an electron configuration for an element based on its position in the periodic table:

- The inner electron configuration for any element is the electron configuration of the noble gas that immediately precedes that element in the periodic table. Represent the inner configuration with the symbol for the noble gas in brackets.
- The outer electrons can be determined from the element's position within a particular block (s, p, d, or f) in the periodic table. Trace the elements between the preceding noble gas and the element of interest, and assign electrons to the appropriate orbitals.
- The highest principal quantum number (highest n value) is equal to the row number of the element in the periodic table.
- For any element containing d electrons, the principal quantum number (n value) of the outermost d electrons is equal to the row number of the element minus 1.

EXAMPLE 9.5 Writing Electron Configurations from the Periodic Table

Write an electron configuration for arsenic based on its position in the periodic table.

SOLUTION

The noble gas that precedes arsenic in the periodic table is argon, so the inner electron configuration is [Ar]. We can obtain the outer electron configuration by

▲ **FIGURE 9.28 Electron configuration of arsenic** Determining the electron configuration for As from its position in the periodic table.

tracing the elements between Ar and As and assigning electrons to the appropriate orbitals. Remember that the highest n value is given by the row number (4 for arsenic). So, we begin with [Ar], then add in the two $4s$ electrons as we trace across the s block, followed by ten $3d$ electrons as we trace across the d block (the n value for d subshells is equal to the row number minus one), and finally the three $4p$ electrons as we trace across the p block to As, which is in the third column of the p block (◄ Figure 9.28).

The electron configuration is:

$$\text{As} \qquad [\text{Ar}]4s^2 3d^{10} 4p^3$$

▶**SKILLBUILDER 9.5 | Writing Electron Configurations from the Periodic Table**

Use the periodic table to determine the electron configuration for tin.

▶**FOR MORE PRACTICE** Example 9.12; Problems 65, 66, 67, 68.

✓ CONCEPTUAL CHECKPOINT 9.4

Which element has the *fewest* valence electrons?

(a) B

(b) Ca

(c) O

(d) K

(e) Ga

9.8 The Explanatory Power of the Quantum-Mechanical Model

At the beginning of this chapter, we learned that the quantum-mechanical model explained the chemical properties of the elements such as the inertness of helium, the reactivity of hydrogen, and the periodic law. We can now see how: *The chemical properties of elements are largely determined by the number of valence electrons they contain. Their properties vary in a periodic fashion because the number of valence electrons is periodic.*

Since elements within a column in the periodic table have the same number of valence electrons, they also have similar chemical properties. The noble gases, for example, all have 8 valence electrons, except for helium, which has 2 (◄ Figure 9.29). Although we don't get into the quantitative (or numerical) aspects of the quantum-mechanical model in this book, calculations show that atoms with 8 valence electrons (or 2 for helium) are particularly low in energy, and therefore stable. The noble gases are indeed chemically stable, and thus relatively inert or nonreactive as accounted for by the quantum model.

Elements with electron configurations close to the noble gases are the most reactive because they can attain noble gas electron configurations by losing or gaining a small number of electrons. Alkali metals (Group 1) are among the most reactive metals since their outer electron configuration (ns^1) is 1 electron beyond a noble gas configuration (◄ Figure 9.30). If they can react to lose the

Noble
gases

| 18 |
| 8A |
| 2 |
| **He** |
| $1s^2$ |
| 10 |
| **Ne** |
| $2s^2 2p^6$ |
| 18 |
| **Ar** |
| $3s^2 3p^6$ |
| 36 |
| **Kr** |
| $4s^2 4p^6$ |
| 54 |
| **Xe** |
| $5s^2 5p^6$ |
| 86 |
| **Rn** |
| $6s^2 6p^6$ |

▲ **FIGURE 9.29 Electron configurations of the noble gases** The noble gases (except for helium) all have 8 valence electrons and completely full outer principal shells.

Alkali
metals

| 1 |
| 1A |
| 3 |
| **Li** |
| $2s^1$ |
| 11 |
| **Na** |
| $3s^1$ |
| 19 |
| **K** |
| $4s^1$ |
| 37 |
| **Rb** |
| $5s^1$ |
| 55 |
| **Cs** |
| $6s^1$ |
| 87 |
| **Fr** |
| $7s^1$ |

◄ **FIGURE 9.30 Electron configurations of the alkali metals** The alkali metals all have ns^1 electron configurations and are therefore 1 electron beyond a noble gas configuration. In their reactions, they tend to lose that electron, forming 1+ ions and attaining a noble gas configuration.

ns^1 electron, they attain a noble gas configuration. This explains why—as we learned in Chapter 4—the Group 1A metals tend to form 1+ cations. As an example, consider the electron configuration of sodium:

$$\text{Na} \quad 1s^2 2s^2 2p^6 3s^1$$

In reactions, sodium loses its $3s$ electron, forming a 1+ ion with the electron configuration of neon.

$$\text{Na}^+ \quad 1s^2 2s^2 2p^6$$

$$\text{Ne} \quad 1s^2 2s^2 2p^6$$

Similarly, alkaline earth metals, with an outer electron configuration of ns^2, also tend to be reactive metals, losing their ns^2 electrons to form 2+ cations (▼ Figure 9.31). For example, consider magnesium:

$$\text{Mg} \quad 1s^2 2s^2 2p^6 3s^2$$

In reactions, magnesium loses its two $3s$ electrons, forming a 2− ion with the electron configuration of neon.

$$\text{Mg}^{2+} \quad 1s^2 2s^2 2p^6$$

On the other side of the periodic table, halogens are among the most reactive nonmetals because of their $ns^2 np^5$ electron configurations (▼ Figure 9.32). They are only one electron away from a noble gas configuration and tend to react to gain that one electron, forming 1− ions. For example, consider fluorine:

$$\text{F} \quad 1s^2 2s^2 2p^5$$

In reactions, fluorine gains an additional $2p$ electron, forming a 1− ion with the electron configuration of neon.

$$\text{F}^- \quad 1s^2 2s^2 2p^6$$

The elements that form predictable ions are shown in ▶ Figure 9.33 (first introduced in Chapter 4). Notice how the charge of these ions reflects their electron configurations—these elements form ions with noble gas electron configurations.

Atoms and/or ions that have the same electron configuration are termed isoelectronic.

Alkaline earth metals

2
2A

| 4 |
| Be |
| $2s^2$ |

| 12 |
| Mg |
| $3s^2$ |

| 20 |
| Ca |
| $4s^2$ |

| 38 |
| Sr |
| $5s^2$ |

| 56 |
| Ba |
| $6s^2$ |

| 88 |
| Ra |
| $7s^2$ |

◀ **FIGURE 9.31   Electron configurations of the alkaline earth metals** The alkaline earth metals all have ns^2 electron configurations and are therefore 2 electrons beyond a noble gas configuration. In their reactions, they tend to lose 2 electrons, forming 2+ ions and attaining a noble gas configuration.

Halogens

17
7A

| 9 |
| F |
| $2s^2 2p^5$ |

| 17 |
| Cl |
| $3s^2 3p^5$ |

| 35 |
| Br |
| $4s^2 4p^5$ |

| 53 |
| I |
| $5s^2 5p^5$ |

| 85 |
| At |
| $6s^2 6p^5$ |

▶ **FIGURE 9.32   Electron configurations of the halogens**   The halogens all have $ns^2 np^5$ electron configurations and are therefore 1 electron short of a noble gas configuration. In their reactions, they tend to gain 1 electron, forming 1− ions and attaining a noble gas configuration.

▲ **FIGURE 9.33   Elements that form predictable ions**

✓ CONCEPTUAL CHECKPOINT 9.5

Below is the electron configuration of calcium:

$$Ca \qquad 1s^2 2s^2 2p^6 3s^2 3p^6 4s^2$$

In its reactions, calcium tends to form the Ca^{2+} ion. Which electrons are lost upon ionization?

(a) the $4s$ electrons

(b) two of the $3p$ electrons

(c) the $3s$ electrons

(d) the $1s$ electrons

9.9 Periodic Trends: Atomic Size, Ionization Energy, and Metallic Character

The quantum-mechanical model also explains other periodic trends such as atomic size, ionization energy, and metallic character. We will examine these one at a time.

ATOMIC SIZE

The **atomic size** of an atom is determined by the distance between the outermost electrons and the nucleus. As we move across a period in the periodic table, we know that electrons occupy orbitals with the same principal quantum number, n. Since the principal quantum number largely determines the size of an orbital, electrons are therefore filling orbitals of approximately the same size, and we might expect atomic size to remain constant across a period. However, with each step across a period, the number of protons in the nucleus is also increasing. This increase in the number of protons results in a greater pull on the electrons from the nucleus, causing atomic size to actually decrease. Therefore:

As you move to the right across a period, or row, in the periodic table, atomic size decreases, as shown in ▶ Figure 9.34 on the next page.

As you move down a column in the periodic table, the highest principal quantum number, n, increases. Since the size of an orbital increases with increasing principal quantum number, the electrons that occupy the outermost orbitals are farther from the nucleus as you move down a column. Therefore:

As you move down a column, or family, in the periodic table, atomic size increases, as shown in Figure 9.34.

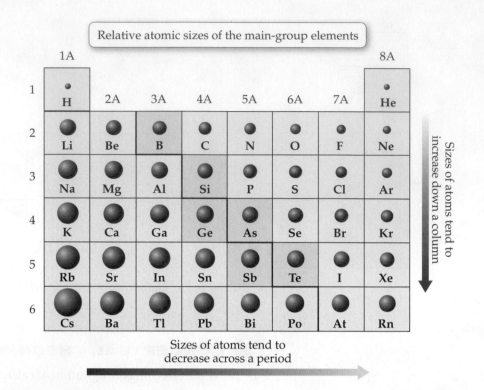

Relative atomic sizes of the main-group elements

Sizes of atoms tend to increase down a column

Sizes of atoms tend to decrease across a period

▶ FIGURE 9.34 **Periodic properties: atomic size** Atomic size decreases as you move to the right across a period and increases as you move down a column in the periodic table.

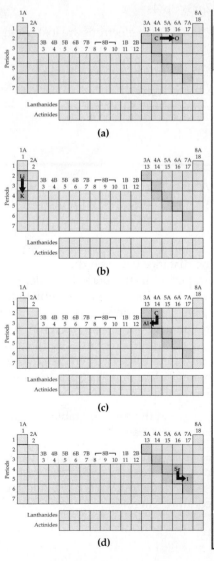

(a)

(b)

(c)

(d)

EXAMPLE 9.6 Atomic Size

Choose the larger atom in each pair.

(a) C or O   **(b)** Li or K   **(c)** C or Al   **(d)** Se or I

SOLUTION

(a) C or O

Carbon atoms are larger than O atoms because, as you trace the path between C and O on the periodic table (see margin), you move to the right within the same period. Atomic size decreases as you go to the right.

(b) Li or K

Potassium atoms are larger than Li atoms because, as you trace the path between Li and K on the periodic table (see margin), you move down a column. Atomic size increases as you go down a column.

(c) C or Al

Aluminum atoms are larger than C atoms because, as you trace the path between C and Al on the periodic table (see margin), you move down a column (atomic size increases) and then to the left across a period (atomic size increases). These effects add together for an overall increase.

(d) Se or I

Based on periodic properties alone, you cannot tell which atom is larger because as you trace the path between Se and I (see margin) you go down a column (atomic size increases) and then to the right across a period (atomic size decreases). These effects tend to cancel one another.

▶**SKILLBUILDER 9.6 | Atomic Size**

Choose the larger atom in each pair.

(a) Pb or Po   **(b)** Rb or Na   **(c)** Sn or Bi   **(d)** F or Se

▶**FOR MORE PRACTICE** Example 9.13a; Problems 81, 82, 83, 84.

CHEMISTRY AND HEALTH
Pumping Ions: Atomic Size and Nerve Impulses

No matter what you are doing at this moment, tiny pumps in each of the trillions of cells that make up your body are hard at work. These pumps, located in the cell membrane, move a number of different ions into and out of the cell. The most important such ions are sodium (Na^+) and potassium (K^+), which happen to be pumped in opposite directions. Sodium ions are pumped *out of cells*, while potassium ions are pumped *into cells*. The result is a *chemical gradient* for each ion: The concentration of sodium is higher outside the cell than within, while exactly the opposite is true for potassium.

The ion pumps within the cell membrane are analogous to water pumps in a high-rise building that pump water against the force of gravity to a tank on the roof. Other structures within the membrane, called ion channels, are like the building's faucets. When they open momentarily, bursts of sodium and potassium ions, driven by their concentration gradients, flow back across the membrane—sodium flowing in and potassium flowing out. These ion pulses are the basis for the transmission of nerve signals in the brain, heart, and throughout the body. Consequently, every move you make or every thought you have is mediated by the flow of these ions.

How do the pumps and channels differentiate between sodium and potassium ions? How do the ion pumps selectively move sodium out of the cell and potassium into

the cell? To answer this question, we must examine the sodium and potassium ions more closely. In what ways do they differ? Both are cations of Group I metals. All Group I metals tend to lose one electron to form cations with 1+ charge, so the magnitude of the charge cannot be the decisive factor. But potassium (atomic number 19) lies directly below sodium in the periodic table (atomic number 11) and based on periodic properties is therefore larger than sodium. The potassium ion has a radius of 133 pm, while the sodium ion has a radius of 95 pm. (Recall from Chapter 2 that $1\ pm = 10^{-12}\ m$.) The pumps and channels within cell membranes are so sensitive that they can distinguish between the sizes of these two ions and selectively allow only one or the other to pass. The result is the transmission of nerve signals that allows you to read this page.

Na^+ K^+

CAN YOU ANSWER THIS? *Other ions, including calcium and magnesium, are also important to nerve signal transmission. Arrange these four ions in order of increasing size: K^+, Na^+, Mg^{2+}, and Ca^{2+}.*

IONIZATION ENERGY

The **ionization energy** of an atom is the energy required to remove an electron from the atom in the gaseous state. For example, the ionization of sodium can be represented with the equation:

$$Na + \text{Ionization energy} \longrightarrow Na^+ + 1e^-$$

Based on what you know about electron configurations, what would you predict about ionization energy trends? Would it take more or less energy to remove an electron from Na than from Cl? We know that Na has an outer electron configuration of $3s^1$ and Cl has an outer electron configuration of $3s^2 3p^5$. Since removing an electron from Na gives it a noble gas configuration—and removing an electron from Cl does not—we would expect sodium to have a lower ionization energy, and that is the case. It is easier to remove an electron from sodium than it is from chlorine. We can generalize this idea in this statement:

> As you move across a period, or row, to the right in the periodic table, ionization energy increases (▶ Figure 9.35 on the next page).

What happens to ionization energy as you move down a column? As we have learned, the principal quantum number, n, increases as you move down a column. Within a given subshell, orbitals with higher principal quantum numbers are larger than orbitals with smaller principal quantum numbers. Consequently, electrons in the outermost principal shell are farther away from the positively charged nucleus—and therefore held less tightly—as you move down a column. This

▶ FIGURE 9.35 Periodic properties: ionization energy Ionization energy increases as you move to the right across a period and decreases as you move down a column in the periodic table.

results in a lower ionization energy (if the electron is held less tightly, it is easier to pull away) as you move down a column. Therefore:

As you move down a column (or family) in the periodic table, ionization energy decreases (Figure 9.35).

Notice that the trends in ionization energy are consistent with the trends in atomic size. Smaller atoms are more difficult to ionize because their electrons are held more tightly. Therefore, as you go across a period, atomic size decreases and ionization energy increases. Similarly, as you go down a column, atomic size increases and ionization energy decreases since electrons are farther from the nucleus and therefore less tightly held.

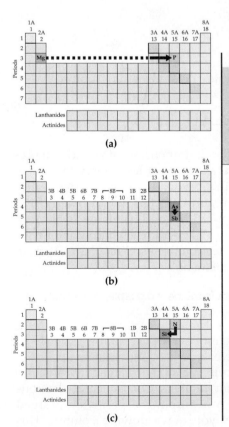

EXAMPLE 9.7 Ionization Energy

Choose the element with the higher ionization energy from each pair.

(a) Mg or P

(b) As or Sb

(c) N or Si

(d) O or Cl

SOLUTION

(a) Mg or P

P has a higher ionization than Mg because, as you trace the path between Mg and P on the periodic table (see margin), you move to the right within the same period. Ionization energy increases as you go to the right.

(b) As or Sb

As has a higher ionization energy than Sb because, as you trace the path between As and Sb on the periodic table (see margin), you move down a column. Ionization energy decreases as you go down a column.

(c) N or Si

N has a higher ionization energy than Si because, as you trace the path between N and Si on the periodic table (see margin), you move down a column (ionization energy decreases) and then to the left across a period (ionization energy decreases). These effects sum together for an overall decrease.

(d)

(d) O or Cl

Based on periodic properties alone, you cannot tell which has a higher ionization energy because, as you trace the path between O and Cl (see margin), you go down a column (ionization energy decreases) and then to the right across a period (ionization energy increases). These effects tend to cancel.

▶**SKILLBUILDER 9.7 | Ionization Energy**

Choose the element with the higher ionization energy from each pair.

(a) Mg or Sr
(b) In or Te
(c) C or P
(d) F or S

▶**FOR MORE PRACTICE**   Example 9.13b; Problems 77, 78, 79, 80.

METALLIC CHARACTER

As we learned in Chapter 4, metals tend to lose electrons in their chemical reactions, while nonmetals tend to gain electrons. As you move across a period in the periodic table, ionization energy increases, which means that electrons are less likely to be lost in chemical reactions. Consequently:

As you move across a period, or row, to the right in the periodic table, **metallic character** decreases (▼ Figure 9.36).

As you move down a column in the periodic table, ionization energy decreases, making electrons more likely to be lost in chemical reactions. Consequently:

As you move down a column, or family, in the periodic table, metallic character increases (Figure 9.36).

These trends, based on the quantum-mechanical model, explain the distribution of metals and nonmetals that we learned in Chapter 4. Metals are found toward the left side of the periodic table and nonmetals (with the exception of hydrogen) toward the upper right.

▲ **FIGURE 9.36 Periodic properties: metallic character** Metallic character decreases as you move to the right across a period and increases as you move down a column in the periodic table.

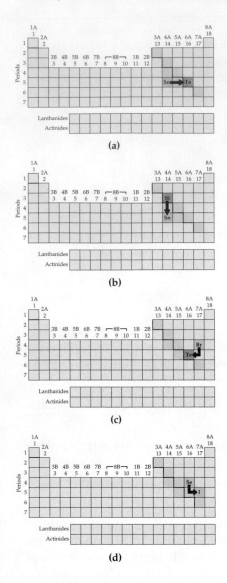

(a)

(b)

(c)

(d)

EXAMPLE 9.8   Metallic Character

Choose the more metallic element from each pair.

(a) Sn or Te

(b) Si or Sn

(c) Br or Te

(d) Se or I

SOLUTION

(a) Sn or Te

Sn is more metallic than Te because, as you trace the path between Sn and Te on the periodic table (see margin), you move to the right within the same period. Metallic character decreases as you go to the right.

(b) Si or Sn

Sn is more metallic than Si because, as you trace the path between Si and Sn on the periodic table (see margin), you move down a column. Metallic character increases as you go down a column.

(c) Br or Te

Te is more metallic than Br because, as you trace the path between Br and Te on the periodic table (see margin), you move down a column (metallic character increases) and then to the left across a period (metallic character increases). These effects add together for an overall increase.

(d) Se or I

Based on periodic properties alone, you cannot tell which is more metallic because as you trace the path between Se and I (see margin), you go down a column (metallic character increases) and then to the right across a period (metallic character decreases). These effects tend to cancel.

▶**SKILLBUILDER 9.8 | Metallic Character**

Choose the more metallic element from each pair.

(a) Ge or In   **(b)** Ga or Sn   **(c)** P or Bi   **(d)** B or N

▶**FOR MORE PRACTICE**   Example 9.13; Problems 85, 86, 87, 88.

CHAPTER IN REVIEW

CHEMICAL PRINCIPLES

Light: Light is electromagnetic radiation, a kind of energy that travels through space at a constant speed of 3.0×10^8 m/s (186,000 mi/s) and exhibits both wavelike and particle-like behavior. Particles of light are called photons. The wave nature of light is characterized by its wavelength, the distance between adjacent crests in the wave. The wavelength of light is inversely proportional to both the frequency—the number of cycles that pass a stationary point in one second—and the energy of a photon. Electromagnetic radiation ranges in wavelength from 10^{-16} m (gamma rays) to 10^6 m (radio waves). In between these lie X-rays, ultraviolet light, visible light, infrared light, and microwaves.

RELEVANCE

Light: Light enables us to see the world. However, we see only visible light, a small sliver in the center of the electromagnetic spectrum. Other forms of electromagnetic radiation are used for cancer therapy, X-ray imaging, night vision, microwave cooking, and communications. Light is also important to many chemical problems. We can learn about the electronic structure of atoms, for example, by examining their interaction with light.

The Bohr Model: The emission spectrum of hydrogen, consisting of bright lines at specific wavelengths, can be explained by the Bohr model for the hydrogen atom. In this model, electrons occupy circular orbits at specific fixed distances from the nucleus. Each orbit is specified by a quantum number (n), which also specifies the orbit's energy. While an electron is in a given orbit, its energy remains constant. When it jumps between orbits, a quantum of energy is absorbed or emitted. Since the difference in energy between orbits is fixed, the energy emitted or absorbed is also fixed. Emitted energy is carried away in the form of a photon of specific wavelength.

The Bohr Model: The Bohr model was a first attempt to explain the bright-line spectra of atoms. While it did predict the spectrum of the hydrogen atom, it failed to predict the spectra of other atoms and was consequently replaced by the quantum-mechanical model.

The Quantum-Mechanical Model: The quantum-mechanical model for the atom describes electron orbitals, which are electron probability maps that show the relative probability of finding an electron in various places surrounding the atomic nucleus. Orbitals are specified with a number (n), called the principal quantum number, and a letter. The principal quantum number ($n = 1, 2, 3 \ldots$) specifies the principal shell, and the letter (s, p, d, or f) specifies the subshell of the orbital. In the hydrogen atom, the energy of orbitals depends only on n. In multi-electron atoms, the energy ordering is $1s$ $2s$ $2p$ $3s$ $3p$ $4s$ $3d$ $4p$ $5s$ $4d$ $5p$ $6s$.

An electron configuration indicates which orbitals are occupied for a particular atom. Orbitals are filled in order of increasing energy and obey the Pauli exclusion principle (each orbital can hold a maximum of two electrons with opposing spins) and Hund's rule (electrons occupy orbitals of identical energy singly before pairing).

The Quantum-Mechanical Model: The quantum-mechanical model changed the way we view nature. Before the quantum-mechanical model, electrons were viewed as small particles, much like any other particle. Electrons were supposed to follow the normal laws of motion, just as a baseball does. However, the electron, with its wavelike properties, does not follow these laws. Instead, electron motion is describable only through probabilistic predictions. Quantum theory single-handedly changed the predictability of nature at its most fundamental level.

The quantum-mechanical model of the atom predicts and explains many of the chemical properties we learned about in earlier chapters.

The Periodic Table: Elements within the same column of the periodic table have similar outer electron configurations and the same number of valence electrons (electrons in the outermost principal shell), and therefore similar chemical properties. The periodic table is divisible into blocks (s block, p block, d block, and f block) in which particular sublevels are filled. As you move across a period to the right in the periodic table, atomic size decreases, ionization energy increases, and metallic character decreases. As you move down a column in the periodic table, atomic size increases, ionization energy decreases, and metallic character increases.

The Periodic Table: The periodic law exists because the number of valence electrons is periodic, and valence electrons determine chemical properties. Quantum theory also predicts that atoms with 8 outershell electrons (or 2 for helium) are particularly stable, thus explaining the inertness of the noble gases. Atoms without noble gas configurations undergo chemical reactions to attain them, explaining the reactivity of the alkali metals and the halogens as well as the tendency of several families to form ions with certain charges.

CHEMICAL SKILLS

EXAMPLES

Predicting Relative Wavelength, Energy, and Frequency of Light (Section 9.3)

- Relative wavelengths can be obtained from Figure 9.4.

- Energy per photon increases with decreasing (shorter) wavelength.

- Frequency increases with decreasing (shorter) wavelength.

EXAMPLE 9.9   Predicting Relative Wavelength, Energy, and Frequency of Light

Which type of light—infrared or ultraviolet—has the longer wavelength? Higher frequency? Higher energy per photon?

SOLUTION

Infrared light has the longer wavelength (Figure 9.4). Ultraviolet light has the higher frequency and the higher energy per photon.

Writing Electron Configurations and Orbital Diagrams (Section 9.6)

To write electron configurations, determine the number of electrons in the atom from the element's atomic number and then follow these rules:

- Electrons occupy orbitals so as to minimize the energy of the atom; therefore, lower-energy orbitals fill before higher-energy orbitals. Orbitals fill in the order: 1s 2s 2p 3s 3p 4s 3d 4p 5s 4d 5p 6s (Figure 9.24). The s subshells hold up to 2 electrons, p subshells hold up to 6, d subshells hold up to 10, and f subshells hold up to 14.

- Orbitals can hold no more than 2 electrons each. When 2 electrons occupy the same orbital, they must have opposing spins.

- When orbitals of identical energy are available, these are first occupied singly with parallel spins rather than in pairs.

EXAMPLE 9.10   Writing Electron Configurations and Orbital Diagrams

Write an electron configuration and orbital diagram (outer electrons only) for germanium.

SOLUTION

Germanium is atomic number 32; therefore, it has 32 electrons.

ELECTRON CONFIGURATION

$$\text{Ge} \quad 1s^2 2s^2 2p^6 3s^2 3p^6 4s^2 3d^{10} 4p^2$$

$$or$$

$$\text{Ge} \quad [\text{Ar}]4s^2 3d^{10} 4p^2$$

ORBITAL DIAGRAM (OUTER ELECTRONS)

Identifying Valence Electrons and Core Electrons (Section 9.7)

- Valence electrons are the electrons in the outermost principal energy shell (the principal shell with the highest principal quantum number).

- Core electrons are those that are not in the outermost principal shell.

EXAMPLE 9.11   Identifying Valence Electrons and Core Electrons

Identify the valence electrons and core electrons in the electron configuration of germanium (given in Example 9.10).

SOLUTION

Writing Electron Configurations for an Element Based on Its Position in the Periodic Table (Section 9.7)

1. The inner electron configuration for any element is the electron configuration of the noble gas that immediately precedes that element in the periodic table. Represent the inner configuration with the symbol for the noble gas in brackets.

2. The outer electrons can be determined from the element's position within a particular block (s, p, d, or f) in the periodic table. Trace the elements between the preceding noble gas and the element of interest and assign electrons to the appropriate orbitals. Figure 9.26 shows the outer electron configuration based on the position of an element in the periodic table.

3. The highest principal quantum number (highest n value) is equal to the row number of the element in the periodic table.

4. The principal quantum number (n value) of the outermost d electrons for any element containing d electrons is equal to the row number of the element minus 1.

EXAMPLE 9.12   Writing Electron Configurations for an Element Based on Its Position in the Periodic Table

Write an electron configuration for iodine based on its position in the periodic table.

SOLUTION

The inner configuration for I is [Kr].

Begin with the [Kr] inner electron configuration. As you trace from Kr to I, add 2 5s electrons, 10 4d electrons, and 5 5p electrons. The overall configuration is:

$$\text{I} \quad [\text{Kr}]5s^2 4d^{10} 5p^5$$

Periodic Trends: Atomic Size, Ionization Energy, and Metallic Character (Section 9.9)

On the periodic table:

- Atomic size decreases as you move to the right and increases as you move down.

- Ionization energy increases as you move to the right and decreases as you move down.

- Metallic character decreases as you move to the right and increases as you move down.

EXAMPLE 9.13 Periodic Trends: Atomic Size, Ionization Energy, and Metallic Character

Arrange Si, In, and S in order of **(a)** increasing atomic size, **(b)** increasing ionization energy, and **(c)** increasing metallic character.

SOLUTION

(a) S, Si, In
(b) In, Si, S
(c) S, Si, In

KEY TERMS

atomic size [9.9]
Bohr model [9.1]
core electrons [9.7]
electromagnetic
 radiation [9.2]
electromagnetic
 spectrum [9.3]
electron configuration [9.6]
electron spin [9.6]
emission spectrum [9.4]

excited state [9.6]
frequency (ν) [9.2]
gamma rays [9.3]
ground state [9.6]
Hund's rule [9.6]
infrared light [9.3]
ionization energy [9.9]
metallic character [9.9]
microwaves [9.3]
orbital [9.5]

orbital diagram [9.6]
Pauli exclusion
 principle [9.6]
photon [9.2]
principal quantum
 number [9.6]
principal shell [9.6]
quantized [9.4]
quantum (plural,
 quanta) [9.4]

quantum-mechanical
 model [9.1]
quantum number [9.4]
radio waves [9.3]
subshell [9.6]
ultraviolet (UV) light [9.3]
valence electron [9.7]
visible light [9.3]
wavelength (λ) [9.2]
X-rays [9.3]

EXERCISES

QUESTIONS

1. When were the Bohr model and the quantum-mechanical model for the atom developed? What purpose do these models serve?
2. What is light? How fast does light travel?
3. What is white light? Colored light?
4. Explain, in terms of absorbed and reflected light, why a blue object appears blue.
5. What is the relationship between the wavelength of light and the amount of energy carried by its photons? How are wavelength and frequency of light related?
6. List some sources of gamma rays.
7. How are X-rays used?
8. Why should excess exposure to gamma rays and X-rays be avoided?
9. Why should excess exposure to ultraviolet light be avoided?
10. What objects emit infrared light? What technology exploits this?
11. Why do microwave ovens heat food, but tend not to heat the dish the food is on?
12. What type of electromagnetic radiation is used in communications devices such as cellular telephones?
13. Describe the Bohr model for the hydrogen atom.
14. What is an emission spectrum? Use the Bohr model to explain why the emission spectrum of the hydrogen atom consists of distinct lines at specific wavelengths.
15. Explain the difference between a Bohr orbit and a quantum-mechanical orbital.
16. What is the difference between the ground state of an atom and an excited state of an atom?
17. Explain how the motion of an electron is different from the motion of a baseball. What is a probability map?
18. Explain why quantum-mechanical orbitals have "fuzzy" boundaries.
19. List the four possible subshells in the quantum-mechanical model, the number of orbitals in each subshell, and the maximum number of electrons that can be contained in each subshell.
20. List all of the quantum-mechanical orbitals through 5s, in the correct energy order for multi-electron atoms.
21. What is the Pauli exclusion principle? Why is it important when writing electron configurations?
22. What is Hund's rule? Why is it important when writing orbital diagrams?
23. Within an electron configuration, what do symbols such as [Ne] and [Kr] represent?
24. Explain the difference between valence electrons and core electrons.

25. Identify each block in the blank periodic table.

(a) *s* block

(b) *p* block

(c) *d* block

(d) *f* block

26. Give some examples of the explanatory power of the quantum-mechanical model.

27. Explain why Group 1 elements tend to form 1+ ions and Group 7 elements tend to form 1− ions.

28. Explain the periodic trends in each chemical property:

(a) ionization energy

(b) atomic size

(c) metallic character

PROBLEMS

WAVELENGTH, ENERGY, AND FREQUENCY OF ELECTROMAGNETIC RADIATION

29. How long does it take light to travel in:

(a) 1.0 ft (report answer in nanoseconds)

(b) 2462 mi, the distance between Los Angeles and New York (report answer in milliseconds)

(c) 4.5 billion km, the average separation between the sun and Neptune (report answer in hours and minutes)

30. How far does light travel in:

(a) 1.0 s

(b) 1.0 day

(c) 1.0 yr

31. Which type of electromagnetic radiation has the longest wavelength?

(a) visible

(b) ultraviolet

(c) infrared

(d) X-ray

32. Which type of electromagnetic radiation has the shortest wavelength?

(a) radio waves

(b) microwaves

(c) infrared

(d) ultraviolet

33. List the types of electromagnetic radiation in order of increasing energy per photon.

(a) radio waves

(b) microwaves

(c) infrared

(d) ultraviolet

34. List the types of electromagnetic radiation in order of decreasing energy per photon.

(a) gamma rays

(b) radio waves

(c) microwaves

(d) visible light

35. List two types of electromagnetic radiation with frequencies higher than visible light.

36. List two types of electromagnetic radiation with frequencies lower than infrared light.

37. List these three types of radiation—infrared, X-ray, and radio waves—in order of:

(a) increasing energy per photon

(b) increasing frequency

(c) increasing wavelength

38. List these three types of electromagnetic radiation—visible, gamma rays, and microwaves—in order of:

(a) decreasing energy per photon

(b) decreasing frequency

(c) decreasing wavelength

THE BOHR MODEL

39. Bohr orbits have fixed _____ and fixed _____.

40. In the Bohr model, what happens when an electron makes a transition between orbits?

41. Two of the emission wavelengths in the hydrogen emission spectrum are 410 nm and 434 nm. One of these is due to the $n = 6$ to $n = 2$ transition, and the other is due to the $n = 5$ to $n = 2$ transition. Which wavelength goes with which transition?

42. Two of the emission wavelengths in the hydrogen emission spectrum are 656 nm and 486 nm. One of these is due to the $n = 4$ to $n = 2$ transition, and the other is due to the $n = 3$ to $n = 2$ transition. Which wavelength goes with which transition?

THE QUANTUM-MECHANICAL MODEL

43. Sketch the $1s$ and $2p$ orbitals. How would the $2s$ and $3p$ orbitals differ from the $1s$ and $2p$ orbitals?

44. Sketch the $3d$ orbitals. How would the $4d$ orbitals differ from the $3d$ orbitals?

45. Which electron is, on average, closer to the nucleus: an electron in a $2s$ orbital or an electron in a $3s$ orbital?

46. Which electron is, on average, farther from the nucleus: an electron in a $3p$ orbital or an electron in a $4p$ orbital?

47. According to the quantum-mechanical model for the hydrogen atom, which electron transition would produce light with longer wavelength: $2p$ to $1s$ or $3p$ to $1s$?

48. According to the quantum-mechanical model for the hydrogen atom, which transition would produce light with longer wavelength: $3p$ to $2s$ or $4p$ to $2s$?

ELECTRON CONFIGURATIONS

49. Write full electron configurations for each element.
  (a) Sr
  (b) Ge
  (c) Li
  (d) Kr

50. Write full electron configurations for each element.
  (a) N
  (b) Mg
  (c) Ar
  (d) Se

51. Write full orbital diagrams and indicate the number of unpaired electrons for each element.
  (a) He
  (b) B
  (c) Li
  (d) N

52. Write full orbital diagrams and indicate the number of unpaired electrons for each element.
  (a) F
  (b) C
  (c) Ne
  (d) Be

53. Write electron configurations for each element. Use the symbol of the previous noble gas in brackets to represent the core electrons.
  (a) Ga
  (b) As
  (c) Rb
  (d) Sn

54. Write electron configurations for each element. Use the symbol of the previous noble gas in brackets to represent the core electrons.
  (a) Te
  (b) Br
  (c) I
  (d) Cs

55. Write electron configurations for each transition metal.
  (a) Zn
  (b) Cu
  (c) Zr
  (d) Fe

56. Write electron configurations for each transition metal.
  (a) Mn
  (b) Ti
  (c) Cd
  (d) V

VALENCE ELECTRONS AND CORE ELECTRONS

57. Write full electron configurations and indicate the valence electrons and the core electrons for each element.
    (a) Kr
    (b) Ge
    (c) Cl
    (d) Sr

58. Write full electron configurations and indicate the valence electrons and the core electrons for each element.
    (a) Sb
    (b) N
    (c) B
    (d) K

59. Write orbital diagrams for the valence electrons and indicate the number of unpaired electrons for each element.
    (a) Br
    (b) Kr
    (c) Na
    (d) In

60. Write orbital diagrams for the valence electrons and indicate the number of unpaired electrons for each element.
    (a) Ne
    (b) I
    (c) Sr
    (d) Ge

61. How many valence electrons are in each element?
    (a) O
    (b) S
    (c) Br
    (d) Rb

62. How many valence electrons are in each element?
    (a) Ba
    (b) Al
    (c) Be
    (d) Se

ELECTRON CONFIGURATIONS AND THE PERIODIC TABLE

63. List the outer electron configuration for each column in the periodic table.
    (a) 1A
    (b) 2A
    (c) 5A
    (d) 7A

64. List the outer electron configuration for each column in the periodic table.
    (a) 3A
    (b) 4A
    (c) 6A
    (d) 8A

65. Use the periodic table to write electron configurations for each element.
    (a) Al
    (b) Be
    (c) In
    (d) Zr

66. Use the periodic table to write electron configurations for each element.
    (a) Tl
    (b) Co
    (c) Ba
    (d) Sb

67. Use the periodic table to write electron configurations for each element.
    (a) Sr
    (b) Y
    (c) Ti
    (d) Te

68. Use the periodic table to write electron configurations for each element.
    (a) Se
    (b) Sn
    (c) Pb
    (d) Cd

69. How many 2*p* electrons are in an atom of each element?

(a) C

(b) N

(c) F

(d) P

70. How many 3*d* electrons are in an atom of each element?

(a) Fe

(b) Zn

(c) K

(d) As

71. List the number of elements in periods 1 and 2 of the periodic table. Why do the two periods have a different number of elements?

72. List the number of elements in periods 3 and 4 of the periodic table. Why do the two periods have a different number of elements?

73. Name an element in the third period (row) of the periodic table with:

(a) 3 valence electrons

(b) a total of 4 3*p* electrons

(c) 6 3*p* electrons

(d) 2 3*s* electrons and no 3*p* electrons

74. Name an element in the fourth period of the periodic table with:

(a) 5 valence electrons

(b) a total of 4 4*p* electrons

(c) a total of 3 3*d* electrons

(d) a complete outer shell

75. Use the periodic table to identify the element with each electron configuration.

(a) $[Ne]3s^23p^5$

(b) $[Ar]4s^23d^{10}4p^1$

(c) $[Ar]4s^23d^6$

(d) $[Kr]5s^1$

76. Use the periodic table to identify the element with each electron configuration.

(a) $[Ne]3s^1$

(b) $[Kr]5s^24d^{10}$

(c) $[Xe]6s^2$

(d) $[Kr]5s^24d^{10}5p^3$

PERIODIC TRENDS

77. Choose the element with the higher ionization energy from each pair.

(a) As or Bi

(b) As or Br

(c) S or I

(d) S or Sb

78. Choose the element with the higher ionization energy from each pair.

(a) Al or In

(b) Cl or Sb

(c) K or Ge

(d) S or Se

79. Arrange the elements in order of increasing ionization energy: Te, Pb, Cl, S, Sn.

80. Arrange the elements in order of increasing ionization energy: Ga, In, F, Si, N.

81. Choose the element with the larger atoms from each pair.

(a) Al or In

(b) Si or N

(c) P or Pb

(d) C or F

82. Choose the element with the larger atoms from each pair.

(a) Sn or Si

(b) Br or Ga

(c) Sn or Bi

(d) Se or Sn

83. Arrange these elements in order of increasing atomic size: Ca, Rb, S, Si, Ge, F.

84. Arrange these elements in order of increasing atomic size: Cs, Sb, S, Pb, Se.

85. Choose the more metallic element from each pair.
    (a) Sr or Sb
    (b) As or Bi
    (c) Cl or O
    (d) S or As

86. Choose the more metallic element from each pair.
    (a) Sb or Pb
    (b) K or Ge
    (c) Ge or Sb
    (d) As or Sn

87. Arrange these elements in order of increasing metallic character: Fr, Sb, In, S, Ba, Se.

88. Arrange these elements in order of increasing metallic character: Sr, N, Si, P, Ga, Al.

CUMULATIVE PROBLEMS

89. What is the maximum number of electrons that can occupy the $n = 3$ quantum shell?

90. What is the maximum number of electrons that can occupy the $n = 4$ quantum shell?

91. Use the electron configurations of the alkaline earth metals to explain why they tend to form 2+ ions.

92. Use the electron configuration of oxygen to explain why it tends to form a 2− ion.

93. Write the electron configuration for each ion. What do all of the electron configurations have in common?
    (a) Ca^{2+} (b) K^+ (c) S^{2-} (d) Br^-

94. Write the electron configuration for each ion. What do all of the electron configurations have in common?
    (a) F^- (b) P^{3-} (c) Li^+ (d) Al^{3+}

95. Examine Figure 4.12, which shows the division of the periodic table into metals, nonmetals, and metalloids. Use what you know about electron configurations to explain these divisions.

96. Examine Figure 4.14, which shows the elements that form predictable ions. Use what you know about electron configurations to explain these trends.

97. Explain what is wrong with each electron configuration and write the correct ground state (or lowest energy) configuration based on the number of electrons.
    (a) $1s^3 2s^3 2p^9$
    (b) $1s^2 2s^2 2p^6 2d^4$
    (c) $1s^2 1p^5$
    (d) $1s^2 2s^2 2p^8 3s^2 3p^1$

98. Explain what is wrong with each electron configuration and write the correct ground state (or lowest energy) configuration based on the number of electrons.
    (a) $1s^4 2s^4 2p^{12}$
    (b) $1s^2 2s^2 2p^6 3s^2 3p^6 3d^{10}$
    (c) $1s^2 2p^6 3s^2$
    (d) $1s^2 2s^2 2p^6 3s^2 3p^6 4s^2 4d^{10} 4p^3$

99. Bromine is a highly reactive liquid, while krypton is an inert gas. Explain this difference based on their electron configurations.

100. Potassium is a highly reactive metal, while argon is an inert gas. Explain this difference based on their electron configurations.

101. Based on periodic trends, which one of these elements would you expect to be most easily oxidized: Ge, K, S, or N?

102. Based on periodic trends, which one of these elements would you expect to be most easily reduced: Ca, Sr, P, or Cl?

103. When an electron makes a transition from the $n = 3$ to the $n = 2$ hydrogen atom Bohr orbit, the energy difference between these two orbits (3.0×10^{-19} J) is emitted as a photon of light. The relationship between the energy of a photon and its wavelength is given by $E = hc/\lambda$, where E is the energy of the photon in J, h is Planck's constant (6.626×10^{-34} J·s), and c is the speed of light (3.00×10^8 m/s). Find the wavelength of light emitted by hydrogen atoms when an electron makes this transition.

104. When an electron makes a transition from the $n = 4$ to the $n = 2$ hydrogen atom Bohr orbit, the energy difference between these two orbits (4.1×10^{-19} J) is emitted as a photon of light. The relationship between the energy of a photon and its wavelength is given by $E = hc/\lambda$, where E is the energy of the photon in J, h is Planck's constant (6.626×10^{-34} J·s), and c is the speed of light (3.00×10^8 m/s). Find the wavelength of light emitted by hydrogen atoms when an electron makes this transition.

105. The distance from the sun to Earth is 1.496×10^8 km. How long does it take light to travel from the sun to Earth?

106. The nearest star is Alpha Centauri, at a distance of 4.3 light-years from Earth. A light-year is the distance that light travels in one year (365 days). How far away, in kilometers, is Alpha Centauri from Earth?

107. In the beginning of this chapter, we learned that the quantum-mechanical model for the atom is the foundation for modern chemical understanding. Explain why this is so.

108. Niels Bohr said, "Anyone who is not shocked by quantum mechanics has not understood it." What do you think he meant by this statement?

109. The wave nature of matter was first proposed by Louis de Broglie, who suggested that the wavelength (λ) of a particle was related to its mass (m) and its velocity (v) by the equation: $\lambda = h/mv$, where h is Planck's constant (6.626×10^{-34} J·s). Calculate the de Broglie wavelength of: (a) a 0.0459-kg golf ball traveling at 95 m/s; (b) an electron traveling at 3.88×10^6 m/s. Can you explain why the wave nature of matter is significant for the electron but not for the golf ball? (*Hint:* Express mass in kilograms.)

110. The particle nature of light was first proposed by Albert Einstein, who suggested that light could be described as a stream of particles called photons. A photon of wavelength λ has an energy (E) given by the equation: $E = hc/\lambda$, where E is the energy of the photon in J, h is Planck's constant (6.626×10^{-34} J·s), and c is the speed of light (3.00×10^8 m/s). Calculate the energy of 1 mol of photons with a wavelength of 632 nm.

111. We learned in this chapter that ionization generally increases as you move from left to right across the periodic table. However, consider the data below, which shows the ionization energies of the period 2 and 3 elements:

Group	Period 2 Elements	Ionization Energy (kJ/mole)	Period 3 Elements	Ionization Energy
1A	Li	520	Na	496
2A	Be	899	Mg	738
3A	B	801	Al	578
4A	C	1086	Si	786
5A	N	1402	P	1012
6A	O	1314	S	1000
7A	F	1681	Cl	1251
8A	Ne	2081	Ar	1521

Notice that the increase is not uniform. In fact, ionization energy actually decreases a bit in going from elements in group 2A to 3A and then again from 5A to 6A. Use what you know about electron configurations to explain why these dips in ionization energy exist.

112. When atoms lose more than one electron, the ionization energy to remove the second electron is always more than the ionization energy to remove the first. Similarly, the ionization energy to remove the third electron is more than the second and so on. However, the increase in ionization energy upon the removal of subsequent electrons is not necessarily uniform. For example, consider the first three ionization energies of magnesium:

First ionization energy	738 kJ/mol
Second ionization energy	1450 kJ/mol
Third ionization energy	7730 kJ/mol

The second ionization energy is roughly twice the first ionization energy, but then the third ionization energy is over five times the second. Use the electron configuration of magnesium to explain why this is so. Would you expect the same behavior in sodium? Why or why not?

HIGHLIGHT PROBLEMS

113. Excessive exposure to sunlight increases the risk of skin cancer because some of the photons have enough energy to break chemical bonds in biological molecules. These bonds require approximately 250–800 kJ/mol of energy to break. The energy of a single photon is given by $E = hc/\lambda$, where E is the energy of the photon in J, h is Planck's constant (6.626×10^{-34} J·s), and c is the speed of light (3.00×10^8 m/s). Determine which kinds of light contain enough energy to break chemical bonds in biological molecules by calculating the total energy in 1 mol of photons for light of each wavelength.

    (a) infrared light (1500 nm)

    (b) visible light (500 nm)

    (c) ultraviolet light (150 nm)

114. The quantum-mechanical model, besides revolutionizing chemistry, shook the philosophical world because of its implications regarding determinism. Determinism is the idea that the outcomes of future events are determined by preceding events. The trajectory of a baseball, for example, is deterministic; that is, its trajectory—and therefore its landing place—is determined by its position, speed, and direction of travel. Before quantum mechanics, most scientists thought that fundamental particles—such as electrons and protons—also behaved deterministically. The implication of this belief was that the entire universe must behave deterministically—the future must be determined by preceding events. Quantum mechanics challenged this reasoning because fundamental particles do not behave deterministically—their future paths are not determined by preceding events. Some scientists struggled with this idea. Einstein himself refused to believe it, stating, "God does not play dice with the universe." Explain what Einstein meant by this statement.

▲ "God does not play dice with the universe."

►ANSWERS TO SKILLBUILDER EXERCISES

Skillbuilder 9.1

(a) blue, green, red

(b) red, green, blue

(c) red, green, blue

Skillbuilder 9.2

(a) Al $1s^2 2s^2 2p^6 3s^2 3p^1$ or

[Ne]$3s^2 3p^1$

(b) Br $1s^2 2s^2 2p^6 3s^2 3p^6 4s^2 3d^{10} 4p^5$ or

[Ar]$4s^2 3d^{10} 4p^5$

(c) Sr $1s^2 2s^2 2p^6 3s^2 3p^6 4s^2 3d^{10} 4p^6 5s^2$ or

[Kr]$5s^2$

Skillbuilder Plus, p. 301 Subtract 1 electron for each unit of positive charge. Add 1 electron for each unit of negative charge.

(a) Al^{3+} $1s^2 2s^2 2p^6$

(b) Cl^- $1s^2 2s^2 2p^6 3s^2 3p^6$

(c) O^{2-} $1s^2 2s^2 2p^6$

Skillbuilder 9.3

Ar

1s   2s   2p   3s   3p

Skillbuilder 9.4

Cl $1s^2 2s^2 2p^6\; 3s^2 3p^5$

10 core electrons   7 valence electrons

Skillbuilder 9.5   [Kr]$5s^2 4d^{10} 5p^2$

Skillbuilder 9.6

(a) Pb

(b) Rb

(c) cannot determine based on periodic properties

(d) Se

Skillbuilder 9.7

(a) Mg

(b) Te

(c) cannot determine based on periodic properties

(d) F

Skillbuilder 9.8

(a) In

(b) cannot determine based on periodic properties

(c) Bi

(d) B

►ANSWERS TO CONCEPTUAL CHECKPOINTS

9.1 (b) Wavelength and energy per photon are inversely related. Since yellow light has a longer wavelength, it has less energy per photon than violet light.

9.2 (c) The higher energy levels are more closely spaced than the lower ones, so the difference in energy between $n = 2$ and $n = 1$ is greater than the difference in energy between $n = 3$ and $n = 2$. The photon emitted when an electron falls from $n = 2$ to $n = 1$ therefore carries more energy, corresponding to radiation with a shorter wavelength and higher frequency.

9.3 (d) Both have 6 electrons in $2p$ orbitals and 6 electrons in $3p$ orbitals.

9.4 (d) The outermost principal shell for K is $n = 4$, which contains only a single valence electron, $4s^1$.

9.5 (a) Calcium loses its 4s electron and attains a noble gas configuration (that of Ar).

CHAPTER 10

Chemical Bonding

"The fascination of a growing science lies in the work of the pioneers at the very borderland of the unknown, but to reach this frontier one must pass over well traveled roads."

GILBERT N. LEWIS (1875–1946)

10.1 Bonding Models and AIDS Drugs

| Proteins are discussed in more detail in Chapter 19.

In 1989, researchers discovered the structure of a molecule called HIV-protease. HIV-protease is a protein (a class of biological molecules) synthesized by the human immunodeficiency virus (HIV), which causes AIDS. HIV-protease is crucial to the virus's ability to replicate itself. Without HIV-protease, HIV could not spread in the human body because the virus could not copy itself, and AIDS would not develop.

With knowledge of the HIV-protease structure, drug companies set out to design a molecule that would disable protease by sticking to the working part of the molecule (called the *active site*). To design such a molecule, researchers used **bonding theories**—models that predict how atoms bond together to form molecules—to simulate how potential drug molecules would interact with the protease molecule. By the early 1990s, these companies had developed several drug molecules that seemed to work. Since these molecules inhibit the action of HIV-protease, they are called *protease inhibitors*. In human trials, protease inhibitors in combination with other drugs have decreased the viral count in HIV-infected individuals to undetectable levels. Although these drugs do not cure AIDS, HIV-infected individuals who regularly take their medication can now expect nearly normal life spans.

Bonding theories are central to chemistry because they predict how atoms bond together to form compounds. They predict what combinations of atoms form compounds and what combinations do not. Bonding theories predict why salt is NaCl and not $NaCl_2$ and why water is H_2O and not H_3O. Bonding theories also explain the shapes of molecules, which in turn determine many of their physical and chemical properties. The bonding theory you will learn in this chapter is called **Lewis theory**, named after G. N. Lewis (1875–1946), the American chemist

◀ The gold-colored structure on the computer screen is a representation of HIV-protease. The molecule shown in the center is Indinavir, a protease inhibitor.

who developed it. In this model, we represent electrons as dots and draw what are called *dot structures* or *Lewis structures* to represent molecules. These structures, which are fairly simple to draw, have tremendous predictive power. It takes just a few minutes to apply Lewis theory to determine whether a particular set of atoms will form a stable molecule and what that molecule might look like. Although modern chemists also use more advanced bonding theories to better predict molecular properties, Lewis theory remains the simplest method for making quick, everyday predictions about molecules.

10.2   Representing Valence Electrons with Dots

As we discussed in Chapter 9, valence electrons are those in the outermost principal shell. Since valence electrons are most important in bonding, Lewis theory focuses on these. In Lewis theory, the valence electrons of main-group elements are represented as dots surrounding the symbol of the element. The result is called a **Lewis structure**, or **dot structure**. For example, the electron configuration of O is:

$$1s^2\,2s^2 2p^4$$

6 valence electrons

and its Lewis structure is:

$$\cdot \ddot{\text{O}} :$$

6 dots representing valence electrons

> Remember, the number of valence electrons for any main-group element (except helium, which has two valence electrons but is in Group 8A) is equal to the group number of the element.

Each dot represents a valence electron. The dots are placed around the element's symbol with a maximum of two dots per side. Although the exact location of dots is not critical, in this book we fill in the dots singly first and then pair them (with the exception of helium, described shortly).

The Lewis structures for all of the period 2 elements are:

$$\text{Li} \cdot \quad \cdot \text{Be} \cdot \quad \cdot \dot{\text{B}} \cdot \quad \cdot \dot{\ddot{\text{C}}} \cdot \quad \cdot \dot{\ddot{\text{N}}} : \quad \cdot \ddot{\text{O}} : \quad : \ddot{\text{F}} : \quad : \ddot{\text{N}} \text{e} :$$

Lewis structures allow us to easily see the number of valence electrons in an atom. Atoms with eight valence electrons—which are particularly stable—are easily identified because they have eight dots, an **octet**.

Helium is somewhat of an exception. Its electron configuration and Lewis structure are:

$$1s^2 \quad \text{He} :$$

The Lewis structure of helium contains two paired dots (a **duet**). For helium, a duet represents a stable electron configuration.

In Lewis theory, a **chemical bond** involves the sharing or transfer of electrons to attain stable electron configurations for the bonding atoms. If the electrons are transferred, the bond is an **ionic bond**. If the electrons are shared, the bond is a **covalent bond**. In either case, the bonding atoms attain stable electron configurations. As we have seen, a stable configuration usually consists of eight electrons in the outermost or valence shell. This observation leads to the **octet rule**:

> In chemical bonding, atoms transfer or share electrons to obtain outer shells with eight electrons.

The octet rule generally applies to all main-group elements except hydrogen and lithium. Each of these elements achieves stability when it has two electrons (a duet) in its outermost shell.

EXAMPLE 10.1   Writing Lewis Structures for Elements

Write the Lewis structure of phosphorus.

Since phosphorus is in Group 5A in the periodic table, it has five valence electrons. Represent these as five dots surrounding the symbol for phosphorus.	**SOLUTION** $\cdot\ddot{P}\!:$

▶**SKILLBUILDER 10.1 | Writing Lewis Structures for Elements**

Write the Lewis structure of Mg.

▶**FOR MORE PRACTICE**   Example 10.12; Problems 25, 26.

✔ **CONCEPTUAL CHECKPOINT 10.1**

Which two elements have the most similar Lewis structures?

(a) C and Si   **(b)** O and P   **(c)** Li and F   **(d)** S and Br

---

10.3   Lewis Structures of Ionic Compounds: Electrons Transferred

Recall from Chapter 5 that when metals bond with nonmetals, electrons are transferred from the metal to the nonmetal. The metal becomes a cation and the nonmetal becomes an anion. The attraction between the cation and the anion results in an ionic compound. In Lewis theory, we represent this by moving electron dots from the metal to the nonmetal. For example, potassium and chlorine have the Lewis structures:

$$K\cdot \quad :\!\ddot{C}l\!:$$

When potassium and chlorine bond, potassium transfers its valence electron to chlorine.

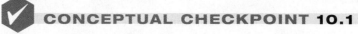

$$K\cdot \quad :\!\ddot{C}l\!: \quad \longrightarrow \quad K^+ \; [:\!\ddot{C}l\!:]^-$$

Recall from Section 4.7 that atoms that lose electrons become positively charged and atoms that gain electrons become negatively charged.	The transfer of the electron gives chlorine an octet (shown as eight dots around chlorine) and leaves potassium with an octet in the previous principal shell, which is now the valence shell. Because the potassium lost an electron, it becomes positively charged, while the chlorine, which gained an electron, becomes negatively charged. The Lewis structure of an anion is usually written within brackets with the charge in the upper right corner (outside the brackets). The positive and negative charges attract one another, forming the compound KCl.

EXAMPLE 10.2   Writing Ionic Lewis Structures

Write the Lewis structure of the compound MgO.

Draw the Lewis structures of magnesium and oxygen by drawing two dots around the symbol for magnesium and six dots around the symbol for oxygen.	**SOLUTION** $\cdot Mg\cdot \quad \cdot\ddot{O}\!:$
In MgO, magnesium loses its two valence electrons, resulting in a 2+ charge, and oxygen gains two electrons, attaining a 2− charge and an octet.	$Mg^{2+} \; [:\!\ddot{O}\!:]^{2-}$

▶**SKILLBUILDER 10.2 | Writing Ionic Lewis Structures**

Write the Lewis structure of the compound NaBr.

▶**FOR MORE PRACTICE**   Example 10.13; Problems 37, 38.

Lewis theory predicts the correct chemical formulas for ionic compounds. For the compound that forms between K and Cl, for example, Lewis theory predicts one potassium cation to every chlorine anion, KCl. As another example, consider the ionic compound formed between sodium and sulfur. The Lewis structures for sodium and sulfur are:

$$Na\cdot \quad \cdot \ddot{\underset{\cdot\cdot}{S}}:$$

Notice that sodium must lose its one valence electron to obtain an octet (in the previous principal shell), while sulfur must gain two electrons to obtain an octet. Consequently, the compound that forms between sodium and sulfur requires two sodium atoms to every one sulfur atom. The Lewis structure is:

$$Na^+ \ [:\ddot{\underset{\cdot\cdot}{S}}:]^{2-} \ Na^+$$

> Recall from Section 5.4 that ionic compounds do not exist as distinct molecules, but rather as part of a large lattice of alternating cations and anions.

The two sodium atoms each lose their one valence electron, while the sulfur atom gains two electrons and obtains an octet. The correct chemical formula is Na_2S.

EXAMPLE 10.3 Using Lewis Theory to Predict the Chemical Formula of an Ionic Compound

Use Lewis theory to predict the formula of the compound that forms between calcium and chlorine.

Draw the Lewis structures of calcium and chlorine by drawing two dots around the symbol for calcium and seven dots around the symbol for chlorine.	**SOLUTION** $\cdot Ca\cdot \quad :\ddot{\underset{\cdot\cdot}{Cl}}:$
Calcium must lose its two valence electrons (to effectively attain an octet in its previous principal shell), while chlorine needs to gain only one electron to obtain an octet. Consequently, the compound that forms between Ca and Cl has two chlorine atoms to every one calcium atom.	$[:\ddot{\underset{\cdot\cdot}{Cl}}:]^- \ Ca^{2+} \ [:\ddot{\underset{\cdot\cdot}{Cl}}:]^-$ The formula is therefore $CaCl_2$.

▶**SKILLBUILDER 10.3 | Using Lewis Theory to Predict the Chemical Formula of an Ionic Compound**

Use Lewis theory to predict the formula of the compound that forms between magnesium and nitrogen.

▶**FOR MORE PRACTICE** Example 10.14; Problems 39, 40, 41, 42.

 CONCEPTUAL CHECKPOINT 10.2

Which nonmetal forms an ionic compound with aluminum with the formula Al_2X_3 (where X represents the nonmetal)?

(a) Cl   **(b)** S   **(c)** N   **(d)** C

10.4 Covalent Lewis Structures: Electrons Shared

Recall from Chapter 5 that when nonmetals bond with other nonmetals, a molecular compound results. Molecular compounds contain covalent bonds in which electrons are shared between atoms rather than transferred. In Lewis theory, we represent covalent bonding by allowing neighboring atoms to share some of their valence electrons in order to attain octets (or duets for hydrogen). For example, hydrogen and oxygen have the Lewis structures:

$$H\cdot \quad \cdot \ddot{\underset{\cdot\cdot}{O}}:$$

In water, hydrogen and oxygen share their electrons so that each hydrogen atom gets a duet and the oxygen atom gets an octet.

$$H:\ddot{\underset{\cdot\cdot}{O}}:H$$

The shared electrons—those that appear in the space between the two atoms—count toward the octets (or duets) of *both of the atoms*.

Duet   Octet   Duet

Electrons that are shared between two atoms are called **bonding pair** electrons, while those that are only on one atom are called **lone pair** (or nonbonding) electrons.

Bonding pair electrons are often represented by dashes to emphasize that they are a chemical bond.

$$H—\ddot{O}—H$$

> Remember that each dash represents a *pair* of shared electrons.

Lewis theory also explains why the halogens form diatomic molecules. Consider the Lewis structure of chlorine.

$$:\ddot{Cl}:$$

If two Cl atoms pair together, they can each attain an octet.

$$:\ddot{Cl}:\ddot{Cl}: \quad or \quad :\ddot{Cl}—\ddot{Cl}:$$

When we examine elemental chlorine, we find that it indeed exists as a diatomic molecule, just as Lewis theory predicts. The same is true for the other halogens.

Similarly, Lewis theory predicts that hydrogen, which has the Lewis structure:

$$H·$$

should exist as H_2. When two hydrogen atoms share their valence electrons, they each get a duet, a stable configuration for hydrogen.

$$H:H \quad or \quad H—H$$

Again, Lewis theory is correct. In nature, elemental hydrogen exists as H_2 molecules.

DOUBLE AND TRIPLE BONDS

In Lewis theory, atoms can share more than one electron pair to attain an octet. For example, we know from Chapter 5 that oxygen exists as the diatomic molecule, O_2. The Lewis structure of an oxygen atom is:

$$·\ddot{O}:$$

If we pair two oxygen atoms together and then try to write a Lewis structure, we do not have enough electrons to give each O atom an octet.

$$:\ddot{O}:\ddot{O}:$$

However, we can convert a lone pair into an additional bonding pair by moving it into the bonding region.

$$:\ddot{O}:\ddot{O}: \quad \longrightarrow \quad :\ddot{O}::\ddot{O}: \quad or \quad :\ddot{O}=\ddot{O}:$$

Each oxygen atom now has an octet because the additional bonding pair counts toward the octet of both oxygen atoms.

Octet ⟶ ⟵ Octet

When two electron pairs are shared between two atoms, the resulting bond is a **double bond**. In general, double bonds are shorter and stronger than single bonds. For example, the distance between oxygen nuclei in an oxygen–oxygen double bond is 121 pm. In a single bond, it is 148 pm.

Atoms can also share three electron pairs. Consider the Lewis structure of N_2. Since each N atom has 5 valence electrons, the Lewis structure for N_2 has 10 electrons. A first attempt at writing the Lewis structure gives:

$$:\!\overset{..}{N}\!:\!\overset{..}{N}\!:$$

As with O_2, we do not have enough electrons to satisfy the octet rule for both N atoms. However, if we convert two additional lone pairs into bonding pairs, each nitrogen atom can get an octet.

$$:\!\overset{\frown\frown}{\underset{..}{N}}\!:\!\overset{..}{N}\!: \quad\longrightarrow\quad :N\!:\!:\!:N: \quad or \quad :N\!\equiv\!N:$$

The resulting bond is a **triple bond**. Triple bonds are even shorter and stronger than double bonds. The distance between nitrogen nuclei in a nitrogen–nitrogen triple bond is 110 pm. In a double bond, the distance is 124 pm. When we examine nitrogen in nature, we find that it indeed exists as a diatomic molecule with a very strong short bond between the two nitrogen atoms. The bond is so strong that it is difficult to break, making N_2 a relatively unreactive molecule.

10.5 Writing Lewis Structures for Covalent Compounds

To write a Lewis structure for a covalent compound, follow these steps:

1. **Write the correct skeletal structure for the molecule.** The skeletal structure shows the relative positions of the atoms and does not include electrons, but it must have the atoms in the correct positions. For example, you could *not* write a Lewis structure for water if you started with the hydrogen atoms next to each other and the oxygen atom at the end (H H O). In nature, oxygen is the central atom, and the hydrogen atoms are **terminal atoms** (at the ends). The correct skeletal structure is H O H.

   The only way to absolutely know the correct skeletal structure for any molecule is to examine its structure in nature. However, we can write likely skeletal structures by remembering two guidelines. First, *hydrogen atoms will always be terminal*. Since hydrogen requires only a duet, it will never be a central atom because central atoms must be able to form at least two bonds and hydrogen can form only one. Second, *many molecules tend to be symmetrical*, so when a molecule contains several atoms of the same type, these tend to be in terminal positions. *This second guideline, however, has many exceptions.* In cases where the skeletal structure is unclear, this text will provide you with the correct skeletal structure.

2. **Calculate the total number of electrons for the Lewis structure by summing the valence electrons of each atom in the molecule.** Remember that the number of valence electrons for any main-group element is equal to its group number in the periodic table. **If you are writing a Lewis structure for a polyatomic ion, the charge of the ion must be considered when calculating the total number of electrons.** Add one electron for each negative charge and subtract one electron for each positive charge.

---

1 pm = 10^{-12} m

When guessing at skeletal structures, put the less metallic elements in terminal positions and the more metallic elements in central positions. Halogens, being among the least metallic elements, are nearly always terminal.

Non-terminal hydrogen atoms exist in some compounds. However, they are rare and beyond our scope.

3. **Distribute the electrons among the atoms, giving octets (or duets for hydrogen) to as many atoms as possible.** Begin by placing two electrons between each pair of atoms. These are the minimal number of bonding electrons. Then distribute the remaining electrons, first to terminal atoms and then to the central atom, giving octets to as many atoms as possible.

4. **If any atoms lack an octet, form double or triple bonds as necessary to give them octets.** Do this by moving lone electron pairs from terminal atoms into the bonding region with the central atom.

A brief version of this procedure is shown in the left column. Two examples of applying it are shown in the center and right columns.

Writing Lewis Structures for Covalent Compounds	**EXAMPLE 10.4** Write the Lewis structure for CO_2.	**EXAMPLE 10.5** Write the Lewis structure for CCl_4.
1. Write the correct skeletal structure for the molecule.	**SOLUTION** Following the symmetry guideline, write: O C O	**SOLUTION** Following the symmetry guideline, write: Cl Cl C Cl Cl
2. Calculate the total number of electrons for the Lewis structure by summing the valence electrons of each atom in the molecule.	Total number of electrons for Lewis structure = $$\begin{pmatrix} \text{\# valence} \\ \text{e}^- \text{ for C} \end{pmatrix} + 2\begin{pmatrix} \text{\# valence} \\ \text{e}^- \text{ for O} \end{pmatrix}$$ $$= 4 + 2(6)$$ $$= 16$$	Total number of electrons for Lewis structure = $$\begin{pmatrix} \text{\# valence} \\ \text{e}^- \text{ for C} \end{pmatrix} + 4\begin{pmatrix} \text{\# valence} \\ \text{e}^- \text{ for Cl} \end{pmatrix}$$ $$= 4 + 4(7)$$ $$= 32$$
3. Distribute the electrons among the atoms, giving octets (or duets for hydrogen) to as many atoms as possible. Begin with the bonding electrons, and then proceed to lone pairs on terminal atoms, and finally to lone pairs on the central atom.	Bonding electrons first. O:C:O (4 of 16 electrons used) Lone pairs on terminal atoms next. :Ö:C:Ö: (16 of 16 electrons used)	Bonding electrons first. Cl Cl:C:Cl Cl (8 of 32 electrons used) Lone pairs on terminal atoms next. :Cl: :Cl:C:Cl: :Cl: (32 of 32 electrons used)
4. If any atoms lack octets, form double or triple bonds as necessary to give them octets.	Move lone pairs from the oxygen atoms to bonding regions to form double bonds. :Ö:C:Ö: ⟶ :O::C::O:	Since all of the atoms have octets, the Lewis structure is complete. ▶**SKILLBUILDER 10.5** Write the Lewis structure for H_2CO.
	▶**SKILLBUILDER 10.4** Write the Lewis structure for CO.	▶**FOR MORE PRACTICE** Example 10.15; Problems 47, 48, 49, 50, 51, 52.

WRITING LEWIS STRUCTURES FOR POLYATOMIC IONS

We write Lewis structures for polyatomic ions by following the same procedure, but we pay special attention to the charge of the ion when calculating the number of electrons for the Lewis structure. Add one electron for each negative charge and subtract one electron for each positive charge. We normally show the Lewis structure for a polyatomic ion within brackets and write the charge of the ion in the upper right corner. For example, suppose we want to write a Lewis structure for the CN^- ion. We begin by writing the skeletal structure.

CN

Next we calculate the total number of electrons for the Lewis structure by summing the number of valence electrons for each atom and adding one for the negative charge.

$$\text{Total number of electrons for Lewis structure} = (\text{\# valence e}^- \text{ in C}) + (\text{\# valence e}^- \text{ in N}) + 1$$

$$= 4 + 5 + 1 \qquad \text{Add one e}^- \text{ to account for 1- charge of ion.}$$

$$= 10$$

We then place two electrons between each pair of atoms

C:N   (2 of 10 electrons used)

and distribute the remaining electrons.

:C̈:N̈:   (10 of 10 electrons used)

Since neither of the atoms has octets, we move two lone pairs into the bonding region to form a triple bond, giving both atoms octets. We also enclose the Lewis structure in brackets and write the charge of the ion in the upper right corner.

[:C:::N:]⁻   or   [:C≡N:]⁻

✓ CONCEPTUAL CHECKPOINT 10.3

The total number of electrons in the Lewis structure of OH^- is:

(a) 6   (b) 7   (c) 8   (d) 9

EXAMPLE 10.6 Writing Lewis Structures for Polyatomic Ions

Write the Lewis structure for the NH_4^+ ion.

Begin by writing the skeletal structure. Hydrogen atoms must be terminal, and following the guideline of symmetry, the nitrogen atom should be in the middle surrounded by four hydrogen atoms.	**SOLUTION** H H N H H
Calculate the total number of electrons for the Lewis structure by summing the number of valence electrons for each atom and subtracting 1 for the positive charge.	$$\text{Total number of electrons for Lewis structure} = 5 + 4 - 1 = 8$$ 4 × (# valence e⁻ in H) # valence e⁻ in N Subtract 1 e⁻ to account for 1+ charge of ion.
Next, place two electrons between each pair of atoms.	(8 of 8 electrons used)

Since the nitrogen atom has an octet and since all of the hydrogen atoms have duets, the placement of electrons is complete. Write the entire Lewis structure in brackets indicating the charge of the ion in the upper right corner.

$$\left[\begin{array}{c} \text{H} \\ \text{H:N:H} \\ \text{H} \end{array}\right]^+ \quad or \quad \left[\begin{array}{c} \text{H} \\ | \\ \text{H}-\text{N}-\text{H} \\ | \\ \text{H} \end{array}\right]^+$$

▶**SKILLBUILDER 10.6 | Writing Lewis Structures for Polyatomic Ions**

Write the Lewis structure for the ClO^- ion.

▶**FOR MORE PRACTICE**   Problems 55bcd, 56abc, 57, 58.

EXCEPTIONS TO THE OCTET RULE

Lewis theory is often correct in its predictions, but exceptions exist. For example, if we try to write the Lewis structure for NO, which has 11 electrons, the best we can do is:

$$:\overset{..}{N}::\overset{..}{O}: \quad or \quad :\overset{.}{N}=\overset{..}{O}:$$

The nitrogen atom does not have an octet, so this is not a great Lewis structure. However, NO exists in nature. Why? As with any simple theory, Lewis theory is not sophisticated enough to be correct every time. It is impossible to write good Lewis structures for molecules with odd numbers of electrons, yet some of these molecules exist in nature. In such cases, we simply write the best Lewis structure that we can. Another significant exception to the octet rule is boron, which tends to form compounds with only six electrons around B, rather than eight. For example, BF_3 and BH_3—both of which exist in nature—lack an octet for B.

$$\begin{array}{c} :\overset{..}{F}: \\ :\overset{..}{F}:B:\overset{..}{F}: \end{array} \qquad \begin{array}{c} \text{H} \\ \text{H}:\overset{..}{B}:\text{H} \end{array}$$

A third type of exception to the octet rule is also common. A number of molecules, such as SF_6 and PCl_5, have more than eight electrons around a central atom in their Lewis structures.

These are often referred to as *expanded octets*. Expanded octets can form for period 3 elements and beyond. Beyond mentioning them, we do not cover expanded octets in this book. In spite of these exceptions, Lewis theory remains a powerful and simple way to understand chemical bonding.

✔ CONCEPTUAL CHECKPOINT 10.4

Which two species have the same number of lone electron pairs in their Lewis structures?

(a) H_2O and H_3O^+

(b) NH_3 and H_3O^+

(c) NH_3 and CH_4

(d) NH_3 and NH_4^+

10.6 Resonance: Equivalent Lewis Structures for the Same Molecule

When writing Lewis structures, we may find that, for some molecules, we can write more than one good Lewis structure. For example, consider writing a Lewis structure for SO_2. We begin with the skeletal structure:

$$O \ S \ O$$

We then sum the valence electrons.

Total number of electrons for Lewis structure

$$= (\# \text{ valence e}^- \text{ in S}) + 2(\# \text{ valence e}^- \text{ in O})$$
$$= 6 + 2(6)$$
$$= 18$$

We next place two electrons between each pair of atoms

$$O:S:O \quad \text{(4 of 18 electrons used)}$$

and then distribute the remaining electrons, first to terminal atoms

$$:\ddot{O}:S:\ddot{O}: \quad \text{(16 of 18 electrons used)}$$

and finally to the central atom.

$$:\ddot{O}:\ddot{S}:\ddot{O}: \quad \text{(18 of 18 electrons used)}$$

Since the central atom lacks an octet, we move one lone pair from an oxygen atom into the bonding region to form a double bond, giving all of the atoms octets.

$$:\ddot{O}::\ddot{S}:\ddot{O}: \quad \text{or} \quad :\ddot{O}=\ddot{S}-\ddot{O}:$$

However, we could have formed the double bond with the other oxygen atom.

$$:\ddot{O}-\ddot{S}=\ddot{O}:$$

These two Lewis structures are equally correct. In cases such as this—where we can write two or more equivalent (or nearly equivalent) Lewis structures for the same molecule—we find that the molecule exists in nature as an average or intermediate between the two Lewis structures. Either *one* of the two Lewis structures for SO_2 would predict that SO_2 should contain two different kinds of bonds (one double bond and one single bond). However, when we examine SO_2 in nature, we find that both of the bonds are equivalent and intermediate in strength and length between a double bond and single bond. We account for this in Lewis theory by representing the molecule with both structures, called **resonance structures**, with a double-headed arrow between them.

$$:\ddot{O}=\ddot{S}-\ddot{O}: \quad \longleftrightarrow \quad :\ddot{O}-\ddot{S}=\ddot{O}:$$

The true structure of SO_2 is intermediate between these two resonance structures.

EXAMPLE 10.7 Writing Resonance Structures

Write the Lewis structure for the NO_3^- ion. Include resonance structures.

Begin by writing the skeletal structure. Applying the guideline of symmetry, make the three oxygen atoms terminal.	**SOLUTION** O $O \quad N \quad O$

Sum the valence electrons (adding one electron to account for the 1− charge) to determine the total number of electrons in the Lewis structure.	Total number of electrons for Lewis structure $= 5 + 3(6) + 1 = 24$ $3 \times$ (# valence e^- in O) # valence e^- in N Add one e^- to account for negative charge of ion.
Place two electrons between each pair of atoms.	O O:N:O (6 of 24 electrons used)
Distribute the remaining electrons, first to terminal atoms.	:Ö: :Ö:N:Ö: (24 of 24 electrons used)
Since there are no electrons remaining to complete the octet of the central atom, form a double bond by moving a lone pair from one of the oxygen atoms into the bonding region with nitrogen. Enclose the structure in brackets and write the charge at the upper right.	$\left[\begin{array}{c} :Ö: \\ :Ö:N::O: \end{array}\right]^-$ or $\left[\begin{array}{c} :Ö: \\ \mid \\ :Ö-N=O: \end{array}\right]^-$
Notice that you could have formed the double bond with either of the other two oxygen atoms.	$\left[\begin{array}{c} :Ö: \\ \mid \\ :O=N-Ö: \end{array}\right]^-$ or $\left[\begin{array}{c} :Ö \\ \parallel \\ :Ö-N-Ö: \end{array}\right]^-$
Since the three Lewis structures are equally correct, write the three structures as resonance structures.	$\left[\begin{array}{c} :Ö: \\ \mid \\ :O=N-Ö: \end{array}\right]^- \longleftrightarrow \left[\begin{array}{c} :Ö \\ \parallel \\ :Ö-N-Ö: \end{array}\right]^- \longleftrightarrow \left[\begin{array}{c} :Ö: \\ \mid \\ :Ö-N=O: \end{array}\right]^-$

▶**SKILLBUILDER 10.7 | Writing Resonance Structures**

Write the Lewis structure for the NO_2^- ion. Include resonance structures.

▶**FOR MORE PRACTICE**   Example 10.16; Problems 55, 56, 57, 58.

10.7 Predicting the Shapes of Molecules

Lewis theory, in combination with **valence shell electron pair repulsion (VSEPR) theory**, can be used to predict the shapes of molecules. VSEPR theory is based on the idea that **electron groups**—lone pairs, single bonds, or multiple bonds—repel each other. This repulsion between the negative charges of electron groups on the central atom determines the geometry of the molecule. For example, consider CO_2, which has the Lewis structure:

$$:Ö=C=Ö:$$

The geometry of CO_2 is determined by the repulsion between the two electron groups (the two double bonds) on the central carbon atom. These two electron groups get as far away from each other as possible, resulting in a bond angle of 180° and a **linear** geometry for CO_2.

180°

CHEMISTRY IN THE ENVIRONMENT

The Lewis Structure of Ozone

Ozone is a form of oxygen in which three oxygen atoms bond together. Its Lewis structure consists of the two resonance structures:

$$:\ddot{O}=\ddot{O}-\ddot{O}: \longleftrightarrow :\ddot{O}-\ddot{O}=\ddot{O}:$$

Compare the Lewis structure of ozone to the Lewis structure of O_2:

$$:\ddot{O}=\ddot{O}:$$

Which molecule, O_3 or O_2, do you think has the stronger oxygen–oxygen bond? If you deduced O_2, you are correct. O_2 has a stronger bond because it is a pure double bond. Ozone, on the other hand, has bonds that are intermediate between single and double, so O_3 has weaker bonds. The effects of this are significant. As discussed in the *Chemistry in the Environment* box in Chapter 6, O_3 shields us from harmful ultraviolet light entering Earth's atmosphere. O_3 is ideally suited to do this because photons at wavelengths of 280–320 nm (the most dangerous components of sunlight to humans) are just strong enough to break ozone's bonds. In the process, the photons are absorbed.

$$:\ddot{O}-\ddot{O}=\ddot{O}: + \text{UV light} \longrightarrow :\ddot{O}=\ddot{O}: + \cdot\ddot{O}\cdot$$

The same wavelengths of UV light, however, do not have sufficient energy to break the stronger double bond of O_2, which is therefore transparent to UV. Consequently, it is important that we continue, and even strengthen, the ban on ozone-depleting compounds.

CAN YOU ANSWER THIS? *Why are these Lewis structures for ozone incorrect?*

$$:\ddot{O}-\ddot{O}-\ddot{O}: \qquad :\ddot{O}=O=\ddot{O}:$$

As another example, consider the molecule H_2CO. Its Lewis structure is:

$$\begin{array}{c} :\ddot{O}: \\ \parallel \\ H-C-H \end{array}$$

This molecule has three electron groups around the central atom. These three electron groups get as far away from each other as possible, resulting in a bond angle of 120° and a **trigonal planar** geometry.

The angles shown here for H_2CO are approximate. The C=O double bond contains more electron density than do C—H single bonds; resulting in a slightly greater repulsion; thus the HCH bond angle is actually 116°, and the HCO bond angles are actually 122°.

A tetrahedron is a geometric shape with four triangular faces.

If a molecule has four electron groups around the central atom, as CH_4, does, it has a **tetrahedral** geometry with bond angles of 109.5°.

CH_4 is shown here with both a ball-and-stick model (left) and a space-filling model (right). Although space-filling models more closely portray molecules, ball-and-stick models are often used to clearly illustrate molecular geometries.

$$\begin{array}{c} H \\ | \\ H-C-H \\ | \\ H \end{array}$$

The mutual repulsion of the four electron groups causes the tetrahedral shape—the tetrahedron allows the maximum separation among the four groups. When we write the structure of CH_4 on paper, it may seem that the molecule should be square planar, with bond angles of 90°. However, in three dimensions the electron groups can get farther away from each other by forming the tetrahedral geometry.

Each of the preceding examples has only bonding groups of electrons around the central atom. What happens in molecules with lone pairs around the central atom? These lone pairs also repel other electron groups. For example, consider the NH_3 molecule:

$$H-\underset{\displaystyle \cdot\cdot}{N}-H$$
$$\overset{\displaystyle H}{|}$$

The four electron groups (one lone pair and three bonding pairs) get as far away from each other possible. If we look only at the electrons, we find that the **electron geometry**—the geometrical arrangement of the electron groups—is tetrahedral.

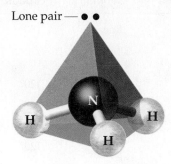

However, the **molecular geometry**—the geometrical arrangement of the atoms—is **trigonal pyramidal**.

Trigonal pyramidal
structure

Notice that, although the electron geometry and the molecular geometry are different, the electron geometry is relevant to the molecular geometry. In other words, the lone pair exerts its influence on the bonding pairs.

Consider one last example, H_2O. Its Lewis structure is:

$$H-\overset{\displaystyle \cdot\cdot}{\underset{\displaystyle \cdot\cdot}{O}}-H$$

Since it has four electron groups, its electron geometry is also tetrahedral.

For reasons we don't cover here, the bond angles in NH_3 and H_2O are actually a few degrees smaller than the ideal tetrahedral angles.

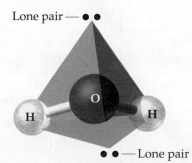

However, its molecular geometry is **bent**.

Bent structure

Table 10.1 summarizes the electron and molecular geometry of a molecule based on the total number of electron groups, the number of bonding groups, and the number of lone pairs.

TABLE 10.1 Electron and Molecular Geometries

Electron Groups*	Bonding Groups	Lone Pairs	Electron Geometry	Angle between Electron Groups**	Molecular Geometry	Example
2	2	0	linear	180°	linear	$:\ddot{O}{=}C{=}\ddot{O}:$
3	3	0	trigonal planar	120°	trigonal planar	$\ddot{O}:$ $H{-}C{-}H$
3	2	1	trigonal planar	120°	bent	$:\ddot{O}{=}\ddot{S}{-}\ddot{O}:$
4	4	0	tetrahedral	109.5°	tetrahedral	$H{-}\underset{\underset{H}{\vert}}{\overset{\overset{H}{\vert}}{C}}{-}H$
4	3	1	tetrahedral	109.5°	trigonal pyramidal	$H{-}\underset{\underset{H}{\vert}}{\ddot{N}}{-}H$
4	2	2	tetrahedral	109.5°	bent	$H{-}\ddot{\underset{..}{O}}{-}H$

*Count only electron groups around the *central* atom. Each of the following is considered one electron group: a lone pair, a single bond, a double bond, and a triple bond.

**Angles listed here are idealized. Actual angles in specific molecules may vary by several degrees.

To determine the geometry of any molecule, use the following procedure. As usual, the steps are in the left column and two examples of applying the steps are in the center and right columns.

Predicting Geometry Using VSEPR Theory	**EXAMPLE 10.8** Predict the electron and molecular geometry of PCl₃.	**EXAMPLE 10.9** Predict the electron and molecular geometry of the [NO₃]⁻ ion.
1. Draw a Lewis structure for the molecule.	**SOLUTION** PCl₃ has 26 electrons. $:\ddot{\underset{..}{C}l:$ $:\ddot{\underset{..}{C}}l:\ddot{P}:\ddot{\underset{..}{C}}l:$	**SOLUTION** [NO₃]⁻ has 24 electrons. $\left[:\ddot{\underset{..}{O}}:\ddot{N}::\ddot{\underset{..}{O}}:\right]^{-}$

2. Determine the total number of electron groups around the central atom. Lone pairs, single bonds, double bonds, and triple bonds each count as one group.	The central atom (P) has four electron groups.	The central atom (N) has three electron groups (the double bond counts as one group).
3. Determine the number of bonding groups and the number of lone pairs around the central atom. These should sum to the result from Step 2. Bonding groups include single bonds, double bonds, and triple bonds.	:C̈l: :C̈l:P̈:C̈l: Lone pair ↗ Three of the four electron groups around P are bonding groups, and one is a lone pair.	[:Ö: :Ö:N::Ö:]⁻ No lone pairs All three of the electron groups around N are bonding groups.
4. Refer to Table 10.1 to determine the electron geometry and molecular geometry.	The electron geometry is tetrahedral (four electron groups), and the molecular geometry—the shape of the molecule—is trigonal pyramidal (four electron groups, three bonding groups, and one lone pair).	The electron geometry is trigonal planar (three electron groups), and the molecular geometry—the shape of the molecule—is trigonal planar (three electron groups, three bonding groups, and no lone pairs).
	▶**SKILLBUILDER 10.8** Predict the molecular geometry of ClNO (N is the central atom).	▶**SKILLBUILDER 10.9** Predict the molecular geometry of the SO_3^{2-} ion.
		▶**FOR MORE PRACTICE** Example 10.17; Problems 65, 66, 69, 70, 73, 74.

✔ CONCEPTUAL CHECKPOINT 10.5

Which condition necessarily leads to a molecular geometry that is identical to the electron geometry?

(a) The presence of a double bond between the central atom and a terminal atom.

(b) The presence of two or more identical terminal atoms bonded to the central atom.

(c) The presence of one or more lone pairs on the central atom.

(d) The absence of any lone pairs on the central atom.

REPRESENTING MOLECULAR GEOMETRIES ON PAPER

Since molecular geometries are three-dimensional, they are often difficult to represent on two-dimensional paper. Many chemists use this notation for bonds to indicate three-dimensional structures on two-dimensional paper.

Straight line	*Hashed lines*	*Wedge*
Bond in plane of paper	Bond projecting into the paper	Bond projecting out of the paper

CHEMISTRY AND HEALTH

Fooled by Molecular Shape

Artificial sweeteners, such as aspartame (Nutrasweet™), taste sweet but have few or no calories. Why? Because taste and caloric value are entirely separate properties of foods. The caloric value of a food depends on the amount of energy released when the food is metabolized. Sucrose (table sugar) is metabolized by oxidation to carbon dioxide and water:

$$C_{12}H_{22}O_{11} + 6\,O_2 \longrightarrow 12\,CO_2 + 11\,H_2O$$
$$\Delta H = -5644 \text{ kJ}$$

When your body metabolizes one mole of sucrose, it obtains 5644 kJ of energy. Some artificial sweeteners, such as saccharin, are not metabolized at all—they just pass through the body unchanged—and therefore have no caloric value. Other artificial sweeteners, such as aspartame, are metabolized but have a much lower caloric content (for a given amount of sweetness) than sucrose.

The taste of a food, however, is independent of its metabolism. The sensation of taste originates in the tongue, where specialized cells called taste cells act as highly sensitive and specific molecular detectors. These cells can distinguish the sugar molecules from the thousands of different types of molecules present in a mouthful of food. The main basis for this discrimination is the molecule's *shape*.

The surface of a taste cell contains specialized protein molecules called taste receptors. Each particular *tastant*—a molecule that we can taste—fits snugly into a special pocket on the taste receptor protein called the *active site*, just as a key fits into a lock (see Section 15.12). For example, a sugar molecule fits only into the active site of the sugar receptor protein called Tlr3. When the sugar molecule (the key) enters the active site (the lock), the different subunits of the Tlr3 protein split apart. This split causes a series of events that result in transmission of a nerve signal, which reaches the brain and registers a sweet taste.

Artificial sweeteners taste sweet because they fit into the receptor pocket that normally binds sucrose. In fact, both aspartame and saccharin bind to the active site in the Tlr3 protein more strongly than sugar does! For this reason, artificial sweeteners are "sweeter than sugar." It takes 200 times as much sucrose as aspartame to trigger the same amount of nerve signal transmission from taste cells.

This type of lock-and-key fit between the active site of a protein and a particular molecule is important not only to taste but to many other biological functions as well. For example, immune response, the sense of smell, and many types of drug action all depend on shape-specific interactions between molecules and proteins. The ability of scientists to determine the shapes of key biological molecules is largely responsible for the revolution in biology that has occurred over the last 50 years.

CAN YOU ANSWER THIS? *Proteins are long-chain molecules in which each link is an amino acid. The simplest amino acid is glycine, which has the structure:*

Determine the geometry around each interior atom in the glycine structure and make a three-dimensional sketch of the molecule.

The major molecular geometries used in this book are shown here using this notation:

| Linear | Trigonal planar | Bent | Tetrahedral | Trigonal pyramidal |

10.8   Electronegativity and Polarity: Why Oil and Water Don't Mix

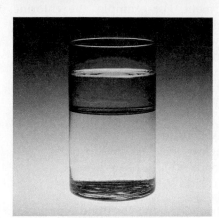

▲ **FIGURE 10.1   Oil and water don't mix**   Question: Why not?

The representation for depicting electron density in this figure is introduced in Section 9.6.

If you combine oil and water in a container, they separate into distinct regions (◄ Figure 10.1). Why? Something about water molecules causes them to bunch together into one region, expelling the oil molecules into a separate region. What is that something? We can begin to understand the answer by examining the Lewis structure of water.

$$H\!-\!\ddot{O}\!-\!H$$

The two bonds between O and H each consist of an electron pair—two electrons shared between the oxygen atom and the hydrogen atom. The oxygen and hydrogen atoms each donate one electron to this electron pair; however, like most children, they don't share them equally. The oxygen atom takes more than its fair share of the electron pair.

ELECTRONEGATIVITY

The ability of an element to attract electrons within a covalent bond is called **electronegativity**. Oxygen is more electronegative than hydrogen, which means that, on average, the shared electrons are more likely to be found near the oxygen atom than near the hydrogen atom. Consider one of the two OH bonds:

Dipole moment

The value of electronegativity is assigned using a relative scale on which fluorine, the most electronegative element, has an electronegativity of 4.0. All other electronegativities are defined relative to fluorine.

Since the electron pair is unequally shared (with oxygen getting the larger share), the oxygen atom has a partial negative charge, symbolized by $\delta-$ (delta minus). The hydrogen atom (getting the smaller share) has a partial positive charge, symbolized by $\delta+$ (delta plus). The result of this uneven electron sharing is a **dipole moment**, a separation of charge within the bond. Covalent bonds that have a dipole moment are called **polar covalent bonds**. The magnitude of the dipole moment, and therefore the degree of polarity of the bond, depend on the electronegativity difference between the two elements in the bond and the length of the bond. For a fixed bond length, the greater the electronegativity difference, the greater the dipole moment and the more polar the bond.

▼ Figure 10.2 shows the relative electronegativities of the elements. Notice that electronegativity increases as you go toward the right across a period in the

▶ **FIGURE 10.2   Electronegativity of the elements**   Linus Pauling introduced the scale shown here. He arbitrarily set the electronegativity of fluorine at 4.0 and computed all other values relative to fluorine.

▲ **FIGURE 10.3 Pure covalent bonding** In Cl₂, the two Cl atoms share the electrons evenly. This is a pure covalent bond.

▲ **FIGURE 10.4 Ionic bonding** In NaCl, Na completely transfers an electron to Cl. This is an ionic bond.

The degree of bond polarity is a continuous function. The guidelines given here are approximate.

▲ **FIGURE 10.5 Polar covalent bonding** In HF, the electrons are shared, but the shared electrons are more likely to be found on F than on H. The bond is polar covalent.

periodic table and decreases as you go down a column in the periodic table. If two elements with identical electronegativities form a covalent bond, they share the electrons equally, and there is no dipole moment. For example, the chlorine molecule, composed of two chlorine atoms (which of course have identical electronegativities), has a pure covalent bond in which electrons are evenly shared (◄ Figure 10.3). The bond has no dipole moment, and the molecule is **nonpolar**.

If there is a large electronegativity difference between the two elements in a bond, such as normally occurs between a metal and a nonmetal, the electron is completely transferred and the bond is ionic. For example, sodium and chlorine form an ionic bond (◄ Figure 10.4).

If there is an intermediate electronegativity difference between the two elements, such as between two different nonmetals, then the bond is polar covalent. For example, HF forms a polar covalent bond (◄ Figure 10.5).

These concepts are summarized in Table 10.2 and ▼ Figure 10.6.

TABLE 10.2 The Effect of Electronegativity Difference on Bond Type

Electronegativity Difference (ΔEN)	Bond Type	Example
zero (0–0.4)	pure covalent	Cl₂
intermediate (0.4–2.0)	polar covalent	HF
large (2.0+)	ionic	NaCl

▲ **FIGURE 10.6 The continuum of bond types** The type of bond (pure covalent, polar covalent, or ionic) is related to the electronegativity difference between the bonded atoms.

EXAMPLE 10.10 Classifying Bonds as Pure Covalent, Polar Covalent, or Ionic

Determine whether the bond formed between each pair of atoms is pure covalent, polar covalent, or ionic.

(a) Sr and F
(b) N and Cl
(c) N and O

SOLUTION

(a) In Figure 10.2, we find the electronegativity of Sr (1.0) and of F (4.0). The electronegativity difference (ΔEN) is:

$$\Delta EN = 4.0 - 1.0 = 3.0$$

Referring to Table 10.2, we classify this bond as ionic.

(b) In Figure 10.2, we find the electronegativity of N (3.0) and of Cl (3.0). The electronegativity difference (ΔEN) is:

$$\Delta EN = 3.0 - 3.0 = 0$$

Referring to Table 10.2, we classify this bond as pure covalent.

(c) In Figure 10.2, we find the electronegativity of N (3.0) and of O (3.5). The electronegativity difference (ΔEN) is:

$$\Delta EN = 3.5 - 3.0 = 0.5$$

Referring to Table 10.2, we classify this bond as polar covalent.

▶SKILLBUILDER 10.10 | **Classifying Bonds as Pure Covalent, Polar Covalent, or Ionic**

Determine whether the bond formed between each pair of atoms is pure covalent, polar covalent, or ionic.

(a) I and I
(b) Cs and Br
(c) P and O

▶**FOR MORE PRACTICE** Problems 81, 82.

POLAR BONDS AND POLAR MOLECULES

Does the presence of one or more polar bonds in a molecule always result in a polar molecule? The answer is no. A **polar molecule** is one with polar bonds that add together—they do not cancel each other—to form a net dipole moment. For diatomic molecules, you can easily tell polar molecules from nonpolar ones. If a diatomic molecule contains a polar bond, then the molecule is polar. However, for molecules with more than two atoms, it is more difficult to tell polar molecules from nonpolar ones because two or more polar bonds may cancel one another. For example, consider carbon dioxide:

$$:\ddot{O}=C=\ddot{O}:$$

Each C=O *bond* is polar because the difference in electronegativity between oxygen and carbon is 1.0. However, since CO_2 has a linear geometry, the dipole moment of one bond completely cancels the dipole moment of the other and the *molecule* is nonpolar. We can understand this with an analogy. Imagine each polar bond to be a rope pulling on the central atom. In CO_2 we can see how the two ropes pulling in opposing directions cancel each other's effect:

$$:\underset{..}{O}=C=\underset{..}{O}:$$

We can also represent polar bonds with arrows (or vectors) that point in the direction of the negative pole and have a plus sign at the positive pole (as just shown for carbon dioxide). If the arrows (or vectors) point in exactly opposing directions as in carbon dioxide, the dipole moments cancel.

Water, on the other hand, has two dipole moments that do not cancel. If we imagine each bond as a rope pulling on oxygen, we see that, because of the angle between the bonds, the pulls of the two ropes do not cancel:

TABLE 10.3 Common Cases of Adding Dipole Moments to Determine Whether a Molecule Is Polar

Nonpolar

Two identical polar bonds pointing in opposite directions will cancel. The molecule is nonpolar.

Nonpolar

Three identical polar bonds at 120° from each other will cancel. The molecule is nonpolar.

Polar

Three polar bonds in a trigonal pyramidal arrangement (109.5°) will not cancel. The molecule is polar.

Polar

Two polar bonds with an angle of less than 180° between them will not cancel. The molecule is polar.

Nonpolar

Four identical polar bonds in a tetrahedral arrangement (109.5° from each other) will cancel. The molecule is nonpolar.

Note: In all cases where the polar bonds cancel, the bonds are assumed to be identical. If one or more of the bonds are different than the other(s), the bonds will not cancel and the molecule is polar.

Consequently, water is a polar molecule. We can use symmetry as a guide to determine whether a molecule containing polar bonds is indeed polar. Highly symmetric molecules tend to be nonpolar even if they have polar bonds because the bond dipole moments (or the pulls of the ropes) tend to cancel. Asymmetric molecules that contain polar bonds tend to be polar because the bond dipole moments (or the pulls of the ropes) tend not to cancel. Table 10.3 summarizes some common cases.

In summary, to determine whether a molecule is polar:

- **Determine whether the molecule contains polar bonds**. A bond is polar if the two bonding atoms have different electronegativities. If there are no polar bonds, the molecule is nonpolar.
- **Determine whether the polar bonds add together to form a net dipole moment**. You must first use VSEPR to determine the geometry of the molecule. Then visualize each bond as a rope pulling on the central atom. Is the molecule highly symmetrical? Do the pulls of the ropes cancel? If so, there is no net dipole moment and the molecule is nonpolar. If the molecule is asymmetrical and the pulls of the rope do not cancel, the molecule is polar.

EXAMPLE 10.11 Determining Whether a Molecule Is Polar

Determine whether NH_3 is polar.

	SOLUTION
Begin by drawing the Lewis structure of NH_3. Since N and H have different electronegativities, the bonds are polar.	
The geometry of NH_3 is trigonal pyramidal (four electron groups, three bonding groups, one lone pair). Draw a three-dimensional picture of NH_3 and imagine each bond as a rope that is being pulled. The pulls of the ropes do not cancel and the molecule is polar.	

NH_3 is polar

▶**SKILLBUILDER 10.11 | Determining Whether a Molecule Is Polar**

Determine whether CH_4 is polar.

▶**FOR MORE PRACTICE**   Example 10.18; Problems 89, 90, 91, 92.

Whether or not a molecule is polar is important because polar molecules tend to behave differently than nonpolar molecules. Water and oil do not mix, for example, because water molecules are polar and the molecules that compose oil are generally nonpolar. Polar molecules interact strongly with other polar molecules because the positive end of one molecule is attracted to the negative end of another, just as the south pole of a magnet is attracted to the north pole of another

EVERYDAY CHEMISTRY

How Soap Works

Imagine eating a greasy cheeseburger with both hands and no napkins. By the end of the meal, your hands are coated with grease and oil. If you try to wash them with only water, they remain greasy. However, if you add a little soap, the grease washes away. Why? As we learned previously, water molecules are polar and the molecules that compose grease and oil are nonpolar. As a result, water and grease repel each other.

The molecules that compose soap, however, have a special structure that allows them to interact strongly with both water and grease. One end of a soap molecule is polar, while the other end is nonpolar.

The polar head of a soap molecule strongly attracts water molecules, while the nonpolar tail strongly attracts grease and oil molecules. Soap is a sort of molecular liaison, one end interacting with water and the other end interacting with grease. Soap therefore allows water and grease to mix, removing the grease from your hands and washing it down the drain.

CAN YOU ANSWER THIS? *Consider this detergent molecule. Which end do you think is polar? Which end is nonpolar?*

$$CH_3(CH_2)_{11}OCH_2CH_2OH$$

Soap molecule

Polar head
attracts water

Nonpolar tail
attracts grease

▶ **FIGURE 10.7 Dipole–dipole attraction** Just as the north pole of one magnet is attracted to the south pole of another, so the positive end of one molecule with a dipole is attracted to the negative end of another molecule with a dipole.

magnet (▲ Figure 10.7). A mixture of polar and nonpolar molecules is similar to a mixture of small magnetic and nonmagnetic particles. The magnetic particles clump together, excluding the nonmagnetic ones and separating into distinct regions (◀ Figure 10.8). Similarly, the polar water molecules attract one another, forming regions from which the nonpolar oil molecules are excluded (▼ Figure 10.9).

▲ **FIGURE 10.8 Magnetic and nonmagnetic particles** Magnetic particles attract one another, excluding nonmagnetic particles. This behavior is analogous to that of polar and nonpolar molecules.

◀ **FIGURE 10.9 Polar and nonpolar molecules** A mixture of polar and nonpolar molecules, like a mixture of magnetic and nonmagnetic particles, separates into distinct regions because the polar molecules attract one another, excluding the nonpolar ones. **Question:** Can you think of some examples of this behavior?

CHAPTER IN REVIEW

CHEMICAL PRINCIPLES

Lewis Theory: Lewis theory is a model for chemical bonding. According to Lewis theory, chemical bonds are formed when atoms transfer valence electrons (ionic bonding) or share valence electrons (covalent bonding) to attain noble gas electron configurations. In Lewis theory, valence electrons are represented as dots surrounding the symbol for an element. When two or more elements bond together, the dots are transferred or shared so that every atom attains eight dots (an octet), or two dots (a duet) in the case of hydrogen.

Molecular Shapes: The shapes of molecules can be predicted by combining Lewis theory with valence shell electron pair repulsion (VSEPR) theory. In this model, electron groups—lone pairs, single bonds, double bonds, and triple bonds—around the central atom repel one another and determine the geometry of the molecule.

Electronegativity and Polarity: Electronegativity refers to the relative ability of elements to attract electrons within a chemical bond. Electronegativity increases as you move to the right across a period in the periodic table and decreases as you move down a column. When two nonmetal atoms of different electronegativities form a covalent bond, the electrons in the bond are not evenly shared and the bond is polar. In diatomic molecules, a polar bond results in a polar molecule. In molecules with more than two atoms, polar bonds may cancel, forming a nonpolar molecule, or they may sum, forming a polar molecule.

RELEVANCE

Lewis Theory: Bonding theories predict what combinations of elements will form stable compounds and can also be used to predict the properties of those compounds. For example, pharmaceutical companies use bonding theories when they are designing drug molecules that must interact with a specific part of a protein molecule.

Molecular Shapes: Molecular shapes determine many of the properties of compounds. Water's bent geometry, for example, causes it to be a liquid at room temperature instead of a gas. It is also the reason ice floats on water and snowflakes have hexagonal patterns.

Electronegativity and Polarity: The polarity of a molecule influences many of its properties such as whether it will be a solid, liquid, or gas at room temperature and whether it will mix with other compounds. Oil and water, for example, do not mix because water is polar while oil is nonpolar.

CHEMICAL SKILLS

Lewis Structures for Elements (Section 10.2)

The Lewis structure of any element is the symbol for the element with the valence electrons represented as dots drawn around the element. The number of valence electrons is equal to the group number of the element (for main-group elements).

EXAMPLES

EXAMPLE 10.12   Lewis Structures for Elements

Draw the Lewis structure of sulfur.

SOLUTION

Since S is in Group 6A, it has six valence electrons. We draw these as dots surrounding its symbol, S.

$\cdot \ddot{\text{S}} \colon$

Writing Lewis Structures of Ionic Compounds (Section 10.3)

In an ionic Lewis structure, the metal loses all of its valence electrons to the nonmetal, which attains an octet. The nonmetal, with its octet, is normally written in brackets with the charge in the upper right corner.

EXAMPLE 10.13 Writing Lewis Structures of Ionic Compounds

Write the Lewis structure for lithium bromide.

SOLUTION

$$Li^+ \; [:\ddot{Br}:]^-$$

Using Lewis Theory to Predict the Chemical Formula of an Ionic Compound (Section 10.3)

To determine the chemical formula of an ionic compound, write the Lewis structures of each of the elements. Then choose the correct number of each type of atom so that the metal atom(s) lose all of their valence electrons and the nonmetal atom(s) attain an octet.

EXAMPLE 10.14 Using Lewis Theory to Predict the Chemical Formula of an Ionic Compound

Use Lewis theory to predict the formula for the compound that forms between potassium and sulfur.

SOLUTION

The Lewis structures of K and S are:

$$K\cdot \qquad \cdot\ddot{S}:$$

Potassium must lose one electron and sulfur must gain two. Consequently, we need two potassium atoms to every sulfur atom. The Lewis structure is:

$$K^+[:\ddot{S}:]^{2-}K^+$$

The correct formula is K_2S.

Writing Lewis Structures for Covalent Compounds (Sections 10.4, 10.5)

To write covalent Lewis structures, follow these steps:

1. **Write the correct skeletal structure for the molecule.** Hydrogen atoms will always be terminal, halogens will usually be terminal, and many molecules tend to be symmetrical.

2. **Calculate the total number of electrons for the Lewis structure by summing the valence electrons of each atom in the molecule.** Remember that the number of valence electrons for any main-group element is equal to its group number in the periodic table. For polyatomic ions, add one electron for each negative charge and subtract one electron for each positive charge.

3. **Distribute the electrons among the atoms, giving octets (or duets for hydrogen) to as many atoms as possible.** Begin by placing two electrons between each pair of atoms. These are the bonding electrons. Then distribute the remaining electrons, first to terminal atoms and then to the central atom.

4. **If any atoms lack an octet, form double or triple bonds as necessary to give them octets.** Do this by moving lone electron pairs from terminal atoms into the bonding region with the central atom.

EXAMPLE 10.15 Writing Lewis Structures for Covalent Compounds

Write the Lewis structure for CS_2.

SOLUTION

S C S

$$
\begin{aligned}
Total\ e^- &= 1 \times (\ \#\ valence\ e^-\ in\ C) \\
&\quad + 2 \times (\ \#\ valence\ e^-\ in\ S) \\
&= 4 + 2(6) \\
&= 16
\end{aligned}
$$

S:C:S   (4 of 16 e^- used)

$:\ddot{S}:C:\ddot{S}:$ (16 of 16 e^- used)

$:\ddot{S}::C::\ddot{S}:$ or $:\ddot{S}=C=\ddot{S}:$

Writing Resonance Structures (Section 10.6)

When you can write two or more equivalent (or nearly equivalent) Lewis structures for a molecule, the true structure is an average between these. Represent this by writing all of the correct structures (called resonance structures) with double-headed arrows between them.

Predicting the Shapes of Molecules (Section 10.7)

To determine the shape of a molecule, follow these steps:

1. **Draw the Lewis structure for the molecule.**

2. **Determine the total number of electron groups around the central atom.** Lone pairs, single bonds, double bonds, and triple bonds each count as one group.

3. **Determine the number of bonding groups and the number of lone pairs around the central atom.** These should sum to the result from Step 2. Bonding groups include single bonds, double bonds, and triple bonds.

4. Refer to Table 10.1 to determine the electron geometry and molecular geometry.

Determining Whether a Molecule Is Polar (Section 10.8)

- **Determine whether the molecule contains polar bonds.** A bond is polar if the two bonding atoms have different electronegativities. If there are no polar bonds, the molecule is nonpolar.

- **Determine whether the polar bonds add together to form a net dipole moment.** Use VSEPR theory to determine the geometry of the molecule. Then visualize each bond as a rope pulling on the central atom. Is the molecule highly symmetrical? Do the pulls of the ropes cancel? If so, there is no net dipole moment and the molecule is nonpolar. If the molecule is asymmetrical and the pulls of the rope do not cancel, the molecule is polar.

EXAMPLE 10.16   Writing Resonance Structures

Write resonance structures for SeO_2.

SOLUTION

We can write a Lewis structure for SeO_2 by following the steps for writing covalent Lewis structures. We find that we can write two equally correct structures, so we draw them both as resonance structures.

$$:\ddot{O}—\ddot{Se}=\ddot{O}: \longleftrightarrow :\ddot{O}=\ddot{Se}—\ddot{O}:$$

EXAMPLE 10.17   Predicting the Shapes of Molecules

Predict the geometry of SeO_2.

SOLUTION

The Lewis structure for SeO_2 (as we saw in Example 10.16) is composed of the following two resonance structures.

$$:\ddot{O}—\ddot{Se}=\ddot{O}: \longleftrightarrow :\ddot{O}=\ddot{Se}—\ddot{O}:$$

Either of the resonance structures will give the same geometry.

Total number of electron groups = 3
Number of bonding groups = 2
Number of lone pairs = 1

Electron geometry = Trigonal planar
Molecular geometry = Bent

EXAMPLE 10.18   Determining Whether a Molecule Is Polar

Determine whether SeO_2 is polar.

SOLUTION

Se and O are nonmetals with different electronegativities (2.4 for Se and 3.5 for O). Therefore, the Se–O bonds are polar.

As we saw in Example 10.17, the geometry of SeO_2 is bent.

The polar bonds do not cancel but rather sum to give a net dipole moment. Therefore the molecule is polar.

KEY TERMS

bent [10.7]
bonding pair [10.4]
bonding theory [10.1]
chemical bond [10.2]
covalent bond [10.2]
dipole moment [10.8]
dot structure [10.2]
double bond [10.4]

duet [10.2]
electron geometry [10.7]
electron group [10.7]
electronegativity [10.8]
ionic bond [10.2]
Lewis structure [10.2]
Lewis theory [10.1]
linear [10.7]

lone pair [10.4]
molecular geometry [10.7]
nonpolar [10.8]
octet [10.2]
octet rule [10.2]
polar covalent bond [10.8]
polar molecule [10.8]
resonance structures [10.6]

terminal atom [10.5]
tetrahedral [10.7]
trigonal planar [10.7]
trigonal pyramidal [10.7]
triple bond [10.4]
valence shell electron pair
    repulsion (VSEPR)
    theory [10.7]

EXERCISES

QUESTIONS

1. Why are bonding theories important? Give some examples of what bonding theories can predict.
2. Write the electron configurations for Ne and Ar. How many valence electrons do they each have?
3. In Lewis theory, what is an octet? What is a duet? What is a chemical bond?
4. What is the difference between ionic bonding and covalent bonding?
5. How can Lewis theory be used to determine the formula of ionic compounds? You may explain this with an example.
6. What is the difference between lone pair and bonding pair electrons?
7. How are double and triple bonds physically different from single bonds?
8. What is the procedure for writing a covalent Lewis structure?
9. How do you determine the number of electrons that go into the Lewis structure of a molecule?
10. How do you determine the number of electrons that go into the Lewis structure of a polyatomic ion?
11. Why does the octet rule have exceptions? Give some examples.
12. What are resonance structures? Why are they necessary?

13. Explain how VSEPR theory predicts the shapes of molecules.
14. If all of the electron groups around a central atom are bonding groups (that is, there are no lone pairs), what is the molecular geometry for:
    (a) two electron groups
    (b) three electron groups
    (c) four electron groups
15. Give the bond angles for each of the geometries in the preceding question.
16. What is the difference between electron geometry and molecular geometry in VSEPR theory?
17. What is electronegativity?
18. What is the most electronegative element on the periodic table?
19. What is a polar covalent bond?
20. What is a dipole moment?
21. What happens when you try to mix a polar liquid with a nonpolar one?
22. If a molecule has polar bonds, will the molecule itself be polar? Why or why not?

PROBLEMS

WRITING LEWIS STRUCTURES FOR ELEMENTS

23. Write an electron configuration for each element and the corresponding Lewis structure. Indicate which electrons in the electron configuration are included in the Lewis structure.
    (a) N
    (b) C
    (c) Cl
    (d) Ar

24. Write an electron configuration for each element and the corresponding Lewis structure. Indicate which electrons in the electron configuration are included in the Lewis structure.
    (a) Li
    (b) P
    (c) F
    (d) Ne

25. Write Lewis structures for each element.
    (a) I
    (b) S
    (c) Ge
    (d) Ca

26. Write Lewis structures for each element.
    (a) Kr
    (b) P
    (c) B
    (d) Na

27. Write a generic Lewis structure for the halogens. Do the halogens tend to gain or lose electrons in chemical reactions? How many?

28. Write a generic Lewis structure for the alkali metals. Do the alkali metals tend to gain or lose electrons in chemical reactions? How many?

29. Write a generic Lewis structure for the alkaline earth metals. Do the alkaline earth metals tend to gain or lose electrons in chemical reactions? How many?

30. Write a generic Lewis structure for the elements in the oxygen family (Group 6A). Do the elements in the oxygen family tend to gain or lose electrons in chemical reactions? How many?

31. Write the Lewis structure for each ion.
    (a) Al^{3+}
    (b) Mg^{2+}
    (c) Se^{2-}
    (d) N^{3-}

32. Write a Lewis structure for each ion.
    (a) Sr^{2+}
    (b) S^{2-}
    (c) Li^+
    (d) Cl^-

33. For each ion, indicate the noble gas that has the same Lewis structure as the ion.
    (a) Br^-
    (b) O^{2-}
    (c) Rb^+
    (d) Ba^{2+}

34. For each ion, indicate the noble gas that has the same Lewis structure as the ion.
    (a) Se^{2-}
    (b) I^-
    (c) Sr^{2+}
    (d) F^-

LEWIS STRUCTURES FOR IONIC COMPOUNDS

35. Determine whether each compound is best represented by an ionic or a covalent Lewis structure.
    (a) SF_6
    (b) $MgCl_2$
    (c) BrCl
    (d) K_2S

36. Determine whether each compounds is best represented by an ionic or a covalent Lewis structure.
    (a) NO
    (b) CO_2
    (c) Rb_2O
    (d) Al_2S_3

37. Write a Lewis structure for each ionic compound.
    (a) NaF
    (b) CaO
    (c) $SrBr_2$
    (d) K_2O

38. Write a Lewis structure for each ionic compound.
    (a) SrO
    (b) Li_2S
    (c) CaI_2
    (d) RbF

39. Use Lewis theory to determine the formula for the compound that forms from:
    (a) Ca and S
    (b) Mg and Br
    (c) Cs and I
    (d) Ca and N

40. Use Lewis theory to determine the formula for the compound that forms from:
    (a) Al and S
    (b) Na and S
    (c) Sr and Se
    (d) Ba and F

41. Draw the Lewis structure for the ionic compound that forms from Mg and:

(a) F

(b) O

(c) N

42. Draw the Lewis structure for the ionic compound that forms from Al and:

(a) F

(b) O

(c) N

43. Determine what is wrong with each ionic Lewis structure and write the correct structure.

(a) $[Cs:]^+ [:\ddot{C}l:]^-$

(b) $Ba^+ [:\ddot{O}:]^-$

(c) $Ca^{2+} [:\ddot{I}:]^-$

44. Determine what is wrong with each ionic Lewis structure and write the correct structure.

(a) $[:\ddot{O}:]^{2-} Na^+ [:\ddot{O}:]^{2-}$

(b) $Mg:\ddot{O}:$

(c) $[Li:]^+ [:\ddot{S}:]^-$

LEWIS STRUCTURES FOR COVALENT COMPOUNDS

45. Use Lewis theory to explain why each element exists as a diatomic molecule:

(a) hydrogen

(b) iodine

(c) nitrogen

(d) oxygen

46. Use Lewis theory to explain why the compound that forms between hydrogen and sulfur has the formula H_2S. Would you expect HS to be stable? H_3S?

47. Write the Lewis structure for each molecule.

(a) PH_3

(b) SCl_2

(c) F_2

(d) HI

48. Write the Lewis structure for each molecule.

(a) CH_4

(b) NF_3

(c) OF_2

(d) H_2O

49. Write the Lewis structure for each molecule.

(a) O_2

(b) CO

(c) HONO (N is central)

(d) SO_2

50. Write the Lewis structure for each molecule.

(a) N_2O (oxygen is terminal)

(b) SiH_4

(c) CI_4

(d) Cl_2CO (carbon is central)

51. Write the Lewis structure for each molecule.

(a) C_2H_2

(b) C_2H_4

(c) N_2H_2

(d) N_2H_4

52. Write the Lewis structure for each molecule.

(a) H_2CO (carbon is central)

(b) H_3COH (carbon and oxygen are both central)

(c) H_3COCH_3 (oxygen is between the two carbon atoms)

(d) H_2O_2

53. Determine what is wrong with each Lewis structure and write the correct structure.

(a) $:\ddot{N}=\ddot{N}:$

(b) $:\ddot{S}-Si-\ddot{S}:$

(c) $H-H-\ddot{O}:$

(d) $:\ddot{I}-N-\ddot{I}:$
$\qquad\quad |$
$\qquad\quad :\ddot{I}:$

54. Determine what is wrong with each Lewis structure and write the correct structure.

(a) $H-H-H-\ddot{N}:$

(b) $:\ddot{C}l=O=\ddot{C}l:$

(c) $\qquad\quad :\ddot{O}:$
$\qquad\qquad |$
$H-C-\ddot{O}-H$

(d) $H=\ddot{B}r:$

55. Write a Lewis structure for each molecule or ion. Include resonance structures if necessary.

(a) SeO_2

(b) CO_3^{2-}

(c) ClO^-

(d) ClO_2^-

56. Write a Lewis structure for each molecule or ion. Include resonance structures if necessary.

(a) ClO_3^-

(b) ClO_4^-

(c) NO_3^-

(d) SO_3

57. Write a Lewis structure for each ion. Include resonance structures if necessary.

(a) PO_4^{3-}

(b) CN^-

(c) NO_2^-

(d) SO_3^{2-}

58. Write a Lewis structure for each ion. Include resonance structures if necessary.

(a) SO_4^{2-}

(b) HSO_4^- (S is central; H is attached to one of the O atoms)

(c) NH_4^+

(d) BrO_2^- (Br is central)

59. Write the Lewis structure for each molecule. These molecules do not follow the octet rule.

(a) BCl_3

(b) NO_2

(c) BH_3

60. Write the Lewis structure for each molecule. These molecules do not follow the octet rule.

(a) BBr_3

(b) NO

PREDICTING THE SHAPES OF MOLECULES

61. Determine the number of electron groups around the central atom for each molecule.

(a) OF_2

(b) NF_3

(c) CS_2

(d) CH_4

62. Determine the number of electron groups around the central atom for each molecule.

(a) CH_2Cl_2

(b) SBr_2

(c) H_2S

(d) PCl_3

63. Determine the number of bonding groups and the number of lone pairs for each of the molecules in Problem 61. The sum of these should equal your answer to Problem 61.

64. Determine the number of bonding groups and the number of lone pairs for each of the molecules in Problem 62. The sum of these should equal your answer to Problem 62.

65. Determine the molecular geometry of each molecule.

(a) CBr_4

(b) H_2CO

(c) CS_2

(d) BH_3

66. Determine the molecular geometry of each molecule.

(a) SiO_2

(b) BF_3

(c) $CFCl_3$ (carbon is central)

(d) H_2CS (carbon is central)

67. Determine the bond angles for each molecule in Problem 65.

68. Determine the bond angles for each molecule in Problem 66.

69. Determine the electron and molecular geometries of each molecule.

(a) N_2O (oxygen is terminal)

(b) SO_2

(c) H_2S

(d) PF_3

70. Determine the electron and molecular geometries of each molecule. (*Hint:* Determine the geometry around each of the two central atoms.)

(a) C_2H_2 (skeletal structure HCCH)

(b) C_2H_4 (skeletal structure H_2CCH_2)

(c) C_2H_6 (skeletal structure H_3CCH_3)

71. Determine the bond angles for each molecule in Problem 69.

72. Determine the bond angles for each molecule in Problem 70.

73. Determine the electron and molecular geometries of each molecule. For molecules with two central atoms, indicate the geometry about each central atom.

(a) N_2

(b) N_2H_2 (skeletal structure HNNH)

(c) N_2H_4 (skeletal structure H_2NNH_2)

74. Determine the electron and molecular geometries of each molecule. For molecules with more than one central atom, indicate the geometry about each central atom.

(a) CH_3OH (skeletal structure H_3COH)

(b) H_3COCH_3 (skeletal structure H_3COCH_3)

(c) H_2O_2 (skeletal structure HOOH)

75. Determine the molecular geometry of each polyatomic ion.

(a) CO_3^{2-}

(b) ClO_2^-

(c) NO_3^-

(d) NH_4^+

76. Determine the molecular geometry of each polyatomic ion.

(a) ClO_4^-

(b) BrO_2^-

(c) NO_2^-

(d) SO_4^{2-}

ELECTRONEGATIVITY AND POLARITY

77. Refer to Figure 10.2 to determine the electronegativity of each element.

(a) Mg

(b) Si

(c) Br

78. Refer to Figure 10.2 to determine the electronegativity of each element.

(a) F

(b) C

(c) S

79. List these elements in order of decreasing electronegativity: Rb, Si, Cl, Ca, Ga.

80. List these elements in order of increasing electronegativity: Ba, N, F, Si, Cs.

81. Refer to Figure 10.2 to find the electronegativity difference between each pair of elements; then refer to Table 10.2 to classify the bonds that occur between them as pure covalent, polar covalent, or ionic.

(a) Mg and Br

(b) Cr and F

(c) Br and Br

(d) Si and O

82. Refer to Figure 10.2 to find the electronegativity difference between each pair of elements; then refer to Table 10.2 to classify the bonds that occur between them as pure covalent, polar covalent, or ionic.

(a) K and Cl

(b) N and N

(c) C and S

(d) C and Cl

83. Arrange these diatomic molecules in order of increasing bond polarity: ICl, HBr, H_2, CO

84. Arrange these diatomic molecules in order of decreasing bond polarity: HCl, NO, F_2, HI

85. Classify each diatomic molecule as polar or nonpolar.
    (a) CO
    (b) O_2
    (c) F_2
    (d) HBr

86. Classify each diatomic molecule as polar or nonpolar.
    (a) I_2
    (b) NO
    (c) HCl
    (d) N_2

87. For each polar molecule in Problem 85 draw the molecule and indicate the positive and negative ends of the dipole moment.

88. For each polar molecule in Problem 86 draw the molecule and indicate the positive and negative ends of the dipole moment.

89. Classify each molecule as polar or nonpolar.
    (a) CS_2
    (b) SO_2
    (c) CH_4
    (d) CH_3Cl

90. Classify each molecule as polar or nonpolar.
    (a) H_2CO
    (b) CH_3OH
    (c) CH_2Cl_2
    (d) CO_2

91. Classify each molecule as polar or nonpolar.
    (a) BH_3
    (b) $CHCl_3$
    (c) C_2H_2
    (d) NH_3

92. Classify each molecule as polar or nonpolar.
    (a) N_2H_2
    (b) H_2O_2
    (c) CF_4
    (d) NO_2

CUMULATIVE PROBLEMS

93. Write electron configurations and Lewis structures for each element. Indicate which of the electrons in the electron configuration are shown in the Lewis structure.
    (a) Ca
    (b) Ga
    (c) As
    (d) I

94. Write electron configurations and Lewis structures for each element. Indicate which of the electrons in the electron configuration are shown in the Lewis structure.
    (a) Rb
    (b) Ge
    (c) Kr
    (d) Se

95. Determine whether each compound is ionic or covalent and write the appropriate Lewis structure.
    (a) K_2S
    (b) CHFO (carbon is central)
    (c) MgSe
    (d) PBr_3

96. Determine whether each compound is ionic or covalent and write the appropriate Lewis structure.
    (a) HCN
    (b) ClF
    (c) MgI_2
    (d) CaS

97. Write the Lewis structure for $OCCl_2$ (carbon is central) and determine whether the molecule is polar. Draw the three-dimensional structure for the molecule.

98. Write the Lewis structure for CH_3COH and determine whether the molecule is polar. Draw the three-dimensional structure for the molecule. The skeletal structure is:

$$
\begin{array}{ccc}
 & H & O \\
H & C & C \quad H \\
 & H &
\end{array}
$$

99. Write the Lewis structure for acetic acid (a component of vinegar), CH_3COOH, and draw the three-dimensional sketch of the molecule. Its skeletal structure is:

$$
\begin{array}{cccc}
 & H & O & \\
H & C & C & O & H \\
 & H & &
\end{array}
$$

100. Write the Lewis structure for benzene, C_6H_6, and draw a three-dimensional sketch of the molecule. The skeletal structure is the ring shown here. (*Hint*: The Lewis structure consists of two resonance structures.)

$$
\begin{array}{ccc}
 & H & \\
 & C & \\
HC & & CH \\
HC & & CH \\
 & C & \\
 & H &
\end{array}
$$

101. Consider the neutralization reaction:

$$HCl(aq) + NaOH(aq) \longrightarrow H_2O(l) + NaCl(aq)$$

Write the reaction showing the Lewis structures of each of the reactants and products.

102. Consider the precipitation reaction.

$$Pb(NO_3)_2(aq) + 2\,LiCl(aq) \longrightarrow PbCl_2(s) + 2\,LiNO_3(aq)$$

Write the reaction showing the Lewis structures of each of the reactants and products.

103. Consider the redox reaction:

$$2\,K(s) + Cl_2(g) \longrightarrow 2\,KCl(s)$$

Draw the Lewis structure for each reactant and product, and determine which reactant was oxidized and which one was reduced.

104. Consider the redox reaction:

$$Ca(s) + Br_2(g) \longrightarrow CaBr_2(s)$$

Draw the Lewis structure for each reactant and product, and determine which reactant was oxidized and which one was reduced.

105. Each compound listed contains both ionic and covalent bonds. Write the ionic Lewis structure for each one, including the covalent structure for the polyatomic ion. Write resonance structures if necessary.

(a) KOH

(b) KNO_3

(c) LiIO

(d) $BaCO_3$

106. Each of the compounds listed contains both ionic and covalent bonds. Write an ionic Lewis structure for each one, including the covalent structure for the polyatomic ion. Write resonance structures if necessary.

(a) $RbIO_2$

(b) $Ca(OH)_2$

(c) NH_4Cl

(d) $Sr(CN)_2$

107. Each molecule listed contains an expanded octet (10 or 12 electrons) around the central atom. Write the Lewis structure for each molecule.

(a) PF_5

(b) SF_4

(c) SeF_4

108. Each molecule listed contains an expanded octet (10 or 12 electrons) around the central atom. Write the Lewis structure for each molecule.

(a) ClF_5

(b) SF_6

(c) IF_5

109. Formic acid is responsible for the sting in biting ants. By mass, formic acid is 26.10% C, 4.38% H, and 69.52% O. The molar mass of formic acid is 46.02 g/mol. Find the molecular formula of formic acid and draw its Lewis structure.

110. Diazomethane has the following composition by mass: 28.57% C, 4.80% H, and 66.64% N. The molar mass of diazomethane is 42.04 g/mol. Find the molecular formula of diazomethane and draw its Lewis structure.

111. Free radicals are molecules that contain an odd number of valence electrons and therefore contain an unpaired electron in their Lewis structure. Write the best possible Lewis structure for the free radical HOO. Does Lewis theory predict that HOO is stable? Predict its geometry.

112. Free radicals (as explained in the previous problem) are molecules that contain an odd number of valence electrons. Write the best possible Lewis structure for the free radical CH_3. Predict its geometry.

HIGHLIGHT PROBLEMS

113. Some theories on aging suggest that free radicals cause a variety of diseases and aging. Free radicals (as explained in Problems 111 and 112) are molecules or ions containing an unpaired electron. As you know from Lewis theory, such molecules are not chemically stable and quickly react with other molecules. Free radicals may attack molecules within the cell, such as DNA, changing them and causing cancer or other diseases. Free radicals may also attack molecules on the surfaces of cells, making them appear foreign to the body's immune system. The immune system then attacks the cell and destroys it, weakening the body. Draw the Lewis structure for each of these free radicals, which have been implicated in theories of aging.

(a) O_2^-

(b) O^-

(c) OH

(d) CH_3OO (unpaired electron on terminal oxygen)

▲ Free radicals, molecules containing unpaired electrons (represented here as X ·), may attack biological molecules such as the DNA molecule depicted here.

114. Free radicals (see Problem 113) are important in many environmentally significant reactions. For example, photochemical smog, which forms as a result of the action of sunlight on air pollutants, is formed in part by these two steps:

$$NO_2 \xrightarrow{\text{UV light}} NO + O$$

$$O + O_2 \longrightarrow O_3$$

The product of this reaction, ozone, is a pollutant in the lower atmosphere. Ozone is an eye and lung irritant and also accelerates the weathering of rubber products. Write Lewis structures for each of the reactants and products in the preceding reactions.

▲ Ozone damages rubber products.

115. Examine the formulas and space-filling models of the molecules shown here. Determine whether the structure is correct and, if not, make a sketch of the correct structure.

(a) H_2Se

(b) CSe_2

(c) PCl_3

(d) CF_2Cl_2

▶ANSWERS TO SKILLBUILDER EXERCISES

Skillbuilder 10.1 $\cdot Mg\cdot$

Skillbuilder 10.2 Na^+ $[:\ddot{\underset{\cdot\cdot}{B}}r:]$

Skillbuilder 10.3 Mg_3N_2

Skillbuilder 10.4 $:C{\equiv}O:$

Skillbuilder 10.5
$$\begin{array}{c} \ddot{O}: \\ \| \\ H-C-H \end{array}$$

Skillbuilder 10.6 $[:\ddot{\underset{\cdot\cdot}{C}}l-\ddot{\underset{\cdot\cdot}{O}}:]^-$

Skillbuilder 10.7 $[:\ddot{O}{=}\ddot{N}{-}\ddot{\underset{\cdot\cdot}{O}}:]^-$ $\longleftrightarrow$ $[:\ddot{\underset{\cdot\cdot}{O}}{-}\ddot{N}{=}\ddot{O}:]^-$

Skillbuilder 10.8 bent

Skillbuilder 10.9 trigonal pyramidal

Skillbuilder 10.10
  (a) pure covalent
  (b) ionic
  (c) polar covalent

Skillbuilder 10.11 CH_4 is nonpolar

▶ANSWERS TO CONCEPTUAL CHECKPOINTS

10.1 (a) C and Si both have four dots in their Lewis structure because they are both in the same column in the periodic table.

10.2 (b) Aluminum must lose its three valence electrons to get an octet. Sulfur must gain two electrons to get an octet. Therefore, two Al atoms are required for every three S atoms.

10.3 (c) The Lewis structure of OH^- has eight electrons; six from oxygen, one from hydrogen, and one from the negative charge.

10.4 (b) Both NH_3 and H_3O^+ have one lone electron pair.

10.5 (d) If there are no lone pairs on the central atom, all of its valence electrons are involved in bonds, so the molecular geometry must be the same as the electron geometry.

CHAPTER 11

Gases

"The generality of men are so accustomed to judge of things by their senses that, because the air is invisible, they ascribe but little to it, and think it but one removed from nothing."

ROBERT BOYLE (1627–1691)

11.1 Extra-Long Straws

Like too many kids, I grew up preferring fast-food restaurants to home cooking. My favorite stunt at the burger restaurant was drinking my orange soda from an extra-long straw that I pieced together from several smaller straws. I would pinch the end of one straw and squeeze it into the end of another. By attaching several straws together in this way, I could put my orange soda on the floor and drink it while standing on my chair (for some reason, my parents did not appreciate my scientific curiosity). I sometimes planned ahead and brought duct tape to the restaurant to form extra-tight seals between adjacent straws. My brother and I would compete to see who could make the longest working straw. Since I was older, I usually won.

I often wondered how long the straw could be if I made perfect seals between the straws. Could I drink my orange soda from a cup on the ground while I sat in my tree house? Could I drink it from the top of a 10-story building? It seemed to me that I could; but I was wrong. Even if the extended straw had perfect seals and rigid walls, and even if I could create a perfect vacuum (the absence of all air), I could never suck my orange soda from a straw longer than about 10.3 m (34 ft). Why?

◀ When you drink from a straw, you remove some of the molecules from inside the straw. This creates a pressure difference between the inside of the straw and the outside of the straw that results in the liquid being pushed up the straw. The pushing is done by molecules in the atmosphere—primarily nitrogen and oxygen—as shown here.

Straws work because sucking creates a *pressure* difference between the inside of the straw and the outside. We define pressure and its units more thoroughly later; for now think of pressure as the force exerted per unit area by gas molecules as they collide with the surfaces around them (▶ Figure 11.1). Just as a ball exerts a force when it bounces against a wall, so a molecule exerts a force when it collides with a surface. The result of many of these collisions is pressure. The total amount of pressure exerted by a gas sample depends on several factors, including the concentration of gas molecules in the sample. On Earth at sea level, the gas molecules in our atmosphere exert an average pressure of $101,325 \text{ N/m}^2$ or, in English units, 14.7 lb/in^2.

Pressure is the result of collisions between gas molecules and the surfaces around them.

Gas molecules

Surface

▲ FIGURE 11.1 **Gas pressure** Pressure is the force exerted by gas molecules as they collide with the surfaces around them.

When you put a straw in a glass of soda, the pressure inside and outside the straw are the same, so the soda does not rise within the straw (▼ Figure 11.2a). When you suck on the straw, you remove some of the air molecules from inside the straw, lowering the number of collisions that occur and therefore lowering the pressure inside the straw (▼ Figure 11.2b). However, the pressure outside the straw remains the same. The result is a pressure differential—the pressure outside the straw becomes greater than the pressure inside of the straw. This greater external pressure pushes the liquid up the straw and into your mouth.

How high can this greater external pressure push the liquid up the straw? If you formed a perfect vacuum within the straw, the pressure outside of the straw at sea level would be enough to push the orange soda (which is mostly water) to a total height of about 10.3 m (▼ Figure 11.3). This is because a 10.3-m column of water exerts the same pressure—101,325 N/m^2 or 14.7 lb/in.^{2}—as do the gas molecules in our atmosphere. In other words, the orange soda would rise up the straw until the pressure exerted by its weight equaled the pressure exerted by the molecules in our atmosphere.

The pressure exerted by the atmosphere is equal to the pressure exerted by a 10.3 m column of water

Pressure lowered

P_{atm}

P_{atm} P_{atm} h

(a) (b)

▲ FIGURE 11.2 **Sipping soda** (a) When a straw is put into a glass of orange soda, the pressure inside and outside the straw is the same, so the liquid levels inside and outside the straw are the same. (b) When a person sucks on the straw, the pressure inside the straw is lowered. The greater pressure on the surface of the liquid outside the straw pushes the liquid up the straw.

Valve

Vacuum

Vacuum pump

Pressure exerted by 10.3 m column of orange soda: 14.7 pounds per square inch

Pressure exerted by our atmosphere: 14.7 pounds per square inch

10.3 m

▲ FIGURE 11.3 **Atmospheric pressure** Even if you formed a perfect vacuum, atmospheric pressure could only push orange soda to a total height of about 10 m. This is because a column of water (or soda) 10.3 m high exerts the same pressure (14.7 lb/in.^{2}) as the gas molecules in our atmosphere.

11.2 Kinetic Molecular Theory: A Model for Gases

In prior chapters we have seen the importance of models or theories in understanding nature. A simple model for understanding the behavior of gases is the **kinetic molecular theory**. This model predicts the correct behavior for most gases under many conditions. Like other models, the kinetic molecular theory is not

Kinetic molecular theory

1. Collection of particles in constant motion

2. No attractions or repulsions between particles; collisions like billiard ball collisions

3. A lot of space between the particles compared to the size of the particles themselves

4. The speed of the particles increases with increasing temperature

▲ **FIGURE 11.4 Simplified representation of an ideal gas** In reality, the spaces between the gas molecules would be larger in relation to the size of the molecules than shown here.

perfect and breaks down under certain conditions. In this book, however, we focus on conditions where it works well.

Kinetic molecular theory makes the following assumptions (▲ Figure 11.4):

1. A gas is a collection of particles (molecules or atoms) in constant, straight-line motion.

2. Gas particles do not attract or repel each other—they do not interact. The particles collide with each other and with the surfaces around them, but they bounce back from these collisions like idealized billiard balls.

3. There is a lot of space between gas particles compared with the size of the particles themselves.

4. The average kinetic energy—energy due to motion—of gas particles is proportional to the temperature of the gas in kelvins. This means that as the temperature increases, the particles move faster and therefore have more energy.

Kinetic molecular theory is consistent with, and indeed predicts, the properties of gases. As described in Section 3.3, gases:

- are compressible
- assume the shape and volume of their container
- have low densities in comparison with liquids and solids

Gases are compressible because the atoms or molecules that compose them have a lot of space between them. By applying external pressure to a gas sample, the atoms or molecules are forced closer together, compressing the gas. Liquids and solids, in contrast, are not compressible because the atoms or molecules composing them are already in close contact—they cannot be forced any closer together. The compressibility of a gas can be seen, for example, by pushing a piston into a cylinder containing a gas. The piston goes down (◄ Figure 11.5) in response to the external pressure. If the cylinder were filled with a liquid or a solid, however, the piston would not move when pushed (◄ Figure 11.6).

Gases assume the shape and volume of their container because gaseous atoms or molecules are in constant, straight-line motion. In contrast to a solid or liquid, whose atoms or molecules interact with one another, the atoms or molecules in a

 Gases are compressible.

Gas

▲ **FIGURE 11.5 Compressibility of gases** Gases are compressible because there is so much empty space between gas particles.

Liquids are not compressible.

Liquid

▲ **FIGURE 11.6 Incompressibility of liquids** Liquids are not compressible because there is so little space between the liquid particles.

▲ **FIGURE 11.7 A gas assumes the shape of its container** Since the attractions between molecules in a gas are negligible, and since the particles are in constant motion, a gas expands to fill the volume of its container.

gas do not interact with one another (or more precisely, their interactions are negligible). They simply move in straight lines, colliding with each other and with the walls of their container. As a result, they fill the entire container, collectively assuming its shape (◄ Figure 11.7).

Gases have a low density in comparison with solids and liquids because there is so much empty space between the atoms or molecules in a gas. For example, if the water in a 350-mL (12-oz) can of soda were converted to steam (gaseous water), the steam would occupy a volume of 595 L (the equivalent of 1700 soda cans).

Convert liquid to gas

(1 can of soda)   (1700 cans of soda)

▲ If all of the water in a 12-oz (350-mL) can of orange soda were converted to gaseous steam (at 1 atm pressure and 100 °C), the steam would occupy a volume equal to 1700 soda cans.

11.3 Pressure: The Result of Constant Molecular Collisions

A prediction of kinetic molecular theory—which we already encountered in explaining how straws work—is the very existence of pressure. **Pressure** is the result of the constant collisions between the atoms or molecules in a gas and the surfaces around them. Because of pressure, we can drink from straws, inflate basketballs, and move air into and out of our lungs. Variation in pressure in Earth's atmosphere creates wind, and changes in pressure help predict weather. Pressure is all around us and even inside us. The pressure exerted by a gas sample is defined as the force per unit area that results from the collisions of gas particles with surrounding surfaces.

$$\text{Pressure} = \frac{\text{Force}}{\text{Area}}$$

The pressure exerted by a gas depends on several factors, including the number of gas particles in a given volume (◄ Figure 11.8). The fewer the gas particles, the lower the pressure. Pressure decreases, for example, with increasing altitude. As we climb a mountain or ascend in an airplane, there are fewer molecules per unit volume in air and the pressure consequently drops. For this reason, most airplane cabins are artificially pressurized (see the *Everyday Chemistry* box on page 365.)

You may feel the effect of a drop in pressure as a pain in your ears. This pain is caused by air-containing cavities within your ear (▶ Figure 11.9). When you climb a mountain, for example, the external pressure (that pressure that surrounds you) drops while the pressure within your ear cavities (the internal pressure) remains the same. This creates an imbalance—the lower external pressure causes your eardrum to bulge outward, causing pain. With time and a yawn or two, the excess air within your ears' cavities escapes, equalizing the internal and external pressure and relieving the pain.

Lower pressure

Higher pressure

◄ **FIGURE 11.8 Pressure** Since pressure is a result of collisions between gas particles and the surfaces around them, the amount of pressure increases when the number of particles in a given volume increases.

▶ FIGURE 11.9 **Pressure imbalance** The pain you feel in your ears upon climbing a mountain or ascending in an airplane is caused by an imbalance of pressure between the cavities inside your ear and the outside air.

Higher pressure

Lower pressure

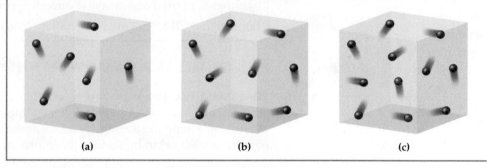

CONCEPTUAL CHECKPOINT 11.1

Which sample of an ideal gas will have the *lowest* pressure? Assume that all of the particles are identical and that the three samples are the same temperature.

(a)   (b)   (c)

PRESSURE UNITS

The simplest unit of pressure is the **atmosphere (atm)**, the average pressure at sea level. The SI unit of pressure is the **pascal (Pa)**, defined as 1 newton (N) per square meter.

$$1 \, Pa = 1 \, N/m^2$$

The pascal is a much smaller unit of pressure; 1 atm is equal to 101,325 Pa.

$$1 \, atm = 101,325 \, Pa$$

The pressure in a fully inflated mountain bike tire is about 6 atm, and the pressure on top of Mount Everest is about 0.31 atm.

A third unit of pressure, the **millimeter of mercury (mm Hg)**, originates from how pressure is measured with a barometer (◀ Figure 11.10). A barometer is an evacuated glass tube whose tip is submerged in a pool of mercury. As described in Section 11.1, liquid is pushed up an evacuated tube by the atmospheric pressure on the liquid's surface. We learned that water is pushed up to a height of 10.3 m by the average pressure at sea level. Mercury, however, with its higher density, is

Vacuum

Glass tube

760 mm
(29.92 in.)

Atmospheric
pressure

Mercury

▲ FIGURE 11.10   **The mercury barometer** Average atmospheric pressure at sea level pushes a column of mercury to a height of 760 mm (29.92 in.).   **Question:** What happens to the height of the mercury column if the external pressure decreases? increases?

Since mercury is 13.5 times as dense as water, it is pushed up 1/13.5 times as high as water by atmospheric pressure.

pushed up to a height of only 0.760 m, or 760 mm, by the average pressure at sea level. This shorter length—0.760 m instead of 10.3 m—makes a column of mercury a convenient way to measure pressure.

In a barometer, the mercury column rises or falls with changes in atmospheric pressure. If the pressure increases, the level of mercury within the column rises. If the pressure decreases, the level of mercury within the column falls. Since 1 atm of pressure pushes a column of mercury to a height of 760 mm, 1 atm and 760 mm Hg are equal.

$$1 \text{ atm} = 760 \text{ mm Hg}$$

A millimeter of mercury is also called a **torr** after Italian physicist Evangelista Torricelli (1608–1647), who invented the barometer.

$$1 \text{ mm Hg} = 1 \text{ torr}$$

Inches of mercury is still a widely used unit in weather reports. You have probably heard a weather forecaster say, "The barometer is 30.07 and rising," meaning that the atmospheric pressure is currently 30.07 in. Hg.

Other common units of pressure include inches of mercury (in. Hg) and **pounds per square inch (psi)**.

$$1 \text{ atm} = 14.7 \text{ psi} \qquad 1 \text{ atm} = 29.92 \text{ in. Hg}$$

All these units are summarized in Table 11.1.

TABLE 11.1 Common Units of Pressure

Unit	Average Air Pressure at Sea Level
pascal (Pa)	101,325 Pa
atmosphere (atm)	1 atm
millimeter of mercury (mm Hg)	760 mm Hg (exact)
torr (torr)	760 torr (exact)
pounds per square inch (psi)	14.7 psi
inches of mercury (in. Hg)	29.92 atm

See Section 2.6.

✔ CONCEPTUAL CHECKPOINT 11.2

A liquid that is about twice as dense as water is used in a barometer. With this barometer, normal atmospheric pressure would be about:

(a) 0.38 m **(b)** 1.52 m **(c)** 5.15 m **(d)** 20.6 m

PRESSURE UNIT CONVERSION

We convert one pressure unit to another in the same way that we converted between other units in Chapter 2. For example, suppose we want to convert 0.311 atm (the approximate average pressure at the top of Mount Everest) to millimeters of mercury. We begin by sorting the information in the problem statement.

GIVEN: 0.311 atm
FIND: mm Hg

SOLUTION MAP
We then strategize by building a solution map that shows how to convert from atm to mm Hg.

$$\frac{760 \text{ mm Hg}}{1 \text{ atm}}$$

RELATIONSHIPS USED

$$1 \text{ atm} = 760 \text{ mm Hg (from Table 11.1)}$$

SOLUTION
The solution begins with the given value (0.311 atm) and converts it to mm Hg.

$$0.311 \text{ atm} \times \frac{760 \text{ mm Hg}}{1 \text{ atm}} = 236 \text{ mm Hg}$$

EVERYDAY CHEMISTRY

Airplane Cabin Pressurization

Most commercial airplanes fly at elevations between 25,000 and 40,000 ft. At these elevations, atmospheric pressure is below 0.50 atm, much less than the normal atmospheric pressure to which our bodies are accustomed. The physiological effects of these lowered pressures—and the correspondingly lowered oxygen levels (see Section 11.9)—include dizziness, headache, shortness of breath, and even unconsciousness. Consequently, commercial airplanes pressurize the air in their cabins. If, for some reason, an airplane cabin should lose its pressurization, passengers are directed to breathe oxygen through an oxygen mask.

Cabin air pressurization is accomplished as part of the cabin's overall air circulation system. As air flows into the plane's jet engines, the large turbines at the front of the engines compress it. Most of this compressed (or pressurized) air exits out the back of the engines, creating the thrust that drives the plane forward. However, some of the pressurized air is directed into the cabin, where it is cooled and mixed with existing cabin air. This air is then circulated through the cabin through the overhead vents. The air leaves the cabin through ducts that direct it into the lower portion of the airplane. About half of this exiting air is mixed with incoming, pressurized air to circulate again. The other half is vented out of the plane through an outflow valve. This valve is adjusted to maintain the desired cabin pressure. Federal regulations require that cabin pressure in commercial airliners be greater than the equivalent of outside air pressure at 8000 ft.

CAN YOU ANSWER THIS? *Atmospheric pressure at elevations of 8000 ft average about 0.72 atm. Convert this pressure to millimeters of mercury, inches of mercury, and pounds per square inch. Would a cabin pressurized at 500 mm Hg meet federal standards?*

◄ Commercial airplane cabins must be pressurized to a pressure greater than the equivalent atmospheric pressure at an elevation of 8000 ft.

EXAMPLE 11.1 Converting between Pressure Units

A high-performance road bicycle tire is inflated to a total pressure of 125 psi. What is this pressure in millimeters of mercury?

SORT You are given a pressure in psi and asked to convert it to mm Hg.	**GIVEN:** 125 psi **FIND:** mm Hg
STRATEGIZE Begin the solution map with the given units of psi. Use the conversion factors to convert first to atm and then to mm Hg.	**SOLUTION MAP** psi → atm → mm Hg $\dfrac{1\text{ atm}}{14.7\text{ psi}}$ $\dfrac{760\text{ mm Hg}}{1\text{ atm}}$ **RELATIONSHIPS USED** 1 atm = 14.7 psi (Table 11.1) 760 mm Hg = 1 atm (Table 11.1)
SOLVE Follow the solution map to solve the problem.	**SOLUTION** $$125\text{ psi} \times \frac{1\text{ atm}}{14.7\text{ psi}} \times \frac{760\text{ mm Hg}}{1\text{ atm}} = 6.46 \times 10^3 \text{ mm Hg}$$
CHECK Check your answer. Are the units correct? Does the answer make physical sense?	The answer has the correct units, mm Hg. The answer is reasonable because the mm Hg is a smaller unit than psi; therefore the value of the pressure in mm Hg should be greater than the value of the same pressure in psi.

►SKILLBUILDER 11.1 | **Converting between Pressure Units**
Convert a pressure of 173 in. Hg into pounds per square inch.

►SKILLBUILDER PLUS
Convert a pressure of 23.8 in. Hg into kilopascals.

►FOR MORE PRACTICE   Example 11.13; Problems 23, 24, 25, 26, 29, 30, 31, 32.

11.4   Boyle's Law: Pressure and Volume

The pressure of a gas sample depends, in part, on its volume. If the temperature and the amount of gas are constant, the pressure of a gas sample *increases* for a *decrease* in volume and *decreases* for an *increase* in volume. A simple hand pump, for example, works on this principle. A hand pump is basically a cylinder equipped with a moveable piston (▼ Figure 11.11). The volume in the cylinder increases when you pull the handle up (the upstroke) and decreases when you push the handle down (the downstroke). On the upstroke, the *increasing* volume causes a *decrease* in the internal pressure (the pressure within the pump's cylinder). This, in turn, draws air into the pump's cylinder through a one-way valve. On the downstroke, the *decreasing* volume causes an *increase* in the internal pressure. This increase forces the air out of the pump, through a different one-way valve, and into the tire or whatever else is being inflated.

The relationships between gas properties—such as the relationship between pressure and volume—are described by gas laws. These laws show how a change in one of these properties affects one or more of the others. The relationship between volume and pressure was discovered by Robert Boyle (1627–1691) and is called **Boyle's law**.

| Boyle's law assumes constant temperature and a constant number of gas particles.

Boyle's law: The volume of a gas and its pressure are inversely proportional.

$$V \propto \frac{1}{P} \quad \propto \text{ means "proportional to"}$$

If two quantities are inversely proportional, then increasing one decreases the other (► Figure 11.12). As we saw for the hand pump, when the volume of a gas sample is decreased, its pressure increases and vice versa. Kinetic molecular theory explains the observed change in pressure. If the volume of a gas sample is decreased, the same number of gas particles is crowded into a smaller volume, causing more collisions with the walls of the container and therefore increasing the pressure (► Figure 11.13).

Scuba divers learn about Boyle's law during certification courses because it explains why ascending too quickly toward the surface is dangerous. For every 10 m of depth that a diver descends in water, he experiences an additional 1 atm of pressure due to the weight of the water above him (► Figure 11.14). The pressure regulator used in scuba diving delivers air at a pressure that matches the external pressure; otherwise the diver

One-way valve

Upstroke
Volume increases
Pressure decreases

Downstroke
Volume decreases
Pressure increases

◄ FIGURE 11.11   **Operation of a hand pump**

▶ **FIGURE 11.12   Volume versus pressure**   (a) A J-tube, such as the one shown here, can be used to measure the volume of a gas at different pressures. Adding mercury to the J-tube causes the pressure on the gas sample to increase and its volume to decrease. **(b)** A plot of the volume of a gas as a function of pressure.

As pressure increases, volume decreases

Added mercury increases pressure on gas sample

As pressure increases, volume decreases

Gas Volume

Gas Volume

(a)

(b)

▶ **FIGURE 11.13   Volume versus pressure: a molecular view**   As the volume of a sample of gas is decreased, the number of collisions between the gas molecules and each square meter of the container increases. This raises the pressure exerted by the gas.

$P = 1$ atm

$V = 1$ L

$P = 2$ atm

$V = 0.5$ L

Depth = 0 m
$P = 1$ atm

Depth = 20 m
$P = 3$ atm

could not inhale the air (see the *Everyday Chemistry* box on page 371). For example, when a diver is at 20 m of depth, the regulator delivers air at a pressure of 3 atm to match the 3 atm of pressure around the diver—1 atm due to normal atmospheric pressure and 2 additional atmospheres due to the weight of the water at 20 m (▶ Figure 11.15).

Suppose that a diver inhaled a lungful of 3-atm air and swam quickly to the surface (where the pressure drops to 1 atm) while holding his breath. What would happen to the volume of air in his lungs? Since the pressure decreases by a factor of 3, the volume of the air in his lungs would increase by a factor of 3, severely damaging his lungs and possibly killing him. Of course, the volume increase in the diver's lungs would be so great that the diver would not be able to hold his breath all the way to the surface—the air would force itself out of his mouth. So the most important rule in diving is *never hold your breath*.

◀ **FIGURE 11.14   Pressure at depth**   For every 10 m of depth, a diver experiences an additional 1 atm of pressure due to the weight of the water surrounding him. At 20 m, the diver experiences a total pressure of 3 atm (1 atm from atmospheric pressure plus an additional 2 atm from the weight of the water).

► **FIGURE 11.15 The dangers of decompression**
(a) A diver at 20 m experiences an external pressure of 3 atm and breathes air pressurized at 3 atm. **(b)** If the diver shoots toward the surface with lungs full of 3 atm air, his lungs will expand as the external pressure drops to 1 atm.

(a)   (b)

Divers must ascend slowly and breathe continuously, allowing the regulator to bring the air pressure in their lungs back to 1 atm by the time they reach the surface.

Boyle's law can be used to calculate the volume of a gas following a pressure change or the pressure of a gas following a volume change *as long as the temperature and the amount of gas remain constant*. For these calculations, we write Boyle's law in a slightly different way.

$$\text{Since } V \; \alpha \; \frac{1}{P}, \text{ then } V = \frac{\text{Constant}}{P}$$

If we multiply both sides by P, we get:

$$PV = \text{Constant}$$

This relationship is true because if the pressure increases, the volume decreases, but the product $P \times V$ is always equal to the same constant. For two different sets of conditions, we can say that

$$P_1V_1 = \text{Constant} = P_2V_2, \text{ or}$$

$$P_1V_1 = P_2V_2$$

where P_1 and V_1 are the initial pressure and volume of the gas, and P_2 and V_2 are the final volume and pressure. For example, suppose we want to calculate the pressure of a gas that was initially at 765 mm Hg and 1.78 L and later compressed to 1.25 L. We first sort the information in the problem.

Based on Boyle's law, and before doing any calculations, do you expect P₂ to be greater than or less than P₁?

GIVEN: $P_1 = 765 \text{ mm Hg}$

$V_1 = 1.78 \text{ L}$

$V_2 = 1.25 \text{ L}$

FIND: P_2

SOLUTION MAP
We then draw a solution map showing how the equation takes us from the given quantities (what we have) to the find quantity (what we want).

$$P_1V_1 = P_2V_2 \qquad \boxed{\text{Equation relating them}}$$

RELATIONSHIPS USED

$$P_1V_1 = P_2V_2 \quad \text{(Boyle's law, presented in this section)}$$

SOLUTION
We then solve the equation for the quantity we are trying to find (P_2).

$$P_1V_1 = P_2V_2$$

$$P_2 = \frac{P_1V_1}{V_2}$$

Lastly, we substitute the numerical values into the equation and calculate the answer.

$$P_2 = \frac{P_1V_1}{V_2} = \frac{(765 \text{ mm Hg})(1.78 \text{ L})}{1.25 \text{ L}}$$

$$= 1.09 \times 10^3 \text{ mm Hg}$$

EXAMPLE 11.2   Boyle's Law

A cylinder equipped with a moveable piston has an applied pressure of 4.0 atm and a volume of 6.0 L. What is the volume of the cylinder if the applied pressure is decreased to 1.0 atm?

SORT You are given an initial pressure, an initial volume, and a final pressure. You are asked to find the final volume.	**GIVEN:** $P_1 = 4.0$ atm $V_1 = 6.0$ L $P_2 = 1.0$ atm **FIND:** V_2
STRATEGIZE Draw a solution map beginning with the given quantities. Boyle's law shows the relationship necessary to get to the find quantity.	**SOLUTION MAP** $\boxed{P_1, V_1, P_2} \longrightarrow \boxed{V_2}$ $P_1V_1 = P_2V_2$ **RELATIONSHIPS USED** $P_1V_1 = P_2V_2$ (Boyle's law, presented in this section)
SOLVE Solve the equation for the quantity you are trying to find (V_2), and then substitute the numerical quantities into the equation to compute the answer.	**SOLUTION** $P_1V_1 = P_2V_2$ $V_2 = \dfrac{V_1P_1}{P_2}$ $= \dfrac{(6.0\text{L})(4.0 \text{ atm})}{1.0 \text{ atm}}$ $= 24$ L
CHECK Check your answer. Are the units correct? Does the answer make physical sense?	The answer has units of volume (L) as expected. The answer is reasonable because we expect the volume to increase as the pressure decreases.

▶SKILLBUILDER 11.2 | Boyle's Law

A snorkeler takes a syringe filled with 16 mL of air from the surface, where the pressure is 1.0 atm, to an unknown depth. The volume of the air in the syringe at this depth is 7.5 mL. What is the pressure at this depth? If the pressure increases by an additional 1 atm for every 10 m of depth, how deep is the snorkeler?

▶FOR MORE PRACTICE   Example 11.14; Problems 33, 34, 35, 36.

✓ CONCEPTUAL CHECKPOINT 11.3

A flask contains a gas sample at pressure x. If the volume of the container triples at constant temperature and constant amount of gas, the pressure becomes:

(a) $3x$

(b) $\frac{1}{3}x$

(c) $9x$

11.5   Charles's Law: Volume and Temperature

Recall from Section 2.9 that Density = Mass/Volume. If the volume increases and the mass remains constant, the density must decrease.

Have you ever noticed that hot air rises? You may have walked upstairs in your house and noticed it getting warmer. Or you may have witnessed a hot-air balloon take flight. The air that fills a hot-air balloon is warmed with a burner, which then causes the balloon to rise in the cooler air around it. Why does hot air rise? Hot air rises because the volume of a gas sample at constant pressure increases with increasing temperature. As long as the amount of gas (and therefore its mass) remains constant, warming it decreases its density because density is mass divided by volume. A lower-density gas floats in a higher-density gas just as wood floats in water.

Suppose you keep the pressure of a gas sample constant and measure its volume at a number of different temperatures. The results of a number of such measurements are shown in ▼ Figure 11.16. The plot reveals the relationship

▲ Heating the air in a balloon makes it expand (Charles's law). As the volume occupied by the hot air increases, its density decreases, allowing the balloon to float in the cooler, denser air that surrounds it.

The extrapolated line could not be measured experimentally because all gases would condense into liquids before −273 °C is reached.

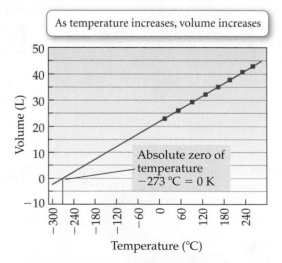

As temperature increases, volume increases

Absolute zero of temperature
−273 °C = 0 K

▲ FIGURE 11.16   Volume versus temperature
The volume of a gas increases linearly with increasing temperature.   Question: How does this graph demonstrate that −273 °C is the coldest possible temperature?

EVERYDAY CHEMISTRY

Extra-long Snorkels

Several episodes of *The Flintstones* featured Fred Flintstone and Barney Rubble snorkeling. Their snorkels, however, were not the modern kind, but long reeds that stretched from the surface of the water down to many meters of depth. Fred and Barney swam around in deep water while breathing air provided to them by these extra-long snorkels. Would this work? Why do people bother with scuba diving equipment if they could simply use 10-m snorkels the way that Fred and Barney did?

When we breathe, we expand the volume of our lungs, lowering the pressure within them (Boyle's law). Air from outside our lungs then flows into them. Extra-long snorkels, such as those used by Fred and Barney, do not work because of the pressure caused by water at depth. A diver at 10 m experiences a pressure of 2 atm that compresses the air in his lungs to a pressure of 2 atm. If the diver had a snorkel that went to the surface—where the air pressure is 1 atm— air would flow out of his lungs, not into them. It would be impossible to breathe.

CAN YOU ANSWER THIS? *Suppose a diver takes a balloon with a volume of 2.5 L from the surface, where the pressure is 1.0 atm, to a depth of 20 m, where the pressure is 3.0 atm. What would happen to the volume of the balloon? What if the end of the balloon was on a long tube that went to the surface and was attached to another balloon, as shown in the drawing below? Which way would air flow as the diver descends?*

◀ Fred and Barney used reeds to breathe air from the surface, even when they were at depth. This would not work because the pressure at depth would push air out of their lungs, preventing them from breathing.

▶ If one end of a long tube with balloons tied on both ends were submerged in water, in which direction would air flow?

Section 3.10 summarizes the three different temperature scales.

between volume and temperature: The volume of a gas increases with increasing temperature. Note also that temperature and volume are *linearly related*. If two variables are linearly related, then plotting one against the other produces a straight line.

We can predict an important property of matter by extending the line on our plot backward from the lowest measured point—a process called *extrapolation*. Our extrapolated line shows that the gas should have a zero volume at −273 °C. Recall from Chapter 3 that −273 °C corresponds to 0 K, the coldest possible temperature. Our extrapolated line shows that below −273 °C, our gas would have a negative volume, which is physically impossible. For this reason, we refer to 0 K as **absolute zero**—colder temperatures do not exist.

The first person to carefully quantify the relationship between the volume of a gas and its temperature was J. A. C. Charles (1746–1823), a French mathematician and physicist. Charles was interested in gases and was among the first people to ascend in a hydrogen-filled balloon. The law he formulated is called **Charles's law**.

Charles's law assumes constant pressure and a constant amount of gas.

Charles's law: The volume (V) of a gas and its Kelvin temperature (T) are directly proportional.

$$V \propto T$$

If two variables are directly proportional, then increasing one by some factor increases the other by the same factor. For example, when the temperature of a gas sample (in kelvins) is doubled, its volume doubles; when the temperature is tripled, its volume triples; and so on. The observed relationship between the temperature and volume of a gas follows from kinetic molecular theory. If the temperature of a gas sample is increased, the gas particles move faster, and if the pressure is to remain constant, the volume must increase (▼ Figure 11.17).

You can experience Charles's law directly by holding a partially inflated balloon over a warm toaster. As the air in the balloon warms, you can feel the balloon expanding. Alternatively, you can put an inflated balloon in the freezer or take it outside on a very cold day (below freezing) and see that it becomes smaller as it cools.

Charles's law can be used to calculate the volume of a gas following a temperature change or the temperature of a gas following a volume change *as long as the pressure and the amount of gas are constant*. For these calculations, we express Charles's law in a different way as follows:

Since $V \propto T$, then $V = \text{Constant} \times T$

If we divide both sides by T, we get:

$$V/T = \text{Constant}$$

If the temperature increases, the volume increases in direct proportion so that the quotient, V/T, is always equal to the same constant. So, for two different measurements, we can say that

$$V_1/T_1 = \text{Constant} = V_2/T_2, \text{ or}$$

$$\frac{V_1}{T_1} = \frac{V_2}{T_2}$$

▲ If you hold a partially inflated balloon over a warm toaster, the balloon will expand as the air within the balloon warms.

As temperature increases, the volume of the balloon increases

Ice water   Boiling water

▲ **FIGURE 11.17 Volume versus temperature: a molecular view** If a balloon is moved from an ice-water bath into a boiling-water bath, the gas molecules inside it move faster due to the increased temperature. If the external pressure remains constant, the molecules will expand the balloon and collectively occupy a larger volume.

where V_1 and T_1 are the initial volume and temperature of the gas and V_2 and T_2 are the final volume and temperature. *All temperatures must be expressed in kelvins.*

For example, suppose we have a 2.37-L sample of a gas at 298 K that is then heated to 354 K with no change in pressure. To determine the final volume of the gas, we begin by sorting the information in the problem statement.

GIVEN: $T_1 = 298$ K

$V_1 = 2.37$ L

$T_2 = 354$ K

FIND: V_2

SOLUTION MAP

We then strategize by building a solution map that shows how the equation takes us from the given quantities to the unknown quantity.

> Based on Charles's law, and before doing any calculations, do you expect V_2 to be greater than or less than V_1?

RELATIONSHIPS USED

$$\frac{V_1}{T_1} = \frac{V_2}{T_2} \text{ (Charles's law, presented in this section)}$$

SOLUTION

We then solve the equation for the quantity we are trying to find (V_2).

$$\frac{V_1}{T_1} = \frac{V_2}{T_2}$$

$$V_2 = \frac{V_1}{T_1} T_2$$

Lastly, we substitute the numerical values into the equation and calculate the answer.

$$V_2 = \frac{V_1}{T_1} T_2$$

$$= \frac{2.37 \text{ L}}{298 \text{ K}} 354 \text{ K}$$

$$= 2.82 \text{ L}$$

EXAMPLE 11.3 Charles's Law

A sample of gas has a volume of 2.80 L at an unknown temperature. When the sample is submerged in ice water at $t = 0\,°C$, its volume decreases to 2.57 L. What was its initial temperature (in kelvins and in Celsius)? Assume a constant pressure. (To distinguish between the two temperature scales, use t for temperature in °C and T for temperature in K.)

SORT You are given an initial volume, a final volume, and a final temperature. You are asked to find the intitial temperature in both kelvins (T_1) and degrees Celsius (t_1).	**GIVEN:** $$V_1 = 2.80 \text{ L}$$ $$V_2 = 2.57$$ $$t_2 = 0\,°C$$ **FIND:** T_1 and t_1
STRATEGIZE Draw a solution map beginning with the given quantities. Charles's law shows the relationship necessary to get to the find quantity.	**SOLUTION MAP** $$\frac{V_1}{T_1} = \frac{V_2}{T_2}$$ **RELATIONSHIPS USED** $$\frac{V_1}{T_1} = \frac{V_2}{T_2} \text{ (Charles's law, presented in this section)}$$
SOLVE Solve the equation for the quantity you are trying to find (T_1). Before you substitute in the numerical values, you must convert the temperature to kelvins. *Remember, gas law problems must always be worked using Kelvin temperatures.* Once you have converted the temperature to kelvins, substitute into the equation to find T_1. Convert the temperature to degrees Celsius to find t_1.	**SOLUTION** $$\frac{V_1}{T_1} = \frac{V_2}{T_2}$$ $$T_1 = \frac{V_1}{V_2} T_2$$ $$T_2 = 0 + 273 = 273 \text{ K}$$ $$T_1 = \frac{V_1}{V_2} T_2$$ $$= \frac{2.80 \cancel{\text{ L}}}{2.57 \cancel{\text{ L}}} 273 \text{ K}$$ $$= 297 \text{ K}$$ $$t_1 = 297 - 273 = 24\,°C$$
CHECK Check your answer. Are the units correct? Does the answer make physical sense?	The answers have the correct units, K and °C. The answer is reasonable because the initial volume was larger than the final volume; therefore the initial temperature must be higher than the final temperature.

▶**SKILLBUILDER 11.3 | Charles's Law**

A gas in a cylinder with a moveable piston has an initial volume of 88.2 mL and is heated from 35 °C to 155 °C. What is the final volume of the gas in milliliters?

▶**FOR MORE PRACTICE** Problems 39, 40, 41, 42.

11.6   The Combined Gas Law: Pressure, Volume, and Temperature

Boyle's law shows how P and V are related at constant temperature, and Charles's law shows how V and T are related at constant pressure. But what if two of these variables change at once? For example, what happens to the volume of a gas if both its pressure and its temperature are changed?

Since volume is inversely proportional to pressure ($V \propto 1/P$) and directly proportional to temperature ($V \propto T$), we can write:

$$V \propto \frac{T}{P} \quad \text{or} \quad \frac{PV}{T} = \text{Constant}$$

For a sample of gases under two different sets of conditions we use the **combined gas law**:

> The combined gas law encompasses both Boyle's law and Charles's law and can be used in place of them. If one physical property (P, V, or T) is constant, it will cancel out of your calculations when you use the combined gas law.

The combined gas law: $\dfrac{P_1 V_1}{T_1} = \dfrac{P_2 V_2}{T_2}$

The combined gas law applies only when the *amount* of gas is constant. The temperature (as with Charles's law) must be expressed in kelvins.

Suppose you carry a cylinder with a moveable piston that has an initial volume of 3.65 L up a mountain. The pressure at the bottom of the mountain is 755 mm Hg, and the temperature is 302 K. The pressure at the top of the mountain is 687 mm Hg, and the temperature is 291 K. What is the volume of the cylinder at the top of the mountain? We begin by sorting the information in the problem statement.

GIVEN:

$$P_1 = 755 \text{ mm Hg} \quad T_2 = 291 \text{ K}$$
$$V_1 = 3.65 \text{ L} \quad\quad\quad P_2 = 687 \text{ mm Hg}$$
$$T_1 = 302 \text{ K}$$

FIND: V_2

SOLUTION MAP

We strategize by building a solution map that shows how the combined gas law equation takes us from the given quantities to the find quantity.

RELATIONSHIPS USED

$$\frac{P_1 V_1}{T_1} = \frac{P_2 V_2}{T_2} \quad \text{(Combined gas law, presented in this section)}$$

SOLUTION

We then solve the equation for the quantity we are trying to find (V_2).

$$\frac{P_1 V_1}{T_1} = \frac{P_2 V_2}{T_2}$$

$$V_2 = \frac{P_1 V_1 T_2}{T_1 P_2}$$

Lastly, we substitute in the appropriate values and calculate the answer.

$$V_2 = \frac{P_1 V_1 T_2}{T_1 P_2}$$

$$= \frac{755 \text{ mm Hg} \times 3.65 \text{ L} \times 291 \text{ K}}{302 \text{ K} \times 687 \text{ mm Hg}}$$

$$= 3.87 \text{ L}$$

EXAMPLE 11.4 The Combined Gas Law

A sample of gas has an initial volume of 158 mL at a pressure of 735 mm Hg and a temperature of 34 °C. If the gas is compressed to a volume of 108 mL and heated to a temperature of 85 °C, what is its final pressure in millimeters of mercury?

SORT You are given an initial pressure, temperature, and volume as well as a final temperature and volume. You are asked to find the final pressure.	**GIVEN:** $P_1 = 735$ mm Hg $t_1 = 34\ °C$ $t_2 = 85\ °C$ $V_1 = 158$ mL $V_2 = 108$ mL **FIND:** P_2
STRATEGIZE Draw a solution map beginning with the given quantities. The combined gas law shows the relationship necessary to get to the find quantity.	**SOLUTION MAP** $\boxed{P_1, T_1, V_1, T_2, V_2} \longrightarrow \boxed{P_2}$ $$\frac{P_1 V_1}{T_1} = \frac{P_2 V_2}{T_2}$$ **RELATIONSHIPS USED** $$\frac{P_1 V_1}{T_1} = \frac{P_2 V_2}{T_2} \quad \text{(Combined gas law, presented in this section)}$$
SOLVE Solve the equation for the quantity you are trying to find (P_2). Before you substitute in the numerical values, you must convert the temperatures to kelvins. Once you have converted the temperature to kelvins, substitute into the equation to find P_2.	**SOLUTION** $$\frac{P_1 V_1}{T_1} = \frac{P_2 V_2}{T_2}$$ $$P_2 = \frac{P_1 V_1 T_2}{T_1 V_2}$$ $$T_1 = 34 + 273 = 307 \text{ K}$$ $$T_2 = 85 + 273 = 358 \text{ K}$$ $$P_2 = \frac{735 \text{ mm Hg} \times 158 \text{ mL} \times 358 \text{ K}}{307 \text{ K} \times 108 \text{ mL}}$$ $$= 1.25 \times 10^3 \text{ mm Hg}$$
CHECK Check your answer. Are the units correct? Does the answer make physical sense?	The answer has the correct units, mm Hg. The answer is reasonable because the decrease in volume and the increase in temperature should result in a pressure that is higher than the initial pressure.

▶ **SKILLBUILDER 11.4 | The Combined Gas Law**

A balloon has a volume of 3.7 L at a pressure of 1.1 atm and a temperature of 30 °C. If the balloon is submerged in water to a depth where the pressure is 4.7 atm and the temperature is 15 °C, what will its volume be (assume that any changes in pressure caused by the skin of the balloon are negligible)?

▶ **FOR MORE PRACTICE** Example 11.15; Problems 51, 52, 53, 54, 55, 56.

✓ **CONCEPTUAL CHECKPOINT 11.4**

A volume of gas is confined to a cylinder with a freely moveable piston at one end. If you apply enough heat to double the Kelvin temperature of the gas,

(a) the pressure and volume will both double.

(b) the pressure will double but the volume will remain the same.

(c) the volume will double but the pressure will remain the same.

(d) the volume will double but the pressure will be halved.

11.7 Avogadro's Law: Volume and Moles

So far, we have learned how V, P, and T are interrelated, but we have considered only a constant amount of a gas. What happens when the amount of gas changes? If we make several measurements of the volume of a gas sample (at constant temperature and pressure) while varying the number of moles in the sample, we get results similar to those shown in ▼ Figure 11.18. We can see that the relationship between volume and number of moles is linear. An extrapolation to zero moles shows a zero volume, as we might expect. This relationship was first stated formally by Amedeo Avogadro (1776–1856) and is called **Avogadro's law**.

> | Avogadro's law assumes constant temperature and pressure.

Avogadro's law: The volume of a gas and the amount of the gas in moles (n) are directly proportional.

$$V \propto n$$

When the amount of gas in a sample is increased, its volume increases in direct proportion, which is yet another prediction of kinetic molecular theory. If the number of gas particles increases at constant pressure and temperature, the particles must occupy more volume.

You experience Avogadro's law when you inflate a balloon, for example. With each exhaled breath, you add more gas particles to the inside of the balloon, increasing its volume (▼ Figure 11.19). Avogadro's law can be used to calculate the volume of a gas following a change in the amount of the gas *as long as the pressure and temperature of the gas are constant*. For these calculations, Avogadro's law is expressed as

> | Since $V \propto n$, then V/n = Constant. If the number of moles increases, then the volume increases in direct proportion so that the quotient, V/n, is always equal to the same constant. Thus, for two different measurements, we can say that
> $$\frac{V_1}{n_1} = \text{Constant} = \frac{V_2}{n_2} \text{ or } \frac{V_1}{n_1} = \frac{V_2}{n_2}.$$

$$\frac{V_1}{n_1} = \frac{V_2}{n_2}$$

where V_1 and n_1 are the initial volume and number of moles of the gas and V_2 and n_2 are the final volume and number of moles. In calculations, Avogadro's law is used in a manner similar to the other gas laws, as shown in Example 11.5.

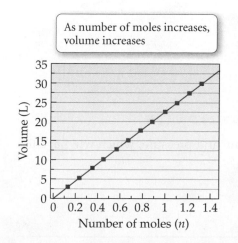

As number of moles increases, volume increases

◀ **FIGURE 11.18   Volume versus number of moles**   The volume of a gas sample increases linearly with the number of moles in the sample.

▶ **FIGURE 11.19   Blow-up**   As you exhale into a balloon, you add gas molecules to the inside of the balloon, increasing its volume.

EXAMPLE 11.5   Avogadro's Law

A 4.8-L sample of helium gas contains 0.22 mol of helium. How many additional moles of helium gas must be added to the sample to obtain a volume of 6.4 L? Assume constant temperature and pressure.

SORT	GIVEN:
You are given an initial volume, an initial number of moles, and a final volume. You are (essentially) asked to find the final number of moles.	$V_1 = 4.8$ L $n_1 = 0.22$ mol $V_2 = 6.4$ L FIND: n_2

STRATEGIZE	SOLUTION MAP
Draw a solution map beginning with the given quantities. Avogadro's law shows the relationship necessary to get to the find quantity.	$V_1, n_1, V_2 \longrightarrow n_2$ $$\dfrac{V_1}{n_1} = \dfrac{V_2}{n_2}$$ RELATIONSHIPS USED $\dfrac{V_1}{n_1} = \dfrac{V_2}{n_2}$ (Avogadro's law, presented in this section)

SOLVE	SOLUTION
Solve the equation for the quantity you are trying to find (n_2) and substitute the appropriate quantities to calculate n_2.	$$\dfrac{V_1}{n_1} = \dfrac{V_2}{n_2}$$ $$n_2 = \dfrac{V_2}{V_1} n_1$$ $$= \dfrac{6.4 \cancel{L}}{4.8 \cancel{L}} 0.22 \text{ mol}$$ $$= 0.29 \text{ mol}$$
Since the balloon already contains 0.22 mol, subtract this quantity from the final number of moles to determine how much must be added.	mol to add $= 0.29 - 0.22 = 0.07$ mol

CHECK	
Check your answer. Are the units correct? Does the answer make physical sense?	The answer has the correct units, moles. The answer is reasonable because the increase in the number of moles is proportional to the given increase in the volume.

▶**SKILLBUILDER 11.5 | Avogadro's Law**

A chemical reaction occurring in a cylinder equipped with a moveable piston produces 0.58 mol of a gaseous product. If the cylinder contained 0.11 mol of gas before the reaction and had an initial volume of 2.1 L, what was its volume after the reaction?

▶**FOR MORE PRACTICE**   Problems 45, 46, 47, 48.

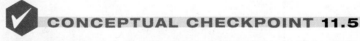

CONCEPTUAL CHECKPOINT 11.5

If each gas sample has the same temperature and pressure, which will have the greatest volume?

(a) 1 g O_2

(b) 1 g Ar

(c) 1 g H_2

11.8   The Ideal Gas Law: Pressure, Volume, Temperature, and Moles

The relationships covered so far can be combined into a single law that encompasses all of them. So far, we know that:

$V \propto \dfrac{1}{P}$ (Boyle's law)

$V \propto T$ (Charles's law)

$V \propto n$ (Avogadro's law)

Combining these three expressions, we get:

$$V \propto \dfrac{nT}{P}$$

The volume of a gas is directly proportional to the number of moles of gas and the temperature of the gas and is inversely proportional to the pressure of the gas. We can replace the proportional sign with an equal sign by adding R, a proportionality constant called the **ideal gas constant**.

$$V = \dfrac{RnT}{P}$$

Rearranging, we get the **ideal gas law**:

The ideal gas law: $PV = nRT$

The value of R, the ideal gas constant, is:

$$R = 0.0821 \ \dfrac{\text{L} \cdot \text{atm}}{\text{mol} \cdot \text{K}}$$

| R can also be expressed in other units, but its numerical value will be different.

The ideal gas law contains within it the simple gas laws that we have learned. For example, recall that Boyle's law states that $V \propto 1/P$ when the amount of gas (n) and the temperature of the gas (T) are kept constant. To derive Boyle's law, we can rearrange the ideal gas law as follows:

$$PV = nRT$$

First, divide both sides by P.

$$V = \dfrac{nRT}{P}$$

Then put the variables that are constant in parentheses.

$$V = (nRT)\dfrac{1}{P}$$

Since n and T are constant in this case and since R is always a constant,

$$V = (\text{Constant}) \times \dfrac{1}{P}$$

which gives us Boyle's law $\left(V \propto \dfrac{1}{P} \right)$.

The ideal gas law also shows how other pairs of variables are related. For example, from Charles's law we know that volume is proportional to temperature at constant pressure and a constant number of moles. But what if we heat a sample of gas at constant *volume* and a constant number of moles? This question applies to the warning labels on aerosol cans such as hair spray or deodorants. These labels warn the user against excessive heating or incineration of the can, even after the contents are used up. Why? An aerosol can that appears empty actually contains a fixed amount of gas trapped in a fixed volume. What would happen if you heated the can? Let's rearrange the ideal gas law to clearly see the relationship between pressure and temperature at constant volume and a constant number of moles.

$$PV = nRT$$

If we divide both sides by V, we get:

$$P = \frac{nRT}{V}$$

$$P = \left(\frac{nR}{V}\right)T$$

The relationship between pressure and temperature is also known as Gay-Lussac's law.

Since n and V are constant and since R is always a constant:

$$P = \text{Constant} \times T$$

As the temperature of a fixed amount of gas in a fixed volume increases, the pressure increases. In an aerosol can, this pressure increase can cause the can to explode, which is why aerosol cans should not be heated or incinerated. Table 11.2 summarizes the relationships between all of the simple gas laws and the ideal gas law.

TABLE 11.2 Relationships between Simple Gas Laws and Ideal Gas Law

Variable Quantities	Constant Quantities	Ideal Gas Law in Form of Variables-constant	Simple Gas Law	Name of Simple Law
V and P	n and T	$PV = nRT$	$P_1V_1 = P_2V_2$	Boyle's law
V and T	n and P	$\frac{V}{T} = \frac{nR}{P}$	$\frac{V_1}{T_1} = \frac{V_2}{T_2}$	Charles's law
P and T	n and V	$\frac{P}{T} = \frac{nR}{V}$	$\frac{P_1}{T_1} = \frac{P_2}{T_2}$	Gay-Lussac's law
P and n	V and T	$\frac{P}{n} = \frac{RT}{V}$	$\frac{P_1}{n_1} = \frac{P_2}{n_2}$	
V and n	T and P	$\frac{V}{n} = \frac{RT}{P}$	$\frac{V_1}{n_1} = \frac{V_2}{n_2}$	Avogadro's law

The ideal gas law can be used to determine the value of any one of the four variables (P, V, n, or T) given the other three. However, each of the quantities in the ideal gas law *must be expressed* in the units within R.

- Pressure **(P)** must be expressed in atmospheres.
- Volume **(V)** must be expressed in liters.
- Amount of gas **(n)** must be expressed in moles.
- Temperature **(T)** must be expressed in kelvins.

For example, suppose we want to know the pressure of 0.18 mol of a gas in a 1.2-L flask at 298 K. We begin by sorting the information in the problem statement.

GIVEN: $n = 0.18$ mol
$V = 1.2$ L
$T = 298$ K

FIND: P

SOLUTION MAP
We then strategize by drawing a solution map that shows how the ideal gas law takes us from the given quantities to the find quantity.

$$PV = nRT$$

RELATIONSHIPS USED

$$PV = nRT \text{ (Ideal gas law, presented in this section)}$$

SOLUTION

We then solve the equation for the quantity we are trying to find (in this case, P).

$$PV = nRT$$

$$P = \frac{nRT}{V}$$

Next we substitute in the numerical values and calculate the answer.

$$P = \frac{0.18 \; \cancel{\text{mol}} \times 0.0821 \; \dfrac{\cancel{\text{L}} \cdot \text{atm}}{\cancel{\text{mol}} \cdot \cancel{\text{K}}} \times 298 \; \cancel{\text{K}}}{1.2 \; \cancel{\text{L}}}$$

$$= 3.7 \text{ atm}$$

Notice that all units cancel except the units of the quantity we need (atm).

EXAMPLE 11.6   The Ideal Gas Law

Calculate the volume occupied by 0.845 mol of nitrogen gas at a pressure of 1.37 atm and a temperature of 315 K.

SORT You are given the number of moles, the pressure, and the temperature of a gas sample. You are asked to find the volume.	**GIVEN:** $n = 0.845$ mol $P = 1.37$ atm $T = 315$ K **FIND:** V
STRATEGIZE Draw a solution map beginning with the given quantities. The ideal gas law shows the relationship necessary to get to the find quantity.	**SOLUTION MAP** $\boxed{n, P, T} \longrightarrow \boxed{V}$ $PV = nRT$ **RELATIONSHIPS USED** $PV = nRT$ (Ideal gas law, presented in this section)
SOLVE Solve the equation for the quantity you are trying to find (V) and substitute the appropriate quantities to compute V.	**SOLUTION** $PV = nRT$ $V = \dfrac{nRT}{P}$ $V = \dfrac{0.845 \; \cancel{\text{mol}} \times 0.0821 \; \dfrac{\text{L} \cdot \cancel{\text{atm}}}{\cancel{\text{mol}} \cdot \cancel{\text{K}}} \times 315 \; \cancel{\text{K}}}{1.37 \; \cancel{\text{atm}}} = 16.0 \text{ L}$
CHECK Check your answer. Are the units correct? Does the answer make physical sense?	The answer has the correct units for volume, liters. The *value* of the answer is a bit more difficult to judge. However, at standard temperature and pressure ($T = 0 \,°C$ or 273.15 K and $P = 1$ atm), 1 mol gas occupies 22.4 L (see Section 11.10). Therefore our answer of 16.0 L seems reasonable for the volume of 0.85 mol of gas under conditions that are not too far from standard temperature and pressure.

▶**SKILLBUILDER 11.6 | The Ideal Gas Law**

An 8.5-L tire is filled with 0.55 mol of gas at a temperature of 305 K. What is the pressure of the gas in the tire?

▶**FOR MORE PRACTICE**   Example 11.16; Problems 59, 60, 61, 62.

If the units given in an ideal gas law problem are different from those of the ideal gas constant (atm, L, mol, and K), you must convert to the correct units before you substitute into the ideal gas equation, as demonstrated in Example 11.7.

EXAMPLE 11.7 The Ideal Gas Law Requiring Unit Conversion

Calculate the number of moles of gas in a basketball inflated to a total pressure of 24.2 psi with a volume of 3.2 L at 25 °C.

SORT	GIVEN:
You are given the pressure, the volume, and the temperature of a gas sample. You are asked to find the number of moles.	$P = 24.2$ psi $V = 3.2$ L $t = 25\,°C$ FIND: n

STRATEGIZE	SOLUTION MAP
Draw a solution map beginning with the given quantities. The ideal gas law shows the relationship necessary to get to the find quantity.	$P, V, T \longrightarrow n$ $PV = nRT$ RELATIONSHIPS USED $PV = nRT$ (Ideal gas law, presented in this section)

SOLVE	SOLUTION
Solve the equation for the quantity you are trying to find (n). Before substituting into the equation, you must convert P and t into the correct units. (Since 1.6462 atm is an intermediate answer, mark the least significant digit, but don't round until the end.) Finally, substitute into the equation to calculate n.	$PV = nRT$ $n = \dfrac{PV}{RT}$ $P = 24.2\ \text{psi} \times \dfrac{1\ \text{atm}}{14.7\ \text{psi}} = 1.6462\ \text{atm}$ $T = t + 273$ $= 25 + 273 = 298\ K$ $n = \dfrac{1.6462\ \text{atm} \times 3.2\ \text{L}}{0.0821\ \dfrac{\text{L}\cdot\text{atm}}{\text{mol}\cdot\text{K}} \times 298\ \text{K}}$ $= 0.22\ \text{mol}$

CHECK	
Check your answer. Are the units correct? Does the answer make physical sense?	The answer has the correct units, moles. The *value* of the answer is a bit more difficult to judge. Again, knowing that at standard temperature and pressure ($T = 0\,°C$ or 273.15 K and $P = 1$ atm), 1 mol of gas occupies 22.4 L can help (see Check step in Example 11.6). A 3.2 L sample of gas at STP would contain about 0.15 mol; therefore at a greater pressure, the sample should contain a bit more than 0.15 mol, which is consistent with the answer.

▶SKILLBUILDER 11.7 | The Ideal Gas Law Requiring Unit Conversion

How much volume does 0.556 mol of gas occupy when its pressure is 715 mm Hg and its temperature is 58 °C?

▶SKILLBUILDER PLUS

Find the pressure in millimeters of mercury of a 0.133-g sample of helium gas at 32 °C and contained in a 648-mL container.

▶FOR MORE PRACTICE Problems 63, 64, 67, 68.

MOLAR MASS OF A GAS FROM THE IDEAL GAS LAW

The ideal gas law can be used in combination with mass measurements to calculate the molar mass of a gas. For example, a sample of gas has a mass of 0.136 g. Its volume is 0.112 L at a temperature of 298 K and a pressure of 1.06 atm. Find its molar mass.

We begin by sorting the information given in the problem.

GIVEN:

$$m = 0.136 \text{ g} \qquad V = 0.112 \text{ L}$$
$$T = 298 \text{ K} \qquad P = 1.06 \text{ atm}$$

FIND: molar mass (g/mol)

SOLUTION MAP

We then strategize by drawing a solution map, which in this case has two parts. In the first part, we use P, V, and T to find the number of moles of gas. In the second part, we use the number of moles of gas and the given mass to find the molar mass.

$$PV = nRT$$

$$\text{Molar mass} = \frac{\text{Mass } (m)}{\text{Moles } (n)}$$

RELATIONSHIPS USED

$$PV = nRT \text{ (Ideal gas law, presented in this section)}$$

$$\text{Molar mass} = \frac{\text{Mass}}{\text{Moles}} \text{ (Definition of molar mass from Section 6.3)}$$

SOLUTION

$$PV = nRT$$

$$n = \frac{PV}{RT}$$

$$= \frac{1.06 \text{ atm} \times 0.112 \text{ L}}{0.0821 \dfrac{\text{L} \cdot \text{atm}}{\text{mol} \cdot \text{K}} \times 298 \text{ K}}$$

$$= 4.8525 \times 10^{-3} \text{ mol}$$

$$\text{Molar mass} = \frac{\text{Mass } (m)}{\text{Moles } (n)}$$

$$= \frac{0.136 \text{ g}}{4.8525 \times 10^{-3} \text{ mol}}$$

$$= 28.0 \text{ g/mol}$$

EXAMPLE 11.8 Molar Mass Using the Ideal Gas Law and Mass Measurement

A sample of gas has a mass of 0.311 g. Its volume is 0.225 L at a temperature of 55 °C and a pressure of 886 mm Hg. Find its molar mass.

SORT You are given the mass, the volume, the temperature, and the pressure of a gas sample. You are asked to find the molar mass of the gas.	**GIVEN:** $m = 0.311$ g $V = 0.225$ L $t = 55$ °C $P = 886$ mm Hg **FIND:** Molar mass (g/mol)
STRATEGIZE In the first part of the solution map, use the ideal gas law to find the number of moles of gas from the other given quantities. In the second part, use the number of moles from the first part, as well as the given mass, to find the molar mass.	**SOLUTION MAP** P, V, T → n $PV = nRT$ n, m → Molar mass $$\text{Molar mass} = \frac{\text{Mass }(m)}{\text{Moles }(n)}$$ **RELATIONSHIPS USED** $PV = nRT$ (Ideal gas law, presented in this section) $\text{Molar mass} = \dfrac{\text{Mass }(m)}{\text{Moles }(n)}$ (Definition of molar mass from Section 6.3)
SOLVE First, solve the ideal gas law for n. Before substituting into the equation, you must convert the pressure to atm and temperature to K. Now, substitute into the equation to calculate n, the number of moles. Finally, use the number of moles just found and the given mass (m) to find the molar mass.	**SOLUTION** $$PV = nRT$$ $$n = \frac{PV}{RT}$$ $$P = 886 \text{ mm Hg} \times \frac{1 \text{ atm}}{760 \text{ mm Hg}} = 1.1658 \text{ atm}$$ $$T = 55 \text{ °C} + 273 = 328 \text{ K}$$ $$n = \frac{1.1658 \text{ atm} \times 0.225 \text{ L}}{0.0821 \dfrac{\text{L} \cdot \text{atm}}{\text{mol} \cdot \text{K}} \times 328 \text{ K}}$$ $$= 9.7406 \times 10^{-3} \text{ mol}$$ $$\text{Molar mass} = \frac{\text{Mass }(m)}{\text{Moles }(n)}$$ $$= \frac{0.311 \text{ g}}{9.7406 \times 10^{-3} \text{ mol}}$$ $$= 31.9 \text{ g/mol}$$
CHECK Check your answer. Are the units correct? Does the answer make physical sense?	The answer has the correct units, g/mol. The answer is reasonable because its value is within the range of molar masses for common compounds.

▶**SKILLBUILDER 11.8** | **Molar Mass Using the Ideal Gas Law and Mass Measurement**

A sample of gas has a mass of 827 mg. Its volume is 0.270 L at a temperature of 88 °C and a pressure of 975 mm Hg. Find its molar mass.

▶**FOR MORE PRACTICE**   Problems 69, 70, 71, 72.

Ideal gas conditions
• High temperature
• Low pressure

Although a complete derivation is beyond the scope of this book, the ideal gas law follows directly from the kinetic molecular theory of gases. Consequently, the ideal gas law holds only under conditions where the kinetic molecular theory holds. The ideal gas law works exactly only for gases that are acting ideally (◀ Figure 11.20), which means that (a) the volume of the gas particles is small compared to the space between them and (b) the forces between the gas particles are not significant. These assumptions break down (▼ Figure 11.21) under conditions of high pressure (because the space between particles is no longer much larger than the size of the particles themselves) or low temperatures (because the gas particles move so slowly that their interactions become significant). For all of the problems encountered in this book, you may assume ideal gas behavior.

• Particle size small compared to space between particles.
• Interactions between particles are insignificant.

▲ **FIGURE 11.20   Conditions for ideal gas behavior**   At high temperatures and low pressures, the assumptions of the kinetic molecular theory apply.

Non-ideal gas conditions
• Low temperature
• High pressure

Intermolecular Interactions

• Particle size significant compared to space between particles.
• Interactions between particles are significant.

▶ **FIGURE 11.21   Conditions for nonideal gas behavior**   At low temperatures and high pressures, the assumptions of the kinetic molecular theory are not valid.

11.9   Mixtures of Gases: Why Deep-Sea Divers Breathe a Mixture of Helium and Oxygen

Many gas samples are not pure but consist of mixtures of gases. The air in our atmosphere, for example, is a mixture containing 78% nitrogen, 21% oxygen, 0.9% argon, 0.04% carbon dioxide (Table 11.3), and a few other gases in smaller amounts.

According to the kinetic molecular theory, each of the components in a gas mixture acts independently of the others. For example, the nitrogen molecules in air exert a certain pressure—78% of the total pressure—that is independent of the

TABLE 11.3   Composition of Dry Air

Gas	Percent by Volume (%)
nitrogen (N_2)	78
oxygen (O_2)	21
argon (Ar)	0.9
carbon dioxide (CO_2)	0.04

Gas mixture (80% He ●, 20% Ne ●)
$P_{tot} = 1.0$ atm
$P_{He} = 0.80$ atm
$P_{Ne} = 0.20$ atm

▲ **FIGURE 11.22   Partial pressures** A gas mixture at a total pressure of 1.0 atm consisting of 80% helium and 20% neon will have a helium partial pressure of 0.80 atm and a neon partial pressure of 0.20 atm.

| The fractional composition is the percent composition divided by 100.

| We can ignore the contribution of CO_2 and other trace gases because they are present in very small amounts.

presence of the other gases in the mixture. Likewise, the oxygen molecules in air exert a certain pressure—21% of the total pressure—that is also independent of the presence of the other gases in the mixture. The pressure due to any individual component in a gas mixture is called the **partial pressure** of that component. The partial pressure of any component is that component's fractional composition times the total pressure of the mixture (◄ Figure 11.22).

Partial pressure of component:

= Fractional composition of component × Total pressure

For example, the partial pressure of nitrogen (P_{N_2}) in air at 1.0 atm is:

$$P_{N_2} = 0.78 \times 1.0 \text{ atm}$$
$$= 0.78 \text{ atm}$$

Similarly, the partial pressure of oxygen in air at 1.0 atm is:

$$P_{O_2} = 0.21 \times 1.0 \text{ atm}$$
$$= 0.21 \text{ atm}$$

The sum of the partial pressures of each of the components in a gas mixture must equal the total pressure, as expressed by **Dalton's law of partial pressures**:

Dalton's law of partial pressures:

$$P_{tot} = P_a + P_b + P_c + \dots$$

where P_{tot} is the total pressure and P_a, P_b, P_c, . . . are the partial pressures of the components.

For 1 atm air:

$$P_{tot} = P_{N_2} + P_{O_2} + P_{Ar}$$
$$P_{tot} = 0.78 \text{ atm} + 0.21 \text{ atm} + 0.01 \text{ atm}$$
$$= 1.00 \text{ atm}$$

EXAMPLE 11.9   Total Pressure and Partial Pressure

A mixture of helium, neon, and argon has a total pressure of 558 mm Hg. If the partial pressure of helium is 341 mm Hg and the partial pressure of neon is 112 mm Hg, what is the partial pressure of argon?

You are given the total pressure of a gas mixture and the partial pressures of two (of its three) components. You are asked to find the partial pressure of the third component.	**GIVEN:** $P_{tot} = 558$ mm Hg $P_{He} = 341$ mm Hg $P_{Ne} = 112$ mm Hg **FIND:** P_{Ar}
To solve this problem, solve Dalton's law for the partial pressure of argon and substitute the correct values to calculate it.	**SOLUTION** $P_{tot} = P_{He} + P_{Ne} + P_{Ar}$ $P_{Ar} = P_{tot} - P_{He} - P_{Ne}$ $= 558$ mm Hg $- 341$ mm Hg $- 112$ mm Hg $= 105$ mm Hg

►**SKILLBUILDER 11.9** | **Total Pressure and Partial Pressure**

A sample of hydrogen gas is mixed with water vapor. The mixture has a total pressure of 745 torr, and the water vapor has a partial pressure of 24 torr. What is the partial pressure of the hydrogen gas?

►**FOR MORE PRACTICE**   Example 11.17; Problems 73, 74, 75, 76.

DEEP-SEA DIVING AND PARTIAL PRESSURE

Our lungs have evolved to breathe oxygen at a partial pressure of $P_{O_2} = 0.21$ atm. If the total pressure decreases—as happens when we climb a mountain, for example—the partial pressure of oxygen also decreases. For example, on top of Mount Everest, where the total pressure is 0.311 atm, the partial pressure of oxygen is only 0.065 atm. As we learned earlier, low oxygen levels can have negative physiological effects, a condition called **hypoxia**, or oxygen starvation. Mild hypoxia causes dizziness, headache, and shortness of breath. Severe hypoxia, which occurs when P_{O_2} drops below 0.1 atm, may cause unconsciousness or even death. For this reason, climbers hoping to make the summit of Mount Everest usually carry oxygen to breathe.

High oxygen levels can also have negative physiological effects. Scuba divers, as we have learned, breathe pressurized air. At 30 m, a scuba diver breathes air at a total pressure of 4.0 atm, making P_{O_2} about 0.84 atm. This increased partial pressure of oxygen causes a higher density of oxygen molecules in the lungs (▼ Figure 11.23), which results in a higher concentration of oxygen in body tissues. When P_{O_2} increases beyond 1.4 atm, the increased oxygen concentration in body tissues causes a condition called **oxygen toxicity**, which is characterized by muscle twitching, tunnel vision, and convulsions (► Figure 11.24). Divers who venture too deep without proper precautions have drowned because of oxygen toxicity.

▲ Mountain climbers on Mount Everest require oxygen because the pressure is so low that the lack of oxygen causes hypoxia, a condition that in severe cases can be fatal.

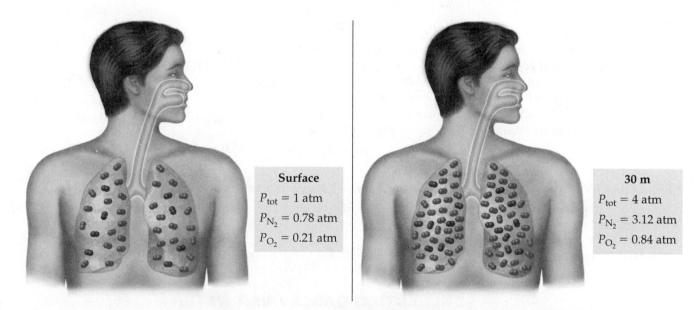

Surface
$P_{tot} = 1$ atm
$P_{N_2} = 0.78$ atm
$P_{O_2} = 0.21$ atm

30 m
$P_{tot} = 4$ atm
$P_{N_2} = 3.12$ atm
$P_{O_2} = 0.84$ atm

▲ **FIGURE 11.23   Too much of a good thing**   When a person is breathing compressed air, there is a larger partial pressure of oxygen in the lungs. A large oxygen partial pressure in the lungs results in a larger amount of oxygen in bodily tissues. When the oxygen partial pressure increases beyond 1.4 atm, oxygen toxicity results. (In this figure, the red molecules are oxygen and the blue ones are nitrogen.)

▶ FIGURE 11.24 **Oxygen partial pressure limits** The partial pressure of oxygen in air at sea level is 0.21 atm. If this pressure drops by 50%, fatal hypoxia can result. High oxygen levels can also be harmful, but only if the partial pressure of oxygen increases by a factor of 7 or more.

A second problem associated with breathing pressurized air is the increase in nitrogen in the lungs. At 30 m a scuba diver breathes nitrogen at P_{N_2} = 3.1 atm, which causes an increase in nitrogen concentration in bodily tissues and fluids. When P_{N_2} increases beyond about 4 atm, a condition called **nitrogen narcosis**, which is also referred to as *rapture of the deep*, results. Divers describe this condition as a feeling of being tipsy. A diver breathing compressed air at 60 m feels as if he has had too much wine.

To avoid oxygen toxicity and nitrogen narcosis, deep-sea divers—those venturing beyond 50 m—breathe specialized mixtures of gases. One common mixture is called heliox, a mixture of helium and oxygen. These mixtures usually contain a smaller percentage of oxygen than would be found in air, thereby lowering the risk of oxygen toxicity. Heliox also contains helium instead of nitrogen, eliminating the risk of nitrogen narcosis.

EXAMPLE 11.10 Partial Pressure, Total Pressure, and Percent Composition

Calculate the partial pressure of oxygen that a diver breathes with a heliox mixture containing 2.0% oxygen at a depth of 100 m where the total pressure is 10.0 atm.

You are given the percent oxygen in the mixture and the total pressure. You are asked to find the partial pressure of oxygen.	**GIVEN:** O_2 percent = 2.0% P_{tot} = 10.0 atm **FIND:** P_{O_2}
The partial pressure of a component in a gas mixture is equal to the fractional composition of the component multiplied by the total pressure. Calculate the fractional composition of O_2 by dividing the percent composition by 100. Calculate the partial pressure of O_2 by multiplying the fractional composition by the total pressure.	**SOLUTION** Partial pressure of component = Fractional composition of component × Total pressure Fractional composition of O_2 = $\dfrac{2.0}{100}$ = 0.020 P_{O_2} = 0.020 × 10.0 atm = 0.20 atm

▶ **SKILLBUILDER 11.10 | Partial Pressure, Total Pressure, and Percent Composition**

A diver breathing heliox with an oxygen composition of 5.0% wants to adjust the total pressure so that P_{O_2} = 0.21 atm. What must the total pressure be?

▶ **FOR MORE PRACTICE** Problems 79, 80, 81, 82.

COLLECTING GASES OVER WATER

When the product of a chemical reaction is gaseous, it is often collected by the displacement of water. For example, suppose the following reaction is used as a source of hydrogen gas.

$$Zn(s) + 2\,HCl(aq) \longrightarrow ZnCl_2(aq) + H_2(g)$$

Hydrogen plus water vapor

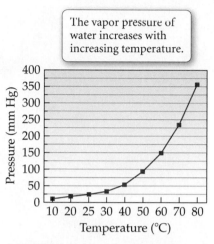

The vapor pressure of water increases with increasing temperature.

▲ **FIGURE 11.25   Vapor pressure**   When a gas from a chemical reaction is collected through water, water molecules become mixed with the gas molecules. The pressure of water vapor in the final mixture is the vapor pressure of water at the temperature at which the gas is collected.

▲ **FIGURE 11.26   Vapor pressure of water as a function of temperature**   Vapor pressure increases with increasing temperature.

Vapor pressure is covered in detail in Chapter 12.

TABLE 11.4   Vapor Pressure of Water versus Temperature

Temperature (°C)	Pressure (mm Hg)
10 °C	9.2
20 °C	17.5
25 °C	23.8
30 °C	31.8
40 °C	55.3
50 °C	92.5
60 °C	149.4
70 °C	233.7
80 °C	355.1

As the hydrogen gas forms, it bubbles through the water and gathers in the collection flask (▲ Figure 11.25). However, the hydrogen gas collected in this way is not pure but is mixed with water vapor because some water molecules evaporate and become mixed with the hydrogen molecules. The partial pressure of water in the mixture depends on the temperature and is called its **vapor pressure** (Table 11.4 and ▲ Figure 11.26). Vapor pressure increases with increasing temperature because higher temperatures cause more water molecules to evaporate.

Suppose we collect the hydrogen gas over water at a total pressure of 758 mm Hg and a temperature of 25 °C. What is the partial pressure of the hydrogen gas? We know that the total pressure is 758 mm Hg and that the partial pressure of water is 23.8 mm Hg (its vapor pressure at 25 °C).

$$P_{tot} = P_{H_2} + P_{H_2O}$$
$$758 \text{ mm Hg} = P_{H_2} + 23.8 \text{ mm Hg}$$

Therefore,

$$P_{H_2} = 758 \text{ mm Hg} - 23.8 \text{ mm Hg}$$
$$= 734 \text{ mm Hg}$$

The partial pressure of the hydrogen in the mixture will be 734 mm Hg.

11.10   Gases in Chemical Reactions

See Section 8.3.

Chapter 8 describes how the coefficients in chemical equations can be used as conversion factors between moles of reactants and moles of products in a chemical reaction. These conversion factors could be used to determine, for example, the amount of product obtained in a chemical reaction based on a given amount of reactant or the amount of one reactant needed to completely react with a given amount of another reactant. The general solution map for these kinds of calculations is

Moles A $\longrightarrow$ Moles B

where A and B are two different substances involved in the reaction and the conversion factor between them comes from the stoichiometric coefficients in the balanced chemical equation.

In reactions involving gaseous reactants or products, the amount of gas is often specified in terms of its volume at a given temperature and pressure. In cases like this, we can use the ideal gas law to convert pressure, volume, and temperature to moles.

$$n = \frac{PV}{RT}$$

We can then use the stoichiometric coefficients to convert to other quantities in the reaction. For example, consider the reaction for the synthesis of ammonia.

$$3\,H_2(g) + N_2(g) \longrightarrow 2\,NH_3(g)$$

How many moles of NH_3 are formed by the complete reaction of 2.5 L of hydrogen at 381 K and 1.32 atm? Assume that there is more than enough N_2.

We begin by sorting the information in the problem statement.

GIVEN:

$V = 2.5\,L$

$T = 381\,K$

$P = 1.32\,atm$ (of H_2)

FIND: mol NH_3

SOLUTION MAP

We strategize by drawing a solution map. The solution map for this problem is similar to the solution maps for other stoichiometric problems (see Sections 8.3 and 8.4). However, we first use the ideal gas law to find mol H_2 from P, V, and T. Then we use the stoichiometric coefficients from the equation to convert mol H_2 to mol NH_3.

RELATIONSHIPS USED

$PV = nRT$ (ideal gas law, Section 11.8)

3 mol H_2 : 2 mol NH_3 (from balanced equation given in problem)

SOLUTION

We first solve the ideal gas equation for n.

$$PV = nRT$$

$$n = \frac{PV}{RT}$$

Then we substitute in the appropriate values.

$$n = \frac{1.32\ \cancel{atm} \times 2.5\ \cancel{L}}{0.0821\ \dfrac{\cancel{L}\cdot\cancel{atm}}{mol\cdot\cancel{K}} \times 381\ \cancel{K}}$$

$$= 0.1055\ mol\ H_2$$

Next, we convert mol H_2 to mol NH_3.

$$0.1055 \text{ mol } H_2 \times \frac{2 \text{ mol } NH_3}{3 \text{ mol } H_2} = 0.070 \text{ mol } NH_3$$

There is enough H_2 to form 0.070 mol NH_3.

EXAMPLE 11.11   Gases in Chemical Reactions

How many liters of oxygen gas form when 294 g of $KClO_3$ completely react in this reaction (which is used in the ignition of fireworks)?

$$2 \, KClO_3(s) \longrightarrow 2 \, KCl(s) + 3 \, O_2(g)$$

Assume that the oxygen gas is collected at $P = 755$ mm Hg and $T = 305$ K.

SORT You are given the mass of a reactant in a chemical reaction. You are asked to find the volume of a gaseous product at a given pressure and temperature.	**GIVEN:** 294 g $KClO_3$ $P = 755$ mm Hg (of oxygen gas) $T = 305$ K **FIND:** Volume of O_2 in liters
STRATEGIZE The solution map has two parts. In the first part, convert from g $KClO_3$ to mol $KClO_3$ and then to mol O_2. In the second part, use mol O_2 as n in the ideal gas law to find the volume of O_2. You will need the molar mass of $KClO_3$ and the stoichiometric relationship between $KClO_3$ and O_2 (from the balanced chemical equation). You will also need the ideal gas law.	**SOLUTION MAP** g $KClO_3$ $\longrightarrow$ mol $KClO_3$ $\longrightarrow$ mol O_2 $\dfrac{1 \text{ mol } KClO_3}{122.6 \text{ g } KClO_3}$ $\quad$ $\dfrac{3 \text{ mol } O_2}{2 \text{ mol } KClO_3}$ n (mol O_2), P, T $\longrightarrow$ V (O_2) $PV = nRT$ **RELATIONSHIPS USED** 1 mol $KClO_3$ = 122.6 g (molar mass of $KClO_3$) 2 mol $KClO_3$: 3 mol O_2 (from balanced equation given in problem) $PV = nRT$ (ideal gas law, Section 11.8)
SOLVE Begin by converting mass $KClO_3$ to mol $KClO_3$ and then to mol O_2. Then solve the ideal gas equation for V. Before substituting the values into this equation, you must convert the pressure to atm. Finally, substitute the given quantities along with the number of moles just calculated to calculate the volume.	**SOLUTION** $294 \text{ g } KClO_3 \times \dfrac{1 \text{ mol } KClO_3}{122.5 \text{ g } KClO_3} \times \dfrac{3 \text{ mol } O_2}{2 \text{ mol } KClO_3} = 3.60 \text{ mol } O_2$ $PV = nRT$ $V = \dfrac{nRT}{P}$ $P = 755 \text{ mm Hg} \times \dfrac{1 \text{ atm}}{760 \text{ mm Hg}} = 0.99342 \text{ atm}$ $V = \dfrac{3.60 \text{ mol} \times 0.0821 \dfrac{L \cdot atm}{mol \cdot K} \times 305 \text{ K}}{0.99342 \text{ atm}}$ $= 90.7$ L
CHECK Check your answer. Are the units correct? Does the answer make physical sense?	The answer has the correct units, liters. The *value* of the answer is a bit more difficult to judge. Again, knowing that at standard temperature and pressure ($T = 0$ °C or 273.15 K and $P = 1$ atm), 1 mol of gas occupies 22.4 L can help (see Check step in Example 11.6). A 90.7 L sample of gas at STP would contain about 4 mol; since we started with a little more than 2 mol $KClO_3$, and since 2 mol $KClO_3$ forms 3 mol O_2, an answer that corresponds to about 4 mol O_2 is reasonable.

▶SKILLBUILDER 11.11 | Gases in Chemical Reactions

In this reaction, 4.58 L of O_2 were formed at 745 mm Hg and 308 K. How many grams of Ag_2O decomposed?

$$2\,Ag_2O(s) \longrightarrow 4\,Ag(s) + O_2(g)$$

▶FOR MORE PRACTICE   Problems 89, 90, 91, 92, 93, 94.

MOLAR VOLUME AT STANDARD TEMPERATURE AND PRESSURE

In Example 11.6 (in the Check step), we saw that the volume occupied by 1 mol of gas at 0 °C (273.15 K) and 1 atm is 22.4 L. These conditions are called **standard temperature and pressure (STP)**, and the volume occupied by 1 mol of gas under these conditions is called the **molar volume** of an ideal gas at STP. Using the ideal gas law, we can confirm that the molar volume at STP is 22.4 L.

$$V = \frac{nRT}{P}$$

$$= \frac{1.00\ \text{mol} \times 0.821\ \dfrac{\text{L} \cdot \text{atm}}{\text{mol} \cdot \text{K}} \times 273\ \text{K}}{1.00\ \text{atm}}$$

$$= 22.4\ \text{L}$$

Under standard conditions, therefore, we can use this ratio as a conversion factor.

| The molar volume of 22.4 L applies only at STP.

1 mol : 22.4 L

One mole of any gas at standard temperature and pressure (STP) occupies 22.4 L.

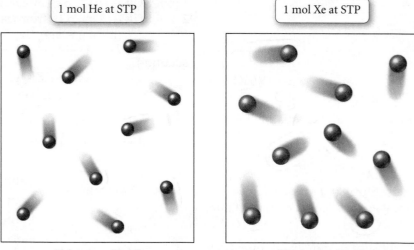

1 mol He at STP

Volume = 22.4 L
Mass = 4.00 g

1 mol Xe at STP

Volume = 22.4 L
Mass = 131.3 g

For example, suppose we wanted to calculate the number of liters of **CO_2** gas that forms at STP when 0.879 moles of **$CaCO_3$** undergoes this reaction:

$$CaCO_3(s) \longrightarrow CaO(s) + CO_2(g)$$

We begin by sorting the information in the problem statement.

GIVEN: 0.879 mol $CaCO_3$
FIND: $CO_2(g)$ in liters

SOLUTION MAP

We strategize by drawing a solution map that shows how to convert from mol $CaCO_3$ to mol CO_2 to L CO_2 using the molar volume at STP.

$$\frac{1 \text{ mol } CO_2}{1 \text{ mol } CaCO_3} \qquad \frac{22.4 \text{ L } CO_2}{1 \text{ mol } CO_2}$$

RELATIONSHIPS USED

1 mol $CaCO_3$: 1 mol CO_2 (from balanced equation given in problem)

1 mol = 22.4 L (at STP)   (molar volume at STP, presented in this section)

SOLUTION

$$0.879 \text{ mol } \cancel{CaCO_3} \times \frac{1 \text{ mol } \cancel{CO_2}}{1 \text{ mol } \cancel{CaCO_3}} \times \frac{22.4 \text{ L } CO_2}{1 \text{ mol } \cancel{CO_2}} = 19.7 \text{ L } CO_2$$

EXAMPLE 11.12   Using Molar Volume in Calculations

How many grams of water form when 1.24 L of H_2 gas at STP completely reacts with O_2?

$$2 H_2(g) + O_2(g) \longrightarrow 2 H_2O(g)$$

SORT You are given the volume of a reactant at STP and asked to find the mass of the product formed.	**GIVEN:** 1.24 L H_2 **FIND:** g H_2O
STRATEGIZE In the solution map, use the molar volume to convert from volume H_2 to mol H_2. Then use the stoichiometric relationship to convert to mol H_2O and finally the molar mass of H_2O to get to mass H_2O.	**SOLUTION MAP** L H_2 → mol H_2 → mol H_2O → g H_2O $\dfrac{1 \text{ mol } H_2}{22.4 \text{ L } H_2} \qquad \dfrac{2 \text{ mol } H_2O}{2 \text{ mol } H_2} \qquad \dfrac{18.02 \text{ g } H_2O}{1 \text{ mol } H_2O}$ **RELATIONSHIPS USED** 1 mol = 22.4 L (molar volume at STP, presented in this section) 2 mol H_2 : 2 mol H_2O (from balanced equation given in problem) 18.02 g H_2O = 1 mol H_2O (molar mass of H_2O)
SOLVE Begin with the volume of H_2 and follow the solution map to arrive at mass H_2O in grams.	**SOLUTION** $1.24 \text{ L } \cancel{H_2} \times \dfrac{1 \text{ mol } \cancel{H_2}}{22.4 \text{ L } \cancel{H_2}} \times \dfrac{2 \text{ mol } \cancel{H_2O}}{2 \text{ mol } \cancel{H_2}} \times \dfrac{18.02 \text{ g } H_2O}{1 \text{ mol } \cancel{H_2O}} = 0.998 \text{ g } H_2O$
CHECK Check your answer. Are the units correct? Does the answer make physical sense?	The answer has the correct units, g H_2O. The answer is reasonable because 1.24 L of a gas is about 0.05 mol of reactant (at STP) and 1 g H_2O is about 0.05 mol product (1 g/18 g/mol ≈ 0.05 mol). Since the reaction produces 1 mol H_2O to 1 mol H_2, we expect the number of moles of H_2O produced to be equal to the number of moles of H_2 that react.

▶ **SKILLBUILDER 11.12 | Using Molar Volume in Calculations**

How many liters of oxygen (at STP) are required to form 10.5 g of H_2O?

$$2 H_2(g) + O_2(g) \longrightarrow 2 H_2O(g)$$

▶ **FOR MORE PRACTICE**   Problems 95, 96, 97, 98.

CHEMISTRY IN THE ENVIRONMENT

Air Pollution

All major cities in the world have polluted air. This pollution comes from a number of sources, including electricity generation, motor vehicles, and industrial waste. While there are many different kinds of air pollutants, some of the major gaseous air pollutants are:

Sulfur dioxide (SO_2)—Sulfur dioxide is emitted primarily as a by-product of electricity generation and industrial metal refining. SO_2 is a lung and eye irritant that affects the respiratory system. SO_2 is also one of the main precursors of acid rain.

Carbon monoxide (CO)—Carbon monoxide is formed by the incomplete combustion of fossil fuels (petroleum, natural gas, and coal). It is emitted mainly by motor vehicles. CO displaces oxygen in the blood and causes the heart and lungs to work harder. At high levels, CO can cause sensory impairment, decreased thinking ability, unconsciousness, and even death.

Ozone (O_3)—Ozone in the upper atmosphere is a normal part of our environment. Upper atmospheric ozone filters out part of the harmful UV light contained in sunlight. Lower-atmospheric or *ground-level* ozone, on the other hand, is a pollutant that results from the action of sunlight on motor vehicle emissions. Ground-level ozone is an eye and lung irritant. Prolonged exposure to ozone has been shown to permanently damage the lungs.

Nitrogen dioxide (NO_2)—Nitrogen dioxide is emitted by motor vehicles and by electricity generation plants. It is an orange-brown gas that causes the dark haze often seen over polluted cities. NO_2 is an eye and lung irritant and a precursor of acid rain.

In the United States, the U.S. Environmental Protection Agency (EPA) has set standards for these pollutants. Beginning in the 1970s, the U.S. Congress passed the Clean Air Act and its amendments, requiring U.S. cities to reduce their pollution and maintain levels below the limits set by the EPA. As a result of this legislation, pollutant levels in U.S. cities have decreased significantly over the last 30 years, even as the number of vehicles has increased. For example, according to the EPA, the levels of all four of the previously mentioned pollutants in major U.S cities decreased during 1980–2008. The amounts of these decreases are shown in Table 11.5.

TABLE 11.5   Changes in Pollutant Levels for Major U.S. Cities, 1980–2008

Pollutant	Change, 1980–2008
SO_2	−71%
CO	−79%
O_3	−25%
NO_2	−46%

(*Source:* U.S. EPA)

Although the levels of pollutants (especially ozone) in many cities are still above what the EPA considers safe, much progress has been made. These trends demonstrate that good legislation can clean up our environment.

CAN YOU ANSWER THIS? *Calculate the amount (in grams) of SO_2 emitted when 1.0 kg of coal containing 4.0% S by mass is completely burned. Under standard conditions, what volume in liters would this SO_2 occupy?*

▲ Air pollution plagues most large cities.

CHEMICAL PRINCIPLES

Kinetic Molecular Theory: The kinetic molecular theory is a model for gases. In this model, gases are composed of widely spaced, noninteracting particles whose average kinetic energy depends on temperature.

Pressure: Pressure is the force per unit area that results from the collision of gas particles with surfaces. The SI unit of pressure is the pascal, but pressure is often expressed in other units such as atmospheres, millimeters of mercury, torr, pounds per square inch, and inches of mercury.

Simple Gas Laws: The simple gas laws show how one of the properties of a gas varies with another. They are:

Volume (*V*) and Pressure (*P*)

$$V \propto \frac{1}{P} \quad \text{(Boyle's law)}$$

Volume (*V*) and Temperature (*T*)

$$V \propto T \quad \text{(Charles's law)}$$

Volume (*V*) and Moles (*n*)

$$V \propto n \quad \text{(Avogadro's law)}$$

The Combined Gas Law: The combined gas law joins Boyle's law and Charles's law.

$$\frac{P_1 V_1}{T_1} = \frac{P_2 V_2}{T_2}$$

The Ideal Gas Law: The ideal gas law combines the four properties of a gas—pressure, volume, temperature, and number of moles—in a single equation showing their interrelatedness.

$$PV = nRT$$

Mixtures of Gases: The pressure due to an individual component in a mixture of gases is its partial pressure and is defined as the fractional composition of the component multiplied by the total pressure:

Partial pressure of component = Fractional composition of component × Total pressure

Dalton's law states that the total pressure of a mixture of gases is equal to the sum of the partial pressures of its components:

$$P_{tot} = P_a + P_b + P_c + \ldots$$

RELEVANCE

Kinetic Molecular Theory: The kinetic molecular theory predicts many of the properties of gases, including their low density in comparison to solids, their compressibility, and their tendency to assume the shape and volume of their container. The kinetic molecular theory also predicts the ideal gas law.

Pressure: Pressure is a fundamental property of a gas. It allows tires to be inflated and makes it possible to drink from straws.

Simple Gas Laws: Each of the simple gas laws allows us to see how two properties of a gas are interrelated. They are also useful in calculating how one of the properties of a gas changes when another does. Boyle's law, for example, can be used to calculate how the volume of a gas will change in response to a change in pressure, or vice versa.

The Combined Gas Law: The combined gas law is used to calculate how a property of a gas (pressure, volume, or temperature) changes when two other properties are changed at the same time.

The Ideal Gas Law: The ideal gas law lets you find any one of the four properties of a gas if you know the other three.

Mixtures of Gases: Since many gases are not pure but mixtures of several components, it is useful to know how each component contributes to the properties of the entire mixture. The concepts of partial pressure are relevant to deep-sea diving, for example, and to collecting gases over water, where water vapor mixes with the gas being collected.

Gases in Chemical Reactions: Stoichiometric calculations involving gases are similar to those that do not involve gases in that the coefficients in a balanced chemical equation provide conversion factors among moles of reactants and products in the reaction. For gases, the amount of a reactant or product is often specified by the volume of reactant or product at a given temperature and pressure. The ideal gas law is then used to convert from these quantities to moles of reactant or product. Alternatively, at standard temperature and pressure, volume can be converted directly to moles with the equality:

$$1 \text{ mol} = 22.4 \text{ L (at STP)}$$

Gases in Chemical Reactions: Reactions involving gases are common in chemistry. For example, many atmospheric reactions—some of which are important to the environment—occur as gaseous reactions.

CHEMICAL SKILLS

EXAMPLES

Pressure Unit Conversion (Section 11.3)

EXAMPLE 11.13 Pressure Unit Conversion

Convert 18.4 in. Hg to torr.

SORT

You are given a pressure in inches of mercury asked to convert the units to torr.

GIVEN: 18.4 in. Hg
FIND: torr

STRATEGIZE

Begin with the quantity you are given and multiply by the appropriate conversion factor(s) to get to the quantity you are trying to find.

SOLUTION MAP

$$\frac{1 \text{ atm}}{29.92 \text{ in. Hg}} \qquad \frac{760 \text{ torr}}{1 \text{ atm}}$$

RELATIONSHIPS USED

$$1 \text{ atm} = 29.92 \text{ in. Hg} \quad \text{(Table 11.1)}$$
$$760 \text{ torr} = 1 \text{ atm} \quad \text{(Table 11.1)}$$

SOLVE

Follow the solution map to solve the problem.

SOLUTION

$$18.4 \text{ in. Hg} \times \frac{1 \text{ atm}}{29.92 \text{ in. Hg}} \times \frac{760 \text{ torr}}{1 \text{ atm}} = 467 \text{ torr}$$

CHECK

Are the units correct? Does the answer make physical sense?

The units (torr) are correct. The value of the answer is reasonable because the torr is a smaller unit than in. Hg; therefore the value of a pressure in torr should be larger than in in. Hg.

Simple Gas Laws (Sections 11.4, 11.5, 11.7)

EXAMPLE 11.14 Simple Gas Laws

A gas has a volume of 5.7 L at a pressure of 3.2 atm. What is its volume at 4.7 atm? (Assume constant temperature.)

SORT

You are given the initial and final pressures and the initial volume. You are asked to find the final volume.

GIVEN:

$$P_1 = 3.2 \text{ atm}$$
$$V_1 = 5.7 \text{ L}$$
$$P_2 = 4.7 \text{ atm}$$

FIND: V_2

STRATEGIZE
Calculations involving the simple gas laws usually consist of finding one of the initial or final conditions given the other initial and final conditions. In this case, use Boyle's law to find V_2 given P_1, V_1, and V_2.

SOLUTION MAP

$$P_1V_1 = P_2V_2$$

RELATIONSHIPS USED

$$P_1V_1 = P_2V_2 \text{ (Boyle's law, Section 11.4)}$$

SOLVE
Solve Boyle's law for V_2 and substitute the correct variables to calculate its value.

SOLUTION

$$P_1V_1 = P_2V_2 \text{ torr}$$

$$V_2 = \frac{P_1}{P_2}V_1$$

$$= \frac{3.2 \text{ atm}}{4.7 \text{ atm}} 5.7 \text{ L}$$

$$= 3.9 \text{ L}$$

CHECK
Are the units correct? Does the answer make physical sense?

The units (L) are correct. The value is reasonable because as the pressure increases the volume should decrease.

The Combined Gas Law (Section 11.6)

EXAMPLE 11.15 The Combined Gas Law

A sample of gas has an initial volume of 2.4 L at a pressure of 855 mm Hg and a temperature of 298 K. If the gas is heated to a temperature of 387 K and expanded to a volume of 4.1 L, what is its final pressure in millimeters of mercury?

SORT
You are given the initial and final volume and temperature, and you are given the initial pressure. You are asked to find the final pressure.

GIVEN:

$$P_1 = 855 \text{ mm Hg}$$

$$V_1 = 2.4 \text{ L}$$

$$T_1 = 298 \text{ K}$$

$$V_2 = 4.1 \text{ L}$$

$$T_2 = 387 \text{ K}$$

FIND: P_2

STRATEGIZE
Problems involving the combined gas law usually consist of finding one of the initial or final conditions given the other initial and final conditions. In this case, use the combined gas law to find the unknown quantity, P_2.

SOLUTION MAP

$$\boxed{P_1, V_1, T_1, V_2, T_2} \longrightarrow \boxed{P_2}$$

$$\frac{P_1V_1}{T_1} = \frac{P_2V_2}{T_2}$$

RELATIONSHIPS USED

$$\frac{P_1V_1}{T_1} = \frac{P_2V_2}{T_2} \quad \text{(Combined gas law, Section 11.6)}$$

SOLVE

Solve the combined gas law for the quantity you are trying to find, in this case P_2, and substitute the known quantities to calculate the value of P_2.

SOLUTION

$$\frac{P_1 V_1}{T_1} = \frac{P_2 V_2}{T_2}$$

$$P_2 = \frac{P_1 V_1 T_2}{T_1 V_2}$$

$$= \frac{855 \text{ mm Hg} \times 2.4 \cancel{L} \times 387 \text{ K}}{298 \cancel{K} \times 4.1 \cancel{L}}$$

$$= 6.5 \times 10^2 \text{ mm Hg}$$

CHECK

Are the units correct? Does the answer make physical sense?

The units, mm Hg, are correct. The value of the answer makes sense because the volume increase was proportionally more than the temperature decrease; therefore you would expect the pressure to decrease.

The Ideal Gas Law (Section 11.8)

EXAMPLE 11.16   The Ideal Gas Law

Calculate the pressure exerted by 1.2 mol of gas in a volume of 28.2 L and at a temperature of 334 K.

SORT

You are given the number of moles of a gas, its volume, and its temperature. You are asked to find its pressure.

GIVEN:

$$n = 1.2 \text{ mol}$$
$$V = 28.2 \text{ L}$$
$$T = 334 \text{ K}$$

FIND: P

SOLUTION MAP

$$PV = nRT$$

RELATIONSHIPS USED

$$PV = nRT \text{ (Ideal gas law, Section 11.8)}$$

STRATEGIZE

Calculations involving the ideal gas law often involve finding one of the four variables (P, V, n, or T) given the other three. In this case, you are asked to find P. Use the given variables and the ideal gas law to arrive at P.

SOLVE

Solve the ideal gas law equation for P and substitute the given variables to calculate P.

SOLUTION

$$PV = nRT$$

$$P = \frac{nRT}{V}$$

$$= \frac{1.2 \cancel{\text{mol}} \times 0.0821 \frac{\cancel{L} \cdot \text{atm}}{\cancel{\text{mol}} \cdot \cancel{K}} \times 334 \cancel{K}}{28.2 \cancel{L}}$$

$$= 1.2 \text{ atm}$$

CHECK

Are the units correct? Does the answer make physical sense?

The units (atm) are correct units for pressure. The *value* of the answer is a bit more difficult to judge. However, knowing that at standard temperature and pressure ($T = 0\ °C$ or 273.15 K and $P = 1$ atm), 1 mol of gas occupies 22.4 L, we can see that the answer is reasonable because we have a bit more than one mole of a gas at a temperature not too far from standard temperature. The volume of the gas is a bit higher than 22.4 L; therefore we might expect the pressure to be a bit higher than 1 atm.

Total Pressure and Partial Pressure (Section 11.9)

You are given the partial pressures of three gases in a mixture and asked to find the total pressure.

Use Dalton's law of partial pressures ($P_{tot} = P_a + P_b + P_c + \ldots$) to solve the problem. Sum the partial pressures to obtain the total pressure.

EXAMPLE 11.17   Total Pressure and Partial Pressure

A mixture of three gases has the partial pressures.

$$P_{CO_2} = 289 \text{ mm Hg}$$

$$P_{O_2} = 342 \text{ mm Hg}$$

$$P_{N_2} = 122 \text{ mm Hg}$$

What is the total pressure of the mixture?

GIVEN:

$$P_{CO_2} = 289 \text{ mm Hg}$$

$$P_{O_2} = 342 \text{ mm Hg}$$

$$P_{N_2} = 122 \text{ mm Hg}$$

FIND: P_{tot}

SOLUTION

$$P_{tot} = P_{CO_2} + P_{O_2} + P_{N_2}$$

$$= 289 \text{ mm Hg} + 342 \text{ mm Hg} + 122 \text{ mm Hg}$$

$$= 753 \text{ mm Hg}$$

KEY TERMS

absolute zero [**11.5**]
atmosphere (atm) [**11.3**]
Avogadro's law [**11.7**]
Boyle's law [**11.4**]
Charles's law [**11.5**]
combined gas
  law [**11.6**]

Dalton's law of partial
  pressures [**11.9**]
hypoxia [**11.9**]
ideal gas constant (R) [**11.8**]
ideal gas law [**11.8**]
kinetic molecular theory
  [**11.2**]

millimeter of mercury
  (mm Hg) [**11.3**]
molar volume [**11.10**]
nitrogen narcosis [**11.9**]
oxygen toxicity [**11.9**]
partial pressure [**11.9**]
pascal (Pa) [**11.3**]

pounds per square inch
  (psi) [**11.3**]
pressure [**11.3**]
standard temperature and
  pressure (STP) [**11.10**]
torr [**11.3**]
vapor pressure [**11.9**]

EXERCISES

QUESTIONS

1. What is pressure?
2. Explain how drinking from a straw works. What causes the drink to go up the straw? Is there an upper limit to how long a straw can theoretically be and still work as a drinking straw?
3. What are the main assumptions of kinetic molecular theory?
4. Describe the main properties of a gas. How are these predicted by kinetic molecular theory?
5. Why do we experience pain in our ears during changes in altitude?
6. What units are used to measure pressure?
7. What is Boyle's law? Explain Boyle's law from the perspective of kinetic molecular theory.

8. Explain why scuba divers should not hold their breath as they ascend to the surface.
9. Why would it be impossible to breathe air through an extra-long snorkel (greater than a couple of meters) while swimming underwater?
10. What is Charles's law? Explain Charles's law from the perspective of kinetic molecular theory.
11. Explain why hot-air balloons float above the ground.
12. What is the combined gas law? When is it useful?
13. What is Avogadro's law? Explain Avogadro's law from the perspective of kinetic molecular theory.
14. What is the ideal gas law? When is it useful?

15. Under what conditions is the ideal gas law most accurate? Under what conditions does the ideal gas law break down? Why?
16. What is partial pressure?
17. What is Dalton's law?
18. Describe hypoxia and oxygen toxicity.
19. Why do deep-sea divers breathe a mixture of helium and oxygen?

20. When a gas is collected over water, is the gas pure? Why or why not?
21. What is vapor pressure?
22. What is standard temperature and pressure (STP)? What is the molar volume of a gas at STP?

PROBLEMS

CONVERTING BETWEEN PRESSURE UNITS

23. Convert each pressure measurement to atm.
    (a) 1277 mm Hg
    (b) 2.38×10^5 Pa
    (c) 127 psi
    (d) 455 torr

24. Convert each pressure measurement to atm.
    (a) 921 torr
    (b) 4.8×10^4 Pa
    (c) 87.5 psi
    (d) 34.22 in. Hg

25. Perform each pressure conversion.
    (a) 2.3 atm to torr
    (b) 4.7×10^{-2} atm to millimeters of mercury
    (c) 24.8 psi to millimeters of mercury
    (d) 32.84 in. Hg to torr

26. Perform each pressure conversion.
    (a) 1.06 atm to millimeters of mercury
    (b) 95,422 Pa to millimeters of mercury
    (c) 22.3 psi to torr
    (d) 35.78 in. Hg to millimeters of mercury

27. Complete the table:

Pascals	Atmospheres	Millimeters of Mercury	Torr	Pounds per Square Inch
882 Pa	_____	6.62 mm Hg	_____	_____
_____	0.558 atm	_____	_____	_____
_____	_____	_____	_____	24.8 psi
_____	_____	_____	764 torr	_____
_____	_____	249 mm Hg	_____	_____

28. Complete the table:

Pascals	Atmospheres	Millimeters of Mercury	Torr	Pounds per Square Inch
_____	1.91 atm	_____	1.45×10^3 torr	_____
1.15×10^4 Pa	_____	_____	_____	_____
_____	_____	_____	721 torr	_____
_____	_____	109 mm Hg	_____	_____
_____	_____	_____	_____	38.9 psi

29. The pressure in Denver, Colorado (5280-ft elevation), averages about 24.9 in. Hg. Convert this pressure to:
    (a) atmospheres
    (b) millimeters of mercury
    (c) pounds per square inch
    (d) pascals

30. The pressure on top of Mount Everest averages about 235 mm Hg. Convert this pressure to:
    (a) torr
    (b) pounds per square inch
    (c) inches of mercury
    (d) atmospheres

31. The North American record for highest recorded barometric pressure is 31.85 in. Hg, set in 1989 in Northway, Alaska. Convert this pressure to:
    (a) millimeters of mercury
    (b) atmospheres
    (c) torr
    (d) kilopascals

32. The world record for lowest pressure (at sea level) was 658 mm Hg, recorded inside Typhoon Ida on September 24, 1958, in the Philippine Sea. Convert this pressure to:
    (a) torr
    (b) atmospheres
    (c) inches of mercury
    (d) pounds per square inch

SIMPLE GAS LAWS

33. A sample of gas has an initial volume of 3.95 L at a pressure of 705 mm Hg. If the volume of the gas is increased to 5.38 L, what will the pressure be? (Assume constant temperature.)

34. A sample of gas has an initial volume of 22.8 L at a pressure of 1.65 atm. If the sample is compressed to a volume of 10.7 L, what will its pressure be? (Assume constant temperature.)

35. A snorkeler with a lung capacity of 6.3 L inhales a lungful of air at the surface, where the pressure is 1.0 atm. The snorkeler then descends to a depth of 25 m, where the pressure increases to 3.5 atm. What is the capacity of the snorkeler's lungs at this depth? (Assume constant temperature.)

36. A scuba diver with a lung capacity of 5.2 L inhales a lungful of air at a depth of 45 m and a pressure of 5.5 atm. If the diver were to ascend to the surface (where the pressure is 1.0 atm) while holding her breath, to what volume would the air in her lungs expand? (Assume constant temperature.)

37. Use Boyle's law to complete the table (assume temperature and number of moles of gas to be constant):

P_1	V_1	P_2	V_2
755 mm Hg	2.85 L	885 mm Hg	_____
_____	1.33 L	4.32 atm	2.88 L
192 mm Hg	382 mL	_____	482 mL
2.11 atm	_____	3.82 atm	125 mL

38. Use Boyle's law to complete the table (assume temperature and number of moles of gas to be constant):

P_1	V_1	P_2	V_2
_____	1.90 L	4.19 atm	1.09 L
755 mm Hg	118 mL	709 mm Hg	_____
2.75 atm	6.75 mL	_____	49.8 mL
343 torr	_____	683 torr	8.79 L

39. A balloon with an initial volume of 3.2 L at a temperature of 299 K is warmed to 376 K. What is its volume at the final temperature?

40. A dramatic classroom demonstration involves cooling a balloon from room temperature (298 K) to liquid nitrogen temperature (77 K). If the initial volume of the balloon is 2.7 L, what will its volume be after it cools?

41. A 48.3-mL sample of gas in a cylinder equipped with a piston is warmed from 22 °C to 87 °C. What is its volume at the final temperature?

42. A syringe containing 1.55 mL of oxygen gas is cooled from 95.3 °C to 0.0 °C. What is the final volume of oxygen gas?

43. Use Charles's law to complete the table (assume pressure and number of moles of gas to be constant):

V_1	T_1	V_2	T_2
1.08 L	25.4 °C	1.33 L	_____
_____	77 K	228 mL	298 K
115 cm^3	_____	119 cm^3	22.4 °C
232 L	18.5 °C	_____	96.2 °C

44. Use Charles's law to complete the table (assume pressure and number of moles of gas to be constant):

V_1	T_1	V_2	T_2
119 L	10.5 °C	_____	112.3 °C
_____	135 K	176 mL	315 K
2.11 L	15.4 °C	2.33 L	_____
15.4 cm^3	_____	19.2 cm^3	10.4 °C

45. A 0.12-mole sample of nitrogen gas occupies a volume of 2.55 L. What is the volume of 0.32 mol of nitrogen gas under the same conditions?

46. A 0.48-mole sample of helium gas occupies a volume of 11.7 L. What is the volume of 0.72 mol of helium gas under the same conditions?

47. A balloon contains 0.128 mol of gas and has a volume of 2.76 L. If an additional 0.073 mol of gas is added to the balloon, what will its final volume be?

48. A cylinder with a moveable piston contains 0.87 mol of gas and has a volume of 334 mL. What will its volume be if an additional 0.22 mol of gas is added to the cylinder?

49. Use Avogadro's law to complete the table (assume pressure and temperature to be constant):

V_1	n_1	V_2	n_2
38.5 mL	1.55×10^{-3} mol	49.4 mL	_____
_____	1.37 mol	26.8 L	4.57 mol
11.2 L	0.628 mol	_____	0.881 mol
422 mL	_____	671 mL	0.0174 mol

50. Use Avogadro's law to complete the table (assume pressure and temperature to be constant):

V_1	n_1	V_2	n_2
25.2 L	5.05 mol	_____	3.03 mol
_____	1.10 mol	414 mL	0.913 mol
8.63 L	0.0018 mol	10.9 L	_____
53 mL	_____	13 mL	2.61×10^{-4} mol

THE COMBINED GAS LAW

51. A sample of gas with an initial volume of 28.4 L at a pressure of 725 mm Hg and a temperature of 305 K is compressed to a volume of 14.8 L and warmed to a temperature of 375 K. What is the final pressure of the gas?

52. A cylinder with a moveable piston contains 218 mL of nitrogen gas at a pressure of 1.32 atm and a temperature of 298 K. What must the final volume be for the pressure of the gas to be 1.55 atm at a temperature of 335 K?

53. A scuba diver takes a 2.8-L balloon from the surface, where the pressure is 1.0 atm and the temperature is 34 °C, to a depth of 25 m, where the pressure is 3.5 atm and the temperature is 18 °C. What is the volume of the balloon at this depth?

54. A bag of potato chips contains 585 mL of air at 25 °C and a pressure of 765 mm Hg. Assuming the bag does not break, what will be its volume at the top of a mountain where the pressure is 442 mm Hg and the temperature is 5.0 °C?

55. A gas sample with a volume of 5.3 L has a pressure of 735 mm Hg at 28 °C. What is the pressure of the sample if the volume remains at 5.3 L but the temperature rises to 86 °C?

56. The total pressure in a 11.7-L automobile tire is 44 psi at 11 °C. How much does the pressure in the tire rise if it warms to a temperature of 37 °C and the volume remains at 11.7 L?

57. Use the combined gas law to complete the table (assume the number of moles of gas to be constant):

P_1	V_1	T_1	P_2	V_2	T_2
1.21 atm	1.58 L	12.2 °C	1.54 atm	_____	32.3 °C
721 torr	141 mL	135 K	801 torr	152 mL	_____
5.51 atm	0.879 L	22.1 °C	_____	1.05 L	38.3 °C

58. Use the combined gas law to complete the table (assume the number of moles of gas to be constant):

P_1	V_1	T_1	P_2	V_2	T_2
1.01 atm	_____	2.7 °C	0.54 atm	0.58 L	42.3 °C
123 torr	41.5 mL	_____	626 torr	36.5 mL	205 K
_____	1.879 L	20.8 °C	0.412 atm	2.05 L	48.1 °C

THE IDEAL GAS LAW

59. What is the volume occupied by 0.255 mol of helium gas at a pressure of 1.25 atm and a temperature of 305 K?

60. What is the pressure in a 20.0-L cylinder filled with 0.683 mol of nitrogen gas at a temperature of 325 K?

61. A cylinder contains 28.5 L of oxygen gas at a pressure of 1.8 atm and a temperature of 298 K. How many moles of gas are in the cylinder?

62. What is the temperature of 0.52 mol of gas at a pressure of 1.3 atm and a volume of 11.8 L?

63. A cylinder contains 11.8 L of air at a total pressure of 43.2 psi and a temperature of 25 °C. How many moles of gas does the cylinder contain?

64. What is the pressure in millimeters of mercury of 0.0115 mol of helium gas with a volume of 214 mL at a temperature of 45 °C?

65. Use the ideal gas law to complete the table:

P	V	n	T
1.05 atm	1.19 L	0.112 mol	_____
112 torr	_____	0.241 mol	304 K
_____	28.5 mL	1.74×10^{-3} mol	25.4 °C
0.559 atm	0.439 L	_____	255 K

66. Use the ideal gas law to complete the table:

P	V	n	T
2.39 atm	1.21 L	_____	205 K
512 torr	_____	0.741 mol	298 K
0.433 atm	0.192 L	0.0131 mol	_____
_____	20.2 mL	5.71×10^{-3} mol	20.4 °C

67. How many moles of gas must be forced into a 3.5-L ball to give it a gauge pressure of 9.4 psi at 25 °C? The gauge pressure is relative to atmospheric pressure. Assume that atmospheric pressure is 14.7 psi so that the total pressure in the ball is 24.1 psi.

68. How many moles of gas must be forced into a 4.8-L tire to give it a gauge pressure of 32.4 psi at 25 °C? The gauge pressure is relative to atmospheric pressure. Assume that atmospheric pressure is 14.7 psi so that the total pressure in the tire is 47.1 psi.

69. An experiment shows that a 248-mL gas sample has a mass of 0.433 g at a pressure of 745 mm Hg and a temperature of 28 °C. What is the molar mass of the gas?

70. An experiment shows that a 113-mL gas sample has a mass of 0.171 g at a pressure of 721 mm Hg and a temperature of 32 °C. What is the molar mass of the gas?

71. A sample of gas has a mass of 38.8 mg. Its volume is 224 mL at a temperature of 55 °C and a pressure of 886 torr. Find the molar mass of the gas.

72. A sample of gas has a mass of 0.555 g. Its volume is 117 mL at a temperature of 85 °C and a pressure of 753 mm Hg. Find the molar mass of the gas.

PARTIAL PRESSURE

73. A gas mixture contains each gas at the indicated partial pressure.

N_2	217 torr
O_2	106 torr
He	248 torr

What is the total pressure of the mixture?

74. A gas mixture contains each gas at the indicated partial pressure.

CO_2	422 mm Hg
Ar	102 mm Hg
O_2	165 mm Hg
H_2	52 mm Hg

What is the total pressure of the mixture?

75. A heliox deep-sea diving mixture delivers an oxygen partial pressure of 0.30 atm when the total pressure is 11.0 atm. What is the partial pressure of helium in this mixture?

76. A mixture of helium, nitrogen, and oxygen has a total pressure of 752 mm Hg. The partial pressures of helium and nitrogen are 234 mm Hg and 197 mm Hg, respectively. What is the partial pressure of oxygen in the mixture?

77. The hydrogen gas formed in a chemical reaction is collected over water at 30 °C at a total pressure of 732 mm Hg. What is the partial pressure of the hydrogen gas collected in this way?

78. The oxygen gas emitted from an aquatic plant during photosynthesis is collected over water at a temperature of 25 °C and a total pressure of 753 torr. What is the partial pressure of the oxygen gas?

79. A gas mixture contains 78% nitrogen and 22% oxygen. If the total pressure is 1.12 atm, what are the partial pressures of each component?

80. An air sample contains 0.038% CO_2. If the total pressure is 758 mm Hg, what is the partial pressure of CO_2?

81. A heliox deep-sea diving mixture contains 4.0% oxygen and 96.0% helium. What is the partial pressure of oxygen when this mixture is delivered at a total pressure of 8.5 atm?

82. A scuba diver breathing normal air descends to 100 m of depth, where the total pressure is 11 atm. What is the partial pressure of oxygen that the diver experiences at this depth? Is the diver in danger of experiencing oxygen toxicity?

MOLAR VOLUME

83. Calculate the volume of each gas sample at STP.
    (a) 22.5 mol Cl_2
    (b) 3.6 mol nitrogen
    (c) 2.2 mol helium
    (d) 27 mol CH_4

84. Calculate the volume of each gas sample at STP.
    (a) 21.2 mol N_2O
    (b) 0.215 mol CO
    (c) 0.364 mol CO_2
    (d) 8.6 mol C_2H_6

85. Calculate the volume of each gas sample at STP.
    (a) 73.9 g N_2
    (b) 42.9 g O_2
    (c) 148 g NO_2
    (d) 245 mg CO_2

86. Calculate the volume of each gas sample at STP.
    (a) 48.9 g He
    (b) 45.2 g Xe
    (c) 48.2 mg Cl_2
    (d) 3.83 kg SO_2

87. Calculate the mass of each gas sample at STP.
    (a) 178 mL CO_2
    (b) 155 mL O_2
    (c) 1.25 L SF_6

88. Calculate the mass of each gas sample at STP.
    (a) 5.82 L NO
    (b) 0.324 L N_2
    (c) 139 cm^3 Ar

GASES IN CHEMICAL REACTIONS

89. Consider the chemical reaction:

    $$C(s) + H_2O(g) \longrightarrow CO(g) + H_2(g)$$

    How many liters of hydrogen gas are formed from the complete reaction of 1.07 mol of C? Assume that the hydrogen gas is collected at a pressure of 1.0 atm and temperature of 315 K.

90. Consider the chemical reaction:

    $$2 H_2O(l) \longrightarrow 2 H_2(g) + O_2(g)$$

    How many moles of H_2O are required to form 1.3 L of O_2 at a temperature of 325 K and a pressure of 0.988 atm?

91. CH_3OH can be synthesized by the reaction:

    $$CO(g) + 2 H_2(g) \longrightarrow CH_3OH(g)$$

    How many liters of H_2 gas, measured at 748 mm Hg and 86 °C, are required to synthesize 0.55 mol of CH_3OH? How many liters of CO gas, measured under the same conditions, are required?

92. Oxygen gas reacts with powdered aluminum according to the reaction:

    $$4 Al(s) + 3 O_2(g) \longrightarrow 2 Al_2O_3(s)$$

    How many liters of O_2 gas, measured at 782 mm Hg and 25 °C, are required to completely react with 2.4 mol of Al?

93. Nitrogen reacts with powdered aluminum according to the reaction:

    $$2 Al(s) + N_2(g) \longrightarrow 2 AlN(s)$$

    How many liters of N_2 gas, measured at 892 torr and 95 °C, are required to completely react with 18.5 g of Al?

94. Sodium reacts with chlorine gas according to the reaction:

    $$2 Na(s) + Cl_2(g) \longrightarrow 2 NaCl(s)$$

    What volume of Cl_2 gas, measured at 687 torr and 35 °C, is required to form 28 g of NaCl?

95. How many grams of NH_3 form when 24.8 L of $H_2(g)$ (measured at STP) reacts with N_2 to form NH_3 according to this reaction?

$$N_2(g) + 3 H_2(g) \longrightarrow 2 NH_3(g)$$

96. Lithium reacts with nitrogen gas according to the reaction:

$$6 Li(s) + N_2(g) \longrightarrow 2 Li_3N(s)$$

How many grams of lithium are required to completely react with 58.5 mL of N_2 gas measured at STP?

97. How many grams of calcium are consumed when 156.8 mL of oxygen gas, measured at STP, reacts with calcium according to this reaction?

$$2 Ca(s) + O_2(g) \longrightarrow 2 CaO(s)$$

98. How many grams of magnesium oxide are formed when 14.8 L of oxygen gas, measured at STP, completely reacts with magnesium metal according to this reaction?

$$2 Mg(s) + O_2(g) \longrightarrow 2 MgO(s)$$

CUMULATIVE PROBLEMS

99. Use the ideal gas law to show that the molar volume of a gas at STP is 22.4 L.

100. Use the ideal gas law to show that 28.0 g of nitrogen gas and 4.00 g of helium gas occupy the same volume at any temperature and pressure.

101. The mass of an evacuated 255-mL flask is 143.187 g. The mass of the flask filled with 267 torr of an unknown gas at 25 °C is 143.289 g. Calculate the molar mass of the unknown gas.

102. A 118-mL flask is evacuated, and its mass is measured as 97.129 g. When the flask is filled with 768 torr of helium gas at 35 °C, it is found to have a mass of 97.171 g. Is the gas pure helium?

103. A gaseous compound containing hydrogen and carbon is decomposed and found to contain 82.66% carbon and 17.34% hydrogen by mass. The mass of 158 mL of the gas, measured at 556 mm Hg and 25 °C, is 0.275 g. What is the molecular formula of the compound?

104. A gaseous compound containing hydrogen and carbon is decomposed and found to contain 85.63% C and 14.37% H by mass. The mass of 258 mL of the gas, measured at STP, is 0.646 g. What is the molecular formula of the compound?

105. The reaction between zinc and hydrochloric acid is carried out as a source of hydrogen gas in the laboratory:

$$Zn(s) + 2 HCl(aq) \longrightarrow ZnCl_2(aq) + H_2(g)$$

If 325 mL of hydrogen gas is collected over water at 25 °C at a total pressure of 748 mm Hg, how many grams of Zn reacted?

106. Consider the reaction:

$$2 NiO(s) \longrightarrow 2 Ni(s) + O_2(g)$$

If O_2 is collected over water at 40 °C and a total pressure of 745 mm Hg, what volume of gas will be collected for the complete reaction of 24.78 g of NiO?

107. How many grams of hydrogen are collected in a reaction where 1.78 L of hydrogen gas is collected over water at a temperature of 40 °C and a total pressure of 748 torr?

108. How many grams of oxygen are collected in a reaction where 235 mL of oxygen gas is collected over water at a temperature of 25 °C and a total pressure of 697 torr?

109. The decomposition of a silver oxide sample forms 15.8 g of $Ag(s)$:

$$2 Ag_2O(s) \longrightarrow 4 Ag(s) + O_2(g)$$

What total volume of gas forms if it is collected over water at a temperature of 25 °C and a total pressure of 752 mm Hg?

110. The following reaction consumes 2.45 kg of $CO(g)$:

$$CO(g) + H_2O(g) \longrightarrow CO_2(g) + H_2(g)$$

How many total liters of gas are formed if the products are collected at STP?

111. When hydrochloric acid is poured over a sample of sodium bicarbonate, 28.2 mL of carbon dioxide gas is produced at a pressure of 0.954 atm and a temperature of 22.7 °C. Write an equation for the gas evolution reaction and determine how much sodium bicarbonate reacted.

112. When hydrochloric acid is poured over potassium sulfide, 42.9 mL of hydrogen sulfide gas is produced at a pressure of 752 torr and a temperature of 25.8 °C. Write an equation for the gas evolution reaction and determine how much potassium sulfide (in grams) reacted.

113. Consider the reaction:

$2 SO_2(g) + O_2(g) \longrightarrow 2 SO_3(g)$

(a) If 285.5 mL of SO_2 is allowed to react with 158.9 mL of O_2 (both measured at STP), what is the limiting reactant and the theoretical yield of SO_3?

(b) If 187.2 mL of SO_3 is collected (measured at STP), what is the percent yield for the reaction?

114. Consider the reaction:

$P_4(s) + 6 H_2(g) \longrightarrow 4 PH_3(g)$

(a) If 88.6 L of $H_2(g)$, measured at STP, is allowed to react with 158.3 g of P_4, what is the limiting reactant?

(b) If 48.3 L of PH_3, measured at STP, forms, what is the percent yield?

115. Consider the reaction for the synthesis of nitric acid:

$3 NO_2(g) + H_2O(l) \longrightarrow 2 HNO_3(aq) + NO(g)$

(a) If 12.8 L of $NO_2(g)$, measured at STP, is allowed to react with 14.9 g of water, find the limiting reagent and the theoretical yield of HNO_3 in grams.

(b) If 14.8 g of HNO_3 forms, what is the percent yield?

116. Consider the reaction for the production of NO_2 from NO:

$2 NO(g) + O_2(g) \longrightarrow 2 NO_2(g)$

(a) If 84.8 L of $O_2(g)$, measured at 35 °C and 632 mm Hg, is allowed to react with 158.2 g of NO, find the limiting reagent.

(b) If 97.3 L of NO_2 forms, measured at 35 °C and 632 mm Hg, what is the percent yield?

117. Ammonium carbonate decomposes upon heating according to the balanced equation:

$(NH_4)_2CO_3(s) \longrightarrow 2 NH_3(g) + CO_2(g) + H_2O(g)$

Calculate the total volume of gas produced at 22 °C and 1.02 atm by the complete decomposition of 11.83 g of ammonium carbonate.

118. Ammonium nitrate decomposes explosively upon heating according to the balanced equation:

$2 NH_4NO_3(s) \longrightarrow 2 N_2(g) + O_2(g) + 4 H_2O(g)$

Calculate the total volume of gas (at 25 °C and 748 mm Hg) produced by the complete decomposition of 1.55 kg of ammonium nitrate.

119. A mixture containing 235 mg of helium and 325 mg of neon has a total pressure of 453 torr. What is the partial pressure of helium in the mixture?

120. A mixture containing 4.33 g of CO_2 and 3.11 g of CH_4 has a total pressure of 1.09 atm. What is the partial pressure of CO_2 in the mixture?

121. Consider the reaction:

$2 SO_2(g) + O_2(g) \longrightarrow 2 SO_3(g)$

A reaction flask initially contains 0.10 atm of SO_2 and 0.10 atm of O_2. What is the total pressure in the flask once the limiting reactant is completely consumed? Assume a constant temperature and volume and a 100% reaction yield.

122. Consider the reaction:

$CO(g) + 2 H_2(g) \longrightarrow CH_3OH(g)$

A reaction flask initially contains 112 torr of CO and 282 torr of H_2. The reaction is allowed to proceed until the pressure stops changing, at which point the total pressure is 196 torr. Determine the percent yield for the reaction. Assume a constant temperature and that no other reactions occur other than the one indicated.

HIGHLIGHT PROBLEMS

123. Which gas sample depicted, all at the same temperature, will have the greatest pressure? Explain.

(a)

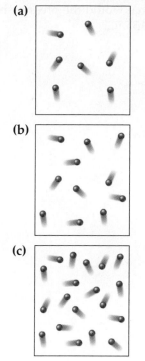

(b)

(c)

124. This image represents a sample of gas at a pressure of 1 atm, a volume of 1 L, and a temperature of 25 °C. Draw a similar picture showing what happens if the volume were reduced to 0.5 L and the temperature increased to 250 °C. What happens to the pressure?

$V = 1.0$ L
$T = 25$ °C
$P = 1.0$ atm

125. Automobile air bags inflate following a serious impact. The impact triggers the chemical reaction:

$$2\,NaN_3(s) \rightarrow 2\,Na(s) + 3\,N_2(g)$$

If an automobile air bag has a volume of 11.8 L, how much NaN_3 in grams is required to fully inflate the air bag upon impact? Assume STP conditions.

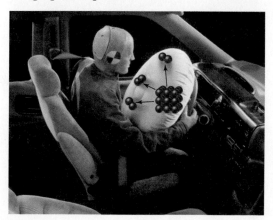

126. Olympic cyclists fill their tires with helium to make them lighter. Calculate the mass of air in an air-filled tire and the mass of helium in a helium-filled tire. What is the mass difference between the two? Assume that the volume of the tire is 855 mL, that it is filled with a total pressure of 125 psi, and that the temperature is 25 °C. Also, assume an average molar mass for air of 28.8 g/mol.

127. In a common classroom demonstration, a balloon is filled with air and submerged into liquid nitrogen. The balloon contracts as the gases within the balloon cool. Suppose the balloon initially contains 2.95 L of air at a temperature of 25.0 °C and a pressure of 0.998 atm. Calculate the expected volume of the balloon upon cooling to −196 °C (the boiling point of liquid nitrogen). When the demonstration is carried out, the actual volume of the balloon decreases to 0.61 L. How well does the observed volume of the balloon compare to your calculated value? Can you explain the difference?

128. Aerosol cans carry clear warnings against incineration because of the high pressures that can develop upon heating. Suppose a can contains a residual amount of gas at a pressure of 755 mm Hg and a temperature of 25 °C. What would the pressure be if the can were heated to 1155 °C?

▶**ANSWERS TO SKILLBUILDER EXERCISES**

Skillbuilder 11.1	85.0 psi	**Skillbuilder 11.7**	16.1 L
Skillbuilder Plus, p. 366	80.6 kPa	**Skillbuilder Plus, p. 382**	977 mm Hg
Skillbuilder 11.2	$P_2 = 2.1$ atm; depth is approximately 11 m	**Skillbuilder 11.8**	70.8 g/mol
		Skillbuilder 11.9	721 torr
Skillbuilder 11.3	123 mL	**Skillbuilder 11.10**	$P_{tot} = 4.2$ atm
Skillbuilder 11.4	0.82 L	**Skillbuilder 11.11**	82.3 g
Skillbuilder 11.5	13 L	**Skillbuilder 11.12**	6.53 L O_2
Skillbuilder 11.6	1.6 atm		

▶**ANSWERS TO CONCEPTUAL CHECKPOINTS**

11.1 (a) Since the particles are all identical, and since (a) has the smallest number of particles per unit volume, it will have the lowest pressure.

11.2 (c) Atmospheric pressure will support a column of water 10.3 m in height. If the liquid in a barometer were twice as dense as water, a column of it would be twice as heavy and the pressure it exerted at its base would be twice as great. Therefore, atmospheric pressure would be able to support a column only half as high.

11.3 (b) Since the volume triples, and since according to Boyle's law the volume and pressure are inversely proportional, the pressure will fall by a factor of 3.

11.4 (c) The piston will move in response to any pressure difference, keeping the pressure of the gas in the cylinder equal to that of the surrounding atmosphere. Thus, we know that the final pressure of the gas will be the same as it was initially. At constant pressure, the volume of the gas will be proportional to the temperature—if the Kelvin temperature doubles, the volume will double.

11.5 (c) Since hydrogen gas has the lowest molar mass of the set, 1 g will have the greatest number of moles and therefore the greatest volume.

CHAPTER 12

Liquids, Solids, and Intermolecular Forces

"It will be found that everything depends on the composition of the forces with which the particles of matter act upon one another; and from these forces ... all phenomena of nature take their origin."

ROGER JOSEPH BOSCOVICH (1711–1787)

12.1 Interactions between Molecules

Bite into a candy bar and taste its sweetness. Drink a cup of strong coffee and experience its bitterness. What causes these flavors? Most tastes originate from interactions between molecules. Certain molecules in coffee, for example, interact with molecular receptors on the surface of specialized cells on the tongue. The receptors are highly specific, recognizing only certain types of molecules. The interaction between the molecule and the receptor triggers a signal that goes to the brain, which we interpret as a bitter taste. Bitter tastes are usually unpleasant because many of the molecules that cause them are poisons. The sensation of bitterness is probably an evolutionary adaptation that helps us avoid these poisons.

The interaction between the molecules in coffee that taste bitter and the taste receptors on the tongue is caused by **intermolecular forces**—attractive forces that exist *between* molecules. Living organisms depend on intermolecular forces not only for taste but also for many other physiological processes. For example, in Chapter 19, we will see how intermolecular forces help determine the shapes of protein molecules—the workhorse molecules in living organisms. Later in this chapter—in the *Chemistry and Health* box in Section 12.6—we learn how intermolecular forces are central to DNA, the inheritable molecules that serve as blueprints for life.

The interactions between bitter molecules in coffee and molecular receptors on the tongue are highly specific. However, less specific intermolecular forces exist between all molecules and atoms. These intermolecular forces are responsible for the very existence of liquids and solids. The state of a sample of matter—solid, liquid, or gas—depends on the magnitude of intermolecular forces relative to the amount of thermal energy in the sample. Recall from Section 3.10 that the molecules and atoms that compose matter are in constant random motion that increases with increasing temperature. The energy associated with this motion is

◄ Flavors are caused by the interactions of molecules in foods or drinks with molecular receptors on the surface of the tongue. This image shows a caffeine molecule, one of the substances responsible for the sometimes bitter flavors in coffee.

called **thermal energy**. The weaker the intermolecular forces relative to thermal energy, the more likely the sample will be gaseous. The stronger the intermolecular forces relative to thermal energy, the more likely the sample will be liquid or solid.

12.2 Properties of Liquids and Solids

We are all familiar with solids and liquids. Water, gasoline, rubbing alcohol, and fingernail-polish remover are all common liquids. Ice, dry ice, and diamond are familiar solids. In contrast to gases—in which molecules or atoms are separated by large distances—the molecules or atoms that compose liquids and solids are in close contact with one another (▼ Figure 12.1).

▶ **FIGURE 12.1 Gas, liquid, and solid states**

Gas   Liquid   Solid

The difference between solids and liquids is in the freedom of movement of the constituent molecules or atoms. In liquids, even though the atoms or molecules are in close contact, they are still free to move around each other. In solids, the atoms or molecules are fixed in their positions, although thermal energy causes them to vibrate about a fixed point. These molecular properties of solids and liquids result in the following macroscopic properties.

Properties of Liquids

- High densities in comparison to gases.
- Indefinite shape; they assume the shape of their container.
- Definite volume; they are not easily compressed.

Properties of Solids

- High densities in comparison to gases.
- Definite shape; they do not assume the shape of their container.
- Definite volume; they are not easily compressed.
- May be crystalline (ordered) or amorphous (disordered).

Table 12.1 summarizes these properties, as well as the properties of gases for comparison.

TABLE 12.1 Properties of the States of Matter

Phase	Density	Shape	Volume	Strength of Intermolecular Forces[a]	Example
gas	low	indefinite	indefinite	weak	carbon dioxide gas (CO_2)
liquid	high	indefinite	definite	moderate	liquid water (H_2O)
solid	high	definite	definite	strong	sugar ($C_{12}H_{22}O_{11}$)

[a]Relative to thermal energy.

Liquids have high densities in comparison to gases because the atoms or molecules that compose liquids are much closer together. The density of liquid water, for example, is 1.0 g/cm^3 (at 25 °C), while the density of gaseous water at 100 °C and 1 atm is 0.59 g/L, or $5.9 \times 10^{-4} \text{ g/cm}^3$. Liquids assume the shape of their containers because the atoms or molecules that compose them are free to flow. When you pour water into a flask, the water flows and assumes the shape of the flask (▼ Figure 12.2). Liquids are not easily compressed because the molecules or atoms that compose them are in close contact—they cannot be pushed closer together.

Like liquids, solids have high densities in comparison to gases because the atoms or molecules that compose solids are also close together. The densities of solids are usually just slightly greater than those of the corresponding liquids. A major exception is water, whose solid (ice) is slightly less dense than liquid water. Solids have a definite shape because, in contrast to liquids or gases, the molecules or atoms that compose solids are fixed in place (▼ Figure 12.3). Each molecule or atom only vibrates about a fixed point. Like liquids, solids have a definite volume and cannot be compressed because the molecules or atoms composing them are in close contact. As described in Section 3.3, solids may be *crystalline*, in which case the atoms or molecules that compose them arrange themselves in a well-ordered, three-dimensional array, or they may be *amorphous*, in which case the atoms or molecules that compose them have no long-range order.

As we will see in Section 12.8, ice is less dense than liquid water because water expands when it freezes due to its unique crystalline structure.

▲ **FIGURE 12.2 A liquid assumes the shape of its container** Because the molecules in liquid water are free to move around each other, they flow and assume the shape of their container.

▲ **FIGURE 12.3 Solids have a definite shape** In a solid such as ice, the molecules are fixed in place. However, they vibrate about fixed points.

12.3 Intermolecular Forces in Action: Surface Tension and Viscosity

The most important manifestation of intermolecular forces is the very existence of liquids and solids. Without intermolecular forces, solids and liquids would not exist and all matter would be gaseous. In liquids, we can observe several other manifestations of intermolecular forces including surface tension and viscosity.

SURFACE TENSION

A fly fisherman delicately casts a small metal hook (with a few feathers and strings attached to make it look like an insect) onto the surface of a moving stream. The hook floats on the surface of the water and attracts trout (▶ Figure 12.4).

Interior molecule interacts with six neighbors.

Surface molecule interacts with only four neighbors.

▲ **FIGURE 12.5   Origin of surface tension**   Molecules at the surface of a liquid interact only with neighbors below them, resulting in a net inward force that creates tension at the surface. This tension tends to minimize the area of the surface and makes it resist penetration.

▲ **FIGURE 12.4   Floating flies** Even though they are denser than water, fly-fishing lures float on the surface of a stream or lake because of surface tension.

▲ **FIGURE 12.6   Surface tension at work**   A paper clip will float on water if it is carefully placed on the surface of the water. It is held up by surface tension.

The hook floats because of **surface tension**, the tendency of liquids to minimize their surface area. This tendency causes liquids to have a sort of "skin" that resists penetration. For the fisherman's hook to sink into the water, the water's surface area would have to increase slightly. The increase is resisted because molecules at the surface interact only with molecules on the interior of the liquid (▲ Figure 12.5). Therefore, the surface molecules are subjected to a net inward force only (which is not balanced by any force from the external side), so the surface is under tension and that tension tends to minimize its area. You can observe surface tension by carefully placing a paper clip on the surface of water (◀ Figure 12.6). The paper clip, even though it is denser than water, will float on the surface of the water. A slight tap on the clip will overcome surface tension and cause the clip to sink. Surface tension increases with increasing intermolecular forces. You can't float a paper clip on gasoline, for example, because the intermolecular forces among the molecules composing gasoline are weaker than the intermolecular forces among water molecules; they are not under as much tension, so they do not form a "skin."

VISCOSITY

Another manifestation of intermolecular forces is **viscosity**, the resistance of a liquid to flow. Liquids that are viscous flow more slowly than liquids that are not viscous. For example, motor oil is more viscous than gasoline, and maple syrup is more viscous than water (◀ Figure 12.7). Viscosity is greater in substances with stronger intermolecular forces because molecules cannot move around each other as freely, hindering flow. Long molecules, such as the hydrocarbons in motor oil, also tend to form viscous liquids because of molecular entanglement (the long chainlike molecules get tangled together).

◀ **FIGURE 12.7   Viscosity**   Maple syrup is more viscous than water because its molecules interact strongly, and so cannot flow past one another easily.

415

EVERYDAY CHEMISTRY

Why Are Water Drops Spherical?

Have you ever seen a close-up photograph of tiny water droplets (▼ Figure 12.8) or carefully watched water in free fall? In both cases, the distorting effects of gravity are diminished, and the water forms nearly perfect spheres. On the space shuttle, the complete absence of gravity results in floating spheres of water (▼ Figure 12.9). Why? Water drops are spherical because of the surface tension caused by the attractive forces between water molecules. Just as gravity pulls matter within a planet or star inward to form a sphere, so intermolecular forces pull water molecules inward to form a sphere. The sphere minimizes the surface-area-to-volume ratio, thereby minimizing the number of molecules at the surface.

A collection of magnetic marbles provides a good physical model of a water drop. Each magnetic marble is like a water molecule, attracted to the marbles around it. If you agitate these marbles slightly, so that they can find their preferred configuration, they tend toward a spherical shape (▼ Figure 12.10) because the attractions between the marbles cause them to minimize the number of marbles at the surface.

▲ **FIGURE 12.8 An almost perfect sphere** If a water droplet is small enough, it will largely be free of the distorting effects of gravity and be almost perfectly spherical.

▲ **FIGURE 12.9 A perfect sphere** In the absence of gravity, as in this picture taken on the space shuttle, water assumes the shape of a sphere.

▲ **FIGURE 12.10 An analogy for surface tension** Magnetic marbles tend to arrange themselves in a spherical shape.

CAN YOU ANSWER THIS? *How does the tendency of a liquid to form spherical drops depend on the strength of intermolecular forces? Do liquids with weaker intermolecular forces have a higher or lower tendency to form spherical drops?*

12.4 Evaporation and Condensation

Leave a glass of water in the open for several days and the water level within the glass slowly drops. Why? The first reason is that water molecules at the surface of the liquid—which experience fewer attractions to neighboring molecules and are therefore held less tightly—can break away from the rest of the liquid. The second reason is that all of the molecules in the liquid have a *distribution of kinetic energy* at

▶ **FIGURE 12.11 Energy distribution** At a given temperature, a sample of molecules or atoms will have a distribution of kinetic energies, as shown here. Only a small fraction of molecules have enough energy to escape. At a higher temperature, the fraction of molecules with enough energy to escape increases.

In evaporation or vaporization, a substance is converted from its liquid state into its gaseous state.

▲ **FIGURE 12.12 Evaporation** Because molecules on the surface of a liquid are held less tightly than those in the interior, the most energetic among them can break away into the gas state in the process called *evaporation*.

Dynamic equilibrium is so named because both condensation and evaporation of individual molecules continue, but at the same rate.

See Table 11.4 for the vapor pressure of water at different temperatures.

any given temperature (▲ Figure 12.11). At any given moment, some molecules in the liquid are moving faster than the average (higher energy), and others are moving more slowly (lower energy). Some of the molecules that are moving faster have enough energy to break free from the surface, resulting in **evaporation** or **vaporization**, a physical change in which a substance is converted from its liquid state to its gaseous state (◀ Figure 12.12).

If you spill the same amount of water (as was in the glass) on a table, it evaporates more quickly, probably within a few hours. Why? The surface area of the spilled water is greater, leaving more molecules susceptible to evaporation. If you warm the glass of water, it also evaporates more quickly because the greater thermal energy causes a greater fraction of molecules to have enough energy to break away from the surface (see Figure 12.11). If you fill the glass with rubbing alcohol instead of water, the liquid again evaporates more quickly because the intermolecular forces between the alcohol molecules are weaker than the intermolecular forces between water molecules. In general, the rate of vaporization increases with:

- Increasing surface area
- Increasing temperature
- Decreasing strength of intermolecular forces

Liquids that evaporate easily are termed **volatile**, while those that do not vaporize easily are termed **nonvolatile**. Rubbing alcohol, for example, is more volatile than water. Motor oil at room temperature is virtually nonvolatile.

If you leave water in a *closed* container, its level remains constant because the molecules that leave the liquid are trapped in the air space above the water. These gaseous molecules bounce off of the walls of the container and eventually hit the surface of the water again and recondense. **Condensation** is a physical change in which a substance is converted from its gaseous state to its liquid state.

Evaporation and condensation are opposites: Evaporation is a liquid turning into a gas, and condensation is a gas turning into a liquid. When liquid water is initially put into a closed container, more evaporation happens than condensation because there are so few gaseous water molecules in the space above the water (▶ Figure 12.13a). However, as the number of gaseous water molecules increases, the rate of condensation also increases (▶ Figure 12.13b). At the point where the rates of condensation and evaporation become equal (▶ Figure 12.13c), **dynamic equilibrium** is reached and the number of gaseous water molecules above the liquid remains constant. The **vapor pressure** of a liquid is the partial pressure of its vapor in dynamic equilibrium with its liquid. For water at 25 °C, the vapor pressure is 23.8 mm Hg. Vapor pressure increases with:

- Increasing temperature
- Decreasing strength of intermolecular forces

Vapor pressure is independent of surface area because an increase in surface area at equilibrium equally affects the rate of evaporation and the rate of condensation.

Evaporation begins to occur.

Evaporation continues, but condensation also begins to occur.

Dynamic equilibrium: rate of evaporation = rate of condensation

(a)   (b)   (c)

▶ **FIGURE 12.13   Evaporation and condensation**   **(a)** When water is first put into a closed container, water molecules begin to evaporate. **(b)** As the number of gaseous molecules increases, some of the molecules begin to collide with the liquid and are recaptured—that is, they recondense into liquid. **(c)** When the rate of evaporation equals the rate of condensation, dynamic equilibrium occurs, and the number of gaseous molecules remains constant.

BOILING

Sometimes you see bubbles begin to form in hot water below 100 °C. These bubbles are dissolved air—not gaseous water—leaving the liquid. Dissolved air comes out of water as you heat it because the solubility of a gas in a liquid decreases with increasing temperature (Section 13.4).

As you increase the temperature of water in an open container, the increasing thermal energy causes molecules to leave the surface and vaporize at a faster and faster rate. At the **boiling point**—the temperature at which the vapor pressure of a liquid is equal to the pressure above it—the thermal energy is enough for molecules within the *interior* of the liquid (not just those at the surface) to break free into the gas phase (▼ Figure 12.14). Water's **normal boiling point**—its boiling point at a pressure of 1 atmosphere—is 100 °C. When a sample of water reaches 100 °C, you can see bubbles form within the liquid. These bubbles are pockets of gaseous water. The bubbles quickly rise to the surface of the liquid, and the water molecules that were in the bubble leave as gaseous water, or steam. Once the boiling point of a liquid is reached, additional heating only causes more rapid boiling; it does not raise the temperature of the liquid above its boiling point (▶ Figure 12.15). A mixture of boiling water *and* steam will always have a temperature of 100 °C (at 1 atm pressure). Only after all the water has been converted to steam can the temperature of the steam rise beyond 100 °C.

▶ **FIGURE 12.14   Boiling**   During boiling, thermal energy is enough to cause water molecules in the interior of the liquid to become gaseous, forming bubbles containing gaseous water molecules.

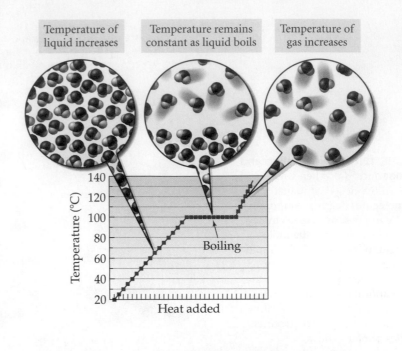

Temperature of liquid increases

Temperature remains constant as liquid boils

Temperature of gas increases

Boiling

Heat added

▶ **FIGURE 12.15 Heating curve during boiling** The temperature of water as it is heated from room temperature through boiling. During boiling, the temperature remains at 100 °C until all the liquid is evaporated.

CONCEPTUAL CHECKPOINT 12.1

The gas over a rapidly boiling pot of water is sampled and analyzed. Which substance do you expect to compose a large fraction of the gas sample?

(a) $H_2(g)$ **(b)** $H_2O(g)$ **(c)** $O_2(g)$ **(d)** $H_2O_2(g)$

ENERGETICS OF EVAPORATION AND CONDENSATION

In an endothermic process, heat is absorbed; in an exothermic process, heat is released.

Evaporation is *endothermic*—when a liquid is converted into a gas, it absorbs heat because energy is required to break molecules away from the rest of the liquid. Imagine a collection of water molecules in the liquid state. As the water evaporates, it cools—typical of endothermic processes—because only the fastest-moving molecules break away, which leaves the slower-moving molecules behind. Under ordinary conditions, the slight decrease in temperature of water as it evaporates is counteracted by thermal energy transfer from the surroundings, which warms the water back up. However, if the evaporating water were thermally isolated from the surroundings, it would continue to cool down as it evaporated.

You can observe the endothermic nature of evaporation by turning off the heat beneath a boiling pot of water; it will quickly stop boiling as the heat lost due to vaporization causes the water to cool below its boiling point. Our bodies use the endothermic nature of evaporation for cooling. When we overheat, we sweat, causing our skin to be covered with liquid water. As this water evaporates it absorbs heat from our bodies, cooling us down. A fan intensifies the cooling effect because it blows newly vaporized water away from the skin, allowing more sweat to vaporize and cause even more cooling. High humidity, on the other hand, slows down evaporation, preventing cooling. When the air already contains high amounts of water vapor, sweat does not evaporate as easily, making our cooling system less efficient.

Condensation, the opposite of evaporation, is *exothermic*—heat is released when a gas condenses to a liquid. If you have ever accidentally put your hand above a steaming kettle, you may have experienced a *steam burn*. As the steam condenses to a liquid on your skin, it releases heat, causing a severe burn. The exothermic nature of condensation is also the reason that winter overnight temperatures in coastal cities, which tend to have water vapor in the air, do not get as low as in deserts, which tend to have dry air. As the air temperature in a coastal city drops, water condenses out of the air, releasing heat and preventing the temperature from dropping further. In deserts, there is little moisture in the air to condense, so the temperature drop is greater.

HEAT OF VAPORIZATION

The amount of heat required to vaporize one mole of liquid is the **heat of vaporization**(ΔH_{vap}). The heat of vaporization of water at its normal boiling point (100 °C) is 40.7 kJ/mole.

$$H_2O(l) \longrightarrow H_2O(g) \quad \Delta H = +40.7 \text{ kJ (at 100 °C)}$$

ΔH is positive because vaporization is endothermic; energy must be added to the water to vaporize it.

The same amount of heat is involved when 1 mol of gas condenses, but *the heat is emitted* rather than absorbed.

$$H_2O(g) \longrightarrow H_2O(l) \quad \Delta H = -40.7 \text{ kJ (at 100 °C)}$$

ΔH is negative because condensation is exothermic; energy is given off as the water condenses.

Different liquids have different heats of vaporization (Table 12.2). Heats of vaporization are also *temperature dependent* (they change with temperature). The higher the temperature, the easier it is to vaporize a given liquid and therefore the lower the heat of vaporization.

TABLE 12.2   Heats of Vaporization of Several Liquids at Their Boiling Points and at 25 °C

Liquid	Chemical Formula	Normal Boiling Point (°C)	Heat of Vaporization (kJ/mol) at Boiling Point	Heat of Vaporization (kJ/mol) at 25 °C
water	H_2O	100.0	40.7	44.0
isopropyl alcohol (rubbing alcohol)	C_3H_8O	82.3	39.9	45.4
acetone	C_3H_6O	56.1	29.1	31.0
diethyl ether	$C_4H_{10}O$	34.5	26.5	27.1

We can use the heat of vaporization of a liquid to calculate the amount of heat energy required to vaporize a given amount of that liquid. To do so, we use the heat of vaporization as a conversion factor between moles of the liquid and the amount of heat required to vaporize it. For example, suppose we want to calculate the amount of heat required to vaporize 25.0 g of water at its boiling point. We begin by sorting the information in the problem statement.

GIVEN: 25.0 g H_2O

FIND: heat (kJ)

SOLUTION MAP

We then strategize by building a solution map that begins with the mass of water and ends with the energy required to vaporize it.

$$\begin{array}{ccc} \boxed{\text{g } H_2O} & \longrightarrow \boxed{\text{mol } H_2O} & \longrightarrow \boxed{\text{kJ}} \end{array}$$

$$\frac{1 \text{ mol } H_2O}{18.02 \text{ g } H_2O} \qquad \frac{40.7 \text{ kJ}}{1 \text{ mol } H_2O}$$

RELATIONSHIPS USED

$$\Delta H_{vap} = 40.7 \text{ kJ/mol at } 100 \text{ °C (Table 12.2)}$$

$$1 \text{ mol } H_2O = 18.02 \text{ g } H_2O \text{ (molar mass of water)}$$

SOLUTION

$$25.0 \text{ g } H_2O \times \frac{1 \text{ mol } H_2O}{18.02 \text{ g } H_2O} \times \frac{40.7 \text{ kJ}}{1 \text{ mol } H_2O} = 56.5 \text{ kJ}$$

EXAMPLE 12.1 Using the Heat of Vaporization in Calculations

Calculate the amount of water in grams that can be vaporized at its boiling point with 155 kJ of heat.

SORT You are given the number of kilojoules of heat energy and asked to find the mass of water that can be vaporized with the given amount of energy.	**GIVEN:** 155 kJ **FIND:** g H_2O
STRATEGIZE Draw the solution map beginning with the energy in kilojoules and converting to moles of water and then to grams of water.	**SOLUTION MAP** $$\boxed{kJ} \longrightarrow \boxed{mol\ H_2O} \longrightarrow \boxed{g\ H_2O}$$ $\dfrac{1\ mol\ H_2O}{40.7\ kJ}$ $\qquad$ $\dfrac{18.02\ g\ H_2O}{1\ mol\ H_2O}$ **RELATIONSHIPS USED** $\Delta H_{vap} = 40.7$ kJ/mol at $100\,°C$ (Table 12.2) 18.02 g $H_2O = 1$ mol H_2O (molar mass of water)
SOLVE Follow the solution map to solve the problem.	**SOLUTION** $155\ kJ \times \dfrac{1\ mol\ H_2O}{40.7\ kJ} \times \dfrac{18.02\ g}{1\ mol\ H_2O} = 68.6\ g$
CHECK Check your answer. Are the units correct? Does the answer make physical sense?	The units (g) are correct. The magnitude of the answer makes sense because each mole of water absorbs about 40 kJ of energy upon vaporization. Therefore 155 kJ should vaporize close to 4 mol of water, which is consistent with the answer (4 mol of water has a mass of about 72 g)

▶ **SKILLBUILDER 12.1 | Using the Heat of Vaporization in Calculations**

Calculate the amount of heat in kilojoules required to vaporize 2.58 kg of water at its boiling point.

▶ **SKILLBUILDER PLUS**

A drop of water weighing 0.48 g condenses on the surface of a 55-g block of aluminum that is initially at 25 °C. If the heat released during condensation goes only toward heating the metal, what is the final temperature in Celsius of the metal block? (The specific heat capacity of aluminum is given in Table 3.4 and is 0.903 J/g °C.)

▶ **FOR MORE PRACTICE** Example 12.7; Problems 47, 48, 49, 50, 51, 52.

12.5 Melting, Freezing, and Sublimation

As the temperature of a solid increases, thermal energy causes the molecules and atoms composing the solid to vibrate faster. At the **melting point**, atoms and molecules have enough thermal energy to overcome the intermolecular forces that hold them at their stationary points, and the solid turns into a liquid. The melting point of ice, for example, is 0 °C. Once the melting point of a solid is reached, additional heating only causes more rapid melting; it does not raise the temperature of the solid above its melting point (▶ Figure 12.16). Only after all of the ice has melted will additional heating raise the temperature of the liquid water past 0 °C. A mixture of water *and* ice will always have a temperature of 0 °C (at 1 atm pressure).

ENERGETICS OF MELTING AND FREEZING

The most common way to cool down a drink is to drop several ice cubes into it. As the ice melts, the drink cools because melting is endothermic—heat is absorbed

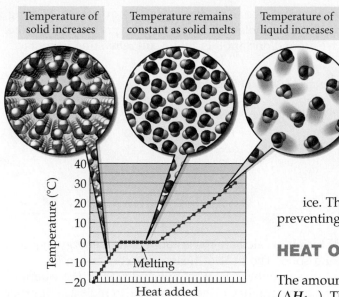

Temperature of solid increases

Temperature remains constant as solid melts

Temperature of liquid increases

▲ **FIGURE 12.16 Heating curve during melting** A graph of the temperature of ice as it is heated from −20 °C to 35 °C. During melting, the temperature of the solid and the liquid remain at 0 °C until the entire solid is melted.

when a solid is converted into a liquid. The melting ice absorbs heat from the liquid in the drink and cools the liquid. Melting is endothermic because energy is required to partially overcome the attractions between molecules in the solid and free them into the liquid state.

Freezing, the opposite of melting, is exothermic— heat is released when a liquid freezes into a solid. For example, as water in your freezer turns into ice, it releases heat, which must be removed by the refrigeration system of the freezer. If the refrigeration system did not remove the heat, the water would not completely freeze into ice. The heat released as it began to freeze would warm the freezer, preventing further freezing.

HEAT OF FUSION

The amount of heat required to melt 1 mol of a solid is the **heat of fusion** (ΔH_{fus}). The heat of fusion for water is 6.02 kJ/mol.

$$H_2O(s) \longrightarrow H_2O(l) \quad \Delta H = +6.02 \text{ kJ}$$

ΔH is positive because melting is endothermic; energy must be added to the ice to melt it.

The same amount of heat is involved when 1 mol of liquid water freezes, but the heat is emitted rather than absorbed.

$$H_2O(l) \longrightarrow H_2O(s) \quad \Delta H = -6.02 \text{ kJ}$$

ΔH is negative because freezing is exothermic; energy is given off as the water freezes.

Different substances have different heats of fusion (Table 12.3 on p. 422). Notice that, in general, the heat of fusion is significantly less than the heat of vaporization. It takes less energy to melt 1 mol of ice than it does to vaporize 1 mol of liquid water. Why? Vaporization requires complete separation of one molecule

▶ When ice melts, water molecules break free from the solid structure and become liquid. As long as ice and water are both present, the temperature will be 0.0 °C.

from another, so the intermolecular forces must be completely overcome. Melting, on the other hand, requires that intermolecular forces be only partially overcome, allowing molecules to move around one another while still remaining in contact.

TABLE 12.3   Heats of Fusion of Several Substances

Liquid	Chemical Formula	Melting Point (°C)	Heat of Fusion (kJ/mol)
water	H_2O	0.00	6.02
isopropyl alcohol (rubbing alcohol)	C_3H_8O	−89.5	5.37
acetone	C_3H_6O	−94.8	5.69
diethyl ether	$C_4H_{10}O$	−116.3	7.27

We can use the heat of fusion to calculate the amount of heat energy required to melt a given amount of a solid. The heat of fusion is a conversion factor between moles of a solid and the amount of heat required to melt it. For example, suppose we want to calculate the amount of heat required to melt 25.0 g of ice (at 0 °C). We first sort the information in the problem statement.

GIVEN: 25.0 g H_2O

FIND: heat (kJ)

SOLUTION MAP

We then draw the solution map, beginning with the mass of water and ending with the energy required to melt it.

$$\boxed{\text{g } H_2O} \longrightarrow \boxed{\text{mol } H_2O} \longrightarrow \boxed{\text{kJ}}$$

$$\frac{1 \text{ mol } H_2O}{18.02 \text{ g } H_2O} \qquad \frac{6.02 \text{ kJ}}{1 \text{ mol } H_2O}$$

RELATIONSHIPS USED

$$\Delta H_{fus} = 6.02 \text{ kJ/mol (Table 12.3)}$$

$$1 \text{ mol } H_2O = 18.02 \text{ g } H_2O \text{ (molar mass of water)}$$

SOLUTION

$$25.0 \text{ g } H_2O \times \frac{1 \text{ mol } H_2O}{18.02 \text{ g } H_2O} \times \frac{6.02 \text{ kJ}}{1 \text{ mol } H_2O} = 8.35 \text{ kJ}$$

EXAMPLE 12.2   Using the Heat of Fusion in Calculations

Calculate the amount of ice in grams that, upon melting (at 0 °C), absorbs 237 kJ of heat.

SORT	
You are given the number of kJ of heat energy and asked to find the mass of ice that absorbs the given amount of energy upon melting.	**GIVEN:** 237 kJ **FIND:** g H_2O (ice)

STRATEGIZE	
Draw the solution map beginning with the energy in kilojoules and converting to moles of water and then to grams of water.	**SOLUTION MAP** $\boxed{\text{kJ}} \longrightarrow \boxed{\text{mol } H_2O} \longrightarrow \boxed{\text{g } H_2O}$ $\frac{1 \text{ mol } H_2O}{6.02 \text{ kJ}} \qquad \frac{18.02 \text{ g}}{1 \text{ mol } H_2O}$

	RELATIONSHIPS USED $\Delta H_{fus} = 6.02 \, \text{kJ/mol}$ (Table 12.3) $1 \, \text{mol H}_2\text{O} = 18.02 \, \text{g H}_2\text{O}$ (molar mass of water)
SOLVE Follow the solution map to solve the problem.	**SOLUTION** $$237 \, \text{kJ} \times \frac{1 \, \text{mol H}_2\text{O}}{6.02 \, \text{kJ}} \times \frac{18.02 \, \text{g}}{1 \, \text{mol H}_2\text{O}} = 709 \, \text{g}$$
CHECK Check your answer. Are the units correct? Does the answer make physical sense?	The units (g) are correct. The magnitude of the answer makes sense because each mole of water absorbs about 6 kJ of energy upon melting. Therefore 237 kJ should melt close to 40 mol of water, which is consistent with the answer (40 mol of water has a mass of about 720 g).

▶**SKILLBUILDER 12.2 | Using the Heat of Fusion in Calculations**

Calculate the amount of heat absorbed when a 15.5-g ice cube melts (at 0 °C).

▶**SKILLBUILDER PLUS**

A 5.6-g ice cube (at 0 °C) is placed into 195 g of water initially at 25 °C. If the heat absorbed for melting the ice comes only from the 195 g of water, what is the temperature change of the 195 g of water?

▶**FOR MORE PRACTICE**   Example 12.8; Problems 55, 56, 57, 58.

✓ **CONCEPTUAL CHECKPOINT 12.2**

This diagram shows a heating curve for ice beginning at −25 °C and ending at 125 °C. Correlate sections i, ii, and iii with the correct states of water.

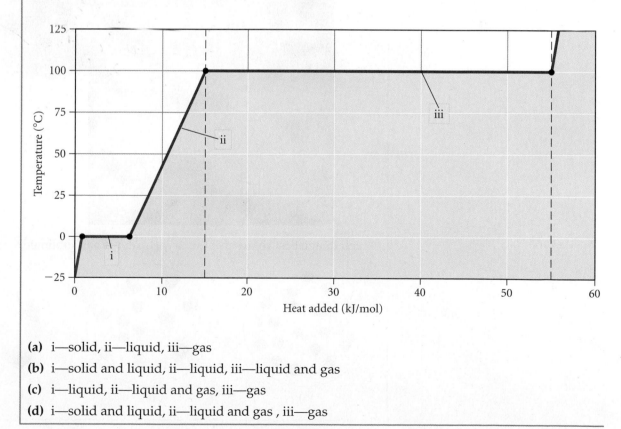

(a) i—solid, ii—liquid, iii—gas

(b) i—solid and liquid, ii—liquid, iii—liquid and gas

(c) i—liquid, ii—liquid and gas, iii—gas

(d) i—solid and liquid, ii—liquid and gas , iii—gas

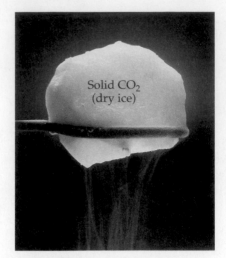

▲ Dry ice is solid carbon dioxide. The solid does not melt but rather sublimes. It transforms directly from solid carbon dioxide to gaseous carbon dioxide.

SUBLIMATION

Sublimation is a physical change in which a substance changes from its solid state directly to its gaseous state. When a substance sublimes, molecules leave the surface of the solid, where they are held less tightly than in the interior, and become gaseous. For example, dry ice, which is solid carbon dioxide, does not melt under atmospheric pressure (at any temperature). At $-78\,°C$, the CO_2 molecules have enough energy to leave the surface of the dry ice and become gaseous. Regular ice will slowly sublime at temperatures below $0\,°C$. You can observe the sublimation of ice in cold climates; ice or snow laying on the ground gradually disappears, even if the temperature remains below $0\,°C$. Similarly, ice cubes left in the freezer for a long time slowly become smaller, even though the freezer is always below $0\,°C$. In both cases, the ice is subliming, turning directly into water vapor.

Ice also sublimes out of frozen foods. You can clearly see this in food that is frozen in an airtight plastic bag for a long time. The ice crystals that form in the bag are water that has sublimed out of the food and redeposited on the surface of the bag. For this reason, food that remains frozen for too long becomes dried out. This can be avoided to some degree by freezing foods to colder temperatures (further below $0\,°C$), a process called deep-freezing. The colder temperature lowers the rate of sublimation and preserves the food longer.

✓ CONCEPTUAL CHECKPOINT 12.3

Solid carbon dioxide (dry ice) can be depicted as follows:

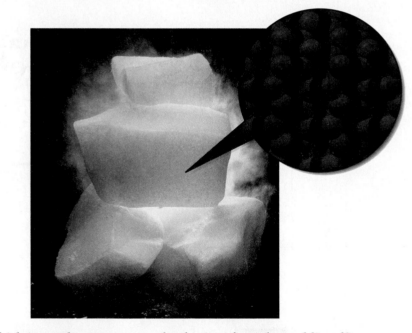

Which image best represents the dry ice after it has sublimed?

(a)   (b)   (c)

12.6 Types of Intermolecular Forces: Dispersion, Dipole–Dipole, and Hydrogen Bonding

The strength of the intermolecular forces between the molecules or atoms that compose a substance determines the state—solid, liquid, or gas—of the substance at room temperature. Strong intermolecular forces tend to result in liquids and solids (with high melting and boiling points). Weak intermolecular forces tend to result in gases (with low melting and boiling points). In this book, we focus on three fundamental types of intermolecular forces. In order of increasing strength, they are the dispersion force, the dipole–dipole force, and the hydrogen bond.

DISPERSION FORCE

| The nature of dispersion forces was first recognized by Fritz W. London (1900–1954), a German-American physicist.

The default intermolecular force, present in all molecules and atoms, is the **dispersion force** (also called the *London force*). Dispersion forces are caused by fluctuations in the electron distribution within molecules or atoms. Since all atoms and molecules have electrons, they all have dispersion forces. The electrons in an atom or a molecule may, at any one instant, be unevenly distributed. For example, imagine a frame-by-frame movie of a helium atom in which each "frame" captures the position of the helium atom's two electrons (▼ Figure 12.17). In any one frame, the electrons are not symmetrically arranged around the nucleus. In Frame 3, for example, helium's two electrons are on the left side of the helium atom. The left side then acquires a slightly negative charge (δ^-). The right side of the atom, which is void of electrons, acquires a slightly positive charge (δ^+).

▶ **FIGURE 12.17 Instantaneous dipoles** Random fluctuations in the electron distribution of a helium atom cause instantaneous dipoles to form.

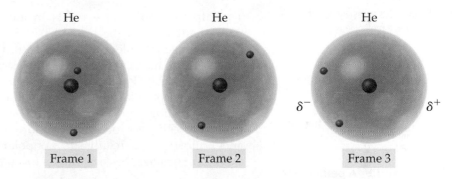

This fleeting charge separation is called an **instantaneous dipole** (or *temporary dipole*). An instantaneous dipole on one helium atom induces an instantaneous dipole on its neighboring atoms because the positive end of the instantaneous dipole attracts electrons in the neighboring atoms (▼ Figure 12.18). The dispersion force occurs as neighboring atoms attract one another—the positive end of one instantaneous dipole attracts the negative end of another. The dipoles responsible for the dispersion force are transient, constantly appearing and disappearing in response to fluctuations in electron clouds.

▶ **FIGURE 12.18 Dispersion force** An instantaneous dipole on any one helium atom induces instantaneous dipoles on neighboring atoms. The neighboring atoms then attract one another. This attraction is called the dispersion force.

| To *polarize* means to form a dipole moment.

The magnitude of the dispersion force depends on how easily the electrons in the atom or molecule can move or *polarize* in response to an instantaneous dipole, which in turn depends on the size of the electron cloud. A larger electron cloud

TABLE 12.4 Noble Gas Boiling Points

Noble Gas	Molar Mass (g/mol)	Boiling Point (K)
He	4.00	4.2 K
Ne	20.18	27 K
Ar	39.95	87 K
Kr	83.80	120 K
Xe	131.29	165 K

results in a greater dispersion force because the electrons are held less tightly by the nucleus and therefore can polarize more easily. If all other variables are constant, the dispersion force increases with increasing molar mass. For example, consider the boiling points of the noble gases displayed in Table 12.4. As the molar mass of the noble gases increase, their boiling points increase. While molar mass alone does not determine the magnitude of the dispersion force, it can be used as a guide when comparing dispersion forces within a family of similar elements or compounds.

EXAMPLE 12.3 Dispersion Forces

Which halogen, Cl_2 or I_2, has the higher boiling point?

SOLUTION

The molar mass of Cl_2 is 70.90 g/mol, and the molar mass of I_2 is 253.80 g/mol. Since I_2 has the higher molar mass, it has stronger dispersion forces and therefore the higher boiling point.

▶SKILLBUILDER 12.3 | Dispersion Forces

Which hydrocarbon, CH_4 or C_2H_6, has the higher boiling point?

▶FOR MORE PRACTICE   Problems 63, 64.

DIPOLE–DIPOLE FORCE

See Section 10.8 to review how to determine whether a molecule is polar.

▲ FIGURE 12.19 A permanent dipole Molecules such as formaldehyde are polar and therefore have a permanent dipole.

The **dipole–dipole force** exists in all polar molecules. Polar molecules have **permanent dipoles** (Section 10.8) that interact with the permanent dipoles of neighboring molecules (◀ Figure 12.19). The positive end of one permanent dipole is attracted to the negative end of another; this attraction is the dipole–dipole force (◀ Figure 12.20). Polar molecules, therefore, have higher melting and boiling points than nonpolar molecules of similar molar mass. Remember that all molecules (including polar ones) have dispersion forces. In addition, polar molecules have dipole–dipole forces. These additional attractive forces raise their melting and boiling points relative to nonpolar molecules of similar molar mass. For example, consider the compounds formaldehyde and ethane:

Name	Formula	Molar mass (g/mol)	Structure	bp (°C)	mp (°C)
Formaldehyde	CH_2O	30.0	$\begin{matrix} O \\ \| \\ H-C-H \end{matrix}$	−19.5	−92
Ethane	C_2H_6	30.1	$\begin{matrix} H \quad H \\ \| \quad \| \\ H-C-C-H \\ \| \quad \| \\ H \quad H \end{matrix}$	−88	−172

Formaldehyde is polar and therefore has a higher melting point and boiling point than nonpolar ethane, even though the two compounds have the same molar mass.

The polarity of molecules composing liquids is also important in determining a liquid's **miscibility**—its ability to mix without separating into two phases. In general, polar liquids are miscible with other polar liquids but are not miscible with nonpolar liquids. For example, water, a polar liquid, is not miscible with pentane (C_5H_{12}), a nonpolar liquid (▶ Figure 12.21). Similarly, water and oil (also nonpolar) do not mix. Consequently, oily hands or oily stains on clothes cannot be washed away with plain water (see Chapter 10, *Everyday Chemistry: How Soap Works*).

▲ FIGURE 12.20 Dipole–dipole attraction The positive end of a polar molecule is attracted to the negative end of its neighbor, giving rise to the dipole–dipole force.

▶ **FIGURE 12.21   Polar and nonpolar compounds   (a)** Pentane, a nonpolar compound, does not mix with water, a polar compound. **(b)** For the same reason, the oil and vinegar (vinegar is largely a water solution of acetic acid) in salad dressing tend to separate into distinct layers. **(c)** An oil spill from a tanker demonstrates dramatically that petroleum and seawater are not miscible.

(a)   (b)   (c)

EXAMPLE 12.4   Dipole–Dipole Forces

Determine whether or not each molecule has dipole–dipole forces.

(a) CO_2 **(b)** CH_2Cl_2 **(c)** CH_4

SOLUTION

A molecule has dipole–dipole forces if it is polar. To determine whether a molecule is polar, you must:

1. determine whether the molecule contains polar bonds, and
2. determine whether the polar bonds add together to form a net dipole moment (Section 10.8).

(a) Since the electronegativities of carbon and oxygen are 2.5 and 3.5, respectively (Figure 10.2), CO_2 has polar bonds. The geometry of CO_2 is linear. Consequently, the polar bonds cancel; the molecule is not polar and does not have dipole–dipole forces.

$O{=}C{=}O$

Nonpolar; no dipole–dipole forces

(b) The electronegativities of C, H, and Cl are 2.5, 2.1, and 3.5, respectively. Consequently, CH_2Cl_2 has two polar bonds (C—Cl) and two bonds that are nearly nonpolar (C—H). The geometry of CH_2Cl_2 is tetrahedral. Since the C—Cl bonds and the C—H bonds are different, they do not cancel, but sum to a net dipole moment. Therefore the molecule is polar and has dipole–dipole forces.

Polar; dipole–dipole forces

(c) Since the electronegativities of C and H are 2.5 and 2.1, respectively, the C—H bonds are nearly nonpolar. In addition, since the geometry of the molecule is tetrahedral, any slight polarities that the bonds might have will cancel. CH_4 is therefore nonpolar and does not have dipole–dipole forces.

Nonpolar; no dipole–dipole forces

▶**SKILLBUILDER 12.4 | Dipole–Dipole Forces**

Determine whether or not each molecule has dipole–dipole forces.

(a) CI_4 **(b)** CH_3Cl **(c)** HCl

▶**FOR MORE PRACTICE**   Problems 59, 60, 61, 62.

HYDROGEN BONDING

Polar molecules containing hydrogen atoms bonded directly to fluorine, oxygen, or nitrogen exhibit an additional intermolecular force called a **hydrogen bond**. HF, NH_3 and H_2O, for example, all undergo hydrogen bonding. A hydrogen bond is a sort of *super* dipole–dipole force. The large electronegativity difference between hydrogen and these electronegative elements, as well as the small size of these atoms (which allows neighboring molecules to get very close to each other), gives rise to a strong attraction between the hydrogen in each of these molecules and the F, O, or N on neighboring molecules. This attraction between a hydrogen atom and an electronegative atom is the hydrogen bond. For example, in HF the hydrogen is strongly attracted to the fluorine on neighboring molecules (◀ Figure 12.22).

Do not confuse hydrogen bonds with chemical bonds. Chemical bonds occur between *individual atoms within a molecule* and are generally much stronger than hydrogen bonds. A hydrogen bond has only 2 to 5% the strength of a typical covalent chemical bond. Hydrogen bonds—like dispersion forces and dipole–dipole forces—are intermolecular forces that occur *between molecules*. In liquid water, for example, the hydrogen bonds are transient, constantly forming, breaking, and re-forming as water molecules move within the liquid. Hydrogen bonds are, however, the strongest of the three intermolecular forces. Substances composed of molecules that form hydrogen bonds have much higher melting and boiling points than you would predict based on molar mass. For example, consider the two compounds, methanol and ethane.

$\delta^+ \delta^-$ $\delta^+ \delta^-$ $\delta^+ \delta^-$

H—F ······ H—F ······ H—F

▲ **FIGURE 12.22 The hydrogen bond** In HF, the hydrogen on each molecule is strongly attracted to the fluorine on its neighbor. The intermolecular attraction of a hydrogen atom to an electronegative atom is called a *hydrogen bond*.

Name	Formula	Molar mass (g/mol)	Structure	bp (°C)	mp (°C)
Methanol	CH_3OH	32.0	H—C—O—H (with H above and H below C)	64.7	−97.8
Ethane	C_2H_6	30.1	H—C—C—H (with H above and below each C)	−88	−172

Since methanol contains hydrogen directly bonded to oxygen, its molecules have hydrogen bonding as an intermolecular force. The hydrogen that is directly bonded to oxygen is strongly attracted to the oxygen on neighboring molecules (◀ Figure 12.23). This strong attraction makes the boiling point of methanol 64.7 °C. Consequently, methanol is a liquid at room temperature. Water is another good example of a molecule with hydrogen bonding as an intermolecular force (▼ Figure 12.24). The boiling point of water (100 °C) is remarkably high for a molecule with such a low molar mass (18.02 g/mol). Hydrogen bonding is important in biological molecules. The shapes of proteins and nucleic acids are largely influenced by hydrogen bonding; for example, the two halves of DNA are held together by hydrogen bonds (see the *Chemistry and Health* box later in this section).

▲ **FIGURE 12.23 Hydrogen bonding in methanol** Since methanol contains hydrogen atoms directly bonded to oxygen, methanol molecules form hydrogen bonds to one another. The hydrogen atom on each methanol molecule is attracted to the oxygen atom of its neighbor.

◀ **FIGURE 12.24 Hydrogen bonding in water** Water molecules form strong hydrogen bonds with one another.

EXAMPLE 12.5   Hydrogen Bonding

One of these compounds is a liquid at room temperature. Which one and why?

Formaldehyde   Fluoromethane   Hydrogen peroxide

SOLUTION

The three compounds have similar molar masses.

formaldehyde	30.03 g/mol
fluoromethane	34.04 g/mol
hydrogen peroxide	34.02 g/mol

Therefore, the strengths of their dispersion forces are similar. All three compounds are also polar, so they have dipole–dipole forces. Hydrogen peroxide, however, is the only compound that also contains H bonded directly to F, O, or N. Therefore it also has hydrogen bonding and is most likely to have the highest boiling point of the three. Since the problem stated that only one of the compounds was a liquid, we can safely assume that hydrogen peroxide is the liquid. Note that although fluoromethane *contains* both H and F, H is not *directly bonded* to F, so fluoromethane does not have hydrogen bonding as an intermolecular force. Similarly, although formaldehyde *contains* both H and O, H is not *directly bonded* to O, so formaldehyde does not have hydrogen bonding either.

> In some cases, hydrogen bonding can occur between one molecule in which H is directly bonded to F, O, or N and another molecule containing an electronegative atom. (See the box, *Chemistry and Health*, in this section for an example.)

►SKILLBUILDER 12.5 | Hydrogen Bonding

Which has the higher boiling point, HF or HCl? Why?

►FOR MORE PRACTICE   Examples 12.9, 12.10; Problems 65, 66, 67, 68, 69, 70.

Table 12.5 summarizes the different types of intermolecular forces. Remember that dispersion forces, the weakest kind of intermolecular force, are present in all

TABLE 12.5   Types of Intermolecular Forces

Type of Force	Relative Strength	Present in	Example
dispersion force (or London force)	weak, but increases with increasing molar mass	all atoms and molecules	H_2 .... ; H_2 H_2
dipole–dipole force	moderate	only polar molecules	HCl δ^+ ...... δ^- δ^+ δ^- ; HCl HCl
hydrogen bond	strong	molecules containing H bonded directly to F, O, or N	HF δ^+ ...... δ^- δ^+ δ^- ; HF HF

CHEMISTRY AND HEALTH

Hydrogen Bonding in DNA

DNA is a long chainlike molecule that acts as a blue-print for living organisms. Copies of DNA are passed from parent to offspring, which is why we inherit traits from our parents. A DNA molecule is composed of thousands of repeating units called *nucleotides* (▶ Figure 12.25). Each nucleotide contains one of four different bases: adenine, thymine, cytosine, and guanine (abbreviated A, T, C, and G, respectively). The order of these bases along DNA encodes the instructions that specify how proteins—the workhorse molecules in living organisms—are made in each cell of the body. Proteins determine virtually all human characteristics, including how we look, how we fight infections, and even how we behave. Consequently, human DNA is a blueprint for how humans are made.

Each time a human cell divides, it must copy the blueprint—which means replicating its DNA. The replicating mechanism is related to the structure of DNA, discovered in 1953 by James Watson and Francis Crick. DNA consists of two complementary strands wrapped around each other in the now famous double helix. Each strand is held to the other by hydrogen bonds that occur between the bases on each strand. DNA replicates because each base (A, T, C, and G) has a complementary partner with which it hydrogen-bonds (▶ Figure 12.26). Adenine (A) hydrogen-bonds with thymine (T), and cytosine (C) hydrogen-bonds with guanine (G). The hydrogen bonds are so specific that each base will pair only with its complementary partner. When a cell is going to divide, the DNA unzips across the hydrogen bonds that run along its length. Then new nucleotides, containing bases complementary to the bases in each half, add along each of the halves, forming hydrogen bonds with their complement. The result is two identical copies of the original DNA (see Chapter 19).

CAN YOU ANSWER THIS? *Why would dispersion forces not work as a way to hold the two halves of DNA together? Why would covalent bonds not work?*

▲ **FIGURE 12.25 The structure of DNA** DNA is composed of repeating units called nucleotides. Each nucleotide is composed of a sugar, a phosphate, and a base.

---

molecules and atoms and increase with increasing molar mass. These forces are weak in small molecules, but they become substantial in molecules with high molar masses. Dipole–dipole forces are present in polar molecules. Hydrogen bonds, the strongest kind of intermolecular force, are present in molecules containing hydrogen bonded directly to fluorine, oxygen, or nitrogen.

✔ CONCEPTUAL CHECKPOINT 12.4

When dry ice sublimes, what forces are overcome?

(a) chemical bonds between carbon atoms and oxygen atoms

(b) hydrogen bonds between carbon dioxide molecules

(c) dispersion forces between carbon dioxide molecules

(d) dipole–dipole forces between carbon dioxide molecules

▲ **FIGURE 12.26 Hydrogen bonding in DNA** The two halves of the DNA double helix are held together by hydrogen bonds.

12.7 Types of Crystalline Solids: Molecular, Ionic, and Atomic

As we learned in Section 12.2, solids may be crystalline (a well-ordered array of atoms or molecules) or amorphous (having no long-range order). Crystalline solids can be divided into three categories—molecular, ionic, and atomic—based on the individual units that compose the solid (▶ Figure 12.27).

MOLECULAR SOLIDS

Molecular solids are solids whose composite units are *molecules*. Ice (solid H_2O) and dry ice (solid CO_2) are examples of molecular solids. Molecular solids are held together by the kinds of intermolecular forces—dispersion forces, dipole–dipole forces, and hydrogen bonding—that we just discussed in Section 12.6. For example, ice is held together by hydrogen bonds, and dry ice is held together by dispersion forces. Molecular solids as a whole tend to have low to moderately low melting points; ice melts at 0 °C and dry ice sublimes at −78.5 °C.

► **FIGURE 12.27 A classification scheme for crystalline solids**

Ice   Table salt   Gold

IONIC SOLIDS

See Section 5.4 for a complete description of the formula unit.

Ionic solids are solids whose composite units are *formula units*, the smallest electrically neutral collection of cations and anions that compose the compound. Table salt (NaCl) and calcium fluoride (CaF_2) are good examples of ionic solids. Ionic solids are held together by electrostatic attractions between cations and anions. For example, in NaCl, the attraction between the Na^+ cation and the Cl^- anion holds the solid lattice together because the lattice is composed of alternating Na^+ cations and Cl^- anions in a three-dimensional array. In other words, the forces that hold ionic solids together are actual ionic bonds. Since ionic bonds are much stronger than any of the intermolecular forces discussed previously, ionic solids tend to have much higher melting points than molecular solids. For example, sodium chloride melts at 801 °C, while carbon disulfide CS_2—a molecular solid with a higher molar mass—melts at −110 °C.

ATOMIC SOLIDS

Atomic solids are solids whose composite units are *individual atoms*. Diamond (C), iron (Fe), and solid xenon (Xe) are good examples of atomic solids. Atomic solids can themselves be divided into three categories—**covalent atomic solids**, **nonbonding atomic solids**, and **metallic atomic solids**—each held together by a different kind of force (► Figure 12.28).

Covalent atomic solids, such as diamond, are held together by covalent bonds. In diamond (► Figure 12.29), each carbon atom forms four covalent bonds to four other carbon atoms in a tetrahedral geometry. This structure extends throughout the entire crystal, so that a diamond crystal can be thought of as a giant molecule held together by these covalent bonds. Since covalent bonds are very strong, covalent atomic solids have high melting points. Diamond is estimated to melt at about 3800 °C.

▶ FIGURE 12.28   **A classification scheme for atomic solids**

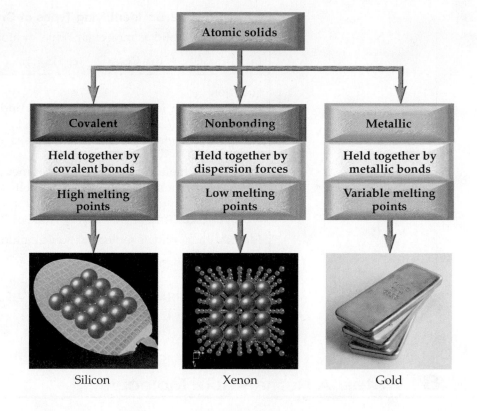

Silicon   Xenon   Gold

Nonbonding atomic solids, such as solid xenon, are held together by relatively weak dispersion forces. Xenon atoms have stable electron configurations and therefore do not form covalent bonds with each other. Consequently, solid xenon, like other nonbonding atomic solids, has a very low melting point (about −112 °C).

Metallic atomic solids, such as iron, have variable melting points. Metals are held together by metallic bonds that, in the simplest model, consist of positively charged ions in a sea of electrons (▼ Figure 12.30). Metallic bonds are of varying strengths, with some metals, such as mercury, having melting points below room temperature, and other metals, such as iron, having relatively high melting points (iron melts at 1809 °C).

▲ FIGURE 12.29   **Diamond: a covalent atomic solid**   In diamond, carbon atoms form covalent bonds in a three-dimensional hexagonal pattern.

▲ FIGURE 12.30   **Structure of a metallic atomic solid**   In the simplest model of a metal, each atom donates one or more electrons to an "electron sea." The metal consists of the metal cations in a negatively charged electron sea.

EXAMPLE 12.6   Identifying Types of Crystalline Solids

Identify each solid as molecular, ionic, or atomic.

(a) $CaCl_2(s)$ **(b)** $Co(s)$ **(c)** $CS_2(s)$

SOLUTION

(a) $CaCl_2$ is an ionic compound (metal and nonmetal) and therefore forms an ionic solid ($CaCl_2$ melts at 772 °C).

(b) Co is a metal and therefore forms a metallic atomic solid (Co melts at 1768 °C).

(c) CS_2 is a molecular compound (nonmetal bonded to a nonmetal) and therefore forms a molecular solid (CS_2 melts at −110 °C).

▶**SKILLBUILDER 12.6 | Identifying Types of Crystalline Solids**

Identify each solid as molecular, ionic, or atomic.

(a) $NH_3(s)$ **(b)** $CaO(s)$ **(c)** $Kr(s)$

▶**FOR MORE PRACTICE**   Problems 73, 74, 75, 76.

12.8   Water: A Remarkable Molecule

▲ **FIGURE 12.31   The water molecule**

| Water reaches its maximum density at 4.0 °C.

Water is easily the most common and important liquid on Earth. It fills our oceans, lakes, and streams. In its solid form, it covers nearly an entire continent (Antarctica), as well as large regions around the North Pole, and caps our tallest mountains. In its gaseous form, it humidifies our air. We drink water, we sweat water, and we excrete bodily wastes dissolved in water. Indeed, the majority of our body mass *is* water. Life is impossible without water, and in most places on Earth where liquid water exists, life exists. Recent evidence of water on Mars—that existed either in the past or exists in the present—has fueled hopes of finding life or evidence of life there. Water is remarkable.

Among liquids, water is unique. It has a low molar mass (18.02 g/mol), yet is a liquid at room temperature. No other compound of similar molar mass even comes close to being a liquid at room temperature. For example, nitrogen (28.02 g/mol) and carbon dioxide (44.01 g/mol) are both gases at room temperature. Water's relatively high boiling point (for its low molar mass) can be understood by examining the structure of the water molecule (◀ Figure 12.31). The bent geometry of the water molecule and the highly polar nature of the O—H bonds result in a molecule with a significant dipole moment. Water's two O—H bonds (hydrogen directly bonded to oxygen) allow water molecules to form strong hydrogen bonds with other water molecules, resulting in a relatively high boiling point. Water's high polarity also allows it to dissolve many other polar and ionic compounds. Consequently, water is the main solvent of living organisms, transporting nutrients and other important compounds throughout the body.

The way water freezes is also unique. Unlike other substances, which contract upon freezing, water expands upon freezing. This seemingly trivial property has significant consequences. For example, because liquid water expands when it freezes, ice is less dense than liquid water. Consequently, ice cubes and icebergs both float. The frozen layer of ice at the surface of a winter lake insulates the water in the lake from further freezing. If this ice layer were to sink, it would kill bottom-dwelling aquatic life and possibly allow the lake to freeze solid, eliminating virtually all aquatic life in the lake.

CHEMISTRY IN THE ENVIRONMENT

Water Pollution

Water quality is critical to human health. Many human diseases—especially in developing nations—are caused by poor water quality. Several kinds of pollutants, including biological contaminants and chemical contaminants, can get into water supplies. Biological contaminants are microorganisms that cause diseases such as hepatitis, cholera, dysentery, and typhoid. Drinking water in developed nations is usually treated to kill microorganisms. Most biological contaminants can be eliminated from untreated water by boiling. Water containing biological contaminants poses an immediate danger to human health and should not be consumed.

Chemical contaminants get into drinking water from sources such as industrial dumping, pesticide and fertilizer use, and household dumping. These contaminants include organic compounds, such as carbon tetrachloride and dioxin, and inorganic elements and compounds such as mercury, lead, and nitrates. Since many chemical contaminants are neither volatile nor alive like biological contaminants, they are *not* eliminated through boiling.

The Environmental Protection Agency (EPA), under the Safe Drinking Water Act of 1974 and its amendments, sets standards that specify the maximum contamination level (MCL) for nearly 100 biological and chemical contaminants in water. Water providers that serve more than 25 people must periodically test the water they deliver to their consumers for these contaminants. If levels exceed the standards set by the EPA, the water provider must notify the consumer and take appropriate measures to remove the contaminant from the water. According to the EPA, if water

▲ Safe drinking water has a major effect on public health and the spread of disease. In many parts of the world, the water supply is unsafe to drink. In the United States the Environmental Protection Agency (EPA) is charged with maintaining water safety.

comes from a provider that serves more than 25 people, it should be safe to consume over a lifetime. If it is not safe to drink for a short period of time, consumers will be notified.

CAN YOU ANSWER THIS? *Suppose a sample of water is contaminated by a nonvolatile contaminant such as lead. Why doesn't boiling eliminate the contaminant?*

The expansion of water upon freezing, however, is one reason that most organisms do not survive freezing. When the water within a cell freezes, it expands and often ruptures the cell, just as water freezing within a pipe bursts the pipe. Many foods, especially those with high water content, do not survive freezing very well either. Have you ever tried, for example, to freeze your own vegetables? Try putting lettuce or spinach in the freezer. When you defrost it, it will be limp and damaged. The frozen food industry gets around this problem by *flash-freezing* vegetables and other foods. In this process, foods are frozen instantaneously, preventing water molecules from settling into their preferred crystalline structure. Consequently, the water does not expand very much, and the food remains largely undamaged.

▶ Lettuce does not survive freezing because the expansion of water upon freezing ruptures the cells within the lettuce leaf.

CHEMICAL PRINCIPLES

Properties of Liquids

- High densities in comparison to gases.
- Indefinite shape; they assume the shape of their container.
- Definite volume; they are not easily compressed.

Properties of Solids

- High densities in comparison to gases.
- Definite shape; they do not assume the shape of their container.
- Definite volume; they are not easily compressed.
- May be crystalline (ordered) or amorphous (disordered).

Manifestations of Intermolecular Forces: Surface tension—the tendency for liquids to minimize their surface area—is a direct result of intermolecular forces. Viscosity—the resistance of liquids to flow—is another result of intermolecular forces. Both surface tension and viscosity increase with greater intermolecular forces.

Evaporation and Condensation: Evaporation or vaporization—an endothermic physical change—is the conversion of a liquid to a gas. Condensation—an exothermic physical change—is the conversion of a gas to a liquid. When the rate of evaporation and condensation in a liquid/gas sample are equal, dynamic equilibrium is reached and the partial pressure of the gas at that point is called its vapor pressure. When the vapor pressure equals the external pressure, the boiling point is reached. At the boiling point, thermal energy causes molecules in the interior of the liquid, as well as those at the surface, to convert to gas, resulting in the bubbling. The heat absorbed or emitted during evaporation and condensation (respectively) can be calculated using the heat of vaporization.

Melting and Freezing: Melting—an endothermic physical change—is the conversion of a solid to a liquid, and freezing—an exothermic physical change—is the conversion of liquid to a solid. The heat absorbed or emitted during melting and freezing (respectively) can be calculated using the heat of fusion.

RELEVANCE

Properties of Liquids: Common liquids include water, acetone (fingernail-polish remover), and rubbing alcohol. Water is the most common and most important liquid on Earth. It is difficult to imagine life without water.

Properties of Solids: Much of the matter we encounter is solid. Common solids include ice, dry ice, and diamond. Understanding the properties of solids involves understanding the particles that compose them and how those particles interact.

Manifestations of Intermolecular Forces: Many insects can walk on water due to surface tension. Water is drawn from roots of trees and up into the leaves because of capillary action, a direct result of intermolecular forces. The viscosity of a liquid is one of its defining properties and is important in applications such as automobile lubrication; the viscosity of a motor oil must be high enough to coat an engine's surfaces, but not so high that it can't flow to remote parts of the engine.

Evaporation and Condensation: Evaporation is the body's natural cooling system. When we get overheated, we sweat; the sweat then evaporates and cools us. Evaporation and condensation both play roles in moderating climate. Humid areas, for example, cool less at night because as the temperature drops, water condenses out of the air, releasing heat and preventing a further temperature drop.

Melting and Freezing: The melting of solid ice is used, for example, to cool drinks when we place ice cubes in them. Since melting is endothermic, it absorbs heat from the liquid and cools it.

Types of Intermolecular Forces: The three main types of intermolecular forces are:

Dispersion forces—Dispersion forces occur between all molecules and atoms due to instantaneous fluctuations in electron charge distribution. The strength of the dispersion force increases with increasing molar mass.

Dipole–dipole forces—Dipole–dipole forces exist between molecules that are polar. Consequently, polar molecules have higher melting and boiling points than nonpolar molecules of similar molar mass.

Hydrogen bonding—Hydrogen bonding exists between molecules that have H bonded directly to F, O, or N. Hydrogen bonds are the strongest of the three intermolecular forces.

Types of Intermolecular Forces: The type of intermolecular force present in a substance determines many of the properties of the substance. The stronger the intermolecular force, for example, the greater the melting and boiling points of the substance. In addition, the miscibility of liquids—their ability to mix without separating—depends on the relative kinds of intermolecular forces present within them. In general, polar liquids are miscible with other polar liquids, but not with nonpolar liquids. Hydrogen bonding is important in many biological molecules such as proteins and DNA.

Types of Crystalline Solids: Crystalline solids can be divided into three categories based on the individual units composing the solid:

Molecular solids—Molecules are the composite units of molecular solids, which are held together by dispersion forces, dipole–dipole forces, or hydrogen bonding.

Ionic solids—Formula units (the smallest electrically neutral collection of cations and anions) are the composite units of ionic solids. They are held together by the electrostatic attractions that occur between cations and anions.

Atomic solids—Atoms are the composite units of atomic solids, which are held together by different forces depending on the particular solid.

Types of Crystalline Solids: Solids have different properties depending on their individual units and the forces that hold those units together. Molecular solids tend to have low melting points. Ionic solids tend to have intermediate to high melting points. Atomic solids have varied melting points, depending on the particular solid.

Water: Water is a unique molecule. Because of its strong hydrogen bonding, water is a liquid at room temperature. Unlike most liquids, water expands when it freezes. In addition, water is highly polar, making it a good solvent for many polar substances.

Water: Water is critical to life. On Earth, wherever there is water, there is life. Water acts as a solvent and transport medium, and virtually all the chemical reactions on which life depends take place in aqueous solution. The expansion of water upon freezing allows life within frozen lakes to survive the winter. The ice on top of the lake acts as insulation, protecting the rest of the lake (and the life within it) from freezing.

CHEMICAL SKILLS

EXAMPLES

Using Heat of Vaporization in Calculations (Section 12.4)

EXAMPLE 12.7 Heat of Vaporization in Calculations

Calculate the amount of heat required to vaporize 84.8 g of water at its boiling point.

SORT

You are given the mass of water and asked to find the amount of heat required to vaporize it.

GIVEN: 84.8 g H_2O

FIND: heat (kJ)

SOLUTION MAP

$$\frac{1 \text{ mol } H_2O}{18.02 \text{ g } H_2O} \qquad \frac{40.7 \text{ kJ}}{1 \text{ mol } H_2O}$$

STRATEGIZE

To calculate the amount of heat required to vaporize a given amount of a substance, first convert the given amount of the substance to moles and then use the heat of vaporization as a conversion factor to get to kilojoules. For vaporization, the heat is always absorbed. For condensation, follow the same procedure, but the heat is always emitted.

RELATIONSHIPS USED

$\Delta H_{vap} = 40.7 \text{ kJ/mol}$ at 100 °C (Table 12.2)

1 mol H_2O = 18.02 g H_2O (molar mass of water)

SOLVE

Follow the solution map to solve the problem.

SOLUTION

$$84.8 \text{ g } H_2O \times \frac{1 \text{ mol } H_2O}{18.02 \text{ g } H_2O} \times \frac{40.7 \text{ kJ}}{1 \text{ mol } H_2O} = 192 \text{ kJ}$$

CHECK

Check your answer. Are the units correct? Does the answer make physical sense?

The units (kJ) are correct. The magnitude of the answer makes sense because each mole of water absorbs about 40 kJ of energy upon vaporization. Therefore 84.8 g (between 4 and 5 mol) should absorb between 160 and 200 kJ upon vaporization.

Using Heat of Fusion in Calculations (Section 12.5)

EXAMPLE 12.8 Using Heat of Fusion in Calculations

Calculate the amount of heat emitted when 12.4 g of water freezes to solid ice.

SORT

You are given the mass of water and asked to find the amount of heat emitted when it freezes.

GIVEN: 12.4 g H_2O

FIND: heat (kJ)

STRATEGIZE

The heat of fusion can be used as a conversion factor between moles of a substance and the amount of heat required to melt it. To calculate the amount of heat required to melt a given amount of a substance, first convert the given amount of the substance to moles and then use the heat of fusion as a conversion factor to get to kilojoules. For melting, the heat is always absorbed. For freezing, follow the same procedure, but the heat is always emitted.

SOLUTION MAP

$$\frac{1 \text{ mol } H_2O}{18.02 \text{ g } H_2O} \qquad \frac{6.02 \text{ kJ}}{1 \text{ mol } H_2O}$$

RELATIONSHIPS USED

$\Delta H_{fus} = 6.02 \text{ kJ/mol}$ (Table 12.3)

1 mol H_2O = 18.02 g H_2O (molar mass of water)

SOLVE

Follow the solution map to solve the problem.

SOLUTION

$$12.4 \text{ g } H_2O \times \frac{1 \text{ mol } H_2O}{18.02 \text{ g } H_2O} \times \frac{6.02 \text{ kJ}}{1 \text{ mol } H_2O} = 4.14 \text{ kJ}$$

The heat emitted is 4.14 kJ.

CHECK

Check your answer. Are the units correct? Does the answer make physical sense?

The units (kJ) are correct. The magnitude of the answer makes sense because each mole of water emits about 6 kJ of energy upon freezing. Therefore 12.2 g (less than 1 mol) should emit less than 6 kJ upon freezing.

Determining the Types of Intermolecular Forces in a Compound (Section 12.6)

All substances exhibit dispersion forces. Polar substances—those whose molecules have polar bonds that add to net dipole moment—also exhibit dipole–dipole forces. Substances whose molecules contain H bonded directly to F, O, or N exhibit hydrogen bonding as well.

EXAMPLE 12.9   Determining the Types of Intermolecular Forces in a Compound

Determine the types of intermolecular forces present in each substance.

(a) N_2
(b) CO
(c) NH_3

SOLUTION

(a) N_2 is nonpolar and therefore has only dispersion forces.
(b) CO is polar and therefore has dipole–dipole forces (in addition to dispersion forces).
(c) NH_3 has hydrogen bonding (in addition to dispersion forces and dipole–dipole forces).

Using Intermolecular Forces to Determine Melting and/or Boiling Points (Section 12.6)

To determine relative boiling points and melting points among compounds, you must evaluate the types of intermolecular forces that each compound exhibits. Dispersion forces are the weakest kind of intermolecular force, but they increase with increasing molar mass. Dipole–dipole forces are stronger than dispersion forces. If two compounds have similar molar mass, but one is polar, it will have higher melting and boiling points. Hydrogen bonds are the strongest type of intermolecular force. Substances that exhibit hydrogen bonding will have much higher boiling and melting points than substances without hydrogen bonding, even if the substance without hydrogen bonding is of higher molar mass.

EXAMPLE 12.10   Using Intermolecular Forces to Determine Melting and/or Boiling Points

Arrange each group of compounds in order of increasing boiling point.

(a) F_2, Cl_2, Br_2
(b) HF, HCl, I IBr

SOLUTION

(a) Since these all have only dispersion forces, and since they are similar substances (all halogens), the strength of the dispersion force will increase with increasing molar mass. Therefore, the correct order is $F_2 < Cl_2 < Br_2$.
(b) Since HF has hydrogen bonding, it has the highest boiling point. Between HCl and HBr, HBr (because of its higher molar mass) has a higher boiling point. Therefore the correct order is $HCl < HBr < HF$.

KEY TERMS

atomic solid [12.7]	heat of fusion (ΔH_{fus}) [12.5]	melting point [12.5]	permanent dipole [12.6]
boiling point [12.4]	heat of vaporization	metallic atomic solid [12.7]	sublimation [12.5]
condensation [12.4]	(ΔH_{vap}) [12.4]	miscibility [12.6]	surface tension [12.3]
covalent atomic solid [12.7]	hydrogen bond [12.6]	molecular solid [12.7]	thermal energy [12.1]
dipole–dipole force [12.6]	instantaneous (temporary)	nonbonding atomic solid	vaporization [12.4]
dispersion force [12.6]	dipole [12.6]	[12.7]	vapor pressure [12.4]
dynamic equilibrium [12.4]	intermolecular forces [12.1]	nonvolatile [12.4]	viscosity [12.3]
evaporation [12.4]	ionic solid [12.7]	normal boiling point [12.4]	volatile [12.4]

EXERCISES

QUESTIONS

1. What are intermolecular forces? Why are intermolecular forces important?
2. Why are water droplets spherical?
3. What determines whether a substance is a solid, liquid, or gas?
4. What are the properties of liquids? Explain the properties of liquids in terms of the molecules or atoms that compose them.
5. What are the properties of solids? Explain the properties of solids in terms of the molecules or atoms that compose them.
6. What is the difference between a crystalline solid and an amorphous solid?
7. What is surface tension? How does it depend on intermolecular forces?
8. What is viscosity? How does it depend on intermolecular forces?
9. What is evaporation? Condensation?
10. Why does a glass of water evaporate more slowly in the glass than if you spilled the same amount of water on a table?
11. Explain the difference between evaporation below the boiling point of a liquid and evaporation at the boiling point of a liquid.
12. What is the boiling point of a liquid? What is the normal boiling point?
13. Acetone evaporates more quickly than water at room temperature. What can you say about the relative strength of the intermolecular forces in the two compounds? Which substance is more volatile?
14. Explain condensation and dynamic equilibrium.
15. What is the vapor pressure of a substance? How does it depend on temperature and strength of intermolecular forces?
16. Explain how sweat cools the body.

17. Explain why a steam burn from gaseous water at 100 °C is worse than a water burn involving the same amount of liquid water at 100 °C.
18. Explain what happens when a liquid boils.
19. Explain why the water in a cup placed in a small ice chest (without a refrigeration mechanism) initially at −5 °C does *not* freeze.
20. Explain how ice cubes cool down beverages.
21. Is the melting of ice endothermic or exothermic? What is the sign of ΔH for the melting of ice? For the freezing of water?
22. Is the boiling of water endothermic or exothermic? What is the sign of ΔH for the boiling of water? For the condensation of steam?
23. What are dispersion forces? How does the strength of dispersion forces relate to molar mass?
24. What are dipole–dipole forces? How can you tell whether a compound has dipole–dipole forces?
25. What is hydrogen bonding? How can you tell whether a compound has hydrogen bonding?
26. What is a molecular solid? What kinds of forces hold molecular solids together?
27. How do the melting points of molecular solids relate to those of other types of solids?
28. What is an ionic solid? What kinds of forces hold ionic solids together?
29. How do the melting points of ionic solids relate to those of other types of solids?
30. What is an atomic solid? What are the properties of atomic solids?
31. In what ways is water unique?
32. How would ice be different if it were denser than water? How would that affect aquatic life in cold-climate lakes?

PROBLEMS

EVAPORATION, CONDENSATION, MELTING, AND FREEZING

33. Which will evaporate more quickly: 55 mL of water in a beaker with a diameter of 4.5 cm or 55 mL of water in a dish with a diameter of 12 cm? Why?

34. Two samples of pure water of equal volume are put into separate dishes and kept at room temperature for several days. The water in the first dish is completely vaporized after 2.8 days, while the water in the second dish takes 8.3 days to completely evaporate. What can you conclude about the two dishes?

35. One milliliter of water is poured onto one hand, and one milliliter of acetone (fingernail-polish remover) is poured onto the other. As they evaporate, they both feel cool. Which one feels cooler and why? (*Hint:* Which substance is more volatile?)

36. Spilling water over your skin on a hot day will cool you down. Spilling vegetable oil over your skin on a hot day will not. Explain the difference.

37. Several ice cubes are placed in a beaker on a lab bench, and their temperature, initially at −5.0 °C, is monitored. Explain what happens to the temperature as a function of time. Make a sketch of how the temperature might change with time. (Assume that the lab is at 25 °C.)

38. Water is put into a beaker and heated with a Bunsen burner. The temperature of the water, initially at 25 °C, is monitored. Explain what happens to the temperature as a function of time. Make a sketch of how the temperature might change with time. (Assume that the Bunsen burner is hot enough to heat the water to its boiling point.)

39. Which causes a more severe burn: spilling 0.50 g of 100 °C water on your hand, or allowing 0.50 g of 100 °C steam to condense on your hand? Why?

40. The nightly winter temperature drop in a seaside town is usually less than that in nearby towns that are farther inland. Explain.

41. When a plastic bag containing a water and ice mixture is placed in an ice chest initially at −8 °C, the temperature of the ice chest goes up. Why?

42. The refrigeration mechanism in a freezer with an automatic ice maker runs extensively each time ice forms from liquid water in the freezer. Why?

43. An ice chest is filled with 3.5 kg of ice at 0 °C. A second ice chest is filled with 3.5 kg of water at 0 °C. After several hours, which ice chest is colder? Why?

44. Why will 50 g of water initially at 0 °C warm more quickly than 50 g of an ice/water mixture initially at 0 °C?

45. In Denver, Colorado, water boils at 95 °C. Explain.

46. At the top of Mount Everest, water boils at 70 °C. Explain.

HEAT OF VAPORIZATION AND HEAT OF FUSION

47. How much heat is required to vaporize 33.8 g of water at 100 °C?

48. How much heat is required to vaporize 43.9 g of acetone at its boiling point?

49. How much heat does your body lose when 2.8 g of sweat evaporates from your skin at 25 °C? (Assume that the sweat is only water.)

50. How much heat does your body lose when 4.86 g of sweat evaporates from your skin at 25 °C? (Assume that the sweat is only water.)

51. How much heat is emitted when 4.25 g of water condenses at 25 °C?

52. How much heat is emitted when 65.6 g of isopropyl alcohol condenses at 25 °C?

53. The human body obtains 835 kJ of energy from a chocolate chip cookie. If this energy were used to vaporize water at 100 °C, how many grams of water could be vaporized? (Assume that the density of water is 1.0 g/mL.)

54. The human body obtains 1078 kJ from a candy bar. If this energy were used to vaporize water at 100 °C, how much water in liters could be vaporized? (Assume that the density of water is 1.0 g/mL.)

55. How much heat is required to melt 37.4 g of ice at 0 °C?

56. How much heat is required to melt 23.9 g of solid diethyl ether (at its melting point)?

57. How much energy is released when 34.2 g of water freezes?

58. How much energy is released when 2.55 kg of diethyl ether freezes?

INTERMOLECULAR FORCES

59. What kinds of intermolecular forces are present in each substance?
    (a) Kr
    (b) N_2
    (c) CO
    (d) HF

60. What kinds of intermolecular forces are present in each substance?
    (a) HCl
    (b) H_2O
    (c) Br_2
    (d) He

61. What kinds of intermolecular forces are present in each substance?
    (a) NCl_3 (trigonal pyramidal)
    (b) NH_3 (trigonal pyramidal)
    (c) SiH_4 (tetrahedral)
    (d) CCl_4 (tetrahedral)

62. What kinds of intermolecular forces present in each substance?
    (a) O_3
    (b) HBr
    (c) CH_3OH
    (d) I_2

63. Which substance has the highest boiling point? Why? *Hint:* They are all nonpolar.
    (a) CH_4
    (b) CH_3CH_3
    (c) $CH_3CH_2CH_3$
    (d) $CH_3CH_2CH_2CH_3$

64. Which noble gas has the highest boiling point? Why?
    (a) Kr
    (b) Xe
    (c) Rn

65. One of these two substances is a liquid at room temperature. Which one and why?
    $$CH_3OH \quad CH_3SH$$

66. One of these two substances is a liquid at room temperature. Which one and why?
    $$CH_3OCH_3 \quad CH_3CH_2OH$$

67. A flask containing a mixture of $NH_3(g)$ and $CH_4(g)$ is cooled. At $-33.3\,°C$ a liquid begins to form in the flask. What is the liquid?

68. Explain why CS_2 is a liquid at room temperature while CO_2 is a gas.

69. Are $CH_3CH_2CH_2CH_2CH_3$ and H_2O miscible?

70. Are CH_3OH and H_2O miscible?

71. Determine whether a homogeneous solution forms when each pair of substances is mixed.
    (a) CCl_4 and H_2O
    (b) Br_2 and CCl_4
    (c) CH_3CH_2OH and H_2O

72. Determine whether a homogeneous solution forms when each pair of substances is mixed.
    (a) $CH_3CH_2CH_2CH_2CH_3$ and $CH_3CH_2CH_2CH_2CH_2CH_3$
    (b) CBr_4 and H_2O
    (c) Cl_2 and H_2O

TYPES OF SOLIDS

73. Identify each solid as molecular, ionic, or atomic.
    (a) Ar(s)
    (b) $H_2O(s)$
    (c) $K_2O(s)$
    (d) Fe(s)

74. Identify each solid as molecular, ionic, or atomic.
    (a) $CaCl_2(s)$
    (b) $CO_2(s)$
    (c) Ni(s)
    (d) $I_2(s)$

75. Identify each solid as molecular, ionic, or atomic.

   (a) $H_2S(s)$

   (b) $KCl(s)$

   (c) $N_2(s)$

   (d) $NI_3(s)$

76. Identify each solid as molecular, ionic, or atomic.

   (a) $SF_6(s)$

   (b) $C(s)$

   (c) $MgCl_2(s)$

   (d) $Ti(s)$

77. Which solid has the highest melting point? Why?

   (a) $Ar(s)$

   (b) $CCl_4(s)$

   (c) $LiCl(s)$

   (d) $CH_3OH(s)$

78. Which solid has the highest melting point? Why?

   (a) C (s, diamond)

   (b) $Kr(s)$

   (c) $NaCl(s)$

   (d) $H_2O(s)$

79. For each pair of solids, determine which one has the higher melting point and explain why.

   (a) $Ti(s)$ and $Ne(s)$

   (b) $H_2O(s)$ and $H_2S(s)$

   (c) $Kr(s)$ and $Xe(s)$

   (d) $NaCl(s)$ and $CH_4(s)$

80. For each pair of solids, determine which one has the higher melting point and explain why.

   (a) $Fe(s)$ and $CCl_4(s)$

   (b) $KCl(s)$ or $HCl(s)$

   (c) $TiO_2(s)$ or $HOOH(s)$

81. List these substances in order of increasing boiling point:

$$H_2O, Ne, NH_3, NaF, SO_2$$

82. List these substances in order of decreasing boiling point:

$$CO_2, Ne, CH_3OH, KF$$

CUMULATIVE PROBLEMS

83. Ice actually has negative caloric content. How much energy, in each of the following units, does your body lose from eating (and therefore melting) 78 g of ice?

   (a) joules

   (b) kilojoules

   (c) calories (1 cal = 4.18 J)

   (d) nutritional Calories or capital "C" Calories (1000 cal = 1 Cal)

84. Ice has negative caloric content. How much energy, in each of the following units, does your body lose from eating (and therefore melting) 145 g of ice?

   (a) joules

   (b) kilojoules

   (c) calories (1 cal = 4.18 J)

   (d) nutritional Calories or capital "C" calories (1000 cal = 1 Cal)

85. An 8.5-g ice cube is placed into 255 g of water. Calculate the temperature change in the water upon the complete melting of the ice. *Hint:* Determine how much heat is absorbed by the melting ice and then use $q = mC\Delta T$ to calculate the temperature change of the 255 g of water.

86. A 14.7-g ice cube is placed into 324 g of water. Calculate the temperature change in the water upon complete melting of the ice. *Hint:* Determine how much heat is absorbed by the melting ice and then use $q = mC\Delta T$ to calculate the temperature change of the 324 g of water.

87. How much ice in grams would have to melt to lower the temperature of 352 mL of water from 25 °C to 0 °C? (Assume that the density of water is 1.0 g/mL.)

88. How much ice in grams would have to melt to lower the temperature of 55.8 g of water from 55.0 °C to 0 °C? (Assume that the density of water is 1.0 g/mL.)

89. How much heat in kilojoules is evolved in converting 1.00 mol of steam at 145 °C to ice at −50.0 °C? The heat capacity of steam is 1.84 J/g °C and that of ice is 2.09 J/g °C.

90. How much heat in kilojoules is required to warm 10.0 g of ice, initially at −10.0 °C, to steam at 110.0 °C. The heat capacity of ice is 2.09 J/g °C and that of steam is 1.84 J/g °C.

91. Draw a Lewis structure for each molecule and determine its molecular geometry. What kind of intermolecular forces are present in each substance?

    (a) H_2Se

    (b) SO_2

    (c) $CHCl_3$

    (d) CO_2

92. Draw a Lewis structure for each molecule and determine its molecular geometry. What kind of intermolecular forces are present in each substance?

    (a) BCl_3 (remember that B is a frequent exception to the octet rule)

    (b) HCOH (carbon is central; each H and O bonded directly to C)

    (c) CS_2

    (d) NCl_3

93. The melting point of ionic solids depends on the magnitude of the electrostatic attractions that hold the solid together. Draw ionic Lewis structures for NaF and MgO. Which do you think has the higher melting point?

94. Draw ionic Lewis structures for KF and CaO. Use the information and the method in the previous problem to predict which of these two ionic solids has the higher melting point.

95. Explain the observed trend in the melting points of the alkyl halides. Why is HF atypical?

Compound	Melting Point
HI	−50.8 °C
HBr	−88.5 °C
HCl	−114.8 °C
HF	−83.1 °C

96. Explain the observed trend in the boiling points of the compounds listed. Why is H_2O atypical?

Compound	Boiling Point
H_2Te	−2 °C
H_2Se	−41.5 °C
H_2S	−60.7 °C
H_2O	+100 °C

97. An ice cube at 0.00 °C with a mass of 23.5 g is placed into 550.0 g of water, initially at 28.0 °C, in an insulated container. Assuming that no heat is lost to the surroundings, what is the temperature of the entire water sample after all of the ice has melted?

98. If 1.10 g of steam at 100.0 °C condenses into 38.5 g of water, initially at 27.0 °C, in an insulated container, what is the final temperature of the entire water sample? Assume no loss of heat into the surroundings.

HIGHLIGHT PROBLEMS

99. Consider the molecular view of water shown here. Pick a molecule in the interior and draw a line to each of its direct neighbors. Pick a molecule near the edge (analogous to a molecule on the surface in three dimensions) and do the same. Which molecule has the most neighbors? Which molecule is more likely to evaporate?

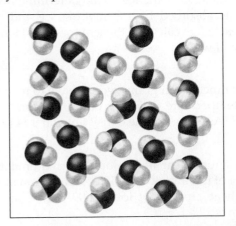

100. Water does not easily remove grease from dirty hands because grease is nonpolar and water is polar; therefore they are immiscible. The addition of soap, however, results in the removal of the grease. Examine the structure of soap shown here and explain how soap works.

Nonpolar tail   Polar head

Sodium stearate
a soap

101. One prediction of global warming is the melting of global ice, which may result in coastal flooding. A criticism of this prediction is that the melting of icebergs does not increase ocean levels any more than the melting of ice in a glass of water increases the level of liquid in the glass.

   (a) Is this a valid criticism? Does the melting of an ice cube in a cup of water raise the level of the liquid in the cup? Why or why not?

      A response to this criticism is that scientists are not worried about rising ocean levels due to melting icebergs; rather, scientists are worried about rising ocean levels due to melting ice sheets that sit on the continent of Antarctica.

   (b) Would the melting of the ice sheets increase ocean levels? Why or why not?

102. Explain why rubbing alcohol feels cold when applied to the skin.

▶ANSWERS TO SKILLBUILDER EXERCISES

Skillbuilder 12.1 5.83×10^3 kJ

Skillbuilder Plus, p. 420   47 °C

Skillbuilder 12.2   5.18 kJ

Skillbuilder Plus, p. 423   −2.3 °C

Skillbuilder 12.3 C_2H_6

Skillbuilder 12.4
   (a) no dipole–dipole forces
   (b) yes, it has dipole–dipole forces
   (c) yes, it has dipole–dipole forces

Skillbuilder 12.5   HF, because it has hydrogen bonding as an intermolecular force

Skillbuilder 12.6
   (a) molecular
   (b) ionic
   (c) atomic

▶ANSWERS TO CONCEPTUAL CHECKPOINTS

12.1 (b) Since boiling is a physical rather than a chemical change, the water molecules (H_2O) undergo no chemical alteration—they merely change from the liquid to the gaseous state.

12.2 (b) the temperature remains flat during a state transition; therefore, since the water starts as ice, the first flat section (i) is the melting of the ice to liquid water. The increasing temperature section (ii) is the water warming to its boiling point. The subsequent flat section (iii) is the boiling of the liquid water to gaseous steam.

12.3 (c) Since sublimation is a physical change, the carbon dioxide molecules do not decompose into other molecules or atoms; they simply change state from the solid to the gaseous state.

12.4 (c) The chemical bonds between carbon and oxygen atoms are not broken by changes of state such as sublimation. Because carbon dioxide contains no hydrogen atoms, it cannot undergo hydrogen bonding, and because the molecule is nonpolar, it does not experience dipole–dipole interactions.

CHAPTER 13

Solutions

"The goal of science is to make sense of the diversity of nature."

JOHN BARROW (B. 1952)

13.1 Tragedy in Cameroon

▲ Cameroon is in West Africa.

◄ Late in the summer of 1986, carbon dioxide bubbled out of Lake Nyos and flowed into the adjacent valley. The carbon dioxide came from the bottom of the lake, where it was held in solution by the pressure of the water above it. When the layers in the lake were disturbed, the carbon dioxide came out of solution due to the decrease in pressure—with lethal consequences.

On August 22, 1986, most people living near Lake Nyos in Cameroon, West Africa, began their day in an ordinary way. Unfortunately, the day ended in tragedy. On that evening, a large cloud of carbon dioxide gas, burped up from the depths of Lake Nyos, killed more than 1700 people and about 3000 head of cattle. Survivors tell of smelling rotten eggs, feeling a warm sensation, and then losing consciousness. Two years before that, a similar tragedy had occurred in Lake Monoun, just 60 miles away, killing 37 people. Today, scientists have taken steps to prevent these lakes from burping again.

Lake Nyos is a water-filled volcanic crater. Some 50 miles beneath the surface of the lake, molten volcanic rock (magma) produces carbon dioxide gas that seeps into the lake through the volcano's plumbing system. The carbon dioxide then mixes with the lake water. However, as we will see later in this chapter, the concentration of a gas (such as carbon dioxide) that can build up in water increases with increasing pressure. The great pressure at the bottom of the deep lake therefore allows the concentration of carbon dioxide to become very high (just as the pressure in a soda can allows the concentration of carbon dioxide in soda to be very high). Over time, the carbon dioxide and water mixture at the bottom of the lake became so concentrated that—either because of the high concentration itself or because of some other natural trigger, such as a landslide—some gaseous carbon dioxide escaped. The rising bubbles disrupted the stratified layers of lake water, causing the highly concentrated carbon dioxide and water mixture at the bottom of the lake to rise, which lowered the pressure on it. The drop in pressure on the mixture released

Carbon dioxide, a colorless and odorless gas, displaced the air in low-lying regions surrounding Lake Nyos, leaving no oxygen for the inhabitants to breathe. The rotten-egg smell is an indication of the presence of additional sulfur-containing gases.

more carbon dioxide bubbles just as the drop in pressure upon opening a soda can releases carbon dioxide bubbles. This in turn caused more churning and more carbon dioxide release. Since carbon dioxide is more dense than air, once freed from the lake, it traveled down the sides of the volcano and into the nearby valley, displacing air and asphyxiating many of the local residents.

In efforts to prevent these events from occurring again—by 2001, carbon dioxide concentrations had already returned to dangerously high levels—scientists built a piping system to slowly vent carbon dioxide from the lake bottom. Since 2001, this system has gradually been releasing the carbon dioxide into the atmosphere, preventing a repeat of the tragedy.

▶ Engineers watch as the carbon dioxide vented from the bottom of Lake Nyos creates a geyser. The controlled release of carbon dioxide from the lake bed is designed to prevent future catastrophes like the one that killed more than 1700 people in 1986.

13.2 Solutions: Homogeneous Mixtures

The carbon dioxide and water mixture at the bottom of Lake Nyos is an example of a **solution**, a homogeneous mixture of two or more substances. Solutions are common—most of the liquids and gases that we encounter every day are actually solutions. When most people think of a solution, they think of a solid dissolved in water. The ocean, for example, is a solution of salt and other solids dissolved in water. Blood plasma (blood that has had blood cells removed from it) is a solution of several solids (as well as some gases) dissolved in water. In addition to these, many other kinds of solutions exist. A solution may be composed of a gas and a liquid (like the carbon dioxide and water of Lake Nyos), a liquid and another liquid, a solid and a gas, or other combinations (see Table 13.1).

Aqueous comes from the Latin word *aqua*, meaning "water."

The most common solutions, however, are those containing a solid, a liquid, or a gas and water. These are *aqueous solutions*—they are critical to life and are the

TABLE 13.1 Common Types of Solutions

Solution Phase	Solute Phase	Solvent Phase	Example
gaseous solutions	gas	gas	air (mainly oxygen and nitrogen)
liquid solutions	gas	liquid	soda water (CO_2 and water)
	liquid	liquid	vodka (ethanol and water)
	solid	liquid	seawater (salt and water)
solid solutions	solid	solid	brass (copper and zinc) and other alloys

In a solid/liquid solution, the liquid is usually considered the solvent, regardless of the relative proportions of the components.

main focus of this chapter. Common examples of aqueous solutions include sugar water and salt water, both solutions of solids and water. Similarly, ethyl alcohol—the alcohol in alcoholic beverages—readily mixes with water to form a solution of a liquid with water, and we have already seen an example of a gas-and-water solution in Lake Nyos.

A solution has at least two components. The majority component is usually called the **solvent**, and the minority component is called the **solute**. In our carbon-dioxide-and-water solution, carbon dioxide is the solute and water is the solvent. In a salt-and-water solution, salt is the solute and water is the solvent. Because water is so abundant on Earth, it is a common solvent. However, other solvents are often used in the laboratory, in industry, and even in the home, especially to form solutions with nonpolar solutes. For example, you may use paint thinner, a nonpolar solvent, to remove grease from a dirty bicycle chain or from ball bearings. The paint thinner dissolves (or forms a solution with) the grease, removing it from the metal.

See Sections 10.8 and 12.6 to review the concept of polarity.

In general, polar solvents dissolve polar or ionic solutes, and nonpolar solvents dissolve nonpolar solutes. This tendency is described by the rule *like dissolves like*. Thus, similar kinds of solvents dissolve similar kinds of solutes. Table 13.2 lists some common polar and nonpolar laboratory solvents.

TABLE 13.2   Common Laboratory Solvents

Common Polar Solvents	Common Nonpolar Solvents
water (H_2O)	hexane (C_6H_{14})
acetone (CH_3COCH_3)	diethyl ether ($CH_3CH_2OCH_2CH_3$)
methyl alcohol (CH_3OH)	toluene (C_7H_8)

✔ **CONCEPTUAL CHECKPOINT 13.1**

Which compound would you expect to be *least* soluble in water?

(a) CCl_4 (b) CH_3Cl (c) H_2S (d) KF

13.3   Solutions of Solids Dissolved in Water: How to Make Rock Candy

We have already seen several examples of solutions of a solid dissolved in water. The ocean, for example, is a solution of salt and other solids dissolved in water. A sweetened cup of coffee is a solution of sugar and other solids dissolved in water. Blood plasma is a solution of several solids (and some gases) dissolved in water. Not all solids, however, dissolve in water. We already know that nonpolar solids—such as lard and shortening—do not dissolve in water. Solids such as calcium carbonate and sand do not dissolve either.

When a solid is put into water, there is competition between the attractive forces that hold the solid together (the solute–solute interactions) and the attractive forces occurring between the water molecules and the particles that compose the solid (the solvent–solute interactions). For example, when sodium chloride is put into water, there is competition between the mutual attraction of Na^+ cations and Cl^- anions and the attraction of Na^+ and Cl^- to water molecules as shown in the margin. For sodium ions, the attraction is between the positive charge of the sodium ion and the negative side of water's dipole moment as shown in

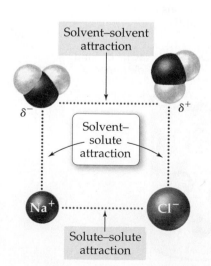

◄ When NaCl is put into water, the attraction between water molecules and Na^+ and Cl^- ions (solvent–solute attraction) overcomes the attraction between Na^+ and Cl^- ions (solute–solute attraction).

▶ **FIGURE 13.1 How a solid dissolves in water** The positive ends of the water dipoles are attracted to the negatively charged Cl^- ions, and the negative ends of the water dipoles are attracted to the positively charged Na^+ ions. The water molecules surround the ions of NaCl and disperse them in the solution.

▲ Figure 13.1 (see Section 10.8 to review dipole moment). For chloride ions, the attraction is between the negative charge of the chloride ion and the positive side of water's dipole moment. In the case of NaCl, the attraction to water wins, and sodium chloride dissolves (▼ Figure 13.2). However, in the case of calcium carbonate ($CaCO_3$), the attraction between Ca^{2+} ions and CO_3^{2-} ions wins and calcium carbonate does not dissolve in water.

SOLUBILITY AND SATURATION

The **solubility** of a compound is defined as the amount of the compound, usually in grams, that dissolves in a certain amount of liquid. For example, the solubility of sodium chloride in water at 25 °C is 36 g NaCl per 100 g water, while the solubility

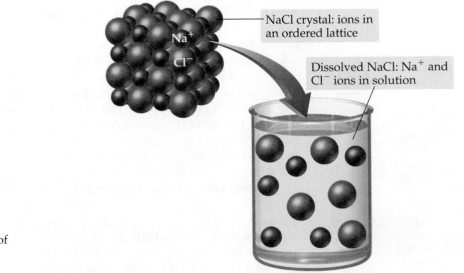

NaCl crystal: ions in an ordered lattice

Dissolved NaCl: Na^+ and Cl^- ions in solution

▶ **FIGURE 13.2 A sodium chloride solution** In a solution of NaCl, the Na^+ and Cl^- ions are dispersed in the water.

of calcium carbonate in water is close to zero. A solution that contains 36 g of NaCl per 100 g water is a *saturated* sodium chloride solution. A **saturated solution** holds the maximum amount of solute under the solution conditions. If additional solute is added to a saturated solution, it will not dissolve. An **unsaturated solution** is holding *less* than the maximum amount of solute. If additional solute is added to an unsaturated solution, it will dissolve. A **supersaturated solution** is one holding *more* than the normal maximum amount of solute. The solute will normally *precipitate* from (or come out of) a supersaturated solution. As the carbon dioxide and water solution rose from the bottom of Lake Nyos, for example, it became supersaturated because of the drop in pressure. The excess gas came out of the solution and rose to the surface of the lake, where it was emitted into the surrounding air.

> Supersaturated solutions can form under special circumstances, such as the sudden release in pressure that occurs in a soda can when it is opened.

▶ A supersaturated solution holds more than the normal maximum amount of solute. In some cases, such as the sodium acetate solution pictured here, a supersaturated solution may be temporarily stable. Any disturbance however, such as dropping in a small piece of solid sodium acetate **(a)**, will cause the solid to come out of solution **(b, c)**.

(a)   (b)   (c)

Chapter 7 introduced the solubility rules (Section 7.5, Table 7.2), which give us a qualitative description of the solubility of ionic solids. Molecular solids may also be soluble in water depending on whether the solid is polar. Table sugar ($C_{12}H_{22}O_{11}$), for example, is polar and soluble in water. Nonpolar solids, such as lard and vegetable shortening, are usually insoluble in water.

ELECTROLYTE SOLUTIONS: DISSOLVED IONIC SOLIDS

A sugar solution (containing a molecular solid) and a salt solution (containing an ionic solid) are very different, as shown in ▼ Figure 13.3. In a salt solution the dissolved particles are ions, while in a sugar solution the dissolved particles are molecules. The ions in the salt solution are mobile charged particles and can therefore conduct electricity. As described in Section 7.5, a solution containing a solute that dissociates into ions is called an **electrolyte solution**. The sugar solution contains dissolved sugar molecules and cannot conduct electricity; it is a **nonelectrolyte solution**. In general, soluble ionic solids form electrolyte solutions, while soluble molecular solids form nonelectrolyte solutions.

> NaCl forms a strong electrolyte solution (Section 7.5). Weak electrolyte solutions are covered in Chapter 14.

Dissolved ions (NaCl)   Dissolved molecules (sugar)

Electrolyte solution   Nonelectrolyte solution

▶ **FIGURE 13.3 Electrolyte and nonelectrolyte solutions** Electrolyte solutions contain dissolved ions (charged particles) and therefore conduct electricity. Nonelectrolyte solutions contain dissolved molecules (neutral particles) and so do not conduct electricity.

► **FIGURE 13.4 Solubility of some ionic solids as a function of temperature**

HOW SOLUBILITY VARIES WITH TEMPERATURE

Have you ever noticed how much easier it is to dissolve sugar in hot tea than in cold tea? The solubility of solids in water can be highly dependent on temperature. In general, the solubility of *solids* in water increases with increasing temperature (▲ Figure 13.4). For example, the solubility of potassium nitrate (KNO_3) at 20 °C is about 30 g KNO_3 per 100 g of water. However, at 50 °C, the solubility rises to 88 g KNO_3 per 100 g of water. A common way to purify a solid is a technique called **recrystallization**. Recrystallization involves putting the solid into water (or some other solvent) at an elevated temperature. Enough solid is added to the solvent to create a saturated solution at the elevated temperature. As the solution cools, the solubility decreases, causing some of the solid to precipitate from solution. If the solution cools slowly, the solid will form crystals as it comes out. The crystalline structure tends to reject impurities, resulting in a purer solid.

ROCK CANDY

▲ Rock candy is composed of sugar crystals that form through recrystallization.

Recrystallization can be used to make rock candy. To make rock candy, prepare a saturated sucrose (table sugar) solution at an elevated temperature. Dangle a string in the solution, and leave it to cool and stand for several days. As the solution cools, it becomes supersaturated and sugar crystals grow on the string. After several days, beautiful and sweet crystals, or "rocks," of sugar cover the string, ready to be admired and eaten.

13.4 Solutions of Gases in Water: How Soda Pop Gets Its Fizz

The water at the bottom of Lake Nyos and a can of soda pop are both examples of solutions in which a gas (carbon dioxide) is dissolved in a liquid (water). Most liquids exposed to air contain some dissolved gases. Lake water and seawater, for example, contain dissolved oxygen necessary for the survival of fish. Our blood contains dissolved nitrogen, oxygen, and carbon dioxide. Even tap water contains dissolved atmospheric gases.

You can see the dissolved gases in ordinary tap water by heating it on a stove. Before the water reaches its boiling point, you will see small bubbles develop in the water. These bubbles are dissolved air (mostly nitrogen and oxygen) coming out of solution. Once the water boils, the bubbling becomes more vigorous—these larger bubbles are composed of water vapor. The dissolved air comes out of solution upon heating because—unlike solids, whose solubility *increases* with increasing

Cold soda pop: carbon dioxide more likely to stay in solution

Warm soda pop: carbon dioxide more likely to bubble out of solution

▲ Warm soda pop fizzes more than cold soda pop because the solubility of the dissolved carbon dioxide decreases with increasing temperature.

temperature—the solubility of gases in water *decreases* with increasing temperature. As the temperature of the water rises, the solubility of the dissolved nitrogen and oxygen decreases and these gases come out of solution, forming small bubbles around the bottom of the pot.

The decrease in the solubility of gases with increasing temperature is the reason that warm soda pop bubbles more than cold soda pop and also the reason that warm soda goes flat faster than cold soda. The carbon dioxide comes out of solution faster (bubbles more) at room temperature than at lower temperature because it is less soluble at room temperature.

The solubility of gases also depends on pressure. The higher the pressure above a liquid, the more soluble the gas is in the liquid (▼ Figure 13.5), a relationship known as **Henry's law**.

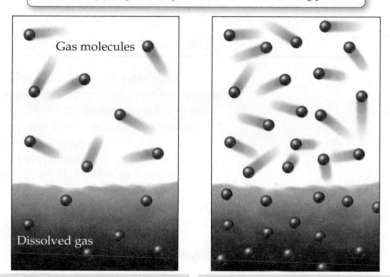

The solubility of a gas in a liquid increases with increasing pressure.

Gas molecules

Dissolved gas

Gas at low pressure over a liquid

Gas at high pressure over a liquid

▶ **FIGURE 13.5   Pressure and solubility**   The higher the pressure above a liquid, the more soluble the gas is in the liquid.

In a can of soda pop and at the bottom of Lake Nyos, carbon dioxide is maintained in solution by high pressure. In soda pop, the pressure is provided by a large amount of carbon dioxide gas that is pumped into the can before sealing it. When the can is opened, the pressure is released and the solubility of carbon dioxide decreases, resulting in bubbling (▼ Figure 13.6). The bubbles are formed by the carbon dioxide gas as it escapes. In Lake Nyos, the pressure is provided by the mass of the lake

▶ **FIGURE 13.6   Pop! Fizz!**   A can of soda pop is pressurized with carbon dioxide. When the can is opened, the pressure is released, lowering the solubility of carbon dioxide in the solution and causing it to come out of solution as bubbles.

CO_2 pressure released

CO_2 under pressure

CO_2 dissolved in solution

CO_2 bubbles out of solution

water itself pushing down on the carbon-dioxide–rich water at the bottom of the lake. When the stratification (or layering) of the lake is disturbed, the pressure on the carbon dioxide solution is lowered and the solubility of carbon dioxide decreases, resulting in the release of excess carbon dioxide gas.

✓ CONCEPTUAL CHECKPOINT 13.2

A solution is saturated in both nitrogen gas (N_2) and potassium chloride (KCl) at 75 °C. What happens when the solution is cooled to room temperature?

(a) Some nitrogen gas bubbles out of solution.

(b) Some potassium chloride precipitates out of solution.

(c) Both (a) and (c).

(d) Nothing happens.

13.5 Specifying Solution Concentration: Mass Percent

As we have seen, the amount of solute in a solution is an important property of the solution. For example, the amount of carbon dioxide in the water at the bottom of Lake Nyos is an important predictor of when the deadly event may repeat itself. A **dilute solution** is one containing small amounts of solute relative to solvent. If the water at the bottom of Lake Nyos were a dilute carbon dioxide solution, it would pose little threat. A **concentrated solution** is one containing large amounts of solute relative to solvent. If the carbon dioxide in the water at the bottom of Lake Nyos becomes concentrated (through the continual feeding of carbon dioxide from magma into the lake), it becomes a large threat. A common way to report solution concentration is *mass percent*.

MASS PERCENT

Also in common use are *parts per million* (ppm), the number of grams of solute per 1 million g of solution, and *parts per billion* (ppb), the number of grams of solute per 1 billion g of solution.

Note that the denominator is the mass of *solution*, not the mass of solvent.

Mass percent is the number of grams of solute per 100 g of solution. So a solution with a concentration of 14% by mass, for example, contains 14 g of solute per 100 g of solution. To calculate mass percent, simply divide the mass of the solute by the mass of the solution (solute *and* solvent) and multiply by 100%.

$$\text{Mass percent} = \frac{\text{Mass solute}}{\text{Mass solute} + \text{Mass solvent}} \times 100\%$$

Suppose you wanted to calculate the mass percent of NaCl in a solution containing 15.3 g of NaCl and 155.0 g of water. Begin by sorting the information in the problem statement.

GIVEN: 15.3 g NaCl
155.0 g H_2O

FIND: mass percent

SOLUTION

To solve this problem, substitute the correct values into the mass percent equation just presented.

$$\text{Mass percent} = \frac{\text{Mass solute}}{\text{Mass solute} + \text{Mass solvent}} \times 100\%$$

$$= \frac{15.3\,\text{g}}{15.3\,\text{g} + 155.0\,\text{g}} \times 100\%$$

$$= \frac{15.3\,\text{g}}{170.3\,\text{g}} \times 100\%$$

$$= 8.98\%$$

The solution is 8.98% NaCl by mass.

EXAMPLE 13.1   Calculating Mass Percent

Calculate the mass percent of a solution containing 27.5 g of ethanol (C_2H_6O) and 175 mL of H_2O. (Assume that the density of water is 1.00 g/mL.)

Begin by setting up the problem. You are given the mass of ethanol and the volume of water and asked to find the mass percent of the solution.	**GIVEN:** 27.5 g C_2H_6O 175 mL H_2O $d_{H_2O} = \dfrac{1.00\,g}{mL}$ **FIND:** mass percent
To find the mass percent, substitute into the equation for mass percent. You need the mass of ethanol and the mass of water. Obtain the mass of water from the volume of water by using the density as a conversion factor.	**SOLUTION** $\text{Mass percent} = \dfrac{\text{Mass solute}}{\text{Mass solute} + \text{Mass solvent}} \times 100\%$ $\text{Mass } H_2O = 175 \text{ mL } H_2O \times \dfrac{1.00\,g}{mL} = 175\,g$
Finally, substitute the correct quantities into the equation and calculate the mass percent.	$\text{Mass percent} = \dfrac{\text{Mass solute}}{\text{Mass solute} + \text{Mass solvent}} \times 100\%$ $= \dfrac{27.5\,g}{27.5\,g + 175\,g}$ $= \dfrac{27.5\,g}{202.5\,g} \times 100\%$ $= 13.6\%$

▶**SKILLBUILDER 13.1 | Calculating Mass Percent**

Calculate the mass percent of a sucrose solution containing 11.3 g of sucrose and 412.1 mL of water. (Assume that the density of water is 1.00 g/mL.)

▶**FOR MORE PRACTICE**   Example 13.11; Problems 41, 42, 43, 44, 45, 46.

USING MASS PERCENT IN CALCULATIONS

We can use the mass percent of a solution as a conversion factor between mass of the solute and mass of the solution. The key to using mass percent in this way is to write it as a fraction.

$$\text{Mass percent} = \frac{\text{g solute}}{100\,\text{g solution}}$$

A solution containing 3.5% sodium chloride, for example, has the following conversion factor.

$$\frac{3.5\,\text{g NaCl}}{100\,\text{g solution}} \qquad \text{converts g solution} \longrightarrow \text{g NaCl}$$

This conversion factor converts from grams of solution to grams of NaCl. If you want to go the other way, simply invert the conversion factor.

$$\frac{100\,\text{g solution}}{3.5\,\text{g NaCl}} \qquad \text{converts g NaCl} \longrightarrow \text{g solution}$$

For example, to use mass percent as a conversion factor, consider a water sample from the bottom of Lake Nyos containing 8.5% carbon dioxide by mass. We can determine how much carbon dioxide in grams is contained in 28.6 L of the water solution. (Assume that the density of the solution is 1.03 g/mL.) We begin by sorting the information in the problem statement.

GIVEN: 8.5% CO_2 by mass

28.6 L solution

$$d = \frac{1.03 \text{ g}}{\text{mL}}$$

FIND: g CO_2

SOLUTION MAP

We strategize by drawing a solution map that begins with L solution and shows the conversion to mL solution and then to g solution using the density. Then we proceed from g solution to g CO_2, using the mass percent (expressed as a fraction) as a conversion factor.

$$\frac{1 \text{ mL}}{10^{-3} \text{ L}} \qquad \frac{1.03 \text{ g}}{\text{mL}} \qquad \frac{8.5 \text{ g } CO_2}{100 \text{ g solution}}$$

RELATIONSHIPS USED

$$\frac{8.5 \text{ g } CO_2}{100 \text{ g solution}} \text{ (given mass percent, written as a fraction)}$$

$$\frac{1.03 \text{ g}}{\text{mL}} \text{ (given density of the solution)}$$

$$1 \text{ mL} = 10^{-3} \text{ L} \quad \text{(from Table 2.2)}$$

SOLUTION

We follow the solution map to calculate the answer.

$$28.6 \text{ L solution} \times \frac{1 \text{ mL}}{10^{-3} \text{ L}} \times \frac{1.03 \text{ g}}{\text{mL}} \times \frac{8.5 \text{ g } CO_2}{100 \text{ g solution}} = 2.5 \times 10^3 \text{ g } CO_2$$

In this example, we used mass percent to convert from a given amount of *solution* to the amount of *solute* present in the solution. In Example 13.2, we use mass percent to convert from a given amount of *solute* to the amount of *solution* containing that solute.

EXAMPLE 13.2 Using Mass Percent in Calculations

A soft drink contains 11.5% sucrose ($C_{12}H_{22}O_{11}$) by mass. What volume of the soft drink solution in milliliters contains 85.2 g of sucrose? (Assume a density of 1.04 g/mL.)

SORT You are given the concentration of sucrose in a soft drink and a mass of sucrose. You are asked to find the volume of the soft drink that contains the given mass of sucrose.	**GIVEN:** 11.5% $C_{12}H_{22}O_{11}$ by mass 85.2 g $C_{12}H_{22}O_{11}$ $d = \dfrac{1.04 \text{ g}}{\text{mL}}$ **FIND:** mL solution (soft drink)
STRATEGIZE Draw a solution map to convert from g solute ($C_{12}H_{22}O_{11}$) to g solution using the mass percent in fractional form as the conversion factor. Convert to mL using the density.	**SOLUTION MAP** $\dfrac{100 \text{ g solution}}{11.5 \text{ g } C_{12}H_{22}O_{11}} \qquad \dfrac{1 \text{ mL}}{1.04 \text{ g}}$

RELATIONSHIPS USED

$$\frac{11.5 \text{ g } C_{12}H_{22}O_{11}}{100 \text{ g solution}} \text{ (given mass percent, written as a fraction)}$$

$$\frac{1.04 \text{ g}}{\text{mL}} \text{ (given density of solution)}$$

SOLVE Follow the solution map to solve the problem.	**SOLUTION** $$85.2 \text{ g } \cancel{C_{12}H_{22}O_{11}} \times \frac{100 \text{ g solution}}{11.5 \text{ g } \cancel{C_{12}H_{22}O_{11}}} \times \frac{1 \text{ mL}}{1.04 \text{ g}} = 712 \text{ mL solution}$$
CHECK Check your answer. Are the units correct? Does the answer make physical sense?	The units (mL solution) are correct. The magnitude of the answer makes sense because each 100 mL of solution contains 11.5 g sucrose; therefore 712 mL should contain a bit more than 77 g, which is close to the given amount of 85.2 g.

▶**SKILLBUILDER 13.2 | Using Mass Percent in Calculations**

How much sucrose ($C_{12}H_{22}O_{11}$) in grams is contained in 355 mL (12 oz) of the soft drink in Example 13.2?

▶**FOR MORE PRACTICE**   Example 13.12; Problems 47, 48, 49, 50, 51, 52.

13.6 Specifying Solution Concentration: Molarity

| Note that molarity is abbreviated with a capital M.

A second way to express solution concentration is **molarity** (M), defined as the number of moles of solute per liter of solution. We calculate the molarity of a solution as follows:

$$\textbf{Molarity (M)} = \frac{\text{Moles solute}}{\text{Liters solution}}$$

Note that molarity is moles of solute per liter of *solution*, not per liter of solvent. To make a solution of a specified molarity, you usually put the solute into a flask and then add water to the desired volume of solution. For example, to make 1.00 L of a 1.00 M NaCl solution, you add 1.00 mol of NaCl to a flask and then add water to make 1.00 L of solution (◀ Figure 13.7). You *do not* combine 1.00 mol of NaCl with 1.00 L of water because that would result in a total volume exceeding 1.00 L and therefore a molarity of less than 1.00 M.

How to prepare a 1.00 molar NaCl solution.

1.00 mol NaCl
(58.44 g)

First add 1.00 mol of NaCl.

Water

Add water until solid is dissolved. Then add additional water until the 1-liter mark is reached.

Mix

A 1.00 molar NaCl solution

◀ **FIGURE 13.7   Making a solution of specific molarity**   To make 1.00 L of a 1.00 M NaCl solution, you add 1.00 mol (58.44 g) of sodium chloride to a flask and then dilute to 1.00 L of total volume.
Question: What would happen if you added 1 L of water to 1 mol of sodium chloride? Would the resulting solution be 1 M?

CHEMISTRY IN THE ENVIRONMENT

The Dirty Dozen

A number of potentially harmful chemicals—such as DDT, dioxin, and polychlorinated biphenyls (PCBs)—can make their way into our water sources from industrial dumping, atmospheric emissions, agriculture, and household dumping. Since crops, livestock, and fish all rely on water, they too can accumulate these chemicals from water. Human consumption of food or water contaminated with these chemicals leads to a number of diseases and adverse health effects such as increased cancer risk, liver damage, or central nervous system damage. Governments around the world have joined forces to ban a number of these chemicals—called persistent organic pollutants or POPs—from production. The original treaty targeted 12 such substances called the dirty dozen (Table 13.3).

A difficult problem posed by these chemicals is their persistence. Once they get into the environment, they stay there for a long time. A second problem is their tendency to undergo *bioamplification*. Because these chemicals are nonpolar, they are stored and concentrated in the fatty tissues of the organisms that consume them. As larger organisms eat smaller ones that have consumed the chemical, the larger organisms consume even more of the stored chemicals. The result is an increase in the concentrations of these chemicals as they move up the food chain. Under the treaty, nearly all intentional production of these chemicals is banned. In the United States, the presence of these contaminants in water

▲ Potentially dangerous chemicals can leak into the environment and contaminate water and food supplies.

TABLE 13.3 The Dirty Dozen

1. aldrin (insecticide)
2. chlordane (insecticide by-product)
3. DDT (insecticide)
4. dieldrin (insecticide)
5. dioxin (industrial by-product)
6. eldrin (insecticide)
7. furan (industrial by-product)
8. heptachlor (insecticide)
9. hexachlorobenzene (fungicide, industrial by-product)
10. mirex (insecticide, fire retardant)
11. polychlorinated biphenyls (PCBs) (electrical insulators)
12. toxaphene (insecticide)

supplies is monitored under supervision of the Environmental Protection Agency (EPA). The EPA has set limits, called maximum contaminant levels (MCLs), for each of the dirty dozen in food and drinking water. Some MCLs for selected compounds in water supplies are listed in Table 13.4.

TABLE 13.4 EPA Maximum Contaminant Level (MCL) for Several "Dirty Dozen" Chemicals

chlordane	0.002 mg/L
dioxin	0.00000003 mg/L
heptachlor	0.0004 mg/L
hexachlorobenzene	0.001 mg/L

Notice the units that the EPA uses to express the concentration of the contaminants: milligrams per liter. This unit is a conversion factor between liters of water consumed and the mass in milligrams of the pollutant. According to the EPA, as long as the contaminant concentrations are below these levels, the water is safe to drink.

CAN YOU ANSWER THIS? *Using what you know about conversion factors, calculate how much of each of the chemicals in Table 13.4 (at their MCL) would be present in 715 L of water, the approximate amount of water consumed by an adult in one year.*

To calculate molarity, divide the number of moles of the solute by the volume of the solution (solute *and* solvent) in liters. For example, to calculate the molarity of a sucrose ($C_{12}H_{22}O_{11}$) solution made with 1.58 mol of sucrose diluted to a total volume of 5.0 L of solution, we begin by sorting the information in the problem statement.

GIVEN: 1.58 mol $C_{12}H_{22}O_{11}$

5.0 L solution

FIND: molarity (M)

SOLUTION

We substitute the correct values into the equation for molarity and calculate the answer.

$$\text{Molarity (M)} = \frac{\text{Moles solute}}{\text{Liters solution}}$$

$$= \frac{1.58 \text{ mol } C_{12}H_{22}O_{11}}{5.0 \text{ L solution}}$$

$$= 0.32 \text{ M}$$

EXAMPLE 13.3   Calculating Molarity

Calculate the molarity of a solution made by putting 15.5 g NaCl into a beaker and adding water to make 1.50 L of NaCl solution.

You are given the mass of sodium chloride (the solute) and the volume of solution. You are asked to find the molarity of the solution.	**GIVEN:** 15.5 g NaCl 1.50 L solution **FIND:** molarity (M)
To calculate molarity, substitute the correct values into the equation and calculate the answer. You must first convert the amount of NaCl from grams to moles using the molar mass of NaCl.	**SOLUTION** $\text{mol NaCl} = 15.5 \text{ g NaCl} \times \dfrac{1 \text{ mol NaCl}}{58.44 \text{ g NaCl}} = 0.2652 \text{ mol NaCl}$ $\text{Molarity (M)} = \dfrac{\text{Moles solute}}{\text{Liters solution}}$ $= \dfrac{0.2652 \text{ mol NaCl}}{1.50 \text{ L solution}}$ $= 0.177 \text{ M}$

▶**SKILLBUILDER 13.3 | Calculating Molarity**

Calculate the molarity of a solution made by putting 55.8 g of $NaNO_3$ into a beaker and diluting to 2.50 L.

▶**FOR MORE PRACTICE**   Example 13.13; Problems 59, 60, 61, 62, 63, 64.

USING MOLARITY IN CALCULATIONS

We can use the molarity of a solution as a conversion factor between moles of the solute and liters of the solution. For example, a 0.500 M NaCl solution contains 0.500 mol NaCl for every liter of solution.

$$\frac{0.500 \text{ mol NaCl}}{\text{L solution}} \qquad \text{converts L solution} \longrightarrow \text{mol NaCl}$$

This conversion factor converts from liters of solution to moles of NaCl. If you want to go the other way, simply invert the conversion factor.

$$\frac{\text{L solution}}{0.500 \text{ mol NaCl}} \qquad \text{converts mol NaCl} \longrightarrow \text{L solution}$$

For example, to determine how many grams of sucrose ($C_{12}H_{22}O_{11}$) are contained in 1.72 L of 0.758 M sucrose solution, begin by sorting the information in the problem statement.

GIVEN: 0.758 M $C_{12}H_{22}O_{11}$

1.72 L solution

FIND: g $C_{12}H_{22}O_{11}$

SOLUTION MAP

We strategize by drawing a solution map that begins with L solution and shows the conversion to moles of sucrose using the molarity, and then the conversion to mass of sucrose using the molar mass.

$$\frac{0.758 \text{ mol } C_{12}H_{22}O_{11}}{\text{L solution}} \qquad \frac{342.34 \text{ g}}{\text{mol}}$$

RELATIONSHIPS USED

$\dfrac{0.758 \text{ mol } C_{12}H_{22}O_{11}}{\text{L solution}}$ (given molarity of solution, written out as a fraction)

$1 \text{ mol } C_{12}H_{22}O_{11} = 342.34 \text{ g}$ (molar mass of sucrose)

SOLUTION

We then follow the solution map to calculate the answer.

$$1.72 \text{ L solution} \times \frac{0.758 \text{ mol } C_{12}H_{22}O_{11}}{\text{L solution}}$$

$$\times \frac{342.34 \text{ g } C_{12}H_{22}O_{11}}{\text{mol } C_{12}H_{22}O_{11}} = 446 \text{ g } C_{12}H_{22}O_{11}$$

In this example, we used molarity to convert from a given amount of *solution* to the amount of *solute* in that solution. In the example that follows, we use molarity to convert from a given amount of *solute* to the amount of *solution* containing that solute.

EXAMPLE 13.4   Using Molarity in Calculations

How many liters of a 0.114 M NaOH solution contains 1.24 mol of NaOH?

SORT You are given the molarity of an NaOH solution and the number of moles of NaOH. You are asked to find the volume of solution that contains the given number of moles.	**GIVEN:** 0.114 M NaOH 1.24 mol NaOH **FIND:** L solution
STRATEGIZE The solution map begins with mol NaOH and shows the conversion to liters of solution using the molarity as a conversion factor.	**SOLUTION MAP** mol NaOH → L solution $\dfrac{1 \text{ L solution}}{0.114 \text{ mol NaOH}}$ **RELATIONSHIP USED** $\dfrac{0.114 \text{ mol NaOH}}{\text{L solution}}$ (given molarity of solution, written out as a fraction)
SOLVE Solve the problem by following the solution map.	**SOLUTION** $1.24 \text{ mol NaOH} \times \dfrac{1 \text{ L solution}}{0.114 \text{ mol NaOH}} = 10.9 \text{ L solution}$

| **CHECK** Check your answer. Are the units correct? Does the answer make physical sense? | The units (L solution) are correct. The magnitude of the answer makes sense because each L of solution contains a little more than 0.10 moles; therefore about 10 L contains a little more than 1 mole. |

▶**SKILLBUILDER 13.4 | Using Molarity in Calculations**

How much of a 0.225 M KCl solution contains 55.8 g of KCl?

▶**FOR MORE PRACTICE** Example 13.14; Problems 65, 66, 67, 68, 69, 70.

ION CONCENTRATIONS

| When an ionic compound dissolves in solution, some of the cations and anions may pair up, so that the actual concentrations of the ions are lower than what you would expect if you assume complete dissociation occurred. | The reported concentration of a solution containing a *molecular* compound usually reflects the concentration of the solute as it actually exists in solution. For example, a 1.0 M glucose ($C_6H_{12}O_6$) solution indicates that the solution contains 1.0 mol of $C_6H_{12}O_6$ per liter of solution. However, the reported concentration of solution containing an *ionic* compound reflects the concentration of the solute *before it is dissolved in solution*. For example, a 1.0 M $CaCl_2$ solution contains 1.0 mol of Ca^{2+} per liter and 2.0 mol of Cl^- per liter. The concentration of the individual ions present in a solution containing an ionic compound can usually be approximated from the overall concentration as shown by the following example. |

EXAMPLE 13.5 Ion Concentration

Determine the molar concentrations of Na^+ and PO_4^{3-} in a 1.50 M Na_3PO_4 solution.

You are given the concentration of an ionic solution and asked to find the concentrations of the component ions.	**GIVEN:** 1.50 M Na_3PO_4 **FIND:** molarity (M) of Na^+ and PO_4^{3-}
A formula unit of Na_3PO_4 contains 3 Na^+ ions (as indicated by the subscript), so the concentration of Na^+ is three times the concentration of Na_3PO_4. Since the same formula unit contains one PO_4^{3-} ion, the concentration of PO_4^{3-} is equal to the concentration of Na_3PO_4.	**SOLUTION** molarity of Na^+ = 3(1.50 M) = 4.50 M molarity of PO_4^{3-} = 1.50 M

▶**SKILLBUILDER 13.5 | Ion Concentration**

Determine the molar concentrations of Ca^{2+} and Cl^- in a 0.75 M $CaCl_2$ solution.

▶**FOR MORE PRACTICE** Problems 77, 78, 79, 80.

✓ **CONCEPTUAL CHECKPOINT 13.3**

A solution is 0.15 M in K_2SO_4. What is the concentration of K^+ in solution?

(a) 0.075 M **(b)** 0.15 M **(c)** 0.30 M **(d)** 0.45 M

13.7 Solution Dilution

| When diluting acids, always add the concentrated acid to the water. *Never add water to concentrated acid solutions.* | To save space in laboratory storerooms, solutions are often stored in concentrated forms called **stock solutions**. For example, hydrochloric acid is typically stored as a 12 M stock solution. However, many lab procedures call for much less concentrated hydrochloric acid solutions, so chemists must dilute the stock solution to the required concentration. This is normally done by diluting a certain amount of the |

stock solution with water. How do we determine how much of the stock solution to use? The easiest way to solve these problems is to use the dilution equation:

$$M_1 V_1 = M_2 V_2$$

where M_1 and V_1 are the molarity and volume of the initial concentrated solution and M_2 and V_2 are the molarity and volume of the final diluted solution. This equation works because the molarity multiplied by the volume gives the number of moles of solute ($M \times V = \text{mol}$), which is the same in both solutions. For example, suppose a laboratory procedure calls for 5.00 L of a 1.50 M KCl solution. How should we prepare this solution from a 12.0 M stock solution? We begin by sorting the information in the problem statement.

| The equation $M_1 V_1 = M_2 V_2$ applies only to solution dilution, NOT to stoichiometry.

GIVEN: $M_1 = 12.0 \text{ M}$

$M_2 = 1.50 \text{ M}$

$V_2 = 5.00 \text{ L}$

FIND: V_1

SOLUTION

We solve the solution dilution equation for V_1 (the volume of the stock solution required for the dilution) and then substitute in the correct values to calculate it.

$$M_1 V_1 = M_2 V_2$$

$$V_1 = \frac{M_2 V_2}{M_1}$$

$$= \frac{1.50 \, \frac{\text{mol}}{\text{L}} \times 5.00 \text{ L}}{12.0 \, \frac{\text{mol}}{\text{L}}}$$

$$= 0.625 \text{ L}$$

We can therefore make the solution by diluting 0.625 L of the stock solution to a total volume of 5.00 L (V_2). The resulting solution will be 1.50 M in KCl (▶ Figure 13.8).

EXAMPLE 13.6 Solution Dilution

To what volume should you dilute 0.100 L of a 15 M NaOH solution to obtain a 1.0 M NaOH solution?

You are given the initial volume and concentration of an NaOH solution and a final concentration. You are asked to find the volume required to dilute the initial solution to the given final concentration.	**GIVEN:** $V_1 = 0.100 \text{ L}$ $M_1 = 15 \text{ M}$ $M_2 = 1.0 \text{ M}$ **FIND:** V_2
Solve the solution dilution equation for V_2 (the volume of the final solution) and substitute the required quantities to calculate V_2. You can make the solution by diluting 0.100 L of the stock solution to a total volume of 1.5 L (V_2). The resulting solution has a concentration of 1.0 M.	**SOLUTION** $M_1 V_1 = M_2 V_2$ $V_2 = \dfrac{M_1 V_1}{M_2}$ $= \dfrac{15 \, \frac{\text{mol}}{\text{L}} \times 0.100 \text{ L}}{1.0 \, \frac{\text{mol}}{\text{L}}}$ $= 1.5 \text{ L}$

▶**SKILLBUILDER 13.6 | Solution Dilution**

How much 6.0 M $NaNO_3$ solution should you use to make 0.585 L of a 1.2 M $NaNO_3$ solution?

▶**FOR MORE PRACTICE**   Example 13.15; Problems 81, 82, 83, 84, 85, 86, 87, 88.

How to make 5.00 L of a 1.50 M KCl solution
from a 12.0 M stock solution.

First add 0.625 L of
12.0 M stock solution

Dilute with water
to total volume
of 5.00 L

1.50 M KCl

▶ **FIGURE 13.8   Making a solution by dilution of a more concentrated solution**

$$M_1 V_1 = M_2 V_2$$

$$\frac{12.0\ \text{mol}}{\cancel{L}} \times 0.625\ \cancel{L} = \frac{1.50\ \text{mol}}{\cancel{L}} \times 5.00\ \cancel{L}$$

$$7.50\ \text{mol} = 7.50\ \text{mol}$$

✔ **CONCEPTUAL CHECKPOINT 13.4**

If 25 g of salt are dissolved in 251 g of water, what is the mass of the resulting solution?

(a) 276 g   **(b)** 251 g   **(c)** 226 g

13.8 Solution Stoichiometry

As we discussed in Chapter 7, many chemical reactions take place in aqueous solutions. Precipitation reactions, neutralization reactions, and gas evolution reactions, for example, all occur in aqueous solutions. Chapter 8 describes how we use the coefficients in chemical equations as conversion factors between moles of reactants and moles of products in stoichiometric calculations. These conversion factors are often used to determine, for example, the amount of product obtained in a chemical reaction based on a given amount of reactant or the amount of one reactant needed to completely react with a given amount of another reactant. The general solution map for these kinds of calculations is:

See Sections 8.2 through 8.4 for a review of reaction stoichiometry.

Moles A ⟶ Moles B

where A and B are two different substances involved in the reaction and the conversion factor between them comes from the stoichiometric coefficients in the balanced chemical equation.

In reactions involving aqueous reactant and products, it is often convenient to specify the amount of reactants or products in terms of their volume and concentration. We can use the volume and concentration to calculate the number of moles of reactants or products, and then use the stoichiometric coefficients to convert to other quantities in the reaction. The general solution map for these kinds of calculations is:

where the conversions between volume and moles are achieved using the molarities of the solutions. For example, consider the reaction for the neutralization of sulfuric acid.

$$H_2SO_4(aq) + 2\ NaOH(aq) \longrightarrow Na_2SO_4(aq) + 2\ H_2O(l)$$

How much 0.125 M NaOH solution do we need to completely neutralize 0.225 L of 0.175 M H_2SO_4 solution? Begin by sorting the information in the problem statement.

GIVEN: 0.225 L H_2SO_4 solution
0.175 M H_2SO_4
0.125 M NaOH

FIND: L NaOH solution

SOLUTION MAP

We strategize by drawing a solution map, which is similar to those for other stoichiometric problems. We first use the volume and molarity of H_2SO_4 solution to get mol H_2SO_4. Then we use the stoichiometric coefficients from the equation to convert mol H_2SO_4 to mol NaOH. Finally, we use the molarity of NaOH to get to L NaOH solution.

$$\frac{0.175\ mol\ H_2SO_4}{L\ H_2SO_4\ solution} \qquad \frac{2\ mol\ NaOH}{1\ mol\ H_2SO_4} \qquad \frac{1\ L\ NaOH\ solution}{0.125\ mol\ NaOH}$$

RELATIONSHIPS USED

$$M(H_2SO_4) = \frac{0.175\ mol\ H_2SO_4}{L\ H_2SO_4\ solution} \quad \text{(given molarity of } H_2SO_4 \text{ solution, written out as a fraction)}$$

$$M(NaOH) = \frac{0.125\ mol\ NaOH}{L\ NaOH\ solution} \quad \text{(given molarity of NaOH solution, written out as a fraction)}$$

1 mol H_2SO_4 : 2 mol NaOH (stoichiometric relationship between H_2SO_4 and NaOH, from balanced chemical equation)

SOLUTION

To solve the problem, we follow the solution map and calculate the answer.

$$0.225\ L\ \cancel{H_2SO_4\ solution} \times \frac{0.175\ \cancel{mol\ H_2SO_4}}{L\ \cancel{H_2SO_4\ solution}} \times \frac{2\ \cancel{mol\ NaOH}}{1\ \cancel{mol\ H_2SO_4}}$$

$$\times \frac{1\ L\ NaOH\ solution}{0.125\ \cancel{mol\ NaOH}} = 0.630\ L\ NaOH\ solution$$

It will take 0.630 L of the NaOH solution to completely neutralize the H_2SO_4.

EXAMPLE 13.7   Solution Stoichiometry

Consider the precipitation reaction:

$$2 \text{ KI}(aq) + \text{Pb(NO}_3)_2(aq) \longrightarrow \text{PbI}_2(s) + 2 \text{ KNO}_3(aq)$$

How much 0.115 M KI solution in liters will completely precipitate the Pb^{2+} in 0.104 L of 0.225 M $Pb(NO_3)_2$ solution?

SORT You are given the concentration of a reactant, KI, in a chemical reaction. You are also given the volume and concentration of a second reactant, $Pb(NO_3)_2$. You are asked to find the volume of the first reactant that completely reacts with the given amount of the second.	**GIVEN:** 0.115 M KI
	
	            0.104 L $Pb(NO_3)_2$ solution
	
	            0.225 M $Pb(NO_3)_2$
	
	FIND: L KI solution

STRATEGIZE The solution map for this problem is similar to the solution maps for other stoichiometric problems. First use the volume and molarity of $Pb(NO_3)_2$ solution to get mol $Pb(NO_3)_2$ Then use the stoichiometric coefficients from the equation to convert mol $Pb(NO_3)_2$ to mol KI. Finally, use mol KI to find L KI solution.	**SOLUTION MAP** **RELATIONSHIPS USED** $\text{M KI} = \dfrac{0.115 \text{ mol KI}}{\text{L KI solution}}$ (given molarity of KI solution, written out as a fraction) $\text{M Pb(NO}_3)_2 = \dfrac{0.225 \text{ mol Pb(NO}_3)_2}{\text{L Pb(NO}_3)_2 \text{ solution}}$ (given molarity of $Pb(NO_3)_2$ solution, written out as a fraction) 2 mol KI $\equiv$ 1 mol $Pb(NO_3)_2$ (stoichiometric relationship between KI and $Pb(NO_3)_2$, from balanced chemical equation)

SOLVE Follow the solution map to solve the problem. Begin with volume of $Pb(NO_3)_2$ solution and cancel units to arrive at volume of KI solution.	**SOLUTION** $0.104 \text{ L Pb(NO}_3)_2 \text{ solution} \times \dfrac{0.225 \text{ mol Pb(NO}_3)_2}{\text{L Pb(NO}_3)_2 \text{ solution}} \times \dfrac{2 \text{ mol KI}}{\text{mol Pb(NO}_3)}$ $\times \dfrac{\text{L KI solution}}{0.115 \text{ mol KI}} = 0.407 \text{ L KI solution}$

CHECK Check your answer. Are the units correct? Does the answer make physical sense?	The units (L KI solution) are correct. The magnitude of the answer makes sense because the lead nitrate solution is about twice as concentrated as the potassium iodide solution and 2 mol of potassium iodide are required to react with 1 mol of lead(II) nitrate. Therefore we would expect the volume of the potassium solution required to completely react with a given volume of the $Pb(NO_3)_2$ solution to be about four times as much.

▶**SKILLBUILDER 13.7 | Solution Stoichiometry**

How many milliliters of 0.112 M Na_2CO_3 will completely react with 27.2 mL of 0.135 M HNO_3 according to the reaction?

$$2 \text{ HNO}_3(aq) + \text{Na}_2\text{CO}_3(aq) \longrightarrow \text{H}_2\text{O}(l) + \text{CO}_2(g) + 2 \text{ NaNO}_3(aq)$$

▶**SKILLBUILDER PLUS**

A 25.0-mL sample of HNO_3 solution requires 35.7 mL of 0.108 M Na_2CO_3 to completely react with all of the HNO_3 in the solution. What is the concentration of the HNO_3 solution?

▶**FOR MORE PRACTICE**   Example 13.16; Problems 89, 90, 91, 92.

13.9 Freezing Point Depression and Boiling Point Elevation: Making Water Freeze Colder and Boil Hotter

▲ Sprinkling salt on icy roads lowers the freezing point of water, so the ice melts even if the temperature is below 0 °C.

Have you ever wondered why salt is added to ice in an ice-cream maker? Or why salt is scattered on icy roads in cold climates? Salt actually lowers the melting point of ice. A salt-and-water solution will remain a liquid even below 0 °C. By adding salt to ice in the ice-cream maker, you form a mixture of ice, salt, and water that can reach a temperature of about −10 °C, cold enough to freeze the cream. On the road, the salt allows the ice to melt, even if the ambient temperature is below freezing.

Adding a nonvolatile solute—one that does not readily evaporate—to a liquid extends the temperature range over which the liquid remains a liquid. The solution has a lower melting point and a higher boiling point than the pure liquid; these effects are called **freezing point depression** and **boiling point elevation**. Freezing point depression and boiling point elevation depend only on the number of solute particles *in solution*, not on the type of solute particles. Properties such as these—which depend on the number of dissolved solute particles and not on the type of solute particles—are called **colligative properties**.

FREEZING POINT DEPRESSION

The freezing point of a solution containing a nonvolatile solute is lower than the freezing point of the pure solvent. For example, antifreeze, added to engine coolant to prevent it from freezing in cold climates, is an aqueous solution of ethylene glycol ($C_2H_6O_2$). The ethylene glycol lowers the freezing point of the aqueous solution. The more concentrated the solution is, the lower the freezing point becomes. For freezing point depression and boiling point elevation, the concentration of the solution is usually expressed in **molality (*m*)**, the number of moles of solute per kilogram of solvent.

Note that molality is abbreviated with a lowercase *m*, while molarity is abbreviated with a capital M.

$$\text{Molality } (m) = \frac{\text{Moles solute}}{\text{Kilograms solvent}}$$

Notice that molality is defined with respect to kilograms of *solvent*, not kilograms of *solution*.

EXAMPLE 13.8 Calculating Molality

Calculate the molality of a solution containing 17.2 g of ethylene glycol ($C_2H_6O_2$) dissolved in 0.500 kg of water.

You are given the mass of ethylene glycol in grams and the mass of the solvent in kilograms. You are asked to find the molality of the resulting solution.	**GIVEN:** 17.2 g $C_2H_6O_2$ 0.500 kg H_2O **FIND:** molality (m)
To calculate molality, substitute the correct values into the equation and calculate the answer. You must first convert the amount of $C_2H_6O_2$ from grams to moles using the molar mass of $C_2H_6O_2$.	**SOLUTION** $\text{mol } C_2H_6O_2 = 17.2 \text{ g } C_2H_6O_2 \times \dfrac{1 \text{ mol } C_2H_6O_2}{62.08 \text{ g } C_2H_6O_2}$ $= 0.2771 \text{ mol } C_2H_6O_2$ $\text{Molality } (m) = \dfrac{\text{Moles solute}}{\text{Kilograms solvent}}$ $= \dfrac{0.2771 \text{ mol } C_2H_6O_2}{0.500 \text{ kg } H_2O}$ $= 0.554 \, m$

▶**SKILLBUILDER 13.8 | Calculating Molality**

Calculate the molality (m) of a sucrose ($C_{12}H_{22}O_{11}$) solution containing 50.4 g sucrose and 0.332 kg of water.

▶**FOR MORE PRACTICE** Example 13.17; Problems 97, 98, 99, 100.

▲ Ethylene glycol is the chief component of antifreeze, which keeps engine coolant from freezing in winter or boiling over in summer.

| The equations for freezing point depression and boiling point elevation given in this section apply only to nonelectrolyte solutions.

| Different solvents have different values of K_f.

✓ CONCEPTUAL CHECKPOINT 13.5

A laboratory procedure calls for a 2.0 molal aqueous solution. A student accidentally makes a 2.0 molar solution. The solution made by the student is:

(a) too concentrated

(b) too dilute

(c) just right

(d) it depends on the molar mass of the solute

With an understanding of molality, we can now quantify freezing point depression. The amount that the freezing point of a solution is lowered by a particular amount of solute is given by the following equation.

The freezing point depression of a solution

$$\Delta T_f = m \times K_f$$

where

- ΔT_f is the change in temperature of the freezing point in °C (from the freezing point of the pure solvent).
- m is the molality of the solution in $\dfrac{\text{mol solute}}{\text{kg solvent}}$.
- K_f is the freezing point depression constant for the solvent.

For water:

$$K_f = 1.86 \, \frac{°\text{C kg solvent}}{\text{mol solute}}$$

Calculating the freezing point of a solution involves substituting into the given equation, as the following example demonstrates.

EXAMPLE 13.9 Freezing Point Depression

Calculate the freezing point of a 1.7 m ethylene glycol solution.

You are given the molality of an aqueous solution and asked to find the freezing point depression. You will need the freezing point depression equation provided in this section.	**GIVEN:** 1.7 m solution **FIND:** ΔT_f
To solve this problem, simply substitute the values into the equation for freezing point depression and calculate ΔT_f.	**SOLUTION** $\Delta T_f = m \times K_f$ $= 1.7 \dfrac{\text{mol solute}}{\text{kg solvent}} \times 1.86 \dfrac{°\text{C kg solvent}}{\text{mol solute}}$ $= 3.2 \, °\text{C}$
The actual freezing point will be the freezing point of pure water (0.00 °C) $- \Delta T_f$.	Freezing point $= 0.00 \, °\text{C} - 3.2 \, °\text{C}$ $= -3.2 \, °\text{C}$

▶**SKILLBUILDER 13.9 | Freezing Point Depression**

Calculate the freezing point of a 2.6 m sucrose solution.

▶**FOR MORE PRACTICE** Example 13.18; Problems 101, 102.

EVERYDAY CHEMISTRY

Antifreeze in Frogs

On the outside, wood frogs (*Rana sylvatica*) look like most other frogs. They are a few inches long and have characteristic greenish-brown skin. However, wood frogs survive cold winters in a remarkable way—they partially freeze. In the frozen state, the frog has no heartbeat, no blood circulation, no breathing, and no brain activity. Within 1 to 2 hours of thawing, however, these vital functions return, and the frog hops off to find food. How is this possible?

Most cold-blooded animals cannot survive freezing temperatures because the water within their cells freezes. As described in Section 12.8, when water freezes, it expands, irreversibly damaging cells. When the wood frog hibernates for the winter, however, it secretes large amounts of glucose into its blood and into the interior of its cells. When the temperature drops below freezing, extracellular bodily fluids, such as those in the abdominal cavity, freeze solid. Fluids within the frogs' cells, however, remain liquid because the high glucose concentration lowers their freezing point. In other words, the concentrated glucose solution within the cells acts as antifreeze, preventing the water within from freezing and allowing the frog to survive.

▲ The wood frog survives cold winters by partially freezing. The fluids in frog cells are protected by a high concentration of glucose that acts as antifreeze, lowering their freezing point so the intercellular fluids remain liquid to temperatures as low as −8 °C.

CAN YOU ANSWER THIS? *The wood frog can survive at body temperatures as low as −8.0 °C. Calculate the molality of a glucose solution ($C_6H_{12}O_6$) required to lower the freezing point of water to −8.0 °C.*

BOILING POINT ELEVATION

The boiling point of a solution containing a nonvolatile solute is higher than the boiling point of the pure solvent. In automobiles, antifreeze not only prevents the freezing of coolant within engine blocks in cold climates, but it also prevents the boiling of engine coolant in hot climates. The amount that the boiling point is raised for solutions is given by the following equation.

The boiling point elevation of a solution:

$$\Delta T_b = m \times K_b$$

where

- ΔT_b is change in temperature of the boiling point in °C (from the boiling point of the pure solvent).
- m is the molality of the solution in $\dfrac{\text{mol solute}}{\text{kg solvent}}$.
- K_b is the boiling point elevation constant for the solvent.

For water:

$$K_b = 0.512 \, \frac{\text{°C kg solvent}}{\text{mol solute}}$$

Different solvents have different values of K_b

The boiling point of solutions is calculated by substituting into the preceding equation as the following example demonstrates.

EXAMPLE 13.10   Boiling Point Elevation

Calculate the boiling point of a 1.7 m ethylene glycol solution.

You are given the molality of an aqueous solution and asked to find the boiling point.	**GIVEN:** 1.7 m solution **FIND:** boiling point
To solve this problem, simply substitute the values into the equation for boiling point elevation and calculate ΔT_b.	**SOLUTION** $\Delta T_b = m \times K_b$ $= 1.7 \, \dfrac{\text{mol solute}}{\text{kg solvent}} \times 0.512 \, \dfrac{°C \, \text{kg solvent}}{\text{mol solute}}$ $= 0.87 °C$
The actual boiling point of the solution will be the boiling point of pure water (100.00 °C) plus ΔT_b.	Boiling point $= 100.00 °C + 0.87 °C$ $= 100.87 °C$

▶**SKILLBUILDER 13.10 | Boiling Point Elevation**

Calculate the boiling point of a 3.5 m glucose solution.

▶**FOR MORE PRACTICE**   Problems 103, 104, 105, 106.

CONCEPTUAL CHECKPOINT 13.6

Which solution has the highest boiling point?

(a) 0.50 M $C_{12}H_{22}O_{11}$ **(b)** 0.50 M $C_6H_{12}O_6$ **(c)** 0.50 M $C_2H_6O_2$

(d) All of these solutions will have the same boiling point.

13.10   Osmosis: Why Drinking Salt Water Causes Dehydration

Humans adrift at sea are surrounded by water, yet drinking that water would only accelerate their dehydration. Why? Salt water causes dehydration because of **osmosis**, the flow of solvent from a less concentrated solution to a more concentrated solution. Solutions containing a high concentration of solute draw solvent from solutions containing a lower concentration of solute. In other words, aqueous solutions with high concentrations of solute, such as seawater, are actually *thirsty solutions*— they draw water away from other, less concentrated solutions, including those in the human body (◀ Figure 13.9).

Direction of water flow

Na^+

Outside of intestine: less concentrated solution

Inside intestine: less concentrated NaCl solution

H_2O

Cl^-

◀ **FIGURE 13.9   Seawater is a *thirsty* solution** As it flows through the stomach and intestine, seawater draws water *out of* bodily tissues, promoting dehydration.

Osmotic Cell: Water flows toward the more concentrated solution

At first

At equilibrium

Osmotic pressure

Semipermeable membrane

Water molecules

Solute particles

▲ **FIGURE 13.10   An osmosis cell**   In an osmosis cell, water flows through a semipermeable membrane from a less concentrated solution into a more concentrated solution. As a result, fluid rises in one side of the tube until the weight of the excess fluid creates enough pressure to stop the flow. This pressure is the osmotic pressure of the solution.

▲ Figure 13.10 shows an osmosis cell. The left side of the cell contains a concentrated saltwater solution, and the right side of the cell contains pure water. A **semipermeable membrane**—a membrane that allows some substances to pass through but not others—separates the two halves of the cell. Through osmosis, water flows from the pure-water side of the cell through the semipermeable membrane into the saltwater side. Over time, the water level on the left side of the cell rises while the water level on the right side of the cell falls. This continues until the pressure created by the weight of the water on the left side is enough to stop the osmotic flow. The pressure required to stop the osmotic flow is the **osmotic pressure** of the solution. Osmotic pressure—like freezing point depression and boiling point elevation—is a colligative property; it depends only on the concentration of the solute particles, not on the type of solute. The more concentrated the solution, the greater its osmotic pressure.

The **membranes** of living cells act as semipermeable membranes. Consequently, if you put a living cell into seawater, it loses water through osmosis and becomes dehydrated. ▼ Figure 13.11 shows red blood cells in solutions of various

▶ **FIGURE 13.11   Red blood cells in solutions of different concentration**   (a) When the solute concentration of the surrounding fluid is equal to that within the cell, there is no net osmotic flow, and the red blood cell exhibits its typical shape. **(b)** When a cell is placed in pure water, osmotic flow of water into the cell causes it to swell up. Eventually it may burst. **(c)** When a cell is placed in a concentrated solution, osmosis draws water out of the cell, distorting its normal shape.

Normal red blood cell

Red blood cell in pure water: water flows into cell

Red blood cell in concentrated solution: water flows out of cell

(a)

(b)

(c)

CHEMISTRY AND HEALTH

Solutions in Medicine

Doctors and others working in health fields often administer solutions to patients. The osmotic pressure of these solutions is controlled for the desired effect on the patient. Solutions having osmotic pressures less than that of bodily fluids are called *hypoosmotic*. These solutions tend to pump water into cells. When a human cell is placed in a hypoosmotic solution—such as pure water—water enters the cell, sometimes causing it to burst (Figure 13.11b). Solutions having osmotic pressures greater than that of bodily fluids are called *hyperosmotic*. These solutions tend to take water out of cells and tissues. When a human cell is placed in a hyperosmotic solution, it typically shrivels as it loses water to the surrounding solution (Figure 13.11c).

Intravenous solutions—those that are administered directly into a patient's veins—must have osmotic pressure equal to that of bodily fluids. These solutions are called *isoosmotic*. When a patient is given an IV in a hospital, the majority of the fluid is usually an isoosmotic saline solution—a solution containing 0.9 g NaCl per 100 mL of solution. In medicine and in other health-related fields, solution concentrations are often reported in units that indicate the mass of the solute in a given volume of solution. Also common is *percent mass to volume*—which is the mass of the solute in grams divided by volume of the solution in milliliters times 100%. In these units, the concentration of an isoosmotic saline solution is 0.9% mass/volume.

CAN YOU ANSWER THIS? *An isoosmotic sucrose* $(C_{12}H_{22}O_{11})$ *solution has a concentration of 0.30 M. Calculate its concentration in percent mass to volume.*

▲ Intravenous fluids consist mostly of isoosmotic saline solutions with an osmotic pressure equal to that of bodily fluids. **Question:** Why would it be dangerous to administer intravenous fluids that do not have an osmotic pressure comparable to that of bodily fluids?

concentrations. The cells in Figure 13.11a, immersed in a solution with the same solute concentration as the cell interior, have the normal red blood cell shape. The cells in Figure 13.11b, in pure water, are swollen. Because the solute concentration within the cells is higher than that of the surrounding fluid, osmosis has pulled water across the membrane *into* the cells. The cells in Figure 13.11c, in a solution more concentrated than the cell interior, are starting to shrivel as osmosis draws water *out of* the cells. Similarly, if you drink seawater, the seawater actually draws water out of your body as it passes through your stomach and intestines. All of that extra water in your intestine promotes dehydration of bodily tissues and diarrhea. Consequently, seawater should never be consumed.

CHAPTER IN REVIEW

CHEMICAL PRINCIPLES

Solutions: A solution is a homogeneous mixture with two or more components. The solvent is the majority component, and the solute is the minority component. Water is the solvent in aqueous solutions.

Solid-and-Liquid Solutions: The solubility—the amount of solute that dissolves in a certain amount of solvent—of solids in liquids increases with increasing temperature. Recrystallization involves dissolving a solid into hot solvent to saturation and then allowing it to cool. As the solution cools, it becomes supersaturated and the solid crystallizes.

Gas-and-Liquid Solutions: The solubility of gases in liquids decreases with increasing temperature but increases with increasing pressure.

Solution Concentration: Solution concentration is used to specify how much of the solute is present in a given amount of solution. Three common ways to express solution concentration are mass percent, molarity, and molality.

$$\text{Mass percent} = \frac{\text{Mass solute}}{\text{Mass solute} + \text{Mass solvent}} \times 100\%$$

$$\text{Molarity (M)} = \frac{\text{Moles solute}}{\text{Liters solution}}$$

$$\text{Molatity } (m) = \frac{\text{Moles solute}}{\text{Kilograms solvent}}$$

Solution Dilution: Solution dilution problems are most conveniently solved using the following equation:

$$M_1 V_1 = M_2 V_2$$

Freezing Point Depression and Boiling Point Elevation: A nonvolatile solute will extend the liquid temperature range of a solution relative to the pure solvent. The freezing point of a solution is lower than the freezing point of the pure solvent, and the boiling point of a solution is higher than the boiling point of the pure solvent. These relationships are quantified by the following equations.

Freezing point depression:
$$\Delta T_f = m \times K_f$$
Boiling point elevation:
$$\Delta T_b = m \times K_b$$

RELEVANCE

Solutions: Solutions are all around us—most of the fluids that we encounter every day are solutions. Common solutions include seawater (solid and liquid), soda pop (gas and liquid), alcoholic spirits such as vodka (liquid and liquid), air (gas and gas), and blood (solid, gas, and liquid).

Solid-and-Liquid Solutions: Solutions of solids dissolved in liquids, such as seawater, coffee, and sugar water, are important both in chemistry and in everyday life. Recrystallization is used extensively in the laboratory to purify solids.

Gas-and-Liquid Solutions: The temperature and pressure dependence of gas solubility is the reason that soda pop fizzes when opened and the reason that warm soda goes flat.

Solution Concentration: Solution concentration is useful in converting between amounts of solute and solution. Mass percent and molarity are the most common concentration units. Molality is used to quantify colligative properties such as freezing point depression and boiling point elevation.

Solution Dilution: Since many solutions are stored in concentrated form, it is often necessary to dilute them to a desired concentration.

Freezing Point Depression and Boiling Point Elevation: Salt is often added to ice in ice-cream makers and is used to melt ice on roads in frigid weather. The salt lowers the freezing point of water, allowing the cream within the ice-cream maker to freeze and the ice on icy roads to melt. Antifreeze is used in the cooling systems of cars both to lower the freezing point of the coolant in winter and to raise its boiling point in summer.

Osmosis: Osmosis is the flow of water from a low-concentration solution to a high-concentration solution through a semipermeable membrane.

Osmosis: Osmosis is the reason drinking seawater causes dehydration. As seawater goes through the stomach and intestines, it draws water away from the body through osmosis, resulting in diarrhea and dehydration. To avoid damage to body tissues, transfused fluids must always be isoosmotic with body fluids. Most transfused fluids consist in whole or part of 0.9% mass/volume saline solution.

CHEMICAL SKILLS

EXAMPLES

Calculating Mass Percent (Section 13.5)

You are given the mass of the solute and the solvent and asked to find the concentration of the solution in mass percent.

To calculate mass percent concentration, divide the mass of the solute by the mass of the solution (solute and solvent) and multiply by 100%.

EXAMPLE 13.11   Calculating Mass Percent

Find the mass percent concentration of a solution containing 19 g of solute and 158 g of solvent.

GIVEN: 19 g solute
158 g solvent

FIND: mass percent

SOLUTION

$$\text{Mass percent} = \frac{\text{mass solute}}{\text{mass solute} + \text{mass solvent}} \times 100\%$$

$$\text{Mass percent} = \frac{19\,g}{19\,g + 158\,g} \times 100\%$$

$$\text{Mass percent} = \frac{19\,g}{177\,g} \times 100\%$$

$$= 11\%$$

Using Mass Percent in Calculations (Section 13.5)

EXAMPLE 13.12   Using Mass Percent in Calculations

How much KCl in grams is in 0.337 L of a 5.80 % mass percent KCl solution? (Assume that the density of the solution is 1.05 g/mL.)

SORT
You are given the volume of a potassium chloride solution and its mass percent concentration. You are asked to find the mass of potassium chloride.

GIVEN: 5.80% KCl by mass
0.337 L solution

$$d = \frac{1.05\,g}{mL}$$

FIND: g KCl

STRATEGIZE
Draw a solution map. Begin with the given volume of the solution in L and convert to mL. Then use the density to find the mass of the solution. Finally, use the mass percent to get to the mass of potassium chloride.

SOLUTION MAP

RELATIONSHIPS USED

$\dfrac{5.80\,\text{g KCl}}{100\,\text{g solution}}$ (given mass percent, written as a fraction)

$\dfrac{1.05\,\text{g}}{\text{mL}}$ (given density)

$1\,\text{mL} = 10^{-3}\,\text{L}\,(\text{Table 2.2})$

SOLVE

Follow your solution map to calculate the answer.

SOLUTION

$$0.337\ \cancel{\text{L solution}} \times \dfrac{1\ \cancel{\text{mL}}}{10^{-3}\ \cancel{\text{L}}} \times \dfrac{1.05\ \cancel{\text{g}}}{\cancel{\text{mL}}} \times \dfrac{5.80\ \text{g KCl}}{100\ \cancel{\text{g solution}}}$$

$$= 20.5\,\text{g KCl}$$

CHECK

Check your answer. Are the units correct? Does the answer make physical sense?

The units (g KCl) are correct. The magnitude of the answer makes sense because 0.337 L is a bit more than 300 g of solution. Each 100 g of solution contains about 6 g KCl, Therefore the answer should be a bit more than 18 g.

Calculating Molarity (Section 13.6)

You are given the number of moles of potassium chloride and the volume of solution. You are asked to find the molarity.

To calculate molarity, divide the number of moles of solute by the volume of the solution in liters.

EXAMPLE 13.13   Calculating Molarity

Calculate the molarity of a KCl solution containing 0.22 mol of KCl in 0.455 L of solution.

GIVEN: 0.22 mol KCl
0.455 L solution

FIND: molarity (M)

SOLUTION

$$\text{Molarity (M)} = \dfrac{0.22\ \text{mol KCl}}{0.455\ \text{L solution}}$$

$$= 0.48\ \text{M}$$

Using Molarity in Calculations (Section 13.6)

SORT

You are given the volume and molarity of a potassium chloride solution and asked to find the mass of potassium chloride contained in the solution.

STRATEGIZE

Draw a solution map beginning with liters of solution and converting to moles of solute using the molarity as a conversion factor. Then convert to grams using the molar mass.

EXAMPLE 13.14   Using Molarity in Calculations

How much KCl in grams is contained in 0.488 L of 1.25 M KCl solution?

GIVEN: 1.25 M KCl
0.488 L solution

FIND: g KCl

SOLUTION MAP

| L KCl solution | mol KCl | g KCl |

$$\dfrac{1.25\ \text{mol KCl}}{\text{L solution}} \qquad \dfrac{74.55\ \text{g KCl}}{\text{mol KCl}}$$

SOLVE

Follow your solution map to compute the answer.

RELATIONSHIPS USED

$$\frac{1.25\,\text{mol KCl}}{\text{L solution}} \quad \text{(given molarity, written as a fraction)}$$

$$1\,\text{mol} = 74.55\,\text{g (molar mass KCl)}$$

SOLUTION

$$0.488\,\cancel{\text{L solution}} \times \frac{1.25\,\cancel{\text{mol KCl}}}{\cancel{\text{L solution}}} \times \frac{74.55\,\text{g KCl}}{\cancel{\text{mol KCl}}}$$

$$= 45.5\,\text{g KCl}$$

CHECK

Check your answer. Are the units correct? Does the answer make physical sense?

The units (g KCl) are correct. The magnitude of the answer makes sense because if each liter of solution contains 1.25 mol, then the given amount of solution (which is about 0.5 L) should contain a bit more than 0.5 mol, which would have a mass that is bit more than about 37 g.

Solution Dilution (Section 13.7)

You are given the initial molarity and final molarity of a solution as well as the final volume. You are asked to find the initial volume.

Most solution dilution problems will use equation $M_1V_1 = M_2V_2$.

Solve the equation for the quantity you are trying to find (in this case, V_1) and then substitute in the correct values to calculate it.

EXAMPLE 13.15   Solution Dilution

How much of an 8.0 M HCl solution should be used to make 0.400 L of a 2.7 M HCl solution?

GIVEN: $M_1 = 8.0\,\text{M}$

$M_2 = 2.7\,\text{M}$

$V_2 = 0.400\,\text{L}$

FIND: V_1

SOLUTION

$$M_1V_1 = M_2V_2$$

$$V_1 = \frac{M_2V_2}{M_1}$$

$$= \frac{2.7\,\dfrac{\text{mol}}{\cancel{\text{L}}} \times 0.400\,\text{L}}{8.0\,\dfrac{\text{mol}}{\cancel{\text{L}}}}$$

$$= 0.14\,\text{L}$$

Solution Stoichiometry (Section 13.8)

EXAMPLE 13.16   Solution Stoichiometry

Consider the reaction:

$$\text{HCl}(aq) + \text{NaOH}(aq) \longrightarrow \text{NaCl}(aq) + \text{H}_2\text{O}(l)$$

How much 0.113 M NaOH solution will completely neutralize 1.25 L of 0.228 M HCl solution?

SORT

You are given the volume and concentration of a hydrochloric acid solution as well as the concentration of a sodium hydroxide solution with which it reacts. You are asked to find the volume of the sodium hydroxide solution that will completely react with the hydrochloric acid.

GIVEN: 1.25 L HCl solution

0.228 M HCl

0.113 M NaOH

FIND: L NaOH solution

STRATEGIZE

Draw a solution map. Use the volume and molarity of HCl to get to mol HCl. Then use the stoichiometric coefficients to convert to mole NaOH. Finally, convert back to volume of NaOH using the molarity of NaOH.

SOLUTION MAP

$$\boxed{\begin{array}{c}\text{L HCl}\\\text{solution}\end{array}} \longrightarrow \boxed{\begin{array}{c}\text{mol}\\\text{HCl}\end{array}} \longrightarrow \boxed{\begin{array}{c}\text{mol}\\\text{NaOH}\end{array}} \longrightarrow \boxed{\begin{array}{c}\text{L NaOH}\\\text{solution}\end{array}}$$

$$\dfrac{0.228\text{ mol HCl}}{\text{L HCl solution}} \qquad \dfrac{1\text{ mol NaOH}}{1\text{ mol HCl}} \qquad \dfrac{1\text{ L NaOH solution}}{0.113\text{ mol NaOH}}$$

RELATIONSHIPS USED

$$M\,(HCl) = \dfrac{0.228\text{ mol HCl}}{\text{L HCl solution}} \begin{array}{l}\text{(given concentration of}\\\text{HCl solution, written}\\\text{as a fraction)}\end{array}$$

$$M\,(NaOH) = \dfrac{0.113\text{ mol NaOH}}{\text{L NaOH solution}} \begin{array}{l}\text{(given concentration}\\\text{of NaOH solution,}\\\text{written as a fraction)}\end{array}$$

1 mol HCl:1 mol NaOH (stoichiometric relationship between HCl and NaOH, from balanced equation)

SOLVE

Follow the solution map to calculate the answer.

SOLUTION

$$1.25\ \cancel{\text{L HCl solution}} \times \dfrac{0.228\ \cancel{\text{mol HCl}}}{\cancel{\text{L HCl solution}}}$$

$$\times \dfrac{1\ \cancel{\text{mol NaOH}}}{1\ \cancel{\text{mol HCl}}} \times \dfrac{\text{L NaOH solution}}{0.113\ \cancel{\text{mol NaOH}}}$$

$$= 2.52\,\text{L NaOH solution}$$

CHECK

Check your answer. Are the units correct? Does the answer make physical sense?

The units (L NaOH solution) are correct. The magnitude of the answer makes sense because the sodium hydroxide solution is about half as concentrated as the hydrochloric acid. Since the reaction stoichiometry is 1:1, the volume of the sodium hydroxide solution should therefore be about twice the volume of the hydrochloric acid.

Calculating Molality (Section 13.9)

You are given the number of moles of sucrose and the mass of water into which it is dissolved. You are asked to find the molality of the resulting solution.

Substitute the correct values into the definition of molality and calculate the answer. If any of the quantities are not in the correct units, convert them into the correct units before substituting into the equation.

EXAMPLE 13.17   Calculating Molality

Calculate the molality of a solution containing 0.183 mol of sucrose dissolved in 1.10 kg of water.

GIVEN: 0.183 mol of sucrose
1.10 kg H_2O

FIND: molality (m)

SOLUTION

$$\text{Molality}\,(m) = \dfrac{\text{Moles solute}}{\text{Kilograms solvent}}$$

$$\text{Molality}\,(m) = \dfrac{0.183\text{ mol sucrose}}{1.10\text{ kg }H_2O}$$

$$= 0.166\ m$$

Freezing Point Depression and Boiling Point Elevation (Section 13.9)

You are given the molality of the solution and asked to find its freezing point.

To find ΔT_f or ΔT_b, simply substitute the values into the equation and calculate the answer.

The freezing point will be the freezing point of pure water $(0.00\,°C) - \Delta T_f$.

EXAMPLE 13.18   Freezing Point Depression and Boiling Point Elevation

Calculate the freezing point of a 2.5 m aqueous sucrose solution.

GIVEN: 2.5 m solution

FIND: ΔT_f

SOLUTION

$$\Delta T_f = m \times K_f$$

$$= 2.5\,\frac{\text{mol solute}}{\text{kg solvent}} \times 1.86\,\frac{°C\ \text{kg solvent}}{\text{mol solute}}$$

$$= 4.7\,°C$$

Freezing point $= 0.00\,°C - 4.7\,°C$

$$= -4.7\,°C$$

KEY TERMS

boiling point elevation
  [13.9]
colligative properties **[13.9]**
concentrated solution **[13.5]**
dilute solution **[13.5]**
electrolyte solution **[13.3]**
freezing point depression
  [13.9]

Henry's law **[13.4]**
mass percent **[13.5]**
molality (m) **[13.9]**
molarity (M) **[13.6]**
nonelectrolyte solution
  [13.3]
osmosis **[13.10]**

osmotic pressure **[13.10]**
recrystallization **[13.3]**
saturated solution **[13.3]**
semipermeable membrane
  [13.10]
solubility **[13.3]**
solute **[13.2]**

solution **[13.2]**
solvent **[13.2]**
stock solution **[13.7]**
supersaturated solution
  [13.3]
unsaturated solution **[13.3]**

EXERCISES

QUESTIONS

1. What is a solution? List some examples.
2. What is an aqueous solution?
3. In a solution, what is the solvent? What is the solute? List some examples.
4. Explain what "like dissolves like" means.
5. What is solubility?
6. Describe what happens when additional solute is added to:
   (a) a saturated solution
   (b) an unsaturated solution
   (c) a supersaturated solution
7. Explain the difference between a strong electrolyte solution and a nonelectrolyte solution. What kinds of solutes form strong electrolyte solutions?
8. How does gas solubility depend on temperature?
9. Explain recrystallization.
10. How is rock candy made?
11. When you heat water on a stove, bubbles form on the bottom of the pot *before* the water boils. What are these bubbles? Why do they form?

12. Explain why warm soda pop goes flat faster than cold soda pop.
13. How does gas solubility depend on pressure? How does this relationship explain why a can of soda pop fizzes when opened.
14. What is the difference between a dilute solution and a concentrated solution?
15. Define the concentration units mass percent and molarity.
16. What is a stock solution?
17. How does the presence of a nonvolatile solute affect the boiling point and melting point of a solution relative to the boiling point and melting point of the pure solvent?
18. What are colligative properties?
19. Define molality.
20. What is osmosis?

21. Two shipwreck survivors were rescued from a life raft. One had drunk seawater while the other had not. The one who had drunk the seawater was more severely dehydrated than the one who did not. Explain.

22. Why are intravenous fluids always isoosmotic saline solutions? What would happen if pure water were administered intravenously?

PROBLEMS

SOLUTIONS

23. Determine whether or not each mixture is a solution.
    (a) sand and water mixture
    (b) oil and water mixture
    (c) salt and water mixture
    (d) sterling silver cup

24. Determine whether or not each mixture is a solution.
    (a) air
    (b) carbon dioxide and water mixture
    (c) a blueberry muffin
    (d) a brass buckle

25. Identify the solute and solvent in each solution.
    (a) salt water
    (b) sugar water
    (c) soda water

26. Identify the solute and solvent in each solution.
    (a) 80-proof vodka (40% ethyl alcohol)
    (b) oxygenated water
    (c) antifreeze (ethylene glycol and water)

27. Pick an appropriate solvent from Table 13.2 to dissolve:
    (a) motor oil (nonpolar)
    (b) sugar (polar)
    (c) lard (nonpolar)
    (d) potassium chloride (ionic)

28. Pick an appropriate solvent from Table 13.2 to dissolve:
    (a) glucose (polar)
    (b) salt (ionic)
    (c) vegetable oil (nonpolar)
    (d) sodium nitrate (ionic)

SOLIDS DISSOLVED IN WATER

29. What are the dissolved particles in a solution containing an ionic solute? What is the name for this kind of solution?

30. What are the dissolved particles in a solution containing a molecular solute? What is the name for this kind of solution?

31. A solution contains 35 g of $NaCl$ per 100 g of water at 25 °C. Is the solution unsaturated, saturated, or supersaturated? (See Figure 13.4.)

32. A solution contains 28 g of KNO_3 per 100 g of water at 25 °C. Is the solution unsaturated, saturated, or supersaturated? (See Figure 13.4.)

33. A KNO_3 solution containing 45 g of KNO_3 per 100 g of water is cooled from 40 °C to 0 °C. What happens during cooling? (See Figure 13.4.)

34. A KCl solution containing 42 g of KCl per 100 g of water is cooled from 60 °C to 0 °C. What happens during cooling? (See Figure 13.4.)

35. Refer to Figure 13.4 to determine whether each of the given amounts of solid will completely dissolve in the given amount of water at the indicated temperature.
    (a) 30.0 g $KClO_3$ in 85.0 g of water at 35 °C
    (b) 65.0 g $NaNO_3$ in 125 g of water at 15 °C
    (c) 32.0 g KCl in 70.0 g of water at 82 °C

36. Refer to Figure 13.4 to determine whether each of the given amounts of solid will completely dissolve in the given amount of water at the indicated temperature.
    (a) 45.0 g $CaCl_2$ in 105 g of water at 5 °C
    (b) 15.0 g $KClO_3$ in 115 g of water at 25 °C
    (c) 50.0 g $Pb(NO_3)_2$ in 95.0 g of water at 10 °C

GASES DISSOLVED IN WATER

37. Some laboratory procedures involving oxygen-sensitive reactants or products call for using preboiled (and then cooled) water. Explain why this is so.

38. A person preparing a fish tank uses preboiled (and then cooled) water to fill it. When the fish is put into the tank, it dies. Explain.

39. Scuba divers breathing air at increased pressure can suffer from nitrogen narcosis—a condition resembling drunkenness—when the partial pressure of nitrogen exceeds about 4 atm. What property of gas/water solutions causes this to happen? How could the diver reverse this effect?

40. Scuba divers breathing air at increased pressure can suffer from oxygen toxicity—too much oxygen in the bloodstream—when the partial pressure of oxygen exceeds about 1.4 atm. What happens to the amount of oxygen in a diver's bloodstream when he or she breathes oxygen at elevated pressures? How can this be reversed?

MASS PERCENT

41. Calculate the concentration of each solution in mass percent.

(a) 41.2 g $C_{12}H_{22}O_{11}$ in 498 g H_2O

(b) 178 mg $C_6H_{12}O_6$ in 4.91 g H_2O

(c) 7.55 g NaCl in 155 g H_2O

42. Calculate the concentration of each solution in mass percent.

(a) 132 g KCl in 598 g H_2O

(b) 22.3 mg KNO_3 in 2.84 g H_2O

(c) 8.72 g C_2H_6O in 76.1 g H_2O

43. A soft drink contains 42 g of sugar in 311 g of H_2O. What is the concentration of sugar in the soft drink in mass percent?

44. A soft drink contains 32 mg of sodium in 309 g of H_2O. What is the concentration of sodium in the soft drink in mass percent?

45. Complete the table:

Mass Solute	Mass Solvent	Mass Solution	Mass Percent
15.5 g	238.1 g	___	___
22.8 g	___	___	12.0%
___	183.3 g	212.1 g	___
___	315.2 g	___	15.3%

46. Complete the table:

Mass Solute	Mass Solvent	Mass Solution	Mass Percent
2.55 g	25.0 g	___	___
___	45.8 g	___	3.8%
1.38 g	___	27.2 g	___
23.7 g	___	___	5.8%

47. Ocean water contains 3.5% NaCl by mass. How much salt can be obtained from 254 g of seawater?

48. A saline solution contains 1.1% NaCl by mass. How much NaCl is present in 96.3 g of this solution?

49. Determine the amount of sucrose in each solution.

(a) 48 g of a solution containing 3.7% sucrose by mass

(b) 103 mg of a solution containing 10.2% sucrose by mass

(c) 3.2 kg of a solution containing 14.3% sucrose by mass

50. Determine the amount of potassium chloride in each solution.

(a) 19.7 g of a solution containing 1.08% KCl by mass

(b) 23.2 kg of a solution containing 18.7% KCl by mass

(c) 38 mg of a solution containing 12% KCl by mass

51. Determine the mass (in g) of each NaCl solution that contains 1.5 g of NaCl.

(a) 0.058% NaCl by mass

(b) 1.46% NaCl by mass

(c) 8.44% NaCl by mass

52. Determine the mass (in g) of each sucrose solution that contains 12 g of sucrose.

(a) 4.1% sucrose by mass

(b) 3.2% sucrose by mass

(c) 12.5% sucrose by mass

53. AgNO$_3$ solutions are often used to plate silver onto other metals. What is the maximum amount of silver in grams that can be plated out of 4.8 L of an AgNO$_3$ solution containing 3.4% Ag by mass? (Assume that the density of the solution is 1.01 g/mL.)

54. A dioxin-contaminated water source contains 0.085% dioxin by mass. How much dioxin is present in 2.5 L of this water? (Assume that the density of the solution is 1.01 g/mL.)

55. Ocean water contains 3.5% NaCl by mass. What mass of ocean water in grams contains 45.8 g of NaCl?

56. A hard water sample contains 0.0085% Ca by mass (in the form of Ca^{2+} ions). What mass of water in grams contains 1.2 g of Ca? (1.2 g of Ca is the recommended daily allowance of calcium for 19- to 24-year-olds.)

57. Lead is a toxic metal that affects the central nervous system. A Pb-contaminated water sample contains 0.0011% Pb by mass. What volume of the water in milliliters contains 115 mg of Pb? (Assume that the density of the solution is 1.0 g/mL.)

58. Benzene is a carcinogenic (cancer-causing) compound. A benzene-contaminated water sample contains 0.000037% benzene by mass. What volume of the water in liters contains 175 mg of benzene? (Assume that the density of the solution is 1.0 g/mL.)

MOLARITY

59. Calculate the molarity of each solution.
  (a) 0.127 mol of sucrose in 655 mL of solution
  (b) 0.205 mol of KNO$_3$ in 0.875 L of solution
  (c) 1.1 mol of KCl in 2.7 L of solution

60. Calculate the molarity of each solution.
  (a) 1.54 mol of LiCl in 22.2 L of solution
  (b) 0.101 mol of LiNO$_3$ in 6.4 L of solution
  (c) 0.0323 mol of glucose in 76.2 mL of solution

61. Calculate the molarity of each solution.
  (a) 22.6 g of C$_{12}$H$_{22}$O$_{11}$ in 0.442 L of solution
  (b) 42.6 g of NaCl in 1.58 L of solution
  (c) 315 mg of C$_6$H$_{12}$O$_6$ in 58.2 mL of solution

62. Calculate the molarity of each solution.
  (a) 33.2 g of KCl in 0.895 L of solution
  (b) 61.3 g of C$_2$H$_6$O in 3.4 L of solution
  (c) 38.2 mg of KI in 112 mL of solution

63. A 205-mL sample of ocean water contains 6.8 g of NaCl. What is the molarity of the solution with respect to NaCl?

64. A 355-mL can of soda pop contains 41 g of sucrose (C$_{12}$H$_{22}$O$_{11}$). What is the molarity of the solution with respect to sucrose?

65. How many moles of NaCl are contained in each solution?
  (a) 1.5 L of a 1.2 M NaCl solution
  (b) 0.448 L of a 0.85 M NaCl solution
  (c) 144 mL of a 1.65 M NaCl solution

66. How many moles of sucrose are contained in each solution?
  (a) 3.4 L of a 0.100 M sucrose solution
  (b) 0.952 L of a 1.88 M sucrose solution
  (c) 21.5 mL of a 0.528 M sucrose solution

67. What volume of each solution contains 0.15 mol of KCl?
  (a) 0.255 M KCl
  (b) 1.8 M KCl
  (c) 0.995 M KCl

68. What volume of each solution contains 0.325 mol of NaI?
  (a) 0.152 M NaI
  (b) 0.982 M NaI
  (c) 1.76 M NaI

69. Complete the table:

Solute	Solute Mass	Mol Solute	Volume Solution	Molarity
KNO_3	22.5 g	_____	125.0 mL	_____
$NaHCO_3$	_____	_____	250.0 mL	0.100 M
$C_{12}H_{22}O_{11}$	55.38 g	_____	_____	0.150 M

70. Complete the table:

Solute	Solute Mass	Mol Solute	Volume Solution	Molarity
$MgSO_4$	0.588 g	_____	25.0 mL	_____
$NaOH$	_____	_____	100.0 mL	1.75 M
CH_3OH	12.5 g	_____	_____	0.500 M

71. Calculate the mass of NaCl in a 35-mL sample of a 1.3 M NaCl solution.

72. Calculate the mass of glucose ($C_6H_{12}O_6$) in a 105-mL sample of a 1.02 M glucose solution.

73. A chemist wants to make 2.5 L of a 0.100 M KCl solution. How much KCl in grams should the chemist use?

74. A laboratory procedure calls for making 500.0 mL of a 1.4 M KNO_3 solution. How much KNO_3 in grams is needed?

75. How many liters of a 0.500 M sucrose ($C_{12}H_{22}O_{11}$) solution contain 1.5 kg of sucrose?

76. What volume of a 0.35 M $Mg(NO_3)_2$ solution contains 87 g of $Mg(NO_3)_2$?

77. Determine the concentration of Cl^- in each aqueous solution. (Assume complete dissociation of each compound.)
  (a) 0.15 M NaCl
  (b) 0.15 M $CuCl_2$
  (c) 0.15 M $AlCl_3$

78. Determine the concentration of NO_3^- in each aqueous solution. (Assume complete dissociation of each compound.)
  (a) 0.10 M KNO_3
  (b) 0.10 M $Ca(NO_3)_2$
  (c) 0.10 M $Cr(NO_3)_3$

79. Determine the concentration of the cation and anion in each aqueous solution. (Assume complete dissociation of each compound.)
  (a) 0.12 M Na_2SO_4
  (b) 0.25 M K_2CO_3
  (c) 0.11 M RbBr

80. Determine the concentration of the cation and anion in each aqueous solution. (Assume complete dissociation of each compound.)
  (a) 0.20 M $SrSO_4$
  (b) 0.15 M $Cr_2(SO_4)_3$
  (c) 0.12 M SrI_2

SOLUTION DILUTION

81. A 122-mL sample of a 1.2 M sucrose solution is diluted to 500.0 mL. What is the molarity of the diluted solution?

82. A 3.5-L sample of a 5.8 M NaCl solution is diluted to 55 L. What is the molarity of the diluted solution?

83. Describe how you would make 2.5 L of a 0.100 M KCl solution from a 5.5 M stock KCl solution.

84. Describe how you would make 500.0 mL of a 0.200 M NaOH solution from a 15.0 M stock NaOH solution.

85. To what volume should you dilute 25 mL of a 12 M stock HCl solution to obtain a 0.500 M HCl solution?

86. To what volume should you dilute 75 mL of a 10.0 M H_2SO_4 solution to obtain a 1.75 M H_2SO_4 solution?

87. How much of a 12.0 M HNO_3 solution should you use to make 850.0 mL of a 0.250 M HNO_3 solution?

88. How much of a 5.0 M sucrose solution should you use to make 85.0 mL of a 0.040 M solution?

SOLUTION STOICHIOMETRY

89. Determine the volume of 0.150 M NaOH solution required to neutralize each sample of hydrochloric acid. The neutralization reaction is:

$$NaOH(aq) + HCl(aq) \longrightarrow H_2O(l) + NaCl(aq)$$

   (a) 25 mL of a 0.150 M HCl solution
   (b) 55 mL of a 0.055 M HCl solution
   (c) 175 mL of a 0.885 M HCl solution

90. Determine the volume of 0.225 M KOH solution required to neutralize each sample of sulfuric acid. The neutralization reaction is:

$$H_2SO_4(aq) + 2\,KOH(aq) \longrightarrow$$
$$K_2SO_4(aq) + 2\,H_2O(l)$$

   (a) 45 mL of 0.225 M H_2SO_4
   (b) 185 mL of 0.125 M H_2SO_4
   (c) 75 mL of 0.100 M H_2SO_4

91. Consider the reaction:

$$2\,K_3PO_4(aq) + 3\,NiCl_2(aq) \longrightarrow$$
$$Ni_3(PO_4)_2(s) + 6\,KCl(aq)$$

   What volume of 0.225 M K_3PO_4 solution is necessary to completely react with 134 mL of 0.0112 M $NiCl_2$?

92. Consider the reaction:

$$K_2S(aq) + Co(NO_3)_2(aq) \longrightarrow 2\,KNO_3(aq) + CoS(s)$$

   What volume of 0.225 M K_2S solution is required to completely react with 175 mL of 0.115 M $Co(NO_3)_2$?

93. A 10.0-mL sample of an unknown H_3PO_4 solution requires 112 mL of 0.100 M KOH to completely react with the H_3PO_4. What was the concentration of the unknown H_3PO_4 solution?

$$H_3PO_4(aq) + 3\,KOH(aq) \longrightarrow 3\,H_2O(l) + K_3PO_4(aq)$$

94. A 25.0-mL sample of an unknown $HClO_4$ solution requires 45.3 mL of 0.101 M NaOH for complete neutralization. What was the concentration of the unknown $HClO_4$ solution? The neutralization reaction is:

$$HClO_4(aq) + NaOH(aq) \longrightarrow H_2O(l) + NaClO_4(aq)$$

95. What is the minimum amount of 6.0 M H_2SO_4 necessary to produce 15.0 g of $H_2(g)$ according to the reaction:

$$2\,Al(s) + 3\,H_2SO_4(aq) \longrightarrow Al_2(SO_4)_3(aq) + 3\,H_2(g)$$

96. What is the molarity of $ZnCl_2(aq)$ that forms when 15.0 g of zinc completely reacts with $CuCl_2(aq)$ according to the following reaction? (Assume a final volume of 175 mL.)

$$Zn(s) + CuCl_2(aq) \longrightarrow ZnCl_2 + Cu(s)$$

MOLALITY, FREEZING POINT DEPRESSION, AND BOILING POINT ELEVATION

97. Calculate the molality of each solution.
   (a) 0.25 mol solute; 0.250 kg solvent
   (b) 0.882 mol solute; 0.225 kg solvent
   (c) 0.012 mol solute; 23.1 g solvent

98. Calculate the molality of each solution.
   (a) 0.455 mol solute; 1.97 kg solvent
   (b) 0.559 mol solute; 1.44 kg solvent
   (c) 0.119 mol solute; 488 g solvent

99. Calculate the molality of a solution containing 12.5 g of ethylene glycol ($C_2H_6O_2$) dissolved in 135 g of water.

100. Calculate the molality of a solution containing 257 g glucose ($C_6H_{12}O_6$) dissolved in 1.62 L of water. (Assume a density of 1.00 g/mL for water.)

101. Calculate the freezing point of a water solution at each concentration.
   (a) 0.85 m
   (b) 1.45 m
   (c) 4.8 m
   (d) 2.35 m

102. Calculate the freezing point of a water solution at each concentration.
   (a) 0.100 m
   (b) 0.469 m
   (c) 1.44 m
   (d) 5.89 m

103. Calculate the boiling point of a water solution at each concentration.

    (a) 0.118 m

    (b) 1.94 m

    (c) 3.88 m

    (d) 2.16 m

104. Calculate the boiling point of a water solution at each concentration.

    (a) 0.225 m

    (b) 2.58 m

    (c) 4.33 m

    (d) 6.77 m

105. A glucose solution contains 55.8 g of glucose ($C_6H_{12}O_6$) in 455 g of water. Calculate the freezing point and boiling point of the solution. (Assume a density of 1.00 g/mL for water.)

106. An ethylene glycol solution contains 21.2 g of ethylene glycol ($C_2H_6O_2$) in 85.4 mL of water. Calculate the freezing point and boiling point of the solution. (Assume a density of 1.00 g/mL for water.)

CUMULATIVE PROBLEMS

107. An NaCl solution is made using 133 g of NaCl and diluting to a total solution volume of 1.00 L. Calculate the molarity and mass percent of the solution. (Assume a density of 1.08 g/mL for the solution.)

108. A KNO$_3$ solution is made using 88.4 g of KNO$_3$ and diluting to a total solution volume of 1.50 L. Calculate the molarity and mass percent of the solution. (Assume a density of 1.05 g/mL for the solution.)

109. A 125-mL sample of an 8.5 M NaCl solution is diluted to 2.5 L. What volume of the diluted solution contains 10.8 g of NaCl?

110. A 45.8-mL sample of a 5.8 M KNO$_3$ solution is diluted to 1.00 L. What volume of the diluted solution contains 15.0 g of KNO$_3$?

111. To what final volume should you dilute 50.0 mL of a 5.00 M KI solution so that 25.0 mL of the diluted solution contains 3.25 g of KI?

112. To what volume should you dilute 125 mL of an 8.00 M CuCl$_2$ solution so that 50.0 mL of the diluted solution contains 5.9 g CuCl$_2$?

113. What is the molarity of an aqueous solution that is 5.88% NaCl by mass? (Assume a density of 1.02 g/mL for the solution.)

114. What is the molarity of an aqueous solution that is 6.75% glucose ($C_6H_{12}O_6$) by mass? (Assume a density of 1.03 g/mL for the solution.)

115. Consider the reaction:

    $$2\,Al(s) + 3\,H_2SO_4(aq) \longrightarrow Al_2(SO_4)_3(aq) + 3\,H_2(g)$$

    What minimum volume of 4.0 M H$_2$SO$_4$ is required to produce 15.0 L of H$_2$ at STP?

116. Consider the reaction:

    $$Mg(s) + 2\,HCl(aq) \longrightarrow MgCl_2(aq) + H_2(g)$$

    What minimum amount of 1.85 M HCl is necessary to produce 28.5 L of H$_2$ at STP?

117. How much of a 1.25 M sodium chloride solution in milliliters is required to completely precipitate all of the silver in 25.0 mL of a 0.45 M silver nitrate solution?

118. How much of a 1.50 M sodium sulfate solution in milliliters is required to completely precipitate all of the barium in 150.0 mL of a 0.250 M barium nitrate solution?

119. Nitric acid is usually purchased in concentrated form with a 70.3% HNO$_3$ concentration by mass and a density of 1.41 g/mL. How much of the concentrated stock solution in milliliters should you use to make 2.5 L of 0.500 M HNO$_3$?

120. Hydrochloric acid is usually purchased in concentrated form with a 37.0% HCl concentration by mass and a density of 1.20 g/mL. How much of the concentrated stock solution in milliliters should you use to make 2.5 L of 0.500 M HCl?

121. An ethylene glycol solution is made using 58.5 g of ethylene glycol ($C_2H_6O_2$) and diluting to a total volume of 500.0 mL. Calculate the freezing point and boiling point of the solution. (Assume a density of 1.09 g/mL for the solution.)

122. A sucrose solution is made using 144 g of sucrose ($C_{12}H_{22}O_{11}$) and diluting to a total volume of 1.00 L. Calculate the freezing point and boiling point of the solution. (Assume a density of 1.06 g/mL for the final solution.)

123. A 250.0-mL sample of a 5.00 M glucose ($C_6H_{12}O_6$) solution is diluted to 1.40 L. What are the freezing and boiling points of the final solution? (Assume a density of 1.06 g/mL for the final solution.)

124. A 135-mL sample of a 10.0 M ethylene glycol ($C_2H_6O_2$) solution is diluted to 1.50 L. What are the freezing and boiling points of the final solution? (Assume a density of 1.05 g/mL for the final solution.)

125. An aqueous solution containing 17.5 g of an unknown molecular (nonelectrolyte) compound in 100.0 g of water has a freezing point of −1.8 °C. Calculate the molar mass of the unknown compound.

126. An aqueous solution containing 35.9 g of an unknown molecular (nonelectrolyte) compound in 150.0 g of water has a freezing point of −1.3 °C. Calculate the molar mass of the unknown compound.

127. What is the boiling point of an aqueous solution that freezes at −6.7 °C?

128. What is the freezing point of an aqueous solution that boils at 102.1 °C?

129. A 125-g sample contains only glucose ($C_6H_{12}O_6$) and sucrose ($C_{12}H_{22}O_{11}$). When the sample is added to 0.500 kg of pure water, the resulting solution has a freezing point of −1.75 °C. What were the masses of glucose and sucrose in the original sample?

130. A 13.03-g sample contains only ethylene glycol ($C_2H_6O_2$) and propylene glycol ($C_3H_8O_2$). When the sample is added to 100.0 g of pure water, the resulting solution has a freezing point of −3.50 °C. What was the percent composition of ethylene glycol and propylene glycol in the original sample?

HIGHLIGHT PROBLEMS

131. Consider the molecular views of osmosis cells. For each cell, determine the direction of water flow.

(a)

(b)

(c)

132. What is wrong with this molecular view of a sodium chloride solution? What would make the picture correct?

133. The Safe Drinking Water Act (SDWA) sets a limit for mercury—a toxin to the central nervous system—at 0.002 mg/L. Water suppliers must periodically test their water to ensure that mercury levels do not exceed 0.002 mg/L. Suppose water became contaminated with mercury at twice the legal limit (0.004 mg/L). How much of this water would have to be consumed to ingest 0.100 g of mercury?

134. Water softeners often replace calcium ions in hard water with sodium ions. Since sodium compounds are soluble, the presence of sodium ions in water does not result in the white, scaly residues caused by calcium ions. However, calcium is more beneficial to human health than sodium. Calcium is a necessary part of the human diet, while high levels of sodium intake are linked to increases in blood pressure. The Food and Drug Administration (FDA) recommends that adults ingest less than 2.4 g of sodium per day. How many liters of softened water, containing a sodium concentration of 0.050% sodium by mass, have to be consumed to exceed the FDA recommendation? (Assume a density of 1.0 g/mL for water.)

◄ Drinking water must be tested for the presence of various pollutants, including mercury compounds that can damage the nervous system.

▶ANSWERS TO SKILLBUILDER EXCERISES

Skillbuilder 13.1	2.67%		**Skillbuilder 13.7**	16.4 mL
Skillbuilder 13.2	42.5 g sucrose		**Skillbuilder Plus, p. 465**	0.308 M
Skillbuilder 13.3	0.263 M		**Skillbuilder 13.8**	0.443 m
Skillbuilder 13.4	3.33 L		**Skillbuilder 13.9**	−4.8 °C
Skillbuilder 13.5	0.75 M Ca^{2+} and 1.5 M Cl^-		**Skillbuilder 13.10**	101.8 °C
Skillbuilder 13.6	0.12 L			

▶ANSWERS TO CONCEPTUAL CHECKPOINTS

13.1 (a) CH_3Cl and H_2S are both polar compounds, and KF is ionic. All three would therefore interact more strongly with water molecules (which are polar) than CCl_4, which is nonpolar.

13.2 (b) Some potassium chloride precipitates out of solution. The solubility of most solids decreases with decreasing temperature. However, the solubility of gases increases with decreasing temperature. Therefore, the nitrogen becomes more soluble and will not bubble out of solution.

13.3 (c) The solution is 0.30 M in K^+ because the compound K_2SO_4 forms two moles of K^+ in solution for each mole of K_2SO_4 that dissolves.

13.4 (a) The mass of a solution is equal to the mass of the solute plus the mass of the solvent. Although the solute seems to disappear, it does not, and its mass becomes part of the mass of the solution, in accordance with the law of conservation of mass.

13.5 (a) A 2.0 m solution would be made by adding 2 mol of solute to 1 kg of solvent. 1 kg of water has a volume of 1 L, but because of the dissolved solute, the final solution would have a volume of slightly *more than* 1 L. A 2.0 M solution, by contrast, would consist of 2 mol of solute in a solution of *exactly* 1 L. Therefore, a 2 M aqueous solution would be slightly more concentrated than a 2 m solution.

13.6 (d) Since boiling point elevation depends only on the *concentration* of the dissolved particles, and not on the *kind* of dissolved particles, all of these solutions have the same boiling point.

CHAPTER 14

Acids and Bases

"The differences between the various acid–base concepts are not concerned with which is 'right,' but which is most convenient to use in a particular situation."

JAMES E. HUHEEY

14.1 Sour Patch Kids and International Spy Movies

Gummy candies have a sweet taste and chewy texture that both children and adults can enjoy. From the original classic gummy bear to the gummy worm to the gummy just-about-any-shape-you-can-imagine, these candies are incredibly popular. A common variation is the *sour* gummy candy, whose best-known incarnation is the Sour Patch Kid. Sour Patch Kids are gummy candies shaped like children and coated with a white powder. When you first put a Sour Patch Kid in your mouth, it tastes incredibly sour. The taste is caused by the white powder coating, a mixture of citric acid and tartaric acid. Like all acids, citric and tartaric acid taste sour.

A number of other foods contain acids as well. The taste of lemons and limes, the bite of sourdough bread, and the tang of a tomato are all caused by acids. Acids are substances that—by one definition that we will elaborate on later—produce H^+ ions in solution. When the citric and tartaric acids from a Sour Patch Kid combine with saliva in your mouth, they produce H^+ ions. Those H^+ ions react with protein molecules on your tongue. The protein molecules then change shape, sending an electrical signal to your brain that you experience as a sour taste (▶ Figure 14.1).

Acids have also been made famous by their use in spy movies. James Bond, for example, often carries an acid-filled gold pen. When Bond is captured and imprisoned—as inevitably happens at least one time in each movie—he squirts some acid out of his pen and onto the iron bars of his cell. The acid quickly dissolves the metal, allowing Bond to escape. Although acids do not dissolve iron bars with the ease depicted in the movies, they do dissolve metals. A small piece of aluminum placed in hydrochloric acid, for example, dissolves away in about

When we say that acids dissolve metals, we mean that acids react with metals in a way that causes them to go into solution as metal cations. Bond's pen is made of gold because gold is one of the few metals that is not dissolved by most acids (see Section 16.5).

◀ Acids are found in many common foods. The molecules shown here are citric acid (upper left), the acid found in lemon and limes; acetic acid (upper right), the acid present in vinegar; and tartaric acid (lower left), one of the acids used to coat sour gummy candies.

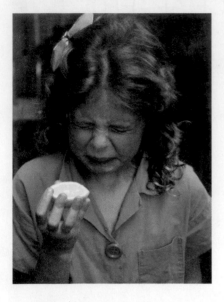

◄ **FIGURE 14.1 Acids taste sour**
When a person eats a sour food, H^+ ions from the acid in the food react with protein molecules in the taste cells of the tongue. This interaction causes the protein molecules to change shape, triggering a nerve impulse to the brain that the person experiences as a sour taste.

► **FIGURE 14.2 Acids dissolve many metals** When aluminum is put into hydrochloric acid, the aluminum dissolves. **Question:** What happens to the aluminum atoms? Where do they go?

10 minutes (▲ Figure 14.2). With enough acid, it would be possible to dissolve the iron bars of a prison cell, but it would take more acid than the amount that fits in a pen.

14.2 Acids: Properties and Examples

| NEVER taste or touch laboratory chemicals.

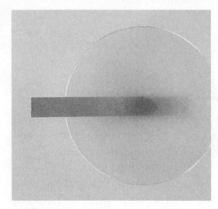

▲ **FIGURE 14.3 Acids turn blue litmus paper red.**

| For a review of naming acids, see Section 5.9.

Acids have the following properties:

- Acids have a sour taste.
- Acids dissolve many metals.
- Acids turn blue litmus paper red.

We have just discussed examples of the sour taste of acids (Sour Patch Kids) and their ability to dissolve metals (spy movies). Acids also turn blue litmus paper red. Litmus paper contains a dye that turns red in acidic solutions (◄ Figure 14.3). In the laboratory, litmus paper is used routinely to test the acidity of solutions.

Some common acids are listed in Table 14.1. Hydrochloric acid is found in most chemistry laboratories. It is used in industry to clean metals, to prepare and process some foods, and to refine metal ores.

HCl

Hydrochloric acid

Hydrochloric acid is also the main component of stomach acid. In the stomach, hydrochloric acid helps break down food and kills harmful bacteria that might enter the body through food. The sour taste sometimes associated with

TABLE 14.1 Some Common Acids

Name	Uses
hydrochloric acid (HCl)	metal cleaning; food preparation; ore refining; main component of stomach acid
sulfuric acid (H_2SO_4)	fertilizer and explosive manufacturing; dye and glue production; automobile batteries
nitric acid (HNO_3)	fertilizer and explosive manufacturing; dye and glue production
acetic acid ($HC_2H_3O_2$)	plastic and rubber manufacturing; food preservation; active component of vinegar
carbonic acid (H_2CO_3)	found in carbonated beverages due to the reaction of carbon dioxide with water
hydrofluoric acid (HF)	metal cleaning; glass frosting and etching

indigestion is caused by the stomach's hydrochloric acid refluxing up into the esophagus (the tube that joins the stomach and the mouth) and throat.

Sulfuric acid—the most widely produced chemical in the United States—and nitric acid are commonly used in the laboratory. In addition, they are used in the manufacture of fertilizers, explosives, dyes, and glue. Sulfuric acid is contained in most automobile batteries.

Annual U.S. production of sulfuric acid exceeds 36 million tons.

Sulfuric acid

Nitric acid

Acetic acid is found in most people's homes as the active component of vinegar. It is also produced in improperly stored wines. The word *vinegar* originates from the French *vin aigre*, which means "sour wine." The presence of vinegar in wines is considered a serious fault, making the wine taste like salad dressing.

Acetic acid

Acetic acid is an example of a **carboxylic acid**, an acid containing the grouping of atoms known as the carboxylic acid group.

Carboxylic acid group

We often find carboxylic acids (covered in more detail in Chapter 18, Section 18.15), in substances derived from living organisms. Other carboxylic acids include citric acid, the main acid in lemons and limes, and malic acid, an acid found in apples, grapes, and wine.

▲ Acetic acid is the active component in vinegar.

$HC_6H_7O_7$ $HC_4H_5O_5$

Citric acid   Malic acid

14.3 Bases: Properties and Examples

NEVER taste or touch laboratory chemicals.

Bases have the following properties:

- Bases have a bitter taste.
- Bases have a slippery feel.
- Bases turn red litmus paper blue.

Coffee is acidic overall, but bases present in coffee—such as caffeine—impart a bitter flavor.

▲ All these consumer products contain bases.

▲ FIGURE 14.4 **Bases turn red litmus paper blue.**

Bases are less common in foods than acids because of their bitter taste. A Sour Patch Kid coated with a base would never sell. Our aversion to the taste of bases is probably an adaptation to protect us against **alkaloids**, organic bases found in plants (see the Chemistry and Health box in Section 14.9). Alkaloids are often poisonous—the active component of hemlock, for example, is the alkaloid coniine—and their bitter taste warns us against eating them. Nonetheless, some foods, such as coffee, contain small amounts of base (caffeine is a base). Many people enjoy the bitterness, but only after acquiring the taste over time.

$C_8H_{17}N$

Coniine

$C_8H_{10}N_4O_2$

Caffeine

Bases feel slippery because they react with oils on your skin to form soaplike substances. Soap itself is basic, and its slippery feel is characteristic of bases. Some household cleaning solutions, such as ammonia, are also basic and have the typical slippery feel of a base. Bases turn red litmus paper blue (◄ Figure 14.4). In the laboratory, litmus paper is routinely used to test the basicity of solutions.

Table 14.2 lists some common bases. Sodium hydroxide and potassium hydroxide are found in most chemistry laboratories. They are also used in processing petroleum and cotton and in soap and plastic manufacturing. Sodium hydroxide is the active ingredient in products such as Drano that work to unclog drains. Sodium bicarbonate can be found in most homes as baking soda and is also an active ingredient in many antacids. When taken as an antacid, sodium bicarbonate neutralizes stomach acid (see Section 14.5), relieving heartburn and sour stomach.

TABLE 14.2 Some Common Bases

Name	Uses
sodium hydroxide (NaOH)	petroleum processing; soap and plastic manufacturing
potassium hydroxide (KOH)	cotton processing; electroplating; soap production
sodium bicarbonate ($NaHCO_3$)*	antacid; ingredient of baking soda; source of CO_2
ammonia (NH_3)	detergent; fertilizer and explosive manufacturing; synthetic fiber production

*Sodium bicarbonate is a salt whose anion (HCO_3^-) is the conjugate base of a weak acid (see Section 14.4) and acts as a base.

14.4 Molecular Definitions of Acids and Bases

We have just seen some of the properties of acids and bases. In this section we examine two different models that explain the molecular basis for acid and base behavior: the Arrhenius model and the Brønsted–Lowry model. The Arrhenius model, which was developed earlier, is more limited in its scope. The Brønsted–Lowry model was developed later and is more broadly applicable.

THE ARRHENIUS DEFINITION

In the 1880s, the Swedish chemist Svante Arrhenius proposed the following molecular definitions of acids and bases.

The Arrhenius definition

Acid—An acid produces H^+ ions in aqueous solution.
Base—A base produces OH^- ions in aqueous solution.

For example, according to the **Arrhenius definition**, HCl is an **Arrhenius acid** because it produces H^+ ions in solution (◀ Figure 14.5).

$$HCl(aq) \longrightarrow H^+(aq) + Cl^-(aq)$$

HCl is a covalent compound and does not contain ions. However, in water it **ionizes** to form $H^+(aq)$ ions and $Cl^-(aq)$ ions. The H^+ ions are highly reactive. In aqueous solution, they bond to water molecules according to the reaction.

The H_3O^+ ion is the **hydronium ion**. In water, H^+ ions *always* associate with H_2O molecules. Chemists often use $H^+(aq)$ and $H_3O^+(aq)$ interchangeably, however, to refer to the same thing—a hydronium ion.

In the molecular formula for an acid, we often write the ionizable hydrogen first. For example, we write the formula for formic acid as follows:

$$HCHO_2$$

Ionizable hydrogen   Non-ionizable hydrogen

The structure of formic acid, however, is not indicated by the molecular formula in the preceding figure. We represent the *structure* of formic acid with its structural formula:

$$\underset{HC-OH}{\overset{O}{\parallel}}$$ Ionizable hydrogen

Notice that the structural formula indicates how the atoms are bonded together; the molecular formula, by contrast, simply indicates the number of each type of atom.

NaOH is an **Arrhenius base** because it produces OH^- ions in solution (◀ Figure 14.6).

$$NaOH(aq) \longrightarrow Na^+(aq) + OH^-(aq)$$

NaOH is an ionic compound and therefore contains Na^+ and OH^- ions. When NaOH is added to water, it **dissociates**, or breaks apart into its component ions.

Under the Arrhenius definition, acids and bases naturally combine to form water, neutralizing each other in the process.

$$H^+(aq) + OH^-(aq) \longrightarrow H_2O(l)$$

THE BRØNSTED–LOWRY DEFINITION

Although the Arrhenius definition of acids and bases works in many cases, it cannot easily explain why some substances act as bases even though they do not contain OH^-. The Arrhenius definition also does not apply to nonaqueous solvents. A second definition of acids and bases, called the **Brønsted–Lowry definition**, introduced in 1923, applies to a wider range of acid–base phenomena. This definition focuses on the *transfer* of H^+ ions in an acid–base reaction. Since an H^+ ion is a proton—a hydrogen atom with its electron taken away—this definition focuses on the idea of a proton donor and a proton acceptor.

$$HCl(aq) \longrightarrow$$
$$H^+(aq) + Cl^-(aq)$$

▲ **FIGURE 14.5 Arrhenius definition of an acid** The Arrhenius definition states that an acid is a substance that produces H^+ ions in solution. These H^+ ions associate with H_2O to form H_3O^+ ions.

$$NaOH(aq) \longrightarrow$$
$$Na^+(aq) + OH^-(aq)$$

▲ **FIGURE 14.6 Arrhenius definition of a base** The Arrhenius definition states that a base is a substance that produces OH^- ions in solution.

Ionic compounds such as NaOH are composed of positive and negative ions. In solution, soluble ionic compounds dissociate into their component ions. Molecular compounds containing an OH group, such as methanol CH_3OH, do not dissociate and therefore do not act as bases.

Brønsted–Lowry definition

Acid—An acid is a proton (H^+ ion) *donor*.
Base—A base is a proton (H^+ ion) *acceptor*.

According to this definition, HCl is a **Brønsted–Lowry acid** because, in solution, it donates a proton to water.

$$HCl(aq) + H_2O(l) \longrightarrow H_3O^+(aq) + Cl^-(aq)$$

This definition more clearly accounts for what happens to the H^+ ion from an acid: it associates with a water molecule to form H_3O^+ (a hydronium ion). The Brønsted–Lowry definition also works well with bases (such as NH_3) that do not inherently contain OH^- ions but that still produce OH^- ions in solution. NH_3 is a **Brønsted–Lowry base** because it accepts a proton from water.

$$NH_3(aq) + H_2O(l) \rightleftharpoons NH_4^+(aq) + OH^-(aq)$$

In the Brønsted–Lowry definition, acids (proton donors) and bases (proton acceptors) always occur together. In the reaction between HCl and H_2O, HCl is the proton donor (acid), and H_2O is the proton acceptor (base).

$$\underset{\substack{\text{Acid}\\ \text{(Proton donor)}}}{HCl(aq)} + \underset{\substack{\text{Base}\\ \text{(Proton acceptor)}}}{H_2O(l)} \longrightarrow H_3O^+(aq) + Cl^-(aq)$$

In the reaction between NH_3 and H_2O, H_2O is the proton donor (acid) and NH_3 is the proton acceptor (base).

$$\underset{\substack{\text{Base}\\ \text{(Proton acceptor)}}}{NH_3(aq)} + \underset{\substack{\text{Acid}\\ \text{(Proton donor)}}}{H_2O(l)} \rightleftharpoons NH_4^+(aq) + OH^-(aq)$$

Notice that under the Brønsted–Lowry definition, some substances—such as water in the previous two equations—can act as acids *or* bases. Substances that can act as acids or bases are **amphoteric**. Notice also what happens when an equation representing Brønsted–Lowry acid–base behavior is reversed.

$$\underset{\substack{\text{Acid}\\ \text{(Proton donor)}}}{NH_4^+(aq)} + \underset{\substack{\text{Base}\\ \text{(Proton acceptor)}}}{OH^-(aq)} \rightleftharpoons NH_3(aq) + H_2O(l)$$

In this reaction, NH_4^+ is the proton donor (acid) and OH^- is the proton acceptor (base). What was the base (NH_3) has become the acid (NH_4^+), and vice versa. NH_4^+ and NH_3 are often referred to as a **conjugate acid–base pair**, two substances related to each other by the transfer of a proton (◀ Figure 14.7). Going back to the original forward reaction, we can identify the conjugate acid–base pairs as follows:

$$\underset{\text{Base}}{NH_3(aq)} + \underset{\text{Acid}}{H_2O(l)} \rightleftharpoons \underset{\substack{\text{Conjugate}\\ \text{acid}}}{NH_4^+(aq)} + \underset{\substack{\text{Conjugate}\\ \text{base}}}{OH^-(aq)}$$

In an acid–base reaction, a base accepts a proton and becomes a conjugate acid. An acid donates a proton and becomes a conjugate base.

Johannes Brønsted, working in Denmark, and Thomas Lowry, working in England, developed the concept of proton transfer in acid–base behavior independently and simultaneously.

The double arrows in this equation indicate that the reaction does not go to completion. We discuss this concept in more detail in Section 14.7.

▲ **FIGURE 14.7 A conjugate acid–base pair** Any two substances related to each other by the transfer of a proton can be considered a conjugate acid–base pair.

EXAMPLE 14.1   Identifying Brønsted–Lowry Acids and Bases and Their Conjugates

In each reaction, identify the Brønsted–Lowry acid, the Brønsted–Lowry base, the conjugate acid, and the conjugate base.

(a) $H_2SO_4(aq) + H_2O(l) \longrightarrow H_3O^+(aq) + HSO_4^-(aq)$
(b) $HCO_3^-(aq) + H_2O(l) \rightleftharpoons H_2CO_3(aq) + OH^-(aq)$

SOLUTION

(a) Since H_2SO_4 donates a proton to H_2O in this reaction, it is the acid (the proton donor). After H_2SO_4 donates the proton, it becomes HSO_4^-, the conjugate base. Since H_2O accepts a proton, it is the base (the proton acceptor). After H_2O accepts the proton, it becomes H_3O^+, the conjugate acid.

(b) Since H_2O donates a proton to HCO_3^- in this reaction, it is the acid (the proton donor). After H_2O donates the proton, it becomes OH^-, the conjugate base. Since HCO_3^- accepts a proton, it is the base (the proton acceptor). After HCO_3^- accepts the proton, it becomes H_2CO_3, the conjugate acid.

▶**SKILLBUILDER 14.1 | Identifying Brønsted–Lowry Acids and Bases and Their Conjugates**

In each reaction, identify the Brønsted–Lowry acid, the Brønsted–Lowry base, the conjugate acid, and the conjugate base.

(a) $C_5H_5N(aq) + H_2O(l) \rightleftharpoons C_5H_5NH^+(aq) + OH^-(aq)$

(b) $HNO_3(aq) + H_2O(l) \longrightarrow NO_3^-(aq) + H_3O^+(aq)$

▶**FOR MORE PRACTICE**   Example 14.11; Problems 35, 36, 37, 38.

✔ **CONCEPTUAL CHECKPOINT 14.1**

Which species is the conjugate base of H_2SO_3?
(a) $H_3SO_3^+$ (b) HSO_3^- (c) SO_3^{2-}

14.5   Reactions of Acids and Bases

NEUTRALIZATION REACTIONS

The reaction between HCl and KOH is also a double-displacement reaction (see Section 7.10).

Neutralization reactions are covered in Section 7.8.

One of the most important reactions of acids and bases is **neutralization**, first introduced in Chapter 7. When an acid and a base are mixed, the $H^+(aq)$ from the acid combines with the $OH^-(aq)$ from the base to form $H_2O(l)$. For example, consider the reaction between hydrochloric acid and potassium hydroxide.

$$\underset{\text{Acid}}{HCl(aq)} + \underset{\text{Base}}{KOH(aq)} \longrightarrow \underset{\text{Water}}{H_2O(l)} + \underset{\text{Salt}}{KCl(aq)}$$

Acid–base reactions generally form water and a **salt**—an ionic compound—that usually remains dissolved in the solution. The salt contains the cation from the base and the anion from the acid.

Ionic compound that contains the cation from the base and the anion from the acid

$$\text{Acid} + \text{Base} \longrightarrow \text{Water} + \text{Salt}$$

Net ionic equations are explained in Section 7.7.

The net ionic equation for many neutralization reactions is:

$$H^+(aq) + OH^-(aq) \longrightarrow H_2O(l)$$

A slightly different but common type of neutralization reaction involves an acid reacting with carbonates or bicarbonates (compounds containing $CO_3{}^{2-}$ or $HCO_3{}^-$). This type of neutralization reaction produces water, gaseous carbon dioxide, and a salt. As an example, consider the reaction of hydrochloric acid and sodium bicarbonate.

$$HCl(aq) + NaHCO_3(aq) \longrightarrow H_2O(l) + CO_2(g) + NaCl(aq)$$

Since this reaction produces gaseous CO_2, it is also called a *gas evolution reaction* (Section 7.8).

$$HCl(aq) + NaHCO_3(aq) \longrightarrow$$

$$H_2O(l) + CO_2(g) + NaCl(aq)$$

▲ The reaction of carbonates or bicarbonates with acids produces water, gaseous carbon dioxide, and a salt.

EXAMPLE 14.2   Writing Equations for Neutralization Reactions

Write a molecular equation for the reaction between aqueous HCl and aqueous $Ca(OH)_2$.

SOLUTION	
First identify the acid and the base and write the skeletal reaction showing the production of water and the salt. The formulas for the ionic compounds in the equation must be charge neutral (see Section 5.5).	$HCl(aq) + Ca(OH)_2(aq) \longrightarrow H_2O(l) + CaCl_2(aq)$
Balance the equation. Notice that $Ca(OH)_2$ contains 2 mol of OH^- for every 1 mol of $Ca(OH)_2$ and will therefore require 2 mol of H^+ to neutralize it.	$2\,HCl(aq) + Ca(OH)_2(aq) \longrightarrow 2\,H_2O(l) + CaCl_2(aq)$

▶SKILLBUILDER 14.2 | Writing Equations for Neutralization Reactions

Write a molecular equation for the reaction that occurs between aqueous H_3PO_4 and aqueous NaOH. *Hint:* H_3PO_4 is a triprotic acid, meaning that 1 mol of H_3PO_4 requires 3 mol of OH^- to completely react with it.

▶FOR MORE PRACTICE   Example 14.12; Problems 43, 44.

ACID REACTIONS

In Section 14.1, we learned that acids dissolve metals, or more precisely, that acids react with metals in a way that causes metals to go into solution. The reaction between an acid and a metal usually produces hydrogen gas and a dissolved salt containing the metal ion as the cation. For example, hydrochloric acid reacts with magnesium metal to form hydrogen gas and magnesium chloride.

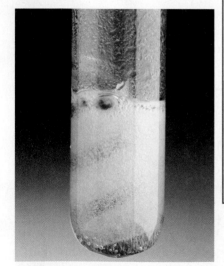

$$2\,HCl(aq) + Mg(s) \longrightarrow$$

$$H_2(g) + MgCl_2(aq)$$

▲ The reaction between an acid and a metal usually produces hydrogen gas and a dissolved salt containing the metal ion.

$$\underset{\text{Acid}}{2\,HCl(aq)} + \underset{\text{Metal}}{Mg(s)} \longrightarrow \underset{\substack{\text{Hydrogen}\\\text{gas}}}{H_2(g)} + \underset{\text{Salt}}{MgCl_2(aq)}$$

Similarly, sulfuric acid reacts with zinc to form hydrogen gas and zinc sulfate.

$$\underset{\text{Acid}}{H_2SO_4(aq)} + \underset{\text{Metal}}{Zn(s)} \longrightarrow \underset{\substack{\text{Hydrogen} \\ \text{gas}}}{H_2(g)} + \underset{\text{Salt}}{ZnSO_4(aq)}$$

It is through reactions such as these that the acid from James Bond's pen in our earlier example dissolves the metal bars that imprison him. For example, if the bars were made of iron and the acid in the pen were hydrochloric acid, the reaction would be:

$$\underset{\text{Acid}}{2\,HCl(aq)} + \underset{\text{Metal}}{Fe(s)} \longrightarrow \underset{\substack{\text{Hydrogen} \\ \text{gas}}}{H_2(g)} + \underset{\text{Salt}}{FeCl_2(aq)}$$

Some metals, however, do not readily react with acids. If the bars that imprisoned James Bond were made of gold, for example, a pen filled with hydrochloric acid would not dissolve the bars. We will discuss the way to determine whether a particular metal dissolves in an acid in Section 16.5.

Acids also react with metal oxides to produce water and a dissolved salt. For example, hydrochloric acid reacts with potassium oxide to form water and potassium chloride.

$$\underset{\text{Acid}}{2\,HCl(aq)} + \underset{\text{Metal oxide}}{K_2O(s)} \longrightarrow \underset{\text{Water}}{H_2O(l)} + \underset{\text{Salt}}{2\,KCl(aq)}$$

Similarly, hydrobromic acid reacts with magnesium oxide to form water and magnesium bromide.

$$\underset{\text{Acid}}{2\,HBr(aq)} + \underset{\text{Metal oxide}}{MgO(s)} \longrightarrow \underset{\text{Water}}{H_2O(l)} + \underset{\text{Salt}}{MgBr_2(aq)}$$

EXAMPLE 14.3   Writing Equations for Acid Reactions

Write an equation for:

(a) The reaction of hydroiodic acid with potassium metal
(b) The reaction of hydrobromic acid with sodium oxide

SOLUTION

(a) The reaction of hydroiodic acid with potassium metal forms hydrogen gas and a salt. The salt contains the ionized form of the metal (K^+) as the cation and the anion of the acid (I^-). Write the skeletal equation and then balance it.	$HI(aq) + K(s) \longrightarrow H_2(g) + KI(aq)$ $2\,HI(aq) + 2\,K(s) \longrightarrow H_2(g) + 2\,KI(aq)$
(b) The reaction of hydrobromic acid with sodium oxide forms water and a salt. The salt contains the cation from the metal oxide (Na^+) and the anion of the acid (Br^-). Write the skeletal equation and then balance it.	$HBr(aq) + Na_2O(s) \longrightarrow H_2O(l) + NaBr(aq)$ $2\,HBr(aq) + Na_2O(s) \longrightarrow H_2O(l) + 2\,NaBr(aq)$

▶SKILLBUILDER 14.3 | **Writing Equations for Acid Reactions**

Write an equation for:

(a) The reaction of hydrochloric acid with strontium metal
(b) The reaction of hydroiodic acid with barium oxide

▶FOR MORE PRACTICE   Example 14.13; Problems 45, 46, 47, 48.

EVERYDAY CHEMISTRY

What Is in My Antacid?

Heartburn, a burning sensation in the lower throat and above the stomach, is caused by the reflux or backflow of stomach acid into the esophagus (the tube that joins the stomach to the throat). In most individuals, this occurs only occasionally, typically after large meals. Physical activity—such as bending, stooping, or lifting—after meals also aggravates heartburn. In some people, the flap between the esophagus and the stomach that normally prevents acid

Drug Facts

Active ingredients (in each 5 mL teaspoon) **Purposes**
Aluminum hydroxide (equivalent to dried gel, USP) 200 mg..........Antacid
Magnesium hydroxide 200 mg...Antacid
Simethicone 20 mg..Antigas

Use relieves:
■ heartburn   ■ acid indigestion   ■ sour stomach
■ upset stomach due to these symptoms   ■ overindulgence in food and drink

Warnings
Ask a doctor before use if you have kidney disease.
Ask a doctor or pharmacist if you are taking a prescription drug. Antacids may interact with certain prescription drugs.
Stop use and ask a doctor if symptoms last more than 2 weeks.
Keep out of reach of children.

Directions ■ shake well   ■ take 2-4 teaspoonfuls between meals, at bedtime, or as directed by a doctor   ■ do not take more than 24 teaspoonfuls in a 24-hour period, or use the maximum dosage for more than 2 weeks

Other information ■ does not meet USP requirements for preservative effectiveness   ■ do not use if breakaway band on plastic cap is broken or missing

reflux becomes damaged, in which case heartburn becomes a regular occurrence.

Drugstores carry many products that either reduce the secretion of stomach acid or neutralize the acid that is produced. Antacids such as Mylanta or Phillips' milk of magnesia contain bases that neutralize the refluxed stomach acid, alleviating heartburn.

CAN YOU ANSWER THIS? *Look at the label of Mylanta shown in the photograph. Can you identify the bases responsible for the antacid action? Write chemical equations showing the reactions of these bases with stomach acid (HCl).*

BASE REACTIONS

The most important base reactions are those in which a base neutralizes an acid (see the beginning of this section). The only other kind of base reaction that we cover in this book is the reaction of sodium hydroxide with aluminum and water.

$$2 \text{ NaOH}(aq) + 2 \text{ Al}(s) + 6 \text{ H}_2\text{O}(l) \longrightarrow 2 \text{ NaAl(OH)}_4(aq) + 3 \text{ H}_2(g)$$

Aluminum is one of the few metals that dissolves in a base. Consequently, it is safe to use NaOH (the main ingredient in many drain-opening products) to unclog your drain as long as your pipes are not made of aluminum, which is generally the case as the use of aluminum pipe is forbidden in most building codes.

14.6 Acid–Base Titration: A Way to Quantify the Amount of Acid or Base in a Solution

The principles we learned in Chapter 13 (Section 13.8) on solution stoichiometry can be applied to a common laboratory procedure called a titration. In a **titration**, a substance in a solution of known concentration is reacted with another substance in a solution of unknown concentration. For example, consider the acid–base reaction between hydrochloric acid and sodium hydroxide:

$$\text{HCl}(aq) + \text{NaOH}(aq) \longrightarrow \text{H}_2\text{O}(l) + \text{NaCl}(aq)$$

The net ionic equation for this reaction is:

$$H^+(aq) + OH^-(aq) \longrightarrow H_2O(l)$$

Suppose you have an HCl solution represented by the molecular diagram at left. (The Cl^- ions and the H_2O molecules not involved in the reaction have been omitted from this representation for clarity.)

In titrating this sample, we slowly add a solution of known OH^- concentration as represented by the following molecular diagrams.

| The OH^- solution also contains Na^+ cations that are not shown in this figure for clarity.

Beginning of titration

Equivalence point

| At the equivalence point, neither reactant is present in excess, and both are limiting. The number of moles of the reactants are related by the reaction stoichiometry (see Chapter 8).

As the OH^- is added, it reacts with and neutralizes the H^+, forming water. At the **equivalence point**—*the point in the titration when the number of moles of OH^- added equals the number of moles of H^+ originally in solution*—the titration is complete. The equivalence point is usually signaled by an **indicator**, a dye whose color depends on the acidity of the solution (▶ Figure 14.8). In most laboratory titrations, the concentration of one of the reactant solutions is unknown, and the concentration of the other is precisely known. By carefully measuring the volume of each solution required to reach the equivalence point, the concentration of the unknown solution can be determined, as demonstrated in Example 14.4.

▶ FIGURE 14.8 **Acid–base titration** In this titration, NaOH is added to an HCl solution. When the NaOH and HCl reach stoichiometric proportions (1 mol of OH$^-$ for every 1 mol of H$^+$), the indicator (phenolphthalein) changes to pink, signaling the equivalence point of the titration. (Phenolphthalein is an indicator that is colorless in acidic solution and pink in basic solution.)

EXAMPLE 14.4 Acid–Base Titration

The titration of 10.00 mL of an HCl solution of unknown concentration requires 12.54 mL of a 0.100 M NaOH solution to reach the equivalence point. What is the concentration of the unknown HCl solution?

SORT	
You are given the volume of an unknown HCl solution and the volume of a known NaOH solution required to titrate the unknown solution. You are asked to find the concentration of the unknown solution.	**GIVEN:** 10.00 mL HCl solution 12.54 mL of a 0.100 M NaOH solution **FIND:** concentration of HCl solution (mol/L)

STRATEGIZE

You must first write the balanced chemical equation for the reaction between the acid and the base (see Example 14.2).

The solution map then has two parts. In the first part, use the volume of NaOH required to reach the equivalence point to calculate the number of moles of HCl in the solution. The final conversion factor comes from the balanced neutralization equation.

In the second part, use the number of moles of HCl and the volume of HCl solution to determine the molarity of the HCl solution.

SOLUTION MAP

$$HCl(aq) + NaOH(aq) \longrightarrow H_2O(l) + NaCl(aq)$$

mL NaOH	→	L NaOH	→	mol NaOH	→	mol HCl

$$\dfrac{1\,L}{1000\,mL} \qquad \dfrac{0.100\,mol\,NaOH}{1\,L\,NaOH} \qquad \dfrac{1\,mol\,HCl}{1\,mol\,NaOH}$$

mol HCl, volume HCl solution	→	molarity

$$M = \dfrac{mol}{L}$$

RELATIONSHIPS USED

1 mol HCl : 1 mol NaOH (from balanced chemical equation)

$$\text{Molarity (M)} = \dfrac{\text{mol solute}}{\text{L solution}} \text{ (definition of molarity, from Section 13.6)}$$

SOLVE

Calculate the moles of HCl in the unknown solution by following the first part of the solution map.

To get the concentration of the solution, divide the number of moles of HCl by the volume of the HCl solution in L. (Note that 10.00 mL is equivalent to 0.01000 L.)

The unknown HCl solution therefore has a concentration of 0.125 M.

SOLUTION

$$12.54\,\cancel{mL\,NaOH} \times \dfrac{1\,\cancel{L}}{1000\,\cancel{mL}} \times \dfrac{0.100\,\cancel{mol\,NaOH}}{1\,\cancel{L\,NaOH}} \times \dfrac{1\,mol\,HCl}{1\,\cancel{mol\,NaOH}}$$

$$= 1.25 \times 10^{-3}\,mol\,HCl$$

$$\text{Molarity} = \dfrac{1.25 \times 10^{-3}\,mol\,HCl}{0.01000\,L} = 0.125\,M$$

CHECK

Check your answer. Are the units correct? Does the answer make physical sense?

The units (M) are correct. The magnitude of the answer makes sense because the reaction has a one-to-one stoichiometry and the volumes of the two solutions are similar; therefore, their concentrations should also be similar.

▶**SKILLBUILDER 14.4** | **Acid–Base Titration**

The titration of a 20.0-mL sample of an H_2SO_4 solution of unknown concentration requires 22.87 mL of a 0.158 M KOH solution to reach the equivalence point. What is the concentration of the unknown H_2SO_4 solution?

▶**FOR MORE PRACTICE**   Example 14.14; Problems 51, 52, 53, 54, 55, 56.

14.7 Strong and Weak Acids and Bases

STRONG ACIDS

Hydrochloric acid (HCl) and hydrofluoric acid (HF) appear to be similar, but there is an important difference between these two acids. HCl is an example of a **strong acid**, one that completely ionizes in solution.

> Single arrow indicates complete ionization

$$HCl(aq) + H_2O(l) \longrightarrow H_3O^+(aq) + Cl^-(aq)$$

| [X] means "molar concentration of X."

| Strong electrolyte solutions were first defined in Section 7.5.

We show the *complete* ionization of HCl with a single arrow pointing to the right in the equation. An HCl solution contains almost no intact HCl; virtually all the HCl has reacted with water to form $H_3O^+(aq)$ and $Cl^-(aq)$ (◀ Figure 14.9). A 1.0 M HCl solution will therefore have an H_3O^+ concentration of 1.0 M. The concentration of H_3O^+ is often abbreviated as $[H_3O^+]$. Using this notation, a 1.0-M HCl solution has $[H_3O^+] = 1.0$ M.

A strong acid is an example of a **strong electrolyte**, a substance whose aqueous solutions are good conductors of electricity (▼ Figure 14.10). Aqueous solutions require the presence of charged particles to conduct electricity. Strong acid solutions are also strong electrolyte solutions because each acid molecule

◀ **FIGURE 14.9   A strong acid**   When HCl dissolves in water, it completely ionizes into H_3O^+ and Cl^- ions. The solution contains no intact HCl.

(a) Pure water   **(b)** HCl solution

▲ **FIGURE 14.10   Conductivity of a strong electrolyte solution**   **(a)** Pure water will not conduct electricity. **(b)** The presence of ions in an HCl solution results in the conduction of electricity, causing the lightbulb to light. Solutions such as these are called strong electrolyte solutions.

ionizes into positive and negative ions. These mobile ions are good conductors of electricity. Pure water is not a good conductor of electricity because it has relatively few charged particles. The danger of using electrical devices—such as a hair dryer—while sitting in the bathtub is that water is seldom pure and often contains dissolved ions. If the device were to come in contact with the water, dangerously high levels of electricity could flow through the water and through your body.

> An ionizable proton is one that becomes an H$^+$ ion in solution.

Table 14.3 lists the six strong acids. The first five acids in the table are **monoprotic acids**, acids containing only one ionizable proton. Sulfuric acid is an example of a **diprotic acid**, an acid that contains two ionizable protons.

TABLE 14.3 Strong Acids

hydrochloric acid (HCl)	nitric acid (HNO$_3$)
hydrobromic acid (HBr)	perchloric acid (HClO$_4$)
hydroiodic acid (HI)	sulfuric acid (H$_2$SO$_4$) *(diprotic)*

WEAK ACIDS

> It is a common mistake to confuse the terms *strong* and *weak acids* with the terms *concentrated* and *dilute acids*. Can you state the difference between these terms?

In contrast to HCl, HF is a **weak acid**, one that does not completely ionize in solution.

Double arrow indicates
partial ionization

$$HF(aq) + H_2O(l) \rightleftharpoons H_3O^+(aq) + F^-(aq)$$

> Calculating exact [H$_3$O$^+$] for weak acids is beyond the scope of this text.

To show that HF does not completely ionize in solution, the equation for its ionization has two opposing arrows, indicating that the reverse reaction occurs to some degree. An HF solution contains a lot of intact HF; it also contains some H$_3$O$^+$(aq) and F$^-$(aq) (▼ Figure 14.11). In other words, a 1.0 M HF solution has [H$_3$O$^+$] < 1.0 M because only some of the HF molecules ionize to form H$_3$O$^+$.

▶ **FIGURE 14.11 A weak acid** When HF dissolves in water, only a fraction of the dissolved molecules ionize into H$_3$O$^+$ and F$^-$ ions. The solution contains many intact HF molecules.

> Notice that the strength of a conjugate base is related to its attraction to H$^+$ in solution.

A weak acid is an example of a **weak electrolyte**, a substance whose aqueous solutions are poor conductors of electricity (▶ Figure 14.12). Weak acid solutions contain few charged particles because only a small fraction of the acid molecules ionize into positive and negative ions.

The degree to which an acid is strong or weak depends in part on the attraction between the anion of the acid (the conjugate base) and the hydrogen ion. Suppose HA is a generic formula for an acid. Then, the degree to which the following reaction proceeds in the forward direction depends in part on the strength of the attraction between H$^+$ and A$^-$.

$$\underset{\text{Acid}}{HA(aq)} + H_2O(l) \longrightarrow H_3O^+(aq) + \underset{\text{Conjugate base}}{A^-(aq)}$$

(a) Pure water

(b) HF solution

▲ **FIGURE 14.12 Conductivity of a weak electrolyte solution** (a) Pure water will not conduct electricity. (b) An HF solution contains some ions, but most of the HF is intact. The light glows only dimly. Solutions such as these are called weak electrolyte solutions.

| Notice that the strength of a conjugate base is related to its attraction to H^+ in solution.

(a) Strong acid

Weak attraction
Complete ionization

(b) Weak acid

Strong attraction
Partial ionization

▲ **FIGURE 14.13 Strong and weak acids** (a) In a strong acid, the attraction between H^+ and A^- is low, resulting in complete ionization. (b) In a weak acid, the attraction between H^+ and A^- is high, resulting in partial ionization.

If the attraction between H^+ and A^- is *weak*, then the reaction favors the forward direction and the acid is *strong* (◄ Figure 14.13a). If the attraction between H^+ and A^- is *strong*, then the reaction favors the reverse direction and the acid is *weak* (Figure 14.13b).

For example, in HCl, the conjugate base (Cl^-) has a relatively weak attraction to H^+, meaning that the reverse reaction does not occur to any significant extent. In HF, on the other hand, the conjugate base (F^-) has a greater attraction to H^+, meaning that the reverse reaction occurs to a significant degree. *In general, the stronger the acid, the weaker the conjugate base and vice versa.* This means that if the forward reaction (that of the acid) has a high tendency to occur, then the reverse reaction (that of the conjugate base) has a low tendency to occur. Table 14.4 lists some common weak acids.

TABLE 14.4 Some Weak Acids

hydrofluoric acid (HF)	sulfurous acid (H_2SO_3) *(diprotic)*
acetic acid ($HC_2H_3O_2$)	carbonic acid (H_2CO_3) *(diprotic)*
formic acid ($HCHO_2$)	phosphoric acid (H_3PO_4) *(triprotic)*

Notice that two of the weak acids in Table 14.4 are diprotic (meaning they have two ionizable protons) and one is triprotic (meaning that it has three ionizable protons). Let us return to sulfuric acid for a moment. Sulfuric acid is a diprotic acid that is strong in its first ionizable proton:

$$H_2SO_4(aq) + H_2O(l) \longrightarrow H_3O^+(aq) + HSO_4^-(aq)$$

but weak in its second ionizable proton.

$$HSO_4^-(aq) + H_2O(l) \rightleftharpoons H_3O^+(aq) + SO_4^{2-}(aq)$$

Sulfurous acid and carbonic acid are weak in both of their ionizable protons, and phosphoric acid is weak in all three of its ionizable protons.

EXAMPLE 14.5 Determining $[H_3O^+]$ in Acid Solutions

What is the H_3O^+ concentration in each solution?

(a) 1.5 M HCl
(b) 3.0 M $HC_2H_3O_2$
(c) 2.5 M HNO_3

SOLUTION

(a) Since HCl is a strong acid, it completely ionizes. The concentration of H_3O^+ will be 1.5 M.

$$[H_3O^+] = 1.5 \text{ M}$$

(b) Since $HC_2H_3O_2$ is a weak acid, it partially ionizes. The calculation of the exact concentration of H_3O^+ is beyond the scope of this text, but we know that it will be less than 3.0 M.

$$[H_3O^+] < 3.0 \text{ M}$$

(c) Since HNO_3 is a strong acid, it completely ionizes. The concentration of H_3O^+ will be 2.5 M.

$$[H_3O^+] = 2.5 \text{ M}$$

▶**SKILLBUILDER 14.5 | Determining $[H_3O^+]$ in Acid Solutions**

What is the H_3O^+ concentration in each solution?

(a) 0.50 M $HCHO_2$
(b) 1.25 M HI
(c) 0.75 M HF

▶**FOR MORE PRACTICE** Example 14.15; Problems 59, 60.

✓ CONCEPTUAL CHECKPOINT 14.2

Examine the molecular views of three different acid solutions shown here. Based on these views, which one of these acids is a weak acid?

(a) (b) (c)

STRONG BASES

In analogy to the definition of a strong acid, a **strong base** is one that completely dissociates in solution. NaOH, for example, is a strong base.

$$NaOH(aq) \longrightarrow Na^+(aq) + OH^-(aq)$$

▶ FIGURE 14.14 **A strong base** When NaOH dissolves in water, it completely dissociates into Na^+ and OH^-. **Question:** The solution contains no intact NaOH. Is NaOH a strong or weak electrolyte?

An NaOH solution contains no intact NaOH—it has all dissociated to form $Na^+(aq)$ and $OH^-(aq)$ (▲ Figure 14.14), and a 1.0 M NaOH solution has $[OH^-] = 1.0$ M and $[Na^+] = 1.0$ M. Some common strong bases are listed in Table 14.5.

TABLE 14.5 Strong Bases

lithium hydroxide (LiOH)	strontium hydroxide ($Sr(OH)_2$)
sodium hydroxide (NaOH)	calcium hydroxide ($Ca(OH)_2$)
potassium hydroxide (KOH)	barium hydroxide ($Ba(OH)_2$)

Unlike diprotic acids, which ionize in two steps, bases containing 2 OH^- ions dissociate in one step.

Some strong bases, such as $Sr(OH)_2$, contain two OH^- ions. These bases completely dissociate, producing two moles of OH^- per mole of base. For example, $Sr(OH)_2$ dissociates as follows:

$$Sr(OH)_2(aq) \longrightarrow Sr^{2+}(aq) + 2\ OH^-(aq)$$

WEAK BASES

A **weak base** is analogous to a weak acid. Unlike strong bases that contain OH^- and dissociate in water, the most common weak bases produce OH^- by accepting a proton from water, ionizing water to form OH^-.

$$B(aq) + H_2O(l) \rightleftharpoons BH^+(aq) + OH^-(aq)$$

In this equation, B is generic for a weak base. Ammonia, for example, ionizes water according to the reaction:

Calculating exact $[OH^-]$ for weak bases is beyond the scope of this text.

$$NH_3(aq) + H_2O(l) \rightleftharpoons NH_4{}^+(aq) + OH^-(aq)$$

The double arrow indicates that the ionization is not complete. An NH_3 solution contains NH_3, $NH_4{}^+$, and OH^- (◀ Figure 14.15). A 1.0 M NH_3 solution has $[OH^-] < 1.0$ M. Table 14.6 lists some common weak bases.

◀ FIGURE 14.15 **A weak base** When NH_3 dissolves in water, it partially ionizes to form $NH_4{}^+$ and OH^-. However, only a fraction of the molecules ionize. Most NH_3 molecules remain as NH_3. **Question:** Is NH_3 a strong or weak electrolyte?

TABLE 14.6   Some Weak Bases

Base	Ionization Reaction
ammonia (NH_3)	$NH_3(aq) + H_2O(l) \rightleftharpoons NH_4^+(aq) + OH^-(aq)$
pyridine (C_5H_5N)	$C_5H_5N(aq) + H_2O(l) \rightleftharpoons C_5H_5NH^+(aq) + OH^-(aq)$
methylamine (CH_3NH_2)	$CH_3NH_2(aq) + H_2O(l) \rightleftharpoons CH_3NH_3^+(aq) + OH^-(aq)$
ethylamine ($C_2H_5NH_2$)	$C_2H_5NH_2(aq) + H_2O(l) \rightleftharpoons C_2H_5NH_3^+(aq) + OH^-(aq)$
bicarbonate ion (HCO_3^-)*	$HCO_3^-(aq) + H_2O(l) \rightleftharpoons H_2CO_3(aq) + OH^-(aq)$

*The bicarbonate ion must occur with a positively charged ion such as Na^+ that serves to balance the charge but does not have any part in the ionization reaction. It is the bicarbonate ion that makes sodium bicarbonate ($NaHCO_3$) basic.

EXAMPLE 14.6   Determining [OH^-] in Base Solutions

What is the OH^- concentration in each solution?

(a) 2.25 M KOH
(b) 0.35 M CH_3NH_2
(c) 0.025 M $Sr(OH)_2$

SOLUTION

(a) Since KOH is a strong base, it completely dissociates into K^+ and OH^- in solution. The concentration of OH^- is 2.25 M.

$$[OH^-] = 2.25 \text{ M}$$

(b) Since CH_3NH_2 is a weak base, it only partially ionizes water. We cannot calculate the exact concentration of OH^-, but we know it is less than 0.35 M.

$$[OH^-] < 0.35 \text{ M}$$

(c) Since $Sr(OH)_2$ is a strong base, it completely dissociates into $Sr^{2+}(aq)$ and $2\ OH^-(aq)$. $Sr(OH)_2$ forms 2 mol of OH^- for every 1 mol of $Sr(OH)_2$. Consequently, the concentration of OH^- is twice the concentration of $Sr(OH)_2$.

$$[OH^-] = 2(0.025 \text{ M}) = 0.050 \text{ M}$$

▶**SKILLBUILDER 14.6 | Determining [OH^-] in Base Solutions**

What is the OH^- concentration in each solution?

(a) 0.055 M $Ba(OH)_2$
(b) 1.05 M C_5H_5N
(c) 0.45 M NaOH

▶**FOR MORE PRACTICE**   Example 14.16; Problems 63, 64.

14.8   Water: Acid and Base in One

We saw earlier that water acts as a base when it reacts with HCl and as an acid when it reacts with NH_3.

Water acting as a base

$$HCl(aq) \ + \ H_2O(l) \longrightarrow H_3O^+(aq) + Cl^-(aq)$$

Acid   Base
(Proton donor)   (Proton acceptor)

Water acting as an acid

$$NH_3(aq) \ + \ H_2O(l) \rightleftharpoons NH_4^+(aq) + OH^-(aq)$$

Base   Acid
(Proton acceptor)   (Proton donor)

Water is *amphoteric*; it can act as either an acid or a base. Even in pure water, water acts as an acid and a base with itself, a process called self-ionization.

Water acting as both an acid and a base

$$H_2O(l) \quad + \quad H_2O(l) \quad \rightleftharpoons \quad H_3O^+(aq) \;+\; OH^-(aq)$$

Acid   Base

(Proton donor)   (Proton acceptor)

In pure water, at 25 °C, the preceding reaction occurs only to a very small extent, resulting in equal and small concentrations of H_3O^+ and OH^-.

$$[H_3O^+] = [OH^-] = 1.0 \times 10^{-7} \text{ M} \quad \text{(in pure water at 25°C)}$$

where $[H_3O^+]$ = the concentration of H_3O^+ in M

and $[OH^-]$ = the concentration of OH^- in M

So all samples of water contain some hydronium ions and some hydroxide ions. The *product* of the concentration of these two ions in aqueous solutions is called the **ion product constant for water (K_w)**.

$$K_w = [H_3O^+][OH^-]$$

| The units of K_w are normally dropped.

We can find the value of K_w at 25 °C by multiplying the hydronium and hydroxide concentrations for pure water listed earlier.

$$K_w = [H_3O^+][OH^-]$$
$$= (1.0 \times 10^{-7})(1.0 \times 10^{-7})$$
$$= (1.0 \times 10^{-7})^2$$
$$= 1.0 \times 10^{-14}$$

The preceding equation holds true for all aqueous solutions at 25 °C. The concentration of H_3O^+ times the concentration of OH^- will be 1.0×10^{-14}. In pure water, since H_2O is the only source of these ions, there is one H_3O^+ ion for every OH^- ion. Consequently, the concentrations of H_3O^+ and OH^- are equal. Such a solution is a **neutral solution**.

| In a neutral solution, $[H_3O^+] = [OH^-]$.

$$[H_3O^+] = [OH^-] = \sqrt{K_w} = 1.0 \times 10^{-7} \text{ M (in pure water)}$$

An **acidic solution** contains an acid that creates additional H_3O^+ ions, causing $[H_3O^+]$ to increase. However, the *ion product constant still applies*.

$$[H_3O^+][OH^-] = K_w = 1.0 \times 10^{-14}$$

| In an acidic solution, $[H_3O^+] > [OH^-]$.

If $[H_3O^+]$ increases, then $[OH^-]$ must decrease for the ion product to remain 1.0×10^{-14}. For example, suppose $[H_3O^+] = 1.0 \times 10^{-3}$ M; then $[OH^-]$ can be found by solving the ion product expression for $[OH^-]$.

$$(1.0 \times 10^{-3})[OH^-] = 1.0 \times 10^{-14}$$

$$[OH^-] = \frac{1.0 \times 10^{-14}}{1.0 \times 10^{-3}} = 1.0 \times 10^{-11} \text{ M}$$

In an acidic solution, $[H_3O^+]$ is greater than 1.0×10^{-7} M, and $[OH^-]$ is less than 1.0×10^{-7} M.

| In a basic solution, $[H_3O^+] < [OH^-]$.

A **basic solution** contains a base that creates additional OH^- ions, causing the $[OH^-]$ to increase and the $[H_3O^+]$ to decrease. For example, suppose $[OH^-] = 1.0 \times 10^{-2}$ M; then $[H_3O^+]$ can be found by solving the ion product expression for $[H_3O^+]$.

$$[H_3O^+](1.0 \times 10^{-2}) = 1.0 \times 10^{-14}$$

$$[H_3O^+] = \frac{1.0 \times 10^{-14}}{1.0 \times 10^{-2}} = 1.0 \times 10^{-12} \text{ M}$$

In a basic solution, $[OH^-]$ is greater than 1.0×10^{-7} M and $[H_3O^+]$ is less than 1.0×10^{-7} M.

▶ FIGURE 14.16 **Acidic and basic solutions**

To summarize, at 25 °C (see ▲ Figure 14.16):

- In a neutral solution, $[H_3O^+] = [OH^-] = 1.0 \times 10^{-7}$ M
- In an acidic solution, $[H_3O^+] > 1.0 \times 10^{-7}$ M $\quad [OH^-] < 1.0 \times 10^{-7}$ M
- In a basic solution, $[H_3O^+] < 1.0 \times 10^{-7}$ M $\quad [OH^-] > 1.0 \times 10^{-7}$ M
- In all aqueous solutions, $[H_3O^+][OH^-] = K_w = 1.0 \times 10^{-14}$

EXAMPLE 14.7 Using K_w in Calculations

Calculate $[OH^-]$ in each solution and determine whether the solution is acidic, basic, or neutral.

(a) $[H_3O^+] = 7.5 \times 10^{-5}$ M
(b) $[H_3O^+] = 1.5 \times 10^{-9}$ M
(c) $[H_3O^+] = 1.0 \times 10^{-7}$ M

To find $[OH^-]$ use the ion product constant, K_w. Substitute the given value for $[H_3O^+]$ and solve the equation for $[OH^-]$. Since $[H_3O^+] > 1.0 \times 10^{-7}$ M and $[OH^-] < 1.0 \times 10^{-7}$ M, the solution is acidic.	**SOLUTION** (a) $[H_3O^+][OH^-] = K_w = 1.0 \times 10^{-14}$ $[7.5 \times 10^{-5}][OH^-] = K_w = 1.0 \times 10^{-14}$ $[OH^-] = \dfrac{1.0 \times 10^{-14}}{7.5 \times 10^{-5}} = 1.3 \times 10^{-10}$ M acidic solution
Substitute the given value for $[H_3O^+]$ into the ion product constant equation and solve the equation for $[OH^-]$. Since $[H_3O^+] < 1.0 \times 10^{-7}$ M and $[OH^-] > 1.0 \times 10^{-7}$ M, the solution is basic.	(b) $[H_3O^+][OH^-] = K_w = 1.0 \times 10^{-14}$ $[1.5 \times 10^{-9}][OH^-] = 1.0 \times 10^{-14}$ $[OH^-] = \dfrac{1.0 \times 10^{-14}}{1.5 \times 10^{-9}} = 6.7 \times 10^{-6}$ M basic solution

Substitute the given value for $[H_3O^+]$ into the ion product constant equation and solve the equation for $[OH^-]$. Since $[H_3O^+] = 1.0 \times 10^{-7}$ M and $[OH^-] = 1.0 \times 10^{-7}$ M, the solution is neutral.

(c) $[H_3O^+][OH^-] = K_w = 1.0 \times 10^{-14}$

$[1.0 \times 10^{-7}][OH^-] = 1.0 \times 10^{-14}$

$$[OH^-] = \frac{1.0 \times 10^{-14}}{1.0 \times 10^{-7}} = 1.0 \times 10^{-7} \text{ M}$$

neutral solution

▶**SKILLBUILDER 14.7 | Using K_w in Calculations**

Calculate $[H_3O^+]$ in each solution and determine whether the solution is acidic, basic, or neutral.

(a) $[OH^-] = 1.5 \times 10^{-2}$ M
(b) $[OH^-] = 1.0 \times 10^{-7}$ M
(c) $[OH^-] = 8.2 \times 10^{-10}$ M

▶**FOR MORE PRACTICE**   Example 14.17; Problems 67, 68, 69, 70.

✓ **CONCEPTUAL CHECKPOINT 14.3**

Which substance would be least likely to act as a base?

(a) H_2O
(b) OH^-
(c) NH_3
(d) NH_4^+

14.9 The pH and pOH Scales: Ways to Express Acidity and Basicity

TABLE 14.7 The pH of Some Common Substances

Substance	pH
gastric (human stomach) acid	1.0–3.0
limes	1.8–2.0
lemons	2.2–2.4
soft drinks	2.0–4.0
plums	2.8–3.0
wine	2.8–3.8
apples	2.9–3.3
peaches	3.4–3.6
cherries	3.2–4.0
beer	4.0–5.0
rainwater (unpolluted)	5.6
human blood	7.3–7.4
egg whites	7.6–8.0
milk of magnesia	10.5
household ammonia	10.5–11.5
4% NaOH solution	14

Notice that an increase of 1 in pH corresponds to a tenfold *decrease* in $[H_3O^+]$.

Chemists have devised a scale based on the hydrogen ion concentration to compactly express the acidity or basicity of solutions. The scale is called the **pH** scale and at 25 °C has these general characteristics:

- pH < 7   *acidic* solution
- pH > 7   *basic* solution
- pH = 7   *neutral* solution

Table 14.7 lists the pH of some common substances. Notice that, as we discussed in Section 14.1, many foods, especially fruits, are acidic and therefore have low pH values. The foods with the lowest pH values are limes and lemons, and they are among the sourest. Relatively few foods, however, are basic.

The pH scale is a **logarithmic scale**; therefore, a change of 1 pH unit corresponds to a tenfold change in H_3O^+ concentration. For example, a lime with a pH of 2.0 is 10 times more acidic than a plum with a pH of 3.0 and 100 times more acidic than a cherry with a pH of 4.0. Each change of 1 in pH scale corresponds to a change of 10 in $[H_3O^+]$ (▶ Figure 14.17).

pH	$[H_3O^+]$	$[H_3O^+]$ Representation
4	10^{-4}	● $\left(\text{Each circle represents } \dfrac{10^{-4}\text{ mol H}^+}{\text{L}}\right)$
3	10^{-3}	●●●●●●●●●●
2	10^{-2}	(grid of circles)

▶ **FIGURE 14.17   The pH scale is a logarithmic scale**   A *decrease* of 1 unit on the pH scale corresponds to an *increase* in $[H_3O^+]$ concentration by a factor of 10. Each circle stands for 10^{-4} mol H^+/L, or 6.022×10^{19} H^+ ions per liter. **Question:** How much of an increase in H_3O^+ concentration corresponds to a decrease of 2 pH units?

CALCULATING pH FROM $[H_3O^+]$

The pH of a solution is defined as the negative of the log of the hydronium ion concentration:

$$pH = -\log[H_3O^+]$$

> Note that pH is defined using the log function (base ten), which is different from the natural log (abbreviated ln).

To calculate pH, you must be able to calculate logarithms. Recall that the log of a number is the exponent to which 10 must be raised to obtain that number, as shown in these examples:

$$\log 10^1 = 1; \ \log 10^2 = 2; \ \log 10^3 = 3$$

$$\log 10^{-1} = -1; \ \log 10^{-2} = -2; \ \log 10^{-3} = -3$$

In the next example, we calculate the log of 1.5×10^{-7}. A solution having an $[H_3O^+] = 1.5 \times 10^{-7}$ M (acidic) has a pH of:

$$
\begin{aligned}
pH &= -\log[H_3O^+] \\
&= -\log(1.5 \times 10^{-7}) \\
&= -(-6.82) \\
&= 6.82
\end{aligned}
$$

Notice that the pH is reported to two decimal places here. This is because only the numbers to the right of the decimal place are significant in a log. Since our original value for the concentration had two significant figures, the log of that number has two decimal places.

> When you take the log of a quantity, the result should have the same number of decimal places as the number of significant figures in the original quantity.

$$\log 1.0 \times 10^{-3} = 3.\overset{\text{2 decimal places}}{00}$$

If the original number had three significant figures, the log would be reported to three decimal places:

$$-\log 1.00 \times 10^{-3} = 3.\overset{\text{3 decimal places}}{000}$$

A solution having $[H_3O^+] = 1.0 \times 10^{-7}$ M (neutral) has a pH of:

$$
\begin{aligned}
pH &= -\log[H_3O^+] \\
&= -\log 1.0 \times 10^{-7} \\
&= -(-7.00) \\
&= 7.00
\end{aligned}
$$

EXAMPLE 14.8   Calculating pH from [H₃O⁺]

Calculate the pH of each solution and indicate whether the solution is acidic or basic.

(a) $[H_3O^+] = 1.8 \times 10^{-4}$ M

(b) $[H_3O^+] = 7.2 \times 10^{-9}$ M

SOLUTION

To calculate pH, substitute the given $[H_3O^+]$ into the pH equation.	**(a)** $pH = -\log[H_3O^+]$ $\quad\quad = -\log 1.8 \times 10^{-4}$ $\quad\quad = -(-3.74)$
Since the pH < 7, this solution is acidic.	$\quad\quad = 3.74$
Again, substitute the given $[H_3O^+]$ into the pH equation.	**(b)** $pH = -\log[H_3O^+]$ $\quad\quad = -\log(7.2 \times 10^{-9})$ $\quad\quad = -(-8.14)$
Since the pH > 7, this solution is basic.	$\quad\quad = 8.14$

▶**SKILLBUILDER 14.8 | Calculating pH from [H₃O⁺]**

Calculate the pH of each solution and indicate whether the solution is acidic or basic.

(a) $[H_3O^+] = 9.5 \times 10^{-9}$ M

(b) $[H_3O^+] = 6.1 \times 10^{-3}$ M

▶**SKILLBUILDER PLUS**

Calculate the pH of a solution with $[OH^-] = 1.3 \times 10^{-2}$ M and indicate whether the solution is acidic or basic. *Hint:* Begin by using K_w to find $[H_3O^+]$.

▶**FOR MORE PRACTICE**   Example 14.18; Problems 73, 74.

CALCULATING [H₃O⁺] FROM pH

Ten raised to the log of a number is equal to that number: $10^{\log x} = x$.	To calculate $[H_3O^+]$ from a pH value, you must *undo* the log. The log can be undone using the inverse log function *(Method 1)* on most calculators or using the 10^x key *(Method 2)*. Both methods do the same thing; the one you use depends on your calculator.

Method 1: Inverse Log Function	Method 2: 10^x Function
$pH = -\log[H_3O^+]$	$pH = -\log[H_3O^+]$
$-pH = \log[H_3O^+]$	$-pH = \log[H_3O^+]$
$\text{invlog}(-pH) = \text{invlog}(\log[H_3O^+])$	$10^{-pH} = 10^{\log[H_3O^+]}$
$\text{invlog}(-pH) = [H_3O^+]$	$10^{-pH} = [H_3O^+]$

The invlog function "undoes" log: $\text{invlog}(\log x) = x$.

The inverse log is sometimes called the antilog.

So, to calculate $[H_3O^+]$ from a pH value, take the inverse log of the negative of the pH value *(Method 1)* or raise 10 to the negative of the pH value *(Method 2)*.

EXAMPLE 14.9   Calculating [H₃O⁺] from pH

Calculate the H_3O^+ concentration for a solution with a pH of 4.80.

SOLUTION

To find the $[H_3O^+]$ from pH, we must undo the log function. Use either Method 1 or Method 2.

Method 1: Inverse Log Function	Method 2: 10ˣ function
$pH = -\log[H_3O^+]$	$pH = -\log[H_3O^+]$
$4.80 = -\log[H_3O^+]$	$4.80 = -\log[H_3O^+]$
$-4.80 = \log[H_3O^+]$	$-4.80 = \log[H_3O^+]$
$\text{invlog}(-4.80) = \text{invlog}(\log[H_3O^+])$	$10^{-4.80} = 10^{\log[H_3O^+]}$
$\text{invlog}(-4.80) = [H_3O^+]$	$10^{-4.80} = [H_3O^+]$
$[H_3O^+] = 1.6 \times 10^{-5}\ M$	$[H_3O^+] = 1.6 \times 10^{-5}\ M$

The number of significant figures in the inverse log of a number is equal to the number of decimal places in the number.

▶**SKILLBUILDER 14.9 | Calculating [H₃O⁺] from pH**

Calculate the H_3O^+ concentration for a solution with a pH of 8.37.

▶**SKILLBUILDER PLUS**

Calculate the OH^- concentration for a solution with a pH of 3.66.

▶**FOR MORE PRACTICE**   Example 14.19; Problems 75, 76.

✔ CONCEPTUAL CHECKPOINT 14.4

Solution A has a pH of 13. Solution B has a pH of 10. The concentration of H_3O^+ in solution B is _____ times that in solution A.

(a) 0.001

(b) $\frac{1}{3}$

(c) 3

(d) 1000

THE pOH SCALE

The **pOH** scale is analogous to the pH scale, but is defined with respect to $[OH^-]$ instead of $[H_3O^+]$.

Notice that *p* is the mathematical function $-\log$; thus, $pX = -\log X$.

$$pOH = -\log[OH^-]$$

A solution having an $[OH^-]$ of $1.0 \times 10^{-3}\,M$ (basic) has a pOH of 3.00. On the pOH scale, a pOH less than 7 is basic and a pOH greater than 7 is acidic. A pOH of 7 is neutral. The $[OH^-]$ concentration can be found from the pOH just as the $[H_3O^+]$ concentration is found from the pH, as shown in the following example.

EXAMPLE 14.10   Calculating [OH⁻] from pOH

Calculate the $[OH^-]$ concentration for a solution with a pOH of 8.55.

SOLUTION

To find the $[OH^-]$ from pOH, we must undo the log function. Use either Method 1 or Method 2.

Method 1: Inverse Log Function	Method 2: 10^x function
$pOH = -\log[OH^-]$	$pOH = -\log[OH^-]$
$8.55 = -\log[OH^-]$	$8.55 = -\log[OH^-]$
$-8.55 = \log[OH^-]$	$-8.55 = \log[OH^-]$
$\text{invlog}(-8.55) = \text{invlog}(\log[OH^-])$	$10^{-8.55} = 10^{\log[OH^-]}$
$\text{invlog}(-8.55) = [OH^-]$	$10^{-8.55} = [OH^-]$
$[OH^-] = 2.8 \times 10^{-9} \text{ M}$	$[OH^-] = 2.8 \times 10^{-9} \text{ M}$

▶**SKILLBUILDER 14.10 | Calculating OH⁻ from pOH**

Calculate the OH^- concentration for a solution with a pOH of 4.25.

▶**SKILLBUILDER PLUS**

Calculate the H_3O^+ concentration for a solution with a pOH of 5.68.

▶**FOR MORE PRACTICE**   Problems 83, 84, 85, 86.

We can derive a relationship between pH and pOH at 25 °C from the expression for K_w.

$$[H_3O^+][OH^-] = 1.0 \times 10^{-14}$$

| $\log (AB) = \log A + \log B$

Taking the log of both sides, we get

$$\log \{[H_3O^+][OH^-]\} = \log (1.0 \times 10^{-14})$$
$$\log[H_3O^+] + \log[OH^-] = -14.00$$
$$-\log[H_3O^+] - \log[OH^-] = 14.00$$
$$pH + pOH = 14.00$$

The sum of pH and pOH is always equal to 14.00 at 25 °C. Therefore, a solution with a pH of 3 has a pOH of 11.

CONCEPTUAL CHECKPOINT 14.5

A solution has a pH of 5. What is the pOH of the solution?

(a) 5   **(b)** 10   **(c)** 14   **(d)** 9

14.10   Buffers: Solutions That Resist pH Change

Most solutions rapidly become more acidic (lower pH) upon addition of an acid or more basic (higher pH) upon addition of a base. A **buffer**, however, resists pH change by neutralizing added acid or added base. Human blood, for example, is a buffer. Acid or base that is added to blood gets neutralized by components within blood, resulting in a nearly constant pH. In healthy individuals, blood pH is between 7.36 and 7.40. If blood pH were to drop below 7.0 or rise above 7.8, death would result.

How does blood maintain such a narrow pH range? Like all buffers, blood contains *significant* amounts of *both a weak acid and its conjugate base*. When additional base is added to blood, the weak acid reacts with the base, neutralizing it. When additional acid is added to blood, the conjugate base reacts with the acid, neutralizing it. In this way, blood maintains a constant pH.

| Buffers can also be composed of a weak base and its conjugate acid.

CHEMISTRY AND HEALTH

Alkaloids

Alkaloids are organic bases that occur naturally in many plants (see Section 14.3) that often have medicinal qualities. Morphine, for example, is a powerful alkaloid drug that occurs in the opium poppy (▶ Figure 14.18) and is used to relieve severe pain. Morphine is an example of a *narcotic*, a drug that dulls the senses and induces sleep. It produces relief from and indifference to pain. Morphine can also produce feelings of euphoria and contentment, which leads to its abuse. Morphine is highly addictive, both psychologically and physically. A person who abuses morphine over long periods of time becomes physically dependent on the drug and suffers severe withdrawal symptoms upon termination of use.

▲ **FIGURE 14.18** **Opium poppy** The opium poppy contains the alkaloids morphine and codeine.

Morphine

Amphetamine is another powerful drug related to the alkaloid ephedrine. Whereas morphine slows down nerve signal transmissions, amphetamine enhances them. Amphetamine is an example of a *stimulant*, a drug that increases alertness and wakefulness. Amphetamine is widely used to treat Attention-Deficit Hyperactivity Disorder (ADHD) and is prescribed under the trade name Adderall. Patients suffering from ADHD find that amphetamine helps them to focus and concentrate more effectively. However, because amphetamine produces alertness and increased stamina, it, too, is often abused.

Amphetamine

Other common alkaloids include caffeine and nicotine, both of which are stimulants. Caffeine is found in the coffee bean, and nicotine is found in tobacco. Although both have some addictive qualities, nicotine is by far the most addictive. A nicotine addiction is among the most difficult to break, as any smoker can attest.

CAN YOU ANSWER THIS? *What part of the amphetamine and morphine molecules makes them bases?*

A simple buffer can be made by mixing both acetic acid ($HC_2H_3O_2$) and its conjugate base, sodium acetate ($NaC_2H_3O_2$) in water (▶ Figure 14.19). (The sodium in sodium acetate is just a spectator ion and does not contribute to buffering action.) Since $HC_2H_3O_2$ is a weak acid and since $C_2H_3O_2^-$ is its conjugate base, a solution containing both of these is a buffer. Note that a weak acid by itself, even though it partially ionizes to form some of its conjugate base, does not contain sufficient base to be a buffer. A buffer must contain *significant* amounts of *both* a weak acid and its conjugate base. Suppose that we add more base, in the form of NaOH, to the buffer solution containing acetic acid and sodium acetate. The acetic acid would neutralize the base according to the reaction:

$$NaOH(aq) + HC_2H_3O_2(aq) \longrightarrow H_2O(l) + NaC_2H_3O_2(aq)$$
$$\text{Base} \qquad\qquad \text{Acid}$$

▲ **FIGURE 14.19   Buffers**   A buffer contains significant amounts of a weak acid and its conjugate base. The acid consumes any added base, and the base consumes any added acid. In this way, a buffer resists pH change.

As long as the amount of NaOH that we add is less than the amount of $HC_2H_3O_2$ in solution, the solution neutralizes the NaOH, and the resulting pH change is small. Suppose, on the other hand, that we add more acid, in the form of HCl, to the solution. Then the conjugate base, $NaC_2H_3O_2$, neutralizes the added HCl according to the reaction:

$$\underset{\text{Acid}}{HCl(aq)} + \underset{\text{Base}}{NaC_2H_3O_2(aq)} \longrightarrow HC_2H_3O_2(aq) + NaCl(aq)$$

As long as the amount of HCl that we add is less than the amount of $NaC_2H_3O_2$ in solution, the solution will neutralize the HCl and the resulting pH change is small.

To summarize:

- Buffers resist pH change.
- Buffers contain significant amounts of both a weak acid and its conjugate base.
- The weak acid neutralizes added base.
- The conjugate base neutralizes added acid.

✔ CONCEPTUAL CHECKPOINT 14.6

Which of the following is a buffer solution?

(a) $H_2SO_4(aq)$ and $H_2SO_3(aq)$

(b) $HF(aq)$ and $NaF(aq)$

(c) $HCl(aq)$ and $NaCl(aq)$

(d) $NaCl(aq)$ and $NaOH(aq)$

CHEMISTRY AND HEALTH

The Danger of Antifreeze

Most types of antifreeze used in cars are solutions of ethylene glycol. Every year, thousands of dogs and cats die from ethylene glycol poisoning because they consume improperly stored antifreeze or antifreeze that has leaked out of a radiator. The antifreeze has a somewhat sweet taste, which attracts a curious dog or cat. Young children are also at risk for ethylene glycol poisoning.

The first stage of ethylene glycol poisoning is a drunken state. Ethylene glycol is an alcohol, and it affects the brain of a dog or cat much as an alcoholic beverage would. Once ethylene glycol begins to metabolize, however, the second and more deadly stage begins. Ethylene glycol is metabolized in the liver into glycolic acid ($HC_2H_3O_3$), which enters the bloodstream. If the original quantities of consumed antifreeze are significant, the glycolic acid overwhelms the blood's natural buffering system, causing blood pH to drop to dangerously low levels. At this point, the cat or dog may begin hyperventilating in an effort to overcome the acidic blood's reduced ability to carry oxygen. If no treatment is administered, the animal will eventually go into a coma and die.

One treatment for ethylene glycol poisoning is the administration of ethyl alcohol (the alcohol found in alcoholic beverages). The liver enzyme that metabolizes ethylene glycol is the same one that metabolizes ethyl alcohol, but it has a higher affinity for ethyl alcohol than for ethylene glycol. Consequently, the enzyme preferentially metabolizes ethyl alcohol, allowing the unmetabolized ethylene glycol to escape through the urine. If administered early, this treatment can save the life of a dog or cat that has consumed ethylene glycol.

CAN YOU ANSWER THIS? *One of the main buffering systems found in blood consists of carbonic acid (H_2CO_3) and bicarbonate ion (HCO_3^-). Write an equation showing how this buffering system could neutralize glycolic acid ($HC_2H_3O_3$) that might enter the blood from ethylene glycol poisoning. Suppose a cat has 0.15 mol of HCO_3^- and 0.15 mol of H_2CO_3 in its bloodstream. How many grams of $HC_2H_3O_3$ could be neutralized before the buffering system in the cat's blood is overwhelmed?*

14.11 Acid Rain: An Environmental Problem Related to Fossil Fuel Combustion

About 90% of U.S. energy comes from fossil fuel combustion. Fossil fuels include petroleum, natural gas, and coal. Some fossil fuels, especially coal, contain significant amounts of sulfur impurities. During combustion, these impurities react with oxygen to form SO_2. In addition, during combustion of any fossil fuel, nitrogen from the air reacts with oxygen to form NO_2. The SO_2 and NO_2 emitted from fossil fuel combustion react with water in the atmosphere to form sulfuric acid and nitric acid.

These equations represent simplified versions of the reactions that actually occur.

$$2\ SO_2 + O_2 + 2\ H_2O \longrightarrow 2\ H_2SO_4$$

$$4\ NO_2 + O_2 + 2\ H_2O \longrightarrow 4\ HNO_3$$

These acids combine with rain to form **acid rain**. In the United States, the problem is greatest in the northeastern portion of the country because many midwestern power plants burn coal. The sulfur and nitrogen oxides produced from coal combustion in the Midwest are carried toward the Northeast by natural air currents, making rain in that portion of the country significantly acidic.

Rain is naturally somewhat acidic because of atmospheric carbon dioxide. Carbon dioxide combines with rainwater to form carbonic acid.

$$CO_2 + H_2O \longrightarrow H_2CO_3$$

However, carbonic acid is a relatively weak acid. Even rain that is saturated with CO_2 has a pH of only about 5.6, which is mildly acidic. However, when nitric acid and sulfuric acid mix with rain, the pH of the rain can fall as low as 4.4 (▶ Figure 14.20). Remember that, because of the logarithmic nature of the pH scale, rain with a pH of 4.4 has an $[H_3O^+]$ about 16 times greater than that of rain with a pH of 5.6. Rain that is this acidic has negative consequences for the environment.

National Atmospheric Deposition Program/National Trends Network
http://nadp.sws.uiuc.edu

▶ **FIGURE 14.20   Acid rain in the United States**   Average pH of precipitation in the United States for 2008.

ACID RAIN DAMAGE

Because acids dissolve metals, acid rain damages metal structures. Bridges, railroads, and even automobiles can be damaged by acid rain. Since acids also react with carbonates (CO_3^{2-}), acid rain damages building materials that contain carbonates, including marble, cement, and limestone. Statues, buildings, and pathways in the Northeast show significant signs of acid rain damage (▼ Figure 14.21).

Acid rain can also accumulate in lakes and rivers and affect aquatic life. In the northeastern United States, more than 2000 lakes and streams have increased acidity levels due to acid rain. Aquatic plants, frogs, salamanders, and some species of fish are sensitive to acid levels and cannot live in the acidified lakes. Trees can also be affected by acid rain because the acid removes nutrients from the soil, making it more difficult for trees to survive.

▶ **FIGURE 14.21   Acid rain damage**   Many monuments and statues, such as this one of George Washington in New York's Washington Square Park, have suffered severe deterioration caused by acid rain. The photo at left was taken in 1935, the one at right some 60 years later. (The statue has recently undergone restoration.)

ACID RAIN LEGISLATION

The Clean Air Act, and its 1990 amendments, have provisions that target acid rain. These provisions force electrical utilities—which are the most significant source of SO_2—to lower their SO_2 emissions gradually over time (▶ Figure 14.22). The decrease in SO_2 emissions has been significant, and the acidity of rain in the Northeast has already stabilized and should decrease in the coming years. Scientists expect most lakes, streams, and forests to recover once the pH of the rain returns to normal levels.

See Section 7.8 for the reaction between acids and carbonates.

	≥ 5.3
	5.2 - 5.3
	5.1 - 5.2
	5.0 - 5.1
	4.9 - 5.0
	4.8 - 4.9
	4.7 - 4.8
	4.6 - 4.7
	4.5 - 4.6
	4.4 - 4.5
	4.3 - 4.4
	< 4.3

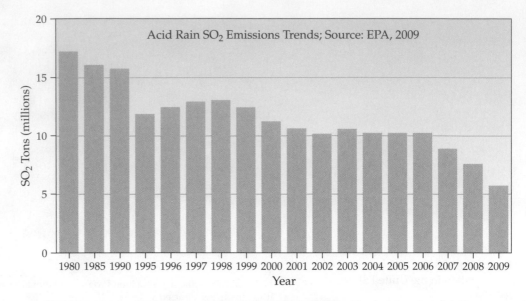

▲ FIGURE 14.22 **Emissions of SO₂ from 1980 to 2009** The height of each bar represents annual SO₂ emissions for the year noted. Under the Clean Air Act and its amendments, SO₂ emissions have been decreasing significantly in the last 30 years.

CHAPTER IN REVIEW

CHEMICAL PRINCIPLES

Acid Properties:
- Acids have a sour taste.
- Acids dissolve many metals.
- Acids turn blue litmus paper red.

Base Properties:
- Bases have a bitter taste.
- Bases have a slippery feel.
- Bases turn red litmus paper blue.

Molecular Definitions of Acids and Bases:
Arrhenius definition

    Acid—substance that produces H^+ ions in solution

    Base—substance that produces OH^- ions in solution

Brønsted–Lowry definition

    Acid—proton donor

    Base—proton acceptor

RELEVANCE

Acid Properties: Acids are responsible for the sour taste in many foods such as lemons, limes, and vinegar. They are also often used in the laboratory and in industry.

Base Properties: Bases are less common in foods, but their presence in some foods—such as coffee and beer— is enjoyed by many as an acquired taste. Bases also have widespread use in the laboratory and in industry.

Molecular Definitions of Acids and Bases: The Arrhenius definition is simpler and easier to use. It also shows how an acid and a base neutralize each other to form water ($H^+ + OH^- \longrightarrow H_2O$). The more generally applicable Brønsted–Lowry definition helps us see that, in water, H^+ ions usually associate with water molecules to form H_3O^+. It also shows how bases that do not contain OH^- ions can still act as bases by accepting a proton from water.

Reactions of Acids and Bases:
Neutralization Reactions In a neutralization reaction, an acid and a base react to form water and a salt.

$$\underset{\text{Acid}}{HCl(aq)} + \underset{\text{Base}}{KOH(aq)} \longrightarrow \underset{\text{Water}}{H_2O(l)} + \underset{\text{Salt}}{KCl(aq)}$$

Acid–Metal Reactions Acids react with many metals to form hydrogen gas and a salt.

$$\underset{\text{Acid}}{2\,HCl(aq)} + \underset{\text{Metal}}{Mg(s)} \longrightarrow \underset{\text{Hydrogen gas}}{H_2(g)} + \underset{\text{Salt}}{MgCl_2(aq)}$$

Acid–Metal Oxide Reactions Acids react with many metal oxides to form water and a salt.

$$\underset{\text{Acid}}{2\,HCl(aq)} + \underset{\text{Metal oxide}}{K_2O(s)} \longrightarrow \underset{\text{Water}}{H_2O(l)} + \underset{\text{Salt}}{2\,KCl(aq)}$$

Reactions of Acids and Bases: Neutralization reactions are common in our everyday lives. Antacids, for example, are bases that react with acids from the stomach to alleviate heartburn and sour stomach.

Acid–metal and acid–metal oxide reactions show the corrosive nature of acids. In both of these reactions, the acid dissolves the metal or the metal oxide. Some of the effects of these kinds of reactions can be seen in the damage to building materials caused by acid rain. Since acids dissolve metals and metal oxides, any building materials composed of these substances are susceptible to acid rain.

Acid–Base Titration: In an acid–base titration, an acid (or base) of known concentration is added to a base (or acid) of unknown concentration. The two reactants are combined until they are in exact stoichiometric proportions (moles of H^+ = moles of OH^-), which marks the equivalence point of the titration. In titration, since you know the moles of H^+ (or OH^-) that you added, you can determine the moles of OH^- (or H^+) in the unknown solution.

Acid–Base Titration: An acid–base titration is a laboratory procedure often used to determine the unknown concentration of an acid or a base.

Strong and Weak Acids and Bases: Strong acids completely ionize, and strong bases completely dissociate in aqueous solutions. For example:

$$HCl(aq) + H_2O(l) \longrightarrow H_3O^+(aq) + Cl^-(aq)$$

$$NaOH(aq) \longrightarrow Na^+(aq) + OH^-(aq)$$

A 1 M HCl solution has $[H_3O^+] = 1$ M and a 1 M NaOH solution has $[OH^-] = 1$ M.

Weak acids only partially ionize in solution. Most weak bases partially ionize water in solution. For example:

$$HF(aq) + H_2O(l) \rightleftharpoons H_3O^+(aq) + F^-(aq)$$

$$NH_3(aq) + H_2O(l) \rightleftharpoons NH_4^+(aq) + OH^-(aq)$$

A 1 M HF solution has $[H_3O^+] < 1$ M, and a 1 M NH_3 solution has $[OH^-] < 1$ M.

Strong and Weak Acids and Bases: Whether an acid is strong or weak depends on the conjugate base: The stronger the conjugate base, the weaker the acid. Since the acidity or basicity of a solution depends on $[H_3O^+]$ and $[OH^-]$, we must know whether an acid is strong or weak to know the degree of acidity or basicity.

Self-Ionization of Water: Water can act as both an acid and a base with itself.

$$\underset{\text{Acid}}{H_2O(l)} + \underset{\text{Base}}{H_2O(l)} \rightleftharpoons H_3O^+(aq) + OH^-(aq)$$

The product of $[H_3O^+]$ and $[OH^-]$ in aqueous solutions will always be equal to the ion product constant, $K_w(10^{-14})$.

$$[H_3O^+][OH^-] = K_w = 1.0 \times 10^{-14}$$

Self-Ionization of Water: The self-ionization of water occurs because aqueous solutions always contain some H_3O^+ and some OH^-. In a neutral solution, the concentrations of these are equal (1.0×10^{-7} M). When an acid is added to water, $[H_3O^+]$ increases and $[OH^-]$ decreases. When a base is added to water, the opposite happens. The ion product constant, however, still equals 1.0×10^{-14}, allowing us to calculate $[H_3O^+]$ given $[OH^-]$ and vice versa.

pH and pOH Scales:

$$pH = -\log[H_3O^+]$$

$$pH > 7 \,(\text{basic})$$

$$pH < 7 \,(\text{acidic})$$

$$pH = 7 \,(\text{neutral})$$

$$pOH = -\log[OH^-]$$

pH and pOH Scales: pH is a convenient way to specify acidity or basicity. Since the pH scale is logarithmic, a change of one on the pH scale corresponds to a tenfold change in the $[H_3O^+]$. The pOH scale, defined with respect to $[OH^-]$ instead of $[H_3O^+]$, is less commonly used.

Buffers: Buffers are solutions containing significant amounts of both a weak acid and its conjugate base. Buffers resist pH change by neutralizing added acid or base.

Buffers: Buffers are important in blood chemistry because blood must stay within a narrow pH range in order to carry oxygen.

Acid Rain: Acid rain is the result of sulfur oxides and nitrogen oxides emitted by fossil fuel combustion. These oxides react with water to form sulfuric acid and nitric acid, which then fall as acid rain.

Acid Rain: Since acids are corrosive, acid rain damages building materials. Because many aquatic plants and animals cannot survive in acidic water, acid rain also affects lakes and rivers, making them too acidic for the survival of some species.

CHEMICAL SKILLS

EXAMPLES

Identifying Brønsted–Lowry Acids and Bases and Their Conjugates (Section 14.4)

The substance that donates the proton is the acid (proton donor) and becomes the conjugate base (as a product). The substance that accepts the proton (proton acceptor) is the base and becomes the conjugate acid (as a product).

EXAMPLE 14.11   Identifying Brønsted–Lowry Acids and Bases and Their Conjugates

Identify the Brønsted–Lowry acid, the Brønsted–Lowry base, the conjugate acid, and the conjugate base in this reaction:

$$HNO_3\,(aq) + H_2O\,(l) \longrightarrow H_3O^+(aq) + NO_3{}^-(aq)$$

SOLUTION

$$\underset{\text{Acid}}{HNO_3(aq)} + \underset{\text{Base}}{H_2O(l)} \longrightarrow \underset{\text{Conjugate acid}}{H_3O^+(aq)} + \underset{\text{Conjugate base}}{NO_3{}^-(aq)}$$

Writing Equations for Neutralization Reactions (Section 14.5)

In a neutralization reaction, an acid and a base usually react to form water and a salt (ionic compound).

$$\text{Acid} + \text{Base} \longrightarrow \text{Water} + \text{Salt}$$

Write the skeletal equation first, making sure to write the formula of the salt so that it is charge-neutral. Then balance the equation.

EXAMPLE 14.12   Writing Equations for Neutralization Reactions

Write a molecular equation for the reaction between aqueous HBr and aqueous $Ca(OH)_2$.

SOLUTION
Skeletal equation:

$$HBr(aq) + Ca(OH)_2(aq) \longrightarrow H_2O(l) + CaBr_2(aq)$$

Balanced equation:

$$2\,HBr(aq) + Ca(OH)_2(aq) \longrightarrow 2\,H_2O(l) + CaBr_2(aq)$$

Writing Equations for the Reactions of Acids with Metals and with Metal Oxides (Section 14.5)

Acids react with many metals to form hydrogen gas and a salt.

$$\text{Acid + Metal} \longrightarrow \text{Hydrogen gas + Salt}$$

Write the skeletal equation first, making sure to write the formula of the salt so that it is charge-neutral. Then balance the equation.

Acids react with many metal oxides to form water and a salt.

$$\text{Acid + Metal oxide} \longrightarrow \text{Water + Salt}$$

Write the skeletal equation first, making sure to write the formula of the salt so that it is charge-neutral. Then balance the equation.

EXAMPLE 14.13 Writing Equations for the Reactions of Acids with Metals and with Metal Oxides

Write equations for the reaction of hydrobromic acid with calcium metal and for the reaction of hydrobromic acid with calcium oxide.

SOLUTION

Skeletal equation:

$$\text{HBr}(aq) + \text{Ca}(s) \longrightarrow \text{H}_2(g) + \text{CaBr}_2(aq)$$

Balanced equation:

$$2\,\text{HBr}(aq) + \text{Ca}(s) \longrightarrow \text{H}_2(g) + \text{CaBr}_2(aq)$$

Skeletal equation:

$$\text{HBr}(aq) + \text{CaO}(s) \longrightarrow \text{H}_2\text{O}(l) + \text{CaBr}_2(aq)$$

Balanced equation:

$$2\,\text{HBr}(aq) + \text{CaO}(s) \longrightarrow \text{H}_2\text{O}(l) + \text{CaBr}_2(aq)$$

Acid–Base Titrations (Section 14.6)

SORT

You are given the volume of a sodium hydroxide solution and the volume and concentration of the sulfuric acid solution required for its titration. You are asked to find the concentration of the sodium hydroxide solution.

STRATEGIZE

Begin by writing the balanced equation for the neutralization reaction (see Example 14.12).

Next draw a solution map. Use the volume and concentration of the known reactant to determine moles of the known reactant. (You have to convert from milliliters to liters first.) Then use the stoichiometric ratio from the balanced equation to get moles of the unknown reactant.

Then add a second part to the solution map indicating how moles and volume can be used to determine molarity.

EXAMPLE 14.14   Acid–Base Titrations

A 15.00-mL sample of a NaOH solution of unknown concentration requires 17.88 mL of a 0.1053 M H_2SO_4 solution to reach the equivalence point in a titration. What is the concentration of the NaOH solution?

GIVEN: 15.00-mL NaOH

   17.88 mL of a 0.1053 M H_2SO_4 solution

FIND: concentration of NaOH solution mol/L

SOLUTION MAP

$$\text{H}_2\text{SO}_4(aq) + 2\,\text{NaOH}(aq) \longrightarrow$$
$$2\,\text{H}_2\text{O}(l) + \text{Na}_2\text{SO}_4(aq)$$

RELATIONSHIPS USED

2 mol NaOH : 1 mol H_2SO_4 (from balanced equation)

$$\text{Molarity(M)} = \frac{\text{mol solute}}{\text{L solution}}$$

SOLVE

Follow the solution map to solve the problem. The first part of the solution gives you moles of the unknown reactant. In the second part of the solution, divide the moles from the first part by the volume to obtain molarity.

CHECK

Check your answer. Are the units correct? Does the answer make physical sense?

SOLUTION

$$17.88 \text{ mL H}_2\text{SO}_4 \times \frac{1 \text{ L}}{1000 \text{ mL}} \times \frac{0.1053 \text{ mol H}_2\text{SO}_4}{\text{L H}_2\text{SO}_4}$$

$$\times \frac{2 \text{ mol NaOH}}{1 \text{ mol H}_2\text{SO}_4} = 3.7655 \times 10^{-3} \text{ mol NaOH}$$

$$M = \frac{\text{mol}}{\text{L}} = \frac{3.7655 \times 10^{-3} \text{ mol NaOH}}{0.01500 \text{ L NaOH}} = 0.2510 \text{ M}$$

The unknown NaOH solution has a concentration of 0.2510 M.

The units (M) are correct. The magnitude of the answer makes sense because the reaction has a two-to-one stoichiometry and the volumes of the two solutions are similar; therefore, the concentration of the NaOH solution must be approximately twice the concentration of the H_2SO_4 solution.

Determining $[H_3O^+]$ in Acid Solutions (Section 14.7)

In a strong acid, $[H_3O^+]$ will be equal to the concentration of the acid. In a weak acid, $[H_3O^+]$ will be less than the concentration of the acid.

EXAMPLE 14.15 Determining $[H_3O^+]$ in Acid Solutions

What is the H_3O^+ concentration in a 0.25 M HCl solution and in a 0.25 M HF solution?

SOLUTION

In the 0.25 M HCl solution (strong acid), $[H_3O^+] =$ 0.25 M. In the 0.25 M HF solution (weak acid), $[H_3O^+] < 0.25$ M.

Determining $[OH^-]$ in Base Solutions (Section 14.7)

In a strong base, $[OH^-]$ is equal to the concentration of the base times the number of hydroxide ions in the base. In a weak base, $[OH^-]$ is less than the concentration of the base.

EXAMPLE 14.16 Determining $[OH^-]$ in Base Solutions

What is the OH^- concentration in a 0.25 M NaOH solution, in a 0.25 M $Sr(OH)_2$ solution, and in a 0.25 M NH_3 solution?

SOLUTION

In the 0.25 M NaOH solution (strong base), $[OH^-] =$ 0.25 M. In the 0.25 M $Sr(OH)_2$ solution (strong base),

$[OH^-] = 0.50$ M. In the 0.25 M NH_3 solution (weak base),

$[OH^-] < 0.25$ M.

Finding the Concentration of $[H_3O^+]$ or $[OH^-]$ from K_w (Section 14.8)

To find $[H_3O^+]$ or $[OH^-]$, use the ion product constant expression.

$$[H_3O^+][OH^-] = 1.0 \times 10^{-14}$$

Substitute the known quantity into the equation ($[H_3O^+]$ or $[OH^-]$) and solve for the unknown quantity.

EXAMPLE 14.17 Finding the Concentration of $[H_3O^+]$ or $[OH^-]$ from K_w

Calculate $[OH^-]$ in a solution with

$$[H_3O^+] = 1.5 \times 10^{-4} \text{ M}.$$

SOLUTION

$$[H_3O^+][OH^-] = 1.0 \times 10^{-14}$$
$$[1.5 \times 10^{-4}][OH^-] = 1.0 \times 10^{-14}$$

$$[OH^-] = \frac{1.0 \times 10^{-14}}{1.5 \times 10^{-4}} = 6.7 \times 10^{-11} \text{ M}$$

Calculating pH from [H₃O⁺] (Section 14.9)

To calculate the pH of a solution from $[H_3O^+]$, simply take the negative log of $[H_3O^+]$.

$$pH = -\log[H_3O^+]$$

EXAMPLE 14.18 Calculating pH from [H₃O⁺]

Calculate the pH of a solution with $[H_3O^+] = 2.4 \times 10^{-5}$ M.

SOLUTION

$$pH = -\log[H_3O^+]$$
$$= -\log(2.4 \times 10^{-5})$$
$$= -(-4.62)$$
$$= 4.62$$

Calculating [H₃O⁺] from pH (Section 14.9)

You can calculate $[H_3O^+]$ from pH by taking the inverse log of the negative of the pH value (Method 1):

$$[H_3O^+] = \text{invlog}(-pH)$$

You can also calculate $[H_3O^+]$ from pH by raising 10 to the negative of the pH (Method 2):

$$[H_3O^+] = 10^{-pH}$$

EXAMPLE 14.19 Calculating [H₃O⁺] from pH

Calculate the $[H_3O^+]$ for a solution with a pH of 6.22.

SOLUTION
Method 1: Inverse Log Function

$$[H_3O^+] = \text{invlog}(-pH)$$
$$= \text{invlog}(-6.22)$$
$$= 6.0 \times 10^{-7}$$

Method 2: 10ˣ Function

$$[H_3O^+] = 10^{-pH}$$
$$= 10^{-6.22}$$
$$= 6.0 \times 10^{-7}$$

KEY TERMS

acid [14.2]
acid rain [14.11]
acidic solution [14.8]
alkaloid [14.3]
amphoteric [14.4]
Arrhenius acid [14.4]
Arrhenius base [14.4]
Arrhenius definition [14.4]
base [14.3]
basic solution [14.8]

Brønsted–Lowry acid [14.4]
Brønsted–Lowry base [14.4]
Brønsted–Lowry definition [14.4]
buffer [14.10]
carboxylic acid [14.2]
conjugate acid–base pair [14.4]
diprotic acid [14.7]
dissociation [14.4]

equivalence point [14.6]
hydronium ion [14.4]
indicator [14.6]
ion product constant for water (K_w) [14.8]
ionize [14.4]
logarithmic scale [14.9]
monoprotic acid [14.7]
neutral solution [14.8]
neutralization [14.5]

pH [14.9]
pOH [14.9]
salt [14.5]
strong acid [14.7]
strong base [14.7]
strong electrolyte [14.7]
titration [14.6]
weak acid [14.7]
weak base [14.7]
weak electrolyte [14.7]

EXERCISES

QUESTIONS

1. What makes tart gummy candies, such as Sour Patch Kids, sour?
2. What are the properties of acids? List some examples of foods that contain acids.
3. What is the main component of stomach acid? Why do we have stomach acid?
4. What are organic acids? List two examples of organic acids.
5. What are the properties of bases? Provide some examples of common substances that contain bases.
6. What are alkaloids?
7. Give the Arrhenius definition of an acid and demonstrate the definition with a chemical equation.
8. Give the Arrhenius definition of a base and demonstrate the definition with a chemical equation.
9. Give the Brønsted–Lowry definitions of acids and bases and demonstrate each with a chemical equation.
10. According to the Brønsted–Lowry definition of acids and bases, what is a conjugate acid–base pair? Provide an example.

11. What is an acid–base neutralization reaction? Provide an example.

12. Provide an example of a reaction between an acid and a metal.

13. List an example of a reaction between an acid and a metal oxide.

14. Name a metal that dissolves in a base and write an equation for the reaction.

15. What is a titration? What is the equivalence point?

16. If a solution contains 0.85 mol of OH^-, how many moles of H^+ would be required to reach the equivalence point in a titration?

17. What is the difference between a strong acid and a weak acid?

18. How is the strength of an acid related to the strength of its conjugate base?

19. What are monoprotic and diprotic acids?

20. What is the difference between a strong base and a weak base?

21. Does pure water contain any H_3O^+ ions? Explain.

22. What happens to $[OH^-]$ in an aqueous solution when $[H_3O^+]$ increases?

23. Give a possible value of $[OH^-]$ and $[H_3O^+]$ in a solution that is:
   (a) acidic   (b) basic   (c) neutral

24. How is pH defined? A change of 1.0 pH unit corresponds to how much of a change in $[H_3O^+]$?

25. How is pOH defined? A change of 2.0 pOH units corresponds to how much of a change in $[OH^-]$?

26. In any aqueous solution at 25 °C, the sum of pH and pOH is 14.0. Explain why this is so.

27. What is a buffer?

28. What are the main components in a buffer?

29. What is the cause of acid rain?

30. Write equations for the chemical reactions by which acid rain forms in the atmosphere.

31. What are the effects of acid rain?

32. How is the problem of acid rain being addressed in the United States?

PROBLEMS

ACID AND BASE DEFINITIONS

33. Identify each substance as an acid or a base and write a chemical equation showing how it is an acid or a base according to the Arrhenius definition.
   (a) $H_2SO_4(aq)$
   (b) $Sr(OH)_2(aq)$
   (c) $HBr(aq)$
   (d) $NaOH(aq)$

34. Identify each substance as an acid or a base and write a chemical equation showing how it is an acid or a base according to the Arrhenius definition.
   (a) $Ca(OH)_2(aq)$
   (b) $HC_2H_3O_2(aq)$
   (c) $KOH(aq)$
   (d) $HNO_3(aq)$

35. For each reaction, identify the Brønsted–Lowry acid, the Brønsted–Lowry base, the conjugate acid, and the conjugate base.
   (a) $HBr(aq) + H_2O(l) \longrightarrow H_3O^+(aq) + Br^-(aq)$
   (b) $NH_3(aq) + H_2O(l) \rightleftharpoons NH_4^+(aq) + OH^-(aq)$
   (c) $HNO_3(aq) + H_2O(l) \longrightarrow$
   $H_3O^+(aq) + NO_3^-(aq)$
   (d) $C_5H_5N(aq) + H_2O(l) \rightleftharpoons$
   $C_5H_5NH^+(aq) + OH^-(aq)$

36. For each reaction, identify the Brønsted–Lowry acid, the Brønsted–Lowry base, the conjugate acid, and the conjugate base.
   (a) $HI(aq) + H_2O(l) \longrightarrow H_3O^+(aq) + I^-(aq)$
   (b) $CH_3NH_2(aq) + H_2O(l) \rightleftharpoons$
   $CH_3NH_3^+(aq) + OH^-(aq)$
   (c) $CO_3^{2-}(aq) + H_2O(l) \rightleftharpoons$
   $HCO_3^-(aq) + OH^-(aq)$
   (d) $H_2CO_3(aq) + H_2O(l) \rightleftharpoons$
   $H_3O^+(aq) + HCO_3^-(aq)$

37. Determine whether each pair is a conjugate acid–base pair.
   (a) NH_3, NH_4^+
   (b) HCl, HBr
   (c) $C_2H_3O_2^-$, $HC_2H_3O_2$
   (d) HCO_3^-, NO_3^-

38. Determine whether each pair is a conjugate acid–base pair
   (a) HI, I^-
   (b) $HCHO_2$, SO_4^{2-}
   (c) PO_4^{3-}, HPO_4^{2-}
   (d) CO_3^{2-}, HCl

39. Write the formula for the conjugate base of each acid.
   (a) HCl (b) H_2SO_3
   (c) $HCHO_2$ (d) HF

40. Write the formula for the conjugate base of each acid.
   (a) HBr (b) H_2CO_3
   (c) $HClO_4$ (d) $HC_2H_3O_2$

41. Write the formula for the conjugate acid of each base.

(a) NH_3

(b) ClO_4^-

(c) HSO_4^-

(d) CO_3^{2-}

42. Write the formula for the conjugate acid of each base.

(a) CH_3NH_2

(b) C_5H_5N

(c) Cl^-

(d) F^-

ACID–BASE REACTIONS

43. Write a neutralization reaction for each acid and base pair.

(a) $HI(aq)$ and $NaOH(aq)$

(b) $HBr(aq)$ and $KOH(aq)$

(c) $HNO_3(aq)$ and $Ba(OH)_2(aq)$

(d) $HClO_4(aq)$ and $Sr(OH)_2(aq)$

44. Write a neutralization reaction for each acid and base pair.

(a) $HF(aq)$ and $Ba(OH)_2(aq)$

(b) $HClO_4(aq)$ and $NaOH(aq)$

(c) $HBr(aq)$ and $Ca(OH)_2(aq)$

(d) $HCl(aq)$ and $KOH(aq)$

45. Write a balanced chemical equation showing how each metal reacts with HBr.

(a) Rb

(b) Mg

(c) Ba

(d) Al

46. Write a balanced chemical equation showing how each metal reacts with HCl.

(a) K

(b) Ca

(c) Na

(d) Sr

47. Write a balanced chemical equation showing how each metal oxide reacts with HI.

(a) MgO

(b) K_2O

(c) Rb_2O

(d) CaO

48. Write a balanced chemical equation showing how each metal oxide reacts with HCl.

(a) SrO

(b) Na_2O

(c) Li_2O

(d) BaO

49. Predict the products of each reaction:

(a) $HClO_4(aq) + Fe_2O_3(s) \longrightarrow$

(b) $H_2SO_4(aq) + Sr(s) \longrightarrow$

(c) $H_3PO_4(aq) + KOH(aq) \longrightarrow$

50. Predict the products of each reaction:

(a) $HI(aq) + Al(s) \longrightarrow$

(b) $H_2SO_4(aq) + TiO_2(s) \longrightarrow$

(c) $H_2CO_3(aq) + LiOH(aq) \longrightarrow$

ACID–BASE TITRATIONS

51. Four solutions of unknown HCl concentration are titrated with solutions of NaOH. The following table lists the volume of each unknown HCl solution, the volume of NaOH solution required to reach the equivalence point, and the concentration of each NaOH solution. Calculate the concentration (in M) of the unknown HCl solution in each case.

HCl Volume (mL)	NaOH Volume (mL)	[NaOH] (M)
(a) 25.00 mL	28.44 mL	0.1231 M
(b) 15.00 mL	21.22 mL	0.0972 M
(c) 20.00 mL	14.88 mL	0.1178 M
(d) 5.00 mL	6.88 mL	0.1325 M

52. Four solutions of unknown NaOH concentration are titrated with solutions of HCl. The following table lists the volume of each unknown NaOH solution, the volume of HCl solution required to reach the equivalence point, and the concentration of each HCl solution. Calculate the concentration (in M) of the unknown NaOH solution in each case.

NaOH Volume (mL)	HCl Volume (mL)	[HCl] (M)
(a) 5.00 mL	9.77 mL	0.1599 M
(b) 15.00 mL	11.34 mL	0.1311 M
(c) 10.00 mL	10.55 mL	0.0889 M
(d) 30.00 mL	36.18 mL	0.1021 M

53. A 25.00-mL sample of an H_2SO_4 solution of unknown concentration is titrated with a 0.1322 M KOH solution. A volume of 41.22 mL of KOH was required to reach the equivalence point. What is the concentration of the unknown H_2SO_4 solution?

54. A 5.00-mL sample of an H_3PO_4 solution of unknown concentration is titrated with a 0.1090 M NaOH solution. A volume of 7.12 mL of the NaOH solution was required to reach the equivalence point. What is the concentration of the unknown H_3PO_4 solution?

55. What volume in milliliters of a 0.121 M sodium hydroxide solution is required to reach the equivalence point in the complete titration of a 10.0-mL sample of 0.102 M sulfuric acid?

56. What volume in milliliters of 0.0985 M sodium hydroxide solution is required to reach the equivalence point in the complete titration of a 15.0-mL sample of 0.124 M phosphoric acid?

STRONG AND WEAK ACIDS AND BASES

57. Classify each acid as strong or weak.
  (a) HCl
  (b) HF
  (c) HBr
  (d) H_2SO_3

58. Classify each acid as strong or weak.
  (a) $HCHO_2$
  (b) H_2SO_4
  (c) HNO_3
  (d) H_2CO_3

59. Determine $[H_3O^+]$ in each acid solution. If the acid is weak, indicate the value that $[H_3O^+]$ is less than.
  (a) 1.7 M HBr
  (b) 1.5 M HNO_3
  (c) 0.38 M H_2CO_3
  (d) 1.75 M $HCHO_2$

60. Determine $[H_3O^+]$ in each acid solution. If the acid is weak, indicate the value that $[H_3O^+]$ is less than.
  (a) 0.125 M $HClO_2$
  (b) 1.25 M H_3PO_4
  (c) 2.77 M HCl
  (d) 0.95 M H_2SO_3

61. Classify each base as strong or weak.
  (a) LiOH
  (b) NH_4OH
  (c) $Ca(OH)_2$
  (d) NH_3

62. Classify each base as strong or weak.
  (a) C_5H_5N
  (b) NaOH
  (c) $Ba(OH)_2$
  (d) KOH

63. Determine $[OH^-]$ in each base solution. If the acid is weak, indicate the value that $[OH^-]$ is less than.
  (a) 0.25 M NaOH
  (b) 0.25 M NH_3
  (c) 0.25 M $Sr(OH)_2$
  (d) 1.25 M KOH

64. Determine $[OH^-]$ in each base solution. If the acid is weak, indicate the value that $[OH^-]$ is less than.
  (a) 2.5 M KOH
  (b) 1.95 M NH_3
  (c) 0.225 M $Ba(OH)_2$
  (d) 1.8 M C_5H_5N

ACIDITY, BASICITY, AND K_W

65. Determine whether each solution is acidic, basic, or neutral.
  (a) $[H_3O^+] = 1 \times 10^{-5}$ M; $[OH^-] = 1 \times 10^{-9}$ M
  (b) $[H_3O^+] = 1 \times 10^{-6}$ M; $[OH^-] = 1 \times 10^{-8}$ M
  (c) $[H_3O^+] = 1 \times 10^{-7}$ M; $[OH^-] = 1 \times 10^{-7}$ M
  (d) $[H_3O^+] = 1 \times 10^{-8}$ M; $[OH^-] = 1 \times 10^{-6}$ M

66. Determine whether each solution is acidic, basic, or neutral.
  (a) $[H_3O^+] = 1 \times 10^{-9}$ M; $[OH^-] = 1 \times 10^{-5}$ M
  (b) $[H_3O^+] = 1 \times 10^{-10}$ M; $[OH^-] = 1 \times 10^{-4}$ M
  (c) $[H_3O^+] = 1 \times 10^{-2}$ M; $[OH^-] = 1 \times 10^{-12}$ M
  (d) $[H_3O^+] = 1 \times 10^{-13}$ M; $[OH^-] = 1 \times 10^{-1}$ M

67. Calculate $[OH^-]$ given $[H_3O^+]$ in each aqueous solution and classify the solution as acidic or basic.

   (a) $[H_3O^+] = 1.5 \times 10^{-9}$ M

   (b) $[H_3O^+] = 9.3 \times 10^{-9}$ M

   (c) $[H_3O^+] = 2.2 \times 10^{-6}$ M

   (d) $[H_3O^+] = 7.4 \times 10^{-4}$ M

68. Calculate $[OH^-]$ given $[H_3O^+]$ in each aqueous solution and classify the solution as acidic or basic.

   (a) $[H_3O^+] = 1.3 \times 10^{-3}$ M

   (b) $[H_3O^+] = 9.1 \times 10^{-12}$ M

   (c) $[H_3O^+] = 5.2 \times 10^{-4}$ M

   (d) $[H_3O^+] = 6.1 \times 10^{-9}$ M

69. Calculate $[H_3O^+]$ given $[OH^-]$ in each aqueous solution and classify each solution as acidic or basic.

   (a) $[OH^-] = 2.7 \times 10^{-12}$ M

   (b) $[OH^-] = 2.5 \times 10^{-2}$ M

   (c) $[OH^-] = 1.1 \times 10^{-10}$ M

   (d) $[OH^-] = 3.3 \times 10^{-4}$ M

70. Calculate $[H_3O^+]$ given $[OH^-]$ in each aqueous solution and classify each solution as acidic or basic.

   (a) $[OH^-] = 2.1 \times 10^{-11}$ M

   (b) $[OH^-] = 7.5 \times 10^{-9}$ M

   (c) $[OH^-] = 2.1 \times 10^{-4}$ M

   (d) $[OH^-] = 1.0 \times 10^{-2}$ M

pH

71. Classify each solution as acidic, basic, or neutral according to its pH value.

   (a) pH = 8.0

   (b) pH = 7.0

   (c) pH = 3.5

   (d) pH = 6.1

72. Classify each solution as acidic, basic, or neutral according to its pH value.

   (a) pH = 4.0

   (b) pH = 3.5

   (c) pH = 13.0

   (d) pH = 0.85

73. Calculate the pH of each solution.

   (a) $[H_3O^+] = 1.7 \times 10^{-8}$ M

   (b) $[H_3O^+] = 1.0 \times 10^{-7}$ M

   (c) $[H_3O^+] = 2.2 \times 10^{-6}$ M

   (d) $[H_3O^+] = 7.4 \times 10^{-4}$ M

74. Calculate the pH of each solution.

   (a) $[H_3O^+] = 2.4 \times 10^{-10}$ M

   (b) $[H_3O^+] = 7.6 \times 10^{-2}$ M

   (c) $[H_3O^+] = 9.2 \times 10^{-13}$ M

   (d) $[H_3O^+] = 3.4 \times 10^{-5}$ M

75. Calculate $[H_3O^+]$ for each solution.

   (a) pH = 8.55

   (b) pH = 11.23

   (c) pH = 2.87

   (d) pH = 1.22

76. Calculate $[H_3O^+]$ for each solution.

   (a) pH = 1.76

   (b) pH = 3.88

   (c) pH = 8.43

   (d) pH = 12.32

77. Calculate the pH of each solution.

   (a) $[OH^-] = 1.9 \times 10^{-7}$ M

   (b) $[OH^-] = 2.6 \times 10^{-8}$ M

   (c) $[OH^-] = 7.2 \times 10^{-11}$ M

   (d) $[OH^-] = 9.5 \times 10^{-2}$ M

78. Calculate the pH of each solution.

   (a) $[OH^-] = 2.8 \times 10^{-11}$ M

   (b) $[OH^-] = 9.6 \times 10^{-3}$ M

   (c) $[OH^-] = 3.8 \times 10^{-12}$ M

   (d) $[OH^-] = 6.4 \times 10^{-4}$ M

79. Calculate $[OH^-]$ for each solution.

   (a) pH = 4.25

   (b) pH = 12.53

   (c) pH = 1.50

   (d) pH = 8.25

80. Calculate $[OH^-]$ for each solution.

   (a) pH = 1.82

   (b) pH = 13.28

   (c) pH = 8.29

   (d) pH = 2.32

81. Calculate the pH of each solution:
    (a) 0.0155 M HBr
    (b) 1.28×10^{-3} M KOH
    (c) 1.89×10^{-3} M HNO_3
    (d) 1.54×10^{-4} M $Sr(OH)_2$

82. Calculate the pH of each solution:
    (a) 1.34×10^{-3} M $HClO_4$
    (b) 0.0211 M NaOH
    (c) 0.0109 M HBr
    (d) 7.02×10^{-5} M $Ba(OH)_2$

pOH

83. Determine the pOH of each solution and classify it as acidic, basic, or neutral.
    (a) $[OH^-] = 1.5 \times 10^{-9}$ M
    (b) $[OH^-] = 7.0 \times 10^{-5}$ M
    (c) $[OH^-] = 1.0 \times 10^{-7}$ M
    (d) $[OH^-] = 8.8 \times 10^{-3}$ M

84. Determine the pOH of each solution and classify it as acidic, basic, or neutral.
    (a) $[OH^-] = 4.5 \times 10^{-2}$ M
    (b) $[OH^-] = 3.1 \times 10^{-12}$ M
    (c) $[OH^-] = 5.4 \times 10^{-5}$ M
    (d) $[OH^-] = 1.2 \times 10^{-2}$ M

85. Determine the pOH of each solution.
    (a) $[H_3O^+] = 1.2 \times 10^{-8}$ M
    (b) $[H_3O^+] = 5.5 \times 10^{-2}$ M
    (c) $[H_3O^+] = 3.9 \times 10^{-9}$ M
    (d) $[OH^-] = 1.88 \times 10^{-13}$ M

86. Determine the pOH of each solution.
    (a) $[H_3O^+] = 8.3 \times 10^{-10}$ M
    (b) $[H_3O^+] = 1.6 \times 10^{-7}$ M
    (c) $[H_3O^+] = 7.3 \times 10^{-2}$ M
    (d) $[OH^-] = 4.32 \times 10^{-4}$ M

87. Determine the pH of each solution and classify it as acidic, basic, or neutral.
    (a) pOH = 8.5
    (b) pOH = 4.2
    (c) pOH = 1.7
    (d) pOH = 7.0

88. Determine the pH of each solution and classify it as acidic, basic, or neutral.
    (a) pOH = 12.5
    (b) pOH = 5.5
    (c) pOH = 0.55
    (d) pOH = 7.98

BUFFERS AND ACID RAIN

89. Determine your location on the map in Figure 14.20. What is the pH of rain where you live? What is the $[H_3O^+]$?

90. Identify the area of the United States with the most acidic rainfall on the map in Figure 14.20. What is the pH of the rain? What is the $[H_3O^+]$?

91. Determine whether or not each mixture is a buffer.
    (a) HCl and HF
    (b) NaOH and NH_3
    (c) HF and NaF
    (d) $HC_2H_3O_2$ and $KC_2H_3O_2$

92. Determine whether or not each mixture is a buffer.
    (a) HBr and NaCl
    (b) $HCHO_2$ and $NaCHO_2$
    (c) HCl and HBr
    (d) KOH and NH_3

93. Write reactions showing how each of the buffers in Problem 91 would neutralize added HCl.

94. Write reactions showing how each of the buffers in Problem 92 would neutralize added NaOH.

95. What substance could you add to each solution to make it a buffer solution?
    (a) 0.100 M $NaC_2H_3O_2$
    (b) 0.500 M H_3PO_4
    (c) 0.200 M $HCHO_2$

96. What substance could you add to each solution to make it a buffer solution?
    (a) 0.050 M $NaHSO_3$
    (b) 0.150 M HF
    (c) 0.200 M $KCHO_2$

CUMULATIVE PROBLEMS

97. How much 0.100 M HCl is required to completely neutralize 20.0 mL of 0.250 M NaOH?

98. How much 0.200 M KOH is required to completely neutralize 25.0 mL of 0.150 M $HClO_4$?

99. What is the minimum volume of 5.0 M HCl required to completely dissolve 10.0 g of magnesium metal?

100. What is the minimum volume of 3.0 M HBr required to completely dissolve 15.0 g of potassium metal?

101. When 18.5 g of $K_2O(s)$ is completely dissolved by HI(aq), how many grams of KI(aq) are formed in solution?

102. When 5.88 g of CaO(s) is completely dissolved by HBr(aq), how many grams of $CaBr_2$(aq) are formed in solution?

103. A 0.125-g sample of a monoprotic acid of unknown molar mass is dissolved in water and titrated with 0.1003 M NaOH. The equivalence point is reached after adding 20.77 mL of base. What is the molar mass of the unknown acid?

104. A 0.105-g sample of a diprotic acid of unknown molar mass is dissolved in water and titrated with 0.1288 M NaOH. The equivalence point is reached after adding 15.2 mL of base. What is the molar mass of the unknown acid?

105. Antacids, such as milk of magnesia, are often taken to reduce the discomfort of acid stomach or heartburn. The recommended dose of milk of magnesia is 1 teaspoon, which contains 400 mg of $Mg(OH)_2$. What volume of HCl solution with a pH of 1.1 can be neutralized by 1 dose of milk of magnesia? (Assume two significant figures in your calculations.)

106. An antacid tablet requires 25.82 mL of 0.200 M HCl to titrate to its equivalence point. What volume in milliliters of stomach acid can be neutralized by the antacid tablet? Assume that stomach acid has a pH of 1.1. (Assume two significant figures in your calculations.)

107. For each $[H_3O^+]$, determine the pH and state whether the solution is acidic or basic.

(a) $[H_3O^+]$ = 0.0025 M

(b) $[H_3O^+]$ = 1.8×10^{-12} M

(c) $[H_3O^+]$ = 9.6×10^{-9} M

(d) $[H_3O^+]$ = 0.0195 M

108. For each $[OH^-]$, determine the pH and state whether the solution is acidic or basic.

(a) $[OH^-]$ = 1.8×10^{-5} M

(b) $[OH^-]$ = 8.9×10^{-12} M

(c) $[OH^-]$ = 3.1×10^{-2} M

(d) $[OH^-]$ = 1.96×10^{-9} M

109. Complete the table. (The first row is completed for you.)

$[H_3O^+]$	$[OH^-]$	pOH	pH	Acidic or Basic
1.0×10^{-4}	1.0×10^{-10}	10.00	4.00	acidic
5.5×10^{-3}	___	___	___	___
___	3.2×10^{-6}	___	___	___
4.8×10^{-9}	___	___	___	___
___	___	___	7.55	___

110. Complete the table. (The first row is completed for you.)

$[H_3O^+]$	$[OH^-]$	pOH	pH	Acidic or Basic
1.0×10^{-8}	1.0×10^{-6}	6.00	8.00	basic
___	___	___	3.55	___
1.7×10^{-9}	___	___	___	___
___	___	___	13.5	___
___	8.6×10^{-11}	___	___	___

111. For each strong acid solution, determine $[H_3O^+]$, $[OH^-]$, and pH.

(a) 0.0088 M $HClO_4$

(b) 1.5×10^{-3} M HBr

(c) 9.77×10^{-4} M HI

(d) 0.0878 M HNO_3

112. For each strong acid solution, determine $[H_3O^+]$, $[OH^-]$, and pH.

(a) 0.0150 M HCl

(b) 1.9×10^{-4} M HI

(c) 0.0226 M HBr

(d) 1.7×10^{-3} M HNO_3

113. For each strong base solution, determine [OH⁻], [H₃O⁺], pH, and pOH.
   (a) 0.15 M NaOH
   (b) 1.5×10^{-3} M Ca(OH)₂
   (c) 4.8×10^{-4} M Sr(OH)₂
   (d) 8.7×10^{-5} M KOH

114. For each strong base solution, determine [OH⁻], [H₃O⁺], pH, and pOH.
   (a) 8.77×10^{-3} M LiOH
   (b) 0.0112 M Ba(OH)₂
   (c) 1.9×10^{-4} M KOH
   (d) 5.0×10^{-4} M Ca(OH)₂

115. As described in Section 14.1, jailed spies on the big screen often use acid stored in a pen to dissolve jail bars and escape. What minimum volume of 12.0 M hydrochloric acid would be required to completely dissolve a 500.0-g iron bar? Would this amount of acid fit into a pen?

116. A popular classroom demonstration consists of filing notches into a new penny and soaking the penny in hydrochloric acid overnight. Since new pennies are made of zinc coated with copper, and since hydrochloric acid dissolves zinc and not copper, the inside of the penny is dissolved by the acid, while the outer copper shell remains. Suppose the penny contains 2.5 g of zinc and is soaked in 20.0 mL of 6.0 M HCl. Calculate the concentration of the HCl solution after all of the zinc has dissolved. *Hint:* The Zn from the penny is oxidized to Zn²⁺.

117. What is the pH of a solution formed by mixing 125.0 mL of 0.0250 M HCl with 75.0 mL of 0.0500 M NaOH?

118. What is the pH of a solution formed by mixing 175.0 mL of 0.0880 M HI with 125.0 mL of 0.0570 M KOH?

119. How many H⁺ (or H₃O⁺) ions are present in one drop (0.050 mL) of pure water at 25 °C?

120. Calculate the number of H⁺ (or H₃O⁺) ions and OH⁻ ions in 1.0 mL of 0.100 M HCl.

121. A 4.00-L base solution contains 0.100 mol total of NaOH and Sr(OH)₂. The pOH of the solution is 1.51. Determine the amounts (in moles) of NaOH and Sr(OH)₂ in the solution.

122. A 1.50-L acid solution contains 0.35 g total of HCl and HBr. The pH of the solution is 2.40. What are the masses of HCl and HBr in the solution?

HIGHLIGHT PROBLEMS

123. Based on the molecular view of each acid solution, determine whether the acid is weak or strong.

(a)   (b)   (c)   (d)

124. Lakes that have been acidified by acid rain can be neutralized by liming, the addition of limestone (CaCO₃). How much limestone in kilograms is required to completely neutralize a 3.8×10^9 L lake with a pH of 5.5?

125. Acid rain over the Great Lakes has a pH of about 4.5. Calculate the [H₃O⁺] of this rain and compare that value to the [H₃O⁺] of rain over the West Coast that has a pH of 5.4. How many times more concentrated is the acid in rain over the Great Lakes?

►ANSWERS TO SKILLBUILDER EXERCISES

Skillbuilder 14.1

(a) $C_5H_5N(aq) + H_2O(l) \rightleftharpoons$
    Base   Acid

$$C_5H_5NH^+(aq) + OH^-(aq)$$
    Conjugate acid   Conjugate base

(b) $HNO_3(aq) + H_2O(l) \longrightarrow H_3O^+(aq) + NO_3^-(aq)$
  Acid   Base   Conjugate acid   Conjugate base

Skillbuilder 14.2

$H_3PO_4(aq) + 3\,NaOH(aq) \longrightarrow 3\,H_2O(l) + Na_3PO_4(aq)$

Skillbuilder 14.3

(a) $2\,HCl(aq) + Sr(s) \longrightarrow H_2(g) + SrCl_2(aq)$

(b) $2\,HI(aq) + BaO(s) \longrightarrow H_2O(l) + BaI_2(aq)$

Skillbuilder 14.4 $9.03 \times 10^{-2}\,M\,H_2SO_4$

Skillbuilder 14.5

(a) $[H_3O^+] < 0.50\,M$

(b) $[H_3O^+] = 1.25\,M$

(c) $[H_3O^+] < 0.75\,M$

Skillbuilder 14.6

(a) $[OH^-] = 0.11\,M$

(b) $[OH^-] < 1.05\,M$

(c) $[OH^-] = 0.45\,M$

Skillbuilder 14.7

(a) $[H_3O^+] = 6.7 \times 10^{-13}\,M$; basic

(b) $[H_3O^+] = 1.0 \times 10^{-7}\,M$; neutral

(c) $[H_3O^+] = 1.2 \times 10^{-5}\,M$; acidic

Skillbuilder 14.8

(a) pH = 8.02; basic

(b) pH = 2.21 acidic

Skillbuilder Plus, p. 509   pH = 12.11; basic

Skillbuilder 14.9 $4.3 \times 10^{-9}\,M$

Skillbuilder Plus, p. 510 $4.6 \times 10^{-11}\,M$

Skillbuilder 14.10 $5.6 \times 10^{-5}\,M$

Skillbuilder Plus, p. 511 $4.8 \times 10^{-9}\,M$

►ANSWERS TO CONCEPTUAL CHECKPOINTS

14.1 (b) The conjugate base of an acid always has one fewer proton and is one charge unit lower (more negative) than the acid.

14.2 (c) Both (a) and (b) show complete ionization and are therefore strong acids. Only the acid depicted in (c) undergoes partial ionization and is therefore a weak acid.

14.3 (d) Each of the others can accept a proton and thus acts as a base. NH_4^+, however, is the conjugate acid of NH_3 and therefore acts as an acid and not as a base.

14.4 (d) Because pH is the *negative* log of the H_3O^+ concentration, a higher pH corresponds to a lower $[H_3O^+]$, and each unit of pH represents a tenfold change in concentration.

14.5 (d) Since the pH is 5, the pOH = 14 − 5 = 9.

14.6 (b) A buffer solution consists of a weak acid and its conjugate base. Of the compounds listed, HF is the only weak acid, and F^- (from NaF in solution) is its conjugate base.

Chemical Equilibrium

"A system is in equilibrium when the forces constituting it are arranged in such a way as to compensate each other, like the two weights pulling at the arms of a pair of scales."

RUDOLF ARNHEIM (1904–2007)

15.1 Life: Controlled Disequilibrium

Have you ever tried to define life? If you have, you know that life is not easy to define. How are living things different from nonliving things? You may try to define living things as those things that can move. But of course many living things do not move—many plants, for example, do not move very much—and some nonliving things, such as glaciers and Earth itself, do move. So motion is neither unique to nor definitive of life. You may try to define living things as those things that can reproduce. But again, many living things, such as mules or sterile humans, cannot reproduce; yet they are alive. In addition, some nonliving things—such as crystals—reproduce (in some sense). So what is unique about living things?

The concept of equilibrium underlies one definition of life. We define *chemical* equilibrium more carefully soon. For now, we can think generally of equilibrium as *sameness and constancy*. When an object is in equilibrium with its surroundings, some property of the object has reached sameness with the surroundings and is no longer changing. For example, a cup of hot water is not in equilibrium with its surroundings with respect to temperature. If left undisturbed, the cup of hot water will slowly cool until it reaches equilibrium with its surroundings. At that point, the temperature of the water is the *same as* that of the surroundings (sameness) and *no longer changes* (constancy).

So equilibrium involves sameness and constancy. Part of a definition for living things, then, is that living things *are not* in equilibrium with their surroundings. Our body temperature, for example, is not the same as the temperature of our

◀ Dynamic equilibrium involves two opposing processes occurring at the same rate. This image draws an analogy between a chemical equilibrium ($N_2O_4 \rightleftharpoons 2NO_2$), in which the two opposing reactions occur at the same rate, and a freeway with traffic moving in opposing directions at the same rate.

surroundings. When we jump into a swimming pool, the pH of our blood does not become the same as the pH of the surrounding water. Living things, even the simplest ones, maintain some measure of *disequilibrium* with their environment.

We must add one more concept, however, to complete our definition of life with respect to equilibrium. A cup of hot water is in disequilibrium with its environment, yet it is not alive. However, the cup of hot water has no control over its disequilibrium and will slowly come to equilibrium with its environment. In contrast, living things—as long as they are alive—maintain and *control* their disequilibrium. Your body temperature, for example, is not only in disequilibrium with your surroundings—it is in controlled disequilibrium. Your body maintains your temperature within a specific range that is not in equilibrium with the surrounding temperature.

So one definition for life is that living things are in *controlled disequilibrium* with their environment. A living thing comes into equilibrium with its surroundings only after it dies. In this chapter, we examine the concept of equilibrium, especially chemical equilibrium—the state that involves sameness and constancy.

15.2 The Rate of a Chemical Reaction

Reaction rates are related to chemical equilibrium because, as we will see in Section 15.3, a chemical system is at equilibrium when the rate of the forward reaction equals the rate of the reverse reaction.

Before we probe more deeply into the concept of chemical equilibrium, we must first understand something about the rates of chemical reactions. The **rate of a chemical reaction**—a measure of how fast the reaction proceeds—is defined as the amount of reactant that changes to product in a given period of time. A reaction with a fast rate proceeds quickly; a large amount of reactant is converted to product in a certain period of time (▶ Figure 15.1a). A reaction with a slow rate proceeds slowly; only a small a small amount of reactant is converted to product in the same period of time (▶ Figure 15.1b).

A reaction rate can also be defined as the amount of a product that forms in a given period of time.

Chemists seek to control reaction rates for many chemical reactions. For example, the space shuttle is propelled by the reaction of hydrogen and oxygen to form water. If the reaction proceeds too slowly, the shuttle will not lift off the ground. If, however, the reaction proceeds too quickly, the shuttle can explode. Reaction rates can be controlled if we understand the factors that influence them.

COLLISION THEORY

According to **collision theory**, chemical reactions occur through collisions between molecules or atoms. For example, consider the gas-phase chemical reaction between $H_2(g)$ and $I_2(g)$ to form $HI(g)$.

$$H_2(g) + I_2(g) \longrightarrow 2\ HI(g)$$

The gas-phase reaction between hydrogen and iodine can proceed by other mechanisms, but this mechanism is valid for the low-temperature thermal reaction.

Whether a collision leads to a reaction also depends on the *orientation* of the colliding molecules, but this topic is beyond the scope of this text.

The reaction begins when an H_2 molecule collides with an I_2 molecule. If the collision occurs with enough energy—that is, if the colliding molecules are moving fast enough—the reaction can proceed to form the products. If the collision occurs with insufficient energy, the reactant molecules (H_2 and I_2) simply bounce off of one another. Since gas-phase molecules have a wide distribution of velocities, collisions occur with a wide distribution of energies. High-energy collisions lead to products, and low-energy collisions do not.

Higher-energy collisions are more likely to lead to products because most chemical reactions have an *activation energy* (or an activation barrier). The activation energy for chemical reactions is discussed in more detail in Section 15.12. For now, we can think of the activation energy as an energy barrier that must be overcome for the reaction to proceed. For example, in the case of H_2 reacting with I_2 to form HI, the product (HI) can begin to form only after the H—H bond and the I—I bond each begin to break. The activation energy is the energy required to begin to break these bonds.

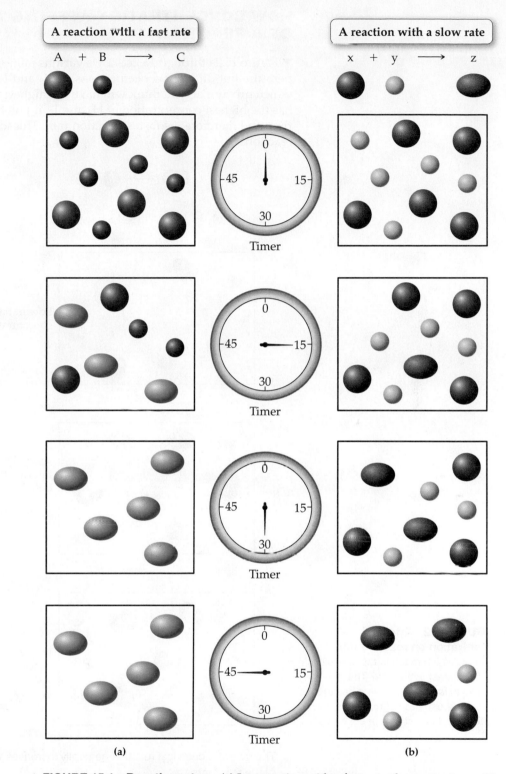

▲ **FIGURE 15.1   Reaction rates   (a)** In a reaction with a fast rate, the reactants react to form products in a short period of time. **(b)** In a reaction with a slow rate, the reactants react to form products over a long period of time.

If molecules react via high-energy collisions, then the factors that influence the rate of a reaction must be the same factors that affect the number of high-energy collisions that occur per unit time. Here, we focus on the two most important factors: the *concentration* of the reacting molecules and the *temperature* of the reaction mixture.

HOW CONCENTRATION AFFECTS THE RATE OF A REACTION

▼ Figure 15.2a through 15.2c shows various mixtures of H_2 and I_2 at the same temperature but different concentrations. If H_2 and I_2 react via collisions to form HI, which mixture do you think will have the highest reaction rate? Since Figure 15.2c has the highest concentration of H_2 and I_2, it will have the most collisions per unit time and therefore the fastest reaction rate. This idea holds true for most chemical reactions.

$$H_2(g) + I_2(g) \longrightarrow 2\,HI(g)$$

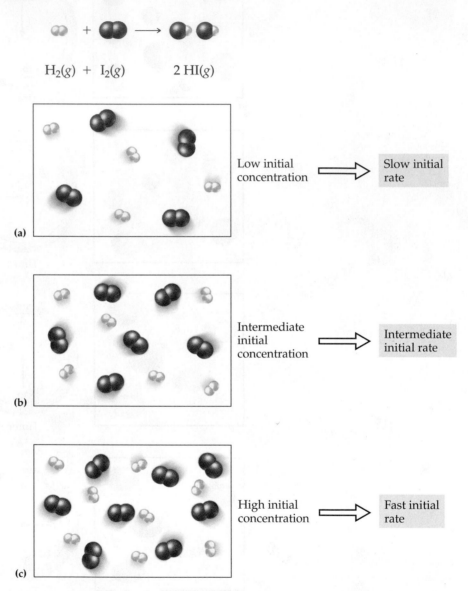

(a) Low initial concentration ⟹ Slow initial rate

(b) Intermediate initial concentration ⟹ Intermediate initial rate

(c) High initial concentration ⟹ Fast initial rate

▶ **FIGURE 15.2 Effect of concentration on reaction rate**
Question: Which reaction mixture will have the fastest initial rate? The mixture in (c) is fastest because it has the highest concentration of reactants and therefore the highest rate of collisions.

The rate of a chemical reaction generally increases with increasing concentration of the reactants.

The exact relationship between increases in concentration and increases in reaction rate varies for different reactions and is beyond the scope of this text. For our purposes, it will suffice to know that for most reactions the reaction rate increases with increasing reactant concentration.

Armed with this knowledge, what can we say about the rate of a reaction as the reaction proceeds? Since reactants turn into products in the course of a reaction, their concentration decreases. Consequently, the reaction rate decreases as well. In other words, as a reaction proceeds, there are fewer reactant molecules (because they have turned into products), and the reaction slows down.

HOW TEMPERATURE AFFECTS THE RATE OF A REACTION

Reaction rates also depend on temperature. ▼ Figure 15.3a through 15.3c show various mixtures of H_2 and I_2 at the same concentration, but different temperatures. Which will have the fastest rate? Raising the temperature makes the molecules move faster (Section 3.10). They therefore experience more collisions per unit time, resulting in a faster reaction rate. In addition, a higher temperature results in more collisions that are (on average) of higher energy. Since it is the high-energy collisions that result in products, this also produces a faster rate. Consequently, Figure 15.3c (which has the highest temperature) has the fastest reaction rate. This relationship holds true for most chemical reactions.

$H_2(g) + I_2(g) \longrightarrow 2 HI(g)$

(a) Low temperature ⟹ Slow rate

(b) Intermediate temperature ⟹ Intermediate rate

(c) High temperature ⟹ Fast rate

▶ **FIGURE 15.3 Effect of temperature on reaction rate** Question: Which reaction mixture will have the fastest initial rate? The mixture in (c) is fastest because it has the highest temperature.

The rate of a chemical reaction generally increases with increasing temperature of the reaction mixture.

The temperature dependence of reaction rates is the reason that cold-blooded animals become more sluggish at lower temperatures. The reactions required for them to think and move become slower, resulting in the sluggish behavior.

To summarize:

- Reaction rates generally increase with increasing reactant concentration.
- Reaction rates generally increase with increasing temperature.
- Reaction rates generally decrease as a reaction proceeds.

✓ **CONCEPTUAL CHECKPOINT 15.1**

In a chemical reaction between two gases, you would expect that increasing the pressure of the gases would probably

(a) increase the reaction rate.

(b) decrease the reaction rate.

(c) not affect the reaction rate.

15.3 The Idea of Dynamic Chemical Equilibrium

What would happen if our reaction between H_2 and I_2 to form HI were able to proceed in both the forward and reverse directions?

$$H_2(g) + I_2(g) \rightleftharpoons 2HI(g)$$

Now, H_2 and I_2 collide and react to form 2 HI molecules, but the 2 HI molecules also collide and react to reform H_2 and I_2. A reaction that can proceed in both the forward and reverse directions is a **reversible reaction**.

Suppose we begin with only H_2 and I_2 in a container (▶ Figure 15.4a). What happens initially? The H_2 and I_2 molecules begin to react to form HI (Figure 15.4b). However, as H_2 and I_2 react, their concentration decreases, which in turn decreases the rate of the forward reaction. At the same time, HI begins to form. As the concentration of HI increases, the reverse reaction begins to occur at an increasingly faster rate because there are more HI collisions with other HI molecules. Eventually, the rate of the reverse reaction (which is increasing) equals the rate of the forward reaction (which is decreasing). At that point, **dynamic equilibrium** is reached (Figure 15.4c and Figure 15.4d).

> **Dynamic equilibrium**—In a chemical reaction, the condition in which the rate of the forward reaction equals the rate of the reverse reaction.

This condition is not static—it is dynamic because the forward and reverse reactions are still occurring but at the same constant rate. When dynamic equilibrium is reached, the concentrations of H_2, I_2, and HI no longer change. They remain the same because the reactants and products are being depleted at the same rate at which they are being formed.

Notice that dynamic equilibrium includes the concepts of sameness and constancy that we discussed in Section 15.1. When dynamic equilibrium is reached, the forward reaction rate is the same as the reverse reaction rate (sameness). Because the reaction rates are the same, the concentrations of the reactants and products no longer change (constancy). However, just because the concentrations of reactants and products no longer change at equilibrium does *not* imply that the concentrations of reactants and products are *equal* to one another at equilibrium. Some reactions reach equilibrium only after most of the reactants have formed products. (Recall strong acids from Chapter 14.) Others reach equilibrium when only a small fraction of the reactants have formed products. (Recall weak acids from Chapter 14.) It depends on the reaction.

A reversible reaction

$$H_2(g) + I_2(g) \rightleftharpoons 2\,HI(g)$$

(a) 8 H_2; 8 I_2; 0 HI

(b) 6 H_2; 6 I_2; 4 HI

Equilibrium

(c) 4 H_2; 4 I_2; 8 HI

(d) 4 H_2; 4 I_2; 8 HI

▶ **FIGURE 15.4   Equilibrium**
When the concentrations of the reactants and products no longer change, equilibrium has been reached.

Narnia is the fictitious world featured in C.S. Lewis's *The Chronicles of Narnia*, and Middle Earth is the fictitious world featured in J.R.R. Tolkien's *The Lord of the Rings*.

We can better understand dynamic equilibrium with a simple analogy. Imagine that Narnia and Middle Earth are two neighboring kingdoms (▶ Figure 15.5). Narnia is overpopulated, and Middle Earth is underpopulated. One day, however, the border between the two kingdoms opens, and people immediately begin to leave Narnia for Middle Earth (call this the forward reaction).

$$\text{Narnia} \longrightarrow \text{Middle Earth} \quad \text{(forward reaction)}$$

Initial

 Represents population

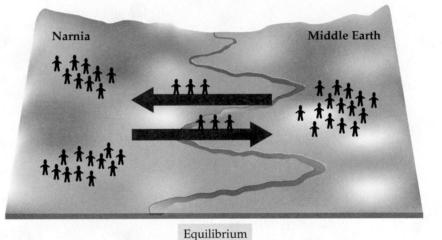

Equilibrium

▶ **FIGURE 15.5 Population analogy for a chemical reaction proceeding to equilibrium**

The population of Narnia decreases as the population of Middle Earth increases. As people leave Narnia, however, the *rate* at which they leave begins to slow down (because Narnia becomes less crowded). On the other hand, as people move into Middle Earth, some decide it was not for them and begin to move back (call this the reverse reaction).

$$\text{Narnia} \longleftarrow \text{Middle Earth} \quad \text{(reverse reaction)}$$

As Middle Earth fills, the rate of people moving back to Narnia accelerates. Eventually, the *rate* of people moving out of Narnia (which has been slowing down as people leave) equals the *rate* of people moving back to Narnia (which has been increasing as Middle Earth gets more crowded). Dynamic equilibrium has been reached.

$$\text{Narnia} \rightleftharpoons \text{Middle Earth}$$

Notice that when the two kingdoms reach dynamic equilibrium, their populations no longer change because the number of people moving out equals the number of people moving in. However, one kingdom—because of its charm, or the character of its leader, or a lower tax rate, or whatever other reason—may have a higher population than the other kingdom, even when dynamic equilibrium is reached.

Similarly, when a chemical reaction reaches dynamic equilibrium, the rate of the forward reaction (analogous to people moving out of Narnia) equals the rate of the reverse reaction (analogous to people moving back into Narnia), and the relative concentrations of reactants and products (analogous to the relative populations of the two kingdoms) become constant. Also, like our two kingdoms, the concentrations of reactants and products will not necessarily be equal at equilibrium, just as the populations of the two kingdoms are not equal at equilibrium.

15.4 The Equilibrium Constant: A Measure of How Far a Reaction Goes

We have just learned that the *concentrations* of reactants and products are not equal at equilibrium; rather, it is the *rates* of the forward and reverse reactions that are equal. But what about the concentrations? What can we know about them? The equilibrium constant (K_{eq}) is a way to quantify the relative concentrations of the reactants and products at equilibrium. Consider the generic chemical reaction:

$$aA + bB \rightleftharpoons cC + dD$$

where A and B are reactants, C and D are products, and a, b, c, and d are the respective stoichiometric coefficients in the chemical equation. The **equilibrium constant (K_{eq})** for the reaction is defined as the ratio—at equilibrium—of the concentrations of the products raised to their stoichiometric coefficients divided by the concentrations of the reactants raised to their stoichiometric coefficients.

$$K_{eq} = \frac{[C]^c [D]^d}{[A]^a [B]^b}$$

Products

Reactants

Notice that the equilibrium constant is a measure of the relative concentrations of reactants and products at equilibrium; the larger the equilibrium constant, the greater the concentration of products relative to reactants at equilibrium.

WRITING EQUILIBRIUM EXPRESSIONS FOR CHEMICAL REACTIONS

To write an equilibrium expression for a chemical reaction, examine the chemical equation and follow the preceding definition. For example, suppose we want to write an equilibrium expression for the reaction:

$$2\,N_2O_5(g) \rightleftharpoons 4\,NO_2(g) + O_2(g)$$

The equilibrium constant is $[NO_2]$ raised to the fourth power multiplied by $[O_2]$ raised to the first power divided by $[N_2O_5]$ raised to the second power.

$$K_{eq} = \frac{[NO_2]^4[O_2]}{[N_2O_5]^2}$$

Notice that the *coefficients* in the chemical equation become the *exponents* in the equilibrium expression.

$$2\,N_2O_5(g) \rightleftharpoons 4\,NO_2(g) + O_2(g)$$

Implied 1

$$K_{eq} = \frac{[NO_2]^4[O_2]}{[N_2O_5]^2}$$

EXAMPLE 15.1   Writing Equilibrium Expressions for Chemical Reactions

Write an equilibrium expression for the chemical equation:

$$CO(g) + 2H_2(g) \rightleftharpoons CH_3OH(g)$$

SOLUTION

The equilibrium expression is the concentration of the products raised to their stoichiometric coefficients divided by the concentration of the reactants raised to their stoichiometric coefficients. Notice that the expression is a ratio of products over reactants. Notice also that the coefficients in the chemical equation are the exponents in the equilibrium expression.

$$K_{eq} = \frac{[CH_3OH]}{[CO][H_2]^2}$$

Product ⟶ [CH₃OH]
Reactants ⟵ [CO][H₂]²

▶**SKILLBUILDER 15.1 | Writing Equilibrium Expressions for Chemical Reactions**

Write an equilibrium expression for the chemical equation:

$$H_2(g) + F_2(g) \rightleftharpoons 2HF(g)$$

▶**FOR MORE PRACTICE**   Example 15.10; Problems 43, 44.

THE SIGNIFICANCE OF THE EQUILIBRIUM CONSTANT

Given this definition of an equilibrium constant, what does it tell us? For example, what does a large equilibrium constant ($K_{eq} \gg 1$) imply about a reaction? It means that the forward reaction is largely favored and that there will be more products than reactants when equilibrium is reached. For example, consider the reaction:

$$H_2(g) + Br_2(g) \rightleftharpoons 2HBr(g) \quad K_{eq} = 1.9 \times 10^{19} \text{ at } 25\,°C$$

The equilibrium constant is large, meaning that at equilibrium the reaction lies far to the right—high concentrations of products, tiny concentrations of reactants (▼ Figure 15.6).

$$H_2(g) + Br_2(g) \rightleftharpoons 2HBr(g)$$

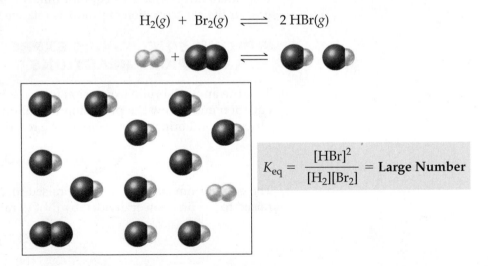

▶ **FIGURE 15.6   The meaning of a large equilibrium constant**   A large equilibrium constant means that there will be a high concentration of products and a low concentration of reactants at equilibrium.

$$K_{eq} = \frac{[HBr]^2}{[H_2][Br_2]} = \textbf{Large Number}$$

Conversely, what does a *small* equilibrium constant ($K_{eq} \ll 1$) mean? It means that the reverse reaction is favored and that there will be more reactants than products when equilibrium is reached. For example, consider the reaction:

$$N_2(g) + O_2(g) \rightleftharpoons 2NO(g) \quad K_{eq} = 4.1 \times 10^{-31} \text{ at } 25\,°C$$

The equilibrium constant is very small, meaning that at equilibrium the reaction lies far to the left—high concentrations of reactants, low concentrations of products

$$N_2(g) + O_2(g) \rightleftharpoons 2\,NO(g)$$

$$K_{eq} = \frac{[NO]^2}{[N_2][O_2]} = \textbf{Small Number}$$

▲ **FIGURE 15.7   The meaning of a small equilibrium constant**   A small equilibrium constant means that there will be a high concentration of reactants and a low concentration of products at equilibrium.

(▲ Figure 15.7). This is fortunate because N_2 and O_2 are the main components of air. If this equilibrium constant were large, much of the N_2 and O_2 in air would react to form NO, a toxic gas.

To summarize:

- $K_{eq} \ll 1$ Reverse reaction is favored; forward reaction does not proceed very far.
- $K_{eq} \approx 1$ Neither direction is favored; forward reaction proceeds about halfway (significant amounts of both reactants and products are present at equilibrium).
- $K_{eq} \gg 1$ Forward reaction is favored; forward reaction proceeds virtually to completion.

The symbol $\approx$ means "approximately equal to."

✓ CONCEPTUAL CHECKPOINT 15.2

Consider a generic chemical reaction in which the reactant, A, simply turns directly into the product, B.

$$A(g) \rightleftharpoons B(g)$$

The reaction is allowed to come to equilibrium at three different temperatures as represented below. At which temperature is the equilibrium constant the largest?

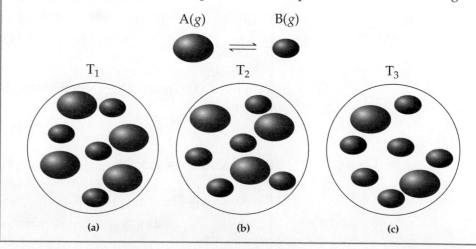

15.5 Heterogeneous Equilibria: The Equilibrium Expression for Reactions Involving a Solid or a Liquid

Consider the chemical reaction:

$$2\,CO(g) \rightleftharpoons CO_2(g) + C(s)$$

We might expect the expression for the equilibrium constant to be:

$$K_{eq} = \frac{[CO_2][C]}{[CO]^2} \quad \text{(incorrect)}$$

However, since carbon is a solid, its concentration is constant—it does not change. Adding more or less carbon to the reaction mixture does not change the concentration of carbon. The concentration of a solid does not change because a solid does not expand to fill its container. The concentration of a solid, therefore, depends only on its density, which (except for slight variations due to temperature) is constant as long as *some solid is present*. Consequently, pure solids—those reactants or products labeled in the chemical equation with an (*s*)—are not included in the equilibrium expression. The correct equilibrium expression is:

> The concentrations of pure solids and pure liquids are excluded from equilibrium expressions because they are constant.

$$K_{eq} = \frac{[CO_2]}{[CO]^2} \quad \text{(correct)}$$

Similarly, the concentration of a pure liquid does not change. Consequently, pure liquids—those reactants or products labeled in the chemical equation with an (*l*)—are also excluded from the equilibrium expression. For example, what is the equilibrium expression for the following reaction?

$$CO_2(g) + H_2O(l) \rightleftharpoons H^+(aq) + HCO_3^-(aq)$$

Since $H_2O(l)$ is pure liquid, it is omitted from the equilibrium expression.

$$K_{eq} = \frac{[H^+][HCO_3^-]}{[CO_2]}$$

EXAMPLE 15.2 Writing Equilibrium Expressions for Reactions Involving a Solid or a Liquid

Write an equilibrium expression for the chemical equation:

$$CaCO_3(s) \rightleftharpoons CaO(s) + CO_2(g)$$

SOLUTION

Since $CaCO_3(s)$ and $CaO(s)$ are both solids, they are omitted from the equilibrium expression.

$$K_{eq} = [CO_2]$$

▶SKILLBUILDER 15.2 | Writing Equilibrium Expressions for Reactions Involving a Solid or a Liquid

Write an equilibrium expression for the chemical equation:

$$4\,HCl(g) + O_2(g) \rightleftharpoons 2\,H_2O(l) + 2\,Cl_2(g)$$

▶FOR MORE PRACTICE Problems 45, 46.

15.6   Calculating and Using Equilibrium Constants

CALCULATING EQUILIBRIUM CONSTANTS

The most direct way to obtain a value for the equilibrium constant of a reaction is to measure the concentrations of the reactants and products in a reaction mixture at equilibrium. For example, consider the reaction:

$$H_2(g) + I_2(g) \rightleftharpoons 2\,HI(g)$$

Suppose a mixture of H_2 and I_2 is allowed to come to equilibrium at 445 °C. The measured equilibrium concentrations are $[H_2] = 0.11\,M$, $[I_2] = 0.11\,M$, and $[HI] = 0.78\,M$. What is the value of the equilibrium constant? We begin by sorting the information in the problem statement.

GIVEN: $[H_2] = 0.11\,M$
$[I_2] = 0.11\,M$
$[HI] = 0.78\,M$

FIND: K_{eq}

SOLUTION
The expression for K_{eq} can be written from the balanced equation.

$$K_{eq} = \frac{[HI]^2}{[H_2][I_2]}$$

To calculate the value of K_{eq}, substitute the correct equilibrium concentrations into the expression for K_{eq}.

$$\begin{aligned}K_{eq} &= \frac{[HI]^2}{[H_2][I_2]} \\ &= \frac{[0.78]^2}{[0.11][0.11]} \\ &= 5.0 \times 10^1\end{aligned}$$

> Equilibrium constants depend on temperature, so temperatures will often be included with equilibrium data. However, the temperature is not a part of the equilibrium expression.

The concentrations within K_{eq} must always be written in moles per liter (M); however, the units are normally dropped in expressing the equilibrium constant.

The *particular concentrations* of reactants and products for a reaction at equilibrium will *not* always be the same for a given reaction; they will depend on the initial concentrations. However, the *equilibrium constant* will always be the same at a given temperature, regardless of the initial concentrations. For example, Table 15.1

> The concentrations in an equilibrium expression should always be in units of molarity (M), but the units themselves are normally dropped.

TABLE 15.1   Initial and Equilibrium Concentrations at 445 °C for the Reaction $H_2(g) + I_2(g) \rightleftharpoons 2\,HI(g)$.

Initial			Equilibrium			Equilibrium Constant
$[H_2]$	$[I_2]$	$[HI]$	$[H_2]$	$[I_2]$	$[HI]$	$K_{eq} = \dfrac{[HI]^2}{[H_2][I_2]}$
0.50	0.50	0.0	0.11	0.11	0.78	$\dfrac{[0.78]^2}{[0.11][0.11]} = 50$
0.0	0.0	0.50	0.055	0.055	0.39	$\dfrac{[0.39]^2}{[0.055][0.055]} = 50$
0.50	0.50	0.50	0.165	0.165	1.17	$\dfrac{[1.17]^2}{[0.165][0.165]} = 50$
1.0	0.5	0.0	0.53	0.033	0.934	$\dfrac{[0.934]^2}{[0.53][0.033]} = 50$

A reaction can approach equilibrium from either direction, depending on the initial concentrations, but its K_{eq} at a given temperature will always be the same.

shows several different equilibrium concentrations of H_2, I_2, and HI, each from a different set of initial concentrations. Notice that the equilibrium constant is always the same, regardless of the initial concentrations. In other words, no matter what the initial concentrations are, the reaction will always go in a direction so that the equilibrium concentrations—when substituted into the equilibrium expression—give the same constant, K_{eq}.

EXAMPLE 15.3 Calculating Equilibrium Constants

Consider the reaction:

$$2\,CH_4(g) \rightleftharpoons C_2H_2(g) + 3\,H_2(g)$$

A mixture of CH_4, C_2H_2, and H_2 is allowed to come to equilibrium at 1700 °C. The measured equilibrium concentrations are $[CH_4] = 0.0203$ M, $[C_2H_2] = 0.0451$ M, and $[H_2] = 0.112$ M. What is the value of the equilibrium constant at this temperature?

You are given the concentrations of the reactants and products of a reaction at equilibrium. You are asked to find the equilibrium constant.	**GIVEN:** $[CH_4] = 0.0203$ M $[C_2H_2] = 0.0451$ M $[H_2] = 0.112$ M **FIND:** K_{eq}
Write the expression for K_{eq} from the balanced equation. To calculate the value of K_{eq}, substitute the correct equilibrium concentrations into the expression for K_{eq}.	**SOLUTION** $K_{eq} = \dfrac{[C_2H_2][H_2]^3}{[CH_4]^2}$ $K_{eq} = \dfrac{[0.0451][0.112]^3}{[0.0203]^2}$ $= 0.154$

▶SKILLBUILDER 15.3 | Calculating Equilibrium Constants

Consider the reaction:

$$CO(g) + 2\,H_2(g) \rightleftharpoons CH_3OH(g)$$

A mixture of CO, H_2, and CH_3OH is allowed to come to equilibrium at 225 °C. The measured equilibrium concentrations are $[CO] = 0.489$ M, $[H_2] = 0.146$ M, and $[CH_3OH] = 0.151$ M. What is the value of the equilibrium constant at this temperature?

▶SKILLBUILDER PLUS

Suppose that the preceding reaction is carried out at a different temperature and that the initial concentrations of the reactants are $[CO] = 0.500$ M and $[H_2] = 1.00$ M. Assuming that there is no product at the beginning of the reaction and that at equilibrium $[CO] = 0.15$ M, find the equilibrium constant at this new temperature. *Hint*: Use the stoichiometric relationships from the balanced equation to find the equilibrium concentrations of H_2 and CH_3OH.

▶FOR MORE PRACTICE Example 15.11; Problems 51, 52, 53, 54, 55, 56.

USING EQUILIBRIUM CONSTANTS IN CALCULATIONS

We can also use the equilibrium constant to calculate the equilibrium concentration of one of the reactants or products, given the equilibrium concentrations of the others. For example, consider the reaction:

$$2\,COF_2(g) \rightleftharpoons CO_2(g) + CF_4(g) \quad K_{eq} = 2.00 \text{ at } 1000\,°C$$

In an equilibrium mixture, the concentration of COF_2 is 0.255 M and the concentration of CF_4 is 0.118 M. What is the equilibrium concentration of CO_2? We begin by sorting the information in the problem statement.

GIVEN: $[COF_2] = 0.255$ M

$[CF_4] = 0.118$ M

$K_{eq} = 2.00$

FIND: $[CO_2]$

SOLUTION MAP

We then draw a solution map showing how the expression for the equilibrium constant provides the equation that gets us from the given quantities to the quantity we are trying to find.

$$K_{eq} = \frac{[CO_2][CF_4]}{[COF_2]^2}$$

SOLUTION

We write the equilibrium expression for the reaction, and then we solve it for the quantity we are trying to find ($[CO_2]$).

$$K_{eq} = \frac{[CO_2][CF_4]}{[COF_2]^2}$$

$$[CO_2] = K_{eq}\frac{[COF_2]^2}{[CF_4]}$$

Now we substitute the appropriate values and calculate $[CO_2]$.

$$[CO_2] = 2.00\frac{[0.255]^2}{[0.118]}$$

$$= 1.10 \text{ M}$$

EXAMPLE 15.4   Using Equilibrium Constants in Calculations

Consider the reaction:

$$H_2(g) + I_2(g) \rightleftharpoons 2\,HI(g) \quad K_{eq} = 69 \text{ at } 340\,°C$$

In an equilibrium mixture, the concentrations of H_2 and I_2 are both 0.020 M. What is the equilibrium concentration of HI?

SORT	**GIVEN:** $[H_2] = [I_2] = 0.020$ M
You are given the equilibrium concentrations of the reactants in a chemical reaction and also the value of the equilibrium constant. You are asked to find the concentration of the product.	$K_{eq} = 69$ **FIND:** $[HI]$

STRATEGIZE

Draw a solution map showing how the equilibrium constant expression gives the relationship between the given concentrations and the concentration you are asked to find.

SOLUTION MAP

$$[H_2], [I_2], K_{eq} \longrightarrow [HI]$$

$$K_{eq} = \frac{[HI]^2}{[H_2][I_2]}$$

SOLVE

Solve the equilibrium expression for [HI] and then substitute in the appropriate values to calculate it.

Since the value of [HI] is squared, you must take the square root of both sides of the equation to solve for [HI] because $\sqrt{[HI]^2} = [HI]$.

SOLUTION

$$K_{eq} = \frac{[HI]^2}{[H_2][I_2]}$$

$$[HI]^2 = K_{eq}[H_2][I_2]$$

$$\sqrt{[HI]^2} = \sqrt{K_{eq}[H_2][I_2]}$$

$$[HI] = \sqrt{69(0.020)(0.020)}$$

$$= 0.17 \text{ M}$$

CHECK

You can check your answer by substituting it back into the expression for K_{eq}.

$$K_{eq} = \frac{[HI]^2}{[H_2][I_2]}$$

$$= \frac{(0.17)^2}{(0.020)(0.020)}$$

$$= 72$$

The calculated value of K_{eq} is about equal to the given value of K_{eq} (which was 69), indicating that your answer is correct. The slight difference is due to rounding error, which is common in problems like these.

▶SKILLBUILDER 15.4 | Using Equilibrium Constants in Calculations

Diatomic iodine (I_2) decomposes at high temperature to form I atoms according to the reaction:

$$I_2(g) \rightleftharpoons 2\,I(g) \quad K_{eq} = 0.011 \text{ at } 1200\,°C$$

In an equilibrium mixture, the concentration of I_2 is 0.10 M. What is the equilibrium concentration of I?

▶FOR MORE PRACTICE   Example 15.12; Problems 57, 58, 59, 60.

✓ **CONCEPTUAL CHECKPOINT 15.3**

When the reaction $A(aq) \rightleftharpoons B(aq) + C(aq)$ is at equilibrium, each of the three compounds has a concentration of 2 M. The equilibrium constant for this reaction is:

(a) 4   **(b)** 2   **(c)** 1   **(d)** 1/2

15.7 Disturbing a Reaction at Equilibrium: Le Châtelier's Principle

| Pronounced "le-sha-te-lyay"

We have seen that a chemical system not in equilibrium tends to go toward equilibrium and that the concentrations of the reactants and products at equilibrium correspond to the equilibrium constant, K_{eq}. What happens, however, when a chemical system already at equilibrium is disturbed? **Le Châtelier's principle** states that the chemical system will respond to minimize the disturbance.

Le Châtelier's principle—When a chemical system at equilibrium is disturbed, the system shifts in a direction that minimizes the disturbance.

In other words, a system at equilibrium tries to maintain that equilibrium—it fights back when disturbed.

We can understand Le Châtelier's principle by returning to our Narnia and Middle Earth analogy. Suppose the populations of Narnia and Middle Earth are at equilibrium. This means that the rate of people moving out of Narnia (and into Middle Earth) is equal to the rate of people moving into Narnia (and out of Middle Earth). It also means that the populations of the two kingdoms are stable. Now imagine disturbing that balance (▶ Figure 15.8). Suppose we add extra people to Middle Earth. What happens? Since Middle Earth suddenly becomes more crowded, the rate of people leaving Middle Earth increases. The net flow of people is *out of Middle Earth and into Narnia*. Notice what happened. We disturbed the equilibrium by adding more people to Middle Earth. The system responded by moving people out of Middle Earth—it shifted in the direction that minimized the disturbance.

On the other hand, what happens if we add extra people to Narnia? Since Narnia suddenly gets more crowded, the rate of people leaving Narnia goes up. The net flow of people is out of Narnia and into Middle Earth. We added people to Narnia,

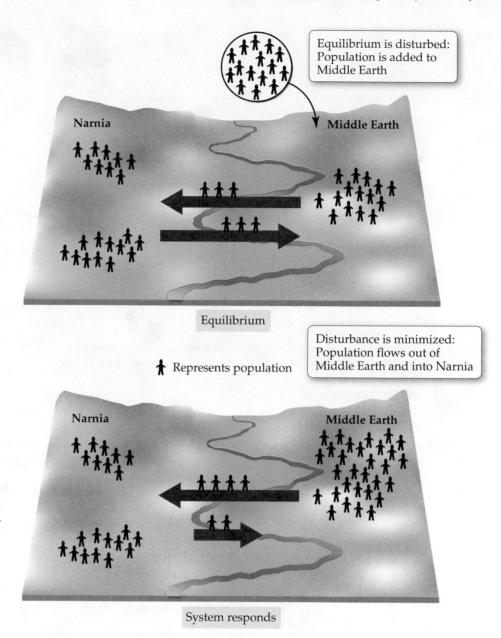

Equilibrium is disturbed:
Population is added to
Middle Earth

Narnia

Middle Earth

Equilibrium

Disturbance is minimized:
Population flows out of
Middle Earth and into Narnia

Represents population

Narnia

Middle Earth

System responds

▶ **FIGURE 15.8 Population analogy for Le Châtelier's principle** When a system at equilibrium is disturbed, it shifts to minimize the disturbance. In this case, adding population to Middle Earth (the disturbance) causes population to move out of Middle Earth (minimizing the disturbance). **Question:** What would happen if you disturbed the equilibrium by taking population out of Middle Earth? In which direction would the population move to minimize the disturbance?

and the system responded by moving people out of Narnia. When systems at equilibrium are disturbed, they react to counter the disturbance. Chemical systems behave similarly. There are several ways to disturb a system in chemical equilibrium. We consider each of these separately in the next three sections of the chapter.

15.8   The Effect of a Concentration Change on Equilibrium

Consider the following reaction at chemical equilibrium:

$$N_2O_4(g) \rightleftharpoons 2\,NO_2(g)$$

Suppose we disturb the equilibrium by adding NO_2 to the equilibrium mixture (▶ Figure 15.9). In other words, we increase the concentration of NO_2. What happens? According to Le Châtelier's principle, the system shifts in a direction to minimize the disturbance. The shift is caused by the increased concentration of NO_2, which in turn increases the rate of the reverse reaction because reaction rates generally increase with increasing concentration (as we discussed in Section 15.2).

$$N_2O_4(g) \rightleftharpoons 2\,NO_2(g)$$

Equilibrium
disturbed

$$N_2O_4(g) \rightleftharpoons 2\,NO_2(g)$$

Add NO₂

$$N_2O_4(g) \rightleftharpoons 2\,NO_2(g)$$

System shifts left

▶ **FIGURE 15.9 Le Châtelier's principle in action I:** When a system at equilibrium is disturbed, it changes to minimize the disturbance. In this case, adding NO_2 (the disturbance) causes the reaction to shift left, consuming NO_2 by forming more N_2O_4.

> When we say that a reaction *shifts to the left* we mean that it proceeds in the reverse direction, consuming products and forming reactants.

The reaction shifts to the left (it proceeds in the reverse direction), consuming some of the added NO_2 and bringing its concentration back down.

$$N_2O_4(g) \rightleftharpoons 2\,NO_2(g)$$

Reaction shifts left   Add NO_2

On the other hand, what happens if we add extra N_2O_4, increasing its concentration? In this case, the rate of the forward reaction increases and the reaction shifts to the right, consuming some of the added N_2O_4 and bringing *its* concentration back down (▶ Figure 15.10).

$$N_2O_4(g) \rightleftharpoons 2\,NO_2(g)$$

Add N_2O_4 Reaction shifts right

> When we say that a reaction *shifts to the right*, we mean that it proceeds in the forward direction, consuming reactants and forming products.

In each case, the system shifts in a direction that minimizes the disturbance.

To summarize, if a chemical system is at equilibrium:

- Increasing the concentration of one or more of the reactants causes the reaction to shift to the right (in the direction of the products).
- Increasing the concentration of one or more of the products causes the reaction to shift to the left (in the direction of the reactants).

EXAMPLE 15.5 The Effect of a Concentration Change on Equilibrium

Consider the following reaction at equilibrium:

$$CaCO_3(s) \rightleftharpoons CaO(s) + CO_2(g)$$

What is the effect of adding additional CO_2 to the reaction mixture? What is the effect of adding additional $CaCO_3$?

SOLUTION

Adding more CO_2 increases the concentration of CO_2 and causes the reaction to shift to the left. Adding more $CaCO_3$ does not increase the concentration of $CaCO_3$ because $CaCO_3$ is a solid and thus has a constant concentration. It is therefore not included in the equilibrium expression and has no effect on the position of the equilibrium.

▶**SKILLBUILDER 15.5 | The Effect of a Concentration Change on Equilibrium**

Consider the following reaction in chemical equilibrium.

$$2\, BrNO(g) \rightleftharpoons 2\, NO(g) + Br_2(g)$$

What is the effect of adding more Br_2 to the reaction mixture? What is the effect of adding more BrNO?

▶**SKILLBUILDER PLUS**

What is the effect of removing some Br_2 from the preceding reaction mixture?

▶**FOR MORE PRACTICE**   Example 15.13a, b; Problems 63, 64, 65, 66.

▶ **FIGURE 15.10   Le Châtelier's principle in action II:**   When a system at equilibrium is disturbed, it changes to minimize the disturbance. In this case, adding N_2O_4 (the disturbance) causes the reaction to shift right, consuming N_2O_4 by producing more NO_2.

$N_2O_4(g) \rightleftharpoons 2\, NO_2(g)$

$N_2O_4(g) \rightleftharpoons 2\, NO_2(g)$
Add N_2O_4

Equilibrium disturbed

$N_2O_4(g) \rightleftharpoons 2\, NO_2(g)$
System shifts right

✔ **CONCEPTUAL CHECKPOINT 15.4**

Consider the equilibrium reaction between carbon monoxide and hydrogen gas to form methanol:

$$CO(g) + 2\, H_2(g) \rightleftharpoons CH_3OH(g)$$

Suppose you have a reaction mixture of these three substances at equilibrium. Which would cause a greater shift toward products: doubling the carbon monoxide concentration or doubling the hydrogen gas concentration?

15.9 The Effect of a Volume Change on Equilibrium

See Section 11.4 for a complete description of Boyle's law.

How does a system in chemical equilibrium respond to a volume change? Remember from Chapter 11 that changing the volume of a gas (or a gas mixture) results in a change in pressure. Remember also that pressure and volume are inversely related: a *decrease* in volume causes an *increase* in pressure, and an *increase* in volume causes a *decrease* in pressure. So, if the volume of a gaseous reaction mixture at chemical equilibrium is changed, the pressure changes and the system will shift in a direction to minimize that change. For example, consider the following reaction at equilibrium in a cylinder equipped with a moveable piston:

$$N_2(g) + 3 H_2(g) \rightleftharpoons 2 NH_3(g)$$

What happens if we push down on the piston, lowering the volume and raising the pressure (▶ Figure 15.11)? How can the chemical system bring the pressure back down? Look carefully at the reaction coefficients in the balanced equation. If the reaction shifts to the right, 4 mol of gas particles (1 mol of N_2 and 3 mol of H_2) are converted to 2 mol of gas particles (2 mol of NH_3). Therefore, as the reaction shifts toward products, the pressure is lowered (because the reaction mixture contains fewer gas particles). So the system then shifts to the right, bringing the pressure back down and minimizing the disturbance.

From the ideal gas law ($PV = nRT$), we can see that lowering the number of moles of a gas (n) results in a lower pressure (P) at constant temperature and volume.

Consider the same reaction mixture at equilibrium again. What happens if, this time, we pull *up* on the piston, *increasing* the volume (▶ Figure 15.12)? The higher volume results in a lower pressure, and the system responds by trying to bring the pressure back up. It can do this by shifting to the left, converting 2 mol of gas particles into 4 mol of gas particles. As the reaction shifts toward reactants, the pressure increases again (because the reaction mixture contains more gas particles) minimizing the disturbance.

To summarize, if a chemical system is at equilibrium:

- Decreasing the volume causes the reaction to shift in the direction that has fewer moles of gas particles.
- Increasing the volume causes the reaction to shift in the direction that has more moles of gas particles.

Notice that if a chemical reaction has an equal number of moles of gas particles on both sides of the chemical equation, a change in volume has no effect. For example, consider the following reaction:

$$H_2(g) + I_2(g) \rightleftharpoons 2 HI(g)$$

Both the left and the right sides of the equation contain 2 mol of gas particles, so a change in volume has no effect on this reaction. In addition, a change in volume has no effect on a reaction that has no gaseous reactants or products.

EXAMPLE 15.6 The Effect of a Volume Change on Equilibrium

Consider the reaction at chemical equilibrium:

$$2 KClO_3(s) \rightleftharpoons 2 KCl(s) + 3 O_2(g)$$

What is the effect of decreasing the volume of the reaction mixture? Increasing the volume of the reaction mixture?

SOLUTION

The chemical equation has 3 mol of gas on the right and 0 mol of gas on the left. Decreasing the volume of the reaction mixture increases the pressure and causes the reaction to shift to the left (toward the side with fewer moles of gas particles). Increasing the volume of the reaction mixture decreases the pressure and causes the reaction to shift to the right (toward the side with more moles of gas particles).

▶**SKILLBUILDER 15.6 | The Effect of a Volume Change on Equilibrium**

Consider the reaction at chemical equilibrium:

$$2\,SO_2(g) + O_2(g) \rightleftharpoons 2\,SO_3(g)$$

What is the effect of decreasing the volume of the reaction mixture? Increasing the volume of the reaction mixture?

▶**FOR MORE PRACTICE**   Example 15.13c; Problems 67, 68, 69, 70.

▲ **FIGURE 15.11   Effect of volume decrease on equilibrium**   When the volume of an equilibrium mixture decreases, the pressure increases. The system responds (to bring the pressure back down) by shifting to the right, the side of the reaction with the fewest moles of gas particles.

▲ **FIGURE 15.12   Effect of volume increase on equilibrium**   When the volume of an equilibrium mixture increases, the pressure decreases. The system responds (to raise the pressure) by shifting to the left, the side of the reaction with the most moles of gas particles.

CHEMISTRY AND HEALTH
How a Developing Fetus Gets Oxygen from Its Mother

Have you ever wondered how a baby in the womb gets oxygen? Unlike you and me, a fetus cannot breathe. Yet like you and me, a fetus needs oxygen. Where does that oxygen come from? In adults, oxygen is absorbed in the lungs and carried in the blood by a protein molecule called hemoglobin, which is abundantly present in red blood cells. Hemoglobin (Hb) reacts with oxygen according to the equilibrium equation:

$$Hb + O_2 \rightleftharpoons HbO_2$$

The equilibrium constant for this reaction is neither large nor small, but intermediate. Consequently, the reaction shifts toward the right or the left, depending on the concentration of oxygen. As blood flows through the lungs, where oxygen concentrations are high, the equilibrium shifts to the right—hemoglobin loads oxygen.

Lung [O_2] high
$$Hb + O_2 \rightleftharpoons HbO_2$$

Reaction shifts right

As blood flows through muscles and organs that are using oxygen (where oxygen concentrations have been depleted), the equilibrium shifts to the left—hemoglobin unloads oxygen.

Muscle [O_2] low
$$Hb + O_2 \rightleftharpoons HbO_2$$

Reaction shifts left

A fetus has its own blood circulatory system. The mother's blood never flows into the fetus's body, and the fetus cannot get any air in the womb. So how does the fetus get oxygen?

The answer lies in fetal hemoglobin (HbF), which is slightly different from adult hemoglobin. Like adult hemoglobin, fetal hemoglobin is in equilibrium with oxygen.

$$HbF + O_2 \rightleftharpoons HbFO_2$$

However, the equilibrium constant for fetal hemoglobin is larger than the equilibrium constant for adult hemoglobin. In other words, fetal hemoglobin will load oxygen at a lower oxygen concentration than adult hemoglobin. So, when the mother's hemoglobin flows through the placenta, it unloads oxygen into the placenta. The baby's blood also flows into the placenta, and even though the baby's blood never mixes with the mother's blood, the fetal hemoglobin within the baby's blood loads the oxygen (that the mother's hemoglobin unloaded) and carries it to the baby. Nature has thus engineered a chemical system where the mother's hemoglobin can in effect *hand off* oxygen to the baby's hemoglobin.

CAN YOU ANSWER THIS? *What would happen if fetal hemoglobin had the same equilibrium constant for the reaction with oxygen as adult hemoglobin?*

▲ A human fetus. **Question:** How does the fetus get oxygen?

✓ CONCEPTUAL CHECKPOINT 15.5

Consider the reaction:

$$H_2(g) + I_2(g) \rightleftharpoons 2\,HI(g)$$

Which change would cause the reaction to shift to the right (toward products)?

(a) decreasing the volume

(b) increasing the volume

(c) increasing the concentration of hydrogen gas

(d) decreasing the concentration of hydrogen gas

15.10 The Effect of a Temperature Change on Equilibrium

According to Le Châtelier's principle, if the temperature of a system at equilibrium is changed, the system should shift in a direction to counter that change. So if the temperature is increased, the reaction should shift in the direction that attempts to lower the temperature and vice versa. Recall from Section 3.9 that energy changes are often associated with chemical reactions. If we want to predict the direction in which a reaction will shift upon a temperature change, we must understand how a shift in the reaction affects the temperature.

In Section 3.9, we classify chemical reactions according to whether they absorb or emit heat energy in the course of the reaction. Recall that an *exothermic reaction* (one with a negative ΔH_{rxn}) emits heat.

Exothermic reaction: $A + B \rightleftharpoons C + D + Heat$

In an exothermic reaction, we can think of heat as a product. Consequently, raising the temperature of an exothermic reaction—think of this as adding heat—causes the reaction to shift left. For example, the reaction of nitrogen with hydrogen to form ammonia is exothermic.

$$N_2(g) + 3\,H_2(g) \rightleftharpoons 2\,NH_3(g) + Heat$$

Reaction shifts left   Add heat

Raising the temperature of an equilibrium mixture of these three gases causes the reaction to shift left, absorbing some of the added heat. Conversely, lowering the temperature of an equilibrium mixture of these three gases causes the reaction to shift right, releasing heat.

$$N_2(g) + 3\,H_2(g) \rightleftharpoons 2\,NH_3(g) + Heat$$

Reaction shifts right   Remove heat

In contrast, an *endothermic reaction* (one with a positive ΔH_{rxn}) absorbs heat.

Endothermic reaction: $A + B + Heat \rightleftharpoons C + D$

In an endothermic reaction, we can think of heat as a reactant. Consequently, raising the temperature (or adding heat) causes an endothermic reaction to shift right. For example, the following reaction is endothermic:

Colorless   Brown
$$N_2O_4(g) + Heat \rightleftharpoons 2\,NO_2(g)$$

Add heat   Reaction shifts right

Raising the temperature of an equilibrium mixture of these two gases causes the reaction to shift right, absorbing some of the added heat. Since N_2O_4 is colorless and NO_2 is brown, we can easily see the effects of changing the temperature of this reaction (▶ Figure 15.13). On the other hand, lowering the temperature of a reaction mixture of these two gases causes the reaction to shift left, releasing heat.

Colorless   Brown
$$N_2O_4(g) + Heat \rightleftharpoons 2\,NO_2(g)$$

Remove   Reaction shifts left
heat

▶ **FIGURE 15.13 Equilibrium as a function of temperature** Since the reaction $N_2O_4(g) \rightleftharpoons 2\,NO_2(g)$ is endothermic, warm temperatures **(a)** cause a shift to the right, toward the production of brown NO_2. Cool temperatures **(b)** cause a shift to the left, to colorless N_2O_4.

(a) Warm: NO_2

(b) Cool: N_2O_4

To summarize:

In an exothermic chemical reaction, heat is a product and:

- Increasing the temperature causes the reaction to shift left (in the direction of the reactants).
- Decreasing the temperature causes the reaction to shift right (in the direction of the products).

In an endothermic chemical reaction, heat is a reactant and:

- Increasing the temperature causes the reaction to shift right (in the direction of the products).
- Decreasing the temperature causes the reaction to shift left (in the direction of the reactants).

EXAMPLE 15.7 The Effect of a Temperature Change on Equilibrium

The following reaction is endothermic.

$$CaCO_3(s) \rightleftharpoons CaO(s) + CO_2(g)$$

What is the effect of increasing the temperature of the reaction mixture? Decreasing the temperature?

SOLUTION

Since the reaction is endothermic, we can think of heat as a reactant.

$$Heat + CaCO_3(s) \rightleftharpoons CaO(s) + CO_2(g)$$

Raising the temperature is adding heat, causing the reaction to shift to the right. Lowering the temperature is removing heat, causing the reaction to shift to the left.

▶**SKILLBUILDER 15.7 | The Effect of a Temperature Change on Equilibrium**

The following reaction is exothermic.

$$2\,SO_2(g) + O_2(g) \rightleftharpoons 2\,SO_3(g)$$

What is the effect of increasing the temperature of the reaction mixture? Decreasing the temperature?

▶**FOR MORE PRACTICE** Example 15.13d; Problems 71, 72, 73, 74.

✓ **CONCEPTUAL CHECKPOINT 15.6**

Consider the endothermic reaction:

$$Cl_2(g) \rightleftharpoons 2\ Cl(g)$$

If a reaction mixture is at equilibrium, which disturbances will increase the amount of product the most?

(a) increasing the temperature and increasing the volume

(b) increasing the temperature and decreasing the volume

(c) decreasing the temperature and increasing the volume

(d) decreasing the temperature and decreasing the volume

15.11 The Solubility-Product Constant

Recall from Section 7.5 that a compound is considered soluble if it dissolves in water and insoluble if it does not. Recall also that, by applying the *solubility rules* (Table 7.2), we can classify many ionic compounds as soluble or insoluble. We can better understand the solubility of an ionic compound with the concept of equilibrium. The process by which an ionic compound dissolves is an equilibrium process. For example, we can represent the dissolving of calcium fluoride in water with the chemical equation:

$$CaF_2(s) \rightleftharpoons Ca^{2+}(aq) + 2\ F^-(aq)$$

The equilibrium expression for a chemical equation that represents the dissolving of an ionic compound is the **solubility-product constant (K_{sp})**. For CaF_2, the solubility-product constant is:

$$K_{sp} = [Ca^{2+}][F^-]^2$$

Notice that, as we discussed in Section 15.5, solids are omitted from the equilibrium expression.

The K_{sp} value is therefore a measure of the solubility of a compound. A large K_{sp} (forward reaction favored) means that the compound is very soluble. A small K_{sp} (reverse reaction favored) means that the compound is not very soluble. Table 15.2 lists the value of K_{sp} for a number of ionic compounds.

TABLE 15.2 Selected Solubility-Product Constants (K_{sp})

Compound	Formula	K_{sp}
barium sulfate	$BaSO_4$	1.07×10^{-10}
calcium carbonate	$CaCO_3$	4.96×10^{-9}
calcium fluoride	CaF_2	1.46×10^{-10}
calcium hydroxide	$Ca(OH)_2$	4.68×10^{-6}
calcium sulfate	$CaSO_4$	7.10×10^{-5}
copper(II) sulfide	CuS	1.27×10^{-36}
iron(II) carbonate	$FeCO_3$	3.07×10^{-11}
iron(II) hydroxide	$Fe(OH)_2$	4.87×10^{-17}
lead(II) chloride	$PbCl_2$	1.17×10^{-5}
lead(II) sulfate	$PbSO_4$	1.82×10^{-8}
lead(II) sulfide	PbS	9.04×10^{-29}
magnesium carbonate	$MgCO_3$	6.82×10^{-6}
magnesium hydroxide	$Mg(OH)_2$	2.06×10^{-13}
silver chloride	$AgCl$	1.77×10^{-10}
silver chromate	Ag_2CrO_4	1.12×10^{-12}
silver iodide	AgI	8.51×10^{-17}

EXAMPLE 15.8 Writing Expressions for K_{sp}

Write expressions for K_{sp} for each ionic compound.

(a) $BaSO_4$　　　(b) $Mn(OH)_2$　　　(c) Ag_2CrO_4

SOLUTION

To write the expression for K_{sp}, first write the chemical reaction showing the solid compound in equilibrium with its dissolved aqueous ions. Then write the equilibrium expression based on this equation.

(a) $BaSO_4(s) \rightleftharpoons Ba^{2+}(aq) + SO_4^{2-}(aq)$

$K_{sp} = [Ba^{2+}][SO_4^{2-}]$

(b) $Mn(OH)_2(s) \rightleftharpoons Mn^{2+}(aq) + 2OH^-(aq)$

$K_{sp} = [Mn^{2+}][OH^-]^2$

(c) $Ag_2CrO_4(s) \rightleftharpoons 2Ag^+(aq) + CrO_4^{2-}(aq)$

$K_{sp} = [Ag^+]^2[CrO_4^{2-}]$

▶**SKILLBUILDER 15.8 | Writing Expressions for K_{sp}**

Write expressions for K_{sp} for each ionic compound.

(a) AgI　　(b) $Ca(OH)_2$

▶**FOR MORE PRACTICE** Example 15.14; Problems 77, 78.

USING K_{sp} TO DETERMINE MOLAR SOLUBILITY

Recall from Section 13.3 that the solubility of a compound is the amount of the compound that dissolves in a certain amount of liquid. The **molar solubility** is the solubility in units of moles per liter. The molar solubility of a compound can be calculated directly from K_{sp}. For example, consider silver chloride.

$$AgCl(s) \rightleftharpoons Ag^+(aq) + Cl^-(aq) \quad K_{sp} = 1.77 \times 10^{-10}$$

How can we find the molar solubility of AgCl from K_{sp}? First, notice that K_{sp} is *not* the molar solubility; it is the solubility-product constant.

Second, notice that the concentration of either Ag^+ or Cl^- at equilibrium will be equal to the amount of AgCl that dissolved. We know this from the relationship of the stoichiometric coefficients in the balanced equation.

$$1 \text{ mol AgCl} : 1 \text{ mol Ag}^+ : 1 \text{ mol Cl}^-$$

Consequently, to find the solubility, we simply need to find $[Ag^+]$ or $[Cl^-]$ at equilibrium. We can do this by writing the expression for the solubility-product constant.

$$K_{sp} = [Ag^+][Cl^-]$$

Because both Ag^+ and Cl^- come from AgCl, their concentrations must be equal. Since the solubility of AgCl is equal to the equilibrium concentration of either dissolved ion, we write:

$$\text{Solubility} = S = [Ag^+] = [Cl^-]$$

Substituting this into the expression for the solubility constant, we get:

$$\begin{aligned} K_{sp} &= [Ag^+][Cl^-] \\ &= S \times S \\ &= S^2 \end{aligned}$$

EVERYDAY CHEMISTRY

Hard Water

Many areas of the United States obtain their water from lakes or reservoirs that have significant concentrations of $CaCO_3$ and $MgCO_3$. These salts dissolve into rainwater as it flows through soils rich in $CaCO_3$ and $MgCO_3$. Water containing these salts is known as hard water. Hard water is not a health hazard because both calcium and magnesium are part of a healthy diet, but their presence in water can be annoying. For example, because of their relatively low solubility-product constants, water can easily become saturated with $CaCO_3$ and $MgCO_3$. A drop of water, for example, becomes saturated with $CaCO_3$ and $MgCO_3$ as it evaporates. A saturated solution such as this precipitates some of its dissolved ions. These precipitates show up as scaly deposits on faucets, sinks, or cookware. Washing cars or dishes with hard water leaves spots of $CaCO_3$ and $MgCO_3$ as these precipitate out of drying drops of water.

◀ Hard water leaves scaly deposits on plumbing fixtures.

CAN YOU ANSWER THIS? *Is the water in your community hard or soft? Use the solubility-product constants from Table 15.2 to calculate the molar solubility of $CaCO_3$ and $MgCO_3$. How many moles of $CaCO_3$ are in 5 L of water that is saturated with $CaCO_3$? How many grams?*

In this text, we limit the calculation of molar solubility to ionic compounds whose chemical formulas have one cation and one anion.

Therefore,

$$S = \sqrt{K_{sp}}$$
$$= \sqrt{1.77 \times 10^{-10}}$$
$$= 1.33 \times 10^{-5}\,M$$

So the molar solubility of AgCl is 1.33×10^{-5} mol/L.

EXAMPLE 15.9 Calculating Molar Solubility from K_{sp}

Calculate the molar solubility of $BaSO_4$.

SOLUTION	
Begin by writing the reaction by which solid $BaSO_4$ dissolves into its constituent aqueous ions.	$BaSO_4(s) \rightleftharpoons Ba^{2+}(aq) + SO_4^{2-}(aq)$
Next, write the expression for K_{sp}.	$K_{sp} = [Ba^{2+}][SO_4^{2-}]$
Define the molar solubility (S) as $[Ba^{2+}]$ or $[SO_4^{2-}]$ at equilibrium.	$S = [Ba^{2+}] = [SO_4^{2-}]$
Substitute S into the equilibrium expression and solve for it.	$K_{sp} = [Ba^{2+}][SO_4^{2-}]$ $= S \times S$ $= S^2$ Therefore $S = \sqrt{K_{sp}}$
Finally, look up the value of K_{sp} in Table 15.2 and calculate S. The molar solubility of $BaSO_4$ is 1.03×10^{-5} mol/L.	$S = \sqrt{K_{sp}}$ $= \sqrt{1.07 \times 10^{-10}}$ $= 1.03 \times 10^{-5}\,M$

▶SKILLBUILDER 15.9 | Calculating Molar Solubility from K_{sp}

Calculate the molar solubility of $CaSO_4$.

▶FOR MORE PRACTICE Example 15.15; Problems 85, 86, 87, 88.

15.12 The Path of a Reaction and the Effect of a Catalyst

In this chapter, we have learned that the equilibrium constant describes the ultimate fate of a chemical reaction. Large equilibrium constants mean that the reaction favors the products. Small equilibrium constants mean that the reaction favors the reactants. But the equilibrium constant by itself does not tell the whole story. For example, consider the reaction between hydrogen gas and oxygen gas to form water:

> Warning: Hydrogen gas is explosive and should never be handled without proper training.

$$2H_2(g) + O_2(g) \rightleftharpoons 2H_2O(g) \quad K_{eq} = 3.2 \times 10^{81} \text{ at } 25\,°C$$

The equilibrium constant is huge, meaning that the forward reaction is heavily favored. Yet you can mix hydrogen and oxygen in a balloon at room temperature, and no reaction occurs. Hydrogen and oxygen peacefully coexist together inside of the balloon and form virtually no water. Why?

To answer this question, we revisit a topic from the beginning of this chapter—*the reaction rate*. At 25 °C, the reaction rate between hydrogen gas and oxygen gas is virtually zero. Even though the equilibrium constant is large, the reaction rate is small and no reaction occurs. The reaction rate between hydrogen and oxygen is slow because the reaction has a large *activation energy*. The **activation energy** (or activation barrier) for a reaction is the energy barrier that must be overcome in order for the reactants to be converted into products. Activation energies exist for most chemical reactions because the original bonds must begin to break before new bonds begin to form, and this requires energy. For example, for H_2 and O_2 to react to form H_2O, the H—H and O=O bonds must begin to break before the new bonds can form. The initial weakening of H_2 and O_2 bonds takes energy—this is the activation energy of the reaction.

> The equilibrium constant describes *how far* a chemical reaction will go. The reaction rate describes *how fast* it will get there.

> The activation energy is sometimes called the *activation barrier*.

HOW ACTIVATION ENERGIES AFFECT REACTION RATES

We can illustrate how activation energies affect reaction rates with a graph showing the energy progress of a reaction (▼ Figure 15.14). We see in the figure that the products have less energy than the reactants, so the reaction is exothermic (it releases energy when it occurs). However, before the reaction can take place, some energy must first be *added*—the energy of the reactants must be raised by an amount that we call the activation energy. The activation energy is a kind of "energy hump" that normally exists between the reactants and products.

▶ **FIGURE 15.14 Activation energy** This plot represents the energy of the reactants and products along the reaction pathway (as the reaction occurs). Notice that the energy of the products is lower than the energy of the reactants, so this is an exothermic reaction. However, notice that the reactants must get over an energy hump—called the *activation energy*—to proceed from reactants to products.

$$2H_2(g) + O_2(g) \rightleftharpoons 2H_2O(g)$$

Activation energy

Energy of reactants

Energy of products

Energy

Reaction pathway

(a) Without catalyst

(b) With catalyst

▲ **FIGURE 15.15   Hill analogy for activation energy**   There are several ways to get these boulders over the hill as fast as possible. **(a)** One way is simply to push them harder—this is analogous to an increase in temperature for a chemical reaction. **(b)** Another way is to find a path that goes *around* the hill—this is analogous to the role of a catalyst for a chemical reaction.

We can understand this concept better by means of a simple analogy—getting a chemical reaction to occur is much like trying to push a bunch of boulders over a hill (▲ Figure 15.15a). We can think of each collision that occurs between reactant molecules as an attempt to roll a boulder over the hill. We can think of a successful collision between two molecules (one that leads to product) as a successful attempt to roll a boulder over the hill and down the other side.

For rolling boulders, the higher the hill is, the harder it will be to get the boulders over the hill, and the fewer the number of boulders that make it over the hill in a given period of time. Similarly, for chemical reactions, the higher the activation energy, the fewer the number of reactant molecules that make it over the barrier, and the slower the reaction rate. In general:

At a given temperature, the higher the activation energy for a chemical reaction, the slower the reaction rate.

Are there any ways to speed up a slow reaction (one with a high activation barrier)? In Section 15.2 we talked of two ways to increase reaction rates. The first way is to increase the concentrations of the reactants, which results in more collisions per unit time. This is analogous to simply pushing more boulders toward the hill in a given period of time. The second way is to increase the temperature. This results in more collisions per unit time, and also in higher energy collisions. Higher-energy collisions are analogous to pushing the boulders harder (with more force), which will result in more boulders making it over the hill per unit time—a faster reaction rate. There is, however, a third way to speed up a slow chemical reaction: by using a *catalyst*.

CATALYSTS LOWER THE ACTIVATION ENERGY

A catalyst does not change the *position* of equilibrium, only *how fast* equilibrium is reached.

Upper-atmospheric ozone forms a shield against harmful ultraviolet light that would otherwise enter Earth's atmosphere. See the *Chemistry in the Environment* box in Chapter 6.

A **catalyst** is a substance that increases the rate of a chemical reaction but is not consumed by the reaction. A catalyst works by lowering the activation energy for the reaction, making it easier for reactants to get over the energy barrier (▶ Figure 15.16). In our boulder analogy, a catalyst creates another path for the boulders to travel—a path with a smaller hill (see Figure 15.15b). For example, consider the noncatalytic destruction of ozone in the upper atmosphere.

$$O_3 + O \longrightarrow 2 O_2$$

▲ **FIGURE 15.16   Function of a catalyst**   A catalyst provides an alternate pathway with a lower activation energy barrier for the reaction.

We have a protective ozone layer because this reaction has a fairly high activation barrier and therefore proceeds at a fairly slow rate. The ozone layer does not rapidly decompose into O_2.

However, the addition of Cl (from synthetic chlorofluorocarbons) to the upper atmosphere has resulted in another pathway by which O_3 can be destroyed. The first step in this pathway—called the *catalytic* destruction of ozone—is the reaction of Cl with O_3 to form ClO and O_2.

$$Cl + O_3 \longrightarrow ClO + O_2$$

This is followed by a second step in which ClO reacts with O, regenerating Cl.

$$ClO + O \longrightarrow Cl + O_2$$

Notice that, if we add the two reactions, the overall reaction is identical to the noncatalytic reaction.

$$\begin{array}{r} \cancel{Cl} + O_3 \longrightarrow \cancel{ClO} + O_2 \\ \cancel{ClO} + O \longrightarrow \cancel{Cl} + O_2 \\ \hline O_3 + O \longrightarrow 2\,O_2 \end{array}$$

However, the activation energies for the two reactions in this pathway are much smaller than for the first, uncatalyzed pathway, and therefore the reaction occurs at a much faster rate. Note that the Cl is not consumed in the overall reaction; this is characteristic of a catalyst.

In the case of the catalytic destruction of ozone, the catalyst speeds up a reaction that we do *not* want to happen. Most of the time, however, catalysts are used to speed up reactions that we *do* want to happen. For example, your car most likely has a catalytic converter in its exhaust system. The catalytic converter contains a catalyst that converts exhaust pollutants (such as carbon monoxide) into less harmful substances (such as carbon dioxide). These reactions occur only with the help of a catalyst because they are too slow to occur otherwise.

The role of catalysis in chemistry cannot be overstated. Without catalysts, chemistry would be a different field. For many reactions, increasing the reaction rate in another way—such as raising the temperature—is simply not feasible. Many reactants are thermally sensitive—increasing the temperature often destroys them. The only way to carry out many reactions is to use catalysts.

A catalyst cannot change the value of K_{eq} for a reaction—it affects only the *rate* of the reaction.

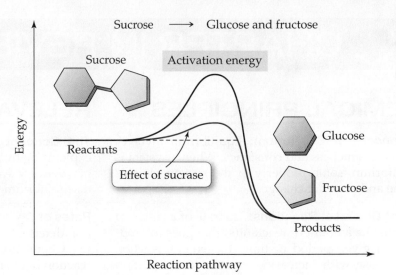

Sucrose ⟶ Glucose and fructose

▶ FIGURE 15.17 **An enzyme catalyst** The enzyme sucrase creates a pathway with a lower activation energy for the conversion of sucrose to glucose and fructose.

▲ FIGURE 15.18 **How an enzyme works** Sucrase has a pocket called the active site where sucrose binds. When a molecule of sucrose enters the active site, the bond between glucose and fructose is weakened, lowering the activation energy for the reaction.

ENZYMES: BIOLOGICAL CATALYSTS

Perhaps the best example of chemical catalysis is found in living organisms. Most of the thousands of reactions that must occur for a living organism to survive would be too slow at normal temperatures. So living organisms use **enzymes**, biological catalysts that increase the rates of biochemical reactions. For example, when we eat sucrose (table sugar), our bodies must break it into two smaller molecules called glucose and fructose. The equilibrium constant for this reaction is large, favoring the products. However, at room temperature, or even at body temperature, the sucrose does not break down into glucose and fructose because the activation energy is high, resulting in a slow reaction rate. In other words, table sugar remains table sugar at room temperature, even though the equilibrium constant for its reaction to glucose and fructose is relatively large (▲ Figure 15.17). In the body, however, an enzyme called *sucrase* catalyzes the conversion of sucrose to glucose and fructose. Sucrase has a pocket—called the active site—into which sucrose snugly fits (like a key into a lock). When sucrose is in the active site, the bond between the glucose and fructose units weakens, lowering the activation energy for the reaction and increasing the reaction rate (◀ Figure 15.18). The reaction can then proceed toward equilibrium—which favors the products—at a much lower temperature.

$$C_{12}H_{22}O_{11} + H_2O \longrightarrow C_6H_{12}O_6 + C_6H_{12}O_6$$
Sucrose — Glucose — Fructose

Not only do enzymes allow otherwise slow reactions to occur at a reasonable rate, they also allow living organisms to have tremendous control over which reactions occur and when. Enzymes are extremely specific—each enzyme catalyzes only a single reaction. So to turn a particular reaction on, a living organism simply needs to produce or activate the correct enzyme to catalyze that reaction.

CHAPTER IN REVIEW

CHEMICAL PRINCIPLES

The Concept of Equilibrium: Equilibrium involves the ideas of sameness and constancy. When a system is in equilibrium, some property of the system remains the same and does not change.

Rates of Chemical Reactions: The rate of a chemical reaction is the amount of reactant(s) that goes to product(s) in a given period of time. In general, reaction rates increase with increasing reactant concentration and increasing temperature. Since reaction rates depend on the concentration of reactants, and since the concentration of reactants decreases as a reaction proceeds, reaction rates usually slow down as a reaction proceeds.

Dynamic Chemical Equilibrium: Dynamic chemical equilibrium occurs when the rate of the forward reaction equals the rate of the reverse reaction.

The Equilibrium Constant: For the generic reaction

$$aA + bB \rightleftharpoons cC + dD$$

the equilibrium constant (K_{eq}) is defined as:

$$K_{eq} = \frac{[C]^c[D]^d}{[A]^a[B]^b}$$

Only the concentrations of gaseous or aqueous reactants and products are included in the equilibrium constant—the concentrations of solid or liquid reactants or products are omitted.

Le Châtelier's Principle: Le Châtelier's principle states that when a chemical system at equilibrium is disturbed, the system shifts in a direction that minimizes the disturbance.

Effect of a Concentration Change on Equilibrium:

- Increasing the concentration of one or more of the *reactants* causes the reaction to shift to the *right*.
- Increasing the concentration of one or more of the *products* causes the reaction to shift to the *left*.

RELEVANCE

The Concept of Equilibrium: The equilibrium concept explains many phenomena such as the human body's oxygen delivery system. Life itself can be defined as controlled disequilibrium with the environment.

Rates of Chemical Reactions: The rate of a chemical reaction determines how fast a reaction will reach its equilibrium. Chemists want to understand the factors that influence reaction rates so that they can control them.

Dynamic Chemical Equilibrium: When dynamic chemical equilibrium is reached, the concentrations of the reactants and products become constant.

The Equilibrium Constant: The equilibrium constant is a measure of how far a reaction will proceed. A large K_{eq} means the forward reaction is favored (lots of products at equilibrium). A small K_{eq} means the reverse reaction is favored (lots of reactants at equilibrium). An intermediate K_{eq} means that there will be significant amounts of both reactants and products at equilibrium.

Le Châtelier's Principle: Le Châtelier's principle helps us predict what happens to a chemical system at equilibrium when the conditions are changed. This allows chemists to modify the conditions of a chemical reaction to obtain a desired result.

Effect of a Concentration Change on Equilibrium: There are many cases when a chemist may want to drive a reaction in one direction or another. For example, suppose a chemist is carrying out a reaction to make a desired compound. The reaction can be pushed to the right by continuously removing the product from the reaction mixture as it forms, thus maximizing the amount of product that can be made.

Effect of a Volume Change on Equilibrium:

- Decreasing the volume causes the reaction to shift in the direction that has *fewer* moles of gas particles.
- Increasing the volume causes the reaction to shift in the direction that has more moles of gas particles.

Effect of a Volume Change on Equilibrium: Like the effect of concentration, the effect of pressure on equilibrium allows a chemist to choose the best conditions under which to carry out a chemical reaction. Some reactions are favored in the forward direction by high pressure (those with fewer moles of gas particles in the products), and others (those with fewer moles of gas particles in the reactants) are favored in the forward direction by low pressure.

Effect of a Temperature Change on Equilibrium:
Exothermic chemical reaction (heat is a product):

- *Increasing* the temperature causes the reaction to shift *left*.
- *Decreasing* the temperature causes the reaction to shift *right*.

Endothermic chemical reaction (heat is a reactant):

- *Increasing* the temperature causes the reaction to shift *right*.
- *Decreasing* the temperature causes the reaction to shift *left*.

Effect of a Temperature Change on Equilibrium: Again, the effect of temperature on a reaction allows chemists to choose conditions that will favor desired reactions. Higher temperatures favor endothermic reactions, while lower temperatures favor exothermic reactions. Most reactions will occur *faster* at higher temperature, so the effect of temperature on the rate, not just on the equilibrium constant, must be considered.

The Solubility-Product Constant, K_{sp}: The solubility-product constant of an ionic compound is the equilibrium constant for the chemical equation that describes the dissolving of the compound.

The Solubility-Product Constant, K_{sp}: The solubility-product constant reflects the solubility of a compound. The greater the solubility-product constant, the greater the solubility of the compound.

Reaction Paths and Catalysts: Most chemical reactions must overcome an energy hump, called the *activation energy*, as they proceed from reactants to products. Increasing the temperature of a reaction mixture increases the fraction of reactant molecules that make it over the energy hump, therefore increasing the rate. A catalyst—a substance that increases the rate of the reaction but is not consumed by it—lowers the activation energy so that it is easier to get over the energy hump *without* increasing the temperature.

Reaction Paths and Catalysts: Catalysts are used in many chemical reactions to increase the rates. Without catalysts, many reactions occur too slowly to be of any value. The thousands of reactions that occur in living organisms are controlled by biological catalysts called *enzymes*.

CHEMICAL SKILLS

EXAMPLES

Writing Equilibrium Expressions for Chemical Reactions (Section 15.5)

Examine the definition of the equilibrium constant in the Chemical Principles section (page 562). To write the equilibrium expression for a reaction, write the concentrations of the products raised to their stoichiometric coefficients divided by the concentrations of the reactants raised to their stoichiometric coefficients. Remember that if the reaction contains reactants or products that are liquids or solids, these are omitted from the equilibrium expression.

EXAMPLE 15.10 Writing Equilibrium Expressions for Chemical Reactions

Write an equilibrium expression for the chemical equation:

$$2\,NO(g) + Br_2(g) \rightleftharpoons 2\,NOBr(g)$$

SOLUTION

$$K_{eq} = \frac{[NOBr]^2}{[NO]^2[Br_2]}$$

Calculating Equilibrium Constants (Section 15.6)

EXAMPLE 15.11   Calculating Equilibrium Constants

An equilibrium mixture of the following reaction had

$$[I] = 0.075 \, M \text{ and } [I_2] = 0.88 \, M.$$

What is the value of the equilibrium constant?

$$I_2(g) \rightleftharpoons 2\,I(g)$$

GIVEN:

$$[I] = 0.075 \, M$$
$$[I_2] = 0.88 \, M$$

FIND: K_{eq}

Begin by setting up the problem in the usual way.

Then write the expression for K_{eq} from the balanced equation. To calculate the value of K_{eq}, substitute the correct equilibrium concentrations into the expression for K_{eq}. The concentrations within K_{eq} should always be written in moles per liter, M. Units are normally dropped in expressing the equilibrium constant so that K_{eq} is unitless.

SOLUTION

$$K_{eq} = \frac{[I]^2}{[I_2]}$$

$$= \frac{[0.075]^2}{[0.88]}$$

$$= 0.0064$$

Using the Equilibrium Constant to Find the Concentration of a Reactant or Product at Equilibrium (Section 15.6)

EXAMPLE 15.12   Using the Equilibrium Constant to Find the Concentration of a Reactant or Product at Equilibrium

Consider the reaction:

$$N_2(g) + 3\,H_2(g) \rightleftharpoons 2\,NH_3(g)$$

$$K_{eq} = 152 \text{ at } 225 \,°C$$

In an equilibrium mixture, $[N_2] = 0.110 \, M$ and $[H_2] = 0.0935 \, M$. What is the equilibrium concentration of NH_3?

SORT

You are given the initial concentrations of nitrogen and hydrogen as well as the equilibrium constant for their reaction to form ammonia. You are asked to find the equilibrium concentration of ammonia.

GIVEN:

$$[N_2] = 0.110 \, M$$

$$[H_2] = 0.0935 \, M$$

$$K_{eq} = 152$$

FIND: $[NH_3]$

STRATEGIZE

Write a solution map that shows how you can use the given concentrations and the equilibrium constant to get to the unknown concentration.

SOLUTION MAP

$$K_{eq} = \frac{[NH_3]^2}{[N_2][H_2]^3}$$

SOLVE

Next, solve the equilibrium expression for the quantity you are trying to find and substitute in the appropriate values to calculate the unknown quantity.

SOLUTION

$$K_{eq} = \frac{[NH_3]^2}{[N_2][H_2]^3}$$

$$[NH_3]^2 = K_{eq}[N_2][H_2]^3$$

$$[NH_3] = \sqrt{K_{eq}[N_2][H_2]^3}$$

$$= \sqrt{(152)(0.110)(0.0935)^3}$$

$$= 0.117 M$$

CHECK

You can check your answer by substituting it back into the expression for K_{eq}.

$$K_{eq} = \frac{[NH_3]^2}{[N_2][H_2]^3}$$

$$= \frac{(0.117)^2}{(0.119)(0.0935)^3}$$

$$= 141$$

The calculated value of K_{eq} is about equal to the given value of K_{eq} (which was 152) indicating that your answer is correct. The slight difference is due to rounding, which is common in problems like these.

Using Le Châtelier's Principle (Sections 15.8, 15.9, 15.10)

To apply Le Châtelier's principle, review the effects of concentration, volume, and temperature in the Chemical Principles section (pages 562–563). For each disturbance, predict how the reaction will change to counter the disturbance.

EXAMPLE 15.13   Using Le Châtelier's Principle

Consider the *endothermic* chemical reaction:

$$C(s) + H_2O(g) \rightleftharpoons CO(g) + H_2(g)$$

Predict the effect of:

(a) increasing [CO]
(b) increasing [H$_2$O]
(c) increasing the reaction volume
(d) increasing the temperature

SOLUTION

(a) Shift left
(b) Shift right
(c) Shift right (more moles of gas on right)
(d) Shift right (heat is a reactant)

Writing an Expression for the Solubility-Product Constant (Section 15.11)

To write the expression for K_{sp}, first write the chemical reaction showing the solid compound in equilibrium with its dissolved aqueous ions. Then write the equilibrium expression as the product of the concentrations of the aqueous ions raised to their stoichiometric coefficients.

EXAMPLE 15.14   Writing an Expression for the Solubility-Product Constant

Write an expression for K_{sp} for PbCl$_2$.

SOLUTION

$$PbCl_2(s) \rightleftharpoons Pb^{2+}(aq) + 2\ Cl^-(aq)$$

$$K_{sp} = [Pb^{2+}][Cl^-]^2$$

Using K_{sp} to Determine Molar Solubility (Section 15.11)

Begin by writing the reaction by which the solid dissolves into its constituent aqueous ions.

Next, write the expression for K_{sp}.

For the problems assigned in this book, the concentration of individual aqueous ions is equal to the solubility, S.

Substitute S into the equilibrium expression and solve the expression for S.

Finally, look up the value of K_{sp} in Table 15.2 and calculate S.

EXAMPLE 15.15 Using K_{sp} to Determine Molar Solubility

Calculate the molar solubility of AgI.

SOLUTION

$$AgI(s) \rightleftharpoons Ag^+(aq) + I^-(aq)$$

$$K_{sp} = [Ag^+][I^-]$$

$$S = [Ag^+] = [I^-]$$

$$K_{sp} = [Ag^+][I^-]$$

$$= S \times S$$

$$= S^2$$

$$S = \sqrt{K_{sp}}$$

$$S = \sqrt{8.51 \times 10^{-17}}$$

$$= 9.22 \times 10^{-9} M$$

KEY TERMS

activation energy [15.12]
catalyst [15.12]
collision theory [15.2]
dynamic equilibrium [15.3]

enzyme [15.12]
equilibrium constant (K_{eq}) [15.4]
Le Châtelier's principle [15.7]

molar solubility [15.11]
rate of a chemical reaction (reaction rate) [15.2]

reversible reaction [15.3]
solubility-product constant (K_{sp}) [15.11]

EXERCISES

QUESTIONS

1. What are the two *general* concepts involved in equilibrium?
2. What is the rate of a chemical reaction? What is the difference between a chemical reaction with a fast rate and one with a slow rate?
3. Why do chemists seek to control reaction rates?
4. How do most chemical reactions occur?
5. What factors influence reaction rates? How?
6. What normally happens to the rate of the forward reaction as a reaction proceeds?
7. What is dynamic chemical equilibrium?
8. Explain how dynamic chemical equilibrium involves the concepts of sameness and constancy.
9. Explain why the concentrations of reactants and products are not necessarily the same at equilibrium.
10. Devise your own analogy—like the Narnia and Middle Earth analogy in the chapter—to explain chemical equilibrium.
11. What is the equilibrium constant? Why is it significant?
12. Write the expression for the equilibrium constant for the following generic chemical equation.

$$aA + bB \rightleftharpoons cC + dD$$

13. What does a small equilibrium constant tell you about a reaction? A large equilibrium constant?
14. Why are solids and liquids omitted from the equilibrium expression?
15. Will the concentrations of reactants and products always be the same in every equilibrium mixture of a particular reaction at a given temperature? Explain.
16. What is Le Châtelier's principle?
17. Apply Le Châtelier's principle to your analogy from Question 12.
18. What is the effect of *increasing* the concentration of a reactant in a reaction mixture at equilibrium?
19. What is the effect of *decreasing* the concentration of a reactant in a reaction mixture at equilibrium?
20. What is the effect of *increasing* the concentration of a product in a reaction mixture at equilibrium?
21. What is the effect of *decreasing* the concentration of a product in a reaction mixture at equilibrium?
22. What is the effect of increasing the pressure of a reaction mixture at equilibrium if the reactant side has fewer moles of gas particles than the product side?

23. What is the effect of increasing the pressure of a reaction mixture at equilibrium if the product side has fewer moles of gas particles than the reactant side?

24. What is the effect of decreasing the pressure of a reaction mixture at equilibrium if the reactant side has fewer moles of gas particles than the product side?

25. What is the effect of decreasing the pressure of a reaction mixture at equilibrium if the product side has fewer moles of gas particles than the reactant side?

26. What is the effect of increasing the temperature of an endothermic reaction mixture at equilibrium? Of decreasing the temperature?

27. What is the effect of increasing the temperature of an exothermic reaction mixture at equilibrium? Of decreasing the temperature?

28. What is the solubility-product constant? What does it signify?

29. Write an expression for the solubility-product constant of $AB_2(s)$. Assume that an ion of B has a charge of -1 (that is, B^-).

30. Write an expression for the solubility-product constant of $A_2B(s)$. Assume that an ion of B has charge of -2 (that is, B^{2-}).

31. What are solubility and molar solubility?

32. What is activation energy for a chemical reaction?

33. Explain why two reactants with a large K_{eq} for a particular reaction might not react immediately when combined.

34. What is the effect of a catalyst on reaction? Why are catalysts so important to chemistry?

35. Does a catalyst affect the value of the equilibrium constant?

36. What are enzymes?

PROBLEMS

THE RATE OF REACTION

37. Two gaseous reactants are allowed to react in a 1-L flask, and the reaction rate is measured. The experiment is repeated with the same amount of each reactant and at the same temperature, but now in a 2-L flask (so the concentration of each reactant is now less). What is likely to happen to the measured reaction rate in the second experiment compared to the first?

38. The rate of phosphorus pentachloride decomposition is measured at a PCl_5 pressure of 0.015 atm and then again at a PCl_5 pressure of 0.30 atm. The temperature is identical in both measurements. Which rate is likely to be faster?

39. The body temperature of cold-blooded animals varies with the ambient temperature. From the point of view of reaction rates, explain why cold-blooded animals are more sluggish at cold temperatures.

40. The rate of a reaction doubles when the temperature increases from 25 °C to 35 °C. Explain why this is so.

41. The initial rate of a chemical reaction was measured, and one of the reactants was found to be reacting at a rate of 0.0011 mol/L s. The reaction was allowed to proceed for 15 minutes, and the rate was measured again. What would you predict about the second measured rate relative to the first?

42. When vinegar is added to a solution of sodium bicarbonate, the mixture immediately begins to bubble furiously. As time passes, however, the bubbling becomes less and less. Explain.

THE EQUILIBRIUM CONSTANT

43. Write an equilibrium expression for each chemical equation.

   (a) $2\,NO_2(g) \rightleftharpoons N_2O_4(g)$

   (b) $2\,BrNO(g) \rightleftharpoons 2\,NO(g) + Br_2(g)$

   (c) $H_2O(g) + CO(g) \rightleftharpoons H_2(g) + CO_2(g)$

   (d) $CH_4(g) + 2\,H_2S(g) \rightleftharpoons CS_2(g) + 4\,H_2(g)$

44. Write an equilibrium expression for each chemical equation.

   (a) $2\,CO(g) + O_2(g) \rightleftharpoons 2\,CO_2(g)$

   (b) $N_2(g) + O_2(g) \rightleftharpoons 2\,NO(g)$

   (c) $SbCl_5(g) \rightleftharpoons SbCl_3(g) + Cl_2(g)$

   (d) $CO(g) + Cl_2(g) \rightleftharpoons COCl_2(g)$

45. Write an equilibrium expression for each chemical equation involving one or more solid or liquid reactants or products.

(a) $PCl_5(g) \rightleftharpoons PCl_3(l) + Cl_2(g)$

(b) $2\ KClO_3(s) \rightleftharpoons 2\ KCl(s) + 3\ O_2(g)$

(c) $HF(aq) + H_2O(l) \rightleftharpoons H_3O^+(aq) + F^-(aq)$

(d) $NH_3(aq) + H_2O(l) \rightleftharpoons NH_4^+(aq) + OH^-(aq)$

46. Write an equilibrium expression for each chemical equation involving one or more solid or liquid reactants or products.

(a) $HCHO_2(aq) + H_2O(l) \rightleftharpoons$
$$H_3O^+(aq) + CHO_2^-(aq)$$

(b) $CO_3{}^{2-}(aq) + H_2O(l) \rightleftharpoons$
$$HCO_3^-(aq) + OH^-(aq)$$

(c) $2\ C(s) + O_2(g) \rightleftharpoons 2\ CO(g)$

(d) $C(s) + CO_2(g) \rightleftharpoons 2\ CO(g)$

---

47. Consider the reaction:

$$2\ H_2S(g) \rightleftharpoons 2\ H_2(g) + S_2(g)$$

Find the mistakes in the equilibrium expression and fix them.

$$K_{eq} = \frac{[H_2][S_2]}{[H_2S]}$$

48. Consider the reaction:

$$CO(g) + Cl_2(g) \rightleftharpoons COCl_2(g)$$

Find the mistake in the equilibrium expression and fix it.

$$K_{eq} = \frac{[CO][Cl_2]}{[COCl_2]}$$

---

49. For each equilibrium constant, indicate whether you would expect an equilibrium reaction mixture to be dominated by reactants, to be dominated by products, or to contain significant amounts of both.

(a) $K_{eq} = 5.2 \times 10^{17}$

(b) $K_{eq} = 1.24$

(c) $K_{eq} = 3.22 \times 10^{-21}$

(d) $K_{eq} = 0.47$

50. For each equilibrium constant, indicate whether you would expect an equilibrium reaction mixture to be dominated by reactants, to be dominated by products, or to contain significant amounts of both.

(a) $K_{eq} = 0.75$

(b) $K_{eq} = 8.5 \times 10^{-7}$

(c) $K_{eq} = 1.4 \times 10^{19}$

(d) $K_{eq} = 4.7 \times 10^{-9}$

CALCULATING AND USING EQUILIBRIUM CONSTANTS

51. Consider the reaction:

$$COCl_2(g) \rightleftharpoons CO(g) + Cl_2(g)$$

An equilibrium mixture of this reaction at a certain temperature was found to have $[COCl_2] = 0.225$ M, $[CO] = 0.105$ M, and $[Cl_2] = 0.0844$ M. What is the value of the equilibrium constant at this temperature?

52. Consider the reaction:

$$CO(g) + 2\ H_2(g) \rightleftharpoons CH_3OH(g)$$

An equilibrium mixture of this reaction at a certain temperature was found to have $[CO] = 0.105$ M, $[H_2] = 0.114$ M, and $[CH_3OH] = 0.185$ M. What is the value of the equilibrium constant at this temperature?

---

53. Consider the reaction:

$$2\ H_2S(g) \rightleftharpoons 2\ H_2(g) + S_2(g)$$

An equilibrium mixture of this reaction at a certain temperature was found to have $[H_2S] = 0.562$ M,

$[H_2] = 2.74 \times 10^{-2}$ M, and $[S_2] = 7.54 \times 10^{-3}$ M.

What is the value of the equilibrium constant at this temperature?

54. Consider the reaction:

$$CO(g) + H_2O(g) \rightleftharpoons CO_2(g) + H_2(g)$$

An equilibrium mixture of this reaction at a certain temperature was found to have $[CO] = 0.0233$ M, $[H_2O] = 0.0115$ M, $[CO_2] = 0.175$ M, and $[H_2] = 0.0274$ M. What is the value of the equilibrium constant at this temperature?

55. Consider the reaction:

$$NH_4HS(s) \rightleftharpoons NH_3(g) + H_2S(g)$$

An equilibrium mixture of this reaction at a certain temperature was found to have $[NH_3] = 0.278\,M$ and $[H_2S] = 0.355\,M$. What is the value of the equilibrium constant at this temperature?

56. Consider the reaction:

$$CaCO_3(s) \rightleftharpoons CaO(s) + CO_2(g)$$

An equilibrium mixture of this reaction at a certain temperature was found to have $[CO_2] = 0.548\,M$. What is the value of the equilibrium constant at this temperature?

57. An equilibrium mixture of the following reaction was found to have $[SbCl_3] = 0.0255\,M$ and $[Cl_2] = 0.135\,M$ at 248 °C. What is the concentration of $SbCl_5$?

$$SbCl_5(g) \rightleftharpoons SbCl_3(g) + Cl_2(g)$$

$$K_{eq} = 4.9 \times 10^{-4} \text{ at } 248\,°C$$

58. An equilibrium mixture of the following reaction was found to have $[I_2] = 0.0205\,M$ at 1200 °C. What is the concentration of I?

$$I_2(g) \rightleftharpoons 2\,I(g)$$

$$K_{eq} = 1.1 \times 10^{-2} \text{ at } 1200\,°C$$

59. An equilibrium mixture of the following reaction was found to have $[I_2] = 0.0112\,M$ and $[Cl_2] = 0.0155\,M$ at 25 °C. What is the concentration of ICl?

$$I_2(g) + Cl_2(g) \rightleftharpoons 2\,ICl(g)$$

$$K_{eq} = 81.9 \text{ at } 25\,°C$$

60. An equilibrium mixture of the following reaction was found to have $[SO_3] = 0.391\,M$ and $[O_2] = 0.125\,M$ at 600 °C. What is the concentration of SO_2?

$$N_2(g) + 3\,H_2(g) \rightleftharpoons 2\,NH_3(g)$$

$$K_{eq} = 4.34 \text{ at } 600\,°C$$

61. Consider the reaction:

$$N_2(g) + 3\,H_2(g) \rightleftharpoons 2\,NH_3(g)$$

Complete the table. Assume that all concentrations are equilibrium concentrations in moles per liter, M.

T (K)	[N₂]	[H₂]	[NH₃]	Keq
500	0.115	0.105	0.439	____
575	0.110	____	0.128	9.6
775	0.120	0.140	____	0.0584

62. Consider the reaction:

$$H_2(g) + I_2(g) \rightleftharpoons 2\,HI(g)$$

Complete the table. Assume that all concentrations are equilibrium concentrations in moles per liter, M.

T (°C)	[H₂]	[I₂]	[HI]	Keq
25	0.355	0.388	0.0922	____
340	____	0.0455	0.387	9.6
445	0.0485	0.0468	____	50.2

LE CHÂTELIER'S PRINCIPLE

63. Consider this reaction at equilibrium.

$$CO(g) + Cl_2(g) \rightleftharpoons COCl_2(g)$$

Predict the effect (shift right, shift left, or no effect) of

(a) adding Cl_2 to the reaction mixture.

(b) adding $COCl_2$ to the reaction mixture.

(c) adding CO to the reaction mixture.

64. Consider this reaction at equilibrium.

$$2\,BrNO(g) \rightleftharpoons 2\,NO(g) + Br_2(g)$$

Predict the effect (shift right, shift left, or no effect) of

(a) adding BrNO to the reaction mixture.

(b) adding NO the reaction mixture.

(c) adding Br_2 to the reaction mixture.

65. Consider this reaction at equilibrium.

$$C(s) + H_2O(g) \rightleftharpoons CO(g) + H_2(g)$$

Predict the effect (shift right, shift left, or no effect) of

(a) adding C to the reaction mixture.

(b) condensing H_2O and removing it from the reaction mixture.

(c) adding CO to the reaction mixture.

(d) removing H_2 from the reaction mixture.

66. Consider this reaction at equilibrium.

$$2\,KClO_3(s) \rightleftharpoons 2\,KCl(s) + 3\,O_2(g)$$

Predict the effect (shift right, shift left, or no effect) of

(a) adding KCl to the reaction mixture.

(b) adding $KClO_3$ to the reaction mixture.

(c) adding O_2 to the reaction mixture.

(d) removing O_2 from the reaction mixture.

67. Consider the effect of a volume change on this reaction at equilibrium.

$$I_2(g) \rightleftharpoons 2I(g)$$

Predict the effect (shift right, shift left, or no effect) of

(a) increasing the reaction volume.

(b) decreasing the reaction volume.

68. Consider the effect of a volume change on this reaction at equilibrium.

$$2H_2S(g) \rightleftharpoons 2H_2(g) + S_2(g)$$

Predict the effect (shift right, shift left, or no effect) of

(a) increasing the reaction volume.

(b) decreasing the reaction volume.

69. Consider the effect of a volume change on this reaction at equilibrium.

$$I_2(g) + Cl_2(g) \rightleftharpoons 2ICl(g)$$

Predict the effect (shift right, shift left, or no effect) of

(a) increasing the reaction volume.

(b) decreasing the reaction volume.

70. Consider the effect of a volume change on this reaction at equilibrium.

$$CO(g) + H_2O(g) \rightleftharpoons CO_2(g) + H_2(g)$$

Predict the effect (shift right, shift left, or no effect) of

(a) increasing the reaction volume.

(b) decreasing the reaction volume.

71. This reaction is endothermic.

$$C(s) + CO_2(g) \rightleftharpoons 2CO(g)$$

Predict the effect (shift right, shift left, or no effect) of

(a) increasing the reaction temperature.

(b) decreasing the reaction temperature.

72. This reaction is endothermic.

$$I_2(g) \rightleftharpoons 2I(g)$$

Predict the effect (shift right, shift left, or no effect) of

(a) increasing the reaction temperature.

(b) decreasing the reaction temperature.

73. This reaction is exothermic.

$$C_6H_{12}O_6(s) + 6O_2(g) \rightleftharpoons 6CO_2(g) + 6H_2O(g)$$

Predict the effect (shift right, shift left, or no effect) of

(a) increasing the reaction temperature.

(b) decreasing the reaction temperature.

74. The following reaction is exothermic.

$$C_2H_4(g) + Br_2(g) \rightleftharpoons C_2H_4Br_2(g)$$

Predict the effect (shift right, shift left, or no effect) of

(a) increasing the reaction temperature.

(b) decreasing the reaction temperature.

75. Coal, which is primarily carbon, can be converted to natural gas, primarily CH_4, by the following exothermic reaction.

$$C(s) + 2H_2(g) \rightleftharpoons CH_4(g)$$

If this reaction mixture is at equilibrium, predict the effect (shift right, shift left, or no effect) of

(a) adding more C to the reaction mixture.

(b) adding more H_2 to the reaction mixture.

(c) raising the temperature of the reaction mixture.

(d) lowering the volume of the reaction mixture.

(e) adding a catalyst to the reaction mixture.

76. Coal can be used to generate hydrogen gas (a potential fuel) by this endothermic reaction.

$$C(s) + H_2O(g) \rightleftharpoons CO(g) + H_2(g)$$

If this reaction mixture is at equilibrium, predict the effect (shift right, shift left, or no effect) of

(a) adding more C to the reaction mixture.

(b) adding more $H_2O(g)$ to the reaction mixture.

(c) raising the temperature of the reaction mixture.

(d) increasing the volume of the reaction mixture.

(e) adding a catalyst to the reaction mixture.

THE SOLUBILITY-PRODUCT CONSTANT

77. For each compound, write an equation showing how the compound dissolves in water and write an expression for K_{sp}.

(a) $CaSO_4$

(b) $AgCl$

(c) CuS

(d) $FeCO_3$

78. For each compound, write an equation showing how the compound dissolves in water and write an expression for K_{sp}.

(a) $Mg(OH)_2$

(b) $FeCO_3$

(c) PbS

(d) $PbSO_4$

79. Determine what is wrong with the following K_{sp} expression for $Fe(OH)_2$ and correct it.

$$K_{sp} = [Fe^{2+}][OH^-]$$

80. Determine what is wrong with the following K_{sp} expression for $Ba(OH)_2$ and correct it.

$$K_{sp} = \frac{[Ba(OH)_2]}{[Ba^{2+}][OH^-]^2}$$

81. A saturated solution of MgF_2 has $[Mg^{2+}] = 2.6 \times 10^{-4}$ M and $[F^-] = 5.2 \times 10^{-4}$ M. What is the value of K_{sp} for MgF_2?

82. A saturated solution of AgI has $[Ag^+] = 9.2 \times 10^{-9}$ M and $[I^-] = 9.2 \times 10^{-9}$ M. What is the value of K_{sp} for AgI?

83. A saturated solution of $PbSO_4$ has $[Pb^{2+}] = 1.35 \times 10^{-4}$ M. What is the concentration of SO_4^{2-}?

84. A saturated solution of $PbCl_2$ has $[Cl^-] = 2.86 \times 10^{-2}$ M. What is the concentration of Pb^{2+}?

85. Calculate the molar solubility of $CaCO_3$.

86. Calculate the molar solubility of PbS.

87. Calculate the molar solubility of $MgCO_3$.

88. Calculate the molar solubility of CuI ($K_{sp} = 1.27 \times 10^{-12}$).

89. Complete the table. Assume that all concentrations are equilibrium concentrations in moles per liter, M.

Compound	[Cation]	[Anion]	K_{sp}
$SrCO_3$	2.4×10^{-5}	2.4×10^{-5}	———
SrF_2	1.0×10^{-3}	———	4.0×10^{-9}
Ag_2CO_3	———	1.3×10^{-4}	8.8×10^{-12}

90. Complete the table. Assume that all concentrations are equilibrium concentrations in moles per liter, M.

Compound	[Cation]	[Anion]	K_{sp}
CdS	3.7×10^{-15}	3.7×10^{-15}	———
BaF_2	———	7.2×10^{-3}	1.9×10^{-7}
Ag_2SO_4	2.8×10^{-2}	———	1.1×10^{-5}

CUMULATIVE PROBLEMS

91. Consider the reaction:

$$Fe^{3+}(aq) + SCN^-(aq) \rightleftharpoons FeSCN^{2+}(aq)$$

A solution is made containing initial $[Fe^{3+}] = 1.0 \times 10^{-3}$ M and initial $[SCN^-] = 8.0 \times 10^{-4}$ M. At equilibrium, $[FeSCN^{2+}] = 1.7 \times 10^{-4}$. Calculate the value of the equilibrium constant. *Hint:* Use the chemical reaction stoichiometry to calculate the equilibrium concentrations of Fe^{3+} and SCN^-.

92. Consider the reaction:

$$SO_2Cl_2(g) \rightleftharpoons SO_2(g) + Cl_2(g)$$

A solution is made containing initial $[SO_2Cl_2] = 0.020$ M. At equilibrium, $[Cl_2] = 1.2 \times 10^{-2}$ M. Calculate the value of the equilibrium constant. *Hint:* Use the chemical reaction stoichiometry to calculate the equilibrium concentrations of SO_2Cl_2 and SO_2.

93. Consider the reaction:

$$CO(g) + Cl_2(g) \rightleftharpoons COCl_2(g)$$
$$K_{eq} = 6.17 \times 10^{-2} \text{ at } 25\,°C$$

A 3.67-L flask containing an equilibrium reaction mixture has $[H_2] = 0.104$ M and $[I_2] = 0.0202$ M. How much HI in grams is in the equilibrium mixture?

94. Consider the reaction:

$$H_2(g) + I_2(g) \rightleftharpoons 2\,HI(g)$$
$$K_{eq} = 2.9 \times 10^{10} \text{ at } 25\,°C$$

A 5.19-L flask containing an equilibrium reaction mixture has $[CO] = 1.8 \times 10^{-6}$ M and $[Cl_2] = 7.3 \times 10^{-7}$ M. How much $COCl_2$ in grams is in the equilibrium mixture?

95. This reaction is exothermic.

$$C_2H_4(g) + Cl_2(g) \rightleftharpoons C_2H_4Cl_2(g)$$

If you were a chemist trying to maximize the amount of $C_2H_4Cl_2$ produced, which of the following might you try? Assume that the reaction mixture reaches equilibrium.

(a) increasing the reaction volume

(b) removing $C_2H_4Cl_2$ from the reaction mixture as it forms

(c) lowering the reaction temperature

(d) adding Cl_2

96. This reaction is endothermic.

$$C_2H_4(g) + I_2(g) \rightleftharpoons C_2H_4I_2(g)$$

If you were a chemist trying to maximize the amount of $C_2H_4I_2$ produced, which of the following might you try? Assume that the reaction mixture reaches equilibrium.

(a) decreasing the reaction volume

(b) removing I_2 from the reaction mixture

(c) raising the reaction temperature

(d) adding C_2H_4 to the reaction mixture

---

97. Calculate the molar solubility of CuS. How many grams of CuS are present in 15.0 L of a saturated CuS solution?

98. Calculate the molar solubility of $FeCO_3$. How many grams of $FeCO_3$ are present in 15.0 L of a saturated $FeCO_3$ solution?

---

99. A sample of tap water is found to be 0.025 M in Ca^{2+}. If 105 mg of Na_2SO_4 is added to 100.0 mL of the tap water, will any $CaSO_4$ precipitate out of solution?

100. If 50.0 mg of Na_2CO_3 are added to 150.0 mL of a solution that is 1.5×10^{-3} M in Mg^{2+}, will any $MgCO_3$ precipitate from the solution?

---

101. The solubility of $CaCrO_4$ at 25 °C is 4.15 g/L. Calculate K_{sp} for $CaCrO_4$.

102. The solubility of nickel(II) carbonate at 25 °C is 0.042 g/L. Calculate K_{sp} for nickel(II) carbonate.

---

103. Consider the reaction:

$$CaCO_3 \rightleftharpoons CaO(s) + CO_2(g)$$

A sample of $CaCO_3$ is placed into a sealed 0.500-L container and heated to 550 K at which the equilibrium constant is 4.1×10^{-4}. When the reaction has come to equilibrium, what mass of solid CaO will be in the container? (Assume that the sample of $CaCO_3$ was large enough that equilibrium could be achieved.)

104. Consider the reaction:

$$NH_4HS(s) \rightleftharpoons NH_3(g) + H_2S(g)$$

A sample of pure NH_4HS is placed in a sealed 2.0-L container and heated to 550 K at which the equilibrium constant is 3.5×10^{-3}. Once the reaction reaches equilibrium, what mass of NH_3 is present in the container? (Assume that the sample of $CaCO_3$ was large enough that equilibrium could be achieved.)

---

105. A 2.55-L solution is 0.115 M in Mg^{2+}. If K_2CO_3 were added to the solution in order to precipitate the magnesium, what minimum mass of K_2CO_3 would be required to get a precipitate?

106. A 75.0-L solution is 0.0251 M in Ca^{2+}. If Na_2SO_4 were added to the solution in order to precipitate the calcium, what minimum mass of Na_2SO_4 would be required to get a precipitate?

HIGHLIGHT PROBLEMS

107. H$_2$ and I$_2$ are combined in a flask and allowed to react according to the reaction:

$$H_2(g) + I_2(g) \rightleftharpoons 2\,HI(g)$$

Examine the following figures (sequential in time) and determine which figure represents the point where equilibrium is reached.

(a)

(b)

(c)

(d)

(e)

(f)

108. Ethene (C$_2$H$_4$) can be halogenated by the reaction:

$$C_2H_4(g) + X_2(g) \rightleftharpoons C_2H_4X_2(g)$$

where X$_2$ can be Cl$_2$, Br$_2$, or I$_2$. Examine the figures representing equilibrium concentrations of this reaction at the same temperature for the three different halogens. Rank the equilibrium constants for these three reactions from largest to smallest.

$$C_2H_4 + Cl_2 \rightleftharpoons C_2H_4Cl_2$$

(a)

$$C_2H_4 + Br_2 \rightleftharpoons C_2H_4Br_2$$

(b)

$$C_2H_4 + I_2 \rightleftharpoons C_2H_4I_2$$

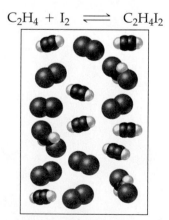

(c)

109. One of the main components of hard water is CaCO$_3$. When hard water evaporates, some of the CaCO$_3$ is left behind as a white mineral deposit. Plumbing fixtures in homes with hard water often acquire these deposits over time. Toilets, for example, may develop these deposits at the water line as the water in the toilet slowly evaporates away. If water is saturated with CaCO$_3$, how much of it has to evaporate to deposit 0.250 g of CaCO$_3$? *Hint:* Begin by using K_{sp} for CaCO$_3$ to determine its solubility.

110. Consider the following generic equilibrium in which a solid reactant is in equilibrium with a gaseous product:

$$A(s) \rightleftharpoons B(g)$$

The following diagrams represent the reaction mixture at the following points: (a) initially; (b) after a short period of time has passed; and (c) at equilibrium.

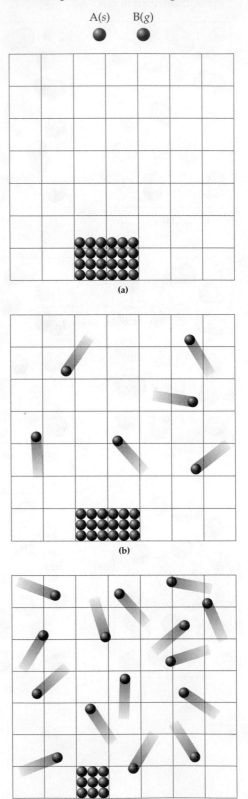

For each diagram, calculate the concentrations of the spheres representing A(s) and B(g). Assume that each block in the grid has an area of 1 cm² and report your answer in units of spheres/cm². (Since the spheres in the solid are not free to move, the solid only occupies the area that it covers. The spheres in the solid, however, are free to move and therefore occupy the entire grid.) What do you notice about the concentrations of A(s) and B(g) in these representations? Write an equilibrium expression for the generic reaction and use the results of your calculations to explain why A(s) is not included in the expression.

▶ANSWERS TO SKILLBUILDER EXERCISES

Skillbuilder 15.1 $K_{eq} = \dfrac{[HF]^2}{[H_2][F_2]}$

Skillbuilder 15.2 $K_{eq} = \dfrac{[Cl_2]^2}{[HCl]^4[O_2]}$

Skillbuilder 15.3 $K_{eq} = 14.5$

Skillbuilder Plus, p. 544 $K_{eq} = 26$

Skillbuilder 15.4 0.033 M

Skillbuilder 15.5 Adding Br_2 causes a shift to the left; adding BrNO causes a shift to the right.

Skillbuilder Plus, p. 549 Removing Br_2 causes a shift to the right

Skillbuilder 15.6 Decreasing volume causes a shift to the right; increasing volume causes a shift to the left.

Skillbuilder 15.7 Increasing the temperature shifts the reaction to the left; decreasing the temperature shifts the reaction to the right.

Skillbuilder 15.8 (a) $K_{sp} = [Ag^+][I^-]$
(b) $K_{sp} = [Ca^{2+}][OH^-]^2$

Skillbuilder 15.9 8.43×10^{-3} M

▶ANSWERS TO CONCEPTUAL CHECKPOINTS

15.1 (a) In accordance with the gas laws (Chapter 11), increasing the pressure would increase the temperature, decrease the volume, or both. Increasing the temperature would increase the reaction rate. Decreasing the volume would increase the concentration of the reactants, which would also increase the reaction rate. Therefore, we would expect that increasing the pressure would speed up the reaction.

15.2 (c) Since the image in (c) has the greatest amount of product, the equilibrium constant must be largest at T_3.

15.3 (b) For this reaction, $K_{eq} = [B][C]/[A] = (2 \times 2)/2 = 2$.

15.4 Doubling the hydrogen concentration leads to a larger shift toward products because the concentration of hydrogen in the denominator is squared.

15.5 (c) Increasing the concentration of a reactant causes the reaction to shift right. For this reaction, changing the volume has no effect because the number of gas particles on both sides of the equation is equal.

15.6 (a) Since the reaction is endothermic, increasing the temperature drives it to the right (toward the product). Since the reaction has 1 mol particles on the left and 2 mol particles on the right, increasing the volume drives it to the right. Therefore, increasing the temperature and increasing the volume will create the greatest amount of product.

CHAPTER 16

Oxidation and Reduction

"We have to learn to understand nature and not merely to observe it and endure what it imposes on us."

JOHN DESMOND BERNAL (1901–1971)

16.1 The End of the Internal Combustion Engine?

▲ **FIGURE 16.1 A fuel-cell car**
The Honda FCX Clarity, a hydrogen-powered, fuel-cell automobile. The only emission is water.

◄ Fuel-cell vehicles (FCVs), such as the one shown here, may someday replace vehicles powered by internal combustion engines. As you can see in this image, FCVs produce only water as exhaust.

It is possible, even likely, that you will see the end of the internal combustion engine within your lifetime. Although it has served us well—powering our airplanes, automobiles, and trains—its time is running out. What will replace it? If our cars don't run on gasoline, what will fuel them? The answers to these questions are not completely settled, but new and better technologies are on the horizon. The most promising of these is the use of **fuel cells** to power electric vehicles. Such whisper-quiet, environmentally friendly supercars are currently available only on a limited basis (◄ Figure 16.1), but they should become more widely available in the years to come.

In 2007, Honda's next-generation FCX Clarity fuel-cell vehicle made its European debut at the Geneva Motorshow. The four-passenger car has a top speed of 100 mph and a range of 270 miles on one tank of fuel. The electric motor is powered by hydrogen, stored as a compressed gas, and its only emission is water—which is so clean you can drink it. The FCX Clarity became available for lease on a limited basis beginning in 2008 and is slated to become available for purchase in 2015. Other automakers have similar prototype models in development. In addition, several cities around the world—including Palm Springs, California; Washington, D.C.; and Vancouver, British Columbia—are currently testing the feasibility of fuel-cell buses in a pilot program.

Fuel-cell technology is possible because of the tendency of some elements to gain electrons from other elements. The most common type of fuel cell—called the hydrogen–oxygen fuel cell—is based on the reaction between hydrogen and oxygen.

$$2\,H_2(g) + O_2(g) \longrightarrow 2\,H_2O(g)$$

In this reaction, hydrogen and oxygen form covalent bonds with one another. Recall from Section 10.2 that a single covalent bond is a shared electron pair.

However, since oxygen is more electronegative than hydrogen (see Section 10.8), the electron pair in a hydrogen–oxygen bond is *unequally* shared, with oxygen getting the larger portion. In effect, oxygen has more electrons in H_2O than in elemental O_2—it has *gained electrons* in the reaction.

In a typical reaction between hydrogen and oxygen, oxygen atoms gain the electrons directly from hydrogen atoms as the reaction proceeds. In a hydrogen–oxygen fuel cell, the same reaction occurs, but the hydrogen and oxygen are separated, forcing the electrons to move through an external wire to get from hydrogen to oxygen. These moving electrons constitute an electrical current, which is used to power the electric motor of a fuel-cell vehicle. In effect, fuel cells use the electron-gaining tendency of oxygen and the electron-losing tendency of hydrogen to force electrons to move through a wire, creating the electricity that powers the car.

Reactions involving the transfer of electrons are called **oxidation–reduction** or **redox reactions**. Besides their application to fuel-cell vehicles, redox reactions are prevalent in nature, in industry, and in many everyday processes. For example, the rusting of iron, the bleaching of hair, and the reactions occurring in batteries all involve redox reactions. Redox reactions are also responsible for providing the energy our bodies need to move, think, and stay alive.

16.2   Oxidation and Reduction: Some Definitions

Consider the following redox reactions:

$$2\,H_2(g) + O_2(g) \longrightarrow 2\,H_2O(g) \text{ (hydrogen–oxygen fuel-cell reaction)}$$

$$4\,Fe(s) + 3\,O_2(g) \longrightarrow 2\,Fe_2O_3(s) \text{ (rusting of iron)}$$

$$CH_4(g) + 2\,O_2(g) \longrightarrow CO_2(g) + 2\,H_2O(g) \text{ (combustion of methane)}$$

What do they all have in common? Each of these reactions involves one or more elements gaining oxygen. In the hydrogen–oxygen fuel-cell reaction, *hydrogen* gains oxygen as it turns into water. In the rusting of iron, *iron* gains oxygen as it turns into iron oxide, the familiar orange substance we call rust (▼ Figure 16.2). In the combustion of methane, *carbon* gains oxygen to form carbon dioxide, producing the brilliant blue flame we see on gas stoves (▼ Figure 16.3). In each

▼ **FIGURE 16.2   Slow oxidation**
Rust is produced by the oxidation of iron to form iron oxide.

▲ **FIGURE 16.3   Rapid oxidation**
The flame on a gas stove results from the oxidation of carbon in natural gas.

case, the substance that gains oxygen is oxidized in the reaction. One definition of *oxidation*—though not the most fundamental one—is simply the *gaining of oxygen.*

Now consider these same three reactions in reverse:

$$2\,H_2O(g) \longrightarrow 2\,H_2(g) + O_2(g)$$

$$2\,Fe_2O_3(s) \longrightarrow 4\,Fe(s) + 3\,O_2(g)$$

$$CO_2(g) + 2\,H_2O(g) \longrightarrow CH_4(g) + 2\,O_2(g)$$

Each of these reactions involves loss of oxygen. In the first reaction, hydrogen loses oxygen; in the second reaction, iron loses oxygen; and in the third reaction, carbon loses oxygen. In each case, the substance that loses oxygen is reduced in the reaction. One definition of *reduction* is simply the *loss of oxygen.*

Redox reactions need not involve oxygen, however. Consider, for example, the similarities between the following two reactions:

$$4\,Li(s) + O_2(g) \longrightarrow 2\,Li_2O(s)$$

$$2\,Li(s) + Cl_2(g) \longrightarrow 2\,LiCl(s)$$

In both cases, lithium (a metal with a strong tendency to lose electrons) reacts with an electronegative nonmetal (which has a tendency to gain electrons). In both cases, lithium atoms lose electrons to become positive ions—lithium is oxidized.

$$Li \longrightarrow Li^+ + e^-$$

The electrons lost by lithium are gained by the nonmetals, which become negative ions—the nonmetals are reduced.

$$O_2 + 4e^- \longrightarrow 2\,O^{2-}$$

$$Cl_2 + 2e^- \longrightarrow 2\,Cl^-$$

A more fundamental definition of **oxidation**, then, is simply the *loss of electrons*, and a more fundamental definition of **reduction** is the *gain of electrons.*

Notice that *oxidation and reduction must occur together.* If one substance loses electrons (oxidation), then another substance must gain electrons (reduction) (◀ Figure 16.4). The substance that is oxidized is called the **reducing agent** because it causes the reduction of the other substance. Similarly, the substance that is reduced is called the **oxidizing agent** because it causes the oxidation of the other substance. For example, consider our hydrogen–oxygen fuel-cell reaction.

$$\underset{\text{Reducing agent}}{2\,H_2(g)} + \underset{\text{Oxidizing agent}}{O_2(g)} \longrightarrow 2\,H_2O(g)$$

In this reaction, hydrogen is oxidized, making it the reducing agent. Oxygen is reduced, making it the oxidizing agent. Substances such as oxygen, which have a strong tendency to attract electrons, are good oxidizing agents—they tend to cause the oxidation of other substances. Substances such as hydrogen, which have a strong tendency to give up electrons, are good reducing agents—they tend to cause the reduction of other substances.

To summarize:

- Oxidation—the loss of electrons
- Reduction—the gain of electrons
- Oxidizing agent—the substance being reduced
- Reducing agent—the substance being oxidized

These definitions of oxidation and reduction are useful because they show the origin of the term *oxidation*, and they allow us to quickly identify reactions involving elemental oxygen as oxidation and reduction reactions. However, as you will see, these definitions are *not* the most fundamental.

In redox reactions between a metal and a nonmetal, the metal is oxidized and the nonmetal is reduced.

Helpful mnemonics: OIL RIG—**O**xidation **I**s **L**oss (of electrons); **R**eduction **I**s **G**ain (of electrons). LEO the lion says GER—**L**ose **E**lectrons **O**xidation; **G**ain **E**lectrons **R**eduction.

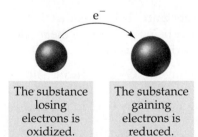

The substance losing electrons is oxidized. | The substance gaining electrons is reduced.

▲ **FIGURE 16.4 Oxidation and reduction** In a redox reaction, one substance loses electrons and another substance gains electrons.

The oxidizing agent may be an element that is itself reduced, or a compound or ion containing an element that is reduced. The reducing agent may be an element that is itself oxidized, or a compound or ion containing an element that is oxidized.

EXAMPLE 16.1 Identifying Oxidation and Reduction

Identify the substance being oxidized and the substance being reduced in each reaction.

(a) $2 Mg(s) + O_2(g) \longrightarrow 2 MgO(s)$

(b) $Fe(s) + Cl_2(g) \longrightarrow FeCl_2(s)$

(c) $Zn(s) + Fe^{2+}(aq) \longrightarrow Zn^{2+}(aq) + Fe(s)$

SOLUTION

(a) In this reaction, magnesium is gaining oxygen and losing electrons to oxygen. Mg is therefore oxidized, and O_2 is reduced.

(b) In this reaction, a metal (Fe) is reacting with an electronegative nonmetal (Cl_2). Fe loses electrons and is therefore oxidized, while Cl_2 gains electrons and is therefore reduced.

(c) In this reaction, electrons are transferred from the Zn to the Fe^{2+}. Zn loses electrons and is oxidized. Fe^{2+} gains electrons and is reduced.

▶**SKILLBUILDER 16.1 | Identifying Oxidation and Reduction**

Identify the substance being oxidized and the substance being reduced in each reaction.

(a) $2 K(s) + Cl_2(g) \longrightarrow 2 KCl(s)$

(b) $2 Al(s) + 3 Sn^{2+}(aq) \longrightarrow 2 Al^{3+}(aq) + 3 Sn(s)$

(c) $C(s) + O_2(g) \longrightarrow CO_2(g)$

▶**FOR MORE PRACTICE** Example 16.11; Problems 33, 34, 35, 36.

EXAMPLE 16.2 Identifying Oxidizing and Reducing Agents

Identify the oxidizing agent and the reducing agent in each reaction.

(a) $2 Mg(s) + O_2(g) \longrightarrow 2 MgO(s)$

(b) $Fe(s) + Cl_2(g) \longrightarrow FeCl_2(s)$

(c) $Zn(s) + Fe^{2+}(aq) \longrightarrow Zn^{2+}(aq) + Fe(s)$

SOLUTION

In the previous example, we identified the substance being oxidized and reduced for these reactions. Recall that the substance being oxidized is the reducing agent, and the substance being reduced is the oxidizing agent.

(a) Mg is oxidized and is therefore the reducing agent; O_2 is reduced and is therefore the oxidizing agent.

(b) Fe is oxidized and is therefore the reducing agent; Cl_2 is reduced and is therefore the oxidizing agent.

(c) Zn is oxidized and is therefore the reducing agent; Fe^{2+} is reduced and is therefore the oxidizing agent.

▶**SKILLBUILDER 16.2 | Identifying Oxidizing and Reducing Agents**

Identify the oxidizing agent and the reducing agent in each reaction.

(a) $2 K(s) + Cl_2(g) \longrightarrow 2 KCl(s)$

(b) $2 Al(s) + 3 Sn^{2+}(aq) \longrightarrow 2 Al^{3+}(aq) + 3 Sn(s)$

(c) $C(s) + O_2(g) \longrightarrow CO_2(g)$

▶**FOR MORE PRACTICE** Example 16.12; Problems 37, 38.

✓ **CONCEPTUAL CHECKPOINT 16.1**

An oxidizing agent:

(a) is always oxidized.

(b) is always reduced.

(c) can either be oxidized or reduced, depending on the reaction.

16.3   Oxidation States: Electron Bookkeeping

For many redox reactions, such as those involving oxygen or other highly electronegative elements, the substances being oxidized and reduced can be identified easily by inspection. For other redox reactions, identification is more difficult. For example, consider the redox reaction between carbon and sulfur.

$$C + 2 S \longrightarrow CS_2$$

What is oxidized here? What is reduced? In order to easily identify oxidation and reduction, chemists have devised a scheme to track electrons and where they go in chemical reactions. In this scheme—which is like bookkeeping for electrons—all shared electrons are assigned to the most electronegative element. Then a number—called the **oxidation state** or the **oxidation number**—is computed for each element based on the number of electrons assigned to it.

> Do not confuse oxidation state with ionic charge. A substance need not be ionic to have an assigned oxidation state.

The procedure just described is a bit cumbersome in practice. However, its main results can be summarized in a series of rules. The easiest way to assign oxidation states is to follow these rules.

Rules for Assigning Oxidation States	Examples

> These rules are hierarchical. If any two rules conflict, follow the rule that is higher on the list.

(1) The oxidation state of an atom in a free element is 0.

Cu   Cl$_2$
0 ox state   0 ox state

(2) The oxidation state of a monoatomic ion is equal to its charge.

Ca^{2+} Cl$^-$
+2 ox state   −1 ox state

(3) The sum of the oxidation states of all atoms in:
  • a neutral molecule or formula unit is 0.

H$_2$O
2(H ox state) + 1(O ox state) = 0

  • an ion is equal to the charge of the ion.

NO$_3^-$
1(N ox state) + 3(O ox state) = −1

(4) In their compounds,
  • Group I metals have an oxidation state of +1.

NaCl
+1 ox state

  • Group II metals have an oxidation state of +2.

CaF$_2$
+2 ox state

(5) In their compounds, nonmetals are assigned oxidation states according to the hierarchical table at left. Entries at the top of the table have priority over entries at the bottom.

Nonmetal	Oxidation State	Example
fluorine	−1	MgF$_2$ −1 ox state
hydrogen	+1	H$_2$O +1 ox state
oxygen	−2	CO$_2$ −2 ox state
Group 7A	−1	CCl$_4$ −1 ox state
Group 6A	−2	H$_2$S −2 ox state
Group 5A	−3	NH$_3$ −3 ox state

EXAMPLE 16.3 Assigning Oxidation States

Assign an oxidation state to each atom in each of the following.

(a) Br_2 (b) K^+ (c) LiF   (d) CO_2 (e) SO_4^{2-} (f) Na_2O_2

Since Br_2 is a free element, the oxidation state of both Br atoms is 0 (Rule 1).	**SOLUTION** (a) Br_2 Br Br 0 0
Since K^+ is a monoatomic ion, the oxidation state of the K^+ ion is +1 (Rule 2).	(b) K^+ K^+ +1
The oxidation state of Li is +1 (Rule 4). The oxidation state of F is −1 (Rule 5). Since this is a neutral compound, the sum of the oxidation states is 0 (Rule 3).	(c) LiF Li F +1 −1 sum: $+1 - 1 = 0$
The oxidation state of oxygen is −2 (Rule 5). The oxidation state of carbon must be deduced from Rule 3, which states that the sum of the oxidation states of all the atoms must be 0. Since there are two oxygen atoms, the oxidation state of O must be multiplied by 2 when calculating the sum.	(d) CO_2 (C ox state) + 2(O ox state) = 0 (C ox state) + 2(−2) = 0 (C ox state) − 4 = 0 C ox state = +4 C O_2 +4 −2 sum: $+4 + 2(-2) = 0$
The oxidation state of oxygen is −2 (Rule 5). The oxidation state of S is expected to be −2 (Rule 5). However, if that were the case, the sum of the oxidation states would not equal the charge of the ion. Since O is higher on the list, it takes priority, and the oxidation state of sulfur is calculated by setting the sum of all of the oxidation states equal to −2 (the charge of the ion).	(e) SO_4^{2-} (S ox state) + 4(O ox state) = −2 (S ox state) + 4(−2) = −2 (Oxygen takes priority over sulfur.) (S ox state) − 8 = −2 S ox state = −2 + 8 S ox state = +6 S O_4^{2-} +6 −2 sum: $+6 + 4(-2) = -2$
The oxidation state of sodium is +1 (Rule 4). The oxidation state of O is expected to be −2 (Rule 5). However, Na takes priority, and we deduce the oxidation state of O by setting the sum of all of the oxidation states equal to 0.	(f) Na_2O_2 2(Na ox state) + 2(O ox state) = 0 2(+1) + 2(O ox state) = 0 (Sodium takes priority over oxygen.) +2 + 2(O ox state) = 0 O ox state = −1 Na_2 O_2 +1 −1 sum: $2(+1) + 2(-1) = 0$

▶SKILLBUILDER 16.3 | Assigning Oxidation States

Assign an oxidation state to each atom in each of the following.

(a) Zn   (b) Cu^{2+} (c) $CaCl_2$ (d) CF_4 (e) NO_2^- (f) SO_3

▶FOR MORE PRACTICE   Example 16.13; Problems 45, 46, 47, 48, 49, 50, 51, 52, 53, 54.

EVERYDAY CHEMISTRY

The Bleaching of Hair

College students, both male and female, with bleached hair are a common sight on most campuses. Many students bleach their own hair with home-bleaching kits available at most drugstores and supermarkets. These kits normally contain hydrogen peroxide (H_2O_2), an excellent oxidizing agent. When applied to hair, hydrogen peroxide oxidizes melanin, the dark pigment that gives hair color. Once melanin is oxidized, it no longer imparts a dark color to hair, leaving the hair with the familiar bleached look.

Hydrogen peroxide also oxidizes other components of hair. For example, the protein molecules in hair contain —SH groups called *thiols*. Thiols are normally slippery (they slide across each other). Hydrogen peroxide oxidizes these thiol groups to sulfonic acid groups (—SO_3H). Sulfonic acid groups are stickier, causing hair to tangle more easily. Consequently, people with heavily bleached hair often use conditioners. Conditioners contain compounds that form thin, lubricating coatings on individual hair shafts. These coatings prevent tangling and make hair softer and more manageable.

CAN YOU ANSWER THIS? *Assign oxidation states to the atoms of H_2O_2. Which atoms in H_2O_2 do you think change oxidation state when H_2O_2 oxidizes hair?*

▲ Hair is often bleached using hydrogen peroxide, a good oxidizing agent.

Now let's return to our original question. What is being oxidized and what is being reduced in the following reaction?

$$C + 2S \longrightarrow CS_2$$

We use the oxidation state rules to assign oxidation states to all elements on both sides of the equation.

$$\underset{0}{C} + \underset{0}{2S} \longrightarrow \underset{+4-2}{CS_2}$$

Carbon went from an oxidation state of 0 to +4. In terms of our electron bookkeeping scheme (the assigned oxidation state), carbon *lost electrons* and was *oxidized*. Sulfur went from an oxidation state of 0 to −2. In terms of our electron bookkeeping scheme, sulfur *gained electrons* and was *reduced*.

In terms of oxidation states, we define oxidation and reduction as follows.

Oxidation—an increase in oxidation state
Reduction—a decrease in oxidation state

EXAMPLE 16.4   Using Oxidation States to Identify Oxidation and Reduction

Use oxidation states to identify the element that is being oxidized and the element that is being reduced in the redox reaction.

$$Ca(s) + 2\,H_2O(l) \longrightarrow Ca(OH)_2(aq) + H_2(g)$$

Assign an oxidation state to each atom in the reaction. Since Ca increased in oxidation state, it was oxidized. Since H decreased in oxidation state, it was reduced. (Note that oxygen has the same oxidation state on both sides of the equation and was therefore neither oxidized nor reduced.)	**SOLUTION**

▶**SKILLBUILDER 16.4 | Using Oxidation States to Identify Oxidation and Reduction**

Use oxidation states to identify the element that is being oxidized and the element that is being reduced in the redox reaction.

$$Sn(s) + 4\,HNO_3(aq) \longrightarrow SnO_2(s) + 4\,NO_2(g) + 2\,H_2O(g)$$

▶**FOR MORE PRACTICE**   Problems 57, 58, 59, 60.

 CONCEPTUAL CHECKPOINT 16.2

In which of the following does nitrogen have the *lowest* oxidation state?

(a) N_2

(b) NO

(c) NO_2

(d) NH_3

16.4   Balancing Redox Equations

In Chapter 7, we learned how to balance chemical equations by inspection. Some redox reactions can be balanced in this way. However, redox reactions occurring in aqueous solutions are usually difficult to balance by inspection and require a special procedure called the *half-reaction method of balancing*. In this procedure, the overall equation is broken down into two **half-reactions**: one for oxidation and one for reduction. The half-reactions are balanced individually and then added together. For example, consider the redox reaction:

$$Al(s) + Ag^+(aq) \longrightarrow Al^{3+}(aq) + Ag(s)$$

We assign oxidation numbers to all atoms to determine what is being oxidized and what is being reduced.

We then divide the reaction into two half-reactions, one for oxidation and one for reduction.

$$\text{Oxidation:} \quad Al(s) \longrightarrow Al^{3+}(aq)$$

$$\text{Reduction:} \quad Ag^+(aq) \longrightarrow Ag(s)$$

We then balance the two half-reactions individually. In this case, the half-reactions are already balanced with respect to mass—the number of each type of atom on both sides of each half-reaction is the same. However, the equations are not balanced with respect to charge—in the oxidation half-reaction, the left side of the equation has 0 charge while the right side has +3 charge, and in the reduction half-reaction, the left side has +1 charge and the right side has 0 charge. We balance the charge of each half-reaction individually by adding the appropriate number of electrons to make the charges on both sides equal.

$$Al(s) \longrightarrow Al^{3+}(aq) + \mathbf{3e^-} \qquad \text{(zero charge on both sides)}$$

$$\mathbf{1e^-} + Ag^+(aq) \longrightarrow Ag(s) \qquad \text{(zero charge on both sides)}$$

Since these half-reactions must occur together, the number of electrons lost in the oxidation half-reaction must equal the number gained in the reduction half-reaction. We equalize these by multiplying one or both half-reactions by appropriate whole numbers to equalize the electrons lost and gained. In this case, we multiply the reduction half-reaction by 3.

$$Al(s) \longrightarrow Al^{3+}(aq) + 3e^-$$

$$3 \times [1e^- + Ag^+(aq) \longrightarrow Ag(s)]$$

We then add the half-reactions together, canceling electrons and other species as necessary.

$$Al(s) \longrightarrow Al^{3+}(aq) + \cancel{3e}$$

$$\cancel{3e} + 3\,Ag^+(aq) \longrightarrow 3\,Ag(s)$$

$$\overline{Al(s) + 3\,Ag^+(aq) \longrightarrow Al^{3+}(aq) + 3\,Ag(s)}$$

Lastly, we verify that the equation is balanced, with respect to both mass and charge as shown in the margin. Notice that the charge need not be zero on both sides of the equation—it just has to be *equal* on both sides. The equation is balanced.

A general procedure for balancing redox reactions is given in the following procedure box. Since aqueous solutions are often acidic or basic, the procedure must account for the presence of H^+ ions or OH^- ions. We cover acidic solutions in the examples that follow and demonstrate how to balance redox reactions in basic solutions in Example 16.8.

Reactants	Products
1 Al	1 Al
3 Ag	3 Ag
+3 charge	+3 charge

Balancing Redox Equations Using the Half-Reaction Method	EXAMPLE 16.5	EXAMPLE 16.6
	Balance the redox reaction.	Balance the redox reaction.
1. Assign oxidation states to all atoms and identify the substances being oxidized and reduced.	**SOLUTION** $Al(s) + Cu^{2+}(aq) \longrightarrow$ $\quad\;0 \qquad\;\; +2$ $Al^{3+}(aq) + Cu(s)$ $\quad\;\;+3 \qquad\quad 0$ Oxidation — Reduction	**SOLUTION** $Fe^{2+}(aq) + MnO_4^-(aq) \longrightarrow$ $\quad+2 \qquad\qquad +7\;-2$ $Fe^{3+}(aq) + Mn^{2+}(aq)$ $\quad\;\;+3 \qquad\qquad +2$ Oxidation — Reduction
2. Separate the overall reaction into two half-reactions, one for oxidation and one for reduction.	**OXIDATION** $Al(s) \longrightarrow Al^{3+}(aq)$ **REDUCTION** $Cu^{2+}(aq) \longrightarrow Cu(s)$	**OXIDATION** $Fe^{2+}(aq) \longrightarrow Fe^{3+}(aq)$ **REDUCTION** $MnO_4^-(aq) \longrightarrow Mn^{2+}(aq)$

3. Balance each half-reaction with respect to mass in the following order.
 	
• Balance all elements other than H and O.
 	
• Balance O by adding H_2O.
 	
• Balance H by adding H^+.
	All elements other than hydrogen and oxygen are balanced, so proceed to the next step.
	
	    No oxygen; proceed to the next step.
	
	    No hydrogen; proceed to the next step.
		All elements other than hydrogen and oxygen are balanced, so proceed to next step.
		
		$$Fe^{2+}(aq) \longrightarrow Fe^{3+}(aq)$$
		$$MnO_4^-(aq) \longrightarrow$$
		$$Mn^{2+}(aq) + \mathbf{4\ H_2O}(l)$$
		$$Fe^{2+}(aq) \longrightarrow Fe^{3+}(aq)$$
		$$\mathbf{8\ H^+}(aq) + MnO_4^-(aq) \longrightarrow$$
		$$Mn^{2+}(aq) + 4\ H_2O(l)$$

4. Balance each half-reaction with respect to charge by adding electrons to the right side of the oxidation half-reaction and the left side of the reduction half-reaction. (The sum of the charges on both sides of each equation should then be equal.)	$$Al(s) \longrightarrow Al^{3+}(aq) + \mathbf{3e^-}$$ $$\mathbf{2e^-} + Cu^{2+}(aq) \longrightarrow Cu(s)$$	$$Fe^{2+}(aq) \longrightarrow Fe^{3+}(aq) + \mathbf{1e^-}$$ $$\mathbf{5e^-} + 8\ H^+(aq) + MnO_4^-(aq) \longrightarrow$$ $$Mn^{2+}(aq) + 4\ H_2O(l)$$
5. Make the number of electrons in both half-reactions equal by multiplying one or both half-reactions by a small whole number.	$$2 \times [Al(s) \longrightarrow Al^{3+}(aq) + 3e^-]$$ $$3 \times [2e^- + Cu^{2+}(aq) \longrightarrow Cu(s)]$$	$$5 \times [Fe^{2+}(aq) \longrightarrow Fe^{3+}(aq) + 1e^-]$$ $$5e^- + 8\ H^+(aq) + MnO_4^-(aq) \longrightarrow$$ $$Mn^{2+}(aq) + 4\ H_2O(l)$$
6. Add the two half-reactions together, canceling electrons and other species as necessary.	$$2\,Al(s) \longrightarrow 2\,Al^{3+}(aq) + \cancel{6e^-}$$ $$\cancel{6e^-} + 3\,Cu^{2+}(aq) \longrightarrow 3\,Cu(s)$$ $$2\,Al(s) + 3\,Cu^{2+}(aq) \longrightarrow$$ $$2\,Al^{3+}(aq) + 3\,Cu(s)$$	$$5\,Fe^{2+}(aq) \longrightarrow 5\,Fe^{3+}(aq) + \cancel{5e^-}$$ $$\cancel{5e^-} + 8\,H^+(aq) + MnO_4^-(aq)$$ $$\longrightarrow Mn^{2+}(aq) + 4\,H_2O(l)$$ $$5\,Fe^{2+}(aq) + 8\,H^+(aq) + MnO_4^-(aq)$$ $$\longrightarrow 5\,Fe^{3+}(aq) + Mn^{2+}(aq) + 4\,H_2O(l)$$
7. Verify that the reaction is balanced with respect to both mass and charge.	<table><tr><td>Reactants</td><td>Products</td></tr><tr><td>2 Al</td><td>2 Al</td></tr><tr><td>3 Cu</td><td>3 Cu</td></tr><tr><td>+6 charge</td><td>+6 charge</td></tr></table>	<table><tr><td>Reactants</td><td>Products</td></tr><tr><td>5 Fe</td><td>5 Fe</td></tr><tr><td>8 H</td><td>8 H</td></tr><tr><td>1 Mn</td><td>1 Mn</td></tr><tr><td>4 O</td><td>4 O</td></tr><tr><td>+17 charge</td><td>+17 charge</td></tr></table>

▶**SKILLBUILDER 16.5**

Balance the following redox reaction occurring in acidic solution.

$$H^+(aq) + Cr(s) \longrightarrow$$
$$H_2(g) + Cr^{3+}(aq)$$

▶**SKILLBUILDER 16.6**

Balance the following redox reaction occurring in acidic solution.

$$Cu(s) + NO_3^-(aq) \longrightarrow$$
$$Cu^{2+}(aq) + NO_2(g)$$

▶**FOR MORE PRACTICE** Example 16.14; Problems 61, 62, 63, 64.

EXAMPLE 16.7   Balancing Redox Reactions

Balance the following redox reaction occurring in acidic solution.

$$I^-(aq) + Cr_2O_7{}^{2-}(aq) \longrightarrow Cr^{3+}(aq) + I_2(s)$$

1. Follow the half-reaction method for balancing redox reactions. Begin by assigning oxidation states.	**SOLUTION**
2. Separate the overall reaction into two half-reactions.	**OXIDATION** $I^-(aq) \longrightarrow I_2(s)$ **REDUCTION** $Cr_2O_7{}^{2-}(aq) \longrightarrow Cr^{3+}(aq)$
3. Balance each half-reaction with respect to mass. • Balance all elements other than H and O. • Balance O by adding H_2O. • Balance H by adding H^+.	$2\,I^-(aq) \longrightarrow I_2(aq)$ $Cr_2O_7{}^{2-}(aq) \longrightarrow 2\,Cr^{3+}(s)$ $2\,I^-(aq) \longrightarrow I_2(s)$ $Cr_2O_7{}^{2-}(aq) \longrightarrow 2\,Cr^{3+}(aq) + \mathbf{7\,H_2O}(l)$ $2\,I^-(aq) \longrightarrow I_2(s)$ $\mathbf{14\,H^+}(aq) + Cr_2O_7{}^{2-}(aq) \longrightarrow 2\,Cr^{3+}(aq) + 7\,H_2O(l)$
4. Balance each half-reaction with respect to charge.	$2\,I^-(aq) \longrightarrow I_2(s) + \mathbf{2e^-}$ $\mathbf{6e^-} + 14\,H^+(aq) + Cr_2O_7{}^{2-}(aq) \longrightarrow 2\,Cr^{3+}(aq) + 7\,H_2O(l)$
5. Make the number of electrons in both half-reactions equal.	$\mathbf{3} \times [\,2\,I^-(aq) \longrightarrow I_2(s) + 2e^-\,]$ $6e^- + 14\,H^+(aq) + Cr_2O_7{}^{2-}(aq) \longrightarrow 2\,Cr^{3+}(aq) + 7\,H_2O(l)$
6. Add the half-reactions together.	$6\,I^-(aq) \longrightarrow 3\,I_2(s) + \cancel{6e^-}$ $\cancel{6e^-} + 14\,H^+(aq) + Cr_2O_7{}^{2-}(aq) \longrightarrow 2\,Cr^{3+}(aq) + 7\,H_2O(l)$ ——————————————————————————— $6\,I^-(aq) + 14\,H^+(aq) + Cr_2O_7{}^{2-}(aq) \longrightarrow 3\,I_2(s) + 2\,Cr^{3+}(aq) + 7\,H_2O(l)$
7. Verify that the reaction is balanced.	**Reactants** **Products** 6 I 6 I 14 H 14 H 2 Cr 2 Cr 7 O 7 O +6 charge +6 charge

▶SKILLBUILDER 16.7 | Balancing Redox Reactions

Balance the following redox reaction occurring in acidic solution.

$$Sn(s) + MnO_4{}^-(aq) \longrightarrow Sn^{2+}(aq) + Mn^{2+}(aq)$$

▶FOR MORE PRACTICE   Problems 65, 66, 67, 68.

When a redox reaction occurs in basic solution, we follow the same general procedure with one addition: the neutralization of H^+ with OH^-. The H^+ and OH^- combine to form water, as shown in the following example.

EXAMPLE 16.8 **Balancing Redox Reactions Occurring in Basic Solution**

Balance the following redox reaction occurring in basic solution.

$$CN^-(aq) + MnO_4^-(aq) \longrightarrow CNO^-(aq) + MnO_2(s) \quad \text{(basic solution)}$$

1. Follow the half-reaction method for balancing redox reactions. Begin by assigning oxidation states.	**SOLUTION** $CN^-(aq) + MnO_4^-(aq) \longrightarrow CNO^-(aq) + MnO_2(s)$ $\;\;\;\;+2\;-3\quad\quad\;+7\;-2\quad\quad\quad\;\;+4\;-3\;-2\quad\;\;+4\;-2$ Reduction / Oxidation
2. Separate the overall reaction into two half-reactions.	**OXIDATION** $CN^-(aq) \longrightarrow CNO^-(aq)$ **REDUCTION** $MnO_4^-(aq) \longrightarrow MnO_2(s)$
3. Balance each half-reaction with respect to mass.	
• Balance all elements other than H and O.	All elements other than H and O are already balanced.
• Balance O by adding H_2O.	$CN^-(aq) + \mathbf{H_2O}(l) \longrightarrow CNO^-(aq)$ $MnO_4^-(aq) \longrightarrow MnO_2(s) + \mathbf{2\,H_2O}(l)$
• Balance H by adding H^+.	$CN^-(aq) + H_2O(l) \longrightarrow CNO^-(aq) + \mathbf{2\,H^+}(aq)$ $MnO_4^-(aq) + \mathbf{4\,H^+}(aq) \longrightarrow MnO_2(s) + 2\,H_2O(l)$
• Neutralize H^+ by adding OH^-. Add the same number of OH^- to each side of the equation (to preserve mass balance).	$CN^-(aq) + H_2O(l) + \mathbf{2\,OH^-}(aq) \longrightarrow CNO^-(aq) + \underbrace{2\,H^+(aq) + \mathbf{2\,OH^-}(aq)}_{2\,H_2O(l)}$ $MnO_4^-(aq) + \underbrace{4\,H^+(aq) + \mathbf{4\,OH^-}(aq)}_{4\,H_2O(l)} \longrightarrow MnO_2(s) + 2\,H_2O(l) + \mathbf{4\,OH^-}(aq)$
• Cancel any water molecules that occur on both sides of a half reaction.	$CN^-(aq) + \cancel{H_2O}(l) + 2\,OH^-(aq) \longrightarrow CNO^-(aq) + 2\,H_2O(l)$ $MnO_4^-(aq) + \overset{2}{\cancel{4}}\,H_2O(l) \longrightarrow MnO_2(s) + \cancel{2\,H_2O}(l) + 4\,OH^-(aq)$
4. Balance each half-reaction with respect to charge.	$CN^-(aq) + 2\,OH^-(aq) \longrightarrow CNO^-(aq) + H_2O(l) + \mathbf{2e^-}$ $\mathbf{3e^-} + MnO_4^-(aq) + 2\,H_2O(l) \longrightarrow MnO_2(s) + 4\,OH^-(aq)$
5. Make the number of electrons in both half-reactions equal.	$\mathbf{3} \times [CN^-(aq) + 2\,OH^-(aq) \longrightarrow CNO^-(aq) + H_2O(l) + 2e^-]$ $\mathbf{2} \times [3e^- + MnO_4^-(aq) + 2\,H_2O(l) \longrightarrow MnO_2(s) + 4\,OH^-(aq)]$
6. Add the half-reactions together and cancel.	$3\,CN^-(aq) + 6\,\cancel{OH^-}(aq) \longrightarrow 3\,CNO^-(aq) + 3\,\cancel{H_2O}(l) + \cancel{6e^-}$ $\cancel{6e^-} + 2\,MnO_4^-(aq) + \overset{}{\cancel{4}}\,H_2O(l) \longrightarrow 2\,MnO_2(s) + \overset{2}{\cancel{8}}\,OH^-(aq)$ $3\,CN^-(aq) + 2\,MnO_4^-(aq) + H_2O(l) \longrightarrow 3\,CNO^-(aq) + 2\,MnO_2(s) + 2\,OH^-(aq)$
7. Verify that the reaction is balanced.	**Reactants** **Products** 3 C 3 C 3 N 3 N 2 Mn 2 Mn 9 O 9 O 2 H 2 H −5 charge −5 charge

▶**SKILLBUILDER 16.8** | **Balancing Redox Reactions**

Balance the following redox reaction occurring in basic solution.

$$H_2O_2(aq) + ClO_2(aq) \longrightarrow ClO_2^-(aq) + O_2(g)$$

▶**FOR MORE PRACTICE** Problems 69, 70.

CHEMISTRY IN THE ENVIRONMENT
Photosynthesis and Respiration: Energy for Life

All living things require energy, and most of that energy comes from the sun. Solar energy reaches Earth in the form of electromagnetic radiation (Chapter 9). This radiation keeps our planet at a temperature that allows life as we know it to flourish. But the wavelengths that make up visible light have an additional and very crucial role to play in the maintenance of life. Plants capture this light and use it to make energy-rich organic molecules such as carbohydrates. (These compounds are discussed more fully in Chapter 19.) Animals get their energy by eating plants or by eating other animals that have eaten plants. So ultimately, virtually all of the energy for life comes from sunlight.

But in chemical terms, how is this energy captured, transferred from organism to organism, and used? *The key reactions in these processes all involve oxidation and reduction.*

Most living things use chemical energy through a process known as *respiration*. In respiration, energy-rich molecules, typified by the sugar glucose, are "burned" in a reaction that can be summarized as follows:

$$\underset{\text{Glucose}}{C_6H_{12}O_6} + \underset{\text{Oxygen}}{6\,O_2} \longrightarrow \underset{\text{Water}}{6\,CO_2} + \text{energy}$$

You can easily see that respiration is a redox reaction. On the simplest level, it is clear that some of the atoms in glucose are gaining oxygen. More precisely, we can use the rules for assigning oxidation states to show that carbon is oxidized from an oxidation number of 0 in glucose to +4 in carbon dioxide.

Respiration is also an exothermic reaction—it releases energy. If you burn glucose in a test tube, the energy is lost as heat. Living things, however, have devised ways of capturing the energy released and using it to power their life processes, such as movement, growth, and the synthesis of other life-sustaining molecules.

Respiration is one half of a larger cycle; the other half is *photosynthesis*. Photosynthesis is the series of reactions by which green plants capture the energy of sunlight and store it as chemical energy in compounds such as glucose. Photosynthesis can be summarized as follows:

$$\underset{\substack{\text{Carbon}\\\text{dioxide}}}{6\,CO_2} + \underset{\text{Water}}{6\,H_2O} + \text{energy (sunlight)} \longrightarrow \underset{\text{Glucose}}{C_6H_{12}O_6} + \underset{\text{Oxygen}}{6\,O_2}$$

This reaction—the exact reverse of respiration—is the ultimate source of the molecules that are oxidized in respiration. And just as the key process in respiration is the oxidation of carbon, the key process in photosynthesis is the reduction of carbon. This reduction is driven by solar energy—energy that is then stored in the resulting glucose molecule. Living things harvest that energy when they "burn" glucose in the respiration half of the cycle.

Thus, oxidation and reduction reactions are at the very center of all life on Earth!

CAN YOU ANSWER THIS? *What is the oxidation state of the oxygen atoms in CO_2, H_2O, and O_2? What does this information tell you about the reactions of photosynthesis and respiration?*

▲ Sunlight, captured by plants in photosynthesis, is the ultimate source of the chemical energy used by nearly all living things on Earth.

16.5 The Activity Series: Predicting Spontaneous Redox Reactions

As we have seen, redox reactions depend on the gaining of electrons by one substance and the losing of electrons by another. Is there a way to predict whether a particular redox reaction will spontaneously occur? Suppose we knew that substance A has a greater tendency to lose electrons than substance B (A is more easily oxidized than B). Then we could predict that if we mix A with cations of B, a redox reaction would occur in which A loses its electrons (A is oxidized) to the cations of B (B cations are reduced). For example, Mg has a greater tendency to lose electrons than Cu. Consequently, if we put solid Mg into a solution containing Cu^{2+} ions, Mg is oxidized and Cu^{2+} is reduced.

$$Mg(s) + Cu^{2+}(aq) \longrightarrow Mg^{2+}(aq) + Cu(s)$$

▶ **FIGURE 16.5 Cu²⁺ oxidizes magnesium** When a magnesium strip is put into a Cu^{2+} solution, the magnesium is oxidized to Mg^{2+} and the copper ion is reduced to $Cu(s)$. Notice the fading of the blue color (due to Cu^{2+} ions) in solution and the appearance of solid copper on the magnesium strip.

Magnesium strip

Cu^{2+} solution

$$Mg(s) + Cu^{2+}(aq) \longrightarrow Mg^{2+}(aq) + Cu(s)$$

We see this as the fading of blue (the color of the Cu^{2+} ions in solution), the dissolving of the solid magnesium, and the appearance of solid copper on the remaining magnesium surface (▲ Figure 16.5). This reaction is spontaneous—it occurs on its own when $Mg(s)$ and $Cu^{2+}(aq)$ come into contact. On the other hand, if we put $Cu(s)$ in a solution containing $Mg^{2+}(aq)$ ions, no reaction occurs (◀ Figure 16.6).

$$Cu(s) + Mg^{2+}(aq) \longrightarrow \text{NO REACTION}$$

No reaction occurs because, as we said previously, Mg atoms have a greater tendency to lose electrons than do Cu atoms; Cu atoms will therefore not lose electrons to Mg^{2+} ions.

Table 16.1 shows the **activity series of metals**. This table lists metals in order of decreasing tendency to lose electrons. The metals at the top of the list have the greatest tendency to lose electrons—they are most easily oxidized and therefore the most reactive. The metals at the bottom of the list have the lowest tendency to lose electrons—they are the most difficult to oxidize and therefore the least reac-

Copper strip

Mg^{2+} solution

▲ **FIGURE 16.6 Mg²⁺ does not oxidize copper** When solid copper is placed in a solution containing Mg^{2+} ions, no reaction occurs. **Question: Why?**

TABLE 16.1 Activity Series of Metals

$Li(s) \longrightarrow Li^{+}(aq) + e^{-}$	Most reactive
$K(s) \longrightarrow K^{+}(aq) + e^{-}$	Most easily oxidized
$Ca(s) \longrightarrow Ca^{2+}(aq) + 2e^{-}$	Strongest tendency to lose electrons
$Na(s) \longrightarrow Na^{+}(aq) + e^{-}$	
$Mg(s) \longrightarrow Mg^{2+}(aq) + 2e^{-}$	
$Al(s) \longrightarrow Al^{3+}(aq) + 3e^{-}$	
$Mn(s) \longrightarrow Mn^{2+}(aq) + 2e^{-}$	
$Zn(s) \longrightarrow Zn^{2+}(aq) + 2e^{-}$	
$Cr(s) \longrightarrow Cr^{3+}(aq) + 3e^{-}$	
$Fe(s) \longrightarrow Fe^{2+}(aq) + 2e^{-}$	
$Ni(s) \longrightarrow Ni^{2+}(aq) + 2e^{-}$	
$Sn(s) \longrightarrow Sn^{2+}(aq) + 2e^{-}$	
$Pb(s) \longrightarrow Pb^{2+}(aq) + 2e^{-}$	
$\mathbf{H_2(g) \longrightarrow 2\,H^{+}(aq) + 2e^{-}}$	
$Cu(s) \longrightarrow Cu^{2+}(aq) + 2e^{-}$	Least reactive
$Ag(s) \longrightarrow Ag^{+}(aq) + e^{-}$	Most difficult to oxidize
$Au(s) \longrightarrow Au^{3+}(aq) + 3e^{-}$	Least tendency to lose electrons

▲ Gold is very low on the activity series. Because it is so difficult to oxidize, it resists the tarnishing and corrosion that more active metals undergo.

tive. It is not a coincidence that the metals used for jewelry, such as silver and gold, are near the bottom of the list. They are among the least reactive metals and therefore do not form compounds easily. Instead, they tend to remain as solid silver and solid gold rather than being oxidized to silver and gold cations by elements in the environment (such as the oxygen in the air).

Each reaction in the activity series is an oxidation half-reaction. The half-reactions at the top are most likely to occur in the *forward* direction, and the half-reactions at the bottom are most likely to occur in the *reverse* direction. Consequently, if we pair a half-reaction from the top of the list with the reverse of a half-reaction from the bottom of the list, we get a spontaneous reaction. More specifically,

> Any half-reaction on the list will be spontaneous when paired with the reverse of any half-reaction below it.

For example, consider the two half-reactions:

$$Mn(s) \longrightarrow Mn^{2+}(aq) + 2e^-$$
$$Ni(s) \longrightarrow Ni^{2+}(aq) + 2e^-$$

The oxidation of Mn is spontaneous when paired with the reduction of Ni^{2+}.

$$Mn(s) \longrightarrow Mn^{2+}(aq) + 2e^-$$
$$\underline{Ni^{2+}(aq) + 2e^- \longrightarrow Ni(s)}$$
$$Mn(s) + Ni^{2+}(aq) \longrightarrow Mn^{2+}(aq) + Ni(s) \quad \text{(spontaneous reaction)}$$

However, if we pair a half-reaction on the list with the reverse of a half-reaction above it, there is no reaction. For example,

$$Mn(s) \longrightarrow Mn^{2+}(aq) + 2e^-$$
$$\underline{Mg^{2+}(aq) + 2e^- \longrightarrow Mg(s)}$$
$$Mn(s) + Mg^{2+}(aq) \longrightarrow \text{NO REACTION}$$

No reaction occurs because Mg has a greater tendency to be oxidized than Mn. Since it is already oxidized in this reaction, nothing else happens.

EXAMPLE 16.9   Predicting Spontaneous Redox Reactions

Are the following redox reactions spontaneous?

(a) $Fe(s) + Mg^{2+}(aq) \longrightarrow Fe^{2+}(aq) + Mg(s)$

(b) $Fe(s) + Pb^{2+}(aq) \longrightarrow Fe^{2+}(aq) + Pb(s)$

SOLUTION

(a) $Fe(s) + Mg^{2+}(aq) \longrightarrow Fe^{2+}(aq) + Mg(s)$

This reaction involves the oxidation of Fe
$$Fe(s) \longrightarrow Fe^{2+}(aq) + 2e^-$$
with the reverse of a half-reaction *above it* in the activity series.
$$Mg^{2+}(aq) + 2e^- \longrightarrow Mg(s)$$
Therefore, the reaction *is not* spontaneous.

(b) $Fe(s) + Pb^{2+}(aq) \longrightarrow Fe^{2+}(aq) + Pb(s)$

This reaction involves the oxidation of Fe
$$Fe(s) \longrightarrow Fe^{2+}(aq) + 2e^-$$
with the reverse of a half-reaction *below it* in the activity series.
$$Pb^{2+}(aq) + 2e^- \longrightarrow Pb(s)$$
Therefore, the reaction *is* spontaneous.

▶**SKILLBUILDER 16.9** | **Predicting Spontaneous Redox Reactions**

Are the following redox reactions spontaneous?

(a) $Zn(s) + Ni^{2+}(aq) \longrightarrow Zn^{2+}(aq) + Ni(s)$

(b) $Zn(s) + Ca^{2+}(aq) \longrightarrow Zn^{2+}(aq) + Ca(s)$

▶**FOR MORE PRACTICE**   Example 16.15; Problems 77, 78.

PREDICTING WHETHER A METAL WILL DISSOLVE IN ACID

In Chapter 14, we learned that acids dissolve metals. Most acids dissolve metals by the reduction of H^+ ions to hydrogen gas and the corresponding oxidation of the metal to its ion. For example, if solid Zn is dropped into hydrochloric acid, the following reaction occurs.

$$Zn(s) \longrightarrow Zn^{2+}(aq) + 2e^-$$
$$2\,H^+(aq) + 2e^- \longrightarrow H_2(g)$$
$$Zn(s) + 2\,H^+(aq) \longrightarrow Zn^{2+}(aq) + H_2(g)$$

We observe the reaction as the dissolving of the zinc and the bubbling of hydrogen gas (◀ Figure 16.7). The zinc is oxidized and the H^+ ions are reduced, dissolving the zinc. Notice that this reaction involves pairing the oxidation half-reaction of Zn with the reverse of a half-reaction below zinc on the activity series (the reduction of H^+). Therefore, this reaction is spontaneous. What happens, however, if we pair the oxidation of Cu with the reduction of H^+? The reaction would not be spontaneous because it involves pairing the oxidation of copper with the reverse of a half-reaction *above it* in the activity series. Consequently, copper does not react with H^+ and will not dissolve in acids such as HCl. In general,

Metals above H_2 on the activity series dissolve in acids, while metals below H_2 do not dissolve in acids.

An important exception to this rule is nitric acid (HNO_3), which through a different reduction half-reaction dissolves some of the metals below H_2 in the activity series.

$Zn(s) + 2\,H^+(aq) \longrightarrow$

$Zn^{2+}(aq) + H_2(g)$

▲ **FIGURE 16.7** **Zinc dissolves in hydrochloric acid**   The zinc metal is oxidized to Zn^{2+} ions, and the H^+ ions are reduced to hydrogen gas.

| Unlike most other acids, nitric acid dissolves some metals below H_2 in the activity series.

EXAMPLE 16.10   **Predicting Whether a Metal Will Dissolve in Acid**

Does Cr dissolve in hydrochloric acid?

SOLUTION

Yes. Since Cr is above H_2 in the activity series, it dissolves in HCl.

▶**SKILLBUILDER 16.10** | **Predicting Whether a Metal Will Dissolve in Acid**

Does Ag dissolve in hydrobromic acid?

▶**FOR MORE PRACTICE**   Problems 83, 84.

✔ **CONCEPTUAL CHECKPOINT 16.3**

It has been suggested that one cause of the decline of the Roman Empire was widespread chronic poisoning. The suspected source was a metal present in the vessels commonly used to store and serve acidic substances such as wine. Which metal would you expect to pose such a danger?

(a) silver

(b) gold

(c) lead

(d) copper

16.6   Batteries: Using Chemistry to Generate Electricity

▲ **FIGURE 16.8   An electrical current**   Electrical current is the flow of electrical charge. In this figure, electrons are flowing through a wire.

The fuel cell that we discussed in the opening section of this chapter is a type of electrochemical cell.

Electrical current is simply the flow of electric charge (◀ Figure 16.8). Electrons flowing through a wire or ions flowing through a solution are both examples of electrical current. Since redox reactions involve the transfer of electrons from one species to another, they can create electrical current. For example, consider the following spontaneous redox reaction.

$$Zn(s) + Cu^{2+}(aq) \longrightarrow Zn^{2+}(aq) + Cu(s)$$

When Zn metal is placed into a Cu^{2+} solution, Zn is oxidized and Cu^{2+} is reduced—electrons are transferred directly from the Zn to the Cu^{2+}. Suppose we separate the reactants and force the electrons to travel through a wire to get from the Zn to the Cu^{2+}. The flowing electrons constitute an electrical current and can be used to do electrical work. This process is normally carried out in an **electrochemical cell**, a device that creates electrical current from a spontaneous redox reaction (or that uses electrical current to drive a nonspontaneous redox reaction). Electrochemical cells that create electrical current from spontaneous reactions are called **voltaic cells** or **galvanic cells**. A *battery* is a voltaic cell that (usually) has been designed for portability.

For example, consider the voltaic cell in ▶ Figure 16.9. In this cell, a solid strip of Zn is placed into a $Zn(NO_3)_2$ solution to form a **half-cell**. Similarly, a solid strip of Cu is placed into a $Cu(NO_3)_2$ solution to form a second half-cell. Then the two half-cells are connected by attaching a wire from the Zn, through a lightbulb or other electrical device, to the copper. The natural tendency of Zn to oxidize and Cu^{2+} to reduce results in the flow of electrons through the wire. The flowing electrons constitute an electrical current that lights the bulb.

In a voltaic cell, the metal strip where oxidation occurs is called the **anode** and is labeled with a negative (−) sign. The metal strip where reduction occurs is called the **cathode** and is labeled with a (+) sign. Electrons flow from the anode to the cathode (away from negative and toward positive).

As electrons flow out of the anode, positive ions form in the oxidation half-cell (Zn^{2+} forms in the preceding example). As electrons flow into the cathode, positive ions deposit as charge-neutral atoms at the reduction half-cell (Cu^{2+} deposits as $Cu(s)$ in the preceding example). However, if this were the only flow of charge, the flow would soon stop as positive charge accumulated at the anode and negative charge at the cathode. The circuit must be completed with a **salt bridge**, an inverted U-shaped tube that joins the two half-cells and contains a strong electrolyte such as KNO_3. The salt bridge allows for the flow of ions that neutralizes the charge imbalance. The negative ions within the salt bridge flow to neutralize the accumulation of positive charge at the anode, and the positive ions flow to neutralize the accumulation of negative charge at the cathode.

The salt bridge completes the circuit—it allows the flow of ions between the two half-cells.

Light bulb

e⁻

e⁻

NO_3^-

K^+

Salt bridge containing $KNO_3(aq)$

Zn(s) anode
−

Cu(s) cathode
+

Glass wool plugs allow ions to pass

Zn^{2+}

Cu^{2+}

$Zn(NO_3)_2(aq)$

$Cu(NO_3)_2(aq)$

Oxidation
$Zn(s) \longrightarrow Zn^{2+}(aq) + 2e^-$

Reduction
$Cu^{2+}(aq) + 2e^- \longrightarrow Cu(s)$

▶ **FIGURE 16.9 A voltaic cell**
Question: Why do electrons flow from left to right in this figure?

In a voltaic cell, electrical voltage is the driving force that causes electrons to flow. A high voltage corresponds to a high driving force, while a low voltage corresponds to a low driving force. We can understand electrical voltage with an analogy. Electrons flowing through a wire are similar to water flowing in a river (▶ Figure 16.10). The *quantity of electrons* that flows through the wire (electrical current) is analogous to the *amount of water* that flows through the river (the river's current). The driving force that causes the electrons to flow through a wire—called *potential difference* or **voltage**—is analogous to the force of gravity that causes water to flow in a river. A high voltage is analogous to a steeply descending streambed (▶ Figure 16.11).

The voltage of a voltaic cell depends on the relative tendencies of the reactants to undergo oxidation and reduction. Combining the oxidation of a metal high on the activity series with the reduction of a metal ion low on the activity series produces a battery with a relatively high voltage. For example, the oxidation of Li(s) combined with the reduction of $Cu^{2+}(aq)$ results in a relatively high voltage. On the other hand, combining the oxidation of a metal on the activity series with the reduction of a metal ion just below it results in a voltaic cell with a relatively low voltage. Combining the oxidation of a metal on the activity series with the reduction of a metal ion above it on the activity series does not produce a voltaic cell at all. For example, you cannot make a voltaic cell by trying to oxidize Cu(s) and reduce $Li^+(aq)$. Such a reaction is not spontaneous and does not produce electrical current.

You can now see why voltaic cells (and batteries) go dead after extended use. As the simple voltaic cell we have just described is used, the zinc electrode dissolves away as zinc is oxidized to zinc ions. Similarly, the Cu^{2+} solution is

High voltage

▲ **FIGURE 16.10 River analogy for electrical current** The flow of electrons through a wire is similar to the flow of water in a river.

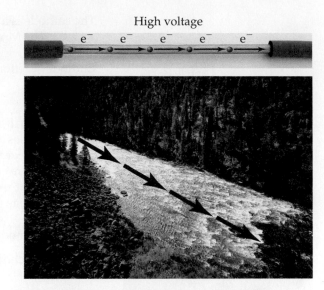

▲ **FIGURE 16.11 River analogy for voltage** A high voltage for electricity is analogous to a steep descent for a river.

depleted of Cu^{2+} ions as they deposit as solid Cu (▼ Figure 16.12). Once the zinc electrode is dissolved and the Cu^{2+} ions are depleted, the cell is dead. Some voltaic cells, such as those employed in rechargeable batteries, can be recharged by running electrical current—from an external source—in the opposite direction. This causes the regeneration of the reactants, allowing repeated use of the battery.

▶ **FIGURE 16.12 Dead voltaic cell** A voltaic cell dies with extended use because the reactants [in this case $Zn(s)$ and $Cu^{2+}(aq)$] become depleted while the products [in this case $Zn^{2+}(aq)$ and $Cu(s)$] accumulate.

$$Zn(s) + Cu^{2+}(aq) \longrightarrow Zn^{2+}(aq) + Cu(s)$$

Carbon rod

Moist paste of MnO_2 and NH_4Cl

Zn case

▲ **FIGURE 16.13 Common dry-cell battery**

DRY-CELL BATTERIES

Flashlight batteries are called **dry cells** because they do not contain large amounts of liquid water. There are several common types of dry-cell batteries. The most inexpensive type of dry cell is composed of a zinc case that acts as the anode (◀ Figure 16.13). The zinc is oxidized according to the following reaction.

Anode reaction: $Zn(s) \longrightarrow Zn^{2+}(aq) + 2e^-$ (oxidation)

The cathode is a carbon rod immersed in a moist paste of MnO_2 that also contains NH_4Cl. The MnO_2 is reduced to Mn_2O_3 according to the following reaction.

Cathode reaction: $2\,MnO_2(s) + 2\,NH_4^+(aq) + 2e^- \longrightarrow$
$$Mn_2O_3(s) + 2\,NH_3(g) + H_2O(l) \quad \text{(reduction)}$$

These two half-reactions produce a voltage of about 1.5 volts. Two or more of these batteries can be connected in series (cathode-to-anode connection) to produce higher voltages.

The more expensive **alkaline batteries** employ slightly different half-reactions that use a base (therefore the name *alkaline*). In an alkaline battery, the reactions are as follows.

Anode reaction: $Zn(s) + 2\,OH^-(aq) \longrightarrow Zn(OH)_2(s) + 2e^-$

Cathode reaction: $2\,MnO_2(s) + 2\,H_2O(l) + 2e^- \longrightarrow$
$$2\,MnO(OH)(s) + 2\,OH^-(aq) \quad \text{(reduction)}$$

Alkaline batteries have a longer working life and a longer shelf life than their nonalkaline counterparts.

LEAD-ACID STORAGE BATTERIES

The batteries in most automobiles are **lead-acid storage batteries**. These batteries consist of six electrochemical cells wired in series (◀ Figure 16.14). Each cell produces 2 volts for a total of 12 volts. The cells each contain a porous lead anode where oxidation occurs according to the following reaction.

Anode reaction: $Pb(s) + SO_4^{2-}(aq) \longrightarrow PbSO_4(s) + 2e^- \quad \text{(oxidation)}$

Each cell also contains a lead(IV) oxide cathode where reduction occurs according to the following reaction.

Cathode reaction: $PbO_2(s) + 4\,H^+(aq) + SO_4^{2-}(aq) + 2e^- \longrightarrow$
$$PbSO_4(s) + 2\,H_2O(l) \quad \text{(reduction)}$$

Both the anode and the cathode are immersed in sulfuric acid (H_2SO_4). As electrical current is drawn from the battery, both the anode and the cathode become coated with $PbSO_4(s)$. If the battery is run for a long time without recharging, too much $PbSO_4(s)$ develops and the battery goes dead. The lead-acid storage battery can be recharged, however, by running electrical current through it in reverse. The electrical current has to come from an external source, such as an alternator in a car. This causes the preceding reactions to occur in reverse, converting the $PbSO_4(s)$ back to $Pb(s)$ and $PbO_2(s)$, recharging the battery.

FUEL CELLS

We discussed fuel cells in Section 16.1. Electric vehicles powered by fuel cells may one day replace internal combustion vehicles. Fuel cells are like batteries, but the reactants are constantly replenished. Normal batteries lose their voltage with use because the reactants are depleted as electrical current is drawn from the battery. In a fuel cell, the reactants—the fuel—constantly flow through the battery, generating electrical current as they undergo a redox reaction.

The most common fuel cell is the hydrogen–oxygen fuel cell (▶ Figure 16.15). In this cell, hydrogen gas flows past the anode (a screen coated with platinum catalyst) and undergoes oxidation.

Anode reaction: $2\,H_2(g) + 4\,OH^-(aq) \longrightarrow 4\,H_2O(l) + 4e^-$

Anode (−): Lead grid packed with porous lead

Anode

Cathode

Electrolyte: A 30% solution of H_2SO_4

Cathode (+): Lead grid packed with PbO_2

▲ FIGURE 16.14 **Lead-acid storage battery** Question: Why do batteries like this become depleted? How are they recharged?

The porosity of the lead anode increases the surface area where electrons can be transferred from the solid lead to the solution.

Electron

Proton

Oxygen atom

Electric circuit

e^-

Hydrogen inlet

Oxygen inlet

H$^+$ OH$^-$

H$_2$

H$_2$O

O$_2$

Electrolyte

Anode	**Cathode**
H$_2(g)$ + 4 OH$^-(aq)$	O$_2(g)$ + 2 H$_2$O(l) + 4e$^-$
$\longrightarrow$ 4 H$_2$O(l) + 4e$^-$	$\longrightarrow$ 4 OH$^-(aq)$

▶ **FIGURE 16.15   Hydrogen–oxygen fuel cell**

Oxygen gas flows past the cathode (a similar screen) and undergoes reduction.

Cathode reaction: O$_2(g)$ + 2 H$_2$O(l) + 4e$^-$ $\longrightarrow$ 4 OH$^-(aq)$

The half-reactions sum to the following overall reaction.

Overall reaction: 2 H$_2(g)$ + O$_2(g)$ $\longrightarrow$ 2 H$_2$O(l)

Notice that the only product is water. In the space shuttle program, hydrogen–oxygen fuel cells provide electricity and astronauts drink the resulting water.

✓ CONCEPTUAL CHECKPOINT 16.4

Suppose you are making a battery composed of a carbon rod inserted into a moist paste of lead(II) ions, which acts as the cathode. You want the metal casing that encloses the battery to act as the anode. Which metal should you use for the casing to achieve the battery with the highest voltage?

(a) Mg   **(b)** Zn   **(c)** Ni

16.7 Electrolysis: Using Electricity to Do Chemistry

In a voltaic cell, a spontaneous redox reaction is used to produce electrical current. In **electrolysis**, electrical current is used to drive an otherwise nonspontaneous redox reaction. An electrochemical cell used for electrolysis is an **electrolytic cell**. We saw that the reaction of hydrogen with oxygen to form water is spontaneous and can be used to produce an electrical current in a fuel cell. By providing electrical current, we can cause the reverse reaction to occur, breaking water into hydrogen and oxygen (▶ Figure 16.16).

2 H$_2(g)$ + O$_2(g)$ $\longrightarrow$ 2 H$_2$O(l)   (spontaneous—produces electrical current; occurs in a voltaic cell)

2 H$_2$O(l) $\longrightarrow$ 2 H$_2(g)$ + O$_2(g)$ (nonspontaneous—consumes electrical current; occurs in an electrolytic cell)

▲ **FIGURE 16.16 Electrolysis of water** As a current passes between the electrodes, liquid water is broken down into hydrogen gas (right tube) and oxygen gas (left tube).

Anode	Cathode
$Ag(s) \longrightarrow Ag^+(aq) + e^-$	$Ag^+(aq) + e^- \longrightarrow Ag(s)$

▲ **FIGURE 16.17 Electrolytic cell for silver plating** Silver is oxidized on the left side of the cell and reduced at the right. As it is reduced, it is deposited on the object to be plated.

One of the problems associated with the widespread adoption of fuel cells is the scarcity of hydrogen. Where is the hydrogen to power these fuel cells going to come from? One possible answer is that the hydrogen can come from water through solar- or wind-powered electrolysis. In other words, a solar- or wind-powered electrolytic cell can be used to make hydrogen from water when the sun is shining or when the wind is blowing. The hydrogen can then be converted back to water to generate electricity when needed. Hydrogen made in this way could also be used to power fuel-cell vehicles.

Electrolysis also has numerous other applications. For example, most metals are found in Earth's crust as metal oxides. Converting them to pure metals requires the reduction of the metal, a nonspontaneous process. Electrolysis can be used to produce these metals. Electrolysis can also be used to plate metals onto other metals. For example, silver can be plated onto another, less expensive metal using the electrolytic cell shown in ▲ Figure 16.17. In this cell, a silver electrode is placed in a solution containing silver ions. An electrical current then causes the oxidation of silver at the anode (replenishing the silver ions in solution) and the reduction of silver ions at the cathode (coating the ordinary metal with solid silver).

$$\text{Anode reaction:} \quad Ag(s) \longrightarrow Ag^+(aq) + e^-$$
$$\text{Cathode reaction:} \quad Ag^+(aq) + e^- \longrightarrow Ag(s)$$

16.8 Corrosion: Undesirable Redox Reactions

▲ Paint can prevent underlying iron from rusting. However, if the paint becomes scratched, the iron will rust at the point of the chip. **Question: Why?**

Corrosion is the oxidation of metals. The most common kind of corrosion is the rusting of iron. A significant part of the iron produced each year goes to replace rusted iron. Rusting is a redox reaction in which iron is oxidized and oxygen is reduced.

Oxidation:	$2\,Fe(s) \longrightarrow 2\,Fe^{2+}(aq) + 4e^-$
Reduction:	$O_2(g) + 2\,H_2O(l) + 4e^- \longrightarrow 4\,OH^-(aq)$
Overall:	$2\,Fe(s) + O_2(g) + 2\,H_2O(l) \longrightarrow 2\,Fe(OH)_2(s)$

The $Fe(OH)_2$ formed in the overall reaction then undergoes several additional reactions to form Fe_2O_3, the familiar orange substance that we call rust. One of the main problems with Fe_2O_3 is that it crumbles off the solid iron below

EVERYDAY CHEMISTRY

The Fuel-Cell Breathalyzer

Police use a device called a breathalyzer to measure the amount of ethyl alcohol (C_2H_5OH) in the bloodstream of a person suspected of driving under the influence of alcohol.

Breathalyzers work because the amount of ethyl alcohol in the breath is proportional to the amount of ethyl alcohol in the bloodstream. One type of breathalyzer employs a fuel cell to measure the amount of alcohol in the breath. The fuel cell consists of two platinum electrodes (▼ Figure 16.18). When a suspect blows into the breathalyzer, any ethyl alcohol in the breath is oxidized to acetic acid at the anode.

Anode: $\underset{\text{Ethyl alcohol}}{C_2H_5OH} + 4\,OH^-(aq) \longrightarrow$

$$\underset{\text{Acetic acid}}{CH_3COOH(aq)} + 3\,H_2O + 4e^-$$

C₂H₅OH

Electrolyte

e⁻

H₂O

OH⁻

e⁻

O₂

Blood alcohol level

Suspect's breath

Anode   Cathode

CH₃COOH   Air intake

▲ **FIGURE 16.18** **Schematic diagram of a fuel-cell breathalyzer**

At the cathode, oxygen is reduced.

Cathode: $O_2(g) + 2\,H_2O(l) + 4e^- \longrightarrow 4\,OH^-(aq)$

The overall reaction is simply the oxidation of ethyl alcohol to acetic acid and water.

Overall: $C_2H_5OH(g) + O_2(g) \longrightarrow CH_3COOH(g) + H_2O$

The amount of electrical current produced depends on the amount of alcohol in the breath. A higher current reveals a higher blood alcohol level. When calibrated correctly, the fuel-cell breathalyzer can precisely measure the blood alcohol level of a suspected drunk driver.

CAN YOU ANSWER THIS? *Assign oxidation states to each element in the reactants and products in the overall equation for the fuel-cell breathalyzer. What element is oxidized, and what element is reduced in the reaction?*

▲ **Fuel-cell breathalyzer** A suspect blows into the top of this device and blood alcohol level is determined.

it, exposing more iron to further rusting. Under the right conditions, an entire piece of iron can rust away.

Iron is not the only metal that undergoes oxidation. Most other metals, such as copper and aluminum, also undergo oxidation. However, the oxides of copper and aluminum do not flake off as iron oxide does. When aluminum oxidizes, the aluminum oxide actually forms a tough clear coating on the underlying metal. This coating protects the underlying metal from further oxidation.

Preventing the iron from rusting is a major industry. The most obvious way to prevent rust is to keep iron dry. Without water, the redox reaction cannot occur. Another way to prevent rust is to coat the iron with a substance that is impervious to water. Cars, for example, are painted and sealed to prevent rust. A scratch in the paint, however, can lead to rusting of the underlying iron.

Rust can also be prevented by placing a *sacrificial electrode* in electrical contact with the iron. The sacrificial electrode must be composed of a metal that is above iron on the activity series. The sacrificial electrode oxidizes in place of the iron, protecting the iron from oxidation. Another way to protect iron from rusting is to coat it with a metal above it in the activity series. Galvanized nails, for example, are coated with a thin layer of zinc. Since zinc is more active than iron, it oxidizes in place of the underlying iron (just like a sacrificial electrode does). The oxide of zinc is not crumbly and remains on the nail as a protective coating.

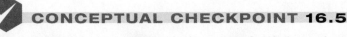 **CONCEPTUAL CHECKPOINT 16.5**

Which metal could NOT be used as a sacrificial electrode to prevent rusting?

(a) Mg   **(b)** Mn   **(c)** Zn   **(d)** Sn

CHAPTER IN REVIEW

CHEMICAL PRINCIPLES

Oxidation and Reduction: Oxidation is:

- the loss of electrons.
- an increase in oxidation state.

Reduction is:

- the gain of electrons.
- a decrease in oxidation state.

Oxidation and reduction reactions always occur together and are sometimes called redox reactions. The substance that is oxidized is the reducing agent, and the substance that is reduced is the oxidizing agent.

Oxidation States: The oxidation state is a fictitious charge assigned to each atom in a compound. It is calculated by assigning all bonding electrons in a compound to the most electronegative element.

The Activity Series: The activity series is a listing of metals from those that are easiest to oxidize to those that are most difficult to oxidize. Any half-reaction in the activity series is spontaneous when paired with the reverse of a half-reaction below it on the list.

RELEVANCE

Oxidation and Reduction: Redox reactions are common in nature, in industry, and in many everyday processes. Batteries use redox reactions to generate electrical current. Our bodies use redox reactions to obtain energy from glucose. In addition, the bleaching of hair, the rusting of iron, and the electroplating of metals all involve redox reactions.

Good oxidizing agents, such as oxygen, hydrogen peroxide, and chlorine, have a strong tendency to gain electrons. Good reducing agents, such as sodium and hydrogen, have a strong tendency to lose electrons.

Oxidation States: Oxidation states help us more easily identify substances being oxidized and reduced in a chemical reaction.

The Activity Series: The activity series allows us to predict whether a redox reaction (involving half-reactions from the series) is spontaneous.

Batteries: In a battery, the reactants of a spontaneous redox reaction are physically separated. As the redox reaction occurs, the transferred electrons are forced to travel through a wire or other external circuit, creating an electrical current that can be used to do electrical work.

Batteries: Batteries are common as portable sources of electrical current. They are used in flashlights, watches, automobiles, and other electrical devices.

Electrolysis: In a battery, a spontaneous redox reaction is used to generate an electrical current. In electrolysis, an electrical current is used to drive a nonspontaneous redox reaction.

Electrolysis: Electrolysis has many applications. For example, electrolysis is used to reduce metal oxides found in Earth's crust to their metals and to plate metals onto other metals.

Corrosion: Corrosion is the oxidation of iron and other metals by atmospheric oxygen. Corrosion can be prevented by keeping the metal dry, sealing it with a protective coating, or depositing a more active metal onto the surface of the metal to be protected.

Corrosion: The most common form of corrosion is the rusting of iron. Since a significant fraction of all iron produced is used to replace rusted iron, the prevention of rust is a major industry.

CHEMICAL SKILLS

EXAMPLES

Identifying Oxidation and Reduction (Section 16.2)

Oxidation can be identified as the gain of oxygen, the loss of electrons, or an increase in oxidation state. Reduction can be identified as the loss of oxygen, the gain of electrons, or a decrease in oxidation state.

When a substance gains oxygen, the substance is oxidized and the oxygen is reduced.

When a metal reacts with an electronegative element, the metal is oxidized and the electronegative element is reduced.

When a metal transfers electrons to a metal ion, the metal is oxidized and the metal ion is reduced.

EXAMPLE 16.11 Identifying Oxidation and Reduction

Determine the substance being oxidized and the substance being reduced in each redox reaction.

(a) $Sn(s) + O_2(g) \longrightarrow SnO_2(s)$
(b) $2\,Na(s) + F_2(g) \longrightarrow 2\,NaF(s)$
(c) $Mg(s) + Cu^{2+}(aq) \longrightarrow Mg^{2+}(aq) + Cu(s)$

SOLUTION

(a) Sn oxidized; O_2 reduced
(b) Na oxidized; F_2 reduced
(c) Mg oxidized; Cu^{2+} reduced

Identifying Oxidizing Agents and Reducing Agents (Section 16.2)

The reducing agent is the substance that is oxidized. The oxidizing agent is the substance that is reduced.

EXAMPLE 16.12 Identifying Oxidizing Agents and Reducing Agents

Identify the oxidizing and reducing agents in each redox reaction.

(a) $Sn(s) + O_2(g) \longrightarrow SnO_2(s)$
(b) $2\,Na(s) + F_2(g) \longrightarrow 2\,NaF(s)$
(c) $Mg(s) + Cu^{2+}(aq) \longrightarrow Mg^{2+}(aq) + Cu(s)$

SOLUTION

(a) Sn is the reducing agent; O_2 is the oxidizing agent.
(b) Na is the reducing agent; F_2 is the oxidizing agent.
(c) Mg is the reducing agent; Cu^{2+} is the oxidizing agent.

Assigning Oxidation States (Section 16.3)

Rules for Assigning Oxidation States

(These rules are hierarchical. If two rules conflict, follow the rule higher on the list.)

1. The oxidation state of an atom in a free element is 0.

2. The oxidation state of a monoatomic ion is equal to its charge.

3. The sum of the oxidation states of all atoms in:
   - a neutral molecule or formula unit is 0.
   - an ion is equal to the charge of the ion.

4. In their compounds,
   - Group I metals have an oxidation state of +1.
   - Group II metals have an oxidation state of +2.

5. In their compounds, nonmetals are assigned oxidation states according to the following hierarchical table.

Fluorine	−1
Hydrogen	+1
Oxygen	−2
Group 7A	−1
Group 6A	−2
Group 5A	−3

Balancing Redox Reactions (Section 16.4)

To balance redox reactions in aqueous acidic solutions, follow this procedure (brief version).

1. Assign oxidation states.

2. Separate the overall reaction into two half-reactions.

3. Balance each half-reaction with respect to mass.
   - Balance all elements other than H and O.

   - Balance O by adding H_2O.

   - Balance H by adding H^+.

4. Balance each half-reaction with respect to charge by adding electrons.

EXAMPLE 16.13 Assigning Oxidation States

Assign an oxidation state to each atom in each compound.

(a) Al

(b) Al^{3+}

(c) N_2O

(d) CO_3^{2-}

SOLUTION

(a) Al (rule 1)
   0

(b) Al^{3+} (rule 2)
   +3

(c) N_2O (rule 5, O takes priority over N)
   +1 −2

(d) CO_3^{2-} (rules 5, 3)
   +4 −2

EXAMPLE 16.14 Balancing Redox Reactions

Balance the following reaction occurring in acidic solution.

$$IO_3^-(aq) + Fe^{2+}(aq) \longrightarrow I_2(s) + Fe^{3+}(aq)$$

SOLUTION

OXIDATION $Fe^{2+}(aq) \longrightarrow Fe^{3+}(aq)$

REDUCTION $IO_3^-(aq) \longrightarrow I_2(s)$

$Fe^{2+}(aq) \longrightarrow Fe^{3+}(aq)$

$2\,IO_3^-(aq) \longrightarrow I_2(s)$

$Fe^{2+}(aq) \longrightarrow Fe^{3+}(aq)$

$2\,IO_3^-(aq) \longrightarrow I_2(s) + \textbf{6 H}_2\textbf{O}(l)$

$Fe^{2+}(aq) \longrightarrow Fe^{3+}(aq)$

$\textbf{12 H}^+(aq) + 2\,IO_3^-(aq) \longrightarrow I_2(s) + 6\,H_2O(l)$

$Fe^{2+}(aq) \longrightarrow Fe^{3+}(aq) + e^-$

$\textbf{10e}^- + 12\,H^+(aq) + 2\,IO_3^-(aq) \longrightarrow I_2(s) + 6\,H_2O(l)$

5. Make the number of electrons in both half-reactions equal.

$$10 \times [Fe^{2+}(aq) \longrightarrow Fe^{3+}(aq) + e^-]$$

$$10e^- + 12\,H^+(aq) + 2\,IO_3^-(aq) \longrightarrow I_2(s) + 6\,H_2O$$

6. Add the two half-reactions together.

$$10\,Fe^{2+}(aq) \longrightarrow 10\,Fe^{3+}(aq) + \cancel{10e^-}$$

$$\cancel{10e^-} + 12\,H^+(aq) + 2\,IO_3^-(aq) \longrightarrow I_2(s) + 6\,H_2O$$

$$\overline{10\,Fe^{2+}(aq) + 12\,H^+(aq) + 2\,IO_3^-(aq) \longrightarrow}$$

$$10\,Fe^{3+}(aq) + I_2(s) + 6\,H_2O$$

7. Verify that the reaction is balanced.

Reactants	Products
10 Fe	10 Fe
12 H	12 H
2 I	2 I
6 O	6 O
+30 charge	+30 charge

Predicting Spontaneous Redox Reactions (Section 16.5)

Any half-reaction in the activity series is spontaneous when paired with the reverse of any half-reaction below it.

EXAMPLE 16.15 Predicting Spontaneous Redox Reactions

Predict whether each redox reaction is spontaneous.

(a) $Cr(s) + 3\,Ag^+(aq) \longrightarrow Cr^{3+}(aq) + 3\,Ag(s)$

(b) $Mn^{2+}(aq) + Fe(s) \longrightarrow Mn(s) + Fe^{2+}(aq)$

SOLUTION

(a) Spontaneous

(b) **Nonspontaneous**

KEY TERMS

activity series of metals **[16.5]**
alkaline battery **[16.6]**
anode **[16.6]**
cathode **[16.6]**
corrosion **[16.8]**
dry cell **[16.6]**

electrical current **[16.6]**
electrochemical cell **[16.6]**
electrolysis **[16.7]**
electrolytic cell **[16.7]**
fuel cell **[16.1]**
half-cell **[16.6]**
half-reaction **[16.4]**

lead-acid storage battery **[16.6]**
oxidation **[16.2]**
oxidation state
    (oxidation number) **[16.3]**
oxidizing agent **[16.2]**

redox (oxidation–reduction) reaction **[16.1]**
reducing agent **[16.2]**
reduction **[16.2]**
salt bridge **[16.6]**
voltage **[16.6]**
voltaic (galvanic) cell **[16.6]**

EXERCISES

QUESTIONS

1. What is a fuel-cell electric vehicle?
2. What is an oxidation–reduction or redox reaction?
3. Define oxidation and reduction with respect to:

    (a) oxygen

    (b) electrons

    (c) oxidation state

4. What is an oxidizing agent? What is a reducing agent?
5. Good oxidation agents have a strong tendency to _____ electrons in reactions.

6. Good reducing agents have a strong tendency to _____ electrons in reactions.
7. What is the oxidation state of a free element? Of a monoatomic ion?
8. For a neutral molecule, the sum of the oxidation states of the individual atoms must add up to _____.
9. For an ion, the sum of the oxidation states of the individual atoms must add up to _____.
10. In their compounds, elements have oxidation states equal to _____. Are there exceptions to this rule? Explain.

11. In a redox reaction, an atom that undergoes an increase in oxidation state is _____. An atom that undergoes a decrease in oxidation state is _____.

12. How does hydrogen peroxide cause hair to change color?

13. When balancing redox equations, the number of electrons lost in the oxidation half-reaction must _____ the number of electrons gained in the reduction half-reaction.

14. When balancing aqueous redox reactions, oxygen is balanced using _____, and hydrogen is balanced using _____.

15. When balancing aqueous redox reactions, charge is balanced using _____.

16. When balancing aqueous redox reactions in basic media, hydrogen ions are neutralized using _____.

17. Are metals at the top of the activity series the most reactive or least reactive?

18. Are metals at the top of the activity series the easiest or hardest to oxidize?

19. Are metals at the bottom of the activity series most likely or least likely to lose electrons?

20. Any half-reaction in the activity series will be spontaneous when paired with the reverse of any half-reaction _____ it.

21. How can you use the activity series to determine whether a metal will dissolve in acids such as HCl or HBr?

22. What is electrical current? Explain how a simple battery creates electrical current.

23. Oxidation occurs at the _____ of an electrochemical cell.

24. Reduction occurs at the _____ of an electrochemical cell.

25. Explain the role of a salt bridge in an electrochemical cell.

26. A high voltage in an electrochemical cell is analogous to _____ in a river.

27. Describe a common dry-cell battery. Include equations for the anode and cathode reactions.

28. Describe a lead-acid storage battery. Include equations for the anode and cathode reactions.

29. Describe a fuel cell. Include equations for the anode and cathode reactions of the hydrogen–oxygen fuel cell.

30. What is electrolysis? Why is it useful?

31. What is corrosion? List reactions for the corrosion of iron.

32. How can rust be prevented?

PROBLEMS

OXIDATION AND REDUCTION

33. What substance is oxidized in each reaction?
    (a) $2 H_2(g) + O_2(g) \longrightarrow 2 H_2O(l)$
    (b) $4 Al(s) + 3 O_2(g) \longrightarrow 2 Al_2O_3(s)$
    (c) $2 Al(s) + 3 Cl_2(g) \longrightarrow 2 AlCl_3(s)$

34. What substance is oxidized in each reaction?
    (a) $2 Zn(s) + O_2(g) \longrightarrow 2 ZnO(s)$
    (b) $CH_4(g) + 2 O_2(g) \longrightarrow CO_2(g) + 2 H_2O(g)$
    (c) $Sr(s) + F_2(g) \longrightarrow SrF_2(s)$

35. For each reaction, identify the substance being oxidized and the substance being reduced.
    (a) $2 Sr(s) + O_2(g) \longrightarrow 2 SrO(s)$
    (b) $Ca(s) + Cl_2(g) \longrightarrow CaCl_2(s)$
    (c) $Ni^{2+}(aq) + Mg(s) \longrightarrow Mg^{2+}(aq) + Ni(s)$

36. For each reaction, identify the substance being oxidized and the substance being reduced.
    (a) $Mg(s) + Br_2(g) \longrightarrow MgBr_2(s)$
    (b) $2 Cr^{3+}(aq) + 3 Mn(s) \longrightarrow 2 Cr(s) + 3 Mn^{2+}(aq)$
    (c) $2 H^+(aq) + Ni(s) \longrightarrow H_2(g) + Ni^{2+}(aq)$

37. For each of the reactions in Problem 35, identify the oxidizing agent and the reducing agent.

38. For each of the reactions in Problem 36, identify the oxidizing agent and the reducing agent.

39. Based on periodic trends, which elements would you expect to be good oxidizing agents?
    (a) potassium
    (b) fluorine
    (c) iron
    (d) chlorine

40. Based on periodic trends, which elements would you expect to be good oxidizing agents?
    (a) oxygen
    (b) bromine
    (c) lithium
    (d) sodium

41. Based on periodic trends, which elements in Problem 39 (in their elemental form) would you expect to be good reducing agents?

42. Based on periodic trends, which elements in Problem 40 (in their elemental form) would you expect to be good reducing agents?

43. For each redox reaction, identify the substance being oxidized, the substance being reduced, the oxidizing agent, and the reducing agent.
    (a) $N_2(g) + O_2(g) \longrightarrow 2\,NO(g)$
    (b) $2\,CO(g) + O_2(g) \longrightarrow 2\,CO_2(g)$
    (c) $SbCl_3(g) + Cl_2(g) \longrightarrow SbCl_5(g)$

44. For each redox reaction, identify the substance being oxidized, the substance being reduced, the oxidizing agent, and the reducing agent.
    (a) $H_2(g) + I_2(g) \longrightarrow 2\,HI(g)$
    (b) $CO(g) + H_2(g) \longrightarrow C(s) + H_2O(g)$
    (c) $2\,Al(s) + 6\,H^+(aq) \longrightarrow 2\,Al^{3+}(aq) + 3\,H_2(g)$

OXIDATION STATES

45. Assign an oxidation state to each element or ion:
    (a) V
    (b) Mg^{2+}
    (c) Cr^{3+}
    (d) O_2

46. Assign an oxidation state to each element or ion:
    (a) Ne
    (b) Br_2
    (c) Cu^+
    (d) Fe^{3+}

47. Assign an oxidation state to each atom in each compound.
    (a) NaCl
    (b) CaF_2
    (c) SO_2
    (d) H_2S

48. Assign an oxidation state to each atom in each compound.
    (a) CH_4
    (b) CH_2Cl_2
    (c) $CuCl_2$
    (d) HI

49. What is the oxidation state of nitrogen in each compound?
    (a) NO
    (b) NO_2
    (c) N_2O

50. What is the oxidation state of Cr in each compound?
    (a) CrO
    (b) CrO_3
    (c) Cr_2O_3

51. Assign an oxidation state to each atom in each polyatomic ion.
    (a) CO_3^{2-}
    (b) OH^-
    (c) NO_3^-
    (d) NO_2^-

52. Assign an oxidation state to each atom in each polyatomic ion.
    (a) CrO_4^{2-}
    (b) $Cr_2O_7^{2-}$
    (c) PO_4^{3-}
    (d) MnO_4^-

53. What is the oxidation state of Cl in each ion?
    (a) ClO^-
    (b) ClO_2^-
    (c) ClO_3^-
    (d) ClO_4^-

54. What is the oxidation state of S in each ion?
    (a) SO_4^{2-}
    (b) SO_3^{2-}
    (c) HSO_3^-
    (d) HSO_4^-

55. Assign an oxidation state to each element in each compound:
    (a) $Cu(NO_3)_2$
    (b) $Sr(OH)_2$
    (c) $K_2Cr_2O_7$
    (d) $NaHCO_3$

56. Assign an oxidation state to each element in each compound:
    (a) Na_3PO_4
    (b) Hg_2S
    (c) $Fe(CN)_3$
    (d) NH_4Cl

57. Assign an oxidation state to each element in each reaction and use the change in oxidation state to determine which element is being oxidized and which element is being reduced.

   (a) $SbCl_5(g) \longrightarrow SbCl_3(g) + Cl_2(g)$

   (b) $CO(g) + Cl_2(g) \longrightarrow COCl_2(g)$

   (c) $2\,NO(g) + Br_2(g) \longrightarrow 2\,BrNO(g)$

   (d) $H_2(g) + CO_2(g) \longrightarrow H_2O(g) + CO(g)$

58. Assign an oxidation state to each element in each reaction and use the change in oxidation state to determine which element is being oxidized and which element is being reduced.

   (a) $CH_4(g) + 2\,H_2S(g) \longrightarrow CS_2(g) + 4\,H_2(g)$

   (b) $2\,H_2S(g) \longrightarrow 2\,H_2(g) + S_2(g)$

   (c) $C_6H_{12}O_6(s) + 6\,O_2(g) \longrightarrow$
   $\qquad\qquad\qquad 6\,CO_2(g) + 6\,H_2O(g)$

   (d) $C_2H_4(g) + Cl_2(g) \longrightarrow C_2H_4Cl_2(g)$

59. Use oxidation states to identify the oxidizing agent and the reducing agent in the following redox reaction.

   $$2\,Na(s) + 2\,H_2O(l) \longrightarrow 2\,NaOH(aq) + H_2(g)$$

60. Use oxidation states to identify the oxidizing agent and the reducing agent in the following redox reaction.

   $$N_2(g) + 3\,H_2(g) \longrightarrow 2\,NH_3(g)$$

BALANCING REDOX REACTIONS

61. Balance each redox reaction using the half-reaction method.

   (a) $K(s) + Cr^{3+}(aq) \longrightarrow Cr(s) + K^+(aq)$

   (b) $Mg(s) + Ag^+(aq) \longrightarrow Mg^{2+}(aq) + Ag(s)$

   (c) $Al(s) + Fe^{2+}(aq) \longrightarrow Al^{3+}(aq) + Fe(s)$

62. Balance each redox reaction using the half-reaction method.

   (a) $Zn(s) + Sn^{2+}(aq) \longrightarrow Zn^{2+}(aq) + Sn(s)$

   (b) $Mg(s) + Cr^{3+}(aq) \longrightarrow Mg^{2+}(aq) + Cr(s)$

   (c) $Al(s) + Ag^+(aq) \longrightarrow Al^{3+}(aq) + Ag(s)$

63. Classify each half-reaction occurring in acidic aqueous solution as an oxidation or a reduction and balance the half-reaction.

   (a) $MnO_4^-(aq) \longrightarrow Mn^{2+}(aq)$

   (b) $Pb^{2+}(aq) \longrightarrow PbO_2(s)$

   (c) $IO_3^-(aq) \longrightarrow I_2(s)$

   (d) $SO_2(g) \longrightarrow SO_4^{2-}(aq)$

64. Classify each half-reaction occurring in acidic aqueous solution as an oxidation or a reduction and balance the half-reaction.

   (a) $S(s) \longrightarrow H_2S(g)$

   (b) $S_2O_8^{2-}(aq) \longrightarrow 2\,SO_4^{2-}(aq)$

   (c) $Cr_2O_7^{2-}(aq) \longrightarrow Cr^{3+}(aq)$

   (d) $NO(g) \longrightarrow NO_3^-(aq)$

65. Balance each redox reaction occurring in acidic aqueous solution. Use the half-reaction method.

   (a) $PbO_2(s) + I^-(aq) \longrightarrow Pb^{2+}(aq) + I_2(s)$

   (b) $SO_3^{2-}(aq) + MnO_4^-(aq) \longrightarrow$
   $\qquad\qquad\qquad SO_4^{2-}(aq) + Mn^{2+}(aq)$

   (c) $S_2O_3^{2-}(aq) + Cl_2(g) \longrightarrow SO_4^{2-}(aq) + Cl^-(aq)$

66. Balance each redox reaction occurring in acidic aqueous solution. Use the half-reaction method.

   (a) $I^-(aq) + NO_2^-(aq) \longrightarrow I_2(s) + NO(g)$

   (b) $BrO_3^-(aq) + N_2H_4(g) \longrightarrow Br^-(aq) + N_2(g)$

   (c) $NO_3^-(aq) + Sn^{2+}(aq) \longrightarrow Sn^{4+}(aq) + NO(g)$

67. Balance each redox reaction occurring in acidic aqueous solution. Use the half-reaction method.

   (a) $ClO_4^-(aq) + Cl^-(aq) \longrightarrow ClO_3^-(aq) + Cl_2(g)$

   (b) $MnO_4^-(aq) + Al(s) \longrightarrow Mn^{2+}(aq) + Al^{3+}(aq)$

   (c) $Br_2(aq) + Sn(s) \longrightarrow Sn^{2+}(aq) + Br^-(aq)$

68. Balance each redox reaction occurring in acidic aqueous solution. Use the half-reaction method.

   (a) $IO_3^-(aq) + SO_2(g) \longrightarrow I_2(s) + SO_4^{2-}(aq)$

   (b) $Sn^{4+}(aq) + H_2(g) \longrightarrow Sn^{2+}(aq) + H^+(aq)$

   (c) $Cr_2O_7^{2-}(aq) + Br^-(aq) \longrightarrow Cr^{3+}(aq) + Br_2(aq)$

69. Balance each redox reaction occurring in basic solution.

   (a) $ClO^-(aq) + Cr(OH)_4^-(aq) \longrightarrow$
   $$CrO_4^{2-}(aq) + Cl^-(aq)$$

   (b) $MnO_4^-(aq) + Br^-(aq) \longrightarrow$
   $$MnO_2(s) + BrO_3^-(aq)$$

70. Balance each redox reaction occurring in basic solution.

   (a) $NO_2^-(aq) + Al(s) \longrightarrow NH_3(g) + AlO_2^-(aq)$

   (b) $Al(s) + MnO_4^-(aq) \longrightarrow$
   $$MnO_2(s) + Al(OH)_4^-(aq)$$

THE ACTIVITY SERIES

71. Which metal has the least tendency to be oxidized?

   (a) Ag

   (b) Na

   (c) Ni

   (d) Pb

72. Which metal has the least tendency to be oxidized?

   (a) Sn

   (b) Mg

   (c) Cu

   (d) Fe

73. Which metal cation has the greatest tendency to be reduced?

   (a) Mn^{2+}

   (b) Cu^{2+}

   (c) K^+

   (d) Ni^{2+}

74. Which metal cation has the greatest tendency to be reduced?

   (a) Pb^{2+}

   (b) Cr^{3+}

   (c) Fe^{2+}

   (d) Sn^{2+}

75. Which metal is the best reducing agent?

   (a) Mn

   (b) Al

   (c) Ni

   (d) Cr

76. Which metal is the best reducing agent?

   (a) Ag

   (b) Mg

   (c) Fe

   (d) Pb

77. Determine whether or not each redox reaction occurs spontaneously in the forward direction.

   (a) $Ni(s) + Zn^{2+}(aq) \longrightarrow Ni^{2+}(aq) + Zn(s)$

   (b) $Ni(s) + Pb^{2+}(aq) \longrightarrow Ni^{2+}(aq) + Pb(s)$

   (c) $Al(s) + 3 Ag^+(aq) \longrightarrow 3 Al^{3+}(aq) + Ag(s)$

   (d) $Pb(s) + Mn^{2+}(aq) \longrightarrow Pb^{2+}(aq) + Mn(s)$

78. Determine whether or not each redox reaction occurs spontaneously in the forward direction.

   (a) $Ca^{2+}(aq) + Zn(s) \longrightarrow Ca(s) + Zn^{2+}(aq)$

   (b) $2 Ag^+(aq) + Ni(s) \longrightarrow 2 Ag(s) + Ni^{2+}(aq)$

   (c) $Fe(s) + Mn^{2+}(aq) \longrightarrow Fe^{2+}(aq) + Mn(s)$

   (d) $2 Al(s) + 3 Pb^{2+}(aq) \longrightarrow 2 Al^{3+}(aq) + 3 Pb(s)$

79. Suppose you wanted to cause Ni^{2+} ions to come out of solution as solid Ni. What metal could you use to accomplish this?

80. Suppose you wanted to cause Pb^{2+} ions to come out of solution as solid Pb. What metal could you use to accomplish this?

81. Which metal in the activity series will reduce Al^{3+} ions but not Na^+ ions?

82. Which metal in the activity series can be oxidized with a Ni^{2+} solution but not with a Cr^{3+} solution?

83. Which metals dissolve in HCl? For those metals that do dissolve, write a balanced redox reaction showing what happens when the metal dissolves.
    (a) Ag
    (b) Fe
    (c) Cu
    (d) Al

84. Which metals dissolve in HCl? For those metals that do dissolve, write a balanced redox reaction showing what happens when the metal dissolves.
    (a) Cr
    (b) Pb
    (c) Au
    (d) Zn

BATTERIES, ELECTROCHEMICAL CELLS, AND ELECTROLYSIS

85. Make a sketch of an electrochemical cell with the following overall reaction. Label the anode, the cathode, and the salt bridge. Indicate the direction of electron flow. *Hint:* When drawing electrochemical cells, the anode is usually drawn on the left side.

$$Mn(s) + Pb^{2+}(aq) \longrightarrow Mn^{2+}(aq) + Pb(s)$$

86. Make a sketch of an electrochemical cell with the following overall reaction. Label the anode, the cathode, and the salt bridge. Indicate the direction of electron flow. *Hint:* When drawing electrochemical cells, the anode is usually drawn on the left side.

$$Mg(s) + Ni^{2+}(aq) \longrightarrow Mg^{2+}(aq) + Ni(s)$$

87. An electrochemical cell has the following reaction occurring at the anode.

$$Zn(s) \longrightarrow Zn^{2+}(aq) + 2e^-$$

Which cathode reaction would produce a battery with the highest voltage?
    (a) $Mg^{2+}(aq) + 2e^- \longrightarrow Mg(s)$
    (b) $Pb^{2+}(aq) + 2e^- \longrightarrow Pb(s)$
    (c) $Cr^{3+}(aq) + 3e^- \longrightarrow Cr(s)$
    (d) $Cu^{2+}(aq) + 2e^- \longrightarrow Cu(s)$

88. An electrochemical cell has the following reaction occurring at the cathode.

$$Ni^{2+}(aq) + 2e^- \longrightarrow Ni(s)$$

Which anode reaction would produce a battery with the highest voltage?
    (a) $Ag(s) \longrightarrow Ag^+(aq) + e^-$
    (b) $Mg(s) \longrightarrow Mg^{2+}(aq) + 2e^-$
    (c) $Cr(s) \longrightarrow Cr^{3+}(aq) + 3e^-$
    (d) $Cu(s) \longrightarrow Cu^{2+}(aq) + 2e^-$

89. Use the half-cell reactions for the alkaline battery to determine the overall reaction that occurs in this battery.

90. Use the half-cell reactions for the lead-acid storage battery to determine the overall reaction that occurs in this battery.

91. Make a sketch of an electrolysis cell that could be used to electroplate copper onto other metal surfaces. Label the anode and the cathode and show the reactions that occur at each.

92. Make a sketch of an electrolysis cell that could be used to electroplate nickel onto other metal surfaces. Label the anode and the cathode and show the reactions that occur at each.

CORROSION

93. Which of the following metals, if coated onto iron, would prevent the corrosion of iron?
    (a) Zn
    (b) Sn
    (c) Mn

94. Which of the following metals, if coated onto iron, would prevent the corrosion of iron?
    (a) Mg
    (b) Cr
    (c) Cu

CUMULATIVE PROBLEMS

95. Determine whether or not each reaction is a redox reaction. For those reactions that are redox reactions, identify the substance being oxidized and the substance being reduced.
    (a) $Zn(s) + CoCl_2(aq) \longrightarrow ZnCl_2(aq) + Co(s)$
    (b) $HI(aq) + NaOH(aq) \longrightarrow H_2O(l) + NaI(aq)$
    (c) $AgNO_3(aq) + NaCl(aq) \longrightarrow$
    $$AgCl(s) + NaNO_3(aq)$$
    (d) $2 K(s) + Br_2(l) \longrightarrow 2 KBr(s)$

96. Determine whether or not each reaction is a redox reaction. For those reactions that are redox reactions, identify the substance being oxidized and the substance being reduced.
    (a) $Pb(NO_3)_2(aq) + 2 LiCl(aq) \longrightarrow$
    $$PbCl_2(s) + 2 LiNO_3(aq)$$
    (b) $2 HBr(aq) + Ca(OH)_2(aq) \longrightarrow$
    $$2 H_2O(l) + CaBr_2(aq)$$
    (c) $2 Al(s) + Fe_2O_3(s) \longrightarrow Al_2O_3(s) + 2 Fe(l)$
    (d) $Na_2O(s) + H_2O(l) \longrightarrow 2 NaOH(aq)$

97. Consider the unbalanced redox reaction:
    $$MnO_4^-(aq) + Zn(s) \longrightarrow Mn^{2+}(aq) + Zn^{2+}(aq)$$
    Balance the equation in acidic solution and determine how much of a 0.500 M KMnO$_4$ solution is required to completely dissolve 2.85 g of Zn.

98. Consider the unbalanced redox reaction:
    $$Cr_2O_7^{2-}(aq) + Cu(s) \longrightarrow Cr^{3+}(aq) + Cu^{2+}$$
    Balance the equation in acidic solution and determine how much of a 0.850 M K$_2$Cr$_2$O$_7$ solution is required to completely dissolve 5.25 g of Cu.

99. If a strip of magnesium metal is dipped into a solution containing silver ions, will a spontaneous reaction occur? If so, write the two half-reactions and the balanced overall equation for the reaction that occurs.

100. If a strip of tin metal is dipped into a solution containing zinc ions, will a spontaneous reaction occur? If so, write the two half-reactions and the balanced overall equation for the reaction that occurs.

101. A 10.0-mL sample of a commercial hydrogen peroxide (H$_2$O$_2$) solution is titrated with 0.0998 M KMnO$_4$. The end point is reached at a volume of 34.81 mL. Find the mass percent of H$_2$O$_2$ in the commercial hydrogen peroxide solution. (Assume a density of 1.00 g/mL for the hydrogen peroxide solution.) The unbalanced redox reaction that occurs in acidic solution during the titration is:
    $$H_2O_2(aq) + MnO_4^-(aq) \longrightarrow O_2(g) + Mn^{2+}(aq)$$

102. A 1.012-g sample of a salt containing Fe^{2+} is titrated with 0.1201 M KMnO$_4$. The end point of the titration is reached at 22.45 mL. Find the mass percent of Fe^{2+} in the sample. The unbalanced redox reaction that occurs in acidic solution during the titration is:
    $$Fe^{2+}(aq) + MnO_4^-(aq) \longrightarrow Fe^{3+}(aq) + Mn^{2+}(aq)$$

103. Silver is electroplated at the cathode of an electrolysis cell by the half-reaction:
    $$Ag^+(aq) + e^- \longrightarrow Ag(s)$$
    How many moles of electrons are required to electroplate 5.8 g of Ag?

104. Gold is electroplated at the cathode of an electrolysis cell by the half-reaction:
    $$Au^{3+}(aq) + 3e^- \longrightarrow Au(s)$$
    How many moles of electrons are required to electroplate 1.40 g of Au?

105. Determine whether HI can dissolve each metal sample. If it can, write a balanced chemical reaction showing how the metal dissolves in HI and determine the minimum amount of 3.5 M HI required to completely dissolve the sample.
    (a) 5.95 g Cr
    (b) 2.15 g Al
    (c) 4.85 g Cu
    (d) 2.42 g Au

106. Determine whether HCl can dissolve each metal sample. If it can, write a balanced chemical reaction showing how the metal dissolves in HCl and determine the minimum amount of 6.0 M HCl required to completely dissolve the sample.
    (a) 5.90 g Ag
    (b) 2.55 g Pb
    (c) 4.83 g Sn
    (d) 1.25 g Mg

107. One drop (assume 0.050 mL) of 6.0 M HCl is placed onto the surface of 0.028-mm-thick aluminum foil. What is the maximum diameter of the hole that will result from the HCl dissolving the aluminum? (Density of aluminum = 2.7 g/cm^3)

108. A graduated cylinder containing 1.00 mL of 12.0 M HCl is accidentally tipped over, and the contents spill onto manganese foil with a thickness of 0.055 mm. Calculate the maximum diameter of the hole that will be dissolved in the foil by the reaction between the manganese and hydrochloric acid. (Density of manganese = 7.47 g/cm^3)

109. The electrolytic cell represented in Figure 16.17 can be used to plate silver onto other metal surfaces. The plating reaction is: $Ag^+(aq) + e^- \longrightarrow Ag(s)$. Notice from the reaction that 1 mol e^- plates out 1 mol Ag(s). Use this stoichiometric relationship to determine how much time is required with an electrical current 0.100 amp to plate out 1.0 g Ag. The amp is a unit of electrical current equivalent to 1 C/s. (*Hint:* Recall that the charge of an electron is 1.60×10^{-19} C.)

110. An electrolytic cell similar to the one represented in Figure 16.17 can be used to plate gold onto other metal surfaces. The plating reaction is: $Au^+(aq) + e^- \longrightarrow Au(s)$. Notice from the reaction that 1 mol e^- plates out 1 mol Au(s). Use this stoichiometric relationship to determine how much time is required with an electrical current of 0.200 amp to plate out 0.400 g Au. The amp is a unit of electrical current equivalent to 1 C/s. (*Hint:* Recall that the charge of an electron is 1.60×10^{-19} C.)

HIGHLIGHT PROBLEMS

111. Consider the molecular views of an Al strip and Cu^{2+} solution. Draw a similar sketch showing what happens to the atoms and ions if the Al strip is submerged in the solution for a few minutes.

Aluminum atoms

Cu^{2+} ions

112. Suppose a fuel-cell generator produces electricity for a house. If each H_2 molecule produces $2e^-$, how many kilograms of hydrogen would be required to generate the electricity needed for a typical house? Assume the home uses about 850 kWh of electricity per month, which corresponds to approximately 2.65×10^4 mol of electrons at the voltage of a fuel cell.

113. Consider the molecular view of an electrochemical cell involving the overall reaction:

$$Zn(s) + Ni^{2+}(aq) \longrightarrow Zn^{2+}(aq) + Ni(s)$$

Draw a similar sketch showing how the cell might appear after it has generated a substantial amount of electrical current.

Anode: $Zn(s) \longrightarrow Zn^{2+}(aq) + 2e^-$ **Cathode:** $Ni^{2+}(aq) + 2e^- \longrightarrow Ni(s)$

▶ANSWERS TO SKILLBUILDER EXERCISES

Skillbuilder 16.1
   (a) K is oxidized, Cl_2 is reduced.
   (b) Al is oxidized, Sn^{2+} is reduced.
   (c) C is oxidized, O_2 is reduced.

Skillbuilder 16.2
   (a) K is the reducing agent; Cl_2 is the oxidizing agent.
   (b) Al is the reducing agent; Sn^{2+} is the oxidizing agent.
   (c) C is the reducing agent; O_2 is the oxidizing agent.

Skillbuilder 16.3
   (a) $\underset{0}{Zn}$
   (b) $\underset{+2}{Cu^{2+}}$
   (c) $\underset{+2-1}{CaCl_2}$
   (d) $\underset{+4-1}{CF_4}$
   (e) $\underset{+3-2}{NO_2^-}$
   (f) $\underset{+6-2}{SO_3}$

Skillbuilder 16.4
   Sn oxidized $(0 \longrightarrow +4)$; N reduced $(+5 \longrightarrow +4)$

Skillbuilder 16.5
   $6 H^+(aq) + 2 Cr(s) \longrightarrow 3 H_2(g) + 2 Cr^{3+}(aq)$

Skillbuilder 16.6
   $Cu(s) + 4 H^+(aq) + 2 NO_3^-(aq) \longrightarrow$
               $Cu^{2+}(aq) + 2 NO_2(g) + 2 H_2O(l)$

Skillbuilder 16.7
   $5 Sn(s) + 16 H^+(aq) + 2 MnO_4^-(aq) \longrightarrow$
            $5 Sn^{2+}(aq) + 2 Mn^{2+}(aq) + 8 H_2O(l)$

Skillbuilder 16.8
   $H_2O_2(aq) + 2 ClO_2(aq) + 2 OH^-(aq) \longrightarrow$
            $O_2(g) + 2 ClO_2^-(aq) + 2 H_2O(l)$

Skillbuilder 16.9   (a) Yes (b) No

Skillbuilder 16.10   No

▶ANSWERS TO CONCEPTUAL CHECKPOINTS

16.1 (b) The oxidizing agent oxidizes another species and is itself always reduced.

16.2 (d) From rule 1, we know that the oxidation state of nitrogen in N_2 is 0. According to rule 3, the sum of the oxidation states of all atoms in a compound = 0. Therefore, by applying rule 5, we can determine that the oxidation state of nitrogen in NO is +2; in NO_2 it is +4; and in NH_3 it is −3.

16.3 (c) Lead is the only one of these metals that lies above hydrogen in the activity series, and therefore the only one that dissolves in an acidic solution.

16.4 (a) Magnesium would lead to the highest voltage because it is highest on the activity series. Of the metals listed, it is most easily oxidized and therefore produces the highest voltage when combined with the reduction of Pb^{2+} ions.

16.5 (d) Tin is the only metal in the list that lies below iron in the activity series. Tin is therefore more difficult to oxidize than iron and cannot prevent the oxidation of the iron.

CHAPTER 17

Radioactivity and Nuclear Chemistry

"Nuclear energy is incomparably greater than the molecular energy which we use today.... What is lacking is the match to set the bonfire alight.... The scientists are looking for this."

WINSTON SPENCER CHURCHILL (1874–1965), IN 1931

17.1 Diagnosing Appendicitis

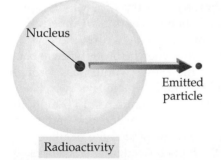

▲ FIGURE 17.1 Radioactivity
Radioactivity is the emission of tiny, energetic particles by the nuclei of certain atoms.

◀ In nuclear medicine, radioactivity is used to obtain clear images of internal organs.

Several years ago I awoke with a dull pain on the lower right side of my belly. The pain worsened over several hours, so I went to the hospital emergency room for evaluation. I was examined by a doctor who said it might be appendicitis, an inflammation of the appendix. The appendix, which has no known function, is a small pouch that extends from the right side of the large intestine. Occasionally, it becomes infected and requires surgical removal.

Patients with appendicitis usually have a high number of white blood cells, so the hospital performed a blood test to determine my white blood cell count. The test was negative—I had a normal white blood cell count. Although my symptoms were consistent with appendicitis, the negative blood test clouded the diagnosis. The doctor gave me the choice of either having my appendix removed (with the chance of it being healthy) or performing an additional test to confirm appendicitis. I chose the additional test.

The additional test involved nuclear medicine, an area of medical practice that uses *radioactivity* to diagnose and treat disease. **Radioactivity** is the emission of tiny, invisible particles by the nuclei of certain atoms (◀ Figure 17.1). Many of these particles can pass right through matter. The atoms that emit these particles are said to be **radioactive**.

To perform the test, antibodies—naturally occurring molecules that fight infection—labeled with radioactive atoms were injected into my bloodstream. Since antibodies attack infection, they migrate to areas of the body where infection is present. If my appendix were infected, the antibodies would accumulate there. After waiting about an hour, I was taken to a room and laid on a table.

▲ **FIGURE 17.2 Nuclear medicine**
In a test for appendicitis, radioactively tagged antibodies are given to the patient. If the patient has an infection in the appendix, the antibodies accumulate there and the emitted radiation can be detected on photographic film.

A photographic film was inserted in a panel above me. Although radioactivity is invisible to the eye, it does expose photographic film. If my appendix were indeed infected, it would now contain a high concentration of the radioactively tagged antibodies, and the film would show a bright spot at the location of my appendix (◄ Figure 17.2). In this test, I—or my appendix to be specific—was the radiation source that would expose film. The test, however, was negative. No radioactivity was emanating from my appendix. It was healthy. After several hours, the pain in my belly subsided and I went home, appendix and all. I never did find out what caused the pain.

Radioactivity is also used to diagnose and treat many other conditions, including cancer, thyroid disease, abnormal kidney and bladder function, and heart disease. These examples from medicine are just a few of the many applications of radioactivity. Naturally occurring radioactivity allows us to estimate the age of fossils and rocks. Radioactivity also led to the discovery of nuclear fission, used for electricity generation and nuclear weapons. In this chapter, we explore radioactivity—how it was discovered, what it is, and how it is used.

17.2 The Discovery of Radioactivity

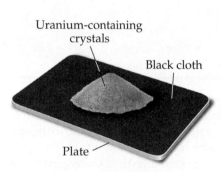

▲ **FIGURE 17.3 Becquerel's experiment**

Element 96 (curium) is named in honor of Marie Curie and her contributions to our understanding of radioactivity.

In 1896, a French scientist named Antoine-Henri Becquerel (1852–1908) discovered radioactivity. Becquerel was not looking for radioactivity at the time. Instead, he was interested in the newly discovered X-rays (see Section 9.3), which were the hot topic of physics research in his time. He hypothesized that X-rays were emitted in conjunction with **phosphorescence**. Phosphorescence is the long-lived *emission* of light that sometimes follows the *absorption* of light by some atoms and molecules. Phosphorescence is probably most familiar to you as the *glow* in glow-in-the-dark toys. After one of these toys is exposed to light, it reemits some of that light, usually at slightly longer wavelengths. If you turn off the lights or put the toy in the dark, you can see the greenish glow of the emitted light. Becquerel hypothesized that the visible greenish glow was associated with the emission of X-rays (which are invisible).

To test his hypothesis, Becquerel placed crystals—composed of potassium uranyl sulfate, a compound known to phosphoresce—on top of a photographic plate wrapped in black cloth (◄ Figure 17.3). He then placed the wrapped plate and the crystals outside to expose them to sunlight. He knew that the crystals phosphoresced because he could see the emitted light when he brought them back into the dark. If the crystals also emitted X-rays, the X-rays would pass through the black cloth and expose the underlying photographic plate. Becquerel performed the experiment several times and always got the same result: the photographic plate showed a bright exposure spot where the crystals had been. Becquerel believed his hypothesis was correct, and he presented the results—that phosphorescence and X-rays were linked—to the French Academy of Sciences.

Becquerel later retracted his results, however, when he discovered that a photographic plate with the same crystals showed a bright exposure spot even when the plate and the crystals were stored in a dark drawer and not exposed to sunlight. Becquerel realized that the crystals themselves were constantly emitting something (independent of whether or not they phosphoresced) that exposed the photographic plate. Becquerel concluded that it was the uranium within the crystals that was the source of the emissions, and he called the emissions *uranic rays*.

Soon after Becquerel's discovery, a young graduate student named Marie Sklodowska Curie (1867–1934), one of the first women in France to attempt doctoral work, decided to pursue the study of uranic rays for her doctoral thesis. Her first task was to determine whether any other substances besides uranium (the heaviest known element at the time) emitted these rays. In her search, Curie discovered two new elements, both of which also emitted uranic rays. Curie named one of her newly discovered elements *polonium* after her home country of Poland. The other element she named *radium* because of the very high amount of radioactivity that it produced. Radium was so radioactive that it gently glowed in the dark and emitted significant amounts of heat. Since it was now clear that these rays were not unique to uranium, Curie changed the name of uranic rays to *radioactivity*. In 1903, Curie received the Nobel Prize in physics—which she shared with Becquerel and her husband, Pierre Curie—for the discovery of radioactivity. In 1911, Curie was awarded a second Nobel Prize, this time in chemistry, for her discovery of the two new elements.

▲ Marie Curie with her two daughters. Irene (right) became a distinguished nuclear physicist in her own right, winning a Nobel Prize in 1935. Eve (left) wrote a highly acclaimed biography of her mother.

▶ In the past, radium was added to some paints that were used on watch dials. The radium made the dial glow.

17.3   Types of Radioactivity: Alpha, Beta, and Gamma Decay

While Curie focused her work on discovering the different kinds of radioactive elements, Ernest Rutherford and others focused on characterizing the radioactivity itself. These scientists found that the emissions were produced by the nuclei of radioactive atoms. These nuclei were unstable and would emit small pieces of themselves in the form of electromagnetic radiation to gain stability. These were the particles that Becquerel and Curie detected. There are several different types of radioactive emissions: alpha (α) rays, beta (β) rays, gamma (γ) rays, and positrons.

In order to understand these different types of radioactivity, we must briefly review the notation to symbolize isotopes that was first introduced in Section 4.8. Recall that any isotope can be represented with the notation:

Remember that the atomic number equals the number of protons and that the mass number equals the number of protons and neutrons.

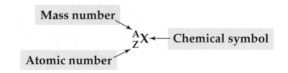

Mass number
$^A_Z X$ ← **Chemical symbol**
Atomic number

For example, the symbol

$$^{21}_{10}\text{Ne}$$

represents the neon isotope containing 10 protons and 11 neutrons. The symbol

$$^{20}_{10}\text{Ne}$$

represents the neon isotope containing 10 protons and 10 neutrons. Remember that many elements have several different isotopes.

The main subatomic particles—protons, neutrons, and electrons—can all be represented with similar notation.

Proton symbol ^1_1p Neutron symbol ^1_0n Electron symbol $^0_{-1}\text{e}$

Alpha decay

α particle $= {}^{4}_{2}\text{He}$

► **FIGURE 17.4 Alpha radiation**
Alpha radiation occurs when an unstable nucleus emits a particle composed of 2 protons and 2 neutrons. **Question:** What happens to the atomic number of an element upon emission of an alpha particle?

Nuclei are unstable when they are too large or contain an unbalanced ratio of neutrons to protons. Small nuclei need about 1 neutron to every proton to be stable, while larger nuclei need about 1.5 neutrons to every proton.

In nuclear chemistry, we are primarily interested in changes within the nucleus; therefore, the 2+ charge that we would normally write for a helium nucleus is omitted for an alpha particle.

The term *nuclide* is used in nuclear chemistry to mean a specific isotope.

ALPHA (α) RADIATION

Alpha (α) radiation occurs when an unstable nucleus emits a small piece of itself consisting of 2 protons and 2 neutrons (▲ Figure 17.4). Since 2 protons and 2 neutrons are identical to a helium-4 nucleus, the symbol for an **alpha (α) particle** is identical to the symbol for helium-4.

alpha (α) particle ${}^{4}_{2}\text{He}$

An α particle

When an atom emits an alpha particle, it becomes a lighter atom. We represent this process with a **nuclear equation**, an equation that represents the changes that occur during radioactivity and other nuclear processes. For example, the nuclear equation for the alpha decay of uranium-238 is:

Parent nuclide — Daughter nuclides

$${}^{238}_{92}\text{U} \longrightarrow {}^{234}_{90}\text{Th} + {}^{4}_{2}\text{He}$$

The original atom is called the **parent nuclide**, and the products are called the **daughter nuclides**. When an element emits an alpha particle, the number of protons in its nucleus changes, transforming it into a different element. In this case, uranium-238 becomes thorium-234. Unlike a chemical reaction, in which elements retain their identity, a nuclear reaction often changes the identity of the elements involved. Like a chemical equation, however, nuclear equations must be balanced.

The sum of the atomic numbers on both sides of a nuclear equation must be equal, and the sum of the mass numbers on both sides must also be equal.

$${}^{238}_{92}\text{U} \longrightarrow {}^{234}_{90}\text{Th} + {}^{4}_{2}\text{He}$$

Left Side	Right Side
Sum of mass numbers = 238	Sum of mass numbers = 234 + 4 = 238
Sum of atomic numbers = 92	Sum of atomic numbers = 90 + 2 = 92

The identity and symbol of the daughter nuclide of any alpha decay can be deduced from the mass and atomic numbers of the parent nuclide. During alpha decay, the mass number decreases by 4 and the atomic number decreases by 2. For example, to write a nuclear equation for the alpha decay of Th-232, we begin with the symbol for Th-232 on the left side of the equation and the symbol for an alpha particle on the right side.

$${}^{232}_{90}\text{Th} \longrightarrow {}^{x}_{y}? + {}^{4}_{2}\text{He}$$

We can deduce the mass number and atomic number of the unknown daughter nuclide because the equation must be balanced.

$$x + 4 = 232; x = 228$$

$$^{232}_{90}\text{Th} \longrightarrow \overbrace{^{x}_{y}?} + {}^{4}_{2}\text{He}$$

$$y + 2 = 90; y = 88$$

Therefore,

$$^{232}_{90}\text{Th} \longrightarrow {}^{228}_{88}? + {}^{4}_{2}\text{He}$$

The atomic number must be 88 and the mass number must be 228. Finally, we can deduce the identity of the daughter nuclide and its symbol from its atomic number. Since the atomic number is 88, the daughter nuclide is radium (Ra).

$$^{232}_{90}\text{Th} \longrightarrow {}^{228}_{88}\text{Ra} + {}^{4}_{2}\text{He}$$

EXAMPLE 17.1   Writing Nuclear Equations for Alpha (α) Decay

Write a nuclear equation for the alpha decay of Ra-224.

SOLUTION

Begin with the symbol for Ra-224 on the left side of the equation and the symbol for an alpha particle on the right side.	$^{224}_{88}\text{Ra} \longrightarrow {}^{x}_{y}? + {}^{4}_{2}\text{He}$
Equalize the sum of the mass numbers and the sum of the atomic numbers on both sides of the equation by writing the appropriate mass number and atomic number for the unknown daughter nuclide.	$^{224}_{88}\text{Ra} \longrightarrow {}^{220}_{86}? + {}^{4}_{2}\text{He}$
Deduce the identity of the unknown daughter nuclide from the atomic number. Since the atomic number is 86, the daughter nuclide is radon (Rn).	$^{224}_{88}\text{Ra} \longrightarrow {}^{220}_{86}\text{Rn} + {}^{4}_{2}\text{He}$

▶SKILLBUILDER 17.1 | Writing Nuclear Equations for Alpha (α) Decay

Write a nuclear equation for the alpha decay of Po-216.

▶FOR MORE PRACTICE   Example 17.6; Problems 59, 60.

Recall that Rutherford used alpha particles to probe the structure of the atom when he discovered the nucleus (Section 4.3).

To ionize means *to create ions* (charged particles).

Alpha radiation is the semi-truck of radioactivity. The alpha particle is by far the most massive of all particles emitted by radioactive nuclei. Consequently, alpha radiation has the most potential to interact with and damage other molecules, including biological ones. Radiation interacts with other molecules and atoms by ionizing them. If radiation ionizes molecules within the cells of living organisms, those molecules become damaged and the cell can die or begin to reproduce abnormally. The ability of radiation to ionize molecules and atoms is called its **ionizing power**. Of all types of radioactivity, alpha radiation has the highest ionizing power.

Because of its large size, alpha radiation also has the lowest **penetrating power**—the ability to penetrate matter. (Imagine a semi-truck trying to get through a traffic jam.) In order for radiation to damage important molecules within living cells, it must penetrate the cell. Alpha radiation does not easily penetrate cells; it can be stopped by a sheet of paper, by clothing, or even by air. Consequently, a low-level alpha emitter kept outside the body is relatively safe. However, if an alpha emitter is ingested or inhaled, it becomes very dangerous because the alpha particles then have direct access to the molecules that compose organs and tissues.

To summarize:

- Alpha particles are composed of 2 protons and 2 neutrons.
- Alpha particles have the symbol ${}^{4}_{2}\text{He}$.
- Alpha particles have a high ionizing power.
- Alpha particles have a low penetrating power.

▶ FIGURE 17.5 **Beta radiation**
Beta radiation occurs when an unstable nucleus emits an electron. As the emission occurs, a neutron turns into a proton. **Question:** What happens to the atomic number of an element upon emission of a beta particle?

BETA (β) RADIATION

Beta (β) radiation occurs when an unstable nucleus emits an electron (▲ Figure 17.5). How does a nucleus, which contains only protons and neutrons, emit an electron? The electron comes from the conversion of a neutron to a proton. In other words, in some unstable nuclei, a neutron changes into a proton and emits an electron in the process.

$$\textbf{Beta decay} \quad \text{Neutron} \longrightarrow \text{Proton} + \text{Electron}$$

The symbol for a **beta (β) particle** in a nuclear equation is:

$$\textbf{Beta } (\boldsymbol{\beta}) \textbf{ particle} \quad {}_{-1}^{0}e \;\bullet$$

Beta radiation is also called beta-minus (β^-) radiation because of its negative charge.

The 0 in the upper left corner reflects the mass number of the electron. The −1 in the lower left corner reflects the charge of the electron, which is equivalent to an atomic number of −1 in a nuclear equation. When an atom emits a beta particle, its atomic number increases by one because it now has an additional proton. For example, the nuclear equation for the beta decay of radium-228 is:

$$^{228}_{88}\text{Ra} \longrightarrow {}^{228}_{89}\text{Ac} + {}^{0}_{-1}e$$

Remember that the mass number is the sum of the number of protons and neutrons. Since electrons have no protons or neutrons, their mass number is zero.

Notice that the nuclear equation is balanced—the sums of the mass numbers on both sides are equal, and the sums of the atomic numbers on both sides are equal.

$$^{228}_{88}\text{Ra} \longrightarrow {}^{228}_{89}\text{Ac} + {}^{0}_{-1}e$$

Left Side	Right Side
Sum of mass numbers = 228	Sum of mass numbers = 228 + 0 = 228
Sum of atomic numbers = 88	Sum of atomic numbers = 89 − 1 = 88

The identity and symbol of the daughter nuclide of any beta decay can be determined in a manner similar to that for alpha decay, as shown in Example 17.2.

EXAMPLE 17.2 Writing Nuclear Equations for Beta (β) Decay

Write a nuclear equation for the beta decay of Bk-249.

SOLUTION

Begin with the symbol for Bk-249 on the left side of the equation and the symbol for a beta particle on the right side.	$^{249}_{97}\text{Bk} \longrightarrow {}^{x}_{y}? + {}^{0}_{-1}e$
Equalize the sum of the mass numbers and the sum of the atomic numbers on both sides of the equation by writing the appropriate mass number and atomic number for the unknown daughter nuclide.	$^{249}_{97}\text{Bk} \longrightarrow {}^{249}_{98}? + {}^{0}_{-1}e$
Deduce the identity of the unknown daughter nuclide from the atomic number. Since the atomic number is 98, the daughter nuclide is californium (Cf).	$^{249}_{97}\text{Bk} \longrightarrow {}^{249}_{98}\text{Cf} + {}^{0}_{-1}e$

▶ **SKILLBUILDER 17.2 | Writing Nuclear Equations for Beta (β) Decay**
Write a nuclear equation for the beta decay of Ac-228.

▶ **SKILLBUILDER PLUS**
Write three nuclear equations to represent the nuclear decay sequence that begins with the alpha decay of U-235 followed by a beta decay of the daughter nuclide and then another alpha decay.

▶ **FOR MORE PRACTICE**   Example 17.7; Problems 61, 62.

Beta radiation is the midsized car of radioactivity. Beta particles are much less massive than alpha particles and consequently have a lower ionizing power. However, because of their smaller size, beta particles have greater penetrating power; a sheet of metal or a thick piece of wood is required to stop them. Consequently, a low-level beta emitter outside the body poses a higher risk than an alpha emitter. Inside the body, however, a beta emitter does less damage than an alpha emitter.

To summarize:

- Beta particles are electrons emitted from atomic nuclei when a neutron changes into a proton.
- Beta particles have the symbol $_{-1}^{0}e$.
- Beta particles have intermediate ionizing power.
- Beta particles have intermediate penetrating power.

GAMMA (γ) RADIATION

See Section 9.3 for a review of electromagnetic radiation.

Gamma (γ) radiation is significantly different from alpha or beta radiation. Gamma radiation is not matter but *electromagnetic radiation:* Gamma rays are high-energy (short-wavelength) photons. The symbol for a **gamma ray** is:

Gamma (γ) ray $_{0}^{0}\gamma$

A gamma ray has no charge and no mass. When a gamma ray is emitted from a radioactive atom, it does not change the mass number or the atomic number of the element. Gamma rays are usually emitted in conjunction with other types of radiation. For example, the alpha emission of U-238 (discussed previously) is also accompanied by the emission of a gamma ray.

$$^{238}_{92}U \longrightarrow \, ^{234}_{90}Th + \, ^{4}_{2}He + \, ^{0}_{0}\gamma$$

Gamma rays are the motorbikes of radioactivity. They have the lowest ionizing power but the highest penetrating power. (Imagine a motorbike zipping through a traffic jam.) Stopping gamma rays requires several inches of lead shielding or thick slabs of concrete.

To summarize:

- Gamma rays are electromagnetic radiation—high-energy, short-wavelength photons.
- Gamma rays have the symbol $_{0}^{0}\gamma$.
- Gamma rays have low ionizing power.
- Gamma rays have high penetrating power.

POSITRON EMISSION

Positron emission can be thought of as a type of beta emission. It is sometimes referred to as beta-plus (β^+) emission.

Positron emission occurs when an unstable nucleus emits a positron (▶ Figure 17.6). A **positron** has the mass of an electron but carries a 1+ charge. In some unstable nuclei, a proton changes into a neutron and emits a positron in the process.

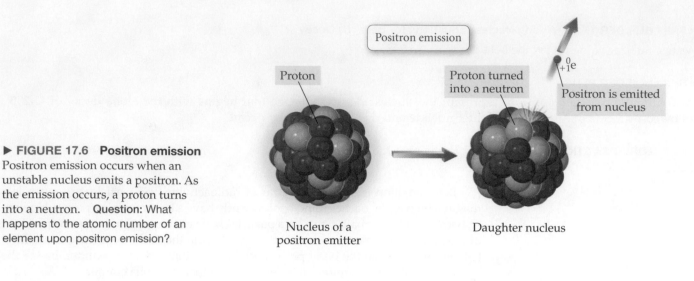

▶ **FIGURE 17.6 Positron emission**
Positron emission occurs when an unstable nucleus emits a positron. As the emission occurs, a proton turns into a neutron.   **Question:** What happens to the atomic number of an element upon positron emission?

Positron emission   Proton $\longrightarrow$ Neutron + Positron

The symbol for a positron in a nuclear equation is:

Positron $^{0}_{+1}e$ ●

The 0 in the upper left corner indicates that a positron has a mass number of 0. The +1 in the lower left corner reflects the charge of the positron, which is equivalent to an atomic number of +1 in a nuclear equation. When an atom emits a positron, its atomic number decreases by 1 because it now has 1 less proton. For example, the nuclear equation for the positron emission of phosphorus-30 is:

$$^{30}_{15}P \longrightarrow {}^{30}_{14}Si + {}^{0}_{+1}e$$

The identity and symbol of the daughter nuclide of any positron emission is determined in a manner similar to alpha and beta decay, as shown in Example 17.3. Positron emission is similar to beta emission in its ionizing and penetrating power. Table 17.1 summarizes the different kinds of radioactivity that we have covered in the chapter.

EXAMPLE 17.3   Writing Nuclear Equations for Positron Emission

Write a nuclear equation for the positron emission of potassium-40.

SOLUTION	
Begin with the symbol for K-40 on the left side of the equation and the symbol for a positron on the right side.	$^{40}_{19}K \longrightarrow {}^{x}_{y}? + {}^{0}_{+1}e$
Equalize the sum of the mass numbers and the sum of the atomic numbers on both sides of the equation by writing the appropriate mass number and atomic number for the unknown daughter nuclide.	$^{40}_{19}K \longrightarrow {}^{40}_{18}? + {}^{0}_{+1}e$
Deduce the identity of the unknown daughter nuclide from the atomic number. Since the atomic number is 18, the daughter nuclide is argon (Ar).	$^{40}_{19}K \longrightarrow {}^{40}_{18}Ar + {}^{0}_{+1}e$

▶**SKILLBUILDER 17.3 | Writing Nuclear Equations for Positron Emission**

Write a nuclear equation for the positron emission of sodium-22.

▶**FOR MORE PRACTICE**   Example 17.8; Problems 63, 64.

TABLE 17.1 Selected Types of Radioactive Decay

Decay Mode	Process	Ionizing Power	Penetrating Power	Example
α	Parent nuclide → Daughter nuclide + α particle ($^4_2$He)	High	Low	$^{238}_{92}U \longrightarrow {}^{234}_{90}Th + {}^4_2He$
β	Neutron; Parent nuclide → Neutron becomes a proton; Daughter nuclide + β particle ($^0_{-1}e$)	Moderate	Moderate	$^{228}_{88}Ra \longrightarrow {}^{228}_{89}Ac + {}^0_{-1}e$
γ	Excited nuclide → Stable nuclide + Photon ($^0_0\gamma$)	Low	High	$^{234}_{90}Th \longrightarrow {}^{234}_{90}Th + {}^0_0\gamma$
Positron emission	Proton; Parent nuclide → Proton becomes a neutron; Daughter nuclide + Positron ($^0_{+1}e$)	Moderate	Moderate	$^{30}_{15}P \longrightarrow {}^{30}_{14}Si + {}^0_{+1}e$

✓ **CONCEPTUAL CHECKPOINT 17.1**

Which kind of radioactive decay changes the mass number of the parent element?

(a) alpha (α) decay

(b) beta (β) decay

(c) gamma (γ) decay

(d) positron decay

17.4 Detecting Radioactivity

The particles emitted from radioactive nuclei contain a large amount of energy and can therefore be easily detected. Radiation detectors detect such particles through their interactions with atoms or molecules. The simplest radiation detectors are pieces of photographic film that become exposed when radiation passes through them. **Film-badge dosimeters**—which consist of photographic film held in a small case that is pinned to clothing—are standard for most people working with or near radioactive substances (▶ Figure 17.7a). These badges are collected and processed (or developed) regularly as a way to monitor exposure. The more exposed the film has become in a given period of time, the more radioactivity to which the person has been exposed.

Radioactivity can be instantly detected with devices such as the **Geiger-Müller counter** (▶ Figure 17.7b). In such an instrument (commonly referred to simply as a Geiger counter), particles emitted by radioactive nuclei pass through an argon-filled chamber. The energetic particles create a trail of ionized argon

▶ FIGURE 17.7 **Detecting radiation** (a) A film badge provides a simple and an inexpensive way of measuring a person's cumulative exposure to radiation. (b) A Geiger counter records the passage of individual energetic particles emitted by radioactive nuclei as they pass through a chamber filled with argon gas. When an argon atom is ionized, the resulting ion is attracted to the anode and the dislodged electron to the cathode, creating a tiny electrical current that can be recorded in various ways.

(a)

(b)

atoms as they pass through the chamber. If the applied voltage is high enough, these newly formed ions produce an electrical signal that can be detected on a meter or turned into an audible click. Each click corresponds to a radioactive particle passing through the argon gas chamber. This clicking is the stereotypical sound most people associate with a radiation detector.

A second type of device commonly used to detect radiation instantly is called a **scintillation counter**. In a scintillation counter, the radioactive emissions pass through a material (such as NaI or CsI) that emits ultraviolet or visible light in response to excitation by energetic particles. This light is detected and turned into an electrical signal that can be read on a meter.

17.5 Natural Radioactivity and Half-Life

Radioactivity is a natural component of our environment. The ground beneath you most likely contains radioactive atoms that emit radiation into the air around you. The food you eat contains a residual amount of radioactive atoms that enter into your body fluids and tissues. Small amounts of radiation from space make it through our atmosphere and constantly bombard Earth. Humans and other living organisms have evolved in this environment and have adapted to survive in it.

One reason for the radioactivity in our environment is the instability of all atomic nuclei beyond atomic number 83 (bismuth). In addition, some isotopes of elements with fewer than 83 protons are also unstable and radioactive.

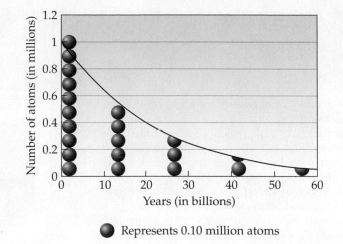

► FIGURE 17.8 **The concept of half-life** A plot of the number of Th-232 atoms in a sample initially containing 1 million atoms as a function of time. Th-232 has a half-life of 14 billion years.

● Represents 0.10 million atoms

| A *decay event* is the emission of radiation by a single radioactive nuclide.

Different radioactive nuclides decay into their daughter nuclides at different rates. Some nuclides decay quickly while others decay slowly. The time it takes for half of the parent nuclides in a radioactive sample to decay to the daughter nuclides is called the **half-life**. Nuclides that decay quickly have short half-lives and are very active (many decay events per unit time), while those that decay slowly have long half-lives and are less active (fewer decay events per unit time). For example, Th-232 is an alpha emitter that decays according to the nuclear reaction:

$$^{232}_{90}\text{Th} \longrightarrow {}^{228}_{88}\text{Ra} + {}^{4}_{2}\text{He}$$

Th-232 has a half-life of 1.4×10^{10} years or 14 billion years. If we start with a sample of Th-232 containing 1 million atoms, the sample would decay to half a million atoms in 14 billion years and then to a quarter of a million in another 14 billion years and so on (▲ Figure 17.8).

$$\begin{array}{ccc} 1 \text{ million} & \frac{1}{2} \text{ million} & \frac{1}{4} \text{ million} \\ \text{Th-232 atoms} \xrightarrow{\text{14 billion years}} & \text{Th-232 atoms} \xrightarrow{\text{14 billion years}} & \text{Th-232 atoms} \end{array}$$

Notice that a radioactive sample *does not* decay to zero atoms in two half-lives—*you can't add two half-lives together to get a "whole" life.* The amount that remains after one half-life is always half of what was present at the start. The amount that remains after two half-lives is a quarter of what was present at the start, and so on.

Some nuclides have short half-lives. Radon-220, for example, has a half-life of approximately 1 minute. If we had a sample of radon-220 that contained 1 million atoms, it would be diminished to $\frac{1}{4}$ million radon-220 atoms in just 2 minutes.

$$\begin{array}{ccc} 1 \text{ million} & \frac{1}{2} \text{ million} & \frac{1}{4} \text{ million} \\ \text{Rn-220 atoms} \xrightarrow{\text{1 minute}} & \text{Rn-220 atoms} \xrightarrow{\text{1 minute}} & \text{Th-232 atoms} \end{array}$$

| Each radioactive nuclide has a unique half-life, which is not affected by physical conditions or chemical environment.

Rn-220 is therefore much more active than Th-232 because it undergoes many more decay events in a given period of time. Some nuclides have even shorter half-lives. Table 17.2 lists several nuclides and their half-lives.

TABLE 17.2 **Selected Nuclides and Their Half-Lives**

Nuclide	Half-Life	Type of Decay
$^{232}_{90}\text{Th}$	1.4×10^{10} yr	alpha
$^{238}_{92}\text{U}$	4.5×10^{9} yr	alpha
$^{14}_{6}\text{C}$	5730 yr	beta
$^{220}_{86}\text{Rn}$	55.6 s	alpha
$^{219}_{90}\text{Th}$	1.05×10^{-6} s	alpha

CHEMISTRY AND HEALTH

Environmental Radon

Radon—a radioactive gas—is one of the products of the radioactive decay series of uranium. Wherever there is uranium in the ground, there is likely to be radon seeping up into the air. Radon and its daughter nuclides (which attach to dust particles) can therefore be inhaled into the lungs, where they decay and increase lung cancer risk. The radioactive decay of radon is by far the single greatest source of human radiation exposure.

Homes built in areas with significant uranium deposits in the ground pose the greatest risk. These homes can accumulate radon levels that are above what the Environmental Protection Agency (EPA) considers safe. Simple test kits are available to test indoor air and determine radon levels. The higher the radon level is, the greater the risk. The risk is even higher for smokers who live in these houses. Excessively high indoor radon levels require the installation of a ventilation system to purge radon from the house. Lower levels can be ventilated by keeping windows and doors open.

CAN YOU ANSWER THIS? *Suppose that a house contains 1.80×10^{-3} mol of radon-222 (which has a half-life of 3.8 days). If no new radon entered the house, how long would it take for the radon to decay to 4.50×10^{-4} mol?*

EPA Map of Radon Zones

LEGEND
Zone 1–Radon > 4 PCi/L
Zone 2–Radon between 2-4 PCi/L
Zone 3–Radon < 2 PCi/L

▲ Map of the United States showing radon levels. Zone 1 counties have the highest levels, and zone 3 counties have the lowest.

EXAMPLE 17.4 Half-Life

How long does it take for a 1.80-mol sample of Th-228 (which has a half-life of 1.9 years) to decay to 0.225 mol?

SOLUTION

It is easiest to draw a table showing the amount of Th-228 as a function of number of half-lives. For each half-life, divide the amount of Th-228 by 2.

Amount of Th-228	Number of Half-Lives	Time in Years
1.80 mol	0	0
0.900 mol	1	1.9
0.450 mol	2	3.8
0.225 mol	3	5.7

It takes three half-lives or 5.7 years for the sample to decay to 0.225 mol.

▶**SKILLBUILDER 17.4 | Half-Life**

A radium-226 sample initially contains 0.112 mol. How much radium-226 is left in the sample after 6400 years? The half-life of radium-226 is 1600 years.

▶**FOR MORE PRACTICE** Example 17.9; Problems 69, 70, 71, 72, 73, 74, 75, 76.

A NATURAL RADIOACTIVE DECAY SERIES

The radioactive elements in our environment are all undergoing radioactive decay. They are always present in our environment because they either have very long half-lives (billions of years) or they are continuously being formed by some

▶ **FIGURE 17.9 Uranium-238 decay series** The red arrows represent alpha decay and the blue arrows represent beta decay.

other process in the environment. In many cases, the daughter nuclide of a radioactive decay is itself radioactive and in turn produces another daughter nuclide that is radioactive and so on, resulting in a radioactive decay series. For example, uranium (atomic number 92) is the heaviest naturally occurring element. It is an alpha emitter that decays to Th-234 with a half-life of 4.47 billion years.

$$^{238}_{92}\text{U} \longrightarrow {}^{234}_{90}\text{Th} + {}^{4}_{2}\text{He}$$

The daughter nuclide, Th-234, is itself radioactive—it is a beta emitter that decays to Pa-234 with a half-life of 24.1 days.

$$^{234}_{90}\text{Th} \longrightarrow {}^{234}_{91}\text{Pa} + {}^{0}_{-1}\text{e}$$

Pa-234 is also radioactive, decaying to U-234 via beta emission with a half-life of 244,500 years. This process continues until it produces Pb-206, which is stable. The entire uranium-238 decay series is shown in ▲ Figure 17.9. In other words, all of the uranium-238 in the environment is slowly decaying away to lead. Since the half-life for the first step in the series is so long, however, there is still plenty of uranium-238 in the environment. All of the other nuclides in the decay series are also present in the environment in varying amounts, depending on their half-lives.

✔️ **CONCEPTUAL CHECKPOINT 17.2**

Suppose that you start with 1 million atoms of a particular radioactive isotope. How many half-lives would be required to reduce the number of undecayed atoms to fewer than 1000?

(a) 10   **(b)** 100   **(c)** 1000   **(d)** 1001

17.6 Radiocarbon Dating: Using Radioactivity to Measure the Age of Fossils and Other Artifacts

Archaeologists, geologists, anthropologists, and other scientists take advantage of the presence of natural radioactivity in our environment to estimate the ages of fossils and artifacts with a technique called **radiocarbon dating**. For example, in

1947, young shepherds searching for a stray goat near the Dead Sea (east of Jerusalem) entered a cave and discovered ancient scrolls stuffed into jars. These scrolls—now named the Dead Sea Scrolls—are 2000-year-old biblical manuscripts, predating other existing manuscripts by almost 1000 years.

The Dead Sea Scrolls—like other ancient artifacts—contain a radioactive signature that reveals their age. This signature results from the presence of carbon-14—which is radioactive—in the environment. Carbon-14 is constantly formed in the upper atmosphere by the neutron bombardment of nitrogen.

$$^{14}_{7}\text{N} + ^{1}_{0}\text{n} \longrightarrow ^{14}_{6}\text{C} + ^{1}_{1}\text{H}$$

Carbon-14 then decays back to nitrogen by beta emission, with a half-life of 5730 years.

$$^{14}_{6}\text{C} \longrightarrow ^{14}_{7}\text{N} + ^{0}_{-1}\text{e}$$

The continuous formation of carbon-14 in the atmosphere and its continuous decay back to nitrogen-14 produces a nearly constant equilibrium concentration of atmospheric carbon-14. That carbon-14 is oxidized to carbon dioxide and then incorporated into plants by photosynthesis. It is also incorporated into animals because animals ultimately depend on plants for food (they either eat plants or eat other animals that eat plants). Consequently, all living organisms contain a residual amount of carbon-14. When a living organism dies, it stops incorporating new carbon-14 into its tissues. The carbon-14 present at the time of death decays with a half-life of 5730 years. Since many artifacts, such as the Dead Sea Scrolls, are made from materials that were once living—such as papyrus, wood, and other plant and animal derivatives—the amount of carbon-14 in these artifacts indicates their age.

The concentration of carbon-14 in all living organisms is the same.

For example, suppose an ancient artifact has a carbon-14 concentration that is 50% of that found in living organisms. How old is the artifact? Since it contains half as much carbon-14 as a living organism, it must be one half-life, or 5730 years old. If the artifact has a carbon-14 concentration that is 25% of that found in living organisms, its age is two half-lives or 11,460 years old. Table 17.3 shows the age of an object based on its carbon-14 content. Ages of less than one half-life, or intermediate between a whole number of half-lives, can also be calculated, but the method for doing so is beyond the scope of this book.

We know that carbon-14 dating is accurate because it can be checked against objects whose ages are known from other methods. For example, old trees can be dated by counting the tree rings within their trunks and by carbon-14 dating. The two methods generally agree to within a few percent, and even this amount of error can be reduced by using the record inherent in the tree rings to calibrate carbon-14 dating. However, carbon-14 dating is not dependable when attempting to date objects that are more than 50,000 years old; the amount of carbon-14 becomes too low to measure.

TABLE 17.3   Age of Object Based on Its Carbon-14 Content

Concentration of C-14 (% Relative to Living Organisms)	Age of Object in Years
100.0	0
50.0	5,730
25.00	11,460
12.50	17,190
6.250	22,920
3.125	28,650
1.563	34,380

CHEMISTRY IN THE MEDIA

The Shroud of Turin

The shroud of Turin—kept in the Cathedral of Turin in Italy—is an old linen cloth that bears the image of a man who appears to have been crucified.

The image becomes clearer if the shroud is photographed and viewed as a negative. Many believe that the shroud is the original burial cloth of Jesus Christ, miraculously imprinted with his image. In 1988, the Roman Catholic Church chose three independent laboratories to perform radiocarbon dating on the shroud. The laboratories took samples from the shroud and measured the carbon-14 content. They all got similar results—the shroud was made from linen originating in about A.D. 1325. Although some have disputed the results, and although no scientific test is 100% reliable, newspapers around the world quickly announced that the shroud could not have been the burial cloth of Jesus.

CAN YOU ANSWER THIS? *An artifact is said to have originated in 3000 B.C. Examination of the C-14 content of the artifacts reveals that the concentration of C-14 is 55% of that found in living organisms. Could the artifact be authentic?*

▲ The Shroud of Turin.

EXAMPLE 17.5 Radiocarbon Dating

A skull believed to belong to an early human being is found to have a carbon-14 content of 3.125% of that found in living organisms. How old is the skull?

SOLUTION

Examine Table 17.3 to determine that a carbon-14 content of 3.125% of that found in living organisms corresponds to an age of 28,650 years.

▶ **SKILLBUILDER 17.5 | Radiocarbon Dating**

An ancient scroll is claimed to have originated from Greek scholars in about 500 B.C. A measure of its carbon-14 content reveals it to contain 100.0% of that found in living organisms. Is the scroll authentic?

▶ **FOR MORE PRACTICE** Example 17.10; Problems 79, 80, 81, 82.

17.7 The Discovery of Fission and the Atomic Bomb

The element with atomic number 100 is named fermium, in honor of Enrico Fermi.

In the mid-1930s Enrico Fermi (1901–1954), an Italian physicist, tried to synthesize a new element by bombarding uranium—the heaviest known element at that time—with neutrons. Fermi hypothesized that if a neutron were incorporated into the nucleus of a uranium atom, the nucleus might undergo beta decay, converting a neutron into a proton. If that happened, a new element, with atomic number 93, would be synthesized for the first time. The nuclear equation for the process is:

$$^{238}_{92}\text{U} + ^{1}_{0}\text{n} \longrightarrow ^{239}_{92}\text{U} \longrightarrow ^{239}_{93}\text{X} + ^{0}_{-1}\text{e}$$

Neutron   Newly synthesized element

Fermi performed the experiment and detected the emission of beta particles. However, his results were inconclusive. Had he synthesized a new element? Fermi never chemically examined the products to determine their composition and therefore could not say with certainty whether he had.

The element with atomic number 109 is named meitnerium, in honor of Lise Meitner.

▲ Lise Meitner in Otto Hahn's Berlin laboratory.

Three researchers in Germany—Lise Meitner (1878–1968), Fritz Strassmann (1902–1980), and Otto Hahn (1879–1968)—repeated Fermi's experiments and then performed careful chemical analysis of the products. What they found in the products—several elements *lighter* than uranium—would change the world forever. On January 6, 1939, Meitner, Strassmann, and Hahn reported that the neutron bombardment of uranium resulted in **nuclear fission**—the splitting of the atom. The nucleus of the neutron-bombarded uranium atom had broken apart into barium, krypton, and other smaller products. They also realized that the process emitted enormous amounts of energy. The following is a nuclear equation for a fission reaction, showing how uranium breaks apart into the daughter nuclides.

$$^{235}_{92}\text{U} + ^{1}_{0}\text{n} \longrightarrow ^{142}_{56}\text{Ba} + ^{91}_{36}\text{Kr} + 3^{1}_{0}\text{n} + \text{Energy}$$

Neutron (n)

Notice that the initial uranium atom is the U-235 isotope, which makes up less than 1% of all naturally occurring uranium., The most abundant uranium isotope, U-238 does not undergo fission. Therefore, the uranium used for fuel in nuclear reactions must be *enriched* in U-235 (it must contain more than the naturally occurring percentage of U-235). Notice also that the process produces three neutrons, which have the potential to initiate fission in three other U-235 atoms.

U.S. scientists quickly realized that uranium enriched with U-235 could undergo a **chain reaction** in which neutrons produced by the fission of one uranium nucleus would induce fission in other uranium nuclei (▼ Figure 17.10).

▲ **FIGURE 17.10 Fission chain reaction** The neutrons produced by the fission of one uranium nucleus induce fission in other uranium nuclei to produce a self-amplifying reaction. **Question:** Why is it necessary that each fission event produce more than one neutron to sustain the chain reaction?

The result would be a self-amplifying reaction capable of producing an enormous amount of energy—an atomic bomb. However, to make a bomb, a **critical mass** of U-235—enough U-235 to produce a self-sustaining reaction—would be necessary. Fearing that Nazi Germany would develop such a bomb, several U.S. scientists persuaded Albert Einstein, the most well-known scientist of the time, to write a letter to President Franklin Roosevelt warning of this possibility. Einstein wrote, "and it is conceivable—though much less certain—that extremely powerful bombs of a new type may thus be constructed. A single bomb of this type, carried by boat and exploded in a port, might very well destroy the whole port together with some of the surrounding territory."

Roosevelt was convinced by Einstein's letter, and in 1941 he assembled the resources to begin the costliest scientific project ever attempted. The top-secret endeavor was called the *Manhattan Project*, and its main goal was to build an atomic bomb before the Germans did. The project was led by physicist J. R. Oppenheimer (1904–1967) and was headquartered at a high-security research facility in Los Alamos, New Mexico. Four years later, on July 16, 1945, the world's first nuclear weapon was successfully detonated at a test site in New Mexico. The first atomic bomb exploded with a force equivalent to 18,000 tons of dynamite. Ironically, the Germans—who had *not* made a successful nuclear bomb—had already been defeated by this time. Instead, the atomic bomb was used on Japan. One bomb was dropped on Hiroshima, and a second bomb was dropped on Nagasaki. Together, the bombs killed approximately 200,000 people and forced Japan to surrender. World War II was over. The atomic age had begun.

▲ The testing of the world's first nuclear bomb at Alamogordo, New Mexico, in 1945.

17.8 Nuclear Power: Using Fission to Generate Electricity

Nuclear reactions, such as fission, generate enormous amounts of energy. In a nuclear bomb, the energy is released all at once. The energy can also be released more slowly and used for other purposes such as electricity generation. In the United States, about 20% of electricity is generated by nuclear fission. In some other countries, as much as 70% of electricity is generated by nuclear fission. To get an idea of the amount of energy released during fission, imagine a hypothetical nuclear-powered car. Suppose the fuel for such a car was a uranium cylinder about the size of a pencil. How often would you have to refuel the car? The energy content of the uranium cylinder would be equivalent to about 1000 twenty-gallon tanks of gasoline. If you refuel your gasoline-powered car once a week, your nuclear-powered car could go 1000 weeks—almost 20 years—before refueling. Imagine a pencil-sized fuel rod lasting for twenty 20 years of driving!

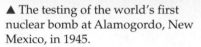

A nuclear-powered car really is hypothetical because the amount of uranium-235 in a pencil-sized cylinder would not be enough to reach a critical mass and produce a self-sustaining reaction.

Similarly, a nuclear-powered electrical plant can produce a lot of electricity with a small amount of fuel. Nuclear power plants generate electricity by using fission to generate heat (▶ Figure 17.11). The heat is used to boil water and create steam, which then turns the turbine on a generator to produce electricity. The fission reaction itself occurs in the nuclear core of the power plant, or *reactor*. The core consists of uranium fuel rods—enriched to about 3.5% U-235—interspersed between retractable neutron-absorbing control rods. When the control rods are fully retracted from the fuel rod assembly, the chain reaction can occur unabated. However, when the control rods are fully inserted into the fuel assembly, they absorb the neutrons that would otherwise induce fission, shutting down the chain reaction.

By inserting or retracting the control rods, the operator can control the rate of fission. If more heat is needed, the control rods are retracted slightly. If the fission reaction begins to get too hot, the control rods are inserted a little more. In this way, the fission reaction is controlled to produce the right amount of heat needed to generate electricity. In case of a power failure, the fuel rods automatically drop into the fuel rod assembly, shutting down the fission reaction.

A typical nuclear power plant generates enough electricity for a city of about 1 million people and uses about 50 kg of fuel per day. In contrast, a coal-burning

▲ Technicians inspect the core of a nuclear reactor, which houses the fuel rods and control rods.

▶ **FIGURE 17.11 Nuclear power**
In a nuclear power plant, fission generates heat that is used to boil water and create steam. The steam then turns a turbine on a generator to produce electricity. Note that the water carrying heat from the reactor core is contained within separate pipes and does not come into direct contact with the steam that drives the turbines.

Reactor cores in the United States are not made of graphite and could not burn in the way that the Chernobyl core did.

power plant uses about 2 million kg of fuel to generate the electricity for that same city. Furthermore, a nuclear power plant generates no air pollution and no greenhouse gases. A coal-burning power plant, on the other hand, emits pollutants such as carbon monoxide, nitrogen oxides, and sulfur oxides. Coal-burning power plants also emit carbon dioxide, a greenhouse gas.

Nuclear power generation, however, is not without problems. Foremost among them is the danger of nuclear accidents. In spite of safety precautions, the fission reaction occurring in a nuclear power plant can overheat. The most famous example of this type of accident occurred in Chernobyl in the former Soviet Union on April 26, 1986. Operators of the plant were performing an experiment designed to reduce maintenance costs. In order to perform the experiment, however, many of the safety features of the reactor core were disabled. The experiment failed, with disastrous results. The nuclear core, composed partly of graphite, overheated and began to burn. The accident directly caused 31 deaths and produced a fire that scattered radioactive debris into the atmosphere, making the surrounding land uninhabitable. The overall death toll from subsequent cancers is undetermined at this time.

Contrary to some public fears, however, a nuclear power plant *cannot* become a nuclear bomb. The uranium fuel used in electricity generation is not sufficiently enriched in U-235 to produce a nuclear detonation. In addition, U.S. nuclear power plants have additional safety features designed to prevent similar accidents. For example, U.S. nuclear power plants have large containment structures designed to contain radioactive debris in the event of an accident.

A second problem associated with nuclear power is waste disposal. Although the amount of nuclear fuel used in electricity generation is small compared to other fuels, the products of the reaction are radioactive and have very long half-lives (thousands of years or more). What do we do with this waste? Currently, in the United States, nuclear waste is stored on site at the nuclear power plants. A permanent disposal site was being developed in Yucca Mountain, Nevada. The site had originally been scheduled to be operational in 2010, and that date was later delayed to 2017. However, the Obama administration determined that the Yucca Mountain site was untenable, and in the spring of 2010, the license application to develop this site was withdrawn. President Obama has also formed a committee (called the Blue Ribbon Commission on America's Nuclear Future) that is charged with developing alternatives to Yucca Mountain.

17.9   Nuclear Fusion: The Power of the Sun

As we have learned, nuclear fission is the splitting of a heavy nucleus to form two or more lighter ones. **Nuclear fusion**, by contrast, is the combination of two light nuclei to form a heavier one. Both fusion and fission emit large amounts of energy. Nuclear fusion is the energy source of stars, including our sun. In stars, hydrogen atoms fuse together to form helium atoms, emitting energy in the process.

Nuclear fusion is also the basis of modern nuclear weapons called hydrogen bombs. A modern hydrogen bomb has up to 1000 times the explosive force of the first atomic bombs. These bombs employ the following fusion reaction.

$$\,^2_1H + \,^3_1H \longrightarrow \,^4_2He + \,^1_0n$$

In this reaction, deuterium (the isotope of hydrogen with one neutron) and tritium (the isotope of hydrogen with two neutrons) combine to form helium-4 and a neutron. Because fusion reactions require two positively charged nuclei (which repel each other) to fuse together, extremely high temperatures are required. In a hydrogen bomb, a small fission bomb is detonated first, providing temperatures high enough for fusion to proceed.

Nuclear fusion has been intensely investigated as a way to produce electricity. Because of the higher energy production—fusion provides about ten times more energy per gram of fuel than fission—and because the products of the reaction are less dangerous than those of fission, fusion holds promise as a future energy source. However, in spite of intense efforts, fusion electricity generation remains elusive. One of the main problems is the high temperature required for fusion to occur—no material can withstand these temperatures. Whether fusion will ever be a viable energy source remains to be seen.

17.10   The Effects of Radiation on Life

Radiation can ionize atoms in biological molecules, thereby initiating reactions that can alter the molecules. When radiation damages important molecules in living cells, problems can develop. The ingestion of radioactive materials—especially alpha and beta emitters—is particularly dangerous because radioactive decay then occurs within the body and can do more damage than external radiation. The effects of radiation are divided into three different types: acute radiation damage, increased cancer risk, and genetic effects.

ACUTE RADIATION DAMAGE

Acute radiation damage results from exposures to large amounts of radiation in a short period of time. The main sources of this kind of exposure are nuclear bombs or exposed nuclear reactor cores. The resulting high levels of radiation kill large numbers of cells. Rapidly dividing cells, such as those in the immune system and the intestinal lining, are most susceptible. Consequently, people exposed to high levels of radiation have weakened immune systems and a lowered ability to absorb nutrients from food. In milder cases, recovery is possible with time. In more extreme cases death results, often from unchecked infection.

INCREASED CANCER RISK

DNA and its function in the body are explained in more detail in Chapter 19.

Lower doses of radiation over extended periods of time can increase cancer risk because radiation can damage DNA, the molecules in cells that carry instructions for cell growth and replication. When the DNA within a cell is damaged, the cell normally dies. Occasionally, however, changes in DNA cause cells to grow abnormally and to become cancerous. These cancerous cells grow into tumors that can spread and, in some cases, cause death. Cancer risk increases with increased radiation exposure. However, cancer is so prevalent and has so many convoluted causes that it is difficult to determine an exact threshold for increased cancer risk from radiation exposure.

GENETIC DEFECTS

Another possible effect of radiation exposure is genetic defects in offspring. If radiation damages the DNA of reproductive cells—such as eggs or sperm—then the offspring that develop from those cells may have genetic abnormalities. Genetic defects of this type have been observed in laboratory animals exposed to high levels of radiation. However, such genetic defects—with a clear causal connection to radiation exposure—have yet to be observed in humans, even in studies of Hiroshima survivors.

MEASURING RADIATION EXPOSURE

Common units of radioactivity include the *curie*, defined as 3.7×10^{10} decay events per second, and the *roentgen*, defined as the amount of radiation that produces 2.58×10^{-4} C of charge per kilogram of air. Human radiation exposure is often reported in a unit called the **rem**. The rem, which stands for *roentgen equivalent man*, is a weighted measure of radiation exposure that accounts for the ionizing power of the different types of radiation. On average, each of us is exposed to approximately one-third of a rem of radiation per year. This radiation comes primarily from natural sources, especially radon, one of the products in the uranium decay series (◀ Figure 17.12). It takes much more radiation than the natural amount to produce measurable health effects in humans. The first measurable effects, a decreased white blood cell count, occur at instantaneous exposures of approximately 20 rem (Table 17.4). Exposures of 100 rem show a definite increase in cancer risk, and exposures of more than 500 rem often result in death.

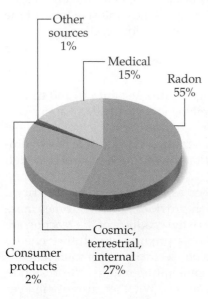

▲ FIGURE 17.12 Radiation exposure by source

TABLE 17.4 Effects of Radiation Exposure

Dose (rem)	Probable Outcome
20–100	decreased white blood cell count; possible increase in cancer risk
100–400	radiation sickness; skin lesions; increase in cancer risk
500	death

17.11 Radioactivity in Medicine

Radioactivity is often perceived as dangerous; however, it is also enormously useful to physicians in the diagnosis and treatment of disease. The use of radioactivity in medicine can be broadly divided into **isotope scanning** and **radiotherapy**.

ISOTOPE SCANNING

An example of isotope scanning was presented in the opening section of this chapter. In isotope scanning, a radioactive isotope is introduced into the body. Then the radiation emitted by the isotope is detected using a photographic film or a scintillation counter. Since different isotopes are taken up by different organs or tissues, isotope scanning has a variety of uses. For example, the radioactive isotope phosphorus-32 is preferentially taken up by cancerous tissue. A cancer patient can be given this isotope to find and identify cancerous tumors. Other isotopes commonly used in medicine include iodine-131, used to diagnose thyroid disorders, and technetium-99, which can produce images of several different internal organs (◀ Figure 17.13).

RADIOTHERAPY

Because radiation kills cells, and because it is particularly effective at killing rapidly dividing cells, it is often used as a therapy for cancer (cancer cells divide more quickly than normal cells). Gamma rays are focused on internal tumors to kill them (▼ Figure 17.14). The gamma-ray beam is usually aimed at the tumor from a number of different angles, maximizing the exposure of the tumor while minimizing the exposure of the healthy tissue around the tumor. (See *Chemistry and Health: Radiation Treatment for Cancer* in Chapter 9.) Nonetheless, cancer patients undergoing radiation therapy usually develop the symptoms of radiation sickness, which include vomiting, skin burns, and hair loss.

Some people wonder how radiation—which is known to cause cancer—can be used to treat cancer. The answer lies in risk analysis. A cancer patient is normally exposed to radiation doses of about 100 rem. Such a dose increases cancer risk by about 1%. However, if the patient has a 100% chance of dying from the cancer that he or she already has, such a risk becomes acceptable, especially since there is often a significant chance of curing the cancer.

▲ **FIGURE 17.13   An isotope scan** Technetium-99 is often used as the radiation source for bone scans such as this one.

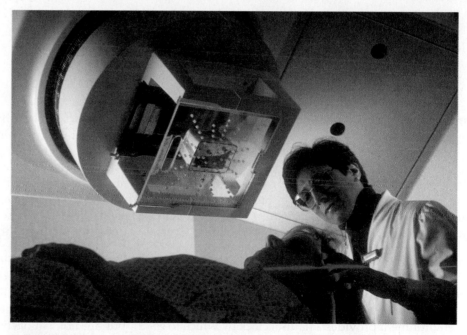

▲ **FIGURE 17.14   Radiotherapy for cancer**   This treatment involves exposing a malignant tumor to gamma rays, typically from radioisotopes such as cobalt-60. The beam is moved in a circular pattern around the tumor to maximize exposure of the cancer cells while minimizing exposure of healthy tissues.

 CONCEPTUAL CHECKPOINT 17.3

Which type of radiation is most likely to be used for isotope scanning?

(a) alpha

(b) beta

(c) gamma

CHAPTER IN REVIEW

CHEMICAL PRINCIPLES

The Nature and Discovery of Radioactivity: Radioactivity is the emission of particles from unstable atomic nuclei. It can be divided into four types.

- **Alpha particles**, composed of 2 protons and 2 neutrons and given the symbol ^4_2He. Alpha particles have a high ionizing power but a low penetrating power.

- **Beta particles**, electrons emitted from atomic nuclei when a neutron changes into a proton. Beta particles have the symbol $^0_{-1}\text{e}$ and have intermediate ionizing power and intermediate penetrating power.

- **Gamma rays**, high-energy, short-wavelength photons. Gamma rays have the symbol $^0_0\gamma$ and have low ionizing power and high penetrating power.

- **Positrons**, emitted from atomic nuclei when protons change into neutrons. Positrons have the same mass as electrons but opposite charge, and so are represented by the symbol $^0_{+1}\text{e}$. They have intermediate ionizing power and intermediate penetrating power.

Detecting Radioactivity: Radioactive emissions carry a large amount of energy and are therefore easily detected. The most common and inexpensive way to detect radioactivity is with photographic film, which is used in film-badge dosimeters to monitor exposure in people working with or near radioactive sources. Detection devices such as a Geiger-Müller counter or a scintillation counter, give instantaneous readings of radiation levels,

Half-Life and Radiocarbon Dating: The half-life of a radioactive nuclide is the time it takes for half of the parent nuclides in a radioactive sample to decay. The presence of radioactive carbon-14 (with a half-life of 5730 years) in the environment provides a natural clock by which to estimate the age of many artifacts and fossils. All living things contain carbon-14. When they die, the carbon-14 decays with its characteristic half-life. A measurement of the amount of carbon-14 remaining in a fossil or are artifact can therefore reveal its age.

Fission, the Atomic Bomb, and Nuclear Power: Fission—the splitting of the atom into smaller fragments—was discovered in 1939. Fission occurs when a U-235 nucleus absorbs a neutron. The nucleus becomes unstable, breaking apart and producing barium, krypton, neutrons, and a lot of energy.

RELEVANCE

The Nature and Discovery of Radioactivity: Radioactivity is a fundamental part of the behavior of some atoms, and it also has many applications. For example, radioactivity is used to diagnose and treat diseases, including cancer, thyroid diseases, abnormal kidney and bladder function, and heart disease. Natural radioactivity is part of our environment and can be used to date ancient objects. The discovery of radioactivity led to the discovery of fission, which in turn led to the development of nuclear bombs and nuclear energy.

Detecting Radioactivity: Since radioactivity is invisible, it must be detected using film or instruments. The detection of radioactivity is important as both a scientific tool and a practical one. Our understanding of what radioactivity is and our continuing research to understand it and its effects on living organisms require the ability to detect it. Our safety in areas where radioactive substances are used also depends on our ability to detect radiation.

Half-Life and Radiocarbon Dating: The half-life of a radioactive nuclide determines the activity of the nuclide and how long it will be radioactive. Nuclides with short half-lives are very active (many decay events per unit time) but are not radioactive for long. Nuclides with long half-lives are less active (fewer decays per unit time) but are radioactive for a long time.

Fission, the Atomic Bomb, and Nuclear Power: The discovery of fission in 1939 changed the world. Within six years, the United States developed and tested fission nuclear bombs, which ended World War II. Fission can also be used to generate electricity. The fission reaction heats water to create steam, which turns the turbine on an electrical generator. Nuclear reactors generate about 20% of the electricity in the United States and up to 70% in some other nations.

Nuclear Fusion: Nuclear fusion is the combination of two light nuclei to form a heavier one. Nuclear fusion is the energy source of stars, including our sun.

Nuclear Fusion: Modern nuclear weapons are fusion bombs with 1000 times the power of the first fission bombs. Nuclear fusion is being explored as a way to generate electricity but has not yet proven successful.

The Effects of Radiation on Life and Nuclear Medicine: Radiation can damage molecules within living cells. Instantaneous high exposure to radiation can lead to radiation sickness and even death. Long-term lower exposure levels can increase cancer risk. However, radiation is also used to attack cancerous tumors and to image internal organs through isotope scanning.

The Effects of Radiation on Life and Nuclear Medicine: Radiation can be used for both good and harm. The destructive effects of radiation can be employed in a nuclear bomb. However, radiation can also be a precise tool in the physician's arsenal against disease.

CHEMICAL SKILLS

EXAMPLES

Writing Nuclear Equations for Alpha Decay (Section 17.3)

Begin with the symbol for the isotope undergoing decay on the left side of the equation and the symbol for an alpha particle on the right side. Leave a space or a question mark for the unknown daughter nuclide.

Equalize the sum of the mass numbers and the sum of the atomic numbers on both sides of the equation by writing the appropriate mass number and atomic number for the unknown daughter nuclide.

Deduce the identity of the unknown daughter nuclide from the atomic number.

EXAMPLE 17.6 Writing Nuclear Equations for Alpha Decay

Write a nuclear equation for the alpha decay of Po-214.

SOLUTION

$$^{214}_{84}\text{Po} \longrightarrow {}^{x}_{y}? + {}^{4}_{2}\text{He}$$

$$^{214}_{84}\text{Po} \longrightarrow {}^{210}_{82}? + {}^{4}_{2}\text{He}$$

$$^{214}_{84}\text{Po} \longrightarrow {}^{210}_{82}\text{Pb} + {}^{4}_{2}\text{He}$$

Writing Nuclear Equations for Beta Decay (Section 17.3)

Begin with the symbol for the isotope undergoing decay on the left side of the equation and the symbol for a beta particle on the right side.

Equalize the sum of the mass numbers and the sum of the atomic numbers on both sides of the equation by writing the appropriate mass number and atomic number for the unknown daughter nuclide.

Deduce the identity of the unknown daughter nuclide from the atomic number.

EXAMPLE 17.7 Writing Nuclear Equations for Beta Decay

Write a nuclear equation for the beta decay of Bi-214.

SOLUTION

$$^{214}_{83}\text{Bi} \longrightarrow {}^{x}_{y}? + {}^{0}_{-1}\text{e}$$

$$^{214}_{83}\text{Bi} \longrightarrow {}^{214}_{84}? + {}^{0}_{-1}\text{e}$$

$$^{214}_{83}\text{Bi} \longrightarrow {}^{214}_{84}\text{Po} + {}^{0}_{-1}\text{e}$$

Writing Nuclear Equations for Positron Decay (Section 17.3)

Begin with the symbol for the isotope undergoing decay on the left side of the equation and the symbol for a positron on the right side.

Equalize the sum of the mass numbers and the sum of the atomic numbers on both sides of the equation by writing the appropriate mass number and atomic number for the unknown daughter nuclide.

Deduce the identity of the unknown daughter nuclide from the atomic number.

EXAMPLE 17.8 Writing Nuclear Equations for Positron Decay

Write a nuclear equation for the positron decay of C-11.

SOLUTION

$$^{11}_{6}\text{C} \longrightarrow {}^{x}_{y}? + {}^{0}_{+1}\text{e}$$

$$^{11}_{6}\text{C} \longrightarrow {}^{11}_{5}? + {}^{0}_{+1}\text{e}$$

$$^{11}_{6}\text{C} \longrightarrow {}^{11}_{5}\text{B} + {}^{0}_{+1}\text{e}$$

Using Half-Life (Section 17.5)

To use half-life to determine the time it takes for a sample to decay to a specified amount or the amount of a sample left after a specified time, draw a table showing the amount of the nuclide as a function of the number of half-lives. For each half-life, divide the amount of parent nuclide by 2.

EXAMPLE 17.9 Using Half-Life

Po-210 is an alpha emitter with a half-life of 138 days. How many grams of Po-210 remain after 552 days if the sample initially contained 5.80 g of Po-210?

SOLUTION

Po-210 (g)	Number of Half-Lives	Time in Days
5.80	0	0
2.90	1	138
1.45	2	276
0.725	3	414
0.363	4	552

The amount of Po-210 left after 552 days is 0.363 g.

Using Carbon-14 Content to Determine the Age of Fossils or Artifacts (Section 17.6)

To determine the age of an artifact or fossil based on its carbon-14 content, you can either examine Table 17.3 or build your own table beginning with 100% carbon-14 (relative to living organisms) and reducing the amount by a factor of one-half for each half-life.

EXAMPLE 17.10 Using Carbon-14 Content to Determine the Age of Fossils or Artifacts

Some wood ashes from a fire pit in the ruins of an ancient village have a carbon-14 content that is 25% of the amount found in living organisms. How old are the ashes and, by implication, the village?

SOLUTION

C-14(%)*	Number of Half-Lives	Time in Years
100	0	0
50.0	1	5730
25.0	2	11,460

*Percent relative to living organisms. The ashes are from wood that was living 11,460 years ago.

KEY TERMS

alpha (α) particle [17.3]
alpha (α) radiation [17.3]
beta (β) particle [17.3]
beta (β) radiation [17.3]
chain reaction [17.7]
critical mass [17.7]
daughter nuclide [17.3]
film-badge dosimeter [17.4]

gamma (γ) radiation [17.3]
gamma ray [17.3]
Geiger-Müller counter [17.4]
half-life [17.5]
ionizing power [17.3]
isotope scanning [17.11]

nuclear equation [17.3]
nuclear fission [17.7]
nuclear fusion [17.9]
parent nuclide [17.3]
penetrating power [17.3]
phosphorescence [17.2]
positron [17.3]

positron emission [17.3]
radioactive [17.1]
radioactivity [17.1]
radiocarbon dating [17.6]
radiotherapy [17.11]
rem [17.10]
scintillation counter [17.4]

EXERCISES

QUESTIONS

1. What is radioactivity? What does it mean for an atom to be radioactive?
2. How was radioactivity first discovered? By whom?
3. What are uranic rays?

4. What role did Marie Sklodowska Curie play in the discovery of radioactivity? How was she acknowledged for her work in radioactivity?

5. Explain what is represented by each symbol in the notation.

$$^A_Z X$$

6. Radioactivity originates from the _____ of radioactive atoms.

7. What is alpha radiation? What is the symbol for an alpha particle?

8. What happens to an atom when it emits an alpha particle?

9. How do the ionizing power and penetrating power of alpha particles compare to other types of radiation?

10. What is beta radiation? What is the symbol for a beta particle?

11. What happens to an atom when it emits a beta particle?

12. How do the ionizing power and penetrating power of beta particles compare to other types of radiation?

13. What is gamma radiation? What is the symbol for a gamma ray?

14. What happens to an atom when it emits a gamma ray?

15. How do the ionizing power and penetrating power of gamma particles compare to other types of radiation?

16. What is positron emission? What is the symbol for a positron?

17. What happens to an atom when it emits a positron?

18. How do the ionizing power and penetrating power of positrons compare to other types of radiation?

19. What is a nuclear equation? What does it mean for a nuclear equation to be balanced?

20. Identify the parent nuclides and daughter nuclides in the nuclear equation. What kind of radioactive decay is involved?

$$^{231}_{91}\text{Pa} \longrightarrow {}^{227}_{89}\text{Ac} + {}^4_2\text{He}$$

21. What is a film-badge dosimeter, and how does it work?

22. How does a Geiger-Müller counter detect radioactivity?

23. Explain how a scintillation counter works.

24. What are some sources of natural radioactivity?

25. Explain the concept of half-life.

26. What is a radioactive decay series?

27. What is the source of radon in our environment? Why is radon problematic?

28. What is the source of carbon-14 in our environment? Why do all living organisms contain a uniform amount of carbon-14?

29. What happens to the carbon-14 in a living organism when it dies? How can this be used to establish how long ago the organism died?

30. How do we know that carbon-14 (or radiocarbon) dating is accurate? What is the age limit for which carbon-14 dating is useful?

31. Explain Fermi's experiment in which he bombarded uranium with neutrons. Include a nuclear equation in your answer.

32. What is nuclear fission? How and by whom was it discovered?

33. Why can nuclear fission be used in a bomb? Include the concept of a chain reaction in your explanation.

34. What is a critical mass?

35. What was the main goal of the Manhattan Project? Who was the project leader?

36. How can nuclear fission be used to generate electricity?

37. Explain the purpose of the control rods in a nuclear reactor core. How do they work?

38. What are the main advantages and problems associated with nuclear electricity generation?

39. Can a nuclear reactor detonate the way a nuclear bomb can? Why or why not?

40. What is nuclear fusion?

41. Do modern nuclear weapons use fission or fusion or both? Explain.

42. Can nuclear fusion be used to generate electricity? What are the advantages of fusion over fission for electricity generation? What are the problems with fusion?

43. How does radiation affect the molecules within living organisms?

44. What is acute radiation damage to living organisms?

45. Explain how radiation can increase cancer risk.

46. Explain how radiation can cause genetic defects. Has this ever been observed in laboratory animals? In humans?

47. What is the main unit of radiation exposure? How much radiation is the average American exposed to per year?

48. Describe the outcomes of radiation exposure at different doses (in rem).

49. Explain the medical use of isotope scanning.

50. How is radioactivity used to treat cancer?

PROBLEMS

ISOTOPIC AND NUCLEAR PARTICLE SYMBOLS

51. Provide a symbol for the isotope of lead that contains 128 neutrons.

52. Provide a symbol for the isotope of bismuth that contains 124 neutrons.

53. How many protons and neutrons are in the nuclide?

$$^{207}_{81}\text{Tl}$$

54. How many protons and neutrons are in the nuclide?

$$^{219}_{86}\text{Rn}$$

55. Identify the particle represented by each symbol as an alpha particle, a beta particle, a gamma ray, a positron, a neutron, or a proton.

    (a) $_{-1}^{0}e$

    (b) $_{0}^{1}n$

    (c) $_{0}^{0}\gamma$

56. Identify the particle represented by each symbol as an alpha particle, a beta particle, a gamma ray, a positron, a neutron, or a proton.

    (a) $_{1}^{1}p$

    (b) $_{2}^{4}He$

    (c) $_{+1}^{0}e$

57. Complete the table:

Chemical Symbol	Atomic Number (Z)	Mass Number (A)	# Protons	# Neutrons
Tc	_____	95	_____	_____
_____	56	128	_____	_____
Eu	_____	_____	_____	82
Fr	_____	_____	_____	136

58. Complete the table:

Chemical Symbol	Atomic Number (Z)	Mass Number (A)	# Protons	# Neutrons
Pd	46	_____	_____	54
Ce	_____	136	_____	_____
_____	84	208	_____	_____
_____	_____	_____	88	138

RADIOACTIVE DECAY

59. Write a nuclear equation for the alpha decay of each nuclide.

    (a) U-234

    (b) Th-230

    (c) Ra-226

    (d) Rn-222

60. Write a nuclear equation for the alpha decay of each nuclide.

    (a) Po-218

    (b) Po-214

    (c) Po-210

    (d) Th-227

61. Write a nuclear equation for the beta decay of each nuclide.

    (a) Pb-214

    (b) Bi-214

    (c) Th-231

    (d) Ac-227

62. Write a nuclear equation for the beta decay of each nuclide.

    (a) Pb-211

    (b) Tl-207

    (c) Th-234

    (d) Pa-234

63. Write a nuclear equation for positron emission by each nuclide.

    (a) C-11

    (b) N-13

    (c) O-15

64. Write a nuclear equation for positron emission by each nuclide.

    (a) Co-55

    (b) Na-22

    (c) F-18

65. Fill in the blanks in the partial decay series.

    $$_{94}^{241}Pu \longrightarrow _{95}^{241}Am + \underline{\quad\quad}$$

    $$_{95}^{241}Am \longrightarrow _{93}^{237}Np + \underline{\quad\quad}$$

    $$_{93}^{237}Np \longrightarrow \underline{\quad\quad} + _{2}^{4}He$$

    $$\underline{\quad\quad} \longrightarrow _{92}^{233}U + _{-1}^{0}e$$

66. Fill in the blanks in the partial decay series.

    $$_{88}^{225}Ra \longrightarrow _{89}^{225}Ac + \underline{\quad\quad}$$

    $$_{89}^{225}Ac \longrightarrow \underline{\quad\quad} + _{2}^{4}He$$

    $$\underline{\quad\quad} \longrightarrow _{85}^{217}At + _{2}^{4}He$$

    $$_{85}^{217}At \longrightarrow \underline{\quad\quad} + _{2}^{4}He$$

67. Write a partial decay series for Th-232 undergoing the sequential decays:

    $\alpha, \beta, \beta, \alpha.$

68. Write a partial decay series for Rn-220 undergoing the sequential decays:

    $\alpha, \alpha, \beta, \alpha.$

HALF-LIFE

69. Suppose you a have a 100,000-atom sample of a radioactive nuclide that decays with a half-life of 2.0 days. How many radioactive atoms are left after 10 days?

70. Iodine-131 is often used in nuclear medicine to obtain images of the thyroid. If you start with 4.0×10^{10} I-131 atoms, how many are left after approximately 1 month? I-131 has a half-life of 8.0 days.

71. A patient is given 0.050 mg of technetium-99m (where m means metastable— an unstable but long-lived state), a radioactive isotope with a half-life of about 6.0 hours. How long until the radioactive isotope decays to 6.3×10^{-3} mg?

72. Radium-223 decays with a half-life of 11.4 days. How long will it take for a 0.240-mol sample of radium to decay to 1.50×10^{-2} mol?

73. One of the nuclides in spent nuclear fuel is U-234, an alpha emitter with a half-life of 2.44×10^5 years. If a spent fuel assembly contains 2.80 kg of U-234, how long will it take for the amount of U-234 to decay to less than 0.10 kg?

74. One of the nuclides in spent nuclear fuel is U-235, an alpha emitter with a half-life of 703 million years. How long would it take for the amount of U-235 to reach one-eighth of its initial amount?

75. A radioactive sample contains 2.45 g of an isotope with a half-life of 3.8 days. How much of the isotope in grams remains after 11.4 days?

76. A 68-mg sample of a radioactive nuclide is administered to a patient to obtain an image of her thyroid. If the nuclide has a half-life of 12 hours, how much of the nuclide remains in the patient after 4.0 days?

77. Each of the following nuclides is used in nuclear medicine. List them in order of most active (largest number of decay events per second) to least active (smallest number of decay events per second).

Nuclide	Half-Life
P-32	14.3 days
Cr-51	27.7 days
Ga-67	78.3 hours
Sr-89	50.5 days

78. Each of the following nuclides is used in nuclear medicine. List them in order of most active (largest number of decay events per second) to least active (smallest number of decay events per second).

Nuclide	Half-Life
Y-90	64.1 hours
Tc-99m	6.02 hours
In-111	2.8 days
I-131	8.0 days

RADIOCARBON DATING

79. A wooden boat discovered just south of the Great Pyramid in Egypt had a carbon-14 content of approximately 50% of that found in living organisms. How old is the boat?

80. A layer of peat buried beneath the glacial sediments from the last ice age had a carbon-14 content of 25% of that found in living organisms. How long ago was this ice age?

81. An ancient skull has a carbon-14 content of 1.563% of that found in living organisms. How old is the skull?

82. A mammoth skeleton has a carbon-14 content of 12.50% of that found in living organisms. When did the mammoth live?

FISSION AND FUSION

83. Write a nuclear reaction for the neutron-induced fission of U-235 to form Xe-144 and Sr-90. How many neutrons are produced in the reaction?

84. Write a nuclear reaction for the neutron-induced fission of U-235 to produce Te-137 and Zr-97. How many neutrons are produced in the reaction?

85. Write a nuclear equation for the fusion of two H-2 atoms to form He-3 and one neutron.

86. Write a nuclear equation for the fusion of H-3 with H-1 to form He-4.

CUMULATIVE PROBLEMS

87. Complete each nuclear equation:

(a) $^{1}_{1}p + ^{9}_{4}Be \longrightarrow$ ____ $+ ^{4}_{2}He$

(b) $^{209}_{83}Bi +$ ____ $\longrightarrow ^{272}_{111}Rg + ^{1}_{0}n$

(c) $^{179}_{74}W + ^{0}_{-1}e \longrightarrow$ ____

88. Complete each nuclear equation:

(a) $^{27}_{13}Al + ^{4}_{2}He \longrightarrow$ ____ $+ ^{1}_{0}n$

(b) ____ $+ ^{1}_{0}n \longrightarrow ^{29}_{14}Si + ^{4}_{2}He$

(c) $^{241}_{95}Am \longrightarrow ^{237}_{93}Np +$ ____

89. A breeder nuclear reactor is a reactor in which U-238 (which does not undergo fission) is converted into Pu-239 (which does undergo fission). The process involves bombardment of U-238 by neutrons to form U-239, which then undergoes two sequential beta decays. Write nuclear equations to represent this process.

90. Write a series of nuclear equations in which Al-27 reacts with a neutron and the product undergoes an alpha decay followed by a beta decay.

91. The fission of U-235 produces 3.2×10^{-11} J/atom. How much energy does it produce per mole of U-235? Per kilogram of U-235?

92. The fusion of deuterium and tritium produces 2.8×10^{-12} J for every atom of deuterium and atom of tritium. How much energy is produced per mole of deuterium and mole of tritium?

93. Bi-210 is a beta emitter with a half-life of 5.0 days. If a sample contains 1.2 g of Bi-210, how many beta emissions occur in 5.0 days?

94. Po-218 is an alpha emitter with a half-life of 3.0 minutes. If a sample contains 55 mg of Po-218, how many alpha emissions occur in 6.0 minutes?

95. If a person living in a high-radon area is exposed to 0.400 rem of radiation from radon per year, and his total exposure is 0.585 rem, what percentage of his total exposure is due to radon?

96. An X-ray technician is exposed to 0.020 rem of radiation at work. If her total exposure is the national average (0.36 rem), what fraction of her exposure is due to on-the-job exposure?

97. Radium-226 (atomic mass 226.03 amu) decays to radon-224, a radioactive gas. The half-life of radium-226 is 1.6×10^{3} years. If a 1.5-g sample of radium-226 decays for 45 days, what volume of radon gas (at 25.0 °C and 1.0 atm) is produced?

98. Consider the fission reaction:

$$^{235}_{92}U + ^{1}_{0}n \longrightarrow ^{142}_{56}Ba + ^{91}_{36}Kr + 3^{1}_{0}n + \text{Energy}$$

What mass of Kr-91 (atomic mass 92.93 amu) is produced by the complete fission of 15 g of U-235 (atomic mass 235.04 amu)?

HIGHLIGHT PROBLEMS

99. Closely examine the diagram representing the alpha decay of sodium-20 and draw the missing nucleus.

100. Closely examine the diagram representing the beta decay of fluorine-21 and draw the missing nucleus.

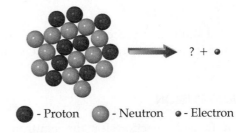

101. Closely examine the diagram representing the positron emission of carbon-10 and draw the missing nucleus.

? + ●

● - Proton   ● - Neutron   ● - Positron

102. A radiometric dating technique uses the decay of U-238 to Pb-206 (the half-life for this process is 4.5 billion years) to determine the age of the oldest rocks on Earth and by implication the age of Earth itself. The oldest uranium-containing rocks on Earth contain approximately equal numbers of uranium atoms and lead atoms. Assuming the rocks were pure uranium when they were formed, how old are the rocks?

►ANSWERS TO SKILLBUILDER EXERCISES

Skillbuilder 17.1 $^{216}_{84}Po \longrightarrow ^{212}_{82}Pb + ^{4}_{2}He$

Skillbuilder 17.2 $^{228}_{89}Ac \longrightarrow ^{228}_{90}Th + ^{0}_{-1}e$

Skillbuilder Plus, p. 619 $^{235}_{92}U \longrightarrow ^{231}_{90}Th + ^{4}_{2}He;$

$^{231}_{90}Th \longrightarrow ^{231}_{91}Pa + ^{0}_{-1}e;$

$^{231}_{91}Pa \longrightarrow ^{227}_{89}Ac + ^{4}_{2}He$

Skillbuilder 17.3 $^{22}_{11}Na \longrightarrow ^{22}_{10}Ne + ^{0}_{+1}e$

Skillbuilder 17.4 7.00×10^{-3} mol

Skillbuilder 17.5   No, the carbon-14 content suggests that the scroll is from very recent times.

►ANSWERS TO CONCEPTUAL CHECKPOINTS

17.1 (a) In alpha decay, the nucleus loses a helium nucleus (2 protons and 2 neutrons), reducing its mass number by 4. The other forms of decay listed involve electrons or positrons, which have negligible mass compared to that of nuclear particles, or gamma-ray photons, which have no mass.

17.2 (a) If you divide 1,000,000 by 2, then divide the remainder by 2, and repeat this process eight more times, you are left with approximately 977 atoms.

17.3 (c) Gamma radiation has the most penetrating power, and therefore it most readily penetrates out of bodily tissues for easier detection.

CHAPTER 18

Organic Chemistry

"The atoms come together in different order and position, like the letters, which, though they are few, yet, by being placed together in different ways, produce innumerable words."

EPICURUS (341–270 B.C.)

18.1 What Do I Smell?

| The smells of substances are not *always* a reliable guide to what is good to eat.

Perfume companies spend millions of dollars trying to produce the most seductive scents. What causes scent? The answer, of course, is molecules. Certain molecules, when they are inhaled, bind with molecular receptors (called olfactory receptors) in our noses. This interaction sends a nerve signal to the brain that we experience as a smell. Some smells, such as that of a flower, are pleasant; other smells, such as that of rotten fish, are unpleasant.

Which molecules cause smell? Many molecules have no scent at all. Nitrogen, oxygen, water, and carbon dioxide molecules, for example, are constantly passing through our noses, yet they produce no smell. Most of the smells that we experience are caused by **organic molecules**, molecules containing carbon combined with several other elements including hydrogen, nitrogen, oxygen, and sulfur. Carbon-containing molecules are responsible for the smells of roses, vanilla, cinnamon, almond, jasmine, body odor, and rotting fish.

When you sprinkle cinnamon onto your French toast, some cinnamaldehyde—an organic compound present in cinnamon—evaporates into the air. You inhale some of the cinnamaldehyde molecules and experience the unique smell of cinnamon. When you walk past a rotting fish on a beach, you inhale triethylamine—an organic compound emitted by the decaying fish—and experience that unique and unpleasant smell. Our reaction to certain smells, positive or negative, is probably an evolutionary adaptation. The pleasant smell of cinnamon tells you that it is good to eat. The unpleasant smell of rotting fish tells you that it has become spoiled and that you should avoid it.

◄ The sweet smell of jasmine is produced by benzyl acetate, an organic compound. When you smell jasmine, benzyl acetate molecules emitted from the flower bind with molecular receptors in your nose, triggering a nerve signal to your brain that you interpret as a sweet smell.

The study of carbon-containing compounds and their reactions is called **organic chemistry**. Besides being prevalent in odors and fragrances, organic compounds are common in foods, drugs, petroleum products, and pesticides. Organic chemistry is also the basis for living organisms. Life has evolved based on carbon-containing compounds, making organic chemistry of utmost importance in understanding living organisms.

$$CH_3-CH_2-N-CH_2-CH_3$$
$$|$$
$$CH_2$$
$$|$$
$$CH_3$$

Triethylamine

▲ Carbon-containing molecules—especially triethylamine—are responsible for the smell of dead fish.

Vanilla   Cinnamon

Vanillin   Cinnamaldehyde

▲ Carbon-containing molecules are responsible for the smell of vanilla beans (vanillin) and cinnamon sticks (cinnamaldehyde).

18.2 Vitalism: The Difference between Organic and Inorganic

By the end of the eighteenth century, chemists had divided compounds into two broad categories: organic and inorganic. Organic compounds came from living things. Sugar—obtained from sugarcane or the sugar beet—is an example of an organic compound. Inorganic compounds, on the other hand, came from Earth. Salt—mined from the ground or extracted from the ocean—is an example of an inorganic compound.

Early chemists realized that organic and inorganic compounds are different, not only in their origin, but also in their properties. Organic compounds are easily decomposed. Sugar, for example, easily decomposes into carbon and water when heated. Think of the last time you burned sugar—the apple-pie filling that dripped in the oven or the sugar in the pan that caramelized a little too much. Inorganic compounds, however, are more difficult to decompose. Salt must be heated to very high temperatures before it decomposes. Even more curious to these early chemists was their inability to synthesize a single organic compound in the laboratory. Many inorganic compounds could be easily synthesized, but organic compounds could not.

The origin and properties of organic compounds led early chemists to postulate that organic compounds are unique to living organisms. They postulated that living organisms employed a **vital force**—a mystical or supernatural power—that allowed them to produce organic compounds. They thought it impossible to produce an organic compound outside of a living organism. This belief—which became known as **vitalism**—was thought to explain why no chemist had succeeded in synthesizing an organic compound in the laboratory.

Organic   Inorganic

▶ Sugar, obtained from sugarcane or sugar beets, is an example of an organic compound. Salt, obtained from a salt mine or from the ocean, is an example of an inorganic compound. **Question:** What are the main differences between organic and inorganic compounds?

An experiment performed in 1828 by a German chemist named Friedrich Wöhler (1800–1882) proved vitalism wrong. Wöhler heated ammonium cyanate (an inorganic compound) and formed urea (an organic compound).

| Vitalism is the belief that living things contain a nonphysical "force" that allows them to synthesize organic compounds.

$$NH_4OCN \xrightarrow{\text{Heat}} H_2NCONH_2$$
Ammonium cyanate   Urea

Urea was a known organic compound that had previously been isolated only from the urine of living organisms. Although it was not realized at the time, Wöhler's simple experiment was a crucial step in opening all of life to scientific investigation. He showed that the compounds that composed living organisms—like all compounds—follow scientific laws and can be studied and understood. Today, known organic compounds number in the millions, and organic chemistry is a vast field that produces substances as diverse as drugs, petroleum, and plastics.

18.3 Carbon: A Versatile Atom

Why did life evolve based on the chemistry of carbon? Why isn't life based on some other element? The answer may not be simple, but we know that life—in order to exist—must have complexity. It is also clear that carbon chemistry is complex. The number of compounds containing carbon is greater than the number of compounds of all the rest of the elements in the periodic table combined. One reason for this complex chemistry is that carbon—with its four valence electrons—can form four covalent bonds. Recall the Lewis structures of carbon and some carbon compounds.

| Lewis structures were first covered in Chapter 10.

As you learn to draw structures for organic compounds, remember that carbon always forms four bonds.

A second reason for carbon's complex chemistry is that carbon, more than any other element, can bond to itself to form chain, branched, and ring structures.

Propane   Isobutane

Cyclohexane

CHEMISTRY IN THE MEDIA

The Origin of Life

The demise of vitalism opened life itself—including its origin—to chemical inquiry. If organic compounds could be made in the laboratory, and if living things were composed of organic compounds, would it be possible to make life in the laboratory? Would it be possible to simulate how life started on Earth?

In 1953, a young scientist named Stanley Miller, working with Harold C. Urey at the University of Chicago, performed an experiment in an attempt to answer this question. Miller re-created the environment of primordial Earth in a flask containing water and certain gases including methane, ammonia, and hydrogen—all believed, at the time, to be components of the early atmosphere. He passed an electrical current through the system to simulate lightning. After several days, Miller analyzed the contents of the flask. What he found made headlines. Not only did the flask contain organic compounds, it contained organic compounds central to life—amino acids. Amino acids, as we will learn in Chapter 19, are the building blocks of biological proteins. Apparently, the foundational compounds of life could be synthesized rather simply under the conditions of early Earth.

Inspired by Miller's results, a number of other scientists set out to understand, and perhaps re-create, life's origin. Some believed that the creation of life in the laboratory was imminent. It has not turned out that way. More than a half century has passed since Miller's seminal experiment, yet we are still struggling to understand how life began. Toward the end of his life, Stanley Miller, who was a professor of chemistry at the University of California at San Diego at the time, said, "The problem of the origin of life has turned out to be much more difficult than I and most people envisioned."

Most scientists investigating the origin of life have a basic hypothesis of how life may have started. A group of molecules developed the ability to copy themselves, but not quite perfectly—some of the copies contained inheritable mistakes. In a few cases, these alterations allowed the molecular "offspring" to replicate even more efficiently. In this way, chemical evolution got its start–producing generation after generation of molecules that slowly got better at copying themselves as they assembled into more complex structures. This process eventually produced a living cell, which is (among other things) a very efficient self-replicating machine.

Although this basic hypothesis is still widely accepted, the details are far from clear. What were these early molecules? How did they form? How did they replicate? Earlier origin-of-life theories proposed that the replicating molecules were primitive forms of the molecules that exist in living organisms today—proteins, RNA, and DNA (see Chapter 19). However, the complexity of these molecules and their inability to replicate independently of one another has caused some researchers to look at other materials that may have been involved in the process, such as clays and sulfur-based compounds. In spite of continuing efforts, no single theory has gained widespread acceptance, and the origin of life continues to be an area of ongoing research.

CAN YOU ANSWER THIS? *How might a belief in vitalism inhibit research into the origin of life?*

▲ An artist's conception of the early Earth. It was this environment that Miller's experiment was designed to simulate.

Molecular geometries and VSEPR theory were covered in Chapter 10.

This versatility allows carbon to serve as the backbone of millions of different chemical compounds—just what is needed for life to exist.

When carbon forms four single bonds, there are four electron groups around it, and VSEPR theory (Section 10.7) predicts a tetrahedral geometry.

Tetrahedral geometry

When carbon forms a double bond and two single bonds, there are three electron groups around each carbon atom, and VSEPR theory predicts a trigonal planar geometry.

Trigonal planar geometry

When carbon forms a triple bond and a single bond (or two double bonds), there are two electron groups around each carbon atom, resulting in a linear geometry.

Linear geometry

18.4 Hydrocarbons: Compounds Containing Only Carbon and Hydrogen

Hydrocarbons—compounds that contain only carbon and hydrogen—are the simplest organic compounds. However, because carbon atoms are so versatile, many different kinds of hydrocarbons exist. Carbon and hydrogen atoms bond together in different numbers and in different ways to form millions of different compounds.

Hydrocarbons are commonly used as fuels. Candle wax, oil, gasoline, liquid propane (LP) gas, and natural gas are all composed of hydrocarbons. Hydrocarbons are also the starting materials in the synthesis of many different consumer products, including fabrics, soaps, dyes, cosmetics, drugs, plastic, and rubber.

The difference between saturated and unsaturated hydrocarbons is explained in Sections 18.5 and 18.8.

As shown in ▼ Figure 18.1, hydrocarbons can be broadly classified into four different types: alkanes, alkenes, alkynes, and aromatic hydrocarbons. Alkanes are saturated hydrocarbons, while alkenes, alkynes, and aromatic hydrocarbons are all unsaturated hydrocarbons. Alkanes, alkenes, and alkynes can be differentiated based on their molecular formulas.

These formulas apply only to open-chain (noncyclical) hydrocarbons.

Alkanes $C_n H_{2n+2}$
Alkenes $C_n H_{2n}$
Alkynes $C_n H_{2n-2}$

▶ **FIGURE 18.1   A flowchart for the classification of hydrocarbons** These formulas apply only to open-chain (noncyclical) hydrocarbons.

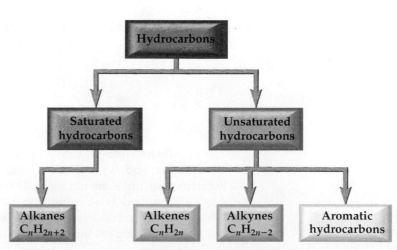

EXAMPLE 18.1 Differentiating between Alkanes, Alkenes, and Alkynes Based on Their Molecular Formulas

Based on the molecular formula, determine whether the noncyclical hydrocarbons are alkanes, alkenes, or alkynes.

(a) C_7H_{14}
(b) $C_{10}H_{22}$
(c) C_3H_4

SOLUTION

(a) C_7H_{14}
The number of carbons is 7; therefore $n = 7$. If $n = 7$, then 14 is $2n$. The molecule must be an alkene.
(b) $C_{10}H_{22}$
The number of carbons is 10; therefore $n = 10$. If $n = 10$, then 22 is $2n + 2$. The molecule must be an alkane.
(c) C_3H_4
The number of carbons is 3; therefore $n = 3$. If $n = 3$, then 4 is $2n - 2$. The molecule must be an alkyne.

▶**SKILLBUILDER 18.1 | Differentiating between Alkanes, Alkenes, and Alkynes Based on Their Molecular Formulas**

Based on the molecular formula, determine whether the noncyclical hydrocarbons are alkanes, alkenes, or alkynes.

(a) C_6H_{12}
(b) C_8H_{14}
(c) C_5H_{12}

▶**FOR MORE PRACTICE** Example 18.9; Problems 33, 34, 35, 36.

18.5 Alkanes: Saturated Hydrocarbons

Alkanes are hydrocarbons containing only single bonds. Alkanes are also called **saturated hydrocarbons** because they are saturated (loaded to capacity) with hydrogen. The simplest alkane is methane, CH_4, the main component of natural gas.

Methane CH_4

| Formula | Structural formula | Space-filling model |

The middle formula above is a **structural formula**, a formula that shows not only the number and type of each atom in a molecule, but the structure as well. Structural formulas are not three-dimensional representations of the molecule—as space-filling models are—but rather two-dimensional representations that show which atoms are bonded together. A structural formula is like a Lewis structure, but it typically depicts bonding electron pairs as dashes and omits lone pairs.

CHEMISTRY IN THE MEDIA
Environmental Problems Associated with Hydrocarbon Combustion

Hydrocarbon fuels are also called **fossil fuels** because they originate from plant and animal life that existed on Earth in prehistoric times. The main types of fossil fuels are natural gas, petroleum, and coal. Fossil fuels are a convenient form of energy because they are relatively cheap, can be easily transported, and burn easily to release large amounts of energy. However, fossil fuels also have several problems associated with their use, including limited supply, smog, acid rain, and global warming.

One of the problems with fossil fuels is that our supply will not last forever. At current rates of consumption, oil and natural gas supplies will be depleted in 40 to 60 years. While there is enough coal to last much longer, it is a dirtier fuel and is less convenient than petroleum and natural gas because it is a solid.

A second problem associated with fossil fuel combustion is smog. Smog results when fossil fuel combustion products are emitted into the air. As we saw in Section 14.11, these include nitrogen oxides (NO and NO_2), sulfur oxides (SO_2 and SO_3), ozone (O_3), and carbon monoxide (CO). These substances make the air above cities brown and dirty. They also irritate the eyes and lungs and put stress on the heart and lungs. Because of good legislation and catalytic converters, however, the level of these pollutants over most cities is decreasing. Even so, in many cities, the levels still exceed what the Environmental Protection Agency (EPA) considers safe.

A third problem associated with fossil fuel combustion is acid rain. The nitrogen oxides and sulfur oxides emitted into air make rain acidic. This acidic rain falls into lakes and streams and makes them acidic as well (see Section 14.11). Some species of aquatic life cannot tolerate the increased acidity and die. Acid rain also affects forests and building materials. Again, good legislation—specifically the Clean Air Amendments of 1990—have addressed acid rain, and sulfur oxide emissions have been decreasing in the United States over the past 20 years. We can expect to see the positive effects of these reductions in lakes and streams in the coming years.

A fourth problem associated with fossil fuel use is climate change, which we discussed in Section 8.1. One of the main products of fossil fuel combustion is carbon dioxide (CO_2). Carbon dioxide is a greenhouse gas; it allows visible light from the sun to enter Earth's atmosphere, but prevents heat (in the form of infrared light) from escaping. In essence, carbon dioxide acts as a blanket, keeping Earth warm. Because of fossil fuel combustion, carbon dioxide levels in our atmosphere have been steadily increasing. This increase is expected to raise Earth's average temperature. Current observations suggest that Earth has already warmed by about 0.6 °C in the last century due to an increase in atmospheric carbon dioxide of about 30%. Computer models suggest that the warming could worsen if carbon dioxide emissions are not curbed.

CAN YOU ANSWER THIS? *Write a balanced equation for the combustion of octane, a component of gasoline. How many moles of carbon dioxide (the main greenhouse gas) are produced for every mole of octane (C_8H_{18}) burned? How many kilograms of carbon dioxide are produced for every kilogram of octane burned?*

◀ Fossil fuels such as gasoline are convenient forms of energy because they are relatively cheap, are easily transported, and release large amounts of energy when burned.

The next simplest alkane is ethane, C_2H_6. To draw the structural formula of ethane, we remove a hydrogen atom from methane and replace it with a methyl ($-CH_3$) group.

Ethane	C_2H_6	
	Formula	Structural formula	Space-filling model

Ethane is a minority component of natural gas.

After ethane, the next simplest alkane is propane, C_3H_8, the main component of LP (liquefied petroleum) gas.

Propane C_3H_8

Formula	Structural formula	Space-filling model

For many organic compounds, it is often useful to write **condensed structural formulas**. A condensed structural formula is a shorthand way to write a structural formula in which you eliminate many or all of the bonds and group like atoms together. For example, the condensed structural formula for propane is:

$$CH_3CH_2CH_3$$

This *does not* mean C—H—H—H—C—H—H—C—H—H—H. No such structure could exist because, as we know, carbon atoms must form four bonds and hydrogen atoms form only one bond. Rather, the condensed structural formula is simply a shorter way to write the true structural formula of propane shown previously.

Next in the series is butane, C_4H_{10}, the main component in lighter fluid.

Butane C_4H_{10}

Formula	Structural formula	Space-filling model

▲ Butane is the primary component of lighter fluid.

Alkanes composed of carbon atoms bonded in a straight chain without any branching—like the ones we have just seen—are called **normal alkanes** or **n-alkanes**. The n-alkanes with three or more carbon atoms have the general structure:

| m = number of —CH$_2$— groups.

$$CH_3(CH_2)_mCH_3$$

Condensed structural formula

Structural formula

As the number of carbon atoms increases in n-alkanes, so does their boiling point. Methane, ethane, propane, and butane are all gases at room temperature, but the next n-alkane in the series, pentane, is a liquid at room temperature.

Alkane	Boiling Point (°C)
methane	−161.5
ethane	−88.6
propane	−42.1
butane	−0.5
pentane	36.0
hexane	68.7
heptane	98.5
octane	125.6

Pentane C_5H_{12}

Formula	Structural formula	Space-filling model

Pentane is a component of gasoline. Table 18.1 summarizes the n-alkanes through decane, which contains 10 carbon atoms. Like pentane, hexane through decane are all components of gasoline. Table 18.2 summarizes the many uses of hydrocarbons.

TABLE 18.1   Alkanes

n	Name	Molecular Formula (C_nH_{2n+2})	Structural Formula	Condensed Structural Formula
1	methane	CH_4		CH_4
2	ethane	C_2H_6		CH_3CH_3
3	propane	C_3H_8		$CH_3CH_2CH_3$
4	n-butane	C_4H_{10}		$CH_3CH_2CH_2CH_3$
5	n-pentane	C_5H_{12}		$CH_3CH_2CH_2CH_2CH_3$
6	n-hexane	C_6H_{14}		$CH_3CH_2CH_2CH_2CH_2CH_3$
7	n-heptane	C_7H_{16}		$CH_3CH_2CH_2CH_2CH_2CH_2CH_3$
8	n-octane	C_8H_{18}		$CH_3CH_2CH_2CH_2CH_2CH_2CH_2CH_3$
9	n-nonane	C_9H_{20}		$CH_3CH_2CH_2CH_2CH_2CH_2CH_2CH_2CH_3$
10	n-decane	$C_{10}H_{22}$		$CH_3CH_2CH_2CH_2CH_2CH_2CH_2CH_2CH_2CH_3$

TABLE 18.2 Uses of Hydrocarbons

Number of Carbon Atoms	State	Major Uses	
1–4	gas	heating fuel, cooking fuel	
5–7	low-boiling liquids	solvents, gasoline	
6–18	liquids	gasoline	
12–24	liquids	jet fuel, portable-stove fuel	
18–50	high-boiling liquids	diesel fuel, lubricants, heating oil	
50+	solids	petroleum jelly, paraffin wax	

EXAMPLE 18.2 Writing Formulas for *n*-Alkanes

Write the structural and condensed structural formula for *n*-octane, C_8H_{18}.

The first step in writing the structural formula is to write out the carbon backbone with eight carbons in it.	**SOLUTION** C—C—C—C—C—C—C—C
The next step is to add H atoms so that all carbons have four bonds.	$$\begin{array}{cccccccc} H & H & H & H & H & H & H & H \\ | & | & | & | & | & | & | & | \\ H-C & -C & -C & -C & -C & -C & -C & -C-H \\ | & | & | & | & | & | & | & | \\ H & H & H & H & H & H & H & H \end{array}$$
To write the condensed structural formula, write the hydrogen atoms bonded to each carbon directly to the right of the carbon atom. Use subscripts to indicate the correct number of hydrogen atoms.	$CH_3CH_2CH_2CH_2CH_2CH_2CH_2CH_3$

▶**SKILLBUILDER 18.2 | Writing Formulas for *n*-Alkanes**

Write the structural and condensed structural formula for C_5H_{12}.

▶**FOR MORE PRACTICE** Problems 37, 38.

18.6 Isomers: Same Formula, Different Structure

In addition to linking together in straight chains to form the *n*-alkanes, carbon atoms can form branched structures called **branched alkanes**. The simplest branched alkane is called isobutane and has the structure:

Isobutane C_4H_{10}

Formula   Structural formula   Space-filling model

Isobutane and butane are **isomers**, molecules with the same molecular formula but different structures. Because of their different structures, they have different properties; indeed, they are different compounds. Isomerism is common in organic compounds. We have seen that butane has two isomers. Pentane (C_5H_{12}) has three isomers, hexane (C_6H_{14}) has five, and decane ($C_{10}H_{22}$) has seventy-five!

EXAMPLE 18.3 Writing Structural Formulas for Isomers

Draw the five isomers for hexane.

SOLUTION

To start, always draw the carbon backbone. The first isomer is the straight-chain isomer, C—C—C—C—C—C. Then, determine the carbon backbone structure of the other isomers by arranging the carbon atoms in four other unique ways.

To finish, fill in all the hydrogen atoms so that each carbon has four bonds.

Notice that

and

are identical to each other because the second structure is just the first one flipped around.

▶**SKILLBUILDER 18.3 | Writing Structural Formulas for Isomers**

Draw the three isomers for pentane.

▶**FOR MORE PRACTICE**   Example 18.10; Problems 39, 40, 41, 42.

CONCEPTUAL CHECKPOINT 18.1

Which two molecules are isomers?

(a) $H_3C-CH-CH_2-CH_2-CH_3$
$\qquad\qquad |$
$\qquad\quad\; CH_3$

(b) $H_3C-CH-\overset{\displaystyle CH_3}{\underset{\displaystyle CH_3}{\overset{|}{\underset{|}{C}}}}-CH_3$
$\qquad\qquad\quad |$
$\qquad\qquad\; CH_3$

(c) $H_3C-CH_2-CH_2-CH_2-CH_3$

(d) $H_3C-CH-CH-CH_2-CH_3$
$\qquad\qquad |\quad\; |$
$\qquad\quad\; CH_3\; CH_3$

18.7 Naming Alkanes

TABLE 18.3 Prefixes for Base Names of Alkane Chains

Number of Carbon Atoms	Prefix
1	*meth-*
2	*eth-*
3	*prop-*
4	*but-*
5	*pent-*
6	*hex-*
7	*hept-*
8	*oct-*
9	*non-*
10	*dec-*

Many organic compounds have common names that can be learned only through familiarity. Because there are so many organic compounds, however, a systematic method of nomenclature is required. In this book, we adopt the nomenclature system recommended by the International Union of Pure and Applied Chemistry (IUPAC), which is used throughout the world. In this system, the base name of an organic compound consists of a prefix and an ending. The prefix is determined by the number of carbon atoms in the **base chain** (the longest continuous chain of carbon atoms). The prefixes for base chains of 1–10 carbon atoms are shown in Table 18.3. Base names for alkanes always have the ending *-ane*. Groups of carbon atoms branching off the base chain are called **alkyl groups** and are named as *substituents*. A **substituent** is an atom or group of atoms that has been *substituted* for a hydrogen atom in an organic compound. Common alkyl groups are shown in Table 18.4.

TABLE 18.4 Common Alkyl Groups

Condensed Structural Formula	Name
$-CH_3$	methyl
$-CH_2CH_3$	ethyl
$-CH_2CH_2CH_3$	propyl
$-CH_2CH_2CH_2CH_3$	butyl
$-CHCH_3$ $\quad\;|$ $\;\; CH_3$	isopropyl
$-CH_2CHCH_3$ $\qquad\;|$ $\quad\;\; CH_3$	isobutyl
$-CHCH_2CH_3$ $\quad\;|$ $\;\; CH_3$	*sec*-butyl
$\quad\; CH_3$ $\quad\;\;|$ $-CCH_3$ $\quad\;\;|$ $\quad\; CH_3$	*tert*-butyl

The rules in Examples 18.4 and 18.5 will allow you to systematically name many alkanes. The rules are presented in the left column, and two examples of how to apply the rules are shown in the center and right columns.

Naming Alkanes	**EXAMPLE 18.4 Naming Alkanes**	**EXAMPLE 18.5 Naming Alkanes**

	Name the following alkane.
	
	$CH_3CH_2CHCH_2CH_2CH_3$
	          |
	          CH_2
	          |
	          CH_3
		Name the following alkane.
		
		$CH_3CHCH_2CHCH_2CH_2CH_3$
		    | |
		    CH_3 CH_2CH_3

1. Count the number of carbon atoms in the longest continuous carbon chain to determine the base name of the compound. Find the prefix corresponding to this number of atoms in Table 18.3 and add the ending *-ane* to form the base name.	**SOLUTION**
	This compound has six carbon atoms in its longest continuous chain.
	
	$CH_3CH_2CHCH_2CH_2CH_3$
	          |
	          CH_2
	          |
	          CH_3
	
	The correct prefix from Table 18.3 is *hex-*. The base name is *hexane*.
		SOLUTION
		This compound has seven carbon atoms in its longest continuous chain.
		
		$CH_3CHCH_2CHCH_2CH_2CH_3$
		    | |
		    CH_3 CH_2CH_3
		
		The correct prefix from Table 18.3 is *hept-*. The base name is *heptane*.

2. Consider every branch from the base chain to be a substituent. Name each substituent according to Table 18.4.	This compound has one substituent named *ethyl*.
	
	$CH_3CH_2CHCH_2CH_2CH_3$
	          |
	          CH_2
	          |
	          CH_3 ← Ethyl
		This compound has one substituent named *methyl* and one named *ethyl*.
		
		$CH_3CHCH_2CHCH_2CH_2CH_3$
		    | |
		    CH_3 CH_2CH_3
		    ↑ ↑
		  Methyl Ethyl

3. Beginning with the end closest to the branching, number the carbon atoms in the base chain and assign a number to each substituent. (If two substituents occur at equal distances from each end, go to the next substituent to determine from which end to start numbering.)	The base chain is numbered as follows:
	
	 1 2 3 4 5 6
	$CH_3CH_2CHCH_2CH_2CH_3$
	          |
	          CH_2
	          |
	          CH_3
	
	The ethyl substituent is assigned the number 3.
		The base chain is numbered as follows:
		
		 1 2 3 4 5 6 7
		$CH_3CHCH_2CHCH_2CH_2CH_3$
		    | |
		    CH_3 CH_2CH_3
		
		The methyl substituent is assigned the number 2, and the ethyl substituent is assigned the number 4.

4. Write the name of the compound in the format:
 	
(subst. number)–
(subst. name) (base name)
 	
If there are two or more substituents, give each one a number and list them alphabetically with hyphens between words and numbers.
	The name of this compound is:
	
	   3-ethylhexane
		The name of this compound is:
		
		   4-ethyl-2-methylheptane
		
		Ethyl is listed before methyl because substituents are listed in alphabetical order.

5. If a compound has two or more identical substituents, designate the number of identical substituents with the prefix *di-* (2), *tri-* (3), or *tetra-* (4) before the substituent's name. Separate the numbers indicating the positions of the substituents relative to each other with a comma. The prefixes are not taken into account when alphabetizing.

Does not apply to this compound.

▶SKILLBUILDER 18.4 | Naming Alkanes

Name the alkane.

CH₃CHCH₃
|
CH₃

Does not apply to this compound.

▶SKILLBUILDER 18.5 | Naming Alkanes

Name the alkane.

CH₃
|
CH₃CHCHCH₂CH₃
|
CH₂CH₃

▶FOR MORE PRACTICE Example 18.11; Problems 43, 44, 45, 46.

EXAMPLE 18.6 Naming Alkanes

Name the alkane.

CH₃CHCH₂CHCH₃
| |
CH₃   CH₃

1. The longest continuous carbon chain has five atoms. Therefore the base name is *pentane*.	**SOLUTION** H₃C—CH—CH₂—CH—CH₃
	              | |
	            CH₃ CH₃

2. This compound has two substituents, both of which are named *methyl*.	CH₃CHCH₂CHCH₃ Methyl→ CH₃ CH₃ ←Methyl

3. Since both substituents are equidistant from the ends, it does not matter which end you start numbering from.	1 2 3 4 5
	CH₃CHCH₂CHCH₃
	       | |
	     CH₃ CH₃

4. and **5.** Since this compound contains two identical substituents, Rule 5 applies and we use the prefix *di-*.	2,4-dimethylpentane

▶SKILLBUILDER 18.6 | Naming Alkanes

Name the alkane.

CH₃
|
CH₃CHCH₂CHCHCH₃
| |
CH₃   CH₃

▶FOR MORE PRACTICE Problems 47, 48, 49, 50.

18.8 Alkenes and Alkynes

Alkenes are hydrocarbons containing at least one double bond between carbon atoms. **Alkynes** are hydrocarbons containing at least one triple bond between carbon atoms. Because of the double or triple bond, alkenes and alkynes have fewer hydrogen atoms than the corresponding alkane and are therefore called **unsaturated hydrocarbons** because they are not loaded to capacity with hydrogen. As we saw earlier, alkenes have the formula C_nH_{2n} and alkynes have the formula C_nH_{2n-2}. The simplest alkene is ethene, C_2H_4, also called *ethylene*.

Ethene or ethylene	C_2H_4		
	Formula	Structural formula	Space-filling model

The geometry about each carbon atom in ethene is trigonal planar, making ethene a flat, rigid molecule. Ethene is a ripening agent in fruit. For example, when a banana within a cluster of bananas begins to ripen, it emits ethene. The ethene then causes other bananas in the cluster to ripen. Banana farmers usually pick bananas green for ease of shipping. When the bananas arrive at their destination, they are often "gassed" with ethene to initiate ripening so that they will be ready to sell. The names and structures of several other alkenes are shown in Table 18.5. Most of them do not have familiar uses other than their presence as minority components of fuels.

▲ Ethene is emitted by ripening bananas. It acts as a chemical messenger, inducing bananas in a bunch to ripen together.

TABLE 18.5 Alkenes

n	Name	Molecular Formula (C_nH_{2n})	Structural Formula	Condensed Structural Formula
2	ethene	C_2H_4		$CH_2{=}CH_2$
3	propene	C_3H_6		$CH_2{=}CHCH_3$
4	1-butene	C_4H_8		$CH_2{=}CHCH_2CH_3$
5	1-pentene	C_5H_{10}		$CH_2{=}CHCH_2CH_2CH_3$
6	1-hexene	C_6H_{12}		$CH_2{=}CHCH_2CH_2CH_2CH_3$

The simplest alkyne is ethyne, C_2H_2, also called acetylene.

Ethyne or acetylene	C_2H_2	$H-C\equiv C-H$	
	Formula	Structural formula	Space-filling model

The geometry about each carbon atom in ethyne is linear, making ethyne a linear molecule. Ethyne (or acetylene) is commonly used as fuel for welding torches. The names and structures of several other alkynes are shown in Table 18.6. Like alkenes, the alkynes do not have familiar uses other than their presence as minority components of gasoline.

TABLE 18.6 Alkynes

n	Name	Molecular Formula (C_nH_{2n-2})	Structural Formula	Condensed Structural Formula
2	ethyne	C_2H_2	$H-C\equiv C-H$	$CH\equiv CH$
3	propyne	C_3H_4	$H-C\equiv C-\overset{\displaystyle H}{\underset{\displaystyle H}{C}}-H$	$CH\equiv CCH_3$
4	1-butyne	C_4H_6	$H-C\equiv C-\overset{\displaystyle H}{\underset{\displaystyle H}{C}}-\overset{\displaystyle H}{\underset{\displaystyle H}{C}}-H$	$CH\equiv CCH_2CH_3$
5	1-pentyne	C_5H_8	$H-C\equiv C-\overset{\displaystyle H}{\underset{\displaystyle H}{C}}-\overset{\displaystyle H}{\underset{\displaystyle H}{C}}-\overset{\displaystyle H}{\underset{\displaystyle H}{C}}-H$	$CH\equiv CCH_2CH_2CH_3$
6	1-hexyne	C_6H_{10}	$H-C\equiv C-\overset{\displaystyle H}{\underset{\displaystyle H}{C}}-\overset{\displaystyle H}{\underset{\displaystyle H}{C}}-\overset{\displaystyle H}{\underset{\displaystyle H}{C}}-\overset{\displaystyle H}{\underset{\displaystyle H}{C}}-H$	$CH\equiv CCH_2CH_2CH_2CH_3$

▲ Acetylene is used as a fuel in welding torches.

NAMING ALKENES AND ALKYNES

Alkenes and alkynes are named in the same way as alkanes with the following exceptions:

- The base chain is the longest continuous carbon chain that *contains the double or triple bond.*
- The base name has the ending *-ene* for alkenes and *-yne* for alkynes.
- The base chain is numbered to *give the double or triple bond the lowest possible number.*
- A number indicating the position of the double or triple bond (lowest possible number) is inserted just before the base name. For example,

$$CH\equiv CCH_2CH_3 \qquad\qquad CH_3CH_2CH=\underset{\underset{\displaystyle CH_3}{|}}{C}CH_3$$

1-Butyne   2-Methyl-2-pentene

EXAMPLE 18.7 Naming Alkenes and Alkynes

Name the following compounds.

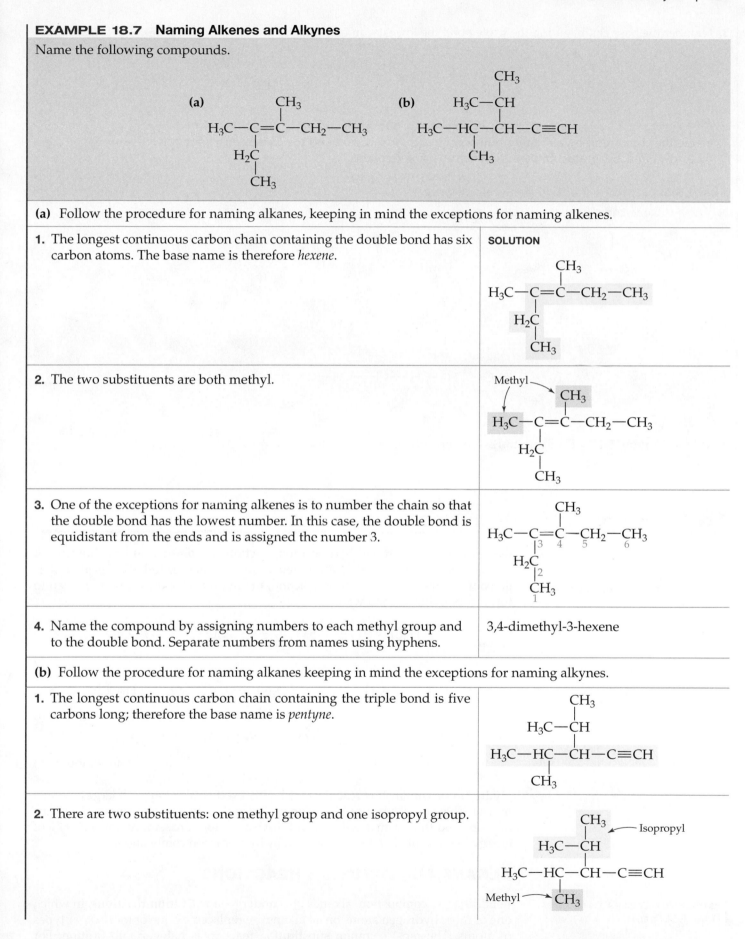

(a) Follow the procedure for naming alkanes, keeping in mind the exceptions for naming alkenes.

1. The longest continuous carbon chain containing the double bond has six carbon atoms. The base name is therefore *hexene*.

SOLUTION

2. The two substituents are both methyl.

3. One of the exceptions for naming alkenes is to number the chain so that the double bond has the lowest number. In this case, the double bond is equidistant from the ends and is assigned the number 3.

4. Name the compound by assigning numbers to each methyl group and to the double bond. Separate numbers from names using hyphens.

3,4-dimethyl-3-hexene

(b) Follow the procedure for naming alkanes keeping in mind the exceptions for naming alkynes.

1. The longest continuous carbon chain containing the triple bond is five carbons long; therefore the base name is *pentyne*.

2. There are two substituents: one methyl group and one isopropyl group.

3. Number the base chain, giving the triple bond the lowest number, 1.

$$\begin{array}{c} CH_3 \\ | \\ H_3C-CH \\ | \\ H_3C-\underset{5}{HC}-\underset{4}{CH}-\underset{3}{CH}-\underset{2}{C}\equiv\underset{1}{CH} \\ | \\ CH_3 \end{array}$$

4. Name the compound by assigning numbers to each substituent and to the triple bond. Separate numbers from names using hyphens.   3-isopropyl-4-methyl-1-pentyne

▶SKILLBUILDER 18.7 | Naming Alkenes and Alkynes

Name the alkene and alkyne.

▶FOR MORE PRACTICE   Example 18.12; Problems 55, 56, 57, 58, 59, 60.

18.9   Hydrocarbon Reactions

One of the most common hydrocarbon reactions is **combustion**, the burning of hydrocarbons in the presence of oxygen. Alkanes, alkenes, and alkynes all undergo combustion. In a combustion reaction, the hydrocarbon reacts with oxygen to form carbon dioxide and water.

Combustion reactions were first covered in Section 7.9.

$$CH_3CH_2CH_3(g) + 5\,O_2(g) \longrightarrow 3\,CO_2(g) + 4\,H_2O(g)$$

(Alkane combustion)

$$CH_2{=}CHCH_2CH_3(g) + 6\,O_2(g) \longrightarrow 4\,CO_2(g) + 4\,H_2O(g)$$

(Alkene combustion)

$$CH{\equiv}CCH_3(g) + 4\,O_2(g) \longrightarrow 3\,CO_2(g) + 2\,H_2O(g)$$

(Alkyne combustion)

Hydrocarbon combustion reactions are highly exothermic: they emit large amounts of heat. This heat can be used to warm homes and buildings, to generate electricity, or to expand the gas in a cylinder and drive a car forward. Approximately 90% of energy in the United States is generated by hydrocarbon combustion.

ALKANE SUBSTITUTION REACTIONS

As discussed in Chapter 4, the halogens include F, Cl, Br, and I.

In addition to combustion, alkanes also undergo **substitution reactions**, in which one or more hydrogen atoms on an alkane are replaced by one or more other types of atoms. The most common substitution reaction is halogen substitution. For example, methane reacts with chlorine gas to form chloromethane.

$$CH_4(g) + Cl_2(g) \xrightarrow[\text{light}]{\text{Heat or}} CH_3Cl(g) + HCl(g)$$

Ethane reacts with chlorine gas to form chloroethane.

$$CH_3CH_3(g) + Cl_2(g) \xrightarrow{\text{Heat or light}} CH_3CH_2Cl(g) + HCl(g)$$

The general form for halogen substitution reactions is:

$$\underset{\text{Alkane}}{R-H} + \underset{\text{Halogen}}{X_2} \longrightarrow \underset{\text{Alkil halide}}{R-X} + \underset{\text{Hydrogen halide}}{HX}$$

In this equation, R represents a hydrocarbon group.

Multiple halogenation reactions can occur because halogens can replace more than one of the hydrogen atoms on an alkane.

ALKENE AND ALKYNE ADDITION REACTIONS

Alkenes and alkynes undergo **addition reactions** in which atoms add across the multiple bond. For example, ethene reacts with chlorine gas to form dichloroethane.

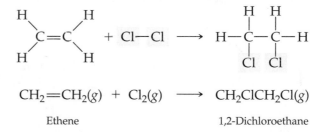

$$\underset{\text{Ethene}}{CH_2{=}CH_2(g)} + Cl_2(g) \longrightarrow \underset{\text{1,2-Dichloroethane}}{CH_2ClCH_2Cl(g)}$$

Notice that the addition of chlorine converts the carbon–carbon double bond into a single bond because each carbon atom now has formed a new bond with a chlorine atom. Alkenes and alkynes can also add hydrogen in **hydrogenation** reactions. For example, in the presence of an appropriate catalyst, propene reacts with hydrogen gas to form propane.

▲ Many foods contain partially hydrogenated vegetable oil. The name means that some of the double bonds in the carbon chains of these molecules have been converted to single bonds by the addition of hydrogen.

$$\underset{\text{Propene}}{CH_3CH{=}CH_2(g)} + H_2(g) \xrightarrow[\text{Catalyst}]{} \underset{\text{Propane}}{CH_3CH_2CH_3(g)}$$

Hydrogenation reactions convert unsaturated hydrocarbons into saturated hydrocarbons. Have you ever read *partially hydrogenated vegetable oil* on a food ingredient label? Vegetable oil is an unsaturated fat—its carbon chains contain double bonds. Unsaturated fats tend to be liquids at room temperature. By means of hydrogenation reactions, hydrogen is added across the double bonds, converting the unsaturated fat into saturated fat, which tends to be solid at room temperature.

More information about fats and oils can be found in Chapter 19.

To summarize:
- All hydrocarbons undergo combustion reactions.
- Alkanes undergo substitution reactions.
- Alkenes and alkynes undergo addition reactions.

18.10 Aromatic Hydrocarbons

As you might imagine, determining the structure of organic compounds has not always been easy. In the mid-1800s chemists were trying to determine the structure of a particularly stable organic compound named benzene that had the formula C_6H_6. In 1865, Friedrich August Kekulé (1829–1896) had a dream in which he envisioned chains of carbon atoms as snakes. The snakes danced before him, and one of them twisted around and bit its tail. Based on that vision, Kekulé proposed the following structure for benzene.

Benzene

This structure shows alternating single and double bonds. When we examine the bond lengths in benzene, however, we find that all of the bonds are of the same length.

The structure of benzene is better represented by the following resonance structures.

The concept of resonance structures was first introduced in Section 10.6.

Recall that the resonance structures indicate that the true structure of benzene is an average between the two structures. In other words, all carbon–carbon bonds in benzene are equivalent and are midway between a single and double bond. The space-filling model of benzene is:

Benzene is often represented with the shorthand notations:

Commonly recognized
representations of benzene

Each point in the hexagon represents a carbon atom with a hydrogen atom attached to it.

The ring structure of benzene occurs in many organic compounds. An atom or group of atoms can be substituted for one or more of the six hydrogen atoms to form substituted benzenes. Two examples of substituted benzenes are chlorobenzene and phenol.

Chlorobenzene   Phenol

Because many compounds containing benzene rings have pleasant aromas, benzene rings are also called **aromatic rings**, and compounds containing them are called *aromatic compounds*. For example, the pleasant smells of cinnamon, vanilla, and jasmine are all caused by aromatic compounds.

NAMING AROMATIC HYDROCARBONS

Monosubstituted benzenes—benzenes in which only one of the hydrogen atoms has been substituted—are often named as derivatives of benzene.

Bromobenzene   Ethylbenzene

The names of monosubstituted benzenes have the general form:

(name of substituent)*benzene*

However, many monosubstituted benzenes also have common names that can be learned only through familiarity.

Toluene   Aniline   Phenol   Styrene

Some substituted benzenes, especially those with large substituents, are named by treating the benzene ring as the substituent. In these cases, the benzene substituent is called a **phenyl group**.

Disubstituted benzenes, benzenes in which two hydrogen atoms have been substituted, are numbered, and the substituents are listed alphabetically. The order of numbering within the ring is then determined by the alphabetical order of the substituents.

1-Chloro-3-ethylbenzene

1-Bromo-2-iodobenzene

When the two substituents are identical, we use the prefix *di-*.

1,2-Dichlorobenzene   1,3-Dichlorobenzene   1,4-Dichlorobenzene

Also in common use—in place of numbering—are the prefixes *ortho-* (1,2 disubstituted), *meta-* (1,3 disubstituted), and *para-* (1,4 disubstituted).

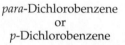

ortho-Dichlorobenzene   *meta*-Dichlorobenzene   *para*-Dichlorobenzene
or   or   or
o-Dichlorobenzene   *m*-Dichlorobenzene   *p*-Dichlorobenzene

EXAMPLE 18.8   Naming Aromatic Compounds

Name the following compound.

SOLUTION

Benzene derivatives are named using the general form (name of substituent) *benzene*. Because this derivative has two substituents, the substituents are numbered and listed alphabetically. The two substituents are *bromo-* and *chloro-*.

Because *bromo-* is first alphabetically, it is assigned the number 1; therefore *chloro-* is assigned the number 2. Following the general form, the name of the compound is 1-bromo-2-chlorobenzene. Alternatively, this compound could be named as *ortho*-bromochlorobenzene.

►**SKILLBUILDER 18.8 | Naming Aromatic Compounds**

Name the following compound.

►**FOR MORE PRACTICE**   Example 18.13; Problems 75, 76, 77, 78, 79, 80, 81, 82.

18.11 Functional Groups

Several families of organic compounds can be thought of as hydrocarbons in which a **functional group**—a characteristic atom or group of atoms—has been inserted into the hydrocarbon. The letter R is often used to represent a hydrocarbon group. If the letter G represents a functional group, then a generic formula for families of organic compounds is:

$$R-G$$

Hydrocarbon group   Functional group

A group of organic compounds with the same functional group forms a **family**. The members of the family of **alcohols** have an —OH functional group and the general formula R—OH. Some specific examples of alcohols are methanol and isopropyl alcohol.

R group CH_3-OH —OH functional group

Methanol

R group CH_3
          |
$CH_3-CH-OH$ —OH functional group

2-propanol or isopropyl alcohol

The insertion of a functional group into a hydrocarbon usually alters the properties of the compound significantly. For example, *methanol*—which can be thought of as methane with an —OH group substituted for one of the hydrogen atoms—is a polar, hydrogen-bonded liquid at room temperature. *Methane*, on the other hand, is a nonpolar gas. While each member of a family is unique, the common functional group within a family gives its members some similarities in both their physical and chemical properties. Table 18.7 lists some common functional groups, their general formulas, and an example of each.

TABLE 18.7 Functional Groups

Family	General Formula	Condensed General Formula	Example	Name
alcohols	R—OH	ROH	CH_3CH_2—OH	ethanol (ethyl alcohol)
ethers	R—O—R	ROR	CH_3—O—CH_3	dimethyl ether
aldehydes	R—$\overset{\overset{\textstyle O}{\|\|}}{C}$—H	RCHO	H_3C—$\overset{\overset{\textstyle O}{\|\|}}{C}$—H	ethanal (acetaldehyde)
ketones	R—$\overset{\overset{\textstyle O}{\|\|}}{C}$—R	RCOR	H_3C—$\overset{\overset{\textstyle O}{\|\|}}{C}$—$CH_3$	propanone (acetone)
carboxylic acids	R—$\overset{\overset{\textstyle O}{\|\|}}{C}$—OH	RCOOH	H_3C—$\overset{\overset{\textstyle O}{\|\|}}{C}$—OH	acetic acid
esters	R—$\overset{\overset{\textstyle O}{\|\|}}{C}$—OR	RCOOR	H_3C—$\overset{\overset{\textstyle O}{\|\|}}{C}$—$OCH_3$	methyl acetate
amines	R—$\overset{\overset{\textstyle R}{\|}}{N}$—R	R_3N	H_3CH_2C—$\overset{\overset{\textstyle H}{\|}}{N}$—H	ethyl amine

18.12   Alcohols

▲ Ethanol is the alcohol in alcoholic beverages.

As previously mentioned, alcohols are organic compounds containing the —OH functional group. They have the general formula R—OH. In addition to methanol and isopropyl alcohol (shown previously), other common alcohols include ethanol and 1-butanol:

$$H_3C—CH_2—OH$$
Ethanol

$$H_3C—CH_2—CH_2—CH_2—OH$$
1-Butanol

NAMING ALCOHOLS

Alcohols are named similarly to alkanes with the following exceptions.

- The base chain is the longest continuous carbon chain that *contains the* —OH *functional group.*
- The base name has the ending -*ol.*
- The base chain is numbered to *give the* —OH *group the lowest possible number.*
- A number indicating the position of the —OH group is inserted just before the base name. For example,

$$CH_3CH_2CH_2CHCH_3$$
|
OH

2-Pentanol

$$CH_2CH_2CHCH_3$$
| |
OH   CH_3

3-Methyl-1-butanol

ABOUT ALCOHOLS

Isopropyl alcohol is the common name for 2-propanol.

Among the most familiar alcohols is ethanol, the alcohol in alcoholic beverages. Ethanol is most commonly formed by the yeast fermentation of sugars, such as glucose, from fruits and grains.

$$C_6H_{12}O_6 \xrightarrow{\text{Yeast}} 2\ CH_3CH_2OH + 2\ CO_2$$
Glucose   Ethanol

Alcoholic beverages contain primarily ethanol and water and a few other components that contribute flavor and color. Beer usually contains 3 to 6% ethanol. Wine contains about 12 to 15% ethanol, and spirits—beverages such as whiskey, rum, or tequila—range from 40 to 80% ethanol, depending on their *proof.* The proof of an alcoholic beverage is twice the percentage of its ethanol content, so an 80-proof whiskey contains 40% ethanol. Ethanol is also used as a gasoline additive because it increases the octane rating of gasoline (which is related to how smoothly the gasoline burns) and fosters its complete combustion, reducing certain pollutants such as carbon monoxide and the precursors of ozone.

Isopropyl alcohol (or 2-propanol) can be purchased at any drugstore as rubbing alcohol. It is commonly used as a disinfectant for wounds and to sterilize medical instruments. Isopropyl alcohol should never be consumed internally, as it is highly toxic. A few ounces of isopropyl alcohol can cause death. A third common alcohol is methanol, also called wood alcohol. Methanol is commonly used as a laboratory solvent and as a fuel additive. Like isopropyl alcohol, methanol is toxic and should never be consumed.

▲ Rubbing alcohol is composed of isopropyl alcohol, or 2-propanol.

18.13 Ethers

Ethers are organic compounds with the general formula R—O—R. The R groups may be the same or different. Some common ethers include dimethyl ether, ethyl methyl ether, and diethyl ether:

H₃C—O—CH₃
Dimethyl ether

H₃C—O—CH₂—CH₃
Ethyl methyl ether

H₃C—CH₂—O—CH₂—CH₃
Diethyl ether

NAMING ETHERS

The IUPAC names for ethers are beyond the scope of this text. Common names for ethers have the format:

   (R group 1)(R group 2) ether

If the two R groups are different, use each of their names (see Table 18.4). If the two R groups are the same, use the prefix *di-*. Some examples include:

H₃C—CH₂—CH₂—O—CH₂—CH₂—CH₃
Dipropyl ether

H₃C—CH₂—O—CH₂—CH₂—CH₃
Ethyl propyl ether

ABOUT ETHERS

The most common ether is diethyl ether. Diethyl ether is frequently used as a laboratory solvent because of its ability to dissolve many organic compounds and because of its low boiling point (34.6 °C). The low boiling point allows for easy removal of the solvent when necessary. Diethyl ether was also used as a general anesthetic for many years. When inhaled, diethyl ether depresses the central nervous system, causing unconsciousness and insensitivity to pain. Its use as an anesthetic, however, has decreased in recent years because other compounds have the same anesthetic effect with fewer side effects (such as nausea).

18.14   Aldehydes and Ketones

The condensed structural formula for aldehydes is RCHO, and for ketones it is RCOR. In ketones the R groups may be the same or different.

Aldehydes and **ketones** have the general formulas:

Both aldehydes and ketones contain a **carbonyl group** (). Ketones have an R group attached to both sides of the carbonyl group, while aldehydes have an R group on one side of the carbonyl group and a hydrogen atom on the other. (An exception is formaldehyde, which is an aldehyde with two H atoms attached to the carbonyl group.)

Methanal or formaldehyde

Other common aldehydes and ketones are shown here.

Ethanal or acetaldehyde   Propanal

Propanone or acetone   Butanone

NAMING ALDEHYDES AND KETONES

Many aldehydes and ketones have common names that can be learned only by becoming familiar with them. Simple aldehydes are systematically named according to the number of carbon atoms in the longest continuous carbon chain that contains the carbonyl group. The base name is formed from the name of the corresponding alkane by dropping the *-e* and adding the ending *-al*.

Simple ketones are systematically named according to the longest continuous carbon chain containing the carbonyl group. The base name is formed from the name of the corresponding alkane by dropping the *-e* and adding the ending *-one*. For ketones, the chain is numbered to give the carbonyl group the lowest possible number (when necessary).

Butanal

Pentanal

2-Pentanone

3-Hexanone

Benzaldehyde

▲ Benzaldehyde is responsible for the smell of almonds.

▲ Fingernail-polish remover is primarily acetone, a ketone.

ABOUT ALDEHYDES AND KETONES

The most familiar aldehyde is probably formaldehyde, shown earlier in this section. Formaldehyde is a gas with a pungent odor. It is often mixed with water to make formalin, a preservative and disinfectant. Formaldehyde is also found in wood smoke, which is one reason that smoking foods preserves them—the formaldehyde kills the bacteria. Aromatic aldehydes, those that also contain an aromatic ring, have pleasant aromas. For example, cinnamaldehyde is the sweet-smelling component of cinnamon, benzaldehyde accounts for the smell of almonds, and vanillin is responsible for the smell of vanilla.

Cinnamaldehyde

Benzaldehyde

Vanillin

The most familiar ketone is acetone, the main component of fingernail-polish remover. Many ketones also have pleasant aromas. For example, 2-heptanone is responsible for the smell of cloves, carvone for the smell of spearmint, and ionone for the smell of raspberries.

2-Heptanone

Carvone

Ionone

Carvone

▲ The smell and taste of spearmint are produced by carvone, an aromatic ketone.

Ionone

▲ Ionone, a ketone, is largely responsible for the smell of raspberries.

18.15 Carboxylic Acids and Esters

The condensed structural formula for carboxylic acids is RCOOH, and for esters it is RCOOR. The R groups in esters may be the same or different.

Carboxylic acids and **esters** have the general formulas:

Carboxylic acid   Ester

Common carboxylic acids and esters are shown here.

Ethanoic acid or acetic acid

Butanoic acid

$$H_3C-CH_2-CH_2-\overset{\overset{\displaystyle O}{\|}}{C}-O-CH_3$$
Methyl butanoate

$$H_3C-CH_2-\overset{\overset{\displaystyle O}{\|}}{C}-O-CH_2CH_3$$
Ethyl propanoate

The generic condensed structural formula for carboxylic acids is RCOOH. Carboxylic acids act as weak acids in solution according to the equation:

$$RCOOH(aq) + H_2O(l) \rightleftharpoons H_3O^+(aq) + RCOO^-(aq)$$

NAMING CARBOXYLIC ACIDS AND ESTERS

Carboxylic acids are systematically named according to the number of carbon atoms in the longest chain containing the —COOH functional group. The base name is formed by dropping the -e from the name of the corresponding alkane and adding the ending -oic acid.

$$H_3C-CH_2-\overset{\overset{\displaystyle O}{\|}}{C}-OH$$
Propanoic acid

$$H_3C-CH_2-CH_2-CH_2-\overset{\overset{\displaystyle O}{\|}}{C}-OH$$
Pentanoic acid

Esters are systematically named as if they were derived from a carboxylic acid by replacing the H on the OH with an alkyl group. The R group from the parent acid forms the base name of the compound. The -ic on the name of the corresponding carboxylic acid is changed to -ate. The R group that replaced the H on the carboxylic acid is named as an alkyl group with the ending -yl. For example:

CH₃CH₂COO**H**
Propano**ic** acid

CH₃CH₂CH₂CH₂COO**H**
Pentano**ic** acid

CH₃CH₂COO*CH₃*
↓
Methyl propano**ate**

CH₃CH₂CH₂CH₂COO*CH₂CH₃*
↓
Ethyl pentano**ate**

ABOUT CARBOXYLIC ACIDS AND ESTERS

Like all acids, carboxylic acids taste sour. The most familiar carboxylic acid is ethanoic acid, which is known by its common name, *acetic acid*. Acetic acid is the active ingredient in vinegar. It can be formed by the oxidation of ethanol, which is why wines left open to air become sour. Some yeasts and bacteria also form acetic acid when they metabolize sugars in bread dough. These are often added to bread dough to make sourdough bread. Other common carboxylic acids include methanoic acid (usually called formic acid), the acid present in bee stings and ant bites; citric acid, the acid present in limes, lemons, and oranges; and lactic acid, the acid that causes muscle soreness after intense exercise.

Acetic acid

▲ Acetic acid, a carboxylic acid, is the active ingredient in vinegar.

Citric acid

▲ Citric acid is responsible for the sour taste of limes and other citrus fruits. **Question:** Can you think of other fruits that might contain citric acid?

Formic or methanoic acid

Lactic acid

Citric acid

Esters are best known for their sweet smells. For example, ethyl butanoate is responsible for the smell and taste of pineapples, and methyl butanoate is responsible for the smell and taste of apples.

$$H_3C-CH_2-CH_2-\overset{\overset{\displaystyle O}{\|}}{C}-O-CH_2-CH_3$$
Ethyl butanoate

$$H_3C-CH_2-CH_2-\overset{\overset{\displaystyle O}{\|}}{C}-O-CH_3$$
Methyl butanoate

Methyl butanoate

▶ Methyl butanoate is an ester found in apples.

Esters are formed from the reaction of a carboxylic acid and an alcohol:

$$\underset{\text{Acid}}{R\overset{\overset{\displaystyle O}{\|}}{C}-OH} + \underset{\text{Alcohol}}{HO-R'} \longrightarrow$$

$$\underset{\text{Ester}}{R-\overset{\overset{\displaystyle O}{\|}}{C}-O-R'} + \underset{\text{Water}}{H_2O}$$

An important example of this reaction is the formation of acetylsalicylic acid (aspirin) from acetic acid and salicylic acid (originally obtained from the bark of the willow tree).

CONCEPTUAL CHECKPOINT 18.2

Which term could *not* be applied to the compound shown?

(a) unsaturated

(b) aromatic

(c) an acid

(d) organic

$$HO-\overset{\overset{O}{\|}}{C}CH_2CH=CHCH_3$$

18.16 Amines

Amines are a class of organic compounds that contain nitrogen. The simplest nitrogen-containing compound is ammonia (NH_3). All other amines are derivatives of ammonia with one or more of the hydrogen atoms replaced by alkyl groups. They are systematically named according to the hydrocarbon groups attached to the nitrogen and given the ending *-amine*.

$$H_3C-CH_2-\underset{\underset{H}{|}}{N}-H$$
Ethylamine

$$H_3C-CH_2-\underset{\underset{CH_3}{|}}{N}-H$$
Ethylmethylamine

Amines are best known for their awful odors. When a living organism dies, bacteria that feast on its proteins emit amines. For example, trimethylamine is a component of the smell of rotten fish, and cadaverine and putrescine are responsible for the smell of decaying animal flesh.

$$H_3C-\underset{\underset{CH_3}{|}}{\overset{\overset{CH_3}{|}}{N}}-CH_3$$
Trimethylamine

$$H_2N-[CH_2]_5-NH_2$$
Cadaverine

$$H_2N-[CH_2]_4-NH_2$$
Putrescine

18.17 Polymers

Polymers are long, chainlike molecules composed of repeating units. The individual repeating units are called **monomers**. In Chapter 19 we will learn about natural polymers such as starches, proteins, and DNA. These natural polymers play important roles in living organisms. In this section, we learn about synthetic polymers. Synthetic polymers compose many frequently encountered plastic products such as PVC tubing, Styrofoam coffee cups, nylon rope, and Plexiglas windows. Polymer materials are common in our everyday lives since they are found in everything from computers to toys to packaging materials. How many things can you think of that are made of plastic?

The simplest synthetic polymer is probably polyethylene. The polyethylene monomer is ethene (also called ethylene).

$$H_2C{=}CH_2 \qquad \textbf{Monomer}$$

Ethene or ethylene

Ethene monomers can be made to react with each other, breaking the double bond between carbons and joining together to make a long polymer chain.

$$\cdots\cdots CH_2{-}CH_2{-}CH_2{-}CH_2{-}CH_2{-}CH_2{-}CH_2{-}CH_2{-}CH_2\cdots\cdots$$

Polymer

Polyethylene

▲ Polyethylene is widely used in containers for beverages.

Polyethylene is the plastic that composes milk jugs, juice containers, and garbage bags. It is an example of an **addition polymer**, a polymer in which the monomers link together without eliminating any atoms.

Substituted polyethylenes comprise an entire class of polymers. For example, polyvinyl chloride (PVC)—the plastic used to make certain kinds of pipes and plumbing fixtures—is composed of monomers in which a chlorine atom has been substituted for one of the hydrogen atoms in ethene.

$$HC{=}CH_2 \atop {|} \atop Cl \qquad \textbf{Monomer}$$

Chloroethene

These monomers react together to form PVC.

$$\cdots CH{-}CH_2{-}CH{-}CH_2{-}CH{-}CH_2{-}CH{-}CH_2{-}CH\cdots \atop {|\qquad\quad |\qquad\quad |\qquad\quad |\qquad\quad |} \atop {Cl\qquad\quad Cl\qquad\quad Cl\qquad\quad Cl\qquad\quad Cl}$$

Polymer

Polyvinyl chloride (PVC)

Table 18.8 shows several other substituted polyethylene polymers.

▶ Polyvinyl chloride (PVC) is used for pipes and plumbing fittings.

TABLE 18.8 Polymers of Commercial Importance

Polymer	Structure	Uses
Addition polymers		
polyethylene	$\left(CH_2-CH_2\right)_n$	films, packaging, bottles
polypropylene	$\left[CH_2-CH \atop \;\;\;\;\;CH_3\right]_n$	kitchenware, fibers, appliances
polystyrene	$\left[CH_2-CH \atop \;\;\;\;\;C_6H_5\right]_n$	packaging, disposable food containers, insulation
polyvinyl chloride	$\left[CH_2-CH \atop \;\;\;\;\;Cl\right]_n$	pipe fittings, clear film for meat packaging
Condensation polymers		
polyurethane	$\left[\underset{O}{\overset{\parallel}{C}}-NH-R-NH-\underset{O}{\overset{\parallel}{C}}-O-R'-O\right]_n$ R, R′ = —CH$_2$—CH$_2$— (for example)	"foam" furniture stuffing, spray-on insulation, automotive parts, footwear, water-protective coatings
polyethylene terephthalate (a polyester)	$\left[O-CH_2-CH_2-O-\underset{O}{\overset{\parallel}{C}}-\bigcirc-\underset{O}{\overset{\parallel}{C}}\right]_n$	tire cord, magnetic tape, apparel, soft-drink bottles
nylon 6,6	$\left[NH-\left(CH_2\right)_6-NH-\underset{O}{\overset{\parallel}{C}}-(CH_2)_4-\underset{O}{\overset{\parallel}{C}}\right]_n$	home furnishings, apparel, carpet fibers, fish line, polymer blends

Some polymers—called **copolymers**—consist of two different kinds of monomers. For example, the monomers that compose nylon 6,6 are hexamethylenediamine and adipic acid. These two monomers add together by eliminating a water molecule for each bond that forms between monomers. Polymers that eliminate an atom or a small group of atoms during polymerization are called **condensation polymers**.

EVERYDAY CHEMISTRY

Kevlar: Stronger Than Steel

In 1965, Stephanie Kwolek, working for DuPont to develop new polymer fibers, noticed an odd cloudy product from a polymerization reaction. Some researchers might have rejected the product, but Kwolek insisted on examining its properties more carefully. The results were astonishing: When the polymer was spun into a fiber, it was stronger than any other fiber known before. Kwolek had discovered Kevlar, a material that is pound for pound five times stronger than steel.

Kevlar is a condensation polymer containing aromatic rings and *amide linkages:*

The amide linkage is the carbonyl (C=O) and nitrogen link between the aromatic rings. The polymeric chains within Kevlar crystallize in a parallel arrangement (like dry spaghetti noodles in a box), with strong cross-linking between neighboring chains due to hydrogen bonding. The hydrogen bonding occurs between the —N—H groups on one chain and the C=O groups on neighboring chains:

This structure is responsible for Kevlar's high strength and its other properties, including chemical resistance and flame resistance.

Today, DuPont sells hundreds of millions of dollars' worth of Kevlar every year. Kevlar is particularly well known for its use in bulletproof vests. With this application alone, Kwolek's discovery has saved thousands of lives. In addition, Kevlar is used to make helmets, radial tires, brake pads, racing sails, suspension bridge cables, skis, and high-performance hiking and camping gear.

CAN YOU ANSWER THIS? *Examine the structure of the Kevlar polymer. Knowing that the polymer is a condensation polymer, draw the structures of the monomers before the condensation reaction.*

▲ The great strength of Kevlar fibers makes this polymer ideal for use in bulletproof vests.

The product that forms between the reaction of two monomers is called a **dimer**. The polymer (nylon 6,6) shown here forms as the dimer continues to add more monomers. Nylon 6,6 and other similar nylons can be drawn into fibers and used to make consumer products such as pantyhose, carpet fibers, and fishing line. Table 18.8 shows other condensation polymers.

Hexamethylenediamine   Adipic acid   Monomers

Dimer

CHAPTER IN REVIEW

CHEMICAL PRINCIPLES

Organic Chemistry: Organic chemistry is the study of carbon-containing compounds and their reactions. Carbon is unique because it can form four bonds and can bond to itself to form chains, branched structures, and ring structures.

Vitalism: Vitalism is the belief that living things contain a mystical force that allows them to produce organic compounds. Vitalism was overthrown when Friedrich Wöhler synthesized urea, an organic compound, in his laboratory.

Hydrocarbons: Hydrocarbons are organic compounds containing only carbon and hydrogen. They can be classified into four different types: alkanes, with the formula C_nH_{2n+2}; alkenes, with the formula C_nH_{2n}; alkynes, with the formula C_nH_{n-2}; and aromatic hydrocarbons, which contain six-carbon-atom ring structures.

Isomers: Isomers are two or more compounds with the same chemical formula but different structures. The more atoms in an organic compound, the greater the number of isomers that exist.

Hydrocarbon Reactions: All hydrocarbons undergo combustion—the reaction of a hydrocarbon with oxygen to form carbon dioxide and water. Alkanes undergo substitution reactions, in which one or more hydrogen atoms on the alkane are replaced by one or more other atoms, usually a halogen. Alkenes and alkynes undergo addition reactions, in which two atoms—often hydrogen or halogens—add across the multiple bond.

RELEVANCE

Organic Chemistry: Organic compounds are prevalent in food, drugs, petroleum products, and pesticides. Organic chemistry is also the basis for living organisms.

Vitalism: The demise of vitalism opened all of life to scientific inquiry and led to our current understanding of organic chemistry.

Hydrocarbons: One of the main uses of hydrocarbons is as fuels. Petroleum and natural gas are both composed of hydrocarbons. Hydrocarbons are also the starting materials in the synthesis of many consumer products, including fabrics, soaps, dyes, cosmetics, drugs, plastics, and rubber.

Isomers: Although isomers may be closely related chemically, they are different compounds with different properties.

Hydrocarbon Reactions: The combustion of hydrocarbons provides about 90% of U.S. energy. We rely on the combustion of hydrocarbons to heat our homes, to propel our automobiles, and to generate electricity. Substitution reactions, especially hydrogenation, are important in the petroleum industry.

Functional Groups: Several families of organic compounds are essentially substituted hydrocarbons in which a functional group has been inserted. The symbol R is convenient shorthand for the hydrocarbon in general formulas. The main families of organic compounds, which are classified according to their functional groups, are:

alcohols	R—OH
ethers	R—O—R
aldehydes	R—COH
ketones	R—CO—R
carboxylic acids	R—COOH
esters	R—COO—R
amines	R_3N

Functional Groups: Organic compounds with similar functional groups form a family of compounds and exhibit similar chemical and physical properties. For example, carboxylic acids behave as acids and therefore have low pH. Many esters, aldehydes, and ketones, especially those that contain an aromatic ring in their R group, tend to have pleasant smells. Amines, on the other hand, tend to have foul smells.

Polymers: Polymers are long, chainlike molecules composed of repeating units called monomers. Polymers that are formed from monomers that attach together without eliminating any atoms or groups of atoms are called addition polymers. Polymers that eliminate water or other atoms during the polymerization process are called condensation polymers. Copolymers are composed of two different kinds of monomers.

Polymers: Many consumer products are composed of polymers. The plastic you use for trash bags and the plastic container you buy water in are both composed of polyethylene. Nylon is used to make clothing and fishing line, and polyvinyl chloride to make plastic pipes and pipe fittings.

CHEMICAL SKILLS

EXAMPLES

Differentiating between Alkanes, Alkenes, and Alkynes Based on Their Molecular Formulas (Section 18.4)
Examine the formula for the molecule in question, and, based on the relative number of carbon and hydrogen atoms, determine whether it is an alkane, alkene, or alkyne.

Alkanes	C_nH_{2n+2}
Alkenes	C_nH_{2n}
Alkynes	C_nH_{n-2}

These formulas apply to open-chain (noncyclical) hydrocarbons only.

EXAMPLE 18.9   Differentiating between Alkanes, Alkenes, and Alkynes Based on Their Molecular Formulas

Identify each noncyclical hydrocarbon as an alkane, alkene, or alkyne.

(a) C_3H_6
(b) C_6H_{14}
(c) $C_{10}H_{18}$

SOLUTION

(a) C_3H_6
  $n = 3, 6 = 2n$; therefore, this is an alkene.
(b) C_6H_{14}
  $n = 6, 14 = 2n + 2$; therefore, this is an alkane.
(c) $C_{10}H_{18}$
  $n = 10, 18 = 2n - 2$; therefore, this is an alkyne.

Writing Structural Formulas for Hydrocarbon Isomers (Section 18.6)

To write structural formulas for a set of hydrocarbon isomers, begin by writing the carbon backbone in as many unique ways as possible. Include double or triple carbon–carbon bonds in your backbone structures.

EXAMPLE 18.10 Writing Structural Formulas for Hydrocarbon Isomers

Write structural formulas for all of the isomers of pentyne (C_5H_8).

SOLUTION

Pentyne has five carbon atoms with one triple bond between two of them. The possible backbone structures are:

$$C\equiv C-C-C-C$$

$$C-C\equiv C-C-C$$

$$C\equiv C-C-C$$
$$\underset{\displaystyle C}{\overset{\displaystyle |}{}}$$

Once you have determined all of the unique backbone structures, add hydrogen atoms to give each carbon atom four bonds.

Then add hydrogen atoms so that each carbon atom forms four bonds.

$$HC\equiv C-CH_2-CH_2-CH_3$$

$$H_3C-C\equiv C-CH_2-CH_3$$

$$HC\equiv C-\underset{\underset{\displaystyle H_3C}{|}}{CH}-CH_3$$

Naming Alkanes (Section 18.7)

For a full review of naming alkanes, see the rules in Section 18.7. The following is a condensed version.

1. Count the number of carbon atoms in the longest continuous carbon chain to determine the base name of the compound. Use the ending *-ane*.

2. Consider every branch from the base chain to be a substituent. Name each substituent according to the number of carbon atoms in the substituent.

3. Beginning with the end closest to the branching, number the base chain and assign a number to each substituent.

4. Write the name of the compound in the format:

   (subst. number)-(subst. name)(base name)

   If there are two or more substituents, give each one a number and list them alphabetically. Words and numbers are separated by hyphens.

5. If a compound has two or more identical substituents, use the prefix *di-* (2), *tri-* (3), or *tetra-* (4) before their name. Separate the numbers indicating their positions relative to each other with a comma.

EXAMPLE 18.11 Naming Alkanes

Name the alkane.

$$\underset{\underset{\displaystyle CH_3 \quad\ CH_3}{|\qquad\quad |}}{CH_3CHCH_2CHCH_3}$$

SOLUTION

The longest continuous chain has five carbon atoms. Therefore, the base name is *pentane*.

$$\overset{\displaystyle 1 \quad\ 2 \quad\ 3 \quad\ 4 \quad 5}{\underset{\underset{\displaystyle CH_3 \quad\ CH_3}{|\qquad\quad |}}{CH_3CHCH_2CHCH_3}}$$

The substituents are both methyl groups.
The substituents are at the 2 and 4 positions.
The name of this compound is 2,4-dimethylpentane.

Naming Alkenes and Alkynes (Section 18.8)

Alkenes and alkynes are named similarly to alkanes with the following exceptions.

- The base chain must contain the double or triple bond.

- The base name has the ending *-ene* for alkenes and *-yne* for alkynes.

- The base chain is numbered to give the double or triple bond the lowest possible number.

- A number indicating the position of the double or triple bond (lowest possible number) is inserted just before the base name.

EXAMPLE 18.12 Naming Alkenes and Alkynes

Name the alkyne.

$$CH_3CH_2C \equiv CCH_3$$
$$|$$
$$CH_3$$

SOLUTION

4-methyl-2-pentyne

Naming Aromatic Compounds (Section 18.10)

The names of substituted benzenes have the following general forms.

Monosubstituted benzenes

(Name of substituent)*benzene*

Compounds in which benzene is better viewed as a substituent

Name the benzene ring as a substituent with the name phenyl.

Disubstituted benzenes

\#-(substituent name)-\#-(substituent name)benzene

- The substituents are listed in alphabetical order.

- The order of numbering within the ring is also determined by the names of the substituents.

- The prefixes *ortho-* (1,2), *meta-* (1,3), and *para-* (1,4) are often used in place of numbers when the substituents are identical.

EXAMPLE 18.13 Naming Aromatic Compounds

Some examples of aromatic compounds and their names are shown here.

Iodobenzene

$$H_3C-CH-CH_2-CH_2-CH_2-CH_3$$

2-Phenylhexane

1-Bromo-3-iodobenzene

meta-Diethylbenzene or
m-Diethylbenzene

KEY TERMS

addition polymer [18.17]	alkyne [18.8]	combustion [18.9]	disubstituted benzene
addition reaction [18.9]	amine [18.16]	condensation polymer	[18.10]
alcohol [18.11]	aromatic ring [18.10]	[18.17]	ester [18.15]
aldehyde [18.14]	base chain [18.7]	condensed structural	ether [18.13]
alkane [18.5]	branched alkane [18.6]	formula [18.5]	family (of organic
alkene [18.8]	carbonyl group [18.14]	copolymer [18.17]	compounds) [18.11]
alkyl group [18.7]	carboxylic acid [18.15]	dimer [18.17]	fossil fuels [18.4]

functional group **[18.11]**
hydrocarbon **[18.4]**
hydrogenation **[18.9]**
isomer **[18.6]**
ketone **[18.14]**
monomer **[18.17]**

monosubstituted benzene **[18.10]**
normal alkane (*n*-alkane) **[18.5]**
organic chemistry **[18.1]**
organic molecule **[18.1]**

phenyl group **[18.10]**
polymer **[18.17]**
saturated hydrocarbon **[18.5]**
structural formula **[18.5]**
substituent **[18.7]**

substitution reaction **[18.9]**
unsaturated hydrocarbon **[18.8]**
vital force **[18.2]**
vitalism **[18.2]**

EXERCISES

QUESTIONS

1. What kinds of molecules are often involved in smell?
2. What is organic chemistry?
3. Explain the difference—as it was viewed at the end of the eighteenth century—between organic and inorganic compounds.
4. What is vitalism? How was vitalism usurped?
5. What is unique about carbon and carbon-based compounds? Why did life evolve around carbon?
6. Describe the geometry about a carbon atom that forms:

   (a) four single bonds

   (b) two single bonds and one double bond

   (c) one single bond and one triple bond
7. What are hydrocarbons? What are the main uses of hydrocarbons?
8. What are the main classifications of hydrocarbons? What are their generic molecular formulas?
9. What is the difference between saturated and unsaturated hydrocarbons?
10. Explain the difference between a molecular formula, a structural formula, and a condensed structural formula.
11. Explain the difference between *n*-alkanes and branched alkanes.
12. What are isomers? List some examples.
13. What are alkenes? How are they different from alkanes?
14. What are alkynes? How are they different from alkanes?
15. What are hydrocarbon combustion reactions? Provide an example.
16. What are alkane substitution reactions? Provide an example.

17. What is an alkene addition reaction? Provide an example.
18. What is an alkyne addition reaction? Provide an example.
19. What is the structure of benzene? What are the different ways in which this structure is represented?
20. What is a functional group? List some examples.
21. What is the generic structure of alcohols? Write the structures of two specific alcohols.
22. List examples of some common alcohols and where you might find them.
23. What is the generic structure of ethers? Write the structures of two specific ethers.
24. Provide an example of a common ether and its main uses.
25. What are the generic structures of aldehydes and ketones? Write the structure of a specific aldehyde and a specific ketone.
26. List some examples of common aldehydes and ketones and where you might find them.
27. What are the generic structures of carboxylic acids and esters? Write the structure of a specific carboxylic acid and a specific ester.
28. List some examples of common carboxylic acids and esters and where you might find them.
29. What is the generic structure of amines? Write the structures of two specific amines.
30. Provide an example of a common amine and where you might find it.
31. Explain what a polymer is and describe the difference between a polymer and a copolymer.
32. Explain the difference between an addition polymer and a condensation polymer.

PROBLEMS

HYDROCARBONS

33. Determine whether or not each compound is a hydrocarbon.

    (a) $C_5H_{12}O$

    (b) NH_3

    (c) C_8H_{16}

    (d) C_2H_6

34. Determine whether or not each compound is a hydrocarbon.

    (a) CH_2O

    (b) CH_4

    (c) C_6H_6

    (d) C_2H_7N

35. Based on the molecular formula, determine whether each compound is an alkane, alkene, or alkyne. (Assume that all the compounds are noncyclic and do not contain more than one multiple bond.)

   (a) C_4H_6

   (b) C_6H_{14}

   (c) C_8H_{14}

   (d) $C_{11}H_{22}$

36. Based on the molecular formula, determine whether each compound is an alkane, alkene, or alkyne. (Assume that all the compounds are noncyclic and do not contain more than one multiple bond.)

   (a) C_3H_6

   (b) C_6H_{10}

   (c) C_3H_8

   (d) C_5H_{10}

ALKANES

37. Write a structural formula and a condensed structural formula for each alkane.

   (a) heptane

   (b) octane

   (c) hexane

   (d) ethane

38. Write a structural formula and a condensed structural formula for each alkane.

   (a) methane

   (b) pentane

   (c) butane

   (d) propane

39. Write structural formulas for each of the two isomers of butane.

40. Write structural formulas for each of the three isomers of pentane.

41. Write structural formulas for any five isomers of octane.

42. Write structural formulas for all of the nine isomers of heptane.

43. Name each alkane.

   (a) $H_3C-CH_2-CH_2-CH_2-CH_3$

   (b) $H_3C-CH_2-CH-CH_3$
                             $|$
                          H_3C

   (c) $H_3C-HC-CH_2-HC-CH_2-CH_3$
                 $|$ $|$
                CH_3 CH_2-CH_3

   (d)
                      H_3C
                      $|$
   $H_3C-CH_2-C-CH_2-CH_3$
                      $|$
                      H_3C

44. Name each alkane.

   (a) $H_3C-CH_2-CH_2-CH_3$

   (b) $H_3C-CH_2-CH_2-CH_2-CH-CH_2-CH_2-CH_3$
                                            $|$
                                   $H_2C-CH_2-CH_3$

   (c) $H_3C-HC-CH_2-HC-CH_2-CH_3$
                 $|$ $|$
                CH_3 CH_2-CH_3

   (d)
              H_3C H_3C
              $|$ $|$
   $H_3C-C-\!-\!-C-CH_3$
              $|$ $|$
              H_3C H_3C

45. Draw a structure for each alkane.

   (a) 2-methylbutane

   (b) 3-ethyl-2-methylhexane

   (c) 3-isopropylheptane

   (d) 2,5-dimethyloctane

46. Draw a structure for each alkane.

   (a) 3-ethylhexane

   (b) 3,3-dimethylpentane

   (c) 3-ethyl-3-methylpentane

   (d) 4,4-diethyloctane

47. Determine what is wrong with the names of each alkane and provide the correct name.

(a) H₂C—CH₂—CH₂—CH₃
|
CH₃

1-Methylbutane

(b) H₃C—CH—CH₂—CH₂—CH₃
|
H₂C—CH₃

2-Ethylpentane

(c) H₃C—CH—CH—CH₂—CH₃
| |
CH₃ CH₃

2-Methyl-3-methyl pentane

48. Determine what is wrong with the names of each alkane and provide the correct name.

(a) H₂C—CH₂—CH₂—CH₃
|
CH₂
|
CH₃

2-Ethyl-butane

(b) CH₂—CH—CH—CH₂—CH₂—CH₂—CH₃
| | |
H₃C H₃C H₃C

1,2,3-Trimethylheptane

(c) H₃C—CH—CH—CH₃
| |
H₃C H₂C—CH₃

2-Ethyl-3-methylbutane

49. Complete the table:

Name	Molecular Formula	Structural Formula	Condensed Structural Formula
2,2,3-trimethyl-pentane	_____	_____	
_____	_____		CH₃CH(CH₃)CH(CH₂CH₂CH₃)CH₂CH₂CH₃
_____	_____	CH₃—C—C—CH₂—CH₂—CH₃ (with CH₃ CH₃ above and CH₃ CH₃ below)	_____
_____	_____	_____	CH₃CH(CH₃)CH(CH₃)CH(CH₂CH₃)₂CH₂CH₃

50. Complete the table:

Name	Molecular Formula	Structural Formula	Condensed Structural Formula
_____	_____	_____	CH₃C(CH₃)₂C(CH₂CH₃)₂CH₂CH₂CH₂CH₂CH₃
_____	_____	CH₃—CH₂—C—C—CH₂—CH₂—CH₂—CH₃ (with CH₃, CH₂, CH₃ branches)	_____
2,2-dimethyl 3-ethylpentane	_____	_____	_____
_____	_____	CH₃—CH₂—CH₂—C(H)—CH₂—CH₂—CH₂—CH₂—CH₃ (with CH₂, CH₂, CH₃ branch)	

ALKENES AND ALKYNES

51. Write a structural formula and a condensed structural formula for any two alkenes.

52. Write a structural formula and a condensed structural formula for any two alkynes.

53. Write structural formulas for all of the possible isomers of *n*-pentene that can be formed by moving the position of the double bond.

54. Write structural formulas for all of the possible isomers of *n*-hexyne that can be formed by moving the position of the triple bond.

55. Name each alkene.

   (a) $H_3C-CH=CH-CH_2-CH_3$

   (b) $H_3C-HC-CH=CH-CH_3$
               |
              CH_3

   (c)
       H_3C
          |
   $H_3C-\overset{}{\underset{}{C}}-CH=CH_2$
         |
        CH_3

   (d)
            H_3C
             |
   $H_3C-HC-CH-CH=CH_2$
           |
         CH_2
         |
         CH_3

56. Name each alkene.

   (a) $H_2C=CH-CH_2-CH_3$

   (b) $H_3C-CH=CH-CH-CH_2-CH_3$
                        |
                     H_2C-CH_3

   (c)
                     CH_3
                     |
   $H_2C=CH-CH-HC-CH_3$
              |
            CH_3

   (d) $H_2C=CH-CH-CH_2-CH_2-CH_3$
                |
          H_3C-CH
               |
             CH_3

57. Name each alkyne.

   (a) $H_3C-C\equiv C-CH_3$

   (b) $H_3C-C\equiv C-HC-CH_3$
                    |
                CH_3

   (c)
               CH_3
               |
   $H_3C-C\equiv C-\overset{}{\underset{}{C}}-CH_2-CH_3$
               |
              CH_3

   (d)

              CH_3
              |
   $HC\equiv C-\overset{}{\underset{}{C}}-CH_2-CH_3$
             |
            CH_2
            |
            CH_3

58. Name each alkyne.

   (a) $HC\equiv C-CH_2-CH_2-CH_3$

   (b) $HC\equiv C-CH-CH_2-CH_2-CH_3$
                  |
                $HC-CH_3$
                 |
                CH_3

   (c) $H_3C-HC-C\equiv C-CH-CH_3$
              | |
            CH_3 H_3C

   (d) $HC\equiv C-C\equiv C-HC-CH_2-CH_3$
                        |
                     CH_2-CH_3

59. Provide correct structures for each compound:

(a) 2-hexene

(b) 3-heptyne

(c) 3-methyl-1-pentyne

(d) 4,4-dimethyl-2-hexene

60. Provide correct structures for each compound:

(a) 3-octyne

(b) 1-pentene

(c) 3,3-dimethyl-1-pentyne

(d) 4-ethyl-3-methyl-2-octene

61. Draw and name all of the possible isomers of $CH_2\!=\!CHCH_2CH_2CH_3$.

62. Draw and name all of the possible alkyne isomers of $CH\!\equiv\!CCH_2CH_2CH_2CH_3$.

63. Complete the table:

Name	Molecular Formula	Structural Formula	Condensed Structural Formula
2,2-dimethyl-3-hexene	___ ___ ___	___	
___	___	$HC\!\equiv\!C\!-\!CH\!-\!CH\!-\!CH_2\!-\!CH_2\!-\!CH_2\!-\!CH_3$ with CH_3 CH_3	$CH_3C(CH_3)_2C(CH_2CH_3)_2C\!\equiv\!CCH_3$ ___
___	___	___	$CH_3C(CH_3)_2C(CH_2CH_2)_2CH\!=\!CHCH_3$

64. Complete the table:

Name	Molecular Formula	Structural Formula	Condensed Structural Formula
3-ethyl-4-methyl-1-heptene	___ ___ ___	___	$CH_3CH\!=\!C(CH_2CH_3)CH_2CH_3$ ___
___	___	$HC\!\equiv\!C\!-\!CH\!-\!CH\!-\!CH_2\!-\!CH_2\!-\!CH_3$ with CH_3 $CH_2CH_2CH_3$	
3,3-dimethyl-4-ethyl-1-hexyne	___	___	___

HYDROCARBON REACTIONS

65. Complete and balance each hydrocarbon combustion reaction.

(a) $CH_3CH_3 + O_2 \longrightarrow$

(b) $CH_2\!=\!CHCH_3 + O_2 \longrightarrow$

(c) $CH\!\equiv\!CH + O_2 \longrightarrow$

66. Complete and balance each hydrocarbon combustion reaction.

(a) $CH_3CH_2CH_2CH_3 + O_2 \longrightarrow$

(b) $CH_2\!=\!CH_2 + O_2 \longrightarrow$

(c) $CH\!\equiv\!CCH_2CH_3 + O_2 \longrightarrow$

67. What are the products of this alkane substitution reaction? (Assume monosubstitution.)

$$CH_4 + Br_2 \longrightarrow$$

68. What are the products of this alkane substitution reaction? (Assume monosubstitution.)

$$CH_3CH_3 + I_2 \longrightarrow$$

69. What are the products of this alkene addition reaction?

$$CH_3CH{=}CHCH_3 + Cl_2 \longrightarrow$$

70. What are the products of this alkene addition reaction?

$$CH_3CHCH{=}CH_2 + Cl_2 \longrightarrow$$
$$\underset{\displaystyle CH_3}{|}$$

71. Complete the hydrogenation reaction:

$$CH_2{=}CH_2 + H_2 \longrightarrow$$

72. Complete the hydrogenation reaction:

$$CH_3CH_2CH_2CH{=}CH_2 + H_2 \longrightarrow$$

AROMATIC HYDROCARBONS

73. Draw the full structural formula that is represented by each shorthand formula.

74. Explain how the two resonance structures *together* represent the true structure of benzene.

75. Name each monosubstituted benzene.

76. Name each monosubstituted benzene.

77. Name each compound in which the benzene ring is treated as a substituent.

(a)

$$H_3C{-}CH_2{-}CH_2{-}CH{-}CH_2{-}CH_2{-}CH_2{-}CH_3$$

(b)

$$H_3C{-}CH_2{-}CH{-}CH{=}CH{-}CH_2{-}CH_3$$

(c)

$$CH_2{-}CH_2{-}CH_2{-}CH_2{-}C{\equiv}C{-}CH_3$$

78. Name each compound in which the benzene ring is treated as a substituent.

(a)

$$H_3C{-}CH_2{-}CH{-}CH{-}CH_2{-}CH_3$$
$$\underset{\displaystyle CH_3}{|}$$

(b)

$$H_3C{-}CH{-}CH{=}CH{-}CH_2{-}CH_2{-}CH_2{-}CH_3$$

(c)

$$H_3C{-}CH{-}CH_2{-}CH{-}CH_2{-}CH_2{-}CH_3$$
$$\underset{\displaystyle H_2C{-}CH_3}{|}$$

79. Name each disubstituted benzene.

(a)

(b) H₂C—CH₃ ... CH₂—CH₃

(c) F ... F

80. Name each disubstituted benzene.

(a)

(b) F ... H₂C—CH₃

(c) I ... I

81. Draw a structure for each compound.
(a) butylbenzene
(b) 1-ethyl-2-iodobenzene
(c) *para*-dimethylbenzene

82. Draw a structure for each compound.
(a) isopropylbenzene
(b) *meta*-dibromobenzene
(c) 1-bromo-4-ethylbenzene

FUNCTIONAL GROUPS

83. Based on its functional group, match the structure on the left with the correct name on the right.

(a) R—C(=O)—H   Ether

(b) R—C(=O)—R   Aldehyde

(c) R—O—R   Amine

(d) R—N(R)—R   Ketone

84. Based on its functional group, match the structure on the left with the correct name on the right.

(a) R—C(=O)—OR   Carboxylic acid

(b) R—C(=O)—OH   Alcohol

(c) R—OH   Ester

(d) R—O—R   Ether

85. For each molecule, identify the functional group and determine the family to which the molecule belongs.

(a) H₃C—CH₂—CH₂—NH—CH₃

(b) H₃C—CH₂—CH₂—CH(=O)

(c)
H₃C—C(CH₃)(H₃C)—C(CH₃)(CH₃)—OH

(d) H₃C—C(CH₃)(CH₃)—O—CH₂—CH₃

86. For each molecule, identify the functional group and determine the family to which the molecule belongs.

(a) H₃C—CH₂—CH₂—C(=O)—CH₃

(b) H₃C—CH₂—C(=O)—O—CH₃

(c) H₃C—HC(OH)—CH₃

(d) H₃C—CH₂—C(=O)—OH

ALCOHOLS

87. Name each alcohol.

(a)

$$H_3C-\overset{\overset{\displaystyle OH}{|}}{HC}-CH_2-CH_3$$

(b)

$$H_2\overset{\overset{\displaystyle OH}{|}}{C}-\underset{\underset{\displaystyle H_3C}{|}}{CH}-CH_3$$

(c) $H_3C-CH_2-CH_2-\underset{\underset{\displaystyle H_2C-CH_3}{|}}{CH}-CH_2-CH_2-OH$

(d)

$$H_3C-CH_2-\overset{\overset{\displaystyle OH}{|}}{\underset{\underset{\displaystyle CH_3}{|}}{C}}-CH_2-CH_3$$

88. Name each alcohol.

(a) $CH_3-CH_2-\underset{\underset{\displaystyle OH}{|}}{CH}-CH_2-CH_2-CH_2-CH_2-CH_3$

(b) $CH_3-\underset{\underset{\displaystyle OH}{|}}{CH}-CH_2-\underset{\underset{\displaystyle CH_3}{|}}{CH}-CH_2-CH_3$

(c)

$$CH_3-\overset{\overset{\displaystyle CH_3}{|}}{\underset{\underset{\displaystyle OH}{|}}{C}}-CH_2-CH_2-CH_3$$

(d) $HO-CH_2-\underset{\underset{\displaystyle CH_3}{|}}{CH}-\underset{\underset{\displaystyle CH_3}{|}}{CH}-CH_2-CH_2-CH_2-CH_3$

89. Draw a structure for each alcohol.
(a) 3-pentanol
(b) 2-methyl-1-butanol
(c) 3-ethyl-2-hexanol
(d) ethanol

90. Draw a structure for each alcohol.
(a) 1-hexanol
(b) 3,4-dimethyl-2-heptanol
(c) 3-propyl-3-octanol
(d) 3,3-diethyl-2,2-dimethyl-1-hexanol

ETHERS

91. For each compound, provide a name if the structure is given, or provide a structure if the name is given.
(a) dibutyl ether
(b) $H_3C-CH_2-O-CH_2-CH_2-CH_3$
(c) $H_3C-CH_2-CH_2-O-CH_2-CH_2-CH_3$
(d) methyl pentyl ether

92. For each compound, provide a name if the structure is given, or provide a structure if the name is given.
(a) ethyl butyl ether
(b) $H_3C-CH_2-O-CH_2-CH_3$
(c) $H_3C-CH_2-CH_2-O-CH_2-CH_2-\underset{\underset{\displaystyle H_3C-CH_2}{|}}{CH_2}$

(d) ethyl hexyl ether

93. For each compound, provide a name if the structure is given, or provide a structure if the name is given.
(a) octanal

(b)

$$H_3C-CH_2-CH_2-\overset{\overset{\displaystyle O}{\|}}{CH}$$

(c)

$$H_3C-CH_2-CH_2-\overset{\overset{\displaystyle O}{\|}}{C}-CH_2-CH_2-CH_3$$

94. For each compound, provide a name if the structure is given, or provide a structure if the name is given.
(a)

$$H_3C-CH_2-CH_2-\overset{\overset{\displaystyle O}{\|}}{C}-CH_3$$

(b) 3-pentanone
(c) propanal
(d)

$$H_3C-CH_2-CH_2-CH_2-CH_2-CH_2-\overset{\overset{\displaystyle O}{\|}}{CH}$$

95. For each compound, provide a name if the structure is given, or provide a structure if the name is given.

(a) octanoic acid

(b)

$$H_3C-\overset{\overset{\displaystyle O}{\|}}{C}-O-CH_3$$

(c) ethyl butanoate

(d)

$$H_3C-CH_2-CH_2-CH_2-CH_2-CH_2-\overset{\overset{\displaystyle O}{\|}}{C}-OH$$

96. For each compound, provide a name if the structure is given, or provide a structure if the name is given.

(a) hexanoic acid

(b)

$$H_3C-CH_2-CH_2-\overset{\overset{\displaystyle O}{\|}}{C}-OH$$

(c)

$$H_3C-CH_2-\overset{\overset{\displaystyle O}{\|}}{C}-O-CH_2-CH_2-CH_3$$

(d) butyl propanoate

AMINES

97. For each compound, provide a name if the structure is given, or provide a structure if the name is given.

(a) diethylamine

(b)

$$\begin{array}{c} CH_3 \\ | \\ CH_2 \\ | \\ H_3C-CH_2-N-CH_2-CH_3 \end{array}$$

(c) $H_3C-CH_2-CH_2-NH-CH_2-CH_2-\overset{\displaystyle CH_2}{\underset{\displaystyle CH_3}{|}}$

98. For each compound, provide a name if the structure is given, or provide a structure if the name is given.

(a) tributylamine

(b) $H_2N-\overset{\displaystyle CH_3}{\underset{}{\overset{|}{H}C}}-CH_3$

(c) ethylmethylamine

POLYMERS

99. Polyisobutylene is an addition polymer formed from the monomer shown here. Draw the structure of the polymer.

$$H_2C=\overset{\displaystyle CH_3}{\underset{\displaystyle CH_3}{\overset{|}{C}}}$$

100. Teflon is an addition polymer formed from the monomer shown here. Draw the structure of the polymer.

101. One kind of polyester is a condensation copolymer formed between terephthalic acid and ethylene glycol. Draw the structure of the dimer and circle the ester functional group. *Hint:* Water (circled) is eliminated when the bond between the monomers forms.

Terephthalic acid   Ethylene glycol

102. Lexan, a polycarbonate, is a condensation copolymer formed between carbonic acid and bisphenol A. Draw the structure of the dimer. *Hint:* Water (circled) is eliminated when the bond between the monomers forms.

Carbonic acid   Bisphenol A

CUMULATIVE PROBLEMS

103. Identify each organic compound as either an alkane, alkene, alkyne, aromatic, alcohol, ether, aldehyde, ketone, carboxylic acid, ester, or amine.

(a) H₃C—CH₂—HC—CH₂—OH
             |
             CH₃

(b) H₃C—HC—CH₂—NH—CH₂—CH—CH₃
        | |
        CH₃   CH₃

(c) H₃C—HC—C—CH—CH₃ (with CH₃ CH₃ above and CH₃ CH₃ below)

(d) H₃C—HC—CH₂—C(=O)—OH (with CH₃ below)

(e) H₃C—HC—CH₂—O—CH₃ (with CH₃ below)

(f) H₃C—HC—C=C—CH₃ (with CH₃ above, CH₃ and CH₃ below)

104. Identify each organic compound as either an alkane, alkene, alkyne, aromatic, alcohol, ether, aldehyde, ketone, carboxylic acid, ester, or amine.

(a) H₂C—CH₃ (attached to benzene ring with CH₃)

(b) H₃C—HC—C≡C—CH₃ (with CH₃ below)

(c) H₃C—HC—CH₂—C(=O)—O—CH₃ (with CH₃ below)

(d) H₃C—CH₂—CH₂—C(=O)—CH—CH₃ (with CH₃ below)

(e) H₃C—CH₂—CH₂—CH(=O)

(f) H₃C—CH—CH₂ (with OH above, CH₃ below)

---

105. Name each compound.

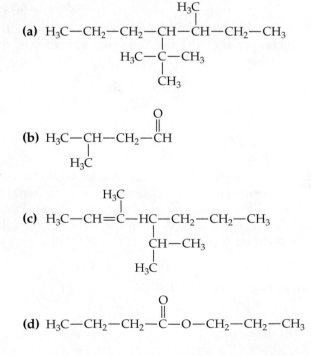

(a) H₃C—CH₂—CH₂—CH—CH—CH₂—CH₃ (with H₃C above, and H₃C—C—CH₃ with CH₃ below)

(b) H₃C—CH—CH₂—CH(=O) (with H₃C below)

(c) H₃C—CH=C—HC—CH₂—CH₂—CH₃ (with H₃C above, CH—CH₃ and H₃C below)

(d) H₃C—CH₂—CH₂—C(=O)—O—CH₂—CH₂—CH₃

106. Name each compound.

(a) H₃C—CH=CH—C—CH—CH₂—CH₃ (with CH₃ CH₃ above, CH₂ and CH₃ below)

(b) benzene ring with F and F

(c) H₃C—CH—CH₂ (with OH above, CH₃ below)

(d) H₃C—CH₂—CH—CH₂—CH—CH₂—CH₂—CH₂ (with CH₃ above, CH—CH₃ and CH₃ below, CH₂ and CH₃ below)

107. For each set of structures, determine whether the two structures are isomers or the same molecule drawn in two different ways.

(a)

$$H_3C-CH-CH-CH_3 \quad\quad H_3C-CH-CH-CH_3$$
$$\;\;\;\;|\quad\;\;| \quad\quad\quad\quad\quad\quad\quad\;\;|$$
$$\;\;CH_3\;\;CH_3 \quad\quad\quad\quad\quad\quad CH_3$$

(b)

$$\quad\quad\quad\quad\quad\quad\quad\quad O\quad\quad\quad\quad\quad\quad\quad\quad O$$
$$\quad\quad\quad\quad\quad\quad\quad\quad\|\quad\quad\quad\quad\quad\quad\quad\quad\|$$
$$H_3C-CH_2-CH_2-CH \quad\quad H_3C-CH-CH$$
$$\quad\quad\quad\quad\quad\quad\quad\quad\quad\quad\quad\quad\quad\quad\quad\quad|$$
$$\quad\quad\quad\quad\quad\quad\quad\quad\quad\quad\quad\quad\quad\quad\quad\;\;CH_3$$

(c)

$$\quad\quad\quad\quad\quad\quad\quad\quad\quad\quad\quad\quad\quad\quad\quad\quad CH_3$$
$$\quad\quad\quad\quad\quad\quad\quad\quad\quad\quad\quad\quad\quad\quad\quad\quad|$$
$$\quad\quad\quad\quad\quad\quad\quad\quad\quad\quad\quad\quad\quad\quad\quad\quad CH_2$$
$$\quad\quad\quad\quad\quad\quad\quad\quad\quad\quad\quad\quad\quad\quad\quad\quad|$$
$$H_3C-CH_2-CH-CH_2-CH_3 \quad\quad H_3C-CH_2-CH$$
$$\quad\quad\quad\quad\quad|\quad\quad\quad\quad\quad\quad\quad\quad\quad\quad\quad|$$
$$\quad\quad\quad\quad CH_3 \quad\quad\quad\quad\quad\quad\quad\quad\quad CH_3$$

108. For each set of structures, determine whether the two structures are isomers or the same molecule drawn in two different ways.

(a)

$$\quad\quad\quad\quad\quad\quad\quad O\quad\quad\quad\quad\quad\quad\quad\quad O$$
$$\quad\quad\quad\quad\quad\quad\quad\|\quad\quad\quad\quad\quad\quad\quad\quad\|$$
$$H_3C-CH_2-C-O-CH_3 \quad\quad H_3C-C-O-CH_2-CH_3$$

(b) $H_3C-C\equiv C-CH_2-CH_3$

$$\quad\quad\quad\quad\quad\quad\quad\quad\quad\quad H_3C-CH_2-C\equiv C-CH_3$$

(c)

$$\quad\quad\quad\quad\quad\quad O\quad\quad\quad\quad\quad\quad\quad O$$
$$\quad\quad\quad\quad\quad\quad\|\quad\quad\quad\quad\quad\quad\quad\|$$
$$\quad\quad H_3C-C-OH \quad\quad HO-C-CH_3$$

109. Complete the equation:

$$CH_2{=}CH_2 + HCl \longrightarrow$$

110. Complete the equation (assume only one addition):

$$CH{\equiv}CH + HI \longrightarrow$$

111. What is the minimum amount of hydrogen gas in grams required to completely hydrogenate 15.5 kg of 2-butene? *Hint:* Begin by writing a balanced equation for the hydrogenation reaction.

112. How many kilograms of CO_2 are produced by the complete combustion of 3.8 kg of *n*-octane? *Hint:* Begin by writing a balanced equation for the combustion of *n*-octane.

113. What minimum volume of oxygen gas (at STP) is required for the complete combustion of 18.9 kg of octane?

114. What minimum volume of hydrogen gas (at STP) is required to fully hydrogenate 15.5 kg of propyne?

HIGHLIGHT PROBLEMS

115. Based on the space-filling model shown, identify the family for each molecule.

Hydrogen   Carbon   Oxygen   Nitrogen

(a)

(b)

(c)

(d)

(e)

(f)

116. In Chapter 13, we learned about the "dirty dozen"—12 chemical compounds that have been targeted by governments around the world to be banned. These compounds are known as persistent organic pollutants (POPs) because once they enter the environment, they remain for long periods of time. Examine the structures of the compounds shown here. What functional groups can you identify within the compounds? (They may have more than one.) What structural features do many of these compounds have in common?

Dioxin—industrial by-product

Furan—industrial by-product

Hexachlorobenzene—fungicide, industrial by-product

DDT—insecticide

▶ANSWERS TO SKILLBUILDER EXERCISES

Skillbuilder 18.1   (a) alkene (b) alkyne (c) alkane

Skillbuilder 18.2

 , $CH_3CH_2CH_2CH_2CH_3$

Skillbuilder 18.3

Skillbuilder 18.4 2-methylpropane	**Skillbuilder 18.8** 1,3-dibromobenzene,
Skillbuilder 18.5 3-ethyl-2-methylpentane	*meta*-dibromobenzene,
Skillbuilder 18.6 2,3,5-trimethylhexane	*m*-dibromobenzene
Skillbuilder 18.7 (a) 4,4-dimethyl-2-pentyne	
(b) 3-ethyl-4,6-dimethyl-1-heptene	

▶ANSWERS TO CONCEPTUAL CHECKPOINTS

18.1 (b) and (d) Isomers have the same formula (in this case C_7H_{16}) but different structures. The compound in (a) has the formula C_6H_{14}, and the compound in (c) has the formula C_5H_{12}. Therefore, (a) and (c) are not isomers of each other, nor are they isomers of (b) and (d).

18.2 (b) The compound shown is organic (carbon-based), unsaturated (it contains a double bond in the carbon chain), and an acid (it includes a —COOH group). However, it does not contain a benzene ring, and so it is not classified as an aromatic compound.

CHAPTER 19

Biochemistry

"Can—and should—life be described in terms of molecules?
For many, such description seems to diminish the beauty of nature.
For others of us, the wonder and beauty of nature are nowhere
more manifest than in the submicroscopic plan of life."

ROBERT A. WEINBERG (1942–)

19.1 The Human Genome Project

◄ The similarities between parents and their children are caused by genes, inheritable blueprints for making organisms. The structure at the bottom of this image is DNA, the molecular basis of genetic information.

In 1990, the U.S. Department of Energy (DOE) and the National Institutes of Health (NIH) embarked on a 15-year project to map the **human genome**, all of the genetic material of a human being. We will define genetic material—and genes—more carefully later. For now, think of genetic material as the *inheritable blueprint* for making organisms. Think of genes as *specific parts* of that blueprint. Each organism has a blueprint unique to itself. The genome of a human, for example, is unique to humans and different from that of other organisms. When organisms reproduce, they pass their genetic material to the next generation.

Within the genetic material of a given species of organism, however, there is variation among individuals. For example, whether you have brown eyes or blue eyes depends on the specific genes for eye color that you inherited from your parents. Many traits—such as physical appearance, intelligence, susceptibility to certain diseases, response to certain drug therapies, and even temperament—are at least partially determined by your specific genes. So understanding the human genome is part of understanding ourselves.

In 2003, the Human Genome Project was completed. The results, which continue to be analyzed, contained some surprises. For example, the mapping of the human genome revealed that humans have only 20,000–25,000 genes. This may seem like a large number, but scientists initially expected more. Before the project, scientists estimated that humans had about 100,000 genes, but that number was clearly too high. The number of genes in humans is not much larger than that in many simpler organisms. The number of genes in a roundworm, for instance, is almost 20,000. Whatever makes humans unique, it is not the number of genes in our genome.

Continuing analysis of the results includes mapping the specific variations between the DNA of different people. Variations called single-nucleotide

Nucleotides are defined in Section 19.7. For now think of them as the units that compose genes.

polymorphisms, or SNPs, are of particular interest. Understanding SNPs can help identify individuals who are susceptible to certain diseases. For example, in the future, a genetic test may reveal that you have the SNPs associated with a certain type of cancer. You can then take preventive steps, or even preventive drug therapy, to avoid actually getting the cancer. Knowledge of SNPs may also allow physicians to tailor drug therapies to match individuals. A genetic test may allow a doctor to give you the drug that is most effective for you.

Analysis of the human genome is also expected to lead to the development of new drugs in two ways. First, an understanding of gene function can lead to smart drug design. Instead of developing drugs by trial and error (the current procedure for many drugs), knowledge of a specific gene will allow scientists to design drugs to carry out a specific function related to that gene. Second, human genes themselves can provide the blueprint for certain types of drugs. For example, interferon, a drug taken by people with multiple sclerosis, is a complex compound normally found in humans. The blueprint for making interferon is in the human genome. Scientists have been able to take this blueprint out of human cells and put it into bacteria, which then synthesize the needed drug. The drug is harvested from bacteria, purified, and given to patients.

Interferon is a protein. Proteins are discussed in Sections 19.5 and 19.6.

The Human Genome Project was possible because of decades of research in **biochemistry**, the study of the chemical substances and processes that occur in plants, animals, and microorganisms. In this chapter, we examine the chemical substances that make life possible and some of the new technology that has resulted from this understanding.

19.2 The Cell and Its Main Chemical Components

The **cell** is the smallest structural unit of living organisms that has the properties traditionally associated with life (▼ Figure 19.1). A cell can be an independent living organism or a building block of a more complex organism. Most cells in higher animals contain a **nucleus**, the control center of the cell and the part of the cell that contains the genetic material. The perimeter of the cell is bound by a **cell membrane** that holds the contents of the cell together. The region between the nucleus and the cell membrane is called the **cytoplasm**. The cytosplasm contains a number of specialized structures that carry out much of the cell's work. The main *chemical* components of the cell can be divided into four classes: *carbohydrates, lipids, proteins,* and *nucleic acids.*

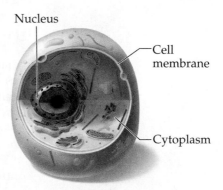

Nucleus

Cell membrane

Cytoplasm

▶ **FIGURE 19.1   A typical animal cell**   The cell is the smallest structural unit of living organisms. The primary genetic material is stored in the nucleus.

19.3 Carbohydrates: Sugar, Starch, and Fiber

As described in Section 18.14, aldehydes have the general structure R—CHO and ketones have the general structure R—CO—R.

Carbohydrates are the primary molecules responsible for short-term energy storage in living organisms. They also form the main structural components of plants. Carbohydrates—as their name, which means carbon and water, implies—often have the general formula $(CH_2O)_n$. Structurally, we identify carbohydrates as aldehydes or ketones containing multiple —OH (hydroxyl) groups. For example, glucose, with the formula $C_6H_{12}O_6$, has the structure:

Glucose

Notice that glucose is an aldehyde (it contains the —CHO group) with —OH groups on most of the carbon atoms. The many —OH groups make glucose soluble in water (and therefore in blood), which is important in glucose's role as the primary fuel of cells. Glucose is easily transported in the bloodstream and is soluble within the aqueous interior of a cell.

MONOSACCHARIDES

Glucose is an example of a **monosaccharide,** a carbohydrate that cannot be broken down into simpler carbohydrates. Monosaccharides such as glucose rearrange in aqueous solution to form ring structures (▼ Figure 19.2).

Glucose (ring form)

▶ FIGURE 19.2 **Rearrangement of glucose from straight-chain to ring form** Question: Can you verify that the straight-chain form and the ring form of glucose are isomers?

Glucose is also an example of a *hexose*, a six-carbon sugar. The general names for monosaccharides have a prefix that depends on the number of carbon atoms, followed by the suffix *-ose*. The most common monosaccharides in living organisms are pentoses and hexoses.

Other monosaccharides in their ring form include fructose and galactose.

Fructose   Galactose

Fructose, also known as fruit sugar, is a hexose found in many fruits and vegetables and is a major component of honey. Galactose, also known as brain sugar, is a hexose usually found combined with other monosaccharides in substances such as lactose (see next section). Galactose is also present within the brain and nervous system of most animals.

DISACCHARIDES

Two monosaccharides can react, eliminating water to form a carbon–oxygen–carbon bond called a **glycosidic linkage** that connects the two rings. The resulting compound is a **disaccharide**, a carbohydrate that can be decomposed into two simpler carbohydrates. For example, glucose and fructose link together to form sucrose, commonly known as table sugar.

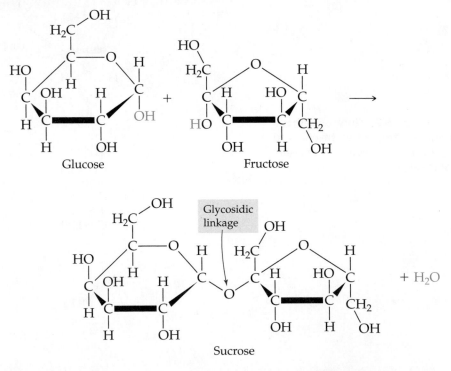

The link between individual monosaccharides is broken during digestion, allowing the individual monosaccharides to pass through the intestinal wall and enter the bloodstream (▶ Figure 19.3).

▲ Fructose is the main sugar in fruit.

▲ Table sugar is composed of sucrose, a disaccharide.

▶ FIGURE 19.3 **Digestion of disaccharides** During digestion, disaccharides are broken down into individual monosaccharide units.

Disaccharides   Monosaccharides

POLYSACCHARIDES

| Polymers were discussed in Section 18.17.

Monosaccharides can link together to form **polysaccharides**, long, chainlike molecules composed of many monosaccharide units. Polysaccharides are a type of *polymer*—chemical compounds (introduced in Section 18.17) composed of repeating structural units in a long chain. Monosaccharides and disaccharides are **simple sugars** or **simple carbohydrates**. Polysaccharides are **complex carbohydrates**. Some common polysacchharides include **starch** and **cellulose**, both of which are composed of repeating glucose units.

Starch

The difference between starch and cellulose is the link between the glucose units. In starch, the oxygen atom joining neighboring glucose units points down (as conventionally drawn) relative to the planes of the rings, a configuration called an *alpha linkage*. In cellulose, the oxygen atoms are roughly parallel with the planes of the rings but pointing slightly up (as conventionally drawn); a configuration called a *beta linkage*. This difference in linkage causes the differences in the properties of starch and cellulose.

Starch is common in potatoes and grains. It is a soft, pliable substance that we can easily chew and swallow. During digestion, the links between individual glucose units are broken, allowing glucose molecules to pass through the intestinal wall and into the bloodstream (▼ Figure 19.4).

▶ FIGURE 19.4 **Digestion of polysaccharides** During digestion, polysaccharides break down into individual monosaccharide units.

Polysaccharides   Monosaccharides

| One form of starch, amylopectin, is also branched, but less so than glycogen.

On the other hand, cellulose—also known as fiber—is a stiffer and more rigid substance. Cellulose is the main structural component of plants. The bonding in cellulose makes it indigestible by humans. When we eat cellulose, it passes right through the intestine, providing bulk to stools and preventing constipation.

A third kind of polysaccharide is **glycogen**. Glycogen has a structure similar to starch, but the chain is highly branched. In animals, excess glucose in the blood is stored as glycogen until it is needed.

EXAMPLE 19.1   Identifying Carbohydrates

Determine which molecules are carbohydrates. Classify each carbohydrate as a monosaccharide, disaccharide, or polysaccharide.

(a)   (b)

(c)   (d)

SOLUTION

You can identify carbohydrates as either an aldehyde or ketone with multiple —OH groups attached or as one or more rings of carbon atoms that include one oxygen atom and also have —OH groups attached to most of the carbon atoms. (a), (b), and (d) are carbohydrates. (a) and (b) are both monosaccharides, and (d) is a disaccharide. (c) is not a carbohydrate because it has only a carboxylic acid group, which is not a characteristic of carbohydrates.

▶ **SKILLBUILDER 19.1 | Identifying Carbohydrates**

Determine which molecules are carbohydrates and classify each carbohydrate as a monosaccharide, disaccharide, or polysaccharide.

(a)   (b)

(c)   (d)

▶ **FOR MORE PRACTICE**   Problems 45, 46, 47, 48.

19.4 Lipids

Lipids are chemical components of the cell that are insoluble in water but soluble in nonpolar solvents. Lipids include fatty acids, fats, oils, phospholipids, glycolipids, and steroids. Insolubility in water makes lipids an ideal structural component of cell membranes. Lipids make up the containers that separate the interior of the cell from its external environment. Lipids are also used for long-term energy storage and for insulation. We all store extra calories from food as lipids, some of us more than others.

FATTY ACIDS

Carboxylic acids were first defined in Section 18.15.

One class of lipids are the **fatty acids**, carboxylic acids with long hydrocarbon tails. The general structure for a fatty acid is:

Fatty acid general structure

where R represents a hydrocarbon chain containing 3 to 19 carbon atoms. Fatty acids differ only in their R group. A common fatty acid is myristic acid, where the R group is $CH_3(CH_2)_{12}-$.

Myristic acid

Myristic acid occurs in butterfat and in coconut oil. Myristic acid is an example of a *saturated* fatty acid—its carbon chain has no double bonds. Other fatty acids—called *monounsaturated* or *polyunsaturated* fatty acids—have one or more double bonds, respectively, in their carbon chains. For example, oleic acid—found in olive oil, peanut oil, and human fat—is an example of a monounsaturated fatty acid.

Oleic acid

The long hydrocarbon tails of fatty acids make them insoluble in water. Table 19.1 lists several different fatty acids and some common sources for each.

TABLE 19.1 Fatty Acids

Saturated Fatty Acids			
Name	Number of Carbon Atoms	Structure	Sources
butyric acid	4	$CH_3CH_2CH_2COOH$	milk fat
capric acid	10	$CH_3(CH_2)_8COOH$	milk fat, whale oil
myristic acid	14	$CH_3(CH_2)_{12}COOH$	butterfat, coconut oil
palmitic acid	16	$CH_3(CH_2)_{14}COOH$	beef fat, butterfat
stearic acid	18	$CH_3(CH_2)_{16}COOH$	beef fat, butterfat

Unsaturated Fatty Acids				
Name	Number of Carbon Atoms	Number of Double Bonds	Structure	Sources
oleic acid	18	1	$CH_3(CH_2)_7CH{=}CH(CH_2)_7COOH$	olive oil, peanut oil
linoleic acid	18	2	$CH_3(CH_2)_4(CH{=}CHCH_2)_2(CH_2)_6COOH$	linseed oil, corn oil
linolenic acid	18	3	$CH_3CH_2(CH{=}CHCH_2)_3(CH_2)_6COOH$	linseed oil, corn oil

FATS AND OILS

Fats and oils are **triglycerides**, triesters composed of glycerol linked to three fatty acids, as shown in the following block diagram.

Esters, first defined in Section 18.15, have the general structure R—COO—R. Glycerol has the following structure.

$$H_2C{-}OH$$
$$HO{-}CH$$
$$H_2C{-}OH$$

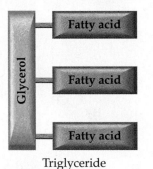

Triglyceride

Triglycerides form by the reaction of glycerol with three fatty acids.

Glycerol 3 fatty acids Triglyceride

The bonds that join the glycerol to the fatty acids are called **ester linkages**. Tristearin—the main component of beef fat—is formed from the reaction of glycerol and three stearic acid molecules.

If the fatty acids in a triglyceride are saturated, the triglyceride is called a **saturated fat** and tends to be solid at room temperature. Lard and many animal fats are examples of saturated fat. On the other hand, if the fatty acids in a triglyceride are unsaturated, the triglyceride is called an **unsaturated fat**, or an *oil*, and tends to be liquid at room temperature. Canola oil, olive oil, and most other vegetable oils are examples of unsaturated fats.

EXAMPLE 19.2   Identifying Triglycerides

Identify the triglycerides and classify each as saturated or unsaturated.

SOLUTION

Triglycerides are readily identified by the three-carbon backbone with long fatty acid tails. Both (b) and (c) are triglycerides. (b) is a saturated fat because it does not have any double bonds in its carbon chains. (c) is an unsaturated fat because it contains double bonds in its carbon chains.

▶**SKILLBUILDER 19.2 | Identifying Triglycerides**

Identify the triglycerides and classify each as a saturated or unsaturated fat.

▶**FOR MORE PRACTICE**   Problems 53, 54.

OTHER LIPIDS

Other lipids found in cells include phospholipids, glycolipids, and steroids. **Phospholipids** have the same basic structure as triglycerides, except that one of the fatty acid groups is replaced with a phosphate group.

Phospholipid

Unlike a fatty acid, which is nonpolar, a phosphate group is polar and often has another polar group attached to it. The phospholipid molecule therefore has a polar section and a nonpolar section. For example, consider the structure of a phosphatidylcholine, a phospholipid found in the cell membranes of higher animals.

Polar head   Nonpolar tails

▲ **FIGURE 19.5   Schematic representation of phospholipids and glycolipids**   The green circle represents the polar part of the molecule, and the tails represent the nonpolar hydrocarbon chains. **Question:** If this molecule were placed in water, how do you think it might orient itself at the surface?

The polar part of the molecule is *hydrophilic* (has a strong affinity for water), while the nonpolar part is *hydrophobic* (avoids water). **Glycolipids** have similar structures and properties. The nonpolar section of a glycolipid is composed of a fatty acid chain and a hydrocarbon chain. The polar section is a sugar molecule such as glucose. Phospholipids and glycolipids are often schematically represented as a circle with two long tails (◀ Figure 19.5). The circle represents the polar hydrophilic part of the molecule, and the tails represent the nonpolar hydrophobic parts. The structure of phospholipids and glycolipids is ideal for constructing cell membranes; the polar parts interact with the aqueous environments of the cell and the nonpolar parts interact with each other. In cell membranes, these lipids form a structure called a **lipid bilayer** (◀ Figure 19.6). Lipid bilayer membranes encapsulate cells and many cellular structures.

Steroids are lipids that contain the following four-ring structure.

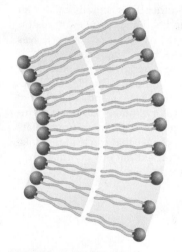

▲ **FIGURE 19.6   Lipid bilayer membrane**   Cell membranes are composed of lipid bilayers, in which phospholipids or glycolipids form a double layer. In this bilayer, the polar heads of the molecules point outward and the nonpolar tails point inward.

Some common steroids include cholesterol, testosterone, and estrogen.

Cholesterol

Testosterone   Estrogen

Although cholesterol has a bad reputation, it serves many important functions in the body. Like phospholipids and glycolipids, cholesterol is part of cell membranes. Cholesterol also serves as a starting material (or precursor) for the body to synthesize other steroids such as testosterone, a principal male hormone, and estrogen, a principal female hormone. Hormones are chemical messengers that regulate many body processes, such as growth and metabolism. They are secreted by specialized tissues and transported in the blood.

CHEMISTRY AND HEALTH

Dietary Fats

Most of the fats and oils in our diet are triglycerides. During digestion, triglycerides are broken down into fatty acids, glycerol, monoglycerides, and diglycerides. These products pass through the intestinal wall and then reassemble into triglycerides before they are absorbed into the blood. This process, however, is slower than the digestion of other food types, and therefore eating fats and oils gives a lasting feeling of fullness.

Triglyceride → Digestion → Glycerol, Diglyceride, Monoglyceride, Fatty acids

▲ The Food and Drug Administration (FDA) recommends that fats and oils provide less than 30% of total caloric intake.

The effect of fats and oils on health has been widely debated. Some diets call for drastic reduction of daily intake of fats and oils; whereas other diets actually call for an increase in fats and oils. The Food and Drug Administration (FDA) recommends that fats and oils compose less than 30% of total caloric intake. However, because fats and oils have a higher caloric content per gram than other food types, it is easy to eat too much of them. The FDA also recommends that of those fats that are consumed, no more than one-third (10% of total caloric intake) should be saturated fats. This is because a diet high in saturated fats increases the risk of artery blockages that can lead to stroke and heart attack. Monounsaturated fats, by contrast, may help protect against these threats.

19.5 Proteins

When most people think of **proteins**, they think of protein sources in their diet such as beef, eggs, poultry, and beans. From a biochemical perspective, however, proteins have a much broader definition. Within living organisms, proteins do much of the work of maintaining life. For example, most of the chemical reactions

Tristearin

CAN YOU ANSWER THIS? *Saturated fats tend to be solid at room temperature, while unsaturated fats tend to be liquid. One reason saturated fats taste good is that they tend to melt in your mouth. Since unsaturated fats are liquid at room temperature, they don't have the same effect. Examine the structures of tristearin, a saturated triglyceride, and trilinolenin, an unsaturated tryglyceride.*

From the structures, determine a reason for why tristearin has a greater tendency to be a solid at room temperature, while trilinolenin has a greater tendency to be a liquid. Hint: Think of the interactions between molecules. Which molecules do you think can interact better with neighboring molecules?

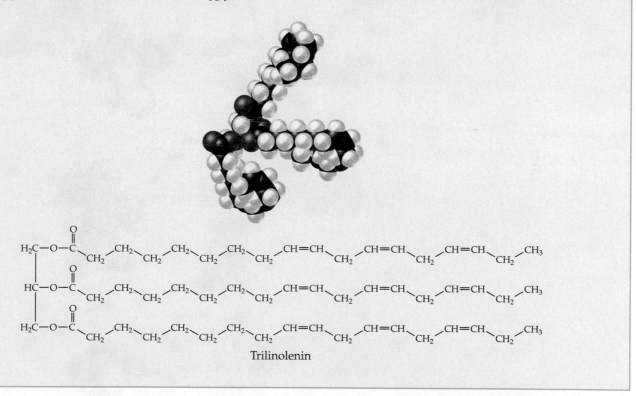

Trilinolenin

See Section 15.12 for a description of catalysts and enzymes.

that occur in living organisms are catalyzed or enabled by proteins. Proteins that act as catalysts are called *enzymes*. Without enzymes, life would be impossible. But acting as enzymes is only one of the many functions of proteins. Proteins are the structural components of muscle, skin, and cartilage. They also transport oxygen in the blood, act as antibodies to fight disease, and function as hormones to regulate metabolic processes. Proteins reign supreme as the working molecules of life.

What are proteins? Proteins are polymers of amino acids. **Amino acids** are molecules containing an amine group, a carboxylic acid group, and an **R group** (also called a *side chain*). The general structure of an amino acid is:

Amine group

Carboxylic acid group

R group

Amino acid general stucture

In a protein, an R group does not necessarily mean a pure alkyl group. See Table 19.2 for common R groups.

Amino acids differ from each other only in their R groups. A simple amino acid is alanine, in which the R group is a methyl ($-CH_3$) group.

R group

Alanine

Other amino acids include serine, $R = -CH_2OH$; aspartic acid, $R = -CH_2COOH$; and lysine, $R = -CH_2(CH_2)_3NH_2$.

Serine

Aspartic acid

Lysine

Notice that the R groups, or side chains, of different amino acids can be very different chemically. Alanine, for example, has a nonpolar side chain while serine has a polar one. Aspartic acid has an acidic side chain while lysine, since it contains nitrogen, has a basic one. When amino acids are strung together to make a protein, these differences determine the structure and properties of the protein. Table 19.2 shows the most common amino acids in proteins.

Amino acids link together because the amine end of one amino acid reacts with the carboxylic acid end of another amino acid.

The resulting bond is a **peptide bond**, and the resulting molecule—two amino acids linked together—is called a **dipeptide**. A dipeptide can link to a third amino acid to form a tripeptide, and so on. Short chains of amino acids are generally called **polypeptides**. Functional proteins usually contain hundreds or even thousands of amino acids joined by peptide bonds.

EXAMPLE 19.3 Peptide Bonds

Show the reaction by which glycine and alanine form a peptide bond.

SOLUTION

Peptide bonds form when the carboxylic end of one amino acid reacts with the amine end of a second amino acid to form a dipeptide and water.

This reaction can also take place between the —NH₂ end of glycine and the —COOH end of alanine, producing a slightly different dipeptide.

▶SKILLBUILDER 19.3 | Peptide Bonds

Show the reaction by which valine and leucine form a peptide bond.

▶FOR MORE PRACTICE   Problems 61, 62, 63, 64.

TABLE 19.2   Common Amino Acids

Glycine (Gly)

Alanine (Ala)

Valine (Val)

Leucine (Leu)

Isoleucine (Ile)

Proline (Pro)

Methionine (Met)

Cysteine (Cys)

Serine (Ser)

Threonine (Thr)

Aspartic acid (Asp)

Glutamic acid (Glu)

Asparagine (Asn)

Glutamine (Glu)

Lysine (Lys)

Arginine (Arg)

Histidine (His)

Phenylalanine (Phe)

Tyrosine (Tyr)

Tryptophan (Trp)

19.6 Protein Structure

When they link together to form proteins, the amino acids interact with one another, causing the protein chain to twist and fold in a very specific way. The exact shape that a protein takes depends on the types of amino acids and their sequence in the protein chain. Different amino acids and different sequences result in different shapes, and these shapes are extremely important.

For example, insulin is a protein that promotes the absorption of glucose out of the blood and into muscle cells where glucose is needed for energy. Insulin recognizes muscle cells because their surfaces contain insulin *receptors*, molecules that fit a specific portion of the insulin protein. If insulin were a different shape, it would not latch onto insulin receptors on muscle cells and therefore would not do its job. So the shape, or *conformation*, of proteins is crucial to their function. We can understand protein structure by exploring it on four levels: primary structure, secondary structure, tertiary structure, and quaternary structure (▼ Figure 19.7).

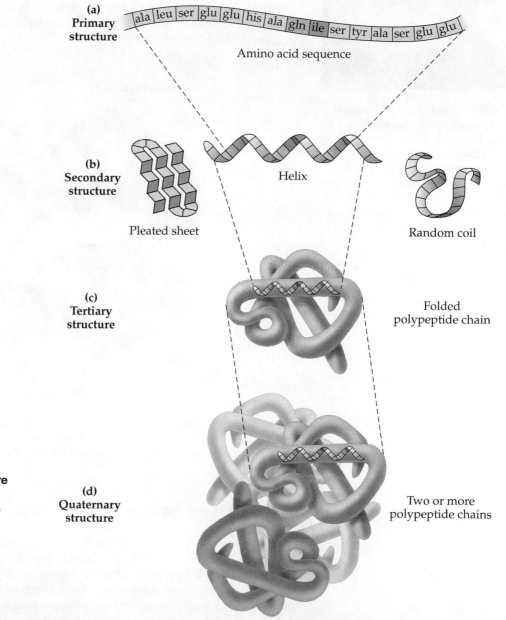

(a) Primary structure
ala leu ser glu glu his ala gln ile ser tyr ala ser glu glu
Amino acid sequence

(b) Secondary structure
Pleated sheet
Helix
Random coil

(c) Tertiary structure
Folded polypeptide chain

(d) Quaternary structure
Two or more polypeptide chains

▶ **FIGURE 19.7   Protein structure**
(a) Primary structure is the amino acid sequence. **(b)** Secondary structure refers to small-scale repeating patterns such as the helix or the pleated sheet. **(c)** Tertiary structure refers to the large-scale bends and folds of the protein. **(d)** Quaternary structure is the arrangement of individual polypeptide chains.

PRIMARY STRUCTURE

The **primary protein structure** is the sequence of amino acids in its chain. Primary structure is maintained by the covalent peptide bonds between individual amino acids. For example, one section of the insulin protein has the sequence:

Gly-Ile-Val-Glu-Gln-Cys-Cys-Ala-Ser-Val-Cys

Each three-letter abbreviation represents an amino acid (see Table 19.2). The first amino acid sequences for proteins were determined in the 1950s. Today, the amino acid sequences for thousands of proteins are known.

Changes in the amino acid sequence of a protein, even minor ones, can have devastating effects on the function of a protein. Hemoglobin, for example, is a protein that transports oxygen in the blood. It is composed of four protein chains, each containing 146 amino acid units (▼ Figure 19.8) for a total of 584 amino acid units. The substitution of glutamic acid for valine in just one position on two of these chains results in the disease known as sickle-cell anemia, in which red blood cells take on a sickle shape that ultimately leads to damage of major organs. In the past, sickle-cell anemia has been fatal, often before age 30—all because of a change in 2 amino acids out of 584.

▲ A sickled red blood cell (center) surrounded by normal red blood cells. The sickled cells, characteristic of sickle-cell anemia, are fragile and easily damaged. They are also more rigid and so tend to become stuck in tiny capillaries, interfering with the flow of blood to tissues and organs.

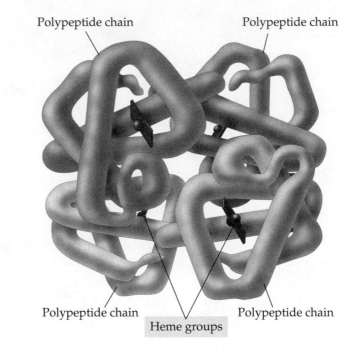

Polypeptide chain   Polypeptide chain

Polypeptide chain   Polypeptide chain

Heme groups

▲ **FIGURE 19.8   Hemoglobin**   Hemoglobin is a protein composed of four chains, each containing 146 amino acid units. Each chain holds a molecule called a *heme*, which contains an iron atom in its center. Oxygen binds at the iron atom.

SECONDARY STRUCTURE

The **secondary protein structure** refers to certain short-range periodic or repeating patterns often found along protein chains. Secondary structure is maintained by interactions between amino acids that are fairly close together in the linear sequence of the protein chain or adjacent to each other on neighboring chains. The most common of these patterns is called the **alpha (α)-helix**, shown in ▶ Figure 19.9. In the α-helix structure, the amino acid chain is wrapped into a tight coil in which the side chains extend outward from the coil. The structure is

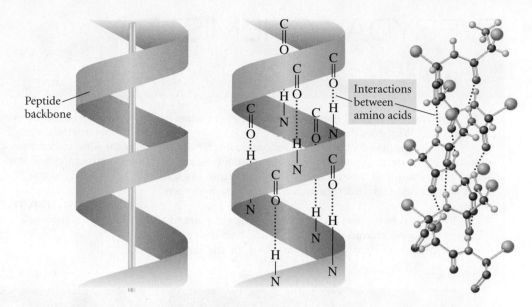

▲ **FIGURE 19.9** **Alpha-helix protein structure** The α-helix is maintained by interactions between the peptide backbones of amino acids that are close to each other in the linear sequence of the protein chain.

maintained by hydrogen-bonding interactions between NH and CO groups along the peptide backbone of the protein. Some proteins—such as keratin, which composes hair—have the α-helix pattern throughout their entire chain. Other proteins have very little or no α-helix pattern in their chain. It depends on the particular protein.

A second common pattern in the secondary structure of proteins is called the **beta (β)-pleated sheet** (▼ Figure 19.10). In this structure, the chain is extended (as opposed to coiled) and forms a zigzag pattern like an accordion pleat. The peptide backbones of the chains interact with one another through hydrogen bonding to maintain the pleated sheet conformation. Some proteins—such as silk—have the β-pleated sheet structure throughout their entire chain. Since its protein chains in the β-pleated sheet are fully extended, silk is inelastic. Many proteins, however, have some sections that are β-pleated sheet, other sections that are α-helix, and still other sections that have less regular patterns called **random coils**.

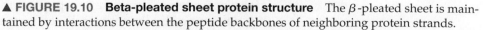

▲ **FIGURE 19.10** **Beta-pleated sheet protein structure** The β-pleated sheet is maintained by interactions between the peptide backbones of neighboring protein strands.

EVERYDAY CHEMISTRY

Why Hair Gets Longer When It Is Wet

Have you ever noticed that your hair gets longer when it is wet? Why does this happen?

Hair is composed of a protein called keratin. The secondary structure of keratin is α-helix throughout, meaning that the protein has a wound-up helical structure. As we learned earlier, this structure is maintained by hydrogen bonding.

Individual hair fibers are composed of several strands of keratin coiled around each other. When hair is dry, the keratin protein is tightly coiled, resulting in the normal length of dry hair. However, when hair becomes wet, water molecules interfere with the hydrogen bonding that maintains the α-helix structure. The result is the relaxation of the α-helix structure and the lengthening of the hair fiber. Completely wet hair is 10 to 12% longer than dry hair.

CAN YOU ANSWER THIS? *When curlers are put into wet hair and the wet hair is allowed to dry, the hair tends to retain the shape of the curler. Can you explain why this happens?*

▲ Wet hair is 10 to 12% longer than dry hair.

TERTIARY STRUCTURE

The **tertiary protein structure** consists of the large-scale bends and folds due to interactions between the R groups of amino acids that are separated by large distances in the linear sequence of the protein chain. These interactions, as shown in ▶ Figure 19.11, include:

- hydrogen bonds
- disulfide linkages (covalent bonds between sulfur atoms on different R groups)
- hydrophobic interactions (attractions between large nonpolar groups)
- salt bridges (acid–base interactions between acidic and basic groups)

Proteins with structural functions—such as keratin, which composes hair, or collagen, which composes tendons and much of the skin—tend to have tertiary structures in which coiled amino acid chains align roughly parallel to one another, forming long, water-insoluble fibers. These kinds of proteins are called **fibrous proteins**. Proteins with nonstructural functions—such as hemoglobin, which carries oxygen, or lysozyme, which fights infections—tend to have tertiary structures in which amino acid chains fold in on themselves, forming water-soluble globules that can travel through the bloodstream. These kinds of proteins are called **globular proteins**. The overall shape of a protein may seem random, but it is not. It is determined by the amino acid sequence and, as we have seen, is critical to its function.

▶ **FIGURE 19.11   Interactions that create tertiary and quaternary structure**   The tertiary structure of a protein is maintained by interactions between the R groups of amino acids that are separated by large distances in the linear sequence of the protein chain. These interactions include hydrogen bonds, disulfide linkages, hydrophobic interactions, and salt bridges. (The figure shows a typical example of each kind of interaction.) The same interactions can also hold different amino acid chains together (quaternary structure).

QUATERNARY STRUCTURE

Many proteins are composed of more than one amino acid chain. As we have seen, for example, hemoglobin is composed of four amino acid chains—each chain is called a *subunit*. The **quaternary protein structure** describes how these subunits fit together. The same kinds of interactions between amino acids maintain quaternary structure and tertiary structure.

To summarize protein structure:

- Primary structure is simply the amino acid sequence. It is maintained by the peptide bonds that hold amino acids together.
- Secondary structure refers to the small-scale repeating patterns often found in proteins. These are maintained by interactions between the peptide backbones of amino acids that are close together in the chain sequence or adjacent to each other on neighboring strands.
- Tertiary structure refers to the large-scale twists and folds within the protein. These are maintained by interactions between the R groups of amino acids that are separated by long distances in the chain sequence.
- Quaternary structure refers to the arrangement of chains (or subunits) in proteins. It is maintained by interactions between amino acids on the individual chains.

✔ CONCEPTUAL CHECKPOINT 19.1

A section of a protein has the amino acid sequence:

ser-gly-glu-phe-ser-ala-leu

This sequence is an example of which level of structure?

(a) primary structure

(b) secondary structure

(c) tertiary structure

(d) quaternary structure

19.7 Nucleic Acids: Molecular Blueprints

We have seen the importance of amino acid sequence in determining protein structure and function. If the amino acid sequence in a protein is incorrect, the protein is unlikely to function properly. How do our bodies constantly synthesize the many thousands of different proteins—each with the correct amino acid sequence—that we need to survive? What ensures that proteins have the correct amino acid sequence? The answer to this question lies in *nucleic acids*. **Nucleic acids** contain a chemical code that specifies the correct amino acid sequences for proteins. Nucleic acids can be divided into two types: deoxyribonucleic acid or **DNA**, which exists primarily in the nucleus of the cell; and ribonucleic acid or **RNA**, which is found throughout the entire interior of the cell.

Like proteins, nucleic acids are polymers. The individual units composing nucleic acids are **nucleotides**. Each nucleotide has three parts: a phosphate, a sugar, and a base (▼ Figure 19.12). In DNA, the sugar is deoxyribose, while in RNA the sugar is ribose.

▶ FIGURE 19.12 **Components of DNA** DNA is a polymer of nucleotides. Each nucleotide has three parts: a sugar group, a phosphate group, and a base. Nucleotides are joined by phosphate linkages.

Nucleotides link together via phosphate linkages to form nucleic acids. Every nucleotide in DNA has the same phosphate and sugar, but can have one of four different bases. In DNA, the four bases are adenine (A), cytosine (C), guanine (G), and thymine (T).

In RNA, the base uracil (U) replaces thymine.

Uracil

The order of bases in a nucleic acid chain specifies the order of amino acids in a protein. However, as there are only four bases and since about 20 different amino acids must be specified, a single base cannot code for a single amino acid. It takes a sequence of three bases—called a **codon**—to code for one amino acid (▼ Figure 19.13). The genetic code—the code that specifies the amino acid that is coded for by a specific codon—was discovered in 1961. It is nearly universal— the same codons specify the same amino acids in nearly all organisms. For example, in DNA the sequence AGT codes for the amino acid serine and the sequence TGA codes for the amino acid threonine. In a rat, a bacterium, or a human—the code is the same.

▲ **FIGURE 19.13   Codons**   A sequence of three nucleotides with their associated bases is called a *codon*. Each codon codes for one amino acid.

A **gene** is a sequence of codons within a DNA molecule that codes for a single protein. Because *proteins* vary in size from 50 to thousands of amino acids, *genes* vary in length from 50 to thousands of codons. For example, egg-white lysozyme is a protein composed of 129 amino acids. The lysozyme gene, then, must contain 129 codons—one for each amino acid in the lysozyme protein. Each codon is like a three-letter word that specifies one amino acid. String the correct number of codons together in the correct sequence, and you have a gene, the instructions for the amino acid sequence in a protein. Genes are contained in structures called **chromosomes**—46 in humans—within the nuclei of cells (▶ Figure 19.14).

Chromosome—a macromolecule of DNA.

Gene—portion of DNA that codes for a single protein.

Codon—sequence of three nucleotides and their associated bases. A codon codes for one amino acid.

T A G

Nucleotide—individual links in the nucleic acid chain. Nucleotides are composed of a sugar group, a phosphate group, and a base.

▶ FIGURE 19.14 **Organization of the genetic material**

✔ **CONCEPTUAL CHECKPOINT 19.2**

The number of DNA bases needed to code for the amino acids in a protein containing 51 amino acids is:

(a) 17

(b) 20

(c) 51

(d) 153

19.8 DNA Structure, DNA Replication, and Protein Synthesis

DNA STRUCTURE

Most of the cells in the human body contain a complete set of instructions (within the DNA in the nucleus) to make—not just the proteins needed by that particular cell—but all of the proteins needed by the human body. However, any one cell does not synthesize every protein specified by the genes of its DNA. Cells synthesize only those proteins that are important to their function. For example, pancreatic cells synthesize insulin and therefore use the insulin gene within their nucleus for the instructions. Pancreatic cells do not, however, synthesize keratin (hair protein), even though the keratin gene is also contained within their nucleus. The cells in the scalp, on the other hand (which also have both insulin and keratin genes in their nuclei), synthesize keratin but not insulin. Cells synthesize only the proteins that are specific to their function.

How do most of the cells in a human get a complete copy of DNA? The answer lies in DNA replication. Cells reproduce by dividing—a parent cell divides into two daughter cells. As it divides, it makes complete copies of its DNA for each daughter cell. The ability of DNA to copy itself is related to its structure. DNA is stored in the nucleus as a double-stranded helix (▼ Figure 19.15). The bases on each DNA strand are directed toward the interior of the helix, where they hydrogen-bond to bases on the other strand. However, the hydrogen bonding between bases is not random. Each base is **complementary**—capable of precise pairing—with only one other base. Adenine (A) hydrogen-bonds only

▲ **FIGURE 19.15** **Structure of the DNA molecule**   DNA has a double-stranded helical structure. Each strand is complementary to the other.

Thymine   Adenine

(a)

Cytosine   Guanine

(b)

▲ **FIGURE 19.16 Complementarity** The complementary nature of DNA is related to the unique way in which the bases interact through hydrogen bonding. Adenine hydrogen-bonds with thymine, and cytosine hydrogen-bonds with guanine.

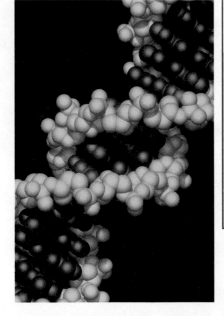

▲ A computer-generated model of the DNA double-helix structure. The yellow atoms are the sugar-phosphate chains, while the blue atoms make up the paired complementary bases.

with thymine (T), and cytosine (C) hydrogen-bonds only with guanine (G) (◄ Figure 19.16). For example, consider a section of DNA containing the following bases.

The complementary strand would then have the following sequence.

The two complementary strands are tightly wrapped into a helical coil, the famous DNA double-helix structure.

EXAMPLE 19.4 Complementary DNA Strand

What is the sequence of the complementary strand for the following DNA strand?

SOLUTION

Draw the complementary strand, remembering that A pairs with T and C pairs with G.

►**SKILLBUILDER 19.4 | Complementary DNA Strand**

What is the sequence of the complementary strand for the following DNA strand?

►**FOR MORE PRACTICE** Problems 71, 72.

DNA REPLICATION

When a cell is about to divide, the DNA within its nucleus unwinds and the hydrogen bonds joining the complementary bases break, forming two single parent strands (► Figure 19.17). With the help of enzymes, a daughter strand complementary to each parent strand—with the correct complementary bases in the correct order—is formed. The hydrogen bonds between the strands then re-form, resulting in two complete copies of the original DNA, one for each daughter cell.

DNA to be replicated

Strands separate

A complement to
each parent strand
is formed.

Two complete copies
of original DNA

▲ **FIGURE 19.17**   **DNA replication**

PROTEIN SYNTHESIS

Humans and animals must synthesize the proteins they need to survive from the
dietary proteins that they eat (the proteins that they eat are not the ones they
need). Dietary protein is split into its constituent amino acids during digestion.
These amino acids are then reconstructed into the correct proteins—those needed
by the particular organism—in the organism's cells. Nucleic acids direct the
process. When a cell needs to make a particular protein, the gene—the section of
the DNA that codes for that specific protein—unravels. The segment of DNA corresponding to the gene acts as a template for the synthesis of a complementary
copy of that gene in the form of another kind of nucleic acid, **messenger RNA** (or
mRNA). The mRNA moves out of the cell's nucleus to a cell structure within the

▲ FIGURE 19.18   Protein synthesis   The mRNA strand that codes for a protein moves through the ribosome. At each codon, the correct amino acid is brought into place and bonds with the previous amino acid.

cytoplasm called a *ribosome*. At the ribosome, protein synthesis occurs. The mRNA chain that codes for the protein moves through the ribosome. As the ribosome "reads" each codon, the corresponding amino acid is brought into place and a peptide bond forms with the previous amino acid (▲ Figure 19.18). As the mRNA moves through the ribosome, the protein (or *polypeptide*) is formed.

To summarize:

- DNA contains the code for the sequence of amino acids in proteins.
- A codon—three nucleotides with their bases—codes for one amino acid.
- DNA strands are composed of four bases, each of which is complementary—capable of precise pairing—with only one other base.
- A gene—a sequence of codons—codes for one protein.
- Chromosomes are molecules of DNA found in the nuclei of cells. Humans have 46 chromosomes.
- When a cell divides, each daughter cell receives a complete copy of the DNA—all 46 chromosomes in humans—within the parent cell's nucleus.
- When a cell synthesizes a protein, the base sequence of the gene that codes for that protein is transferred to mRNA. The mRNA then moves out to a ribosome, where the amino acids are linked in the correct sequence to synthesize the protein. The general sequence is

$$\text{DNA} \longrightarrow \text{RNA} \longrightarrow \text{Protein}$$

✔ CONCEPTUAL CHECKPOINT 19.3

Which biological molecules are *not* polymers?

(a) proteins

(b) steroids

(c) nucleic acids

(d) polysaccharides

CHEMISTRY AND HEALTH

Drugs for Diabetes

Diabetes is a disease in which a person's body does not make enough insulin, the substance that promotes the absorption of sugar from the blood. Consequently, diabetics have high blood sugar levels, which can—over time—lead to a number of complications, including kidney failure, heart attacks, strokes, blindness, and nerve damage. One treatment for diabetes is the injection of insulin, which can help manage blood sugar levels and reduce the risk of these complications. Insulin is a human protein and cannot be easily synthesized in the laboratory. So where do diabetics get life-saving insulin? For many years, the primary source was animals, particularly pigs and cattle. Although animal insulin worked to lower blood sugar levels, some patients could not tolerate it.

Today, diabetics inject *human* insulin. Where does it come from? Its source is one of the success stories of

biotechnology. Scientists were able to remove the gene for insulin from a sample of healthy human cells. They inserted that gene into bacteria, which incorporated the gene into their genome. When the bacteria reproduced, they passed on exact copies of the gene to their offspring. The result was a colony of bacteria that all contained the human insulin gene. Even more amazing, the chemical machinery within the bacteria *expressed* the gene—meaning the bacteria synthesized the human insulin that the gene codes for. Today insulin made in this way is harvested from the cell cultures and bottled for distribution to diabetics. Millions of diabetics manage their disease with human insulin made in this way.

CAN YOU ANSWER THIS?   *Can all drugs be made in this way? What kinds of drugs can be made with these techniques?*

CHAPTER IN REVIEW

CHEMICAL PRINCIPLES

The Cell: The cell is the smallest structural unit of a living organism that has the properties normally associated with life. The main chemical components of the cell can be divided into four categories:

• Carbohydrates

• Lipids

• Proteins

• Nucleic acids

Carbohydrates: Carbohydrates are aldehydes or ketones containing multiple —OH groups. Monosaccharides, such as glucose and fructose, are carbohydrates that cannot be broken down into simpler carbohydrates. Disaccharides, such as sucrose and lactose, are two monosaccharides linked together by glycoside linkages. Polysaccharides, such as starch and cellulose, consist of many monosaccharides linked together. Polysaccharides are also called complex carbohydrates.

Lipids: Lipids are chemical components of the cell that are insoluble in water but soluble in nonpolar solvents. Important lipids include fatty acids, triglycerides, phospholipids, glycolipids, and steroids.

RELEVANCE

The Cell: Some living organisms, such as bacteria, are composed of a single cell. In more complex organisms, cells are the building blocks that compose organs, which together compose the organism.

Carbohydrates: Living organisms use carbohydrates for short-term energy storage. Complex carbohydrates also form the main structural components of plants. Carbohydrates are a major part of our diet. Table sugar (or sucrose), for example, is a disaccharide. Starch and fiber (also called cellulose) are polysaccharides.

Lipids: Lipids compose many of the structural components of cells. Lipids are also used for long-term energy storage and for insulation. In the diet, saturated fats—triglycerides containing no double bonds in their carbon chains—are more likely to increase the risk of stroke and heart attack than unsaturated fats.

Proteins: Proteins are polymers of amino acids. Amino acids are molecules composed of an amine group on one end and a carboxylic acid on the other. Between these two groups is a central carbon atom that has an R group (also called a *side chain*) attached. Amino acids link together by means of peptide bonds, formed by the reaction between the amine end of one amino acid and the carboxylic acid of another. Functional proteins are composed of hundreds or thousands of amino acids.

Proteins: Proteins are the working molecules of life. As biological catalysts (called *enzymes*), they enable thousands of chemical reactions. Proteins also compose the structural elements of muscle, skin, and cartilage; transport oxygen in the blood; act as antibodies to fight disease; and function as hormones to regulate metabolic processes.

Protein Structure: Primary protein structure is the linear amino acid sequence in the protein chain. It is maintained by the peptide bonds that hold amino acids together.

Secondary structure refers to the small-scale repeating patterns often found in proteins. These are maintained by interactions between the peptide backbones of amino acids that are close together in the chain sequence or on neighboring strands.

Tertiary structure refers to the large-scale twists and folds within the protein. These are maintained by interactions between R groups of amino acids that are separated by long distances in the chain sequence.

Quaternary structure refers to the arrangement of chains in proteins. Quaternary structure is maintained by interactions between amino acids on the individual chains.

Protein Structure: The structure of proteins is critical to their function. The shapes of proteins largely determine how they interact with other molecular structures to do their job. That structure depends on the sequence of amino acids within the protein chain and how those amino acids interact with one another.

Nucleic Acids, DNA Replication, and Protein Synthesis: Nucleic acids, including DNA and RNA, are polymers of nucleotides. In DNA, each nucleotide contains one of four bases: adenine (A), cytosine (C), thymine (T), and guanine (G). The order of these bases contains a code that specifies the amino acid sequence in proteins. A codon, a sequence of three bases, codes for an amino acid. A gene, a sequence of hundreds to thousands of codons, codes for a protein. Genes are contained in cellular structures called chromosomes.

Complete copies of DNA are transferred from parent cells to daughter cells via DNA replication. In this process, the two complementary strands of DNA within a cell unravel, and two new strands that complement the original strands are synthesized. In this way, two complete copies of the DNA are made, one for each daughter cell.

When a cell synthesizes a protein, the base sequence of the gene that codes for that protein is transferred to mRNA. The mRNA then moves out to a ribosome, where the amino acids are linked in the correct sequence to synthesize the protein. The general sequence is:

$$DNA \longrightarrow RNA \longrightarrow Protein$$

Nucleic Acids and DNA Replication: Since DNA contains the instructions for making proteins and since proteins are the working molecules of life, our DNA determines a great deal of who and what we are. Humans have basically the same body parts, organs, and metabolic processes because most of our DNA is the same. The differences between humans are at least partly caused by slight differences in their DNA. In recent years, scientists have learned how to manipulate and change DNA, and in this way they can manipulate and change the organisms that result from that DNA.

KEY TERMS

alpha (α)-helix [19.6]
amino acid [19.5]
beta (β)-pleated sheet [19.6]
biochemistry [19.1]
carbohydrate [19.3]
cell [19.2]
cell membrane [19.2]
cellulose [19.3]
chromosome [19.7]
codon [19.7]
complementary base [19.8]
complex carbohydrate [19.3]

cytoplasm [19.2]
dipeptide [19.5]
disaccharide [19.3]
DNA [19.7]
ester linkage [19.4]
fatty acid [19.4]
fibrous protein [19.6]
gene [19.7]
globular protein [19.6]
glycogen [19.3]
glycolipid [19.4]
glycosidic linkage [19.3]
human genome [19.1]
lipid [19.4]
lipid bilayer [19.4]

messenger RNA (mRNA) [19.8]
monosaccharide [19.3]
nucleic acid [19.7]
nucleotide [19.7]
nucleus (of a cell) [19.2]
peptide bond [19.5]
phospholipid [19.4]
polypeptide [19.5]
polysaccharide [19.3]
primary protein structure [19.6]
protein [19.5]
quaternary protein structure [19.6]

random coil [19.6]
R group (side chain) [19.5]
RNA [19.7]
saturated fat [19.4]
secondary protein structure [19.6]
simple carbohydrate [19.3]
simple sugar [19.3]
starch [19.3]
steroid [19.4]
tertiary protein structure [19.6]
triglyceride [19.4]
unsaturated fat [19.4]

EXERCISES

QUESTIONS

1. What is the Human Genome Project? What is one of the surprising findings of the Human Genome Project?
2. What are some of the expected benefits of the Human Genome Project?
3. Explain what a cell is and list its main chemical components.
4. What are carbohydrates? What functions do carbohydrates serve in living organisms?
5. Is glucose soluble in water? How can you tell? Why is this important?
6. Explain the differences between a monosaccharide, a disaccharide, and a polysaccharide.
7. What happens to disaccharides and polysaccharides during digestion?
8. What is the difference between a simple sugar and a complex carbohydrate?
9. What is the difference between starch and cellulose? How does this difference affect digestion?
10. What are lipids? What are the main functions of lipids?
11. What are fatty acids? Draw the general structure of a fatty acid.
12. What is the difference between a saturated fatty acid and an unsaturated fatty acid?
13. What is a triglyceride? Draw the general structure of a triglyceride.
14. What is the difference between a saturated fat and an unsaturated fat, in terms of both structure and properties?
15. What are phospholipids and glycolipids? What properties do they have in common?
16. What are the main functions of phospholipids and glycolipids in the body?
17. What are steroids? What are some of the functions of steroids in the body?
18. What are proteins?
19. What are the main functions of proteins within living organisms?
20. What are amino acids? Draw a general structure for amino acids.
21. How do amino acids differ from one another?
22. What is a peptide bond?
23. Use two generic amino acids to show how a peptide bond forms.
24. What determines the shape of a protein? Why is the shape of a protein so important?
25. What does primary protein structure refer to? What kinds of interactions maintain primary protein structure?
26. What does secondary protein structure refer to? What kinds of interactions maintain secondary protein structure?
27. What does tertiary protein structure refer to? What kinds of interactions maintain tertiary protein structure?
28. What does quaternary protein structure refer to? What kinds of interactions maintain quaternary protein structure?
29. Explain the α-helix structure and the β-pleated sheet structure.
30. What are nucleic acids?
31. What is the main function of nucleic acids?
32. What are the two main types of nucleic acids?
33. What four different bases occur within DNA?
34. What is a codon?
35. What is the genetic code?
36. Is the genetic code different for different organisms?

37. What is a gene?

38. If a protein contains 300 amino acids, about how many nucleotides are in the gene that codes for it?

39. What are chromosomes?

40. Do most cells in the human body contain genes for all of the proteins that a human needs?

41. Do most cells in the human body synthesize all the proteins that they have genes for?

42. Describe how DNA replication occurs.

43. List the complementary base of:

  (a) adenine (A)

  (b) thymine (T)

  (c) cytosine (C)

  (d) guanine (G)

44. Describe how protein synthesis occurs.

PROBLEMS

CARBOHYDRATES

45. Determine whether or not each molecule is a carbohydrate. If it is, classify it as a monosaccharide, a disaccharide, or a trisaccharide.

46. Determine whether or not each molecule is a carbohydrate. If it is, classify it as a monosaccharide, a disaccharide, or a trisaccharide.

47. Classify each carbohydrate as a triose, tetrose, pentose, and so on.

(a)

(b)

(c)

(d)

48. Classify each carbohydrate as a triose, tetrose, pentose, and so on.

(a)

(b)

(c) HO

(d)

---

49. Draw the structure of glucose in both its straight chain and its ring form.

50. Draw the structure of fructose in its ring form. Are fructose and glucose isomers?

---

51. Draw the structure of sucrose. Label the glucose and fructose rings in this disaccharide.

52. Draw the structure of lactose. Label the glucose and galactose rings in this disaccharide.

LIPIDS

53. Determine whether or not each molecule is a lipid. If the molecule is a lipid, determine the kind of lipid. If it is a fatty acid or a triglyceride, classify it as saturated or unsaturated.

(a)

(b)

(c)

(d)

54. Determine whether or not each molecule is a lipid. If the molecule is a lipid, determine the kind of lipid. If it is a fatty acid or a triglyceride, classify it as saturated or unsaturated.

(a)

(b)

(c)

(d)

55. Sketch the block diagram for a triglyceride.

56. Sketch the block diagram for a phospholipid. How are phospholipids different from triglycerides?

57. Draw the structure of the triglyceride that would form from the reaction of myristic acid with glycerol. Would you expect this triglyceride to be a fat or an oil?

58. Draw the structure of the triglyceride that would form from the reaction of oleic acid with glycerol. Would you expect this triglyceride to be a fat or an oil?

AMINO ACIDS AND PROTEINS

59. Determine whether or not each molecule is an amino acid.

(a)

$$H_3C-CH_2-CH_2-\overset{\overset{\displaystyle O}{\|}}{C}-OH$$

(b)

(c)

$$H_2N-\overset{\overset{\displaystyle O}{\|}}{C}-CH_2-CH_3$$

(d)

60. Determine whether or not each molecule is an amino acid.

(a)

$$H_2N-\overset{\overset{\displaystyle H}{|}}{\underset{\underset{\displaystyle H}{|}}{C}}-\overset{\overset{\displaystyle O}{\|}}{C}-OH$$

(b)

(c)

(d)

61. Show the reaction by which isoleucine and serine form a peptide bond.

62. Show the reaction by which valine and lysine form a peptide bond.

63. Draw a structure for each tripeptide.
 (a) leu-ala-gly
 (b) val-thr-lys
 (c) gly-phe-ser

64. Draw a structure for each tripeptide.
 (a) thr-glu-leu
 (b) glu-tyr-ser
 (c) ala-ser-val

65. A cysteine on a protein strand forms a covalent bond with another cysteine that is 35 amino acid units away. The resulting fold in the protein is an example of what kind of structure? (primary, secondary, tertiary, or quaternary)

66. An amino acid on a protein strand hydrogen-bonds to another amino acid that is four amino acid units away. The next amino acid on the chain does the same, hydrogen bonding to an amino acid that is four amino acids away from it. This pattern repeats itself over a significant part of the protein chain. The resulting pattern in the protein is an example of what kind of structure? (primary, secondary, tertiary, or quaternary)

67. The following is the amino acid sequence in one section of a protein. It represents what kind of structure? (primary, secondary, tertiary, or quaternary)

-lys-glu-thr-ala-ala-ala-lys-phe-glu-

68. A particular protein is composed of two individual chains of amino acids. The way these two chains fit together is an example of what kind of structure? (primary, secondary, tertiary, or quaternary)

NUCLEIC ACIDS

69. Determine whether or not each structure is that of a nucleotide. For those that are nucleotides, identify A, T, C, or G.

(a)

(b)

(c)

(d)

70. Determine whether each structure is that of a nucleotide. For those that are nucleotides, identify the base as A, T, C, or G.

(a)

(b)

(c)

(d)

71. Draw the complementary strand of this DNA strand.

72. Draw the complementary strand of this DNA strand.

73. In a step-by-step fashion, show how this section of DNA would replicate to form two copies.

74. In a step-by-step fashion, show how this section of DNA would replicate to form two copies.

CUMULATIVE PROBLEMS

75. Match each linkage with the correct class of biochemicals.

  (a) glycoside linkage

  (b) peptide bonds

  (c) ester linkage

- proteins
- triglycerides
- carbohydrates

76. Match each monomer with the correct class of biopolymers.

  (a) nucleotide

  (b) saccharide

  (c) amino acid

- protein
- DNA
- starch

77. Match each biochemical with the correct function in living organisms.

  (a) glucose

  (b) DNA

  (c) phospholipids

  (d) triglycerides

- compose cell membranes
- long-term energy storage
- short-term energy storage
- blueprint for proteins

78. Match each biochemical with the correct function in living organisms.

  (a) proteins

  (b) cellulose

  (c) RNA

- act as enzymes (among other things)
- involved in protein synthesis
- structural components of plants

79. Match each term with its correct meaning.

  (a) codon

  (b) gene

  (c) human genome

  (d) chromosome

- codes for a single protein
- codes for a single amino acid
- all of the genetic material of a human
- structure that contains genes

80. Match each term with its correct meaning.

  (a) pentose

  (b) dipeptide

  (c) diglyceride

  (d) fatty acid

- a carboxylic acid with a long hydrocarbon R group
- two amino acids joined by a peptide bond
- a glycerol molecule with two fatty acids attached
- a five-carbon sugar

81. The amino acid glycine has the condensed structural formula:

$$NH_2CH_2COOH$$

Determine the VSEPR geometry about each internal atom and make a three-dimensional sketch of the molecule.

82. The amino acid serine has the condensed structural formula:

$$NH_2CH(CH_2OH)COOH$$

Determine the VSEPR geometry about each internal atom and make a three-dimensional sketch of the molecule.

83. Since amino acids are asymmetrical, a peptide with amino acids in a certain order is different from a peptide with the amino acids in the reverse order. For example, gly-val is different from val-gly. Draw the structures of these two dipeptides and show how they are different.

84. Use the abbreviations to write the sequences for all the possible polypeptides that can be made from the three amino acids: thr, gly, ala.

85. Determining the amino acid sequence in a protein usually involves treating the protein with various reagents that break up the protein into smaller fragments that can be individually sequenced. Treating a particular 11-amino-acid polypeptide with one reagent produced the fragments:

    trp-glu-val, gly-arg, ala-ser-phe-gly-asn-lys

    Treating the same polypeptide with a different reagent produced the following fragments:

    gly-asn-lys-trp, glu-val, gly-arg-ala-ser-phe

    What is the amino acid sequence of the polypeptide?

86. Treating a particular polypeptide with one reagent (as described in the previous problem) produced the following fragments:

    asp-thr-ala-trp, gly-glu-ser-lys, trp-arg

    Treating the same polypeptide with a different reagent produced the following fragments:

    thr-ala-trp, gly-glu, ser-lys-trp-arg-asp

    What is the amino acid sequence of the polypeptide?

87. The insulin protein contains 51 amino acids. How many DNA base pairs are required to code for all the amino acids in insulin?

88. A hemoglobin subunit contains 146 amino acids. How many DNA base pairs are required to code for all the amino acids in the subunit?

89. A solution is made by dissolving 23.88 mg of a protein in water and diluting to a total volume of 20.0 mL. The osmotic pressure of the solution is 3.66 torr at 25 °C. What is the molar mass of the protein? (The osmotic pressure of a solution is given by the following equation: Osmotic Pressure = $M \times R \times T$, where M is the molarity of the solution, R is the gas constant, and T is temperature in K.)

90. A solution is made by dissolving 28.65 mg of a protein in water and diluting to a total volume of 25.0 mL. The osmotic pressure of the solution was 4.55 torr at 25 °C. What is the molar mass of the protein? (The osmotic pressure of a solution is given by the equation: Osmotic Pressure = $M \times R \times T$, where M is the molarity of the solution, R is the gas constant, and T is temperature in K.)

HIGHLIGHT PROBLEMS

91. One way to fight viral infections is to prevent viruses from replicating their DNA. Without DNA replication, the virus cannot multiply. Some viral drug therapies cause the introduction of *fake* nucleotides into cells. When the virus uses one of these fake nucleotides in an attempt to replicate its DNA, the fake nucleotide doesn't work and DNA replication is halted. For example, AZT, a drug used to fight the human immunodeficiency virus (HIV) that causes AIDS, results in the introduction of the following fake thymine-containing nucleotide into cells.

Thymine

Actual thymine-containing nucleotide

Fake nucleotide that results from taking
azidothymidine (AZT)

Examine the structures of the real nucleotide and the
AZT fake nucleotide. Propose a mechanism for how
this fake nucleotide might halt DNA replication.

92. Sickle-cell anemia is caused by a genetic defect that
substitutes valine for glutamic acid at one position
in two of the four chains of the hemoglobin protein.

The result is a decrease in the water solubility of
hemoglobin. Examine the structures of valine and
glutamic acid and explain why this might be so.

►ANSWERS TO SKILLBUILDER EXERCISES

Skillbuilder 19.1   (b) monosaccharide; (d) disaccharide

Skillbuilder 19.2   (b) unsaturated fat; (d) saturated fat

Skillbuilder 19.3

Skillbuilder 19.4   Answer: G   G   T   A   A   C   C

►ANSWERS TO CONCEPTUAL CHECKPOINTS

19.1 (a) The amino acid sequence is an example of primary
structure.

19.2 (d) Each of the 51 amino acids is coded for by a single
codon. A codon consists of three nucleotides, each
containing one base.

19.3 (b) Proteins are polymers of amino acids; nucleic acids
are polymers of nucleotides; polysaccharides are
polymers of monosaccharides. Steroids, however,
are not chains of any repeating unit.

APPENDIX: Mathematics Review

Basic Algebra

In chemistry, you often have to solve an equation for a particular variable. For example, suppose you want to solve the following equation for V.

$$PV = nRT$$

To solve an equation for a particular variable, you must isolate that variable on one side of the equation. The rest of the variables or numbers will then be on the other side of the equation. To solve the above equation for V, simply divide both sides by P.

$$\frac{PV}{P} = \frac{nRT}{P}$$

$$V = \frac{nRT}{P}$$

The Ps cancel, and we are left with an expression for V. For another example, consider solving the following equation for °F.

$$°C = \frac{(°F - 32)}{1.8}$$

First, eliminate the 1.8 in the denominator of the right side by multiplying both sides by 1.8.

$$(1.8) \, °C = \frac{(°F - 32)}{1.8} (1.8)$$

$$(1.8) \, °C = (°F - 32)$$

Then eliminate the -32 on the right by adding 32 to both sides.

$$(1.8) \, °C + 32 = (°F - 32) + 32$$

$$(1.8) \, °C + 32 = °F$$

We are now left with an expression for °F.

In general, solve equations by following these guidelines:

- Cancel numbers or symbols in the denominator (bottom part of a fraction) by multiplying by the number or symbol to be canceled.
- Cancel numbers or symbols in the numerator (upper part of a fraction) by dividing by the number or symbol to be canceled.
- Eliminate numbers or symbols that are added by subtracting the same number or symbol.
- Eliminate numbers or symbols that are subtracted by adding the same number or symbol.
- Whether you add, subtract, multiply, or divide, **always perform the same operation for both sides of a mathematical equation**. (Otherwise, the two sides will no longer be equal.)

For a final example, solve the following equation for x.

$$\frac{67x - y + 3}{6} = 2z$$

Cancel the 6 in the denominator by multiplying both sides by 6.

$$(6)\frac{67x - y + 3}{6} = (6)2z$$

$$67x - y + 3 = 12z$$

Eliminate the $+3$ by subtracting 3 from both sides.

$$67x - y + 3 - 3 = 12z - 3$$

$$67x - y = 12z - 3$$

Eliminate the $-y$ by adding y to both sides.

$$67x - y + y = 12z - 3 + y$$

$$67x = 12z - 3 + y$$

Cancel the 67 by dividing both sides by 67.

$$\frac{67x}{67} = \frac{12z - 3 + y}{67}$$

$$x = \frac{12z - 3 + y}{67}$$

FOR PRACTICE Using Algebra to Solve Equations

Solve each of the following for the indicated variable.

(a) $P_1V_1 = P_2V_2$; solve for V_2

(b) $\dfrac{V_1}{T_1} = \dfrac{V_2}{T_2}$; solve for T_1

(c) $PV = nRT$; solve for n

(d) $K = {}^\circ C + 273$; solve for ${}^\circ C$

(e) $\dfrac{3x + 7}{2} = y$; solve for x

(f) $\dfrac{32}{y + 3} = 8$; solve for y

ANSWERS

(a) $V_2 = \dfrac{P_1V_1}{P_2}$

(b) $T_1 = \dfrac{V_1T_2}{V_2}$

(c) $n = \dfrac{PV}{RT}$

(d) ${}^\circ C = K - 273$

(e) $x = \dfrac{2y - 7}{3}$

(f) $y = 1$

Mathematical Operations with Scientific Notation

Writing numbers in scientific notation is covered in detail in Section 2.2. Briefly, a number written in scientific notation consists of a **decimal part**, a number that is usually between 1 and 10, and an **exponential part**, 10 raised to an **exponent**, n.

Each of the following numbers is written in both scientific and decimal notation.

$$1.0 \times 10^5 = 100{,}000 \qquad 1.0 \times 10^{-6} = 0.000001$$

$$6.7 \times 10^3 = 6700 \qquad 6.7 \times 10^{-3} = 0.0067$$

MULTIPLICATION AND DIVISION

To multiply numbers expressed in scientific notation, multiply the decimal parts and add the exponents.

$$(A \times 10^m)(B \times 10^n) = (A \times B) \times 10^{m+n}$$

To divide numbers expressed in scientific notation, divide the decimal parts and subtract the exponent in the denominator from the exponent in the numerator.

$$\frac{(A \times 10^m)}{(B \times 10^n)} = \left(\frac{A}{B}\right) \times 10^{m-n}$$

Consider the following example involving multiplication.

$$(3.5 \times 10^4)(1.8 \times 10^6) = (3.5 \times 1.8) \times 10^{4+6}$$
$$= 6.3 \times 10^{10}$$

Consider the following example involving division.

$$\frac{(5.6 \times 10^7)}{(1.4 \times 10^3)} = \left(\frac{5.6}{1.4}\right) \times 10^{7-3}$$
$$= 4.0 \times 10^4$$

ADDITION AND SUBTRACTION

To add or subtract numbers expressed in scientific notation, rewrite all the numbers so that they have the same exponent, and then add or subtract the decimal parts of the numbers. The exponents remained unchanged.

$$A \times 10^n$$
$$\pm B \times 10^n$$
$$\overline{(A \pm B) \times 10^n}$$

Notice that the numbers *must have* the same exponent.

Consider the following example involving addition.

$$4.82 \times 10^7$$
$$+3.4 \times 10^6$$

First, express both numbers with the same exponent. In this case, we rewrite the lower number and perform the addition as follows:

$$4.82 \times 10^7$$
$$+0.34 \times 10^7$$
$$\overline{5.16 \times 10^7}$$

Consider the following example involving subtraction.

$$7.33 \times 10^5$$
$$-1.9 \times 10^4$$

First, express both numbers with the same exponent. In this case, we rewrite the lower number and perform the subtraction as follows:

$$
\begin{array}{r}
7.33 \times 10^5 \\
-0.19 \times 10^5 \\
\hline
7.14 \times 10^5
\end{array}
$$

FOR PRACTICE **Mathematical Operations with Scientific Notation**

Perform each of the following operations.

(a) $(2.1 \times 10^7)(9.3 \times 10^5)$

(b) $(5.58 \times 10^{12})(7.84 \times 10^{-8})$

(c) $\dfrac{(1.5 \times 10^{14})}{(5.9 \times 10^8)}$

(d) $\dfrac{(2.69 \times 10^7)}{(8.44 \times 10^{11})}$

(e) $\begin{array}{r} 1.823 \times 10^9 \\ +1.11 \times 10^7 \\ \hline \end{array}$

(f) $\begin{array}{r} 3.32 \times 10^{-5} \\ +3.400 \times 10^{-7} \\ \hline \end{array}$

(g) $\begin{array}{r} 6.893 \times 10^9 \\ -2.44 \times 10^8 \\ \hline \end{array}$

(h) $\begin{array}{r} 1.74 \times 10^4 \\ -2.9 \times 10^3 \\ \hline \end{array}$

ANSWERS

(a) 2.0×10^{13}

(b) 4.37×10^5

(c) 2.5×10^5

(d) 3.19×10^{-5}

(e) 1.834×10^9

(f) 3.35×10^{-5}

(g) 6.649×10^9

(h) 1.45×10^4

Logarithms

The logarithm (or log) of a number is the exponent to which 10 must be raised to obtain that number. For example, the log of 100 is 2 because 10 must be raised to the 2nd power to get 100. Similarly, the log of 1000 is 3 because 10 must be raised to the 3rd power to get 1000. The logs of several multiples of 10 are shown on the following page.

$$\log 10 = 1$$
$$\log 100 = 2$$
$$\log 1000 = 3$$
$$\log 10{,}000 = 4$$

Because $10^0 = 1$ by definition, $\log 1 = 0$.

The log of a number smaller than 1 is negative because 10 must be raised to a negative exponent to get a number smaller than 1. For example, the log of 0.01 is -2 because 10 must be raised to the power of -2 to get 0.01. Similarly, the log of 0.001 is -3 because 10 must be raised to the power of -3 to get 0.001. The logs of several fractional numbers are as follows.

$$\log 0.1 = -1$$
$$\log 0.01 = -2$$
$$\log 0.001 = -3$$
$$\log 0.0001 = -4$$

The logs of numbers that are not multiples of 10 can be computed on your calculator. See your calculator manual for specific instructions.

INVERSE LOGARITHMS

The inverse logarithm or invlog function (sometimes called antilog) is exactly the opposite of the log function. For example, the log of 100 is 2 and the inverse log of 2 is 100. The log function and the invlog function undo one another.

$$\log 1000 = 3$$

$$\text{invlog } 3 = 1000$$

$$\text{invlog } (\log 1000) = 1000$$

The inverse log of a number is simply 10 raised to that number.

$$\text{invlog } x = 10^x$$

$$\text{invlog } 3 = 10^3 = 1000$$

The inverse logs of numbers can be computed on your calculator. See your calculator manual for specific instructions.

FOR PRACTICE   Logarithms and Inverse Logarithms

Perform each of the following operations.

(a) $\log 1.0 \times 10^5$

(b) $\log 59$

(c) $\log 1.0 \times 10^{-5}$

(d) $\log 0.068$

(e) invlog 7.0

(f) invlog 1.44

(g) invlog −6.0

(h) invlog −0.250

(i) invlog (log 88)

ANSWERS

(a) 5.00

(b) 1.77

(c) −5.00

(d) −1.17

(e) 1×10^7

(f) 28

(g) 1×10^{-6}

(h) 0.56

(i) 88

absolute zero The coldest temperature possible. Absolute zero (0 K or −273 °C or −459 °F) is the temperature at which molecular motion stops. Lower temperatures do not exist.

acid A molecular compound that dissolves in solution to form H^+ ions. Acids have the ability to dissolve some metals and will turn litmus paper red.

acid rain Acidic precipitation in the form of rain; created when fossil fuels are burned, which releases SO_2 and NO_2, which then react with water in the atmosphere to form sulfuric acid and nitric acid.

acid–base reaction A reaction that forms water and typically a salt.

acidic solution A solution containing a concentration of H_3O^+ ions greater than 1.0×10^{-7} M (pH < 7).

activation energy The amount of energy that must be absorbed by reactants before a reaction can occur; an energy hump that normally exists between the reactants and products.

activity series of metals A listing of metals (and hydrogen) in order of decreasing activity, decreasing ability to oxidize, and decreasing tendency to lose electrons.

actual yield The amount of product actually produced by a chemical reaction.

addition polymer A polymer formed by addition of monomers to one another without elimination of any atoms.

alcohol An organic compound containing an —OH functional group bonded to a carbon atom and having the general formula ROH.

aldehyde An organic compound with the general formula RCHO.

alkali metals The Group 1A elements, which are highly reactive metals.

alkaline battery A dry cell employing half-reactions that use a base.

alkaline earth metals The Group 2A elements, which are fairly reactive metals.

alkaloids Organic compounds that are typically found in plants and act as bases.

alkanes Hydrocarbons in which all carbon atoms are connected by single bonds. Noncyclic alkanes have the general formula C_nH_{2n+2}.

alkene A hydrocarbon that contains at least one double bond between carbon atoms. Noncyclic alkenes have the general formula C_nH_{2n}.

alkyl group In an organic molecule, any group containing only singly bonded carbon atoms and hydrogen atoms.

alkyne A hydrocarbon that contains at least one triple bond between carbon atoms. Noncyclic alkynes have the general formula C_nH_{2n-2}.

alpha particle A particle consisting of two protons and two neutrons (a helium nucleus), represented by the symbol ^4_2He.

alpha (α) radiation Radiation emitted by an unstable nucleus, consisting of alpha particles.

alpha (α)-helix The most common secondary protein structure. The amino acid chain is wrapped into a tight coil from which the side chains extend outward. The structure is maintained by hydrogen bonding interactions between NH and CO groups along the peptide backbone of the protein.

amine An organic compound that contains nitrogen and has the general formula NR_3, where R may be an alkyl group or a hydrogen atom.

amino acid A molecule containing an amine group, a carboxylic acid group, and an R group (also called a side chain). Amino acids are the building blocks of proteins.

amorphous A type of solid matter in which atoms or molecules do not have long-range order (e.g., glass and plastic).

amphoteric In Brønsted–Lowry terminology, able to act as either an acid or a base.

anion A negatively charged ion.

anode The electrode where oxidation occurs in an electrochemical cell.

aqueous solution A homogeneous mixture of a substance with water.

aromatic ring A ring of carbon atoms containing alternating single and double bonds; another name for the benzene ring.

Arrhenius acid A substance that produces H^+ ions in aqueous solution.

Arrhenius base A substance that produces OH^- ions in aqueous solution.

atmosphere (atm) The average pressure at sea level, 101,325 Pa (760 mmHg).

atom The smallest identifiable unit of an element.

atomic element An element that exists in nature with single atoms as the basic unit.

atomic mass A weighted average of the masses of each naturally occurring isotope of an element; atomic mass is the average mass of the atoms of an element.

atomic mass unit (amu) The unit commonly used to express the masses of protons, neutrons, and nuclei. 1 amu = 1.66×10^{-24} g.

atomic number (Z) The number of protons in the nucleus of an atom.

atomic size The size of atom, which is determined by how far the outermost electrons are from the nucleus.

atomic solid A solid whose component units are individual atoms (e.g., diamond, C; iron, Fe).

atomic theory A theory stating that all matter is composed of tiny particles called atoms.

Avogadro's law A law stating that the volume (V) of a gas and the amount of the gas in moles (n) are directly proportional.

Avogadro's number The number of entities in a mole, 6.022×10^{23}.

balanced equation A chemical equation in which the numbers of each type of atom on both sides of the equation are equal.

base A molecular compound that dissolves in solution to form OH^- ions. Bases have a slippery feel and turn litmus paper blue.

base chain The longest continuous chain of carbon atoms in an organic compound.

basic solution A solution containing a concentration of OH^- ions greater than 1.0×10^{-7} M (pH > 7).

bent The molecular geometry in which 3 atoms are not in a straight line. This geometry occurs when the central atoms contain 4 electron groups (2 bonding and 2 nonbonding) or 3 electron groups (2 bonding and 1 nonbonding).

benzene (C_6H_6) A particularly stable organic compound consisting of six carbon atoms joined by alternating single and double bonds in a ring structure.

beta particle A form of radiation consisting of an energetic electron and represented by the symbol $_{-1}^{0}e$.

beta (β) radiation Energetic electrons emitted by an unstable nucleus.

beta (β)-pleated sheet A common pattern in the secondary structure of proteins. The protein chain is extended in a zigzag pattern, and the peptide backbones of adjacent strands interact with one another through hydrogen bonding to form sheets.

binary acid An acid containing only hydrogen and a nonmetal.

binary compound A compound containing only two different kinds of elements.

biochemistry The study of the chemical substances and processes that occur in living organisms.

Bohr model A model for the atom in which electrons travel around the nucleus in circular orbits at specific, fixed distances from the nucleus.

boiling point The temperature at which the vapor pressure of a liquid is equal to the pressure above it.

boiling point elevation The increase in the boiling point of a solution caused by the presence of the solute.

bonding pair Electrons that are shared between two atoms in a chemical bond.

bonding theory A model that predicts how atoms bond together to form molecules.

Boyle's law A law maintaining that the volume (V) of a gas and its pressure (P) are inversely proportional.

branched alkane An alkane composed of carbon atoms bonded in chains containing branches.

Brønsted–Lowry acid A proton (H^+ ion) donor.

Brønsted–Lowry base A proton (H^+ ion) acceptor.

buffer A solution that resists pH change by neutralizing added acid or added base.

Calorie (Cal) An energy unit equivalent to 1000 little-c calories.

calorie (cal) The amount of energy required to raise the temperature of 1 g of water by 1 °C.

carbohydrates Polyhydroxyl aldehydes or ketones or their derivatives, containing multiple —OH groups and often having the general formula $(CH_2O)_n$.

carbonyl group A carbon atom double bonded to an oxygen atom.

carboxylic acid An organic compound with the general formula $RCOOH$.

catalyst A substance that increases the rate of a chemical reaction but is not consumed by the reaction.

cathode The electrode where reduction occurs in an electrochemical cell.

cation A positively charged ion.

cell The smallest structural unit of living organisms that has the properties associated with life.

cell membrane The structure that bounds the cell and holds the contents of the cell together.

cellulose A common polysaccharide composed of repeating glucose units linked together.

Celsius (°C) scale A temperature scale often used by scientists. On this scale, water freezes at 0 °C and boils at 100 °C at 1 atm pressure. Room temperature is approximately 22 °C.

chain reaction A self-sustaining chemical or nuclear reaction yielding energy or products that cause further reactions of the same kind.

charge A fundamental property of protons and electrons. Charged particles experience forces such that like charges repel and unlike charges attract.

Charles's law A law stating that the volume (V) of a gas and its temperature (T) expressed in kelvins are directly proportional.

chemical bond The sharing or transfer of electrons to attain stable electron configurations among the bonding atoms.

chemical change A change in which matter changes its composition.

chemical energy The energy associated with chemical changes.

chemical equation An equation that represents a chemical reaction; the reactants are on the left side of the equation and the products are on the right side.

chemical formula A way to represent a compound. At a minimum, the chemical formula indicates the elements present in the compound and the relative number of atoms of each element.

chemical properties Properties that a substance can display only through changing its composition.

chemical reaction The process by which one or more substances transform into different substances via a chemical change. Chemical reactions often emit or absorb energy.

chemical symbol A one- or two-letter abbreviation for an element. Chemical symbols are listed directly below the atomic number in the periodic table.

chemistry The science that seeks to understand the behavior of matter by studying what atoms and molecules do.

chromosome A biological structure containing genes, located within the nucleus of a cell.

codon A sequence of three bases in a nucleic acid that codes for one amino acid.

colligative properties Physical properties of solutions that depend on the number of solute particles present but not the type of solute particles.

collision theory A theory of reaction rates stating that effective collisions between reactant molecules must take place in order for the reaction to occur.

color change One type of evidence of a chemical reaction, involving the change in color of a substance after a reaction.

combined gas law A law that combines Boyle's law and Charles's law; it is used to calculate how a property of a gas (P, V, or T) changes when two other properties are changed at the same time.

combustion reaction A reaction in which a substance reacts with oxygen, emitting heat and forming one or more oxygen-containing compounds.

complementary base In DNA, a base capable of precise pairing with a specific other DNA base.

complete ionic equation A chemical equation showing all the species as they are actually present in solution.

complex carbohydrate A carbohydrate composed of many repeating saccharide units.

compound A substance composed of two or more elements in fixed, definite proportions.

compressible Able to occupy a smaller volume when subjected to increased pressure. Gases are compressible because, in the gas phase, atoms or molecules are widely separated.

concentrated solution A solution containing large amounts of solute.

condensation A physical change in which a substance is converted from its gaseous form to its liquid form.

condensation polymer A class of polymers that expel atoms, usually water, during their formation or polymerization.

condensed structural formula A shorthand way of writing a structural formula.

conjugate acid–base pair In Brønsted–Lowry terminology, two substances related to each other by the transfer of a proton.

conservation of energy, law of A law stating that energy can be neither created nor destroyed. The total amount of energy is constant and cannot change; it can only be transferred from one object to another or converted from one form to another.

conservation of mass, law of A law stating that in a chemical reaction, matter is neither created nor destroyed.

constant composition, law of A law stating that all samples of a given compound have the same proportions of their constituent elements.

conversion factor A factor used to convert between two separate units; a conversion factor is constructed from any two quantities known to be equivalent.

copolymers Polymers that are composed of two different kinds of monomers and result in chains composed of alternating units rather than a single repeating unit.

core electrons The electrons that are not in the outermost principal shell of an atom.

corrosion The oxidation of metals (e.g., rusting of iron).

covalent atomic solid An atomic solid, such as diamond, that is held together by covalent bonds.

covalent bond The bond that results when two nonmetals combine in a chemical reaction. In a covalent bond, the atoms share their electrons.

critical mass The mass of uranium or plutonium required for a nuclear reaction to be self-sustaining.

crystalline A type of solid matter with atoms or molecules arranged in a well-ordered, three-dimensional array with long-range, repeating order (e.g., salt and diamond).

cytoplasm In a cell, the region between the nucleus and the cell membrane.

Dalton's law of partial pressure A law stating that the sum of the partial pressures of each component in a gas mixture equals the total pressure.

daughter nuclide The nuclide product of a nuclear decay.

decimal part One part of a number expressed in scientific notation.

decomposition A reaction in which a complex substance decomposes to form simpler substances; $AB \longrightarrow A + B$.

density (d) A fundamental property of materials that relates mass and volume and differs from one substance to another. The units of density are those of mass divided by volume, most commonly expressed in g/cm^3, g/mL, or g/L.

derived unit A unit formed from the combination of other units.

dilute solution A solution containing small amounts of solute.

dimer A molecule formed by the joining together of two smaller molecules.

dipeptide Two amino acids linked together via a peptide bond.

dipole moment A measure of the separation of charge in a bond or in a molecule.

diprotic acid An acid containing two ionizable protons.

disaccharide A carbohydrate that can be decomposed into two simpler carbohydrates.

dispersion force The intermolecular force present in all molecules and atoms. Dispersion forces are caused by fluctuations in the electron distribution within molecules or atoms.

displacement A reaction in which one element displaces another in a compound; $A + BC \longrightarrow AC + B$.

dissociation In aqueous solution, the process by which a solid ionic compound separates into its ions.

disubstituted benzene A benzene in which two hydrogen atoms have been replaced by an atom or group of atoms.

DNA (deoxyribonucleic acid) Long chainlike molecules that occur in the nucleus of cells and act as blueprints for the construction of proteins.

dot structure A drawing that represents the valence electrons in atoms as dots; it shows a chemical bond as the sharing or transfer of electron dots.

double bond The bond that exists when two electron pairs are shared between two atoms. In general, double bonds are shorter and stronger than single bonds.

double displacement A reaction in which two elements or groups of elements in two different compounds exchange places to form two new compounds; $AB + CD \longrightarrow AD + CB$.

dry cell An ordinary battery (voltaic cell); it does not contain large amounts of liquid water.

duet The name for the two electrons corresponding to a stable Lewis structure in hydrogen and helium.

dynamic equilibrium In a chemical reaction, the condition in which the rate of the forward reaction equals the rate of the reverse reaction.

electrical current The flow of electric charge—for example, electrons flowing through a wire or ions through a solution.

electrical energy   Energy associated with the flow of electric charge.

electrochemical cell   A device that creates electrical current from a redox reaction.

electrolysis   A process in which electrical current is used to drive an otherwise nonspontaneous redox reaction.

electrolytic cell   An electrochemical cell used for electrolysis.

electromagnetic radiation   A type of energy that travels through space at a constant speed of 3.0×10^8 m/s (186,000 miles/s) and exhibits both wavelike and particlelike behavior. Light is a form of electromagnetic radiation.

electromagnetic spectrum   A spectrum that includes all wavelengths of electromagnetic radiation.

electron   A negatively charged particle that occupies most of the atom's volume but contributes almost none of its mass.

electron configuration   A representation that shows the occupation of orbitals by electrons for a particular element.

electron geometry   The geometrical arrangement of the electron groups in a molecule.

electron group   A general term for a lone pair, single bond, or multiple bond in a molecule.

electron spin   A fundamental property of all electrons that causes them to have magnetic fields associated with them. The spin of an electron can either be oriented up $\left(+\frac{1}{2}\right)$ or down $\left(-\frac{1}{2}\right)$.

electronegativity   The ability of an element to attract electrons within a covalent bond.

element   A substance that cannot be broken down into simpler substances.

emission spectrum   A spectrum associated with the emission of electromagnetic radiation by elements or compounds.

empirical formula   A formula for a compound that gives the smallest whole-number ratio of each type of atom.

empirical formula molar mass   The sum of the molar masses of all the atoms in an empirical formula.

endothermic   Describes a process that absorbs heat energy.

endothermic reaction   A chemical reaction that absorbs energy from the surroundings.

energy   The capacity to do work.

English system   A unit system commonly used in the United States.

enzymes   Biological catalysts that increase the rates of biochemical reactions; enzymes are abundant in living organisms.

equilibrium constant (K_{eq})   The ratio, at equilibrium, of the concentrations of the products raised to their stoichiometric coefficients divided by the concentrations of the reactants raised to their stoichiometric coefficients.

equivalence point   The point in a reaction at which the reactants are in exact stoichiometric proportions.

equivalent   The stoichiometric proportions of elements and compounds in a chemical equation.

ester   An organic compound with the general formula $RCOOR$.

ester linkage   A type of bond with the general structure —COO—. Ester linkages join glycerol to fatty acids.

ether   An organic compound with the general formula ROR.

evaporation   A process in which molecules of a liquid, undergoing constant random motion, acquire enough energy to overcome attractions to neighbors and enter the gas phase.

excited state   An unstable state for an atom or a molecule in which energy has been absorbed but not reemitted, raising an electron from the ground state into a higher energy orbital.

exothermic   Describes a process that releases heat energy.

exothermic reaction   A chemical reaction that releases energy to the surroundings.

experiment   A procedure that attempts to measure observable predictions to test a theory or law.

exponent   A number that represents the number of times a term is multiplied by itself. For example, in 2^4 the exponent is 4 and represents $2 \times 2 \times 2 \times 2$.

exponential part   One part of a number expressed in scientific notation; it represents the number of places the decimal point has moved.

Fahrenheit (°F) scale   The temperature scale that is most familiar in the United States; water freezes at 32 °F and boils at 212 °F at 1 atm prssure.

family (of elements)   A group of elements that have similar outer electron configurations and therefore similar properties. Families occur in vertical columns in the periodic table.

family (of organic compounds)   A group of organic compounds with the same functional group.

fatty acid   A type of lipid consisting of a carboxylic acid with a long hydrocarbon tail.

film badge dosimeter   Badges used to measure radiation exposure, consisting of photographic film held in a small case that is pinned to clothing.

fission, nuclear   The process by which a heavy nucleus is split into nuclei of smaller masses and energy is emitted.

formula mass   The average mass of the molecules (or formula units) that compose a compound.

formula unit   The basic unit of ionic compounds; the smallest electrically neutral collection of cations and anions that compose the compound.

freezing point depression   The decrease in the freezing point of a solvent caused by the presence of a solute.

frequency   The number of wave cycles or crests that pass through a stationary point in one second.

fuel cell   A voltaic cell in which the reactants are constantly replenished.

functional group   A set of atoms that characterize a family of organic compounds.

fusion, nuclear   The combination of light atomic nuclei to form heavier ones with emission of large amounts of energy.

galvanic (voltaic) cell   An electrochemical cell that spontaneously produces electrical current.

gamma radiation   High-energy, short-wavelength electromagnetic radiation emitted by an atomic nucleus.

gamma rays   The shortest-wavelength, most energetic form of electromagnetic radiation. Gamma ray photons are represented by the symbol $_0^0\gamma$.

gas   A state of matter in which atoms or molecules are widely separated and free to move relative to one another.

gas-evolution reaction   A reaction that occurs in solution and forms a gas as one of the products.

gas formation   One type of evidence of a chemical reaction, a gas forms when two substances are mixed together.

Geiger-Müller counter A radioactivity detector consisting of a chamber filled with argon gas that discharges electrical signals when high-energy particles pass through it.

gene A sequence of codons within a DNA molecule that codes for a single protein. Genes vary in length from hundreds to thousands of codons.

genetic material The inheritable blueprint for making organisms.

glycogen A type of polysaccharide; it has a structure similar to that of starch, but the chain is highly branched.

glycolipid A biological molecule composed of a nonpolar fatty acid and hydrocarbon chain and a polar section composed of a sugar molecule such as glucose.

glycoside linkage The link between monosaccharides in a polysaccharide.

ground state The state of an atom or molecule in which the electrons occupy the lowest possible energy orbitals available.

group (of elements) Elements that have similar outer electron configurations and therefore similar properties. Groups occur in vertical columns in the periodic table.

half-cell A compartment in which the oxidation or reduction half-reaction occurs in a galvanic or voltaic cell.

half-life The time it takes for one-half of the parent nuclides in a radioactive sample to decay to the daughter nuclides.

half-reaction Either the oxidation part or the reduction part of a redox reaction.

halogens The Group 7A elements, which are very reactive nonmetals.

heat absorption One type of evidence of a chemical reaction, involving the intake of energy.

heat capacity The quantity of heat energy required to change the temperature of a given amount of a substance by 1 °C.

heat emission One type of evidence of a chemical reaction, involving the evolution of thermal energy.

heat of fusion The amount of heat required to melt one mole of a solid at its melting point with no change in temperature.

heat of vaporization The amount of heat required to vaporize one mole of a liquid at its boiling point with no change in temperature.

heterogeneous mixture A mixture, such as oil and water, that has two or more regions with different compositions.

homogeneous mixture A mixture, such as salt water, that has the same composition throughout.

human genome All of the genetic material of a human being; the total DNA of a human cell.

Hund's rule A rule stating that when filling orbitals of equal energy, electrons will occupy empty orbitals singly before pairing with other electrons.

hydrocarbon A compound that contains only carbon and hydrogen atoms.

hydrogen bond A strong dipole–dipole interaction between molecules containing hydrogen directly bonded to a small, highly electronegative atom, such as N, O, or F.

hydrogenation The chemical addition of hydrogen to a compound.

hydronium ion The H_3O^+ ion. Chemists often use $H^+(aq)$ and $H_3O^+(aq)$ interchangeably to mean the same thing—a hydronium ion.

hypothesis A theory or law before it has become well established; a tentative explanation for an observation or a scientific problem that can be tested by further investigation.

hypoxia A shortage of oxygen in the tissues of the body.

ideal gas law A law that combines the four properties of a gas—pressure (P), volume (V), temperature (T), and number of moles (n) in a single equation showing their interrelatedness: $PV = nRT$ (R = ideal gas constant).

indicator A substance that changes color with acidity level, often used to detect the endpoint of a titration.

infrared (IR) light The fraction of the electromagnetic spectrum between visible light and microwaves. Infrared light is invisible to the human eye.

insoluble Not soluble in water.

instantaneous dipole A type of intermolecular force resulting from transient shifts in electron density within an atom or molecule.

intermolecular forces Attractive forces that exist between molecules.

International System (SI) The standard set of units for science measurements, based on the metric system.

ion An atom (or group of atoms) that has gained or lost one or more electrons, so that it has an electric charge.

ion product constant (K_w) The product of the H_3O^+ ion concentration and the OH^- ion concentration in an aqueous solution. At room temperature, $K_w = 1.0 \times 10^{-14}$.

ionic bond The bond that results when a metal and a nonmetal combine in a chemical reaction. In an ionic bond, the metal transfers one or more electrons to the nonmetal.

ionic compound A compound formed between a metal and one or more nonmetals.

ionic solid A solid compound composed of metals and nonmetals joined by ionic bonds.

ionization The forming of ions.

ionization energy The energy required to remove an electron from an atom in the gaseous state.

ionizing power The ability of radiation to ionize other molecules and atoms.

isomers Molecules with the same molecular formula but different structures.

isoosmotic Describes solutions having equal osmotic pressure.

isotope scanning The use of radioactive isotopes to identify disease in the body.

isotopes Atoms with the same number of protons but different numbers of neutrons.

Kelvin (K) scale The temperature scale that assigns 0 K to the coldest temperature possible, absolute zero (−273 °C or −459 °F), the temperature at which molecular motion stops. The size of the kelvin is identical to that of the Celsius degree.

ketone An organic compound with the general formula RCOR.

kilogram (kg) The SI standard unit of mass.

kilowatt-hour (kWh) A unit of energy equal to 3.6 million joules.

kinetic energy Energy associated with the motion of an object.

kinetic molecular theory A simple model for gases that predicts the behavior of most gases under many conditions.

Le Châtelier's principle A principle stating that when a chemical system at equilibrium is disturbed, the system shifts in a direction that minimizes the disturbance.

lead-acid storage battery An automobile battery consisting of six electrochemical cells wired in series. Each cell produces 2 volts for a total of 12 volts.

Lewis structure A drawing that represents chemical bonds between atoms as shared or transferred electrons; the valence electrons of atoms are represented as dots.

Lewis theory A simple theory for chemical bonding involving diagrams showing bonds between atoms as lines or dots. In this theory, atoms bond together to obtain stable octets (8 valence electrons).

light emission One type of evidence of a chemical reaction, involving the giving off of electromagnetic radiation.

limiting reactant The reactant that determines the amount of product formed in a chemical reaction.

linear Describes the molecular geometry of a molecule containing two electron groups (two bonding groups and no lone pairs).

linearly related A relationship between two variables such that, when they are plotted one against the other, the graph produced is a straight line.

lipid A cellular component that is insoluble in water but soluble in nonpolar solvents.

lipid bilayer A structure formed by lipids in the cell membrane.

liquid A state of matter in which atoms or molecules are packed close to each other (about as closely as in a solid) but are free to move around and by each other.

logarithmic scale A scale involving logarithms. A logarithm entails an exponent that indicates the power to which a number is raised to produce a given number (e.g., the logarithm of 100 to the base 10 is 2).

lone pair Electrons that are only on one atom in a Lewis structure.

main-group elements Groups 1A–8A on the periodic table. These groups have properties that tend to be predictable based on their position in the periodic table.

mass A measure of the quantity of matter within an object.

mass number (A) The sum of the number of neutrons and protons in an atom.

mass percent composition (or mass percent) The percentage, by mass, of each element in a compound.

matter Anything that occupies space and has mass. Matter exists in three different states: solid, liquid, and gas.

melting point The temperature at which a solid turns into a liquid.

messenger RNA (mRNA) Long chainlike molecules that act as blueprints for the construction of proteins.

metallic atomic solid An atomic solid, such as iron, which is held together by metallic bonds that, in the simplest model, consist of positively charged ions in a sea of electrons.

metallic character The properties typical of a metal, especially the tendency to lose electrons in chemical reactions. Elements become more metallic as you move from right to left across the periodic table.

metalloids Those elements that fall along the boundary between the metals and the nonmetals in the periodic table; their properties are intermediate between those of metals and those of nonmetals.

metals Elements that tend to lose electrons in chemical reactions. They are found at the left side and in the center of the periodic table.

meter (m) The SI standard unit of length.

metric system The unit system commonly used throughout most of the world.

microwaves The part of the electromagnetic spectrum between the infrared region and the radio wave region. Microwaves are efficiently absorbed by water molecules and can therefore be used to heat water-containing substances.

millimeter of mercury (mmHg) A unit of pressure that originates from the method used to measure pressure with a barometer. Also called a *torr*.

miscibility The ability of two liquids to mix without separating into two phases, or the ability of one liquid to mix with (dissolve in) another liquid.

mixture A substance composed of two or more different types of atoms or molecules combined in variable proportions.

molality (m) A common unit of solution concentration, defined as the number of moles of solute per kilogram of solvent.

molar mass The mass of one mole of atoms of an element or one mole of molecules (or formula units) for a compound. An element's molar mass in grams per mole is numerically equivalent to the element's atomic mass in amu.

molar solubility The solubility of a substance in units of moles per liter (mol/L).

molar volume The volume occupied by one mole of gas. Under standard temperature and pressure conditions the molar volume of ideal gas is 22.5 L.

molarity (M) A common unit of solution concentration, defined as the number of moles of solute per liter of solution.

mole Avogadro's number (6.022×10^{23}) of particles—especially, of atoms, ions, or molecules. A mole of any element has a mass in grams that is numerically equivalent to its atomic mass in amu.

molecular compound A compound formed from two or more nonmetals. Molecular compounds have distinct molecules as their simplest identifiable units.

molecular element An element that does not normally exist in nature with single atoms as the basic unit. These elements usually exist as diatomic molecules—2 atoms of that element bonded together—as their basic units.

molecular equation A chemical equation showing the complete, neutral formulas for every compound in a reaction.

molecular formula A formula for a compound that gives the specific number of each type of atom in a molecule.

molecular geometry The geometrical arrangement of the atoms in a molecule.

molecular solid A solid whose composite units are molecules.

molecule Two or more atoms joined in a specific arrangement by chemical bonds. A molecule is the smallest identifiable unit of a molecular compound.

monomer An individual repeating unit that makes up a polymer.

monoprotic acid An acid containing only one ionizable proton.

monosaccharide A carbohydrate that cannot be decomposed into simpler carbohydrates.

monosubstituted benzene A benzene in which one of the hydrogen atoms has been replaced by another atom or group of atoms.

net ionic equation An equation that shows only the species that actually participate in a reaction.

neutral solution A solution in which the concentrations of H_3O^+ and OH^- are equal (pH = 7).

neutralization A reaction that takes place when an acid and a base are mixed; the $H^+(aq)$ from the acid combines with the $OH^-(aq)$ from the base to form $H_2O(l)$.

neutron A nuclear particle with no electrical charge and nearly the same mass as a proton.

nitrogen narcosis An increase in nitrogen concentration in bodily tissues and fluids that results in feelings of drunkenness.

noble gases The Group 8A elements, which are chemically unreactive.

nonbonding atomic solid An atomic solid that is held together by relatively weak dispersion forces.

nonelectrolyte solution A solution containing a solute that dissolves as molecules; therefore, the solution does not conduct electricity.

nonmetals Elements that tend to gain electrons in chemical reactions. They are found at the upper right side of the periodic table.

nonpolar molecule A molecule that does not have a net dipole moment.

nonvolatile Describes a compound that does not vaporize easily.

normal alkane (or *n*-alkane) An alkane composed of carbon atoms bonded in a straight chain with no branches.

normal boiling point The boiling point of a liquid at a pressure of 1 atmosphere.

nuclear equation An equation that represents the changes that occur during radioactivity and other nuclear processes.

nuclear radiation The energetic particles emitted from the nucleus of an atom when it is undergoing a nuclear process.

nuclear theory of the atom A theory stating that most of the atom's mass and all of its positive charge is contained in a small, dense nucleus. Most of the volume of the atom is empty space occupied by negatively charged electrons.

nucleic acids Biological molecules, such as deoxyribonucleic acid (DNA) and ribonucleic acid (RNA), that store and transmit genetic information.

nucleotide An individual unit of a nucleic acid. Nucleic acids are polymers of nucleotides.

nucleus (of a cell) The part of the cell that contains the genetic material.

nucleus (of an atom) The small core containing most of the atom's mass and all of its positive charge. The nucleus is made up of protons and neutrons.

observation Often the first step in the scientific method. An observation must measure or describe something about the physical world.

octet The number of electrons, eight, around atoms with stable Lewis structures.

octet rule A rule that states that an atom will give up, accept, or share electrons in order to achieve a filled outer electron shell, which usually consists of 8 electrons.

orbital The region around the nucleus of an atom where an electron is most likely to be found.

orbital diagram An electron configuration in which electrons are represented as arrows in boxes corresponding to orbitals of a particular atom.

organic chemistry The study of carbon-containing compounds and their reactions.

organic molecule A molecule whose main structural component is carbon.

osmosis The flow of solvent from a lower-concentration solution through a semipermeable membrane to a higher-concentration solution.

osmotic pressure The pressure produced on the surface of a semipermeable membrane by osmosis or the pressure required to stop osmotic flow.

oxidation The gain of oxygen, the loss of hydrogen, or the loss of electrons (the most fundamental definition).

oxidation state (or oxidation number) A number that can be used as an aid in writing formulas and balancing equations. It is computed for each element based on the number of electrons assigned to it in a scheme where the most electronegative element is assigned all of the bonding electrons.

oxidation–reduction (redox) reaction A reaction in which electrons are transferred from one substance to another.

oxidizing agent In a redox reaction, the substance being reduced. Oxidizing agents tend to gain electrons easily.

oxyacid An acid containing hydrogen, a nonmetal, and oxygen.

oxyanion An anion containing oxygen. Most polyatomic ions are oxyanions.

oxygen toxicity The result of increased oxygen concentration in bodily tissues.

parent nuclide The original nuclide in a nuclear decay.

partial pressure The pressure due to any individual component in a gas mixture.

pascal (Pa) The SI unit of pressure, defined as 1 newton per square meter.

Pauli exclusion principle A principle stating that no more than two electrons can occupy an orbital and that the two electrons must have opposite spins.

penetrating power The ability of a radioactive particle to penetrate matter.

peptide bond The bond between the amine end of one amino acid and the carboxylic acid end of another. Amino acids link together via peptide bonds to form proteins.

percent natural abundance The percentage amount of each isotope of an element in a naturally occurring sample of the element.

percent yield In a chemical reaction, the percentage of the theoretical yield that was actually attained.

period A horizontal row of the periodic table.

periodic law A law that states that when the elements are arranged in order of increasing relative mass, certain sets of properties recur periodically.

periodic table An arrangement of the elements in which atomic number increases from left to right and elements with similar properties fall in columns called families or groups.

permanent dipole A separation of charge resulting from the unequal sharing of electrons between atoms.

pH scale A scale used to quantify acidity or basicity. A pH of 7 is neutral; a pH lower than 7 is acidic, and a pH greater than 7 is basic. The pH is defined as follows: $pH = -\log[H_3O^+]$.

phenyl group The term for a benzene ring when other substituents are attached to it.

phospholipid A lipid with the same basic structure as a triglyceride, except that one of the fatty acid groups is replaced with a phosphate group.

phosphorescence The slow, long-lived emission of light that sometimes follows the absorption of light by some atoms and molecules.

photon A particle of light or a packet of light energy.

physical change A change in which matter does not change its composition, even though its appearance might change.

physical properties Those properties that a substance displays without changing its composition.

polar covalent bond A covalent bond between atoms of different electronegativities. Polar covalent bonds have a dipole moment.

polar molecule A molecule with polar bonds that add together to create a net dipole moment.

polyatomic ion An ion composed of a group of atoms with an overall charge.

polymer A molecule with many similar units, called monomers, bonded together in a long chain.

polypeptide A short chain of amino acids joined by peptide bonds.

polysaccharide A long, chainlike molecule composed of many linked monosaccharide units. Polysaccharides are polymers of monosaccharides.

positron A nuclear particle that has the mass of an electron but carries a +1 charge.

positron emission Expulsion of a positron from an unstable atomic nucleus. In positron emission, a proton is transformed into a neutron.

potential energy The energy of a body that is associated with its position or the arrangement of its parts.

precipitate An insoluble product formed through the reaction of two solutions containing soluble compounds.

precipitation reaction A reaction that forms a solid or precipitate when two aqueous solutions are mixed.

prefix multipliers Prefixes used by the SI system with the standard units. These multipliers change the value of the unit by powers of 10.

pressure The force exerted per unit area by gaseous molecules as they collide with the surfaces around them.

primary protein structure The sequence of amino acids in a protein's chain. Primary protein structure is maintained by the covalent peptide bonds between individual amino acids.

principal quantum number A number that indicates the shell that an electron occupies.

principal shell The shell indicated by the principal quantum number.

products The final substances produced in a chemical reaction; represented on the right side of a chemical equation.

properties The characteristics we use to distinguish one substance from another.

protein A biological molecule composed of a long chain of amino acids joined by peptide bonds. In living organisms, proteins serve many varied and important functions.

proton A positively charged nuclear particle. A proton's mass is approximately 1 amu.

pure substance A substance composed of only one type of atom or molecule.

quantification The assigning of a number to an observation so as to specify a quantity or property precisely.

quantum (*pl.* quanta) The precise amount of energy possessed by a photon; the difference in energy between two atomic orbitals.

quantum number (*n*) An integer that specifies the energy of an orbital. The higher the quantum number n, the greater the distance between the electron and the nucleus and the higher its energy.

quantum-mechanical model The foundation of modern chemistry; explains how electrons exist in atoms, and how they affect the chemical and physical properties of elements.

quaternary structure In a protein, the way that individual chains fit together to compose the protein. Quaternary structure is maintained by interactions between the R groups of amino acids on the different chains.

R group (side chain) An organic group attached to the central carbon atom of an amino acid.

radio waves The longest wavelength and least energetic form of electromagnetic radiation.

radioactive Describes a substance that emits tiny, invisible, energetic particles from the nuclei of its component atoms.

radioactivity The emission of tiny, invisible, energetic particles from the unstable nuclei of atoms. Many of these particles can penetrate matter.

radiocarbon dating A technique used to estimate the age of fossils and artifacts through the measurement of natural radioactivity of carbon atoms in the environment.

radiotherapy Treatment of disease with radiation, such as the use of gamma rays to kill rapidly dividing cancer cells.

random coil The name given to an irregular pattern of a secondary protein structure.

rate of a chemical reaction (reaction rate) The amount of reactant that changes to product in a given period of time. Also defined as the amount of a product that forms in a given period of time.

reactants The initial substances in a chemical reaction, represented on the left side of a chemical equation.

recrystallization A technique used to purify a solid; involves dissolving the solid in a solvent at high temperature, creating a saturated solution, then cooling the solution to cause the crystallization of the solid.

reducing agent In a redox reaction, the substance being oxidized. Reducing agents tend to lose electrons easily.

reduction The loss of oxygen, the gain of hydrogen, or the gain of electrons (the most fundamental definition).

rem Stands for *roentgen equivalent man;* a weighted measure of radiation exposure that accounts for the ionizing power of the different types of radiation.

resonance structures Two or more Lewis structures that are necessary to describe the bonding in a molecule or ion.

reversible reaction A reaction that is able to proceed in both the forward and reverse directions.

RNA (ribonucleic acid) Long chainlike molecules that occur throughout cells and take part in the construction of proteins.

salt An ionic compound that usually remains dissolved in a solution after an acid–base reaction has occurred.

salt bridge An inverted, U-shaped tube containing a strong electrolyte; completes the circuit in an electrochemical cell by allowing the flow of ions between the two half-cells.

saturated fat A triglyceride composed of saturated fatty acids. Saturated fat tends to be solid at room temperature.

saturated hydrocarbon A hydrocarbon that contains no double or triple bonds between the carbon atoms.

saturated solution A solution that holds the maximum amount of solute under the solution conditions. If additional solute is added to a saturated solution, it will not dissolve.

scientific law A statement that summarizes past observations and predicts future ones. Scientific laws are usually formulated from a series of related observations.

scientific method The way that scientists learn about the natural world. The scientific method involves observations, laws, hypotheses, theories, and experimentation.

scientific notation A system used to write very big or very small numbers, often containing many zeros, more compactly and precisely. A number written in scientific notation consists of a decimal part and an exponential part (10 raised to a particular exponent).

scintillation counter A device used to detect radioactivity in which energetic particles traverse a material that emits ultraviolet or visible light when excited by their passage. The light is detected and turned into an electrical signal.

second (s) The SI standard unit of time.

secondary protein structure Short-range periodic or repeating patterns often found in proteins. Secondary protein structure is maintained by interactions between amino acids that are fairly close together in the linear sequence of the protein chain or adjacent to each other on neighboring chains.

semiconductor A compound or element exhibiting intermediate electrical conductivity that can be changed and controlled.

semipermeable membrane A membrane that selectively allows some substances to pass through but not others.

SI units The most convenient system of units for science measurements, based on the metric system. The set of standard units agreed on by scientists throughout the world.

significant digits (figures) The non-place-holding digits in a reported measurement; they represent the precision of a measured quantity.

simple carbohydrate (simple sugar) A monosaccharide or disaccharide.

single bond A chemical bond in which one electron pair is shared between two atoms.

solid A state of matter in which atoms or molecules are packed close to each other in fixed locations.

solid formation One type of evidence of a chemical reaction, involving the formation of a solid.

solubility The amount of a compound, usually in grams, that will dissolve in a certain amount of solvent.

solubility rules A set of empirical rules used to determine whether an ionic compound is soluble.

solubility-product constant (K_{sp}) The equilibrium expression for a chemical equation that represents the dissolving of an ionic compound in solution.

soluble Dissolves in solution.

solute The minority component of a solution.

solution A homogeneous mixture of two or more substances.

solvent The majority component of a solution.

specific heat capacity (or specific heat) The heat capacity of a substance in joules per gram degree celsius (J/g °C).

spectator ions Ions that do not participate in a reaction; they appear unchanged on both sides of a chemical equation.

standard temperature and pressure (STP) Conditions often assumed in calculations involving gases: $T = 0$ °C (273 K) and $P = 1$ atm.

starch A common polysaccharide composed of repeating glucose units.

states of matter The three forms in which matter can exist: solid, liquid, and gas.

steroid A biological compound containing a 17-carbon 4-ring system.

stock solution A concentrated form in which solutions are often stored.

stoichiometry The numerical relationships among chemical quantities in a balanced chemical equation. Stoichiometry allows us to predict the amounts of products that form in a chemical reaction based on the amounts of reactants.

strong acid An acid that completely ionizes in solution.

strong base A base that completely dissociates in solution.

strong electrolyte A substance whose aqueous solutions are good conductors of electricity.

strong electrolyte solution A solution containing a solute that dissociates into ions; therefore, a solution that conducts electricity well.

structural formula A two-dimensional representation of molecules that not only shows the number and type of atoms, but also how the atoms are bonded together.

sublimation A physical change in which a substance is converted from its solid form directly into its gaseous form.

subshell In quantum mechanics, specifies the shape of the orbital and is represented by different letters (s, p, d, f).

substituent An atom or a group of atoms that has been substituted for a hydrogen atom in an organic compound.

substitution reaction A reaction in which one or more atoms are replaced by one or more different atoms.

supersaturated solution A solution holding more than the normal maximum amount of solute.

surface tension The tendency of liquids to minimize their surface area, resulting in a "skin" on the surface of the liquid.

synthesis A reaction in which simpler substances combine to form more complex substances; A + B $\longrightarrow$ AB.

temporary dipole A type of intermolecular force resulting from transient shifts in electron density within an atom or molecule.

terminal atom An atom that is located at the end of a molecule or chain.

tertiary structure A protein's structure that consists of the large-scale bends and folds due to interactions between the R groups of amino acids that are separated by large distances in the linear sequence of the protein chain.

tetrahedral The molecular geometry of a molecule containing four electron groups (four bonding groups and no lone pairs).

theoretical yield The maximum amount of product that can be made in a chemical reaction based on the amount of limiting reactant.

theory A proposed explanation for observations and laws. A theory presents a model of the way nature works and predicts behavior that extends well beyond the observations and laws from which it was formed.

titration A laboratory procedure used to determine the amount of a substance in solution. In a titration, a reactant in a solution of known concentration is reacted with another reactant in a solution of unknown concentration until the reaction reaches the endpoint.

torr A unit of pressure named after the Italian physicist Evangelista Torricelli; also called a millimeter of mercury.

transition metals The elements in the middle of the periodic table whose properties tend to be less predictable based simply on their position in the periodic table. Transition metals lose electrons in their chemical reactions, but do not necessarily acquire noble gas configurations.

triglyceride A fat or oil; a tryglyceride is a tri-ester composed of glycerol with three fatty acids attached.

trigonal planar The molecular geometry of a molecule containing three electron groups, three bonding groups, and no lone pairs.

trigonal pyramidal The molecular geometry of a molecule containing four electron groups, three bonding groups, and one lone pair.

triple bond A chemical bond consisting of three electron pairs shared between two atoms. In general, triple bonds are shorter and stronger than double bonds.

Type I compounds Compounds containing metals that always form cations with the same charge.

Type II compounds Compounds containing metals that can form cations with different charges.

ultraviolet (UV) light The fraction of the electromagnetic spectrum between the visible region and the X-ray region. UV light is invisible to the human eye.

units Previously agreed-on quantities used to report experimental measurements. Units are vital in chemistry.

unsaturated fat (or oil) A triglyceride composed of unsaturated fatty acids. Unsaturated fats tend to be liquids at room temperature.

unsaturated hydrocarbon A hydrocarbon that contains one or more double or triple bonds between its carbon atoms.

unsaturated solution A solution holding less than the maximum possible amount of solute under the solution conditions.

valence electrons The electrons in the outermost principal shell of an atom; they are involved in chemical bonding.

valence shell electron pair repulsion (VSEPR) A theory that allows prediction of the shapes of molecules based on the idea that electrons—either as lone pairs or as bonding pairs—repel one another.

vapor pressure The partial pressure of a vapor in dynamic equilibrium with its liquid.

vaporization The phase transition between a liquid and a gas.

viscosity The resistance of a liquid to flow; manifestation of intermolecular forces.

visible light The fraction of the electromagnetic spectrum that is visible to the human eye, bounded by wavelengths of 400 nm (violet) and 780 nm (red).

vital force A mystical or supernatural power that, it was once believed, was possessed only by living organisms and allowed them to produce organic compounds.

vitalism The belief that living things contain a nonphysical "force" that allows them to synthesize organic compounds.

volatile Tending to vaporize easily.

voltage The potential difference between two electrodes; the driving force that causes electrons to flow.

volume A measure of space. Any unit of length, when cubed, becomes a unit of volume.

wavelength The distance between adjacent wave crests in a wave.

weak acid An acid that does not completely ionize in solution.

weak base A base that does not completely dissociate in solution.

weak electrolyte A substance whose aqueous solutions are poor conductors of electricity.

X-rays The portion of the electromagnetic spectrum between the ultraviolet (UV) region and the gamma-ray region.

Note: Answers in the Questions section are written as briefly as possible. Student answers may vary and still be correct.

CHAPTER 1
QUESTIONS

1. Soda fizzes due to the interactions between carbon dioxide and water under high pressure. At room temperature, carbon dioxide is a gas and water is a liquid. Through the use of pressure, the makers of soda force the carbon dioxide gas to dissolve in the water. When the can is sealed, the solution remains mixed. When the can is opened, the pressure is released and the carbon dioxide molecules escape in bubbles of gas.

3. Chemists study molecules and interactions at the molecular level to learn about and explain macroscopic events. Chemists attempt to explain why ordinary things are as they are.

5. Chemistry is the science that seeks to understand what matter does by studying what atoms and molecules do.

7. The scientific method is the way chemists investigate the chemical world. The first step consists of observing the natural world. Later observations can be combined to create a scientific law, which summarizes and predicts behavior. Theories are models that strive to explain the cause of the observed phenomenon. Theories are tested through experiment. When a theory is not well established, it is sometimes referred to as a hypothesis.

9. A law is simply a general statement that summarizes and predicts observed behavior. Theories seek to explain the causes of observed behavior.

11. To say "It is just a theory" makes it seem as if theories are easily discardable. However, many theories are very well established and are as close to truth as we get in science. Established theories are backed up with years of experimental evidence, and they are the pinnacle of scientific understanding.

13. The atomic theory states that all matter is composed of small, indestructible particles called atoms. John Dalton formulated this theory.

PROBLEMS

15. Carbon dioxide contains one carbon atom and two oxygen atoms. Water contains one oxygen atom and two hydrogen atoms.

17. **a.** observation   **b.** theory
     c. law   **d.** observation

19. Mass (g) Volume (L) Ratio (g/L)
22.5 1.6 14
35.8 2.55 14.0
70.2 5.00 14.0
98.5 7.01 14.1
The ratio of mass to volume is constant.

21. **a.** All atoms contain a degree of chemical reactivity. The larger the size of an atom, the higher the chemical reactivity of that atom.

b. There are many correct answers. One example is: Conceivably, when the size of an atom is increased, the surface area of the atom is also increased; an atom with a greater surface area is more likely to react chemically.

CHAPTER 2
QUESTIONS

1. Without units, the results are unclear and it is hard to keep track of what each separate measurement entails.

3. Often scientists work with very large or very small numbers that contain a lot of zeros. Scientific notation allows these numbers to be written more compactly, and the information is more organized.

5. Zeros count as significant digits when they are interior zeros (zeros between two numbers) and when they are trailing zeros (zeros after a decimal point). Zeros are **not** significant digits when they are leading zeros, which are zeros to the left of the first nonzero number.

7. For calculations involving only multiplication and division, the result carries the same number of significant figures as the factor with the fewest significant figures.

9. In calculations involving both multiplication/division and addition/subtraction, do the steps in parentheses first; determine the correct number of significant figures in the intermediate answer; then do the remaining steps.

11. The basic SI unit of length is the meter. The kilogram is the SI unit of mass. Lastly, the second is the SI unit of time.

13. For measuring a Frisbee, the unit would be the meter and the prefix multiplier would be *centi*-. The final measurement would be in centimeters.

15. **a.** 2.42 cm   **b.** 1.79 cm
     c. 21.58 cm   **d.** 21.85 cm

17. Units act as a guide in the calculation and are able to show if the calculation is off track. The units must be followed in the calculation, so that the answer is correctly written and understood.

19. A conversion factor is a quantity used to relate two separate units. They are constructed from any two quantities known to be equivalent.

21. The conversion factor is $\dfrac{1 \text{ ft}}{12 \text{ in.}}$. For a feet-to-inches conversion, the conversion factor must be inverted $\left(\dfrac{12 \text{ in.}}{1 \text{ ft}} \right)$.

23. a. Sort the information into the **given** information (the starting point for the problem) and the **find** information (the end point).

b. Create a solution map to get from the given information to the information you are trying to find. This will likely include conversion factors or equations.

c. Follow the solution map to solve the problem. Carry out mathematical operations and cancel units as needed.

d. Ask, does this answer make physical sense? Are the units correct? Is the number of significant figures correct?

25. The solution map for converting grams to pounds is:

$$\frac{1 \text{ lb}}{453.59 \text{ g}}$$

27. The solution map for converting meters to feet is:

$$\frac{100 \text{ cm}}{1 \text{ m}} \qquad \frac{1 \text{ ft}}{30.48 \text{ cm}}$$

29. The density of a substance is the ratio of its mass to its volume. Density is a fundamental property of materials and differs from one substance to another. Density can be used to relate two separate units, thus working as a conversion factor. Density is a conversion factor between mass and volume.

PROBLEMS

31. a. 3.6756×10^7 **b.** 1.288×10^6
c. 1.949×10^7 **d.** 5.32×10^5

33. a. 7.461×10^{-11} m   **b.** 1.58×10^{-5} mi
c. 6.32×10^{-7} m   **d.** 1.5×10^{-5} m

35. a. 602,200,000,000,000,000,000,000
b. 0.00000000000000000016 C
c. 299,000,000 m/s
d. 344 m/s

37. a. 32,200,000   **b.** 0.0072
c. 118,000,000,000   **d.** 0.00000943

39. 2,000,000,000 2×10^9
1,211,000,000 1.211×10^9
0.000874 8.74×10^{-4}
320,000,000,000 3.2×10^{11}

41. a. 54.9 mL   **b.** 48.7 °C
c. 46.83 °C   **d.** 64 mL

43. a. 0.005050   **b.** 0.000000000000000060
c. 220,103   **d.** 0.00108

45. a. 4   **b.** 4
c. 6   **d.** 5

47. a. correct   **b.** 3
c. 7   **d.** correct

49. a. 256.0   **b.** 0.0004893
c. 2.901×10^{-4} **d.** 2.231×10^6

51. a. 2.3   **b.** 2.4
c. 2.3   **d.** 2.4

53. a. 42.3   **b.** correct
c. correct   **d.** 0.0456

55. 8.32466   8.325   8.3   8
84.57225   84.57   85 8×10^1
132.5512   132.6 1.3×10^2 1×10^2

57. a. 0.054   **b.** 0.619
c. 1.2×10^8 **d.** 6.6

59. a. 4.22×10^3 **b.** correct
c. 3.9969   **d.** correct

61. a. 110.6   **b.** 41.4
c. 183.3   **d.** 1.22

63. a. correct   **b.** 1.0982
c. correct   **d.** 3.53

65. a. 3.9×10^3 **b.** 632
c. 8.93×10^4 **d.** 6.34

67. a. 3.15×10^3 **b.** correct
c. correct   **d.** correct

69. a. 3.55×10^3 g   **b.** 8.944 m
c. 4.598×10^{-3} kg   **d.** 18.7 mL

71. a. 0.588 L   **b.** 34.1 μg
c. 10.1 ns   **d.** 2.19×10^{-12} m

73. a. 57.2 cm   **b.** 38.4 m
c. 0.754 km   **d.** 61 mm

75. a. 15.7 in   **b.** 91.2 ft
c. 6.21 mi   **d.** 8478 lb

77. 5.08×10^8 m 5.08×10^5 km   508 Mm
5.08×10^{-1} Gm 5.08×10^{-4} Tm
2.7976×10^{10} m 2.7976×10^7 km   27,976 Mm
2.7976×10^1 Gm 2.7976×10^{-2} Tm
1.77×10^{12} m 1.77×10^9 km 1.77×10^6 Mm
1.77×10^3 Gm   1.77 Tm
1.5×10^8 m 1.5×10^5 km 1.5×10^2 Mm
0.15 Gm 1.5×10^{-4} Tm
4.23×10^{11} m 4.23×10^8 km 4.23×10^5 Mm
423 Gm   0.423 Tm

79. a. 2.255×10^7 kg   **b.** 2.255×10^4 Mg
c. 2.255×10^{13} mg   **d.** 2.255×10^4 metric tons

81. 1.5×10^3 g

83. 5.0×10^1 min

85. 4.7×10^3 cm^3

87. a. 1.0×10^6 m^2 **b.** 1.0×10^{-6} m^3
c. 1.0×10^{-9} m^3

89. **a.** $6.2 \times 10^5 \, \text{pm}^3$ **b.** $6.2 \times 10^{-4} \, \text{nm}^3$
     c. $6.2 \times 10^{-1} \, \text{Å}^3$

91. **a.** $2.15 \times 10^{-4} \, \text{km}^2$ **b.** $2.15 \times 10^4 \, \text{dm}^2$
     c. $2.15 \times 10^6 \, \text{cm}^2$

93. $1.49 \times 10^6 \, \text{mi}^2$

95. $11.4 \, \text{g/cm}^3$, lead

97. $1.26 \, \text{g/cm}^3$

99. Yes, the density of the crown is $19.3 \, \text{g/cm}^3$.

101. **a.** $4.30 \times 10^2 \, \text{g}$ **b.** $3.12 \, \text{L}$

103. **a.** $3.38 \times 10^4 \, \text{g}$ (gold); $5.25 \times 10^3 \, \text{g}$ (sand)

     b. Yes, the mass of the bag of sand is different from the mass of the gold vase; thus, the weight-sensitive pedestal will sound the alarm.

105. $10.6 \, \text{g/cm}^3$

107. $2.7 \times 10^3 \, \dfrac{\text{kg}}{\text{m}^3}$

109. $2.5 \times 10^5 \, \text{lbs}$

111. $1.19 \times 10^5 \, \text{kg}$

113. $18 \, \text{km/L}$

115. $768 \, \text{mi}$

117. Metal A is denser than metal B.

119. $2.26 \, \text{g/cm}^3$

121. $108 \, \text{km}$; $47.2 \, \text{km}$

123. $9.1 \times 10^{10} \, \text{g/cm}^3$

CHAPTER 3
QUESTIONS

1. Matter is defined as anything that occupies space and possesses mass. It can be thought of as the physical material that makes up the universe.

3. The three states of matter are solid, liquid, and gas.

5. In a crystalline solid, the atoms/molecules are arranged in geometric patterns with repeating order. In amorphous solids, the atoms/molecules do not have long-range order.

7. The atoms/molecules in gases are not in contact with each other and are free to move relative to one another. The spacing between separate atoms/molecules is very far apart. A gas has no fixed volume or shape; rather, it assumes both the shape and the volume of the container it occupies.

9. A mixture is two or more pure substances combined in variable proportions.

11. Pure substances are those composed of only one type of atom or molecule.

13. A mixture is formed when two or more pure substances are mixed together; however, a new substance is not formed. A compound is formed when two or more elements are bonded together and form a new substance.

15. In a physical change, the composition of the substance does not change, even though its appearance might change. However, in a chemical change, the substance undergoes a change in its composition.

17. Energy is defined as the capacity to do work.

19. Kinetic energy is the energy associated with the motion of an object. Potential energy is the energy associated with the position or composition of an object.

21. Three common units for energy are joules, calories, and kilowatt-hour.

23. An endothermic reaction is one that absorbs energy from the surroundings. The products have more energy than the reactants in an endothermic reaction.

25. Heat is the transfer of thermal energy caused by a temperature difference, whereas temperature is a measure of the thermal energy of matter.

27. Heat capacity is the quantity of heat energy required to change the temperature of a given amount of the substance by $1 \, °\text{C}$.

29. $°\text{F} = \dfrac{9}{5}(°\text{C}) + 32$

PROBLEMS

31. **a.** element   **b.** element
     c. compound   **d.** compound

33. **a.** homogeneous   **b.** heterogeneous
     c. homogeneous   **d.** homogeneous

35. **a.** pure substance-element
     b. mixture-homogeneous
     c. mixture-heterogeneous
     d. mixture-heterogeneous

37. **a.** chemical   **b.** physical
     c. physical   **d.** chemical

39. physical–colorless; odorless; gas at room temperature; one liter has a mass of 1.260 g under standard conditions; mixes with acetone; chemical–flammable; polymerizes to form polyethylene

41. **a.** chemical   **b.** physical
     c. chemical   **d.** chemical

43. **a.** physical   **b.** chemical

45. $2.10 \times 10^2 \, \text{kg}$

47. **a.** Yes   **b.** No

49. $15.1 \, \text{g of water}$

51. **a.** $2.46 \times 10^3 \, \text{J}$ **b.** $4.16 \times 10^{-3} \, \text{Cal}$
     c. $32.0 \, \text{Cal}$ **d.** $2.35 \times 10^5 \, \text{J}$

53. **a.** $9.0 \times 10^7 \, \text{J}$ **b.** $0.249 \, \text{Cal}$
     c. $1.31 \times 10^{-4} \, \text{kWh}$ **d.** $1.1 \times 10^4 \, \text{cal}$

55.

J	cal	Cal	kWh
225	53.8	5.38×10^{-2}	6.25×10^{-5}
3.44×10^6	8.21×10^5	8.21×10^2	9.54×10^{-1}
1.06×10^9	2.54×10^8	2.54×10^5	295
6.49×10^5	1.55×10^5	155	1.80×10^{-1}

57. $3.697 \times 10^9 \, \text{J}$

59. $8 \times 10^2 \, \text{kJ}$; 17 days

61. Exothermic.

63. **a.** exothermic, $-\Delta H$ **b.** endothermic, $+\Delta H$
    c. exothermic, $-\Delta H$

65. **a.** $1.00 \times 10^2 \,°C$ **b.** $-3.2 \times 10^2 \,°F$
    c. 298 K   **d.** $3.10 \times 10^2 \,K$

67. $-62 \,°C$, 211 K

69. 159 K, $-173 \,°F$

71. $-75.5 \,°F$

73.

0.0 K	$-459.4 \,°F$	$-273.0 \,°C$
301 K	$82.5 \,°F$	$28.1 \,°C$
282 K	$47 \,°F$	$8.5 \,°C$

75. $9.0 \times 10^3 \,J$

77. $8.7 \times 10^5 \,J$

79. $58 \,°C$

81. $31 \,°C$

83. $1.0 \times 10^1 \,°C$

85. $0.24 \,J/g°C$; silver

87. $2.2 \,J/g°C$

89. When warm drinks are placed into the ice, they release heat, which then melts the ice. The prechilled drinks, on the other hand, are already cold, so they do not release much heat.

91. $49 \,°C$

93. 70.2 J

95. $1.7 \times 10^4 \,kJ$

97. $67 \,°C$

99. 6.0 kWh

101. 22 g of fuel

103. 78 g

105. $27.2 \,°C$

107. $5.96 \times 10^5 \,kJ$; \$25

109. $-40°$

111. **a.** pure substance   **b.** pure substance
    c. pure substance   **d.** mixture

113. physical change

115. Small temperature changes in the ocean have a great impact on global weather because of the high heat capacity of water.

117. **a.** Sacramento is farther inland than San Francisco, so Sacramento is not as close to the ocean. The ocean water has a high heat capacity and will be able to keep San Francisco cooler in the hot days of summer. However, Sacramento is away from the ocean in a valley, so it will experience high temperatures in the summer.

b. San Francisco is located right next to the ocean, so the high heat capacity of the seawater keeps the temperature in the city from dropping. In the winter, the ocean actually helps to keep the city warmer, compared to an inland city like Sacramento.

CHAPTER 4
QUESTIONS

1. Democritus theorized that matter was ultimately composed of small, indivisible particles called atoms. Upon dividing matter, one would find tiny, indestructible atoms.

3. Rutherford's gold foil experiment involved sending positively charged alpha-particles through a thin sheet of gold foil and detecting if there was any deflection of the particles. He found that most passed straight through, yet some particles showed some deflection. This result contradicts the plum-pudding model of the atom because the plum-pudding model does not explain the deflection of the alpha-particles.

5.

Particle	Mass (kg)	Mass (amu)	Charge
Proton	1.67262×10^{-27}	1	+1
Neutron	1.67493×10^{-27}	1	0
Electron	0.00091×10^{-27}	0.00055	-1

7. Matter is usually charge-neutral due to protons and electrons having opposite charges. If matter were not charge neutral, many unnatural things would occur, such as objects repelling or attracting each other.

9. A chemical symbol is a unique one- or two-letter abbreviation for an element. It is listed below the atomic number for that element on the periodic table.

11. Mendeleev noticed that many patterns were evident when elements were organized by increasing mass; from this observation he formulated the periodic law. He also organized the elements based on this law and created the basis for the periodic table being used today.

13. The periodic table is organized by listing the elements in order of increasing atomic number.

15. Nonmetals have varied properties (solid, liquid, or gas at room temperature); however, as a whole they tend to be poor conductors of heat and electricity, and they all tend to gain electrons when they undergo chemical changes. They are located toward the upper right side of the periodic table.

17. Each column within the main group elements in the periodic table is labeled as a family or group of elements. The elements within a group usually have similar chemical properties.

19. An ion is an atom or group of atoms that has lost or gained electrons and has become charged.

21. **a.** ion charge = +1   **b.** ion charge = +2
    c. ion charge = +3   **d.** ion charge = -2
    e. ion charge = -1

23. The percent natural abundance of isotopes is the relative amount of each different isotope in a naturally occurring sample of a given element.

25. Isotopes are noted in this manner: $^A_Z X$. X represents the chemical symbol, A represents the mass number, and Z represents the atomic number.

PROBLEMS

27. a. Correct.

b. False; different elements contain different types of atoms according to Dalton.

c. False; one cannot have 1.5 hydrogen atoms; combinations must be in simple, whole-number ratios.

d. Correct.

29. a. Correct.

b. False; most of the volume of the atom is empty space occupied by tiny, negatively charged electrons.

c. False, the number of negatively charged particles outside the nucleus equals the number of positively charged particles inside the nucleus.

d. False, the majority of the mass of an atom is found in the nucleus.

31. Solid matter seems to have no empty space within it because electromagnetic forces hold the atoms in a tight arrangement and the variation in density is too small to perceive with our eyes.

33. a, b, d

35. b, d

37. approximately 1.8×10^3 electrons

39. 5.4×10^{-4} g

41. a. 87   **b.** 36

c. 91   **d.** 32

e. 13

43. a. 18   **b.** 50

c. 54   **d.** 8

e. 81

45. a. C, 6   **b.** N, 7

c. Na, 11   **d.** K, 19

e. Cu, 29

47. a. manganese, 25   **b.** silver, 47

c. gold, 79   **d.** lead, 82

e. sulfur, 16

49.

Element Name	Element Symbol	Atomic Number
Gold	Au	79
Tin	Sn	50
Arsenic	As	33
Copper	Cu	29
Iron	Fe	26
Mercury	Hg	80

51. a. metal   **b.** metal

c. nonmetal   **d.** nonmetal

e. metalloid

53. a, d, e

55. a, b

57. c, d

59. b, e

61. a. halogen   **b.** noble gas

c. halogen   **d.** neither

e. noble gas

63. a. 6A   **b.** 3A

c. 4A   **d.** 4A

e. 5A

65. b, oxygen; it is in the same group or family.

67. b, chlorine and fluorine; they are in the same family or group.

69.

Chemical Symbol	Group Number	Group Name	Metal or Nonmetal
K	1A	Alkali Metals	Metals
Br	7A	Halogens	Nonmetal
Sr	2A	Alkaline Earth	Metal
He	8A	Noble Gas	Nonmetal
Ar	8A	Noble Gas	Nonmetal

71. a. e^- **b.** O^{2-}

c. $2e^-$ **d.** Cl^-

73. a. 2−   **b.** 3+

c. 4+   **d.** 1−

75. a. 11 protons, 10 electrons

b. 56 protons, 54 electrons

c. 8 protons, 10 electrons

d. 27 protons, 24 electrons

77. a. False; Ti^{2+} has 22 protons and 20 electrons.

b. True

c. False; Mg^{2+} has 12 protons and 10 electrons

d. True

79. a. Rb^+ **b.** K^+

c. Al^{3+} **d.** O^{2-}

81. a. 3 electrons lost   **b.** 1 electron lost

c. 1 electron gained   **d.** 2 electrons gained

83.

Symbol	Ion Commonly Formed	Number of Electrons in Ion	Number of Protons in Ion
Te	Te^{2-}	54	52
In	In^{3+}	46	49
Sr	Sr^{2+}	36	38
Mg	Mg^{2+}	10	12
Cl	Cl^-	18	17

85. a. Z = 1, A = 3   **b.** Z = 24, A = 52

c. Z = 20, A = 42   **d.** Z = 73, A = 182

87. a. $^{16}_8 O$ **b.** $^{19}_9 F$

c. $^{23}_{11} Na$ **d.** $^{27}_{13} Al$

89. a. $^{60}_{27} Co$ **b.** $^{22}_{10} Ne$

c. $^{131}_{53} I$ **d.** $^{244}_{94} Pu$

91. a. 11 protons, 12 neutrons
b. 88 protons, 178 neutrons
c. 82 protons, 126 neutrons
d. 7 protons, 7 neutrons

93. 6 protons, 8 neutrons, $^{14}_{6}C$

95. 85.47 amu

97. a. 49.31% **b.** 78.91 amu

99. 121.8 amu, Sb

101. 7.8×10^{17} electrons

103. $4.2 \times 10^{-45}\,m^3$; $6.2 \times 10^{-31}\,m^3$; $6.7 \times 10^{-13}\,\%$

105.

Number Symbol	Number of Protons	Number of Neutrons	A (Mass Number)	Natural Abundance
Sr-84 or $^{84}_{38}Sr$	38	46	84	0.56%
Sr-86 or $^{86}_{38}Sr$	38	48	86	9.86%
Sr-87 or $^{87}_{38}Sr$	38	49	87	7.00%
Sr-88 or $^{88}_{38}Sr$	38	50	88	82.58%

Atomic mass of Sr = 87.62 amu

107.

Symbol	Z	A	Number of Protons	Number of Electrons	Number of Neutrons	Charge
Zn^+	30	64	30	29	34	1+
Mn^{3+}	25	55	25	22	30	3+
P	15	31	15	15	16	0
O^{2-}	8	16	8	10	8	2−
S^{2-}	16	34	16	18	18	2−

109. 153 amu, 52.2%

111. The atomic theory and nuclear model of the atom are both theories because they attempt to provide a broader understanding and model behavior of chemical systems.

113. Atomic mass is measured as the mean value of masses of all isotopes in a sample. In the case of fluorine, only the 19.00 amu isotope is naturally occurring. In the case of chlorine, about 76% of naturally occurring atoms are 35 amu, and 24% are 37 amu.

115. 69.3% Cu-63, 30.7% Cu-65

117. a. Nt − 304 = 72%; Nt − 305 = 4%; Nt − 306 = 24%

b.

CHAPTER 5
QUESTIONS

1. Yes; when elements combine with other elements, a compound is created. Each compound is unique and contains properties different from those of the elements that compose it.

3. The law of constant composition states that all samples of a given compound have the same proportions of their constituent elements. Joseph Proust formulated this law.

5. The more metallic element is generally listed first in a chemical formula.

7. The empirical formula gives the relative number of atoms of each element in a compound. The molecular formula gives the actual number of atoms of each element in a molecule of the compound.

9. An atomic element is one that exists in nature with a single atom as the basic unit. A molecular element is one that exists as a diatomic molecule as the basic unit. Molecular elements include H_2, N_2, O_2, F_2, Cl_2, Br_2, and I_2.

11. The systematic name can be directly derived by looking at the compound's formula. The common name for a compound acts like a nickname and can only be learned through familiarity.

13. The block that contains the elements for Type II compounds is known as the transition metals.

15. The basic form for the names of Type II ionic compounds is to have the name of the metal cation first, followed by the charge of the metal cation (in parentheses, using Roman numerals), and finally the base name of the nonmetal anion with -ide attached to the end.

17. For compounds containing a polyatomic anion, the name of the cation is first, followed by the name for the polyatomic anion. Also, if the compound contains both a polyatomic cation and a polyatomic anion, one would just use the names of both polyatomic ions.

19. The form for naming molecular compounds is to have the first element preceded by a prefix to indicate the number of atoms present. This is then followed by the second element with its corresponding prefix and -ide placed on the end of the second element.

21. To correctly name a binary acid, one must begin the first word with *hydro-*, which is followed by the base name of the nonmetal plus -ic added on the end. Finally, the word *acid* follows the first word.

23. To name an acid with oxyanions ending with -ite, one must take the base name of the oxyanion and attach -ous to it; the word *acid* follows this.

PROBLEMS

25. Yes; the ratios of sodium to chlorine in both samples were equal.

27. $2.06 \times 10^3\,g$

29.

	Mass N_2O	Mass N	Mass O
Sample A	2.85	1.82	1.03
Sample B	4.55	2.91	1.64
Sample C	3.74	2.39	1.35
Sample D	1.74	1.11	0.63

31. NI_3

33. a. Fe_3O_4 **b.** PCl_3
c. PCl_5 **d.** Ag_2O

35. a. 4 **b.** 4
c. 6 **d.** 4

37. a. magnesium, 1; chlorine, 2
b. sodium, 1; nitrogen, 1; oxygen, 3
c. calcium, 1; nitrogen, 2; oxygen, 4
d. strontium, 1; oxygen, 2; hydrogen, 2

39.

Formula	Number of $C_2H_3O_2$	Number of C Atoms	Number of H Atoms	Number of O Atoms	Number of Metal Atoms
$Mg(C_2H_3O_2)_2$	2	4	6	4	1
$NaC_2H_3O_2$	1	2	3	2	1
$Cr_2(C_2H_3O_2)_4$	4	8	12	8	2

41. a. CH_3 **b.** NO_2
   c. C_2H_3O **d.** NH_3

43. a. molecular **b.** atomic
   c. atomic **d.** molecular

45. a. molecular **b.** ionic
   c. ionic **d.** molecular

47. helium $\longrightarrow$ single atoms
   $CCl_4 \longrightarrow$ molecules
   $K_2SO_4 \longrightarrow$ formula units
   bromine $\longrightarrow$ diatomic molecules

49. a. formula units **b.** single atoms
   c. molecules **d.** molecules

51. a. ionic; forms only one type of ion
   b. molecular
   c. molecular
   d. ionic; forms more that one type of ion

53. a. Na_2S **b.** SrO
   c. Al_2S_3 **d.** $MgCl_2$

55. a. $KC_2H_3O_2$ **b.** K_2CrO_4
   c. K_3PO_4 **d.** KCN

57. a. Li_3N, Li_2O, LiF **b.** Ba_3N_2, BaO, BaF_2
   c. AlN, Al_2O_3, AlF_3

59. a. cesium chloride **b.** strontium bromide
   c. potassium oxide **d.** lithium fluoride

61. a. chromium(II) chloride **b.** chromium(III) chloride
   c. tin(IV) oxide **d.** lead(II) iodide

63. a. forms more than one type of ion, chromium(III) oxide
   b. forms only one type of ion, sodium iodide
   c. forms only one type of ion, calcium bromide
   d. forms more than one type of ion, tin(II) oxide

65. a. barium nitrate **b.** lead(II) acetate
   c. ammonium iodide **d.** potassium chlorate
   e. cobalt(II) sulfate **f.** sodium perchlorate

67. a. hypobromite ion **b.** bromite ion
   c. bromate ion **d.** perbromate ion

69. a. $CuBr_2$ **b.** $AgNO_3$
   c. KOH **d.** Na_2SO_4
   e. $KHSO_4$ **f.** $NaHCO_3$

71. a. sulfur dioxide
   b. nitrogen triiodide
   c. bromine pentafluoride
   d. nitrogen monoxide
   e. tetranitrogen tetraselenide

73. a. CO **b.** S_2F_4
   c. Cl_2O **d.** PF_5
   e. BBr_3 **f.** P_2S_5

75. a. PBr_5 phosphorus pentabromide
   b. P_2O_3 diphosphorus trioxide
   c. SF_4 sulfur tetraflouride
   d. correct

77. a. oxyacid, nitrous acid, nitrite
   b. binary acid, hydroiodic acid
   c. oxyacid, sulfuric acid, sulfate
   d. oxyacid, nitric acid, nitrate

79. a. hypochlorous acid **b.** chlorous acid
   c. chloric acid **d.** perchloric acid

81. a. H_3PO_4 **b.** HBr
   c. H_2SO_3

83. a. 63.02 amu **b.** 199.88 amu
   c. 153.81 amu **d.** 211.64 amu

85. PBr_3, Ag_2O, PtO_2, $Al(NO_3)_3$

87. a. CH_4 **b.** SO_3
   c. NO_2

89. a. 12 **b.** 4
   c. 12 **d.** 7

91. a. 8 **b.** 12
   c. 12

93.

Formula	Type	Name
N_2H_4	molecular	dinitrogen tetrahydride
KCl	ionic	potassium chloride
H_2CrO_4	acid	chromic acid
$Co(CN)_3$	ionic	cobalt(III) cyanide

95. a. calcium nitrite
   b. potassium oxide
   c. phosphorus trichloride
   d. correct
   e. potassium iodite

97. a. $Sn(SO_4)_2$ 310.9 amu **b.** HNO_2 47.02 amu
   c. $NaHCO_3$ 84.01 amu **d.** PF_5 125.97 amu

99. a. platinum(IV) oxide 227.08 amu
   b. dinitrogen pentoxide 108.02 amu
   c. aluminum chlorate 277.33 amu
   d. phosphorus pentabromide 430.47 amu

101. C_2H_4

103. 10 different isotopes can exist. 151.88 amu, 152.88 amu, 153.88 amu, 154.88 amu, 155.88 amu, 156.88 amu, 157.88 amu, 158.88 amu, 159.88 amu, and 160.88 amu.

105. a. molecular element **b.** atomic element
   c. ionic compound **d.** molecular compound

107. a. $NaOCl$; $NaOH$ **b.** $Al(OH)_3$; $Mg(OH)_2$
   c. $CaCO_3$
   d. $NaHCO_3$, $Ca_3(PO_4)_2$, $NaAl(SO_4)_2$

CHAPTER 6
QUESTIONS

1. Chemical composition lets us determine how much of a particular element is contained within a particular compound.

3. There are 6.022×10^{23} atoms in 1 mole of atoms.

5. One mole of any element has a mass equal to its atomic mass in grams.

7. **a.** 30.97 g **b.** 195.08 g
   c. 12.01 g **d.** 52.00 g

9. Each element has a different atomic mass number. So, the subscripts that represent mole ratios cannot be used to represent the ratios of grams of a compound. The grams per mole of one element always differ from the grams per mole of a different element.

11. **a.** $11.19 \text{ g H} \equiv 100 \text{ g H}_2\text{O}$
    b. $53.29 \text{ g O} \equiv 100 \text{ g fructose}$
    c. $84.12 \text{ g C} \equiv 100 \text{ g octane}$
    d. $52.14 \text{ g C} \equiv 100 \text{ g ethanol}$

13. The empirical formula gives the smallest whole-number ratio of each type of atom. The molecular formula gives the specific number of each type of atom in the molecule. The molecular formula is always a multiple of the empirical formula.

15. The empirical formula mass of a compound is the sum of the masses of all the atoms in the empirical formula.

PROBLEMS

17. 3.5×10^{24} atoms

19. **a.** 2.0×10^{24} atoms **b.** 5.8×10^{21} atoms.
    c. 1.38×10^{25} atoms **d.** 1.29×10^{23} atoms

21.
Element	Moles	Number of Atoms
Ne	0.552	3.32×10^{23}
Ar	5.40	3.25×10^{24}
Xe	1.78	1.07×10^{24}
He	1.79×10^{-4}	1.08×10^{20}

23. **a.** 72.7 dozen **b.** 6.06 gross
    c. 1.74 reams **d.** 1.45×10^{-21} moles

25. 0.321 mol

27. 28.6 g

29. **a.** 2.05×10^{-2} mol **b.** 0.623 mol
    c. 0.401 mol **d.** 3.21×10^{-3} mol

31.
Element	Moles	Mass
Ne	1.11	22.5 g
Ar	0.117	4.67 g
Xe	7.62	1.00 kg
He	1.44×10^{-4}	5.76×10^{-4} g

33. 8.07×10^{18} atoms

35. 8.44×10^{22} atoms

37. **a.** 1.16×10^{23} atoms **b.** 2.81×10^{23} atoms
    c. 2.46×10^{22} atoms **d.** 7.43×10^{23} atoms

39. 1.9×10^{21} atoms

41. 1.61×10^{25} atoms

43.
Element	Mass	Moles	Number of Atoms
Na	38.5 mg	1.67×10^{-3}	1.01×10^{21}
C	13.5 g	1.12	6.74×10^{23}
V	1.81×10^{-20} g	3.55×10^{-22}	214
Hg	1.44 kg	7.18	4.32×10^{24}

45. **b**

47. **a.** 0.654 mol **b.** 1.22 mol
    c. 96.6 mol **d.** 1.76×10^{-5} mol

49.
Compound	Mass	Moles	Molecules
H_2O	112 kg	6.22×10^3	3.74×10^{27}
N_2O	6.33 g	0.144	8.66×10^{22}
SO_2	156	2.44	1.47×10^{24}
CH_2Cl_2	5.46	0.0643	3.87×10^{22}

51. 6.20×10^{21} molecules

53. **a.** 1.2×10^{23} molecules **b.** 1.21×10^{24} molecules
    c. 3.5×10^{23} molecules **d.** 6.4×10^{22} molecules

55. 0.10 mg

57. $\$6.022 \times 10^{21}$ total. $\$9.1 \times 10^{11}$ per person. Each person would be a billionaire.

59. 5.4 mol Cl

61. d, 3 mol O

63. **a.** 2.5 mol C **b.** 0.230 mol C
    c. 22.7 mol C **d.** 201 mol C

65. **a.** 2 moles H per mole of molecules; 8 H atoms present
    b. 4 moles H per mole of molecules; 20 H atoms present
    c. 3 moles H per mole of molecules; 9 H atoms present

67. **a.** 22.3 g **b.** 29.4 g
    c. 21.6 g **d.** 12.9 g

69. **a.** 1.4×10^3 kg **b.** 1.4×10^3 kg
    c. 2.1×10^3 kg

71. 84.8% Sr

73. 36.1% Ca; 63.9% Cl

75. 10.7 g

77. 6.6 mg

79. **a.** 63.65% **b.** 46.68%
    c. 30.45% **d.** 25.94%

81. **a.** 39.99% C; 6.73% H; 53.28% O
    b. 26.09% C; 4.39% H; 69.52% O
    c. 60.93% C; 15.37% H; 23.69% N
    d. 54.48% C; 13.74% H; 31.78% N

83. Fe_3O_4, 72.36% Fe; Fe_2O_3, 69.94% Fe; $FeCO_3$, 48.20% Fe; magnetite

85. NO_2

87. **a.** NiI_2 **b.** $SeBr_4$
    c. $BeSO_4$

89. C_2H_6N

91. **a.** C_3H_6O **b.** $C_5H_{10}O_2$
   c. $C_9H_{10}O_2$

93. P_2O_3

95. NCl_3

97. C_4H_8

99. **a.** C_6Cl_6 **b.** C_2HCl_3
   c. $C_6H_3Cl_3$

101. 2.43×10^{23} atoms

103. 2×10^{21} molecules

105.

Substance	Mass	Moles	Number of Particles
Ar	0.018 g	4.5×10^{-4}	2.7×10^{20}
NO_2	8.33×10^{-3} g	1.81×10^{-4}	1.09×10^{20}
K	22.4 mg	5.73×10^{-4}	3.45×10^{20}
C_8H_{18}	3.76 kg	32.9	1.98×10^{25}

107. **a.** CuI_2: 20.03% Cu; 79.97% I
   b. $NaNO_3$: 27.05% Na; 16.48% N; 56.47% O
   c. $PbSO_4$: 68.32% Pb; 10.57% S; 21.10% O
   d. CaF_2: 51.33% Ca; 48.67% F

109. 1.8×10^3 kg rock

111. 59 kg Cl

113. 1.1×10^2 g H

115.

Formula	Molar Mass	%C (by mass)	%H (by mass)
C_2H_4	28.06	85.60%	14.40%
C_4H_{10}	58.12	82.66%	17.34%
C_4H_8	56.12	85.60%	14.40%
C_3H_8	44.09	81.71%	18.29%

117. $C_4H_6O_2$

119. $C_{10}H_{14}N_2$

121. 70.4% KBr, 29.6% KI

123. 29.6 g SO_2

125. 2.66 kg Fe

127. **a.** 1×10^{57} atoms per star
   b. 1×10^{68} atoms per galaxy
   c. 1×10^{79} atoms in the universe

129. $C_{16}H_{10}$

CHAPTER 7
QUESTIONS

1. A chemical reaction is the change of one or more substances into different substances, for example, burning wood, rusting iron, and protein synthesis.

3. The main evidence of a chemical reaction includes a color change, the formation of a solid, the formation of a gas, the emission of light, and the emission or absorption of heat.

5. **a.** gas   **b.** liquid
   c. solid   **d.** aqueous

7. **a.** reactants: 4 Ag, 2 O, 1 C products: 4 Ag, 2 O, 1 C balanced: yes

b. reactants: 1 Pb, 2 N, 6 O, 2 Na, 2 Cl products: 1 Pb, 2 N, 6 O, 2 Na, 2 Cl balanced: yes

c. reactants: 3 C, 8 H, 2 O products: 3 C, 8 H, 10 O balanced: no

9. If a compound dissolves in water, then it is soluble. If it does not dissolve in water, it is insoluble.

11. When ionic compounds containing polyatomic ions dissolve in water, the polyatomic ions usually dissolve as intact units.

13. The solubility rules are a set of empirical rules for ionic compounds that were deduced from observations on many compounds. The rules help us determine whether particular compounds will be soluble or insoluble.

15. The precipitate will always be insoluble; it is the solid that forms upon mixing two aqueous solutions.

17. Acid–base reactions involve an acid and a base reacting to form water and an ionic compound. An example is the reaction between hydrobromic acid and sodium hydroxide: $HBr + NaOH \longrightarrow H_2O + NaBr$

19. Gas evolution reactions are reactions that evolve a gas. An example is the reaction between hydrochloric acid and sodium bicarbonate: $HCl + NaHCO_3 \longrightarrow H_2O + CO_2 + NaCl$

21. Combustion reactions are a type of redox reaction and are characterized by the exothermic reaction of a substance with O_2. An example is the reaction between methane and oxygen: $CH_4 + 2O_2 \longrightarrow CO_2 + 2H_2O$

23. A synthesis reaction combines simpler substances to form more complex substances. An example is the reaction between elemental potassium and chloride: $2K + Cl_2 \longrightarrow 2KCl$. A decomposition decomposes a more complex substance into simpler substances. An example is the decomposition of water: $2H_2O \longrightarrow 2H_2 + O_2$

PROBLEMS

25. **a.** Yes; there is a color change showing a chemical reaction

b. No; the state of the compound changes, but no chemical reaction takes place.

c. Yes; there is a formation of a solid in a previously clear solution.

d. Yes; there is a formation of a gas when the yeast was added to the solution.

27. Yes; a chemical reaction has occurred, for the presence of the bubbles is evidence for the formation of a gas.

29. Yes; a chemical reaction has occurred. We know this due to the color change of the hair.

31. Placing a subscript 2 after H_2O would change the compound from water to hydrogen peroxide (H_2O_2). To balance chemical reactions, one must add coefficients, not subscripts.

$2H_2O(l) \longrightarrow 2H_2(g) + O_2(g)$

33. **a.** $PbS + 2HCl \longrightarrow PbCl_2 + H_2S$
   b. $CO + 3H_2 \longrightarrow CH_4 + H_2O$
   c. $Fe_2O_3 + 3H_2 \longrightarrow 2Fe + 3H_2O$
   d. $4NH_3 + 5O_2 \longrightarrow 4NO + 6H_2O$

35. a. $Mg(s) + 2 CuNO_3(aq) \longrightarrow 2 Cu(s) + Mg(NO_3)_2(aq)$

b. $2 N_2O_5(g) \longrightarrow 4 NO_2(g) + O_2(g)$

c. $Ca(s) + 2 HNO_3(aq) \longrightarrow H_2(g) + Ca(NO_3)_2(aq)$

d. $2 CH_3OH(l) + 3 O_2(g) \longrightarrow 2 CO_2(g) + 4 H_2O(g)$

37. $2 H_2(g) + O_2(g) \longrightarrow 2 H_2O(l)$;

$Cl(g) + O_3(g) \longrightarrow ClO(g) + O_2(g)$

39. $2 Na(s) + 2 H_2O(l) \longrightarrow H_2(g) + 2 NaOH(aq)$

41. $2 SO_2(g) + O_2(g) + 2 H_2O(l) \longrightarrow 2 H_2SO_4(aq)$

43. $V_2O_5(s) + 2 H_2(g) \longrightarrow V_2O_3(s) + 2 H_2O(l)$

45. $C_{12}H_{22}O_{11}(aq) + H_2O(l) \longrightarrow 4 CO_2(g) + 4 C_2H_5OH(aq)$

47. a. $Na_2S(aq) + Cu(NO_3)_2(aq) \longrightarrow$
$$2 NaNO_3(aq) + CuS(s)$$

b. $4 HCl(aq) + O_2(g) \longrightarrow 2 H_2O(l) + 2 Cl_2(g)$

c. $2 H_2(g) + O_2(g) \longrightarrow 2 H_2O(l)$

d. $FeS(s) + 2 HCl(aq) \longrightarrow FeCl_2(aq) + H_2S(g)$

49. a. $BaO_2(s) + H_2SO_4(aq) \longrightarrow BaSO_4(s) + H_2O_2(aq)$

b. $2 Co(NO_3)_3(aq) + 3 (NH_4)_2S(aq) \longrightarrow$
$$Co_2S_3(s) + 6 NH_4NO_3(aq)$$

c. $Li_2O(s) + H_2O(l) \longrightarrow 2 LiOH(aq)$

d. $Hg_2(C_2H_3O_2)_2(aq) + 2 KCl(aq) \longrightarrow$
$$Hg_2Cl_2(s) + 2 KC_2H_3O_2(aq)$$

51. a. $2 Rb(s) + 2 H_2O(l) \longrightarrow 2 RbOH(aq) + H_2(g)$

b. Equation is balanced

c. $2 NiS(s) + 3 O_2(g) \longrightarrow 2 NiO(s) + 2 SO_2(g)$

d. $3 PbO(s) + 2 NH_3(g) \longrightarrow 3 Pb(s) + N_2(g) + 3 H_2O(l)$

53. $C_6H_{12}O_6(aq) + 6 O_2(g) \longrightarrow 6 CO_2(g) + 6 H_2O(l)$

55. $2 NO(g) + 2 CO(g) \longrightarrow N_2(g) + 2 CO_2(g)$

57. a. soluble; $Na^+, C_2H_3O_2^-$

b. soluble; Sn^{2+}, NO_3^-

c. insoluble

d. soluble; Na^+, PO_4^{3-}

59. $AgCl; BaSO_4; CuCO_3; Fe_2S_3$

61.

Soluble	Insoluble
K_2S	Hg_2I_2
BaS	$Cu_3(PO_4)_2$
NH_4Cl	MgS
Na_2CO_3	$CaSO_4$
K_2SO_4	$PbSO_4$
SrS	$PbCl_2$
Li_2S	Hg_2Cl_2

63. a. NO REACTION

b. $K_2SO_4(aq) + BaBr_2(aq) \longrightarrow BaSO_4(s) + 2 KBr(aq)$

c. $2 NaCl(aq) + Hg_2(C_2H_3O_2)_2(aq) \longrightarrow$
$$Hg_2Cl_2(s) + 2 NaC_2H_3O_2(aq)$$

d. NO REACTION

65. a. $Na_2CO_3(aq) + Pb(NO_3)_2(aq) \longrightarrow$
$$PbCO_3(s) + 2 NaNO_3(aq)$$

b. $K_2SO_4(aq) + Pb(C_2H_3O_2)_2(aq) \longrightarrow$
$$PbSO_4(s) + 2 KC_2H_3O_2$$

c. $Cu(NO_3)_2(aq) + BaS(aq) \longrightarrow CuS(s) + Ba(NO_3)_2(aq)$

d. NO REACTION

67. a. correct

b. NO REACTION

c. correct

d. $Pb(NO_3)_2(aq) + 2 LiCl(aq) \longrightarrow$
$$PbCl_2(s) + 2 LiNO_3(aq)$$

69. K^+, NO_3^-

71. a. $Ag^+(aq) + NO_3^-(aq) + K^+(aq) + Cl^-(aq) \longrightarrow$
$$AgCl(s) + K^+(aq) + NO_3^-(aq)$$

$Ag^+(aq) + Cl^-(aq) \longrightarrow AgCl(s)$

b. $Ca^{2+}(aq) + S^{2-}(aq) + Cu^{2+}(aq) + 2 Cl^-(aq) \longrightarrow$
$$CuS(s) + Ca^{2+}(aq) + 2 Cl^-(aq)$$

$Cu^{2+}(aq) + S^{2-}(aq) \longrightarrow CuS(s)$

c. $Na^+(aq) + OH^-(aq) + H^+(aq) + NO_3^-(aq) \longrightarrow$
$$H_2O(l) + Na^+(aq) + NO_3^-(aq)$$

$H^+(aq) + OH^-(aq) \longrightarrow H_2O(l)$

d. $6 K^+(aq) + 2 PO_4^{3-}(aq) + 3 Ni^{2+}(aq) + 6 Cl^-(aq) \longrightarrow$
$$Ni_3(PO_4)_2(s) + 6 K^+(aq) + 6 Cl^-(aq)$$

$3 Ni^{2+}(aq) + 2 PO_4^{3-}(aq) \longrightarrow Ni_3(PO_4)_2(s)$

73. $Hg_2^{2+}(aq) + 2 NO_3^-(aq) + 2 Na^+(aq) + 2 Cl^-(aq) \longrightarrow$
$$Hg_2Cl_2(s) + 2 Na^+(aq) + 2 NO_3^-(aq)$$

$Hg_2^{2+}(aq) + 2 Cl^-(aq) \longrightarrow Hg_2Cl_2(s)$

75. a. $2 Na^+(aq) + CO_3^{2-}(aq) + Pb^{2+}(aq) + 2 NO_3^-(aq)$
$$\longrightarrow PbCO_3(s) + 2 Na^+(aq) + 2 NO_3^-(aq)$$

$Pb^{2+}(aq) + CO_3^{2-}(aq) \longrightarrow PbCO_3(s)$

b. $2 K^+(aq) + SO_4^{2-}(aq) + Pb^{2+}(aq) + 2 CH_3CO_2^-(aq)$
$$\longrightarrow PbSO_4(s) + 2 K^+(aq) + 2 CH_3CO_2^-(aq)$$

$Pb^{2+}(aq) + SO_4^{2-}(aq) \longrightarrow PbSO_4(s)$

c. $Cu^{2+}(aq) + 2 NO_3^-(aq) + Ba^{2+}(aq) + S^{2-}(aq) \longrightarrow$
$$CuS(s) + Ba^{2+}(aq) + 2 NO_3^-(aq)$$

$Cu^{2+}(aq) + S^{2-}(aq) \longrightarrow CuS(s)$

d. NO REACTION

77. $HCl(aq) + KOH(aq) \longrightarrow H_2O(l) + KCl(aq)$

$H^+(aq) + OH^-(aq) \longrightarrow H_2O(l)$

79. a. $2 HCl(aq) + Ba(OH)_2(aq) \longrightarrow 2 H_2O(l) + BaCl_2(aq)$

b. $H_2SO_4(aq) + 2 KOH(aq) \longrightarrow 2 H_2O(l) + K_2SO_4(aq)$

c. $HClO_4(aq) + NaOH(aq) \longrightarrow H_2O(l) + NaClO_4(aq)$

81. a. $HBr(aq) + NaHCO_3(aq) \longrightarrow$
$$H_2O(l) + CO_2(g) + NaBr(aq)$$

b. $NH_4I(aq) + KOH(aq) \longrightarrow H_2O(l) + NH_3(g) + KI(aq)$

c. $2 HNO_3(aq) + K_2SO_3(aq) \longrightarrow$
$$H_2O(l) + SO_2(g) + 2 KNO_3(aq)$$

d. $2 HI(aq) + Li_2S(aq) \longrightarrow H_2S(g) + 2 LiI(aq)$

83. b and d are redox reactions; a and c are not.

85. a. $2 C_2H_6(g) + 7 O_2(g) \longrightarrow 4 CO_2(g) + 6 H_2O(g)$

b. $2 Ca(s) + O_2(g) \longrightarrow 2 CaO(s)$

c. $2 C_3H_8O(l) + 9 O_2(g) \longrightarrow 6 CO_2(g) + 8 H_2O(g)$

d. $2 C_4H_{10}S(l) + 15 O_2(g) \longrightarrow$
$$8 CO_2(g) + 10 H_2O(g) + 2 SO_2(g)$$

87. a. $2 \, Ag(s) + Br_2(g) \longrightarrow 2 \, AgBr(s)$

  b. $2 \, K(s) + Br_2(g) \longrightarrow 2 \, KBr(s)$

  c. $2 \, Al(s) + 3 \, Br_2(g) \longrightarrow 2 \, AlBr_3(s)$

  d. $Ca(s) + Br_2(g) \longrightarrow CaBr_2(s)$

89. a. double displacement

  b. synthesis or combination

  c. single displacement

  d. decomposition

91. a. synthesis

  b. decomposition

  c. synthesis

93. a. $2 \, Na^+(aq) + 2 \, I^-(aq) + Hg_2^{\,+}(aq) + 2 \, NO_3^-(aq)$

  $\longrightarrow Hg_2I_2(s) + 2 \, Na^+(aq) + 2 \, NO_3^-(aq)$

  b. $2 \, H^+(aq) + 2 \, ClO_4^-(aq) + Ba^{2+}(aq) + 2 \, OH^-(aq)$

  $\longrightarrow 2 \, H_2O(l) + Ba^{2+}(aq) + 2 \, ClO_4^-(aq)$

  $H^+(aq) + OH^-(aq) \longrightarrow H_2O(s)$

  c. NO REACTION

  d. $2 \, H^+(aq) + 2 \, Cl^-(aq) + 2 \, Li^+(aq) + CO_3^{2-}(aq) \longrightarrow$

  $H_2O(l) + CO_2(g) + 2 \, Li^+(aq) + 2 \, Cl^-(aq)$

  $2 \, H^+(aq) + CO_3^{2-}(aq) \longrightarrow H_2O(l) + CO_2(g)$

95. a. NO REACTION

  b. NO REACTION

  c. $K^+(aq) + HSO_3^-(aq) + H^+(aq) + NO_3^-(aq) \longrightarrow$

  $H_2O(l) + SO_2(g) + K^+(aq) + NO_3^-(aq)$

  $H^+(aq) + HSO_3^-(aq) \longrightarrow H_2O(l) + SO_2(g)$

  d. $Mn^{3+}(aq) + 3 \, Cl^-(aq) + 3 \, K^+(aq) + PO_4^{3-}(aq) \longrightarrow$

  $MnPO_4(s) + 3 \, K^+(aq) + 3 \, Cl^-(aq)$

  $Mn^{3+}(aq) + PO_4^{3-}(aq) \longrightarrow MnPO_4(s)$

97. a. acid–base; $KOH(aq) + HC_2H_3O_2(aq) \longrightarrow$

  $H_2O(l) + KC_2H_3O_2(aq)$

  b. gas evolution; $2 \, HBr(aq) + K_2CO_3(aq) \longrightarrow$

  $H_2O(l) + CO_2(g) + 2 \, KBr(aq)$

  c. synthesis; $2 \, H_2(g) + O_2(g) \longrightarrow 2 \, H_2O(l)$

  d. precipitation; $2 \, NH_4Cl(aq) + Pb(NO_3)_2(aq) \longrightarrow$

  $PbCl_2(s) + 2 \, NH_4NO_3(aq)$

99. a. oxidation–reduction; single displacement

  b. gas evolution; acid–base

  c. gas evolution; double displacement

  d. precipitation; double displacement

101. $3 \, CaCl_2(aq) + 2 \, Na_3PO_4(aq) \longrightarrow$

  $Ca_3(PO_4)_2(s) + 6 \, NaCl(aq)$

  $3 \, Ca^{2+}(aq) + 6 \, Cl^-(aq) + 6 \, Na^+(aq) + 2 \, PO_4^{3-}(aq)$

  $\longrightarrow Ca_3(PO_4)_2(s) + 6 \, Na^+(aq) + 6 \, Cl^-(aq)$

  $3 \, Ca^{2+}(aq) + 2 \, PO_4^{3-}(aq) \longrightarrow Ca_3(PO_4)_2(s)$

  $3 \, Mg(NO_3)_2(aq) + 2 \, Na_3PO_4(aq) \longrightarrow$

  $Mg_3(PO_4)_2(s) + 6 \, NaNO_3(aq)$

  $3 \, Mg^{2+}(aq) + 6 \, NO_3^-(aq) + 6 \, Na^+(aq) + 2 \, PO_4^{3-}(aq)$

  $\longrightarrow Mg_3(PO_4)_2(s) + 6 \, Na^+(aq) + 6 \, NO_3^-(aq)$

  $3 \, Mg^{2+}(aq) + 2 \, PO_4^{3-}(aq) \longrightarrow Mg_3(PO_4)_2(s)$

103. *Correct answers may vary; representative correct answers are:

  a. addition of a solution containing SO_4^{2-};
  $Pb^{2+}(aq) + SO_4^{2-}(aq) \longrightarrow PbSO_4(s)$

  b. addition of a solution containing SO_4^{2-};
  $Ca^{2+}(aq) + SO_4^{2-}(aq) \longrightarrow CaSO_4(s)$

  c. addition of a solution containing SO_4^{2-};
  $Ba^{2+}(aq) + SO_4^{2-}(aq) \longrightarrow BaSO_4(s)$

  d. addition of a solution containing Cl^-;
  $Hg_2^{\,2+}(aq) + 2 \, Cl^-(aq) \longrightarrow Hg_2Cl_2(s)$

105. Ca^{2+} and Cu^{2+} were present in the original solution.

  1st: $Ca^{2+}(aq) + SO_4^{2-}(aq) \longrightarrow CaSO_4(s)$

  2nd: $Cu^{2+}(aq) + CO_3^{2-}(aq) \longrightarrow CuCO_3(s)$

107. 0.168 mol Ca; 6.73 g Ca

109. 0.00128 mol NaCl, 0.0750 g NaCl

111. a. chemical

  b. physical

CHAPTER 8

QUESTIONS

1. Reaction stoichiometry is very important to chemistry. It gives us a numerical relationship between the reactants and products that allows chemists to plan and carry out chemical reactions to obtain products in the desired quantities.

For example, how much CO_2 is produced when a given amount of C_8H_{10} is burned?

How much $H_2(g)$ is produced when a given amount of water decomposes?

3. 1 mol $Cl_2 \equiv$ 2 mol NaCl

5. mass A $\longrightarrow$ moles A $\longrightarrow$ moles B $\longrightarrow$

mass B (A = reactant, B = product)

7. The limiting reactant is the reactant that limits the amount of product in a chemical reaction.

9. The actual yield is the amount of product actually produced by a chemical reaction. The percent yield is the percentage of the theoretical yield that was actually attained.

11. d

13. The enthalpy of reaction is the total amount of heat generated or absorbed by a particular chemical reaction. The quantity is important because it quantifies the change in heat for the chemical reaction. It is useful for determining the necessary starting conditions and predicting the outcome of various reactions.

PROBLEMS

15. a. 2 mol C   **b.** 1 mol C

  c. 3 mol C   **d.** 1.5 mol C

17. a. 2.6 mol NO_2 **b.** 11.6 mol NO_2

  c. 8.90×10^3 mol NO_2 **d.** 2.012×10^{-3} mol NO_2

19. c

21. a. 3.50 mol HCl   **b.** 3.50 mol H_2O

  c. 0.875 mol Na_2O_2 **d.** 1.17 mol SO_3

23. a. 2.4 mol $PbO(s)$, 2.4 mol $SO_2(g)$
   b. 1.6 mol $PbO(s)$, 1.6 mol $SO_2(g)$
   c. 5.3 mol $PbO(s)$, 5.3 mol $SO_2(g)$
   d. 3.5 mol $PbO(s)$, 3.5 mol $SO_2(g)$

25.

mol N_2H_4	mol N_2O_4	mol N_2	mol H_2O
4	2	6	8
6	3	9	12
4	2	6	8
11	5.5	16.5	22
3	1.5	4.5	6
8.26	4.13	12.4	16.5

27. $2 C_4H_{10}(g) + 13 O_2(g) \longrightarrow$
                 $8 CO_2(g) + 10 H_2O(g)$; 32 mol O_2

29. a. $Pb(s) + 2 AgNO_3(aq) \longrightarrow Pb(NO_3)_2(aq) + 2 Ag(s)$
   b. 19 mol $AgNO_3$ **c.** 56.8 mol Ag

31. a. 0.157 g O_2 **b.** 0.500 g O_2
   c. 114 g O_2 **d.** 2.86×10^{-4} g O_2

33. a. 4.0 g NaCl   **b.** 4.3 g $CaCO_3$
   c. 4.0 g MgO   **d.** 3.1 g NaOH

35. a. 8.9 g Al_2O_3, 9.7 g Fe   **b.** 3.0 g Al_2O_3, 3.3 g Fe

37.

Mass CH_4	Mass O_2	Mass CO_2	Mass H_2O
0.645 g	2.57 g	1.77 g	1.45 g
22.32 g	89.00 g	61.20 g	50.09 g
5.044 g	20.11 g	13.83 g	11.32 g
1.07 g	4.28 g	2.94 g	2.41 g
3.18 kg	12.7 kg	8.72 kg	7.14 kg
8.57×10^2 kg	3.42×10^3 kg	2.35×10^3 kg	1.92×10^3 kg

39. a. 2.3 g HCl   **b.** 4.3 g HNO_3
   c. 2.2 g H_2SO_4

41. 123 g H_2SO_4, 2.53 g H_2

43. a. 2 mol A   **b.** 1.8 mol A
   c. 4 mol B   **d.** 40 mol B

45. a. 1.5 mol C   **b.** 3 mol C
   c. 3 mol C   **d.** 96 mol C

47. a. 1 mol K   **b.** 1.8 mol K
   c. 1 mol Cl_2 **d.** 14.6 mol K

49. a. 1.3 mol MnO_3 **b.** 4.8 mol MnO_3
   c. 0.107 mol MnO_3 **d.** 27.5 mol MnO_3

51. 3 mol A, 0 mol B, 4 mol C

53. a. 2 Cl_2 **b.** 3 Cl_2
   c. 2 Cl_2

55. a. 1.0 g F_2 **b.** 10.5 g Li
   c. 6.79×10^3 g F_2

57. a. 1.3 g $AlCl_3$ **b.** 24.8 g $AlCl_3$
   c. 2.17 g $AlCl_3$

59. 74.6%

61. CaO; 25.7 g $CaCO_3$; 75.5%

63. O_2; 5.07 g NiO; 95.9%

65. Pb^{2+}; 262.7 g $PbCl_2$; 96.09%

67. TiO_2: 0 g, C: 7.0 g, Ti: 5.99 g, CO: 7.00 g

69. a. exothermic, $-\Delta H$ **b.** endothermic, $+\Delta H$
   c. exothermic, $-\Delta H$

71. a. 55 kJ   **b.** 110 kJ
   c. 28 kJ   **d.** 55 kJ

73. 4.78×10^3 kJ

75. 34.9 g C_8H_{18}

77. N_2

79. 0.152 g Ba^{2+}

81. 1.5 g HCl

83. 3.1 kg CO_2

85. 4.7 g Na_3PO_4

87. 469 g Zn

89. $2 NH_4NO_3(s) \longrightarrow$
         $2 N_2(g) + O_2(g) + 4 H_2O(l)$; 2.00×10^2 g O_2

91. salicylic acid ($C_7H_6O_3$); 2.71 g $C_9H_8O_4$; 74.1%

93. NH_3; 120 kg (CH_4N_2O); 72.9%

95. 2.4 mg $C_4H_6O_4S_2$

97. 1.0×10^3 g CO_2

99. b; the loudest explosion will occur when the ratio is 2 hydrogen to 1 oxygen, for that is the ratio that occurs in water.

101. 2.8×10^{13} kg CO_2 per year; 1.1×10^2 years

CHAPTER 9
QUESTIONS

1. Both the Bohr model and the quantum-mechanical model for the atom were developed in the early 1900s. These models serve to explain how electrons are arranged within the atomic structure and how the electrons affect the chemical and physical properties of each element.

3. White light contains a spectrum of wavelengths and therefore a spectrum of color. Colored light is produced by a single wavelength and is therefore a single color.

5. Energy carried per photon is greater for shorter wavelengths than for longer wavelengths. Wavelength and frequency are inversely related—the shorter the wavelength, the higher the frequency.

7. X-rays pass through many substances that block visible light and are therefore used to image bones and organs.

9. Ultraviolet light contains enough energy to damage biological molecules, and excessive exposure increases the risk of skin cancer and cataracts.

11. Microwaves can only heat things containing water, and therefore the food, which contains water, becomes hot, but the plate does not.

13. The Bohr model is a representation for the atom in which electrons travel around the nucleus in circular orbits with a fixed energy at specific, fixed distances from the nucleus.

15. The Bohr orbit describes the path of an electron as an orbit or trajectory (a specified path). A quantum-mechanical orbital describes the path of an electron using a probability map.

17. The e^- has wave particle duality, which means the path of an electron is not predictable. The motion of a baseball is predictable. A probability map shows a statistical, reproducible pattern of where the electron is located.

19. The subshells are s (1 orbital, which contains a maximum of 2 electrons); p (3 orbitals, which contain a maximum of 6 electrons); d (5 orbitals, which contain a maximum of 10 electrons); and f (7 orbitals, which contain a maximum of 14 electrons).

21. The Pauli exclusion principle states that separate orbitals may hold no more than 2 electrons, and when 2 electrons are present in a single orbital, they must have opposite spins. When writing electron configurations, the principle means that no box can have more than 2 arrows, and the arrows will point in opposite directions.

23. [Ne] represents $1s^2 2s^2 2p^6$.
[Kr] represents $1s^2 2s^2 2p^6 3s^2 3p^6 4s^2 3d^{10} 4p^6$.

25.

27. Group 1 elements form 1+ ions because they lose one valence electron in the outer s shell to obtain a noble gas configuration. Group 7 elements form 1− ions because they gain an electron to fill their outer p orbital to obtain a noble gas configuration.

PROBLEMS

29. a. 1.0 ns

    b. 13.21 ms

    c. 4 hrs 10 min

31. infrared

33. radiowaves < microwaves < infrared < ultraviolet

35. gamma, ultraviolet, or X-rays

37. a. radio waves < infrared < X-rays

    b. radio waves < infrared < X-rays

    c. X-rays < infrared < radio waves

39. energies, distances

41. $n = 6 \longrightarrow n = 2$: 410 nm

    $n = 5 \longrightarrow n = 2$: 434 nm

43.

The 2s and 3p orbitals are bigger than the 1s and 2p orbitals.

45. Electron in the 2s orbital

47. $2p \longrightarrow 1s$

49. a. $1s^2 2p^2 2p^6 3s^2 3p^6 4s^2 3d^{10} 4p^6 5s^2$

    b. $1s^2 2s^2 2p^6 3s^2 3p^6 4s^2 3d^{10} 4p^2$

    c. $1s^2 2s^1$

    d. $1s^2 2s^2 2p^6 3s^2 3p^6 4s^2 3d^{10} 4p^6$

51.

53. a. $[\text{Ar}]4s^2 3d^{10} 4p^1$ **b.** $[\text{Ar}]4s^2 3d^{10} 4p^3$

    c. $[\text{Kr}]5s^1$ **d.** $[\text{Kr}]5s^2 4d^{10} 5p^2$

55. a. $[\text{Ar}]4s^2 3d^{10}$ **b.** $[\text{Ar}]4s^1 3d^{10}$

    c. $[\text{Kr}]5s^2 4d^2$ **d.** $[\text{Ar}]4s^2 3d^6$

57. Valence electrons are underlined

    a. $1s^2\,2s^2\,2p^6\,3s^2\,3p^6\,\underline{4s^2}\,3d^{10}\,\underline{4p^6}$

    b. $1s^2\,2s^2\,2p^6\,3s^2\,3p^6\,\underline{4s^2}\,3d^{10}\,\underline{4p^2}$

    c. $1s^2\,2s^2\,2p^6\,\underline{3s^2}\,\underline{3p^5}$

    d. $1s^2\,2s^2\,2p^6\,3s^2\,3p^6\,4s^2\,3d^{10}\,4p^6\,\underline{5s^2}$

59.

61. a. 6   **b.** 6

    c. 7   **d.** 1

63. a. ns^1 **b.** ns^2

    c. $ns^2 np^3$ **d.** $ns^2 np^5$

65. a. $[\text{Ne}]3s^2 3p^1$ **b.** $[\text{He}]2s^2$

    c. $[\text{Kr}]5s^2 4d^{10} 5p^1$ **d.** $[\text{Kr}]5s^2 4d^2$

67. a. $[\text{Kr}]5s^2$ **b.** $[\text{Kr}]5s^2 4d^1$

    c. $[\text{Ar}]4s^2 3d^2$

    d. $[\text{Kr}]5s^2 4d^{10} 5p^4$

69. a. 2   **b.** 3

    c. 5   **d.** 6

71. Period 1 has two elements. Period 2 has eight elements. The number of subshells is equal to the principal quantum number. For Period 1, $n = 1$ and the s subshell contains only two elements. For Period 2, $n = 2$ and contains s and p subshells that have a total of 8 elements.

73. **a.** Al   **b.** S
     c. Ar   **d.** Mg

75. **a.** Cl   **b.** Ga
     c. Fe   **d.** Rb

77. **a.** As   **b.** Br
     c. cannot tell   **d.** S

79. Pb < Sn < Te < S < Cl

81. **a.** In   **b.** Si
     c. Pb   **d.** C

83. F < S < Si < Ge < Ca < Rb

85. **a.** Sr   **b.** Bi
     c. cannot tell   **d.** As

87. S < Se < Sb < In < Ba < Fr

89. $18 \, e^-$

91. Alkaline earth metals have the general electron configuration of ns^2. If they lose their two s electrons, they will obtain the stable electron configuration of a noble gas. This loss of electrons will give the metal a +2 charge.

93. **a.** $1s^2 2s^2 2p^6 3s^2 3p^6$
b. $1s^2 2s^2 2p^6 3s^2 3p^6$
c. $1s^2 2s^2 2p^6 3s^2 3p^6$
d. $1s^2 2s^2 2p^6 3s^2 3p^6 4s^2 3d^{10} 4p^6$
They all have noble gas electron configurations.

95. Metals tend to form positive ions because they tend to lose electrons. Elements on the left side of the periodic table have only a few extra electrons, which they will lose to gain a noble gas configuration. Metalloids tend to be elements with 3 to 5 valence electrons; they could lose or gain electrons to obtain a noble gas configuration. Nonmetals tend to gain electrons to fill their almost full valence shell, so they tend to form negative ions and are on the right side of the table.

97. **a.** Can only have 2 in the s shell and 6 in the p shell: $1s^2 2s^2 2p^6 3s^2 3p^3$.
b. There is no $2d$ subshell: $1s^2 2s^2 2p^6 3s^2 3p^2$.
c. There is no $1p$ subshell: $1s^2 2s^2 2p^3$.
d. Can only have 6 in the p shell: $1s^2 2s^2 2p^6 3s^2 3p^3$.

99. Bromine is highly reactive because it reacts quickly to gain an electron and obtain a stable valence shell. Krypton is a noble gas because it already has a stable valence shell.

101. K

103. 660 nm

105. 8 min, 19 sec

107. The quantum-mechanical model provided the ability to understand and predict chemical bonding, which is the basic level of understanding of matter and how it interacts. This model was critical in the areas of lasers, computers, semiconductors, and drug design. The quantum-mechanical model for the atom is considered the foundation of modern chemistry.

109. **a.** 1.5×10^{-34} m   **b.** 1.88×10^{-10} m
Electrons have wave-particle duality, whereas golf balls do not.

111. The ionization energy dips at column 3A because removing an electron from one of those atoms leaves the atom with a fairly stable, filled s-orbital as its valence shell. For the group 6A elements, special stability occurs when those elements lose an electron and achieve a half-filled p-orbital as their valence shell.

113. Ultraviolet light is the only one of these three types of light that contains enough energy to break chemical bonds in biological molecules.

CHAPTER 10
QUESTIONS

1. Bonding theories predict how atoms bond together to form molecules, and they also predict what combinations of atoms form molecules and what combinations do not. Likewise, bonding theories explain the shapes of molecules, which in turn determine many of their physical and chemical properties.

3. Atoms with eight valence electrons are particularly stable and are said to have an octet. Atoms such as hydrogen, helium, lithium, and beryllium are exceptions to the octet rule as they achieve stability when their outermost shell contains two electrons—a duet. A chemical bond is the sharing or transfer of electrons to attain stable electron configurations among the bonding atoms.

5. The Lewis structure for potassium has 1 valence electron, whereas the Lewis structure for monatomic chlorine has 7 valence electrons. From these structures we can determine that if potassium gives up its one valence electron to chlorine, K^+ and Cl^- are formed; therefore the formula must be KCl.

7. Double and triple bonds are shorter and stronger than single bonds.

9. You determine the number of electrons that go into the Lewis structure of a molecule by summing the valence electrons of each atom in the molecule.

11. The octet rule is not sophisticated enough to be correct every time. For example, some molecules that exist in nature have an odd number of valence electrons and thus will not have octets on all their constituent atoms. Some elements tend to form compounds in nature in which they have more (sulfur) or less (boron) than 8 valence electrons.

13. VSEPR theory predicts the shape of molecules using the idea that electron groups repel each other.

15. **a.** 180°   **b.** 120°
     c. 109.5°

17. Electronegativity is the ability of an element to attract electrons within a covalent bond.

19. A polar covalent bond is a covalent bond that has a dipole moment.

21. If a polar liquid and a nonpolar liquid are mixed they will separate into distinct regions because the polar molecules will be attracted to one another and will exclude the nonpolar molecules.

PROBLEMS

23. **a.** $1s^2 2s^2 2p^3$, :N̈·

   b. $1s^2 2s^2 2p^2$, ·Ċ·

   c. $1s^2 2s^2 2p^6 3s^2 3p^5$, :Cl̈·

   d. $1s^2 2s^2 2p^6 3s^2 3p^6$, :Är̈:

25. **a.** :Ï: **b.** ·S̈:

   c. ·Ġe·   **d.** ·Ca·

27. :Ẍ: Halogens tend to gain one electron in a chemical reaction.

29. M: Alkaline earth metals tend to lose two electrons in a chemical reaction.

31. **a.** Al^{3+} **b.** Mg^{2+}

   c. $[:S̈e:]^{2-}$ **d.** $[:N̈:]^{3-}$

33. **a.** Kr   **b.** Ne

   c. Kr   **d.** Xe

35. **a.** covalent   **b.** ionic

   c. covalent   **d.** ionic

37. **a.** $Na^+[:F̈:]^-$ **b.** $Ca^{2+}[:Ö:]^{2-}$

   c. $[:B̈r:]^- Sr^{2+}[:B̈r:]^-$ **d.** $K^+[:Ö:]^{2-}K^+$

39. **a.** CaS   **b.** $MgBr_2$

   c. CsI   **d.** Ca_3N_2

41. **a.** $[:F̈:]^- Mg^{2+}[:F̈:]^-$ **b.** $Mg^{2+}[:Ö:]^{2-}$

   c. $Mg^{2+}[:N̈:]^{3-} Mg^{2+}[:N̈:]^{3-} Mg^{2+}$

43. **a.** $Cs^+[:Cl̈:]^-$ **b.** $Ba^{2+}[:Ö:]^{2-}$

   c. $[:Ï:]^- Ca^{2+}[:Ï:]^-$

45. **a.** Hydrogen exists as a diatomic molecule because two hydrogen molecules achieve a stable duet when they share their electrons and form a single covalent bond.

   b. Iodine achieves a stable octet when two atoms share electrons and form a single bond.

   c. Nitrogen achieves a stable octet when two atoms share electrons and form a triple bond.

   d. Oxygen achieves a stable octet when two atoms share electrons and form a double bond.

47. **a.**
```
   H—P̈—H
     |
     H
```
   b. :Cl̈—S̈—Cl̈:

   c. :F̈—F̈:   **d.** H—Ï:

49. **a.** Ö=Ö   **b.** :C≡O:

   c. H—Ö—N̈=Ö   **d.** :Ö=S̈—Ö:

51. **a.** H—C≡C—H   **b.**
```
   H—C=C—H
     |   |
     H   H
```
   c. H—N̈=N̈—H   **d.**
```
   H—N̈—N̈—H
     |   |
     H   H
```

53. **a.** :N≡N: **b.** S̈=Si=S̈

   c. H—Ö—H   **d.**
```
   :Ï—N̈—Ï:
        |
       :Ï:
```

55. **a.** :Ö=S̈e—Ö: ⟷ :Ö—S̈e=Ö:

   b.

$$\left[\ddot{O}=C-\ddot{O}:\atop\;\;\;\;\;\;:\ddot{O}:\right]^{2-} \!\!\longleftrightarrow \left[:\ddot{O}-C=\ddot{O}\atop\;\;\;\;\;\;:\ddot{O}:\right]^{2-} \!\!\longleftrightarrow \left[:\ddot{O}-C-\ddot{O}:\atop\;\;\;\;\;\;\;\;\|\atop\;\;\;\;\;\;\;\;:\ddot{O}\right]^{2-}$$

   c. $[:Cl̈—Ö:]^-$ **d.** $[:Ö—Cl̈—Ö:]^-$

57. **a.**
$$\left[\;\;\;\;\;\;:\ddot{O}:\atop:\ddot{O}-P-\ddot{O}:\atop\;\;\;\;\;\;:\ddot{O}:\right]^{3-}$$
   b. $[:C≡N:]^-$

   c. $[:Ö=N̈—Ö:]^- ⟷ [:Ö—N̈=Ö:]^-$

   d.
$$\left[:\ddot{O}-S-\ddot{O}:\atop\;\;\;\;\;\;:\ddot{O}:\right]^{2-}$$

59. **a.**
```
   :Cl̈—B—Cl̈:
        |
       :Cl̈:
```
   b. Ö=N̈—Ö: ⟷ :Ö—N̈=Ö

   c.
```
   H—B—H
     |
     H
```

61. **a.** 4   **b.** 4

   c. 2   **d.** 4

63. **a.** 2 bonding groups, 2 lone pairs

   b. 3 bonding groups, 1 lone pair

   c. 2 bonding groups, 0 lone pairs

   d. 4 bonding groups, 0 lone pairs

65. **a.** tetrahedral   **b.** trigonal planar

   c. linear   **d.** trigonal planar

67. **a.** 109.5°   **b.** 120°

   c. 180°   **d.** 120°

69. **a.** linear, linear

   b. trigonal planar, bent

   c. tetrahedral, bent

   d. tetrahedral, trigonal pyramidal

71. **a.** 180°   **b.** 120°

   c. 109.5°   **d.** 109.5°

73. **a.** linear, linear

   b. trigonal planar, bent (about both nitrogen atoms)

   c. tetrahedral, trigonal pyramidal (about both nitrogen atoms)

75. **a.** trigonal planar   **b.** bent

   c. trigonal planar   **d.** tetrahedral

77. **a.** 1.2   **b.** 1.8

   c. 2.8

79. Cl > Si > Ga > Ca > Rb

81. **a.** polar covalent   **b.** ionic
    c. pure covalent   **d.** polar covalent

83. $H_2 <$ ICl $<$ HBr $<$ CO

85. **a.** polar   **b.** nonpolar
    c. nonpolar   **d.** polar

87. **a.** $(+)$:C≡O:$(-)$
    b. nonpolar
    c. nonpolar
    d. $(+)$H—B̈r:$(-)$

89. **a.** nonpolar   **b.** polar
    c. nonpolar   **d.** polar

91. **a.** nonpolar   **b.** polar
    c. nonpolar   **d.** polar

93. **a.** $1s^2 2s^2 2p^6 3s^2 3p^6 \underline{4s^2}$, Ca: (underlined electrons are the ones included)
    b. $1s^2 2s^2 2p^3 3s^2 3p^6 \underline{4s^2} 3d^{10} \underline{4p^1}$, Ġa:
    c. $[Ar]\underline{4s^2} 3d^{10} \underline{4p^3}$, ·Äs:
    d. $[Kr]\underline{5s^2} 4d^{10} \underline{5p^5}$, :Ï:

95. **a.** ionic, $K^+ \left[:\ddot{\underset{..}{S}}:\right]^{2-} K^+$

    b. covalent,
    c. ionic, $Mg^{2+} \left[:\ddot{\underset{..}{Se}}:\right]^{2-}$
    d. covalent,

97. :C̈l—C—C̈l: polar,

99.

101. H:C̈l: + $Na^+ [:\ddot{O}:H]^-$ ⟶ H:Ö:H + $Na^+ [:\ddot{\underset{..}{Cl}}:]^-$

103. K·, :C̈l—C̈l:, $K^+ [:\ddot{\underset{..}{Cl}}:]^-$, Cl reduced, K oxidized

105. **a.** $K^+ [:\ddot{O}:H]^-$
    b.
$$K^+ \left[:\ddot{O}=N:\ddot{\underset{..}{O}}:\right]^- \longleftrightarrow K^+ \left[:\ddot{\underset{..}{O}}:N:\ddot{\underset{..}{O}}:\right]^- \longleftrightarrow K^+ \left[:\ddot{\underset{..}{O}}:N=\ddot{O}:\right]^-$$
    c. $Li^+ [:\ddot{\underset{..}{I}}:\ddot{\underset{..}{O}}:]^-$
    d.
$$Ba^{2+} \left[\ddot{\underset{..}{O}}\underset{:\ddot{\underset{..}{O}}:\; :\ddot{\underset{..}{O}}:}{\overset{\;}{\underset{C}{\|}}}\right]^{2-} \longleftrightarrow Ba^{2+} \left[\underset{:\ddot{\underset{..}{O}}:\; :\ddot{\underset{..}{O}}:}{\overset{:\ddot{\underset{..}{O}}:}{C}}\right]^{2-} \longleftrightarrow Ba^{2+} \left[\underset{:\ddot{\underset{..}{O}}:\; :\ddot{\underset{..}{O}}:}{\overset{:\ddot{\underset{..}{O}}:}{C}}\right]^{2-}$$

107. **a.**   **b.**
    c.

109. CH_2O_2 or
$$\underset{\quad}{\overset{:\ddot{O}:}{\overset{\|}{H—C—\ddot{O}—H}}}$$

111. H—Ö—Ö· HOO is not stable because one oxygen atom does not have an octet. The geometry for HOO is *bent*.

113. **a.** $[:\ddot{O}—\ddot{\underset{..}{O}}:]^-$ **b.** $[:\ddot{\underset{..}{O}}·]^-$
    c. ·Ö—H
    d. $H—\overset{H}{\underset{H}{C}}—\ddot{O}—\ddot{O}·$

115. **a.** The structure has 2 bonding electron pairs and 2 lone pairs. The Lewis structure is analogous to that of water, and the molecular geometry is bent.
    b. Correct
    c. The structure has 3 bonding electron pairs and 1 lone pair. The Lewis structure is analogous to that of NH_3, and the geometry is trigonal pyramidal.
    d. Correct

CHAPTER 11
QUESTIONS

1. Pressure is the push (or force) exerted per unit area by gaseous molecules as they collide with the surfaces around them.

3. The kinetic molecular theory makes four main assumptions. The first is that the gas is a collection of molecules in constant motion. Secondly, there is no attraction or repulsion between the particles and collisions are perfectly elastic. Thirdly, there is a lot of space in between the particles relative to the particle size. Lastly, the speed of the particles increases with temperature.

5. The pain we experience in our ears during a change in altitude is due to a pressure difference between the cavities inside of our ears and the surrounding air.

7. Boyle's Law states that the volume of a gas and its pressure are inversely proportional. This relationship can be explained by the kinetic molecular theory. If the volume of a sample is decreased, the same number of particles are crowded into a smaller space, causing more collisions with the walls of the container. This causes the pressure to increase.

9. When an individual is more than a couple of meters underwater, the air pressure in the lungs is greater than the air pressure at the water's surface. If a snorkel were used, it would move the air from the lungs to the surface, making it very difficult to breathe.

11. Increasing the temperature of the air in the balloon causes it to expand. As the volume of the air increases, the density decreases allowing it to float in the cooler, more dense air surrounding it.

13. Avogadro's law states that the volume of a gas is directly proportional to the amount of gas in moles. Kinetic molecular theory predicts that if the number of gas particles increases at a constant pressure and temperature, the volume increases.

15. The ideal gas law is most accurate when the volume of gas particles is small compared to the space between them. It is also accurate when the forces between particles are not important. The ideal gas law breaks down at high pressures and low temperatures. This breakdown occurs because the gases are no longer acting according to the kinetic molecular theory.

17. Dalton's law states that the sum of the partial pressures in a gas mixture must equal the total pressure. $P_{tot} = P_A + P_B + P_C + \ldots$

19. Deep-sea divers breathe helium with oxygen because helium, unlike nitrogen, does not have physiological effects under high-pressure conditions. The oxygen concentration in the mixture is low to avoid oxygen toxicity.

21. Vapor pressure is the partial pressure of a gas above its liquid. Partial pressure increases with increasing temperature.

PROBLEMS

23. **a.** 1.680 atm   **b.** 2.35 atm
    c. 8.64 atm   **d.** .599 atm

25. **a.** 1.7×10^3 torr   **b.** 36 mmHg
    c. 1.28×10^3 mmHg   **d.** 834.2 torr

27.

Pascals	Atmospheres	mmHg	Torr	PSI
882	0.00871	6.62	6.62	0.128
5.65×10^4	0.558	424	424	8.20
1.71×10^5	1.69	1.28×10^3	1.28×10^3	24.8
1.02×10^5	1.01	764	764	14.8
3.32×10^4	0.328	249	249	4.82

29. **a.** 0.832 atm   **b.** 632 mmHg
    c. 12.2 psi   **d.** 8.43×10^4 Pa

31. **a.** 809.0 mmHg   **b.** 1.065 atm
    c. 809.0 torr   **d.** 107.9 kPa

33. 518 mmHg

35. 1.8 L

37.

P_1	V_1	P_2	V_2
755 mmHg	2.85 L	885 mmHg	2.43 L
9.35 atm	1.33 L	4.32 atm	2.88 L
192 mmHg	382 mL	152 mmHg	482 mL
2.11 atm	226 mL	3.82 atm	125 mL

39. 4.0 L

41. 58.9 mL

43.

V_1	T_1	V_2	T_2
1.08 L	25.4 °C	1.33 L	94.5 °C
58.9 mL	77 K	228 mL	298 K
115 cm^3	12.5 °C	119 cm^3	22.4 °C
232 L	18.5 °C	294 L	96.2 °C

45. 6.8 L

47. 4.33 L

49.

V_1	n_1	V_2	n_2
38.5 mL	1.55×10^{-3}	49.4 mL	1.99×10^{-3}
8.03 L	1.37	26.8 L	4.57
11.2 L	0.628	15.7 L	0.881
422 mL	0.0109	671 mL	0.0174

51. 1.71×10^3 mmHg

53. 0.76 L

55. 877 mmHg

57.

P_1	V_1	T_1	P_2	V_2	T_2
121 atm	1.58 L	12.2 °C	1.54 torr	1.33 L	32.3 °C
721 torr	141 mL	135 K	801 torr	152 mL	162 K
5.51 atm	0.879 L	22.1 °C	4.87 atm	1.05 L	38.3 °C

59. 5.11 L

61. 2.1 mol

63. 1.42 mol

65.

P	V	n	T
1.05 atm	1.19 L	0.112 mol	136 K
112 torr	40.8 L	0.241 mol	304 K
1.50 atm	28.5 mL	1.74×10^{-3} mol	25.4 °C
0.559 atm	0.439 L	0.0117 mol	255 K

67. 0.23 mol

69. 44.0 g/mol

71. 4.00 g/mol

73. 571 torr

75. 10.7 atm

77. 7.00×10^2 mmHg

79. 0.87 atm N_2; 0.25 atm O_2

81. 0.34 atm

83. **a.** 504 L   **b.** 81 L
    c. 49 L   **d.** 6.0×10^2

85. **a.** 59.1 L   **b.** 30.0 L
    c. 72.1 L   **d.** 0.125 L

87. **a.** 0.350 g   **b.** 0.221 g
    c. 8.15 g

89. 28 L

91. 33 L H_2; 16 L CO

93. 8.82 L

95. 12.6 g

97. 0.5611 g

99. $V = \dfrac{nRT}{P}$

$$= \dfrac{1.00 \text{ mol} \left(0.0821 \dfrac{\text{L} \cdot \text{atm}}{\text{mol} \cdot \text{K}} \right) (273 \text{ K})}{1.00 \text{ atm}} = 22.4 \text{ L}$$

101. 27.8 g/mol

103. C_4H_{10}

105. 0.828 g

107. 0.128 g

109. 0.935 L

111. $HCl + NaHCO_3 \longrightarrow CO_2 + H_2O + NaCl$;

1.11×10^{-3} mol

113. **a.** SO_2, 0.0127 mol **b.** 65.6%

115. **a.** NO_2, 24.0 g **b.** 61.6%

117. 11.7 L

119. 356 torr

121. 0.15 atm

123. **c.** From the ideal gas law, we see that pressure is directly proportional to the number of moles of gas per unit volume (n/V). The gas in (c) contains the greatest concentration of particles and thus has the highest pressure.

125. 22.8 g

127. $V_2 = 0.76$ L, actual volume is 0.61 L. Difference is due to the fact that the ideal gas law is not ideal, especially at low temperatures.

CHAPTER 12
QUESTIONS

1. Intermolecular forces are attractive forces that occur between molecules. Intermolecular forces are what living organisms depend on for many physiological processes. Intermolecular forces are also responsible for the existence of liquids and solids.

3. The magnitude of intermolecular forces relative to the amount of thermal energy in the sample determines the state of the matter.

5. Properties of solids:

   a. Solids have high densities in comparison to gases.

   b. Solids have a definite shape.

   c. Solids have a definite volume.

   d. Solids may be crystalline or amorphous.

7. Surface tension is the tendency of liquids to minimize their surface area. Molecules at the surface have few neighbors to interact with via intermolecular forces.

9. Evaporation is a physical change in which a substance is converted from its liquid form to its gaseous form. Condensation is a physical change in which a substance is converted from its gaseous form to its liquid form.

11. Evaporation below the boiling point occurs because molecules on the surface of the liquid experience fewer attractions to the neighboring molecules and can therefore break away. At the boiling point, evaporation occurs faster because more of the molecules have sufficient thermal energy to break away (including internal molecules).

13. Acetone has weaker intermolecular forces than water. Acetone is more volatile than water.

15. Vapor pressure is the partial pressure of a gas in dynamic equilibrium with its liquid. It increases with increasing temperature, and it also increases with decreasing strength of intermolecular forces.

17. A steam burn is worse than a water burn at the same temperature (100 °C), because when the steam condenses on the skin, it releases large amounts of additional heat.

19. As the first molecules freeze, they release heat, making it harder for other molecules to freeze without the aid of a refrigeration mechanism, which would draw heat out.

21. The melting of ice is endothermic. ΔH for melting is positive (+), whereas ΔH for freezing is negative (−).

23. Dispersion forces are the default intermolecular force present in all molecules and atoms. Dispersion forces are caused by fluctuations in the electron distribution within molecules or atoms. Dispersion forces are the weakest type of intermolecular force and increase with increasing molar mass.

25. Hydrogen bonding is an intermolecular force and is sort of a super dipole–dipole force. Hydrogen bonding occurs in compounds containing hydrogen atoms bonded directly to fluorine, oxygen, or nitrogen.

27. Molecular solids as a whole tend to have low to moderately low melting points relative to other types of solids; however, strong molecular forces can increase their melting points relative to each other.

29. Ionic solids tend to have much higher melting points relative to the melting points of other types of solids.

31. Water is unique for a couple of reasons. Water has a low molar mass, yet it is still liquid at room temperature and has a relatively high boiling point. Unlike other substances, which contract upon freezing, water expands upon freezing.

PROBLEMS

33. The 55 mL of water in a dish with a diameter of 12 cm will evaporate more quickly because it has a larger surface area.

35. Acetone feels cooler while evaporating from one's hand, for it is more volatile than water and evaporates much faster.

37. The ice's temperature will increase from −5 °C to 0 °C, where it will then stay constant while the ice completely melts. After the melting process is complete, the water will continue to rise steadily in temperature until it reaches room temperature (25 °C).

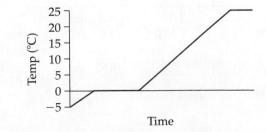

39. Allowing 0.50 g of 100 °C steam to condense on your hand would cause a more severe burn than spilling 0.50 g of 100 °C water on your hand. When the steam condenses on the skin, it releases large amounts of additional heat.

41. The −8 °C ice chest will cause the water in the watery bag of ice to freeze, which is an exothermic process. The freezing process of the water will release heat, and the temperature of the ice chest will increase.

43. The ice chest filled with ice at 0 °C will be colder after a couple of hours than the ice chest filled with water at 0 °C. This is because the ice can absorb additional heat as it melts.

45. Denver is a mile above sea level and thus has a lower air pressure. Due to this low pressure, the point at which the vapor pressure of water equals the external pressure will occur at a lower temperature.

47. 76.3 kJ

49. 6.8 kJ

51. 10.4 kJ

53. 3.70×10^2 g

55. 12.5 kJ

57. 11.4 kJ

59. **a.** dispersion

b. dispersion

c. dispersion, dipole–dipole

d. dispersion, dipole–dipole, hydrogen bonding

61. **a.** dispersion, dipole–dipole

b. dispersion, dipole–dipole, hydrogen bonding

c. dispersion

d. dispersion

63. d, because it has the highest molecular weight

65. CH_3OH, due to strong hydrogen bonding

67. NH_3, because it has the ability to form hydrogen bonds

69. These two substances are not miscible, for H_2O is polar and $CH_3CH_2CH_2CH_2CH_3$ is nonpolar.

71. **a.** no   **b.** yes

c. yes

73. **a.** atomic   **b.** molecular

c. ionic   **d.** atomic

75. **a.** molecular   **b.** ionic

c. molecular   **d.** molecular

77. **c.** $LiCl(s)$; it is an ionic solid and possesses ionic bonds resulting in a higher melting point.

79. **a.** $Ti(s)$; Ti is a covalent atomic solid, and Ne is a nonbonding atomic solid.

b. $H_2O(s)$; while both are molecular solids, water has strong hydrogen bonding.

c. $Xe(s)$; both are nonbonding atomic solids, but Xe has a higher molar mass.

d. $NaCl(s)$; $NaCl(s)$ is an ionic solid, and $CH_4(s)$ is a molecular solid.

81. $Ne < SO_2 < NH_3 < H_2O < NaF$

83. **a.** 2.6×10^4 J   **b.** 26 kJ

c. 6.2×10^3 cal   **d.** 6.2 Cal

85. 2.7 °C

87. 1.1×10^2 g

89. 57.5 kJ

91. **a.** H—$\overset{\displaystyle ..}{\underset{\displaystyle ..}{Se}}$—H Bent: Dispersion and dipole–dipole

b. $\ddot{O}$=$\ddot{S}$—$\ddot{O}$: Bent: Dispersion and dipole–dipole

c. H—$\overset{\displaystyle :\ddot{Cl}:}{\underset{\displaystyle :\ddot{Cl}:}{C}}$—$\ddot{Cl}$: Tetrahedral: Dispersion and dipole–dipole

d. :$\ddot{O}$=C=$\ddot{O}$: Linear: Dispersion

93. $Na^+\left[:\ddot{F}:\right]^-$; $Mg^{2+}\left[:\ddot{O}:\right]^{2-}$; MgO has a higher melting point, because the magnitude of the charges on the ions is greater.

95. As the molecular weight increases from Cl to I, the greater the London dispersion forces present, which will increase the boiling point as observed. However, HF is the only compound listed that has the ability to form hydrogen bonds, which explains the anomaly in the trend.

97. 23.6 °C

99. The molecule in the interior has the most neighbors. The molecule on the surface is more likely to evaporate. In three dimensions, the molecules that make up the surface area have the fewest neighbors and are more likely to evaporate than the molecules in the interior.

101. **a.** Yes, if the focus is on the melting of icebergs. No, the melting of an ice cube in a cup of water will *not* raise the level of the liquid in the cup; when the ice cube melts, the volume of the water created will be less than the volume of the initial ice cube.

b. Yes, for the ice sheets that sit on the continent of Antarctica are above sea level, and if some melted, the water created would be added to the ocean without decreasing the amount of ice below the ocean's surface.

CHAPTER 13
QUESTIONS

1. A solution is a homogeneous mixture of two or more substances. Some examples are air, seawater, soda water, and brass.

3. In a solution, the solvent is the majority component of the mixture, and the solute is the minority component. For example, in a seawater solution, the water is the solvent, and the salt content is the solute.

5. Solubility is the amount of the compound, usually in grams, that will dissolve in a specified amount of solvent.

7. In solutions with solids, soluble ionic solids form strong electrolyte solutions, while soluble molecular solids form nonelectrolyte solutions. Strong electrolyte solutions are solutions containing solutes that dissociate into ions, for example, $BaCl_2$ and NaOH.

9. Recrystallization is a common way to purify a solid. In recrystallization, enough solid is put into high-temperature water until a saturated solution is created. Then the solution cools slowly, and crystals result from the solution. The crystalline structure tends to reject impurities, resulting in a purer solid.

11. The bubbles formed on the bottom of a pot of heated water (before boiling) are dissolved air coming out of the solution. These gases come out of solution because the solubility of the dissolved nitrogen and oxygen decreases as the temperature of the water rises.

13. The solubility of gases increases with increasing pressure. When a soda can is opened, the pressure is lowered, decreasing the solubility of carbon dioxide. This causes bubbles of carbon dioxide to come out of the solution.

15. Mass percent is the number of grams of solute per 100 grams of solution. Molarity is defined as the number of moles of solute per liter of solution.

17. The boiling point of a solution containing a nonvolatile solute is higher than the boiling point of the pure solvent. The melting point of the solution, however, is lower.

19. Molality is a common unit of concentration of a solution expressed as number of moles of solute per kilogram of solvent.

21. Water tends to move from lower concentrations to higher concentrations, and when the salt water is being passed through the human body, the salt content draws the water out of the body, causing dehydration.

PROBLEMS

23. c and d are solutions

25. **a.** solute: salt, solvent: water

    b. solute: sugar, solvent: water

    c. solute: CO_2, solvent: water

27. **a.** hexane   **b.** water

    c. ethyl ether   **d.** water

29. ions, strong electrolyte solution

31. unsaturated

33. recrystallization

35. **a.** no

    b. yes

    c. yes

37. At room temperature water contains some dissolved oxygen gas; however, the boiling of the water will remove dissolved gases.

39. Under higher pressure, the gas (nitrogen) will be more easily dissolved in the blood. To reverse this process, the diver should ascend to relieve the pressure.

41. **a.** 7.64%

    b. 3.50%

    c. 4.64%

43. 12%

45.

Mass Solute	Mass Solvent	Mass Solution	Mass%
15.5	238.1	253.6	6.11%
22.8	167.2	190.0	12.0%
28.8	183.3	212.1	13.6%
56.9	315.2	372.1	15.3%

47. 8.9 g NaCl

49. **a.** 1.8 g   **b.** 10.5 mg

    c. 0.46 kg

51. **a.** 2.6 kg   **b.** 1.0×10^2 g

    c. 18 g

53. 1.6×10^2 g

55. 1.3×10^3 g

57. 11 L

59. **a.** 0.194 M   **b.** 0.234 M

    c. 0.41 M

61. **a.** 0.149 M   **b.** 0.461 M

    c. 3.00×10^{-2} M

63. 0.57 M

65. **a.** 1.8 mol   **b.** 0.38 mol

    c. 0.238 mol

67. **a.** 0.59 L   **b.** 0.083 L

    c. 0.15 L

69.

Solute	Mass Solute	Mol Solute	Volume Solution	Molarity
KNO_3	22.5 g	0.223	125 mL	1.78 M
$NaHCO_3$	2.10 g	0.0250	250.0 mL	0.100 M
$C_{12}H_{22}O_{11}$	55.38 g	0.162	1.08 L	0.150 M

71. 2.7 g

73. 19 g

75. 8.8 L

77. **a.** 0.15 M   **b.** 0.30 M

    c. 0.45 M

79. **a.** 0.24 M Na^+, 0.12 M SO_4^{2-}

    b. 0.50 M K^+, 0.25 M CO_3^{2-}

    c. 0.11 M Rb^+, 0.11 M Br^-

81. 0.29 M

83. Dilute 0.045 L of the stock solution to 2.5 L.

85. 6.0×10^2 mL

87. 17.7 mL

89. **a.** 0.025 L   **b.** 0.020 L

    c. 1.03 L

91. 4.45 mL

93. 0.373 M

95. 1.2 L

97. **a.** 1.0 m   **b.** 3.92 m

    c. 0.52 m

99. 1.49 m

101. **a.** $-1.6 \,^{\circ}C$ **b.** $-2.70 \,^{\circ}C$

    c. $-8.9 \,^{\circ}C$ **d.** $-4.37 \,^{\circ}C$

103. **a.** $100.060 \,^{\circ}C$ **b.** $100.993 \,^{\circ}C$

    c. $101.99 \,^{\circ}C$ **d.** $101.11 \,^{\circ}C$

105. $-1.27 \,^{\circ}C$, $100.348 \,^{\circ}C$

107. 2.28 M, 12.3%

109. 0.43 L

111. 319 mL

113. 1.03 M

115. 0.17 L

117. 9.0 mL

119. 8.0×10^2 mL

121. $-3.60\,°C$, $100.992\,°C$

123. $-1.86\,°C$, $100.508\,°C$

125. 1.8×10^2 g/mol

127. $101.8\,°C$

129. 39.8 g glucose, 85.2 g sucrose

131. **a.** Water will flow from left to right.

  b. Water will flow from right to left.

  c. Water won't flow between the two.

133. 3×10^4 L

CHAPTER 14

QUESTIONS

1. Sour gummy candies are coated with a white powder that is a mixture of citric acid and tartaric acid. The combination of these two acids creates the sour taste.

3. The main component of stomach acid is hydrochloric acid. Its purpose is to help break down food and kill harmful bacteria.

5. The properties of bases are bitter taste, slippery feel, and the ability to turn red litmus paper blue. Some common substances that contain bases are ammonia, Drano, baking soda, and antacids.

7. The Arrhenius definition of an acid is a substance that produces H^+ ions in aqueous solution. An example:

$$HCl(aq) \longrightarrow H^+(aq) + Cl^-(aq)$$

9. The Brønsted–Lowry definition states that an acid is a proton donor and a base is a proton acceptor. The following is an example of a chemical equation demonstrating this definition:

$$\underset{\text{acid}}{HCl(aq)} + \underset{\text{base}}{H_2O(l)} \longrightarrow H_3O^+(aq) + Cl^-(aq)$$

11. An acid–base neutralization reaction occurs when an acid and a base are mixed and the $H^+(aq)$ from the acid combines with the $OH^-(aq)$ from the base to form $H_2O(l)$. An example follows.

$$HCl(aq) + KOH(aq) \longrightarrow H_2O(l) + KCl(aq)$$

13. $2\,HCl(aq) + K_2O(s) \longrightarrow H_2O(l) + 2\,KCl(aq)$

15. A titration is a laboratory procedure in which a reactant in a solution of known concentration is reacted with another reactant in a solution of unknown concentration until the reaction has reached the equivalence point. The equivalence point is the point at which the reactants are in exact stoichiometric proportions.

17. A strong acid is one that will completely dissociate in solution, while a weak acid does not completely dissociate in solution.

19. Monoprotic acids (such as HCl) contain only one hydrogen ion that will dissociate in solution, while diprotic acids (such as H_2SO_4) contain two hydrogen ions that will dissociate in solution.

21. Yes, pure water contains H_3O^+ ions. Through self-ionization, water acts as an acid and a base with itself; water is amphoteric.

23. **a.** $[H_3O^+] > 1.0 \times 10^{-7}\,M$; $[OH^-] < 1.0 \times 10^{-7}\,M$

  b. $[H_3O^+] < 1.0 \times 10^{-7}\,M$; $[OH^-] > 1.0 \times 10^{-7}\,M$

  c. $[H_3O^+] = 1.0 \times 10^{-7}\,M$; $[OH^-] = 1.0 \times 10^{-7}\,M$

25. The pOH of a solution is the negative base-10 logarithm of the concentration of OH^- ions ($-\log[OH^-]$). A change of 2.0 pOH units corresponds to a 100-fold change in $[OH^-]$.

27. A buffer is a solution that resists pH change by neutralizing added acid or added base.

29. The cause of acid rain is the formation of SO_2, NO, and NO_2 during the combustion of fossil fuels.

31. Acid rain damages structures made out of metal, marble, cement, and limestone, as well as harming and possibly killing aquatic life and trees.

PROBLEMS

33. **a.** acid; $H_2SO_4(aq) \longrightarrow H^+(aq) + HSO_4^-(aq)$

  b. base; $Sr(OH)_2(aq) \longrightarrow Sr^{2+}(aq) + 2\,OH^-(aq)$

  c. acid; $HBr(aq) \longrightarrow H^+(aq) + Br^-(aq)$

  d. base; $NaOH(aq) \longrightarrow Na^+(aq) + OH^-(aq)$

35.

B-L Acid	B-L Base	Conjugate Acid	Conjugate Base
a. HBr	H_2O	H_3O^+	Br^-
b. H_2O	NH_3	NH_4^+	OH^-
c. HNO_3	H_2O	H_3O^+	NO_3^-
d. H_2O	C_5H_5N	$C_5H_5NH^+$	OH^-

37. a, c

39. **a.** Cl^- **b.** HSO_3^-

  c. CHO_2^- **d.** F^-

41. **a.** NH_4^+ **b.** $HClO_4$

  c. H_2SO_4 **d.** HCO_3^-

43. **a.** $HI(aq) + NaOH(aq) \longrightarrow H_2O(l) + NaI(aq)$

  b. $HBr(aq) + KOH(aq) \longrightarrow H_2O(l) + KBr(aq)$

  c. $2\,HNO_3(aq) + Ba(OH)_2(aq) \longrightarrow$
  $$2\,H_2O(l) + Ba(NO_3)_2(aq)$$

  d. $2\,HClO_4(aq) + Sr(OH)_2(aq) \longrightarrow$
  $$2\,H_2O(l) + Sr(ClO_4)_2(aq)$$

45. **a.** $2\,HBr(aq) + 2\,Rb(s) \longrightarrow 2\,RbBr(aq) + H_2(g)$

  b. $2\,HBr(aq) + Mg(s) \longrightarrow MgBr_2(aq) + H_2(g)$

  c. $2\,HBr(aq) + 2\,Ba(s) \longrightarrow 2\,BaBr(aq) + H_2(g)$

  d. $6\,HBr(aq) + 2\,Al(s) \longrightarrow 2\,AlBr_3(aq) + 3\,H_2(g)$

47. **a.** $MgO(aq) + 2\,HI(aq) \longrightarrow H_2O(l) + MgI_2(aq)$

  b. $K_2O(aq) + 2\,HI(aq) \longrightarrow H_2O(l) + 2\,KI(aq)$

  c. $Rb_2O(aq) + 2\,HI(aq) \longrightarrow H_2O(l) + 2\,RbI(aq)$

  d. $CaO(aq) + 2\,HI(aq) \longrightarrow H_2O(l) + CaI_2(aq)$

49. **a.** $6\,HClO_4(aq) + Fe_2O_3(s) \longrightarrow$
  $$2\,Fe(ClO_4)_3(aq) + 3\,H_2O(l)$$

  b. $H_2SO_4(aq) + Sr(s) \longrightarrow SrSO_4(s) + H_2(g)$

  c. $H_3PO_4(aq) + 3\,KOH(aq) \longrightarrow 3\,H_2O(l) + K_3PO_4(aq)$

51. **a.** 0.1400 M **b.** 0.138 M

  c. 0.08764 M **d.** 0.182 M

53. 0.1090 M H_2SO_4

55. 16.9 mL

57. **a.** strong   **b.** weak
     c. strong   **d.** weak

59. **a.** $[H_3O^+] = 1.7$ M   **b.** $[H_3O^+] = 1.5$ M
     c. $[H_3O^+] < 0.38$ M   **d.** $[H_3O^+] < 1.75$ M

61. **a.** strong   **b.** weak
     c. strong   **d.** weak

63. **a.** $[OH^-] = 0.25$ M   **b.** $[OH^-] < 0.25$ M
     c. $[OH^-] = 0.50$ M   **d.** $[OH^-] = 1.25$ M

65. **a.** acidic   **b.** acidic
     c. neutral   **d.** basic

67. **a.** 6.7×10^{-6} M, basic   **b.** 1.1×10^{-6} M, basic
     c. 4.5×10^{-9} M, acidic   **d.** 1.4×10^{-11} M, acidic

69. **a.** 3.7×10^{-3} M, acidic   **b.** 4.0×10^{-13} M, basic
     c. 9.1×10^{-5} M, acidic   **d.** 3.0×10^{-11} M, basic

71. **a.** basic   **b.** neutral
     c. acidic   **d.** acidic

73. **a.** 7.77   **b.** 7.00
     c. 5.66   **d.** 3.13

75. **a.** 2.8×10^{-9} M   **b.** 5.9×10^{-12} M
     c. 1.3×10^{-3} M   **d.** 6.0×10^{-2} M

77. **a.** 7.28   **b.** 6.42
     c. 3.86   **d.** 12.98

79. **a.** 1.8×10^{-10} M   **b.** 3.4×10^{-2} M
     c. 3.2×10^{-13} M   **d.** 1.8×10^{-6} M

81. **a.** 1.810   **b.** 11.107
     c. 2.724   **d.** 10.489

83. **a.** pOH = 8.82, acidic   **b.** pOH = 4.15, basic
     c. pOH = 7.00, neutral   **d.** pOH = 2.06, basic

85. **a.** pOH = 6.08, basic   **b.** pOH = 12.74, acidic
     c. pOH = 5.59, basic   **d.** pOH = 1.274, basic

87. **a.** pH = 5.5, acidic   **b.** pH = 9.8, basic
     c. pH = 12.3, basic   **d.** pH = 7.0, neutral

89. various answers

91. c and d are buffers

93. $HCl(aq) + NaF(aq) \longrightarrow HF(aq) + NaCl(aq)$

    $HCl(aq) + KC_2H_3O_2(aq) \longrightarrow HC_2H_3O_2(aq) + KCl(aq)$

95. **a.** $HC_2H_3O_2$ **b.** NaH_2PO_4
     c. NaCHOO

97. 50.0 mL

99. 0.16 L

101. 65.2 g

103. 60.0 g/mol

105. 0.17 L

107. **a.** 2.60, acidic
     b. 11.75, basic
     c. 8.02, basic
     d. 1.710, acidic

109.

$[H_3O^+]$	$[OH^-]$	pOH	pH	Acidic or Basic
1.0×10^{-4}	1.0×10^{-10}	10.00	4.00	acidic
5.5×10^{-3}	1.8×10^{-12}	11.74	2.26	acidic
3.1×10^{-9}	3.2×10^{-6}	5.50	8.50	basic
4.8×10^{-9}	2.1×10^{-6}	5.68	8.32	basic
2.8×10^{-8}	3.5×10^{-7}	6.45	7.55	basic

111. **a.** $[H_3O^+] = 0.0088$ M
       $[OH^-] = 1.1 \times 10^{-12}$ M
       pH = 2.06

     b. $[H_3O^+] = 1.5 \times 10^{-3}$ M
       $[OH^-] = 6.7 \times 10^{-12}$ M
       pH = 2.82

     c. $[H_3O^+] = 9.77 \times 10^{-4}$ M
       $[OH^-] = 1.02 \times 10^{-11}$ M
       pH = 3.010

     d. $[H_3O^+] = 0.0878$ M
       $[OH^-] = 1.14 \times 10^{-13}$ M
       pH = 1.057

113. **a.** $[OH^-] = 0.15$ M
       $[H_3O^+] = 6.7 \times 10^{-14}$ M
       pH = 13.18

     b. $[OH^-] = 3.0 \times 10^{-3}$ M
       $[H_3O^+] = 3.3 \times 10^{-12}$ M
       pH = 11.48

     c. $[OH^-] = 9.6 \times 10^{-4}$ M
       $[H_3O^+] = 1.0 \times 10^{-11}$ M
       pH = 10.98

     d. $[OH^-] = 8.7 \times 10^{-5}$ M
       $[H_3O^+] = 1.1 \times 10^{-10}$ M
       pH = 9.94

115. 1.49 L

117. 11.495

119. 3.0×10^{12} H^+ ions

121. 0.024 mol $Sr(OH)_2$, 0.076 mol NaOH

123. **a.** weak
     b. strong
     c. weak
     d. strong

125. approximately 8 times more concentrated

CHAPTER 15
QUESTIONS

1. The two general concepts involved in equilibrium are sameness and changelessness.

3. By controlling reaction rates, chemists can control the amount of a product that forms in a given period of time and have control of the outcome.

5. The two factors that influence reaction rates are concentration and temperature. The rate of a reaction increases with increasing concentration. The rate of a reaction increases with increasing temperature.

7. In a chemical reaction, dynamic equilibrium is the condition in which the rate of the forward reaction equals the rate of the reverse reaction.

9. Because the rate of the forward and reverse reactions is the same at equilibrium, the relative concentrations of reactants and products become constant.

11. The equilibrium constant is a measure of how far a reaction goes; it is significant because it is a way to quantify the concentrations of the reactants and products at equilibrium.

13. A small equilibrium constant shows that a reverse reaction is favored and that when equilibrium is reached, there will be more reactants than products. A large equilibrium constant shows that a forward reaction is favored and that when equilibrium is reached, there will be more products than reactants.

15. No, the particular concentrations of reactants and products at equilibrium will not always be the same for a given reaction—they will depend on the initial concentrations.

17. Various answers depending on the answer for Question 12.

19. Decreasing the concentration of a reactant in a reaction mixture at equilibrium causes the reaction to shift to the left.

21. Decreasing the concentration of a product in a reaction mixture at equilibrium causes the reaction to shift to the right.

23. Increasing the pressure of a reaction mixture at equilibrium if the product side has fewer moles of gas particles than the reactant side causes the reaction to shift to the right.

25. Decreasing the pressure of a reaction mixture at equilibrium if the product side has fewer moles of gas particles than the reactant side causes the reaction to shift to the left.

27. Increasing the temperature of an exothermic reaction mixture at equilibrium causes the reaction to shift left, absorbing some of the added heat. Decreasing the temperature of an exothermic reaction mixture at equilibrium causes the reaction to shift right, releasing heat.

29. $K_{sp} = [A^{2+}][B^-]^2$

31. The solubility of a compound is the amount of the compound that dissolves in a certain amount of liquid, and the molar solubility is the solubility in units of moles per liter.

33. Two reactants with a large K_{eq} for a particular reaction might not react immediately when combined because of a large activation energy, which is an energy hump that normally exists between the reactants and products. The activation energy must be overcome before the system will undergo a reaction.

35. No, a catalyst does not affect the value of the equilibrium constant; it simply lowers the activation energy and increases the rate of a chemical reaction.

PROBLEMS

37. Rate would decrease because the effective concentration of the reactants has been decreased, which lowers the rate of a reaction.

39. Reaction rates tend to decrease with decreasing temperature, so all life processes (chemical reactions) would have decreased rates.

41. The rate would be lower because the concentration of reactants decreases as they are consumed in the reaction.

43. **a.** $K_{eq} = \dfrac{[N_2O_4]}{[NO_2]^2}$ **b.** $K_{eq} = \dfrac{[NO]^2[Br_2]}{[BrNO]^2}$

    c. $K_{eq} = \dfrac{[H_2][CO_2]}{[H_2O][CO]}$ **d.** $K_{eq} = \dfrac{[CS_2][H_2]^4}{[CH_4][H_2S]^2}$

45. **a.** $K_{eq} = \dfrac{[Cl_2]}{[PCl_5]}$ **b.** $K_{eq} = [O_2]^3$

    c. $K_{eq} = \dfrac{[H_3O^+][F^-]}{[HF]}$ **d.** $K_{eq} = \dfrac{[NH_4^+][OH^-]}{[NH_3]}$

47. $K_{eq} = \dfrac{[H_2]^2[S_2]}{[H_2S]^2}$

49. **a.** products   **b.** both

    c. reactants   **d.** both

51. 0.0394

53. 1.79×10^{-5}

55. 0.0987

57. 7.0 M

59. 0.119 M

61.

$T(K)$	$[N_2]$	$[H_2]$	$[NH_3]$	K_{eq}
500	0.115	0.105	0.439	1.45×10^3
575	0.110	0.25	0.128	9.6
775	0.120	0.140	0.00439	0.0584

63. **a.** shift right   **b.** shift left

    c. shift right

65. **a.** no effect   **b.** shift left

    c. shift left   **d.** shift right

67. **a.** shift right   **b.** shift left

69. **a.** no effect   **b.** no effect

71. **a.** shift right   **b.** shift left

73. **a.** shift left   **b.** shift right

75. **a.** no effect   **b.** shift right

    c. shift left   **d.** shift right

    e. no effect

77. **a.** $CaSO_4(s) \longrightarrow Ca^{2+}(aq) + SO_4^{2-}(aq)$

       $K_{sp} = [Ca^{2+}][SO_4^{2-}]$

    b. $AgCl(s) \longrightarrow Ag^+(aq) + Cl^-(aq)$

       $K_{sp} = [Ag^+][Cl^-]$

    c. $CuS(s) \longrightarrow Cu^{2+}(aq) + S^{2-}(aq)$

       $K_{sp} = [Cu^{2+}][S^{2-}]$

    d. $FeCO_3(s) \longrightarrow Fe^{2+}(aq) + CO_3^{2-}(aq)$

       $K_{sp} = [Fe^{2+}][CO_3^{2-}]$

79. $K_{sp} = [Fe^{2+}][OH^-]^2$

81. 7.0×10^{-11}

83. 1.35×10^{-4} M

85. 7.04×10^{-5} M

87. 2.61×10^{-3} M

89.

Compound	[Cation]	[Anion]	K_{sp}
$SrCO_3$	2.4×10^{-5}	2.4×10^{-5}	5.8×10^{-10}
SrF_2	1.0×10^{-3}	2.0×10^{-3}	4.0×10^{-9}
Ag_2CO_3	2.6×10^{-4}	1.3×10^{-4}	8.8×10^{-12}

91. 3.3×10^2

93. 5.34 g

95. b, c, d

97. 1.13×10^{-18} M, 1.62×10^{-15} g

99. Yes

101. 7.07×10^{-4}

103. 1.2×10^{-2} g

105. 0.021 g K_2CO_3

107. e

109. 35.5 L

CHAPTER 16
QUESTIONS

1. A fuel-cell electric vehicle is an automobile running on an electric motor that is powered by hydrogen. The fuel cells use the electron-gaining tendency of oxygen and the electron-losing tendency of hydrogen to force electrons to move through a wire, creating the electricity that powers the car.

3. **a.** Oxidation is the gaining of oxygen, and reduction is the losing of oxygen.

   b. Oxidation is the loss of electrons, and reduction is the gain of electrons.

   c. Oxidation is an increase in oxidation state, and reduction is a decrease in oxidation state.

5. gain

7. The oxidization state of a free element is zero. The oxidization state of a monoatomic ion equals its charge.

9. For an ion, the sum of the oxidation states of the individual atoms must add up to *the charge of the ion*.

11. In a redox reaction, an atom that undergoes an increase in oxidation state is *oxidized*. An atom that undergoes a decrease in oxidation state is *reduced*.

13. When balancing redox equations, the number of electrons lost in the oxidation half-reaction must *equal* the number of electrons gained in the reduction half-reaction.

15. When balancing aqueous redox reactions, charge is balanced using *electrons*.

17. The metals at the top of the activity series are the most reactive.

19. The metals at the bottom of the activity series are least likely to lose electrons.

21. If the metal is listed above H_2 on the activity series, it will dissolve in acids such as HCl or HBr.

23. Oxidation occurs at the *anode* of an electrochemical cell.

25. The salt bridge joins the two half-cells or completes the circuit; it allows the flow of ions between the two half-cells.

27. The common dry cell battery does not contain large amounts of liquid water and is composed of a zinc case that acts as the anode. The cathode is a carbon rod immersed in a moist paste of MnO_2 that also contains NH_4Cl. The anode and cathode reactions that occur produce a voltage of about 1.5 volts.

   anode reaction:

   $Zn(s) \longrightarrow Zn^{2+}(aq) + 2\,e^-$

   cathode reaction:

   $2\,MnO_2(s) + 2\,NH_2^+(aq) + 2\,e^- \longrightarrow$
   $$2\,Mn_2O_3(s) + 2\,NH_3(g) + H_2O(l)$$

29. Fuel cells are like batteries, but the reactants are constantly replenished. The reactants constantly flow through the battery, generating electrical current as they undergo a redox reaction.

   anode reaction:

   $2\,H_2(g) + 4\,OH^-(aq) \longrightarrow 4\,H_2O(g) + 4\,e^-$

   cathode reaction:

   $O_2(g) + 2\,H_2O(l) + 4\,e^- \longrightarrow 4\,OH^-(aq)$

31. Corrosion is the oxidation of metals; the most common example is rusting of iron.

   oxidation:

   $2\,Fe(s) \longrightarrow 2\,Fe^{2+}(aq) + 4\,e^-$

   reduction:

   $O_2(g) + 2\,H_2O(l) + 4\,e^- \longrightarrow 4\,OH^-(aq)$

   overall:

   $2\,Fe(s) + O_2(g) + 2\,H_2O(l) \longrightarrow 2\,Fe(OH)_2(s)$

PROBLEMS

33. **a.** H_2 **b.** Al

   c. Al

35. **a.** Sr is oxidized, O_2 is reduced.

   b. Ca is oxidized, Cl_2 is reduced.

   c. Mg is oxidized, Ni^{2+} is reduced.

37. **a.** Sr is the reducing agent; O_2 is the oxidizing agent.

   b. Ca is the reducing agent; Cl_2 is the oxidizing agent.

   c. Mg is the reducing agent; Ni^{2+} is the oxidizing agent.

39. b (F_2), d (Cl_2)

41. a (K), c (Fe)

43. **a.** N_2 is oxidized and is the reducing agent.

   O_2 is reduced and is the oxidizing agent.

   b. C is oxidized and is the reducing agent.

   O_2 is reduced and is the oxidizing agent.

   c. Sb is oxidized and is the reducing agent.

   Cl_2 is reduced and is the oxidizing agent.

45. a. 0   **b.** +2
    c. +3   **d.** 0

47. a. Na: +1; Cl: −1   **b.** Ca: +2; F: −1
    c. S: +4; O: −2   **d.** H: +1; S: −2

49. a. +2   **b.** +4
    c. +1

51. a. C: +4; O: −2   **b.** O: −2; H: +1
    c. N: +5; O: −2   **d.** N: +3; O: −2

53. a. +1   **b.** +3
    c. +5   **d.** +7

55. a. Cu, +2; N, +5; O, −2
    b. Sr, +2; O, −2; H, +1
    c. K, +1; O, −2; Cr, +6
    d. Na, +1; H, +1; O, −2; C, +4

57. a. Sb + 5 $\longrightarrow$ +3, reduced
       Cl − 1 $\longrightarrow$ 0, oxidized
    b. C + 2 $\longrightarrow$ +4, oxidized
       Cl 0 $\longrightarrow$ −1, reduced
    c. N + 2 $\longrightarrow$ +3, oxidized
       Br 0 $\longrightarrow$ −1, reduced
    d. H 0 $\longrightarrow$ +1, oxidized
       C + 4 $\longrightarrow$ +2, reduced

59. Na is the reducing agent.
    H is the oxidizing agent.

61. a. $3 \text{ K}(s) + \text{Cr}^{3+}(aq) \longrightarrow \text{Cr}(s) + 3 \text{ K}^+(aq)$
    b. $\text{Mg}(s) + 2 \text{ Ag}^+(aq) \longrightarrow \text{Mg}^{2+}(aq) + 2 \text{ Ag}(s)$
    c. $2 \text{ Al}(s) + 3 \text{ Fe}^{2+}(aq) \longrightarrow 2 \text{ Al}^{3+}(aq) + 3 \text{ Fe}(s)$

63. a. reduction, $5 \text{ e}^- + \text{MnO}_4^-(aq) + 8 \text{ H}^+(aq) \longrightarrow$
                $\text{Mn}^{2+}(aq) + 4 \text{ H}_2\text{O}(l)$
    b. oxidation, $2 \text{ H}_2\text{O}(l) + \text{Pb}^{2+}(aq) \longrightarrow$
                $\text{PbO}_2(s) + 4 \text{ H}^+(aq) + 2\text{e}^-$
    c. reduction, $10 \text{ e}^- + 2 \text{ IO}_3^-(aq) + 12 \text{ H}^+(aq) \longrightarrow$
                $\text{I}_2(s) + 6 \text{ H}_2\text{O}(l)$
    d. oxidation, $\text{SO}_2(g) + 2 \text{ H}_2\text{O}(l) \longrightarrow$
                $\text{SO}_4^{2-}(aq) + 4 \text{ H}^+(aq) + 2 \text{ e}^-$

65. a. $\text{PbO}_2(s) + 4 \text{ H}^+(aq) + 2 \text{ I}^-(aq) \longrightarrow$
                $\text{I}_2(s) + \text{Pb}^{2+}(aq) + 2 \text{ H}_2\text{O}(l)$
    b. $5 \text{ SO}_3^{2-}(aq) + 6 \text{ H}^+(aq) + 2 \text{ MnO}_4^-(aq) \longrightarrow$
                $5 \text{ SO}_4^{2-}(aq) + 2 \text{ Mn}^{2+}(aq) + 3 \text{ H}_2\text{O}(l)$
    c. $\text{S}_2\text{O}_3^{2-}(aq) + 4 \text{ Cl}_2(g) + 5 \text{ H}_2\text{O}(l) \longrightarrow$
                $2 \text{ SO}_4^{2-}(aq) + 8 \text{ Cl}^-(aq) + 10 \text{ H}^+(aq)$

67. a. $\text{ClO}_4^-(aq) + 2 \text{ H}^+(aq) + 2 \text{ Cl}^-(aq) \longrightarrow$
                $\text{ClO}_3^-(aq) + \text{Cl}_2(aq) + \text{H}_2\text{O}(l)$
    b. $3 \text{ MnO}_4^-(aq) + 24 \text{ H}^+(aq) + 5 \text{ Al}(s) \longrightarrow$
                $3 \text{ Mn}^{2+}(aq) + 5 \text{ Al}^{3+}(aq) + 12 \text{ H}_2\text{O}(l)$
    c. $\text{Br}_2(aq) + \text{Sn}(s) \longrightarrow \text{Sn}^{2+}(aq) + 2 \text{ Br}^-(aq)$

69. a. $3 \text{ ClO}^-(aq) + 2 \text{ Cr(OH)}_4^-(aq) + 2 \text{ OH}^-(aq) \rightarrow$
                $3 \text{ Cl}^-(aq) + 2 \text{ CrO}_4^{2-}(aq) + 5 \text{ H}_2\text{O}(l)$
    b. $2 \text{ MnO}_4^-(aq) + \text{Br}(aq) + \text{H}_2\text{O}(l) \longrightarrow$
                $2 \text{ MnO}_2(s) + \text{BrO}_3^-(aq) + 2 \text{ OH}^-(aq)$

71. a, Ag

73. b, Cu^{2+}

75. b, Al

77. b and c occur spontaneously in the forward direction

79. Fe, Cr, Zn, Mn, Al, Mg, Na, Ca, K, Li

81. Mg

83. a. no reaction
    b. $2 \text{ HCl}(aq) + \text{Fe}(s) \longrightarrow \text{H}_2(g) + \text{FeCl}_2(aq)$
    c. no reaction
    d. $6 \text{ HCl}(aq) + 2 \text{ Al}(s) \longrightarrow 3 \text{ H}_2(g) + 2 \text{ AlCl}_3(aq)$

85.

87. d

89. $\text{Zn}(s) + 2 \text{ MnO}_2(s) + 2 \text{ H}_2\text{O}(l) \longrightarrow$
                $\text{Zn(OH)}_2(s) + 2 \text{ MnO(OH)}(s)$

91.

93. a, Zn; c, Mn

95. a. redox; Zn is oxidized; Co is reduced.
    b. not redox
    c. not redox
    d. redox; K is oxidized; Br is reduced.

97. $16 \text{ H}^+(aq) + 2 \text{ MnO}_4^-(aq) + 5 \text{ Zn}(s) \longrightarrow$
       $2 \text{ Mn}^{2+}(aq) + 5 \text{ Zn}^{2+}(aq) + 8 \text{ H}_2\text{O}(l)$; 34.9 mL

99. Yes, the reaction will occur spontaneously.
    $\text{Mg}(s) \longrightarrow \text{Mg}^{2+}(aq) + 2 \text{ e}^-$
    $\text{Ag}^+(aq) + \text{e}^- \longrightarrow \text{Ag}(s)$
    $2 \text{ Ag}^+(aq) + \text{Mg}(s) \longrightarrow \text{Mg}^{2+}(aq) + 2 \text{ Ag}(s)$

101. 2.95%

103. 0.054 mol

105. a. $2 \text{ Cr}(s) + 6 \text{ HI}(aq) \longrightarrow$
           $2 \text{ Cr}^{3+}(aq) + 6 \text{ I}^-(aq) + 3 \text{ H}_2(g)$, 98 mL HI
    b. $2 \text{ Al}(s) + 6 \text{ HI}(aq) \longrightarrow$
           $2 \text{ Al}^{3+}(aq) + 6 \text{ I}^-(aq) + 3 \text{ H}_2(g)$, 68 mL HI
    c. no
    d. no

107. 0.67 cm

109. 8.9×10^3 s or 2.5 hrs

111.

Aluminum atoms

Al^{3+} ions

Copper atoms

Cu^{2+} ions

113. Many of the Zn atoms on the electrode would become Zn^{2+} ions in solution. Many Ni^{2+} ions in solution would become Ni atoms on the electrode.

CHAPTER 17
QUESTIONS

1. Radioactivity is the emission of tiny, invisible particles by disintegration of atomic nuclei. Many of these particles can pass right through matter. Atoms that emit these particles are radioactive.

3. Uranic rays were the name given by Henri Becquerel to the radiation emitted by crystals that contained uranium.

5. X: chemical symbol, used to identify the element.

A: mass number, which is the sum of the number of protons and number of neutrons in the nucleus.

Z: atomic number, which is the number of protons in the nucleus.

7. Alpha radiation occurs when an unstable nucleus emits a small piece of itself composed of 2 protons and 2 neutrons. The symbol for an alpha particle is $^4_2$He.

9. Alpha particles have high ionizing power and low penetrating power compared to beta and gamma particles.

11. When an atom emits a beta particle, its atomic number increases by one because it now has an additional proton. The mass of an atom does not change as a result of beta emission.

13. Gamma radiation is electromagnetic radiation, and the symbol for a gamma ray is $^0_0\gamma$.

15. Gamma particles have low ionizing power and high penetrating power compared to alpha and beta particles.

17. When an atom emits a positron, its atomic number decreases by one because it now has one less proton. The mass of an atom does not change when it emits a positron.

19. A nuclear equation represents the changes that occur during radioactivity and other nuclear processes. For a nuclear equation to be balanced, the sum of the atomic numbers on both sides of the equation must be equal, and the sum of the mass numbers on both sides of the equation must be equal.

21. A film-badge dosimeter is a badge that consists of photographic film held in a small case that is pinned to clothing. It is used to monitor a person's exposure to radiation. The more exposed the film has become in a given period of time, the more the person has been exposed to radioactivity.

23. In a scintillation counter, the radioactive particles pass through a material that emits ultraviolet or visible light in response to excitation by radioactive particles. The light is detected and turned into an electrical signal.

25. The half-life is the time it takes for one-half of the parent nuclides in a radioactive sample to decay to the daughter nuclides. One can relate the half-life of objects to find their radioactive decay rates.

27. The decaying of uranium in the ground is the source of radon in our environment. Radon increases the risk of lung cancer because it is a gas that can be inhaled.

29. When an organism dies, it stops incorporating carbon-14 into its tissues, and the amount present at its death will decay with a half-life of 5730 years. Using this information, one can determine the age of the organism by the amount of carbon-14 still present in the remains.

31. Fermi believed that if a neutron could be incorporated into the nucleus of an atom, the nucleus might undergo beta decay and convert a neutron into a proton. The nuclear equation for this process is:

$$^{238}_{92}U + {}^1_0n \longrightarrow {}^{239}_{93}U \longrightarrow {}^{239}_{93}X + {}^{0}_{-1}e$$

33. Fission can be used in a bomb because it is a self amplifying reaction (fission of one atom induces fission of another) which can produce great amounts of energy.

35. The main goal of the Manhattan Project was to build an atomic bomb before the Germans did. Its project leader was J.R. Oppenheimer.

37. In nuclear reactors control rods control the amount of fission that can occur. When the control rods are inserted into the fuel assembly, they absorb neutrons, preventing them from inducing fission in the fuel rods.

39. No, a nuclear reactor cannot detonate the way a nuclear bomb can, because the uranium fuel used in electricity generation is not sufficiently enriched in U-235 to produce a nuclear detonation.

41. Modern nuclear weapons use both fission and fusion. In the hydrogen bomb, a small fission bomb is detonated first to create a high enough temperature for the fusion reaction to proceed.

43. Radiation can affect the molecules in living organisms by ionizing them.

45. Lower doses of radiation over extended periods of time can increase cancer risk by damaging DNA. Occasionally a change in DNA can cause cells to grow abnormally and to become cancerous.

47. The main unit of radiation exposure is the rem, which stands for *roentgen equivalent man*. The average American is exposed to 1/3 of a rem of radiation per year.

49. Isotope scanning can be used in the medical community to detect and identify cancerous tumors. Likewise, isotope scanning can produce necessary images of several different internal organs.

PROBLEMS

51. $^{128}_{82}Pb$

53. 81 protons, 126 neutrons

55. **a.** beta particle   **b.** neutron

    c. gamma ray

57.

Chemical Symbol	Atomic Number (Z)	Mass Number (A)	Number of Protons	Number of Neutrons
Tc	<u>43</u>	95	<u>43</u>	<u>52</u>
Ba	56	128	<u>56</u>	<u>72</u>
Eu	63	<u>145</u>	63	82
Fr	87	<u>223</u>	87	136

59. **a.** $^{234}_{92}U \longrightarrow ^{230}_{90}Th + ^{4}_{2}He$

b. $^{230}_{90}Th \longrightarrow ^{226}_{88}Ra + ^{4}_{2}He$

c. $^{226}_{88}Ra \longrightarrow ^{222}_{86}Rn + ^{4}_{2}He$

d. $^{222}_{86}Rn \longrightarrow ^{218}_{84}Po + ^{4}_{2}He$

61. **a.** $^{214}_{82}Pb \longrightarrow ^{214}_{83}Bi + ^{0}_{-1}e$

b. $^{214}_{83}Bi \longrightarrow ^{214}_{84}Po + ^{0}_{-1}e$

c. $^{231}_{90}Th \longrightarrow ^{231}_{91}Pa + ^{0}_{-1}e$

d. $^{227}_{89}Ac \longrightarrow ^{227}_{90}Th + ^{0}_{-1}e$

63. **a.** $^{11}_{6}C \longrightarrow ^{11}_{5}B + ^{0}_{+1}e$

b. $^{13}_{7}N \longrightarrow ^{13}_{6}C + ^{0}_{+1}e$

c. $^{15}_{8}O \longrightarrow ^{15}_{7}N + ^{0}_{+1}e$

65. $^{241}_{94}Pu \longrightarrow ^{241}_{95}Am + ^{0}_{-1}e$

$^{241}_{95}Am \longrightarrow ^{237}_{93}Np + ^{4}_{2}He$

$^{237}_{93}Np \longrightarrow ^{233}_{91}Pa + ^{4}_{2}He$

$^{233}_{91}Pa \longrightarrow ^{233}_{92}U + ^{0}_{-1}e$

67. $^{232}_{90}Th \longrightarrow ^{228}_{88}Ra + ^{4}_{2}He$

$^{228}_{88}Ra \longrightarrow ^{228}_{89}Ac + ^{0}_{-1}e$

$^{228}_{89}Ac \longrightarrow ^{228}_{90}Th + ^{0}_{-1}e$

$^{228}_{90}Th \longrightarrow ^{224}_{88}Ra + ^{4}_{2}He$

69. 3.1×10^3

71. 18 hrs

73. 1.2×10^6 yrs

75. 0.31 g

77. Ga-67 > P-32 > Cr-51 > Sr-89

79. 5,730 yrs

81. 34,380 yrs

83. $^{235}_{92}U + ^{1}_{0}n \longrightarrow ^{144}_{54}Xe + ^{90}_{38}Sr + 2\,^{1}_{0}n$; 2 neutrons

85. $^{2}_{1}H + ^{2}_{1}H \longrightarrow ^{3}_{2}He + ^{1}_{0}n$

87. **a.** $^{6}_{3}Li$

b. $^{64}_{28}Ni$

c. $^{179}_{73}Ta$

89. $^{238}_{92}U + ^{1}_{0}n \longrightarrow ^{239}_{92}U$; $^{239}_{92}U \longrightarrow ^{0}_{-1}e + ^{239}_{93}Np$;

$^{239}_{93}NP \longrightarrow ^{0}_{-1}e + ^{239}_{94}Pu$

91. per mole $= 1.9 \times 10^{13}$ J

per kg $= 8.1 \times 10^{13}$ J

93. $1.7 \times 10^{21} \beta$ emissions

95. 68.4%

97. 8.7×10^{-6} L

99. nucleus with 9 protons and 7 neutrons

101. nucleus with 5 protons and 5 neutrons

CHAPTER 18

QUESTIONS

1. Organic molecules are often involved in smell.

3. At the end of the eighteenth century, it was believed that organic compounds came from living things and were easily decomposed, while inorganic compounds came from the earth and were more difficult to decompose. A final difference is that many inorganic compounds could be easily synthesized, but organic compounds could not be.

5. Carbon is unique because it has four valence electrons and can form covalent bonds. Also it can combine with itself to form chains, branches, and rings.

7. Hydrocarbons are compounds that contain only carbon and hydrogen. Hydrocarbons are commonly used for fuels and are also the starting materials in the synthesis of many different consumer products.

9. A saturated hydrocarbon contains no double or triple bonds between the carbon atoms; however, an unsaturated hydrocarbon contains one or more double or triple bonds between the carbon atoms.

11. Alkanes composed of carbon atoms bonded in a straight chain with no branches are called *n*-alkanes. Alkanes composed of carbon atoms forming branched structures are called branched alkanes.

13. Alkenes are hydrocarbons containing at least one double bond between carbon atoms, whereas alkanes contain only single bonds.

15. Hydrocarbon combustion reactions involve the burning of hydrocarbons in the presence of oxygen. An example is:

$$CH_2{=}CHCH_2CH_3(g) + 6\,O_2(g) \longrightarrow$$
$$4\,CO_2(g) + 4\,H_2O(g)$$

17. An alkene addition reaction occurs when atoms add across the double bond. For example:

$$CH_2{=}CH_2(g) + Cl_2(g) \longrightarrow CH_2ClCH_2Cl(g)$$

19. The structure of benzene is 6 carbon atoms connected together in a circle, with each bonded to a hydrogen atom. Here are two ways that benzene is often represented.

21. The generic structure of alcohols is $R{-}OH$. The structure of methanol is CH_3OH, and the structure of ethanol is CH_3CH_2OH.

23. The generic structure of ethers is $R{-}O{-}R$. The structure of dimethyl ether is CH_3OCH_3, and the structure of diethyl ether is $CH_3CH_2OCH_2CH_3$.

25. The generic structure of an aldehyde is $R{-}CHO$; for example, propanal is CH_3CH_2CHO. The generic structure of a ketone is $R{-}CO{-}R$; for example, acetone is CH_3COCH_3.

27. The generic structure of a carboxylic acid is $R{-}COOH$; for example, acetic acid is CH_3COOH. The generic structure of an ester is $R{-}COO{-}R$; for example, ethyl propanoate is $CH_3CH_2COOCH_2CH_3$.

29. An amine is an organic compound containing nitrogen, the structure being NR_x. Two examples are methylamine, CH_3NH_2, and ethylamine, $CH_3CH_2NH_2$.

31. A polymer is a long chainlike molecule composed of repeating units called *monomers*. A copolymer consists of two different kinds of monomers.

PROBLEMS

33. c and d are hydrocarbons

35. a. alkyne   **b.** alkane
    c. alkyne   **d.** alkene

37.

39. $H_3C-CH_2-CH_2-CH_3$ $H_3C-CH-CH_3$
                                        CH_3

41.

43. a. *n*-pentane   **b.** 2-methylbutane
    c. 4-ethyl-2-methylhexane
    d. 3,3-dimethylpentane

45. a. $H_3C-CH-CH_2-CH_3$
                   CH_3

b.
$$CH_3$$
$$|$$
$$CH_2$$
$$|$$
$$H_3C-CH-CH-CH_2-CH_2-CH_3$$
$$|$$
$$CH_3$$

c.
$$H_3C-CH_2-CH-CH_2-CH_2-CH_2-CH_3$$
$$|$$
$$HC-CH_3$$
$$|$$
$$CH_3$$

d.
$$H_3C-CH-CH_2-CH_2-CH-CH_2-CH_2-CH_3$$
$$|\qquad\qquad\qquad |$$
$$CH_3\qquad\qquad CH_3$$

47. a. *n*-pentane **b.** 3-methylhexane
    c. 2,3-dimethylpentane

49.

Name	Molecular Formula	Structural Formula	Condensed Structural Formula
2,2,3-trimethylpentane	C_8H_{18}		$CH_3C(CH_3)_2CH(CH_3)CH_2CH_3$
2-methyl-3-propylhexane	$C_{10}H_{22}$		$CH_3CH(CH_3)CH(CH_2CH_2CH_3)CH_2CH_2CH_3$
2,2,3,3-tetramethylhexane	$C_{10}H_{22}$		$CH_3C(CH_3)_2C(CH_3)_2CH_2CH_2CH_3$
4,4-diethyl 2,3-dimethylhexane	$C_{12}H_{26}$		$CH_3CH(CH_3)CH(CH_3)CH(CH_2CH_3)_2CH_2CH_3$

51.

$CH_2{=}CH_2$ $CH_2{=}CHCH_3$

53. $H_2C{-}CH{-}CH_2{-}CH_2{-}CH_3$

$H_3C{-}CH{=}CH{-}CH_2{-}CH_3$

55. a. 2-pentene
   b. 4-methyl-2-pentene
   c. 3,3-dimethyl-1-butene
   d. 3,4-dimethyl-1-hexene

57. a. 2-butyne
   b. 4-methyl-2-pentyne
   c. 4,4-dimethyl-2-hexyne
   d. 3-ethyl-3-methyl-1-pentyne

59. a. $H_3C{-}CH{=}CH{-}CH_2{-}CH_2{-}CH_3$

   b. $H_3C{-}CH_2{-}C{\equiv}C{-}CH_2{-}CH_2{-}CH_3$

   c.

   d.

61. $H_2C{=}CH{-}CH_2{-}CH_2{-}CH_3$ 1-pentene

$H_3C{-}CH{=}CH{-}CH_2{-}CH_3$ 2-pentene

  2-methyl-1-butene

  3-methyl-1-butene

  2-methyl-2-butene

63.

Name	Molecular Formula	Structural Formula	Condensed Structural Formula
2,2-dimethyl-3-hexene	C_8H_{16}		$CH_3C(CH_3)_2CH=CHCH_2CH_3$
4,4-diethyl-5,5-dimethyl-2-hexyne	$C_{12}H_{22}$		$CH_3C(CH_3)_2C(CH_2CH_3)_2C\equiv CCH_3$
3,4-dimethyl-1-octyne	$C_{10}H_{18}$		$CH\equiv CCH(CH_3)CH(CH_3)CH_2CH_2CH_2CH_3$
4,4-diethyl-5,5-dimethyl-2-hexene	$C_{12}H_{24}$		$CH_3C(CH_3)_2C(CH_2CH_3)_2CH=CHCH_3$

65. **a.** $CH_3CH_3(g) + 7\,O_2(g) \longrightarrow 4\,CO_2(g) + 6\,H_2O(g)$

b. $2\,CH_2=CHCH_3(g) + 9\,O_2(g) \longrightarrow$
$$6\,CO_2(g) + 6\,H_2O(g)$$

c. $2\,CH\equiv CH(g) + 5\,O_2(g) \longrightarrow 4\,CO_2(g) + 2\,H_2O(g)$

67. $CH_4(g) + Br_2(g) \longrightarrow CH_3Br(g) + HBr(g)$

69. $CH_3CH=CHCH_3(g) + Cl_2(g) \longrightarrow$
$$CH_3CHClCHClCH_3(g)$$

71. $CH_2=CH_2(g) + H_2(g) \longrightarrow CH_3CH_3(g)$

73.

75. **a.** fluorobenzene

b. isopropylbenzene

c. ethylbenzene

77. **a.** 4-phenyloctane

b. 5-phenyl-3-heptene

c. 7-phenyl-2-heptyne

79. **a.** 1-bromo-2-chlorobenzene

b. 1,2-diethylbenzene or orthodiethylbenzene

c. 1,3-difluorobenzene or metadifluorobenzene

81. **a.**

b.

c.

83. **a.** ether = $R-O-R$

b. aldehyde =

c. amine =

d. ketone =

85. **a.** $H_3C-CH_2-CH_2-NH$ amine
CH_3

b. $H_3C-CH_2-CH_2-CH$ aldehyde

c. alcohol

d. ether

87. **a.** 2-butanol

b. 2-methyl-1-propanol

c. 3-ethyl-1-hexanol

d. 3-methyl-3-pentanol

89. **a.** $CH_3-CH_2-CH-CH_2-CH_3$ (with OH)

b. $CH_2-CH-CH_2-CH_3$ (with OH, CH_3)

c.

$$CH_3-CH-CH-CH_2-CH_2-CH_3$$
with OH and CH$_2$ substituents, and CH$_3$ on the CH$_2$

d. CH_3-CH_2-OH

91. a. $CH_3-CH_2-CH_2-CH_2-O-CH_2-CH_2-CH_2-CH_3$

b. ethyl propyl ether

c. dipropyl ether

d. $CH_3-O-CH_2-CH_2-CH_2-CH_2-CH_3$

93. a.

$$CH_3-CH_2-CH_2-CH_2-CH_2-CH_2-CH_2-CH\overset{O}{\|}$$

b. butanal

c. 4-heptanone

d.

$$CH_3-CH_2-\overset{O}{\underset{\|}{C}}-CH_2-CH_2-CH_3$$

95. a.

$$CH_3-CH_2-CH_2-CH_2-CH_2-CH_2-CH_2-\overset{O}{\underset{\|}{C}}-OH$$

b. methyl ethanoate

c.

$$H_3C-CH_2-CH_2-\overset{O}{\underset{\|}{C}}-O-CH_2-CH_3$$

d. heptanoic acid

97. a.

$$CH_3-CH_2-\underset{\underset{H}{|}}{N}-CH_2-CH_3$$

b. triethylamine

c. butylpropylamine

99.

$$\left[\!\!\begin{array}{c} CH_3 \\ | \\ CH_2-C \\ | \\ CH_3 \end{array}\!\!\right]_n$$

101.

$$\underset{HO}{\overset{O}{\underset{\|}{C}}}-\!\!\bigcirc\!\!-\overset{O}{\underset{\|}{C}}-O-CH_2-CH_2-OH$$

103. a. alcohol b. amine

c. alkane d. carboxylic acid

e. ether f. alkene

105. a. 3-methyl-4-tert-butylheptane

b. 3-methyl butanal

c. 4-isopropyl-3-methyl-2-heptene

d. propyl butanoate

107. a. same molecule b. isomers

c. same molecule

109. $CH_2{=}CH_2 + HCl \longrightarrow CH_3CH_2Cl$

111. 558 g H_2

113. 4.63×10^4 L

115. a. alcohol b. amine

c. carboxylic acid d. ester

e. alkane f. ether

CHAPTER 19

QUESTIONS

1. The human genome project is a 15-year project to map all of the genetic material of a human being. Scientists were surprised to discover that humans have only 20,000–25,000 genes, which is not much more than the number of genes in many simpler organisms.

3. A cell is the smallest structural unit of life. The main chemical components of a cell are carbohydrates, lipids, proteins, and nucleic acids.

5. Glucose is soluble in water due to its many —OH groups. This is important because glucose is the primary fuel of cells and can be easily transported in the blood-stream.

7. During digestion the links in disaccharides and polysac-charides are broken, allowing individual monosaccha-rides to pass through the intestinal wall and enter the bloodstream.

9. Starch and cellulose are both polysaccharides, but the bond between saccharide units is slightly different. Con-sequently, humans can digest starch and use it for energy, whereas cellulose cannot be digested and passes directly through humans.

11. Fatty acids are carboxylic acids with long hydrocarbon tails. The general structure of a fatty acid is $R-\overset{O}{\underset{\|}{C}}-OH$, where R is 3 to 19 carbon atoms long.

13. A triglyceride is a triester composed of glycerol with three fatty acids attached as shown here.

15. A phospholipid is similar to a triglyceride except instead of one of the fatty acid groups it has a phosphate group. A glycolipid has a nonpolar section made of a fatty acid chain and hydrocarbon chain, and a polar section made of a sugar molecule. Phospholipids and glycolipids both have a polar and nonpolar section.

17. Steroids are lipids that contain a four-ring structure. Cho-lesterol is a steroid that is part of cell membranes and also serves as a starting material for the body to synthesize other steroids. Also, steroids serve as male and female hormones in the body.

19. Proteins serve as catalysts; structural units of muscle, skin, and cartilage; transporters of oxygen; disease-fighting antibodies; and as hormones.

21. Amino acids differ from each other only in their R group or side chain.

23.

25. Primary protein structure refers to the sequence of amino acids in the protein's chain. Primary protein structure is maintained by the covalent peptide bonds between individual amino acids.

27. Tertiary protein structure refers to the large-scale twists and folds within the protein. These are maintained by interactions between the R groups of amino acids that are separated by long distances in the chain sequence.

29. In the α-helix structure, the amino acid chain is wound into a tight coil by hydrogen bonding between C=O and N—H groups at different locations along the backbone. The side chains extend outward. In the β-pleated sheet structure, the amino acid chain doubles back on itself repeatedly in a zig-zag pattern, with adjacent sections held together by hydrogen bonding between C=O and N—H groups along the backbone. The resulting structure is an undulating sheet with the side chains extending above and below it.

31. Nucleic acids contain a chemical code that specifies the correct amino acid sequences for proteins.

33. The four different bases that occur within DNA are adenine (A), cytosine (C), guanine (G), and thymine (T).

35. The genetic code is the code that links a specific codon to an amino acid.

37. A gene is a sequence of codons within a DNA molecule that codes for a single protein. Genes vary in length from 50 to thousands of codons.

39. Chromosomes located within the nuclei of cells are structures containing genes.

41. No; most cells in the human body only synthesize proteins that are important to their function.

43. **a.** The complementary base of adenine (A) is thymine (T).

b. The complementary base of thymine (T) is adenine (A).

c. The complementary base of cytosine (C) is guanine (G).

d. The complementary base of guanine (G) is cytosine (C).

PROBLEMS

45. **a.** monosaccharide
   b. not a carbohydrate
   c. not a carbohydrate
   d. disaccharide

47. **a.** hexose   **b.** tetrose
   c. pentose   **d.** tetrose

49.

51.

Glucose   Fructose

53. **a.** fatty acid, saturated   **b.** steroid
   c. triglyceride, unsaturated
   d. not a lipid

55.

```
G ——[ Fatty acid ]
l
y
c ——[ Fatty acid ]
e
r
o
l ——[ Fatty acid ]
```
Triglyceride

57.

H_2C—O—...

59. b, d

61.

Isoleucine   Serine

63. a.

b.

c.

65. tertiary

67. primary

69. a. nucleotide, G

  b. not a nucleotide

  c. not a nucleotide

  d. not a nucleotide

71.

73.

75. a. glycoside linkage—carbohydrates

  b. peptide bonds—proteins

  c. ester linkage—triglycerides

77. a. glucose—short-term energy storage

  b. DNA—blueprint for proteins

  c. phospholipids—compose cell membranes

  d. triglycerides—long-term energy storage

79. a. codon—codes for a single amino acid

  b. gene—codes for a single protein

  c. genome—all of the genetic material of an organism

  d. chromosome—structure that contains genes

81. Nitrogen: tetrahedral electron geometry, trigonal pyramidal molecular geometry

  1st Carbon: tetrahedral electron geometry, tetrahedral molecular geometry

  2nd Carbon: trigonal planar electron geometry, trigonal planar molecular geometry

83.

gly-val   val-gly

The difference lies in the end groups. In gly-val, glycine has the amine end and valine has the carboxyl end. For val-gly, the reverse is true.

85. gly-arg-ala-ser-phe-gly-asn-lys-trp-glu-val

87. 153 base pairs

89. 6.07×10^3 g/mol

91. The actual thymine-containing nucleotide uses the —OH end to bond and replicate; however, with the fake nucleotide having a nitrogen-based end instead, the possibility of replication is halted.

Chapter 1 Page 2: Richard Megna/Fundamental Photographs. Page 3 (Tl): Maxwellartandphoto.com. Page 3 (Tr): De Visu/Shutterstock. Page 3 center: Jaroslav74/Shutterstock. Page 3 bottom: Kevin Fleming/Corbis. Page 5 (Tr): Jacques-Louis David (French, 1748-1825), "Antoine-Laurent Lavoisier (1743-1794) and His Wife (Marie-Anne-Pierrette Paulze, 1758-1836)", 1788; Oil on canvas, 102-1/4 in. x 76-5/8 in. (259.7 x 194.6 cm). Page 5 (Tl): Corbis. Page 5 (B): IBM Research, Almaden Research Center. Page 7: bonniej/Istockphoto.com. Page 6: SPL/Photo Researchers. **Chapter 2** Page 12: Getty. Page 14 (T): Maxwellartandphoto.com. Page 14 (C): Paul Silverman/Fundamental Photographs. Page 14 (Bl & R): Richard Megna/Fundamental Photographs. Page 15 (T): tzara/Istockphoto.com. Page 15 (B): Richard Megna/Fundamental Photographs. Page 18: NASA. Page 22: Richard Megna/Fundamental Photographs. Page 23 (Tl): National Institute of Standards and Technology. Page 23 (Tc): Bureau International des Poids et Mesures. Page 23 (Tr): National Institute of Standards and Technology. Page 23 (B): Richard Megna/Fundamental Photographs. Page 24: David Mager/Pearson Education. Page 32: Maxwellartandphoto.com. Page 34: Litespeed Titanium. Page 40: Richard Megna/Fundamental Photographs. Page 52 (L): NASA. Page 52 (R): NASA Ozone Watch. **Chapter 3** Page 56 (Tl): Getty Images. Page 56 (Tr): Ken Liska. Page 56 (Bl): IBM Research, Almaden Research Center. Page 56 (Br): Driscoll, Youngquist, & Baldeschwieler, Caltech/SPL/Photo Researchers. Page 57 (Tl): Marek CECH/Shutterstock. Page 57 (Tc): Getty Images. Page 57 (Tr): Getty Images. Page 57 (B): Natural History Museum, London/Alamy. Page 59 (Tl): Shutterstock. Page 60: Getty Images. Page 59 (Tr): Getty Images. Page 59 (B1, 2): Richard Megna/Fundamental Photographs. Page 59 (B3): DenisLarkin/Istockphoto.com. Page 59 (B4): Kip Peticolas/Fundamental Photographs. Page 61: istockphoto.com. Page 63: Richard Megna/Fundamental Photographs. Page 64: Etienne du Preez/Shutterstock. Page 66: JLGutierrez/istockphoto.com. Page 74, 75: Getty Images. Page 90 (T): International Research Institute for Climate and Society. Page 90 (B): peeterv/istockphoto.com. **Chapter 4** Page 94 (B): Library of Congress. Page 94 (Tl): Getty Images. Page 94 (Tr): Corbis. Page 95: IBM Communications Media Relations. Page 97: Maxwellartandphoto.com. Page 99 (T): Getty Images. Page 99 (Bl): Photos.com/Jupiter Images. Page 99 (Br): Getty Images. Page 101 (T): Charles D. Winters/Photo Researchers. Page 101 (B): Library of Congress. Page 102: Stamp from the private collection of Professor C. M. Lang, photography by Gary J. Shulfer, University of Wisconsin, Stevens Point. "Russia: #3607 (1969)"; Scott Standard Postage Stamp Catalogue, Scott Pub. Co., Sidney, Ohio. Page 103: Corbis RF. Page 104 (T): Getty Images. Page 104 (B): Graeme Dawes/Shutterstock. Page 105 (Tlt): Richard Megna/Fundamental Photographs. Page 105 (Tlb): sciencephotos/Alamy. Page 105 (Bl): Richard Megna/Fundamental Photographs. Page 105 (Trt): Charles D. Winters/Photo Researchers. Page 105 (Trb): Pearson Education. Page 109: Getty Images. Page 112: U.S. Department of Energy. **Chapter 5** Page 127 (Bl): Pearson Education/PH College. Page 127 (Br NIP): Charles D. Winters/Photo Researchers. Page 128 (T): Diane Diederich/Istockphoto.com. Page 128 (T Bl): Getty Images. Page 128 (T Br): Mike Kemp/Rubberball/Alamy. Page 132 (NIP): Istockphoto.com. Page 133 (C): Harry Taylor/Dorling Kindersly Media Library. Page 133 (B NIP): Charles D. Winters/Photo Researchers. Page 134 (T): Ken Karp/Omni-Photo Communication. Page 134 (B): Paul Silverman/Fundamental Photographs. Page 141: Maxwellartandphoto.com. Page 146 (L): Photoshot/Alamy. Page 146

(R): Nickos/Istockphoto.com. Page 160 (L): Charles D. Winters/Photo Researchers. Page 160 (R): Richard Megna/Fundamental Photographs. Page 161 (T): Richard Megna/Fundamental Photographs. Page 161 (Bl): Getty Images. Page 161 (Br): Ken Eward/Photo Researchers. Page 162: Maxwellartandphoto.com. **Chapter 6** Page 166 (Tl): Richard Kittenberger/Shutterstock. Page 166 (Tr): Joseph P. Sinnot. Page 166 (C & B): Richard Megna/Fundamental Photographs. Page 167 (T): Maxwellartandphoto.com. Page 167 (B): Getty Images. Page 175, 176: Getty Images. Page 169 (T): Maxwellartandphoto.com. Page 169 (B): Richard Megna/Fundamental Photographs. Page 179: Tom Pantages. Page 181: Maxwellartandphoto.com. Page 184: Charles D. Winters/Photo Researchers. Page 202 (Bc & L): Getty Images. Page 202 (Br): NASA. Page 203: NASA Johnson Space Center. **Chapter 7** Page 206: Maxwellartandphoto.com. Page 207 (Bl): Maxwellartandphoto.com. Page 207 (Tl): Richard Megna/Fundamental Photographs. Page 207 (Tr): Charles D. Winters/Photo Researchers. Page 207 (Bl): Tom Bochsler/Pearson Education/PH College. Page 208 (Tl): Magnascan/Istockphoto.com. Page 208 (B1, 2, 3 from left): Richard Megna/Fundamental Photographs. Page 208 (B4): Jeff Daly/Fundamental Photographs. Page 208 (B5): ayazad/Shutterstock. Page 211: Getty Images. Page 215: Photodisc. Page 218: Richard Megna/Fundamental Photographs. Page 223: Maxwellartandphoto.com. Page 224: Fundamental Photographs. Page 225: Maxwellartandphoto.com. Page 226 (L): Renn SMinkey/Creative Digital Visions/Pearson. Page 226 (R): Tom McCarthy/PhotoEdit. Pages 229, 230, 246: Richard Megna/Fundamental Photographs. **Chapter 8** Pages 250, 251, 254, 258: Maxwellartandphoto.com. Page 265: Richard Megna/Fundamental Photographs. Page 282 (L): Getty Images. Page 282 (R): Katrina Leigh/Shutterstock. **Chapter 9** Page 285: Associated Press. Page 286 (L): Burgess Companies. Page 286 (R): American Institute of Physics/Emilio Segre Visual Archives. Page 287: Martin Barraud/Getty Images. Page 288: zooropa/Shutterstock. Page 289: [copyright] Sierra Pacific Innovations. All rights reserved. SPI CORP, www.x20.org. Page 290: Larry Mulvehill/Photo Researchers. Page 291 (Tl): Getty Images. Page 291 (B): Richard Megna/Fundamental Photographs. Page 295 (R): Joe Sohm/Photo Researchers. Page 295 (L): Associated Press. Page 322 (L): Getty Images. Page 322 (R): Library of Congress. **Chapter 10** Page 341: Kip Peticolas/Fundamental Photographs. Page 340: Maxwellartandphoto.com. Page 345: Richard Megna/Fundamental Photographs. Page 356: Tom Grundy/Shutterstock. **Chapter 11** Page 370: Getty Images. Page 365: Jerry Driendl/Getty Images. Page 372: Richard Megna/Fundamental Photographs. Page 377: Matka Wariatka/Shutterstock. Page 387: Hiro Kuraoka/Corbis. Page 394: Getty Images. Page 407: Don Johnston/Getty Images. Page 408: Tom Bochsler/Pearson Education/PH College. **Chapter 12** Page 413: Getty Images. Page 414 (T NIP): Anthony West/Corbis. Page 414 (C): Richard Megna/Fundamental Photographs. Page 414 (B): LauriPatterson/Istockphoto.com. Page 415 (Tl): lostbear/Istockphoto.com. Page 415 (Bl): NASA/Johnson Space Center. Page 415 (Br): Richard Megna/Fundamental Photographs. Page 417 (T): Richard Megna/Fundamental Photographs. Page 417 (B): Magnascan/Istockphoto.com. Page 421: Richard Megna/Fundamental Photographs. Page 424 (B): Ken Karp/Omni-photo communications. Page 424 (T): Charles Winters/Photo Researchers. Page 427 (L): Richard Megna/Fundamental Photographs. Page 427 (C): Richard

Haynes/Prentice Hall School Division. Page 427 (R): John Caps III/Associated Press. Page 432 (L): KieselUndStein/Istockphoto.com. Page 432 (C): Matejay/Istockphoto.com. Page 432 (R): Shutterstock. Page 433 (L): Getty Images. Page 433 (C): Dr. Mark J. Winter. Page 433 (R): Shutterstock. Page 433 (B): Maciej Toporowicz/Getty Images. Page 435 (B): Kristen Brochmann/Fundamental Photographs. Page 435 (T): Lisa F. Young/Shutterstock. **Chapter 13** Page 448: University of Savoie. Page 451: Richard Megna/Fundamental Photographs. Page 452: bodhihill/Istockphoto.com. Page 453: Maxwellartandphoto.com. Page 458: Don B. Stevenson/Alamy. Page 466: Dennis McDonald/PhotoEdit. Page 467: jane/Istockphoto.com. Page 468: Masonjar /Shutterstock. Page 471 (L): David Kamm/Alamy. Page 471 (R): Al Dodge. Page 485: Ken Lax/Photo Researchers. Page 470: Sam Singer/Pearson Education. **Chapter 14** Page 488 (Tl): George Goodwin/SuperStock. Page 488 (Tr, B): Richard Megna/Fundamental Photographs. Page 489: Ivaylo Ivanov/Shutterstock. Page 490 (T): Maxwellartandphoto.com. Page 490 (B): Richard Megna/Fundamental Photographs. Page 494 (T): Paul Silverman/Fundamental Photographs. Page 494 (B): Richard Megna/Fundamental Photographs. Page 496: Maxwellartandphoto.com. Page 498: Richard Megna/Fundamental Photographs. Page 499: Richard Megna/Fundamental Photographs. Page 501: Richard Megna/Fundamental Photographs. Page 512: Vera Bogaerts/Shutterstock. Page 515: NYC Parks Photo Archive/Fundamental Photographs. Page 515: Spencer Platt/Getty Images. **Chapter 15** Page 552: Claude Edelmann/Petit Format/Photo Researchers. Page 554: Richard Megna/Fundamental Photographs. Page 557: Alamy Images. **Chapter 16** Page 577: American Honda Motor Co. Inc. Page 578: Ron Zmiri/Shutterstock. Page 578: Getty Images. Page 579: Pearson Education/PH College. Page 583: John-Francis Bourke/Digital Vision/Getty Images. Page 589: Yellowj/Shutterstock. Page 590: Richard Megna/Fundamental Photographs. Page 591: Chris Cheadle/Getty Images. Page 592: Richard Megna/Fundamental Photographs. Page 595 (L): Kim Steele/Getty Images. Page 595 (R): Edmond Van Hoorick/Getty Images. Page 598 (T): Charles Winters/Photo Researchers. Page 598

(B): Angus Beare/Dorling Kindersly Media Library. Page 599: Draeger Safety, Inc. Page 610: Tony McNicol/Alamy. **Chapter 17** Page 614: Southern Illinois State University/Photo Researchers. Page 615 (L): Getty Images. Page 615 (R): Richard Megna/Fundamental Photographs. Page 622 (T): Hank Morgan/Photo Researchers. Page 622 (B): Scott Peterson/Liaison/Getty Images. Page 628: Associated Press. Page 627: "Otto Hahn, A Scientific Autobiography", Charles Scribner's Sons, New York, Page 1966, courtesy AIP Emilio Segre Visual Archives. Page 629 (T): Getty Images. Page 629 (B): H. Mark Weidman Photography/Alamy. Page 630: Istockphoto.com. Page 633 (T): Roger Tully/Getty Images. Page 633 (B): Peter Arnold/Alamy. **Chapter 18** Page 644 (L): Alex L. Fradkin/Getty Images. Page 644 (R): Emma Lee/Getty Images. Page 645 (T): Ryan McVey/Getty Images. Page 645 (T): C Squared Studios/Getty Images. Page 646: Science Photo Library/Photo Researchers. Page 649: jdjana/Istockphoto.com. Page 648: David Toase/Getty Images. Page 652 (1 from top): ayazad/Shutterstock. Page 652 (2 from top): Maxwellartandphoto.com. Page 652 (3 from top): David Buffington/Getty Images. Page 652 (4 from top): PhotoEdit. Page 652 (5 from top): Getty Images. Page 652 (6 from top): David Toase/Getty Images. Page 657: Melissa Brandes/Shutterstock. Page 658: Arthur S. Aubry/Getty Images. Page 661: Maxwellartandphoto.com. Page 666 (T): Todd Taulman/Shutterstock. Page 666 (B): Maxwellartandphoto.com. Page 669 (L): Spencer Jones/Getty Images. Page 669 (R): Maxwellartandphoto.com. Page 670 (L): GrigoryL/Shutterstock. Page 670 (R): C Squared Studios/Getty Images. Page 671: Getty Images. Page 672 (T): Getty Images. Page 672 (B): Lawrence Lawry/Getty Images. Page 674: Mary Kay Denney/PhotoEdit. Page 675: Maxwellartandphoto.com. Page 676: Professor Nivaldo Jose Tro. **Chapter 19** Page 694: Huntstock/Subjects/Getty Images. Page 698 (T): Spencer Jones/Getty Images. Page 698 (B): Geostock/Getty Images. Page 706: Maxwellartandphoto.com. Page 713: Oliver Meckes & Nicole Ottawa/Photo Researchers. Page 714: Professor Nivaldo Jose Tro. Page 720: Kenneth Eward/Photo Researchers.

INDEX

Vital force, 644
Vitalism, 644–645
Volatile, 65
   liquids, 416
Voltage, 594
Voltaic cell, 593–595
Volume, 24
   Avogadro's law, 377–378
   Boyle's law, 366–370
   changes and effects on equilibrium,
      550–551
   Charles's law, 370–374
   combined gas law, 375–377
   ideal gas law, 379–385
   moles and, 377–378
   pressure and, 366–370
   pressure, temperature and, 375–377
   pressure, temperature, moles and,
      379–385
   temperature and, 370–374

W

Water
   as an acid, 504–507
   adding salt to, 466
   as a base, 504–507
   boiling, 417–418, 434
   chemical formula for, 129–130
   collecting gases over, 388–389
   composition of, 128–129
   density of, 34
   drops, shape of, 415
   fluoridation of drinking, 183
   freezing, 434, 435
   hard, 557
   heat capacity of, 74
   heat of fusion, 422
   heat of vaporization, 419
   hydrogen bonding, 428
   ion product constant for, 505–507
   molecules, 434–435
   as a polar solvent, 449
   pollution, 435
   solutions of gases in, 452–454
   solutions of solids dissolved in,
      449–452
   why oil doesn't mix with, 341
Watson, James, 430
Wavelength, 287–288

Weak acids, 500–502
Weak bases, 503–504
Weak electrolytes, 500
White light
   components of, 287–288
   spectrum, 292
Wilson, Robert, 18
Winkler, Clemens, 102
Wöhler, Friedrich, 645
Wood alcohol, 666
Work, 66

X

Xenon, 104
   as an atomic solid, 432, 433
X-rays, 289

Z

Z (atomic number), 100
Zinc, 137
   oxidation of, 592

Fundamental Physical Constants

Atomic mass unit	$1\text{ amu} = 1.660539 \times 10^{-27}\text{ kg}$
	$1\text{ g} = 6.022142 \times 10^{23}\text{ amu}$
Avogadro's number	$N_A = 6.022142 \times 10^{23}/\text{mol}$
Electron charge	$e = 1.602176 \times 10^{-19}\text{ C}$
Gas constant	$R = 8.314472\text{ J}/(\text{mol} \cdot \text{K})$
	$= 0.0820582\text{ (L} \cdot \text{atm})/(\text{mol} \cdot \text{K})$
Mass of electron	$m_e = 5.485799 \times 10^{-4}\text{ amu}$
	$= 9.109382 \times 10^{-31}\text{ kg}$
Mass of neutron	$m_n = 1.008665\text{ amu}$
	$= 1.674927 \times 10^{-27}\text{ kg}$
Mass of proton	$m_p = 1.007276\text{ amu}$
	$= 1.672622 \times 10^{-27}\text{ kg}$
Pi	$\pi = 3.1415926536$
Planck's constant	$h = 6.626069 \times 10^{-34}\text{ J} \cdot \text{s}$
Speed of light in vacuum	$c = 2.99792458 \times 10^{8}\text{ m/s}$

Useful Geometric Formulas

Perimeter of a rectangle $= 2l + 2w$

Circumference of a circle $= 2\pi r$

Area of a triangle $= (1/2)(\text{base} \times \text{height})$

Area of a circle $= \pi r^2$

Surface area of a sphere $= 4\pi r^2$

Volume of a sphere $= (4/3)\pi r^3$

Volume of a cylinder or prism $= \text{area of base} \times \text{height}$

Important Conversion Factors

Length: SI unit = meter (m)

- $1\text{ m} = 39.37\text{ in.}$
- $1\text{ in.} = 2.54\text{ cm (exactly)}$
- $1\text{ mile} = 5280\text{ ft} = 1.609\text{ km}$
- $1\text{ angstrom (Å)} = 10^{-10}\text{ m}$

Volume: SI unit = cubic meter (m^3)

- $1\text{ L} = 1000\text{ cm}^3 = 1.057\text{ qt (U.S.)}$
- $1\text{ gal (U.S.)} = 4\text{ qt} = 8\text{ pt}$
  - $= 128\text{ fluid ounces}$
  - $= 3.785\text{ L}$

Mass: SI unit = kilogram (kg)

- $1\text{ kg} = 2.205\text{ lb}$
- $1\text{ lb} = 16\text{ oz} = 453.6\text{ g}$
- $1\text{ ton} = 2000\text{ lb}$
- $1\text{ metric ton} = 1000\text{ kg} = 1.103\text{ tons}$
- $1\text{ g} = 6.022 \times 10^{23}\text{ atomic mass units (amu)}$

Pressure: SI unit = pascal (Pa)

- $1\text{ Pa} = 1\text{ N/m}^2$
- $1\text{ bar} = 10^5\text{ Pa}$
- $1\text{ atm} = 1.01325 \times 10^5\text{ Pa (exactly)}$
  - $= 1.01325\text{ bar}$
  - $= 760\text{ mmHg}$
  - $= 760\text{ torr (exactly)}$

Energy: SI unit = joule (J)

- $1\text{ J} = 1\text{ N} \cdot \text{m}$
- $1\text{ cal} = 4.184\text{ J (exactly)}$
- $1\text{ L} \cdot \text{atm} = 101.33\text{ J}$

Temperature: SI unit = kelvin (K)

- $K = {}^\circ C + 273.15$
- ${}^\circ C = (5/9)({}^\circ F - 32^\circ)$
- ${}^\circ F = (9/5)({}^\circ C) + 32^\circ$